FISH NEUROENDOCRINOLOGY

This is Volume 28 in the

FISH PHYSIOLOGY series
Edited by Anthony P. Farrell and Colin J. Brauner
Honorary Editors: William S. Hoar and David J. Randall

A complete list of books in this series appears at the end of the volume

FISH NEUROENDOCRINOLOGY

Edited by

Dr. NICHOLAS J. BERNIER
Department of Integrative Biology
University of Guelph
Guelph, Ontario
Canada

Dr. GLEN VAN DER KRAAK
Department of Integrative Biology
University of Guelph
Guelph, Ontario
Canada

Dr. ANTHONY P. FARRELL
Faculty of Land and Food Systems & Department of Zoology
The University of British Columbia
Vancouver, British Columbia
Canada

Dr. COLIN J. BRAUNER
Department of Zoology
The University of British Columbia
Vancouver, British Columbia
Canada

AMSTERDAM • BOSTON • HEIDELBERG • LONDON
NEW YORK • OXFORD • PARIS • SAN DIEGO
SAN FRANCISCO • SINGAPORE • SYDNEY • TOKYO
Academic Press is an imprint of Elsevier

Academic Press is an imprint of Elsevier
84 Theobald's Road, London WC1X 8RR, UK
Radarweg 29, PO Box 211, 1000 AE Amsterdam, The Netherlands
30 Corporate Drive, Suite 400, Burlington, MA 01803, USA
525 B Street, Suite 1900, San Diego, CA 92101-4495, USA

First edition 2009

British Library Cataloging in Publication Data
A catalog record for this book is available from the British Library

Library of Congress Cataloging in Publication Data
A catalog record for this book is available from the Library of Congress

ISBN: 978-0-12-374631-3

For information on all Academic Press publications
visit our website at www.elsevierdirect.com

Printed and bound in

09 10 11 12 10 9 8 7 6 5 4 3 2 1

CONTENTS

CONTRIBUTORS ix

PREFACE xi

ABBREVIATIONS xiii

Section I. Anatomical Neuroendocrinology **1**

1. Neuroendocrine Systems of the Fish Brain
Jose Miguel Cerdá-Reverter and Luis Fabián Canosa

1. Introduction 4
2. Cytoarchitecture of the Main Hypophysiotropic Territories 9
3. Hypophysiotropic Territories in the Fish Brain: Tract-Tracing Studies 19
4. Central Neurohormones 22
5. Hypophysiotropic Peptides 30
6. Hypothalamic Neurotransmitters 51
7. Concluding Remarks 54

References 55

2. Endocrine Targets of the Hypothalamus and Pituitary
Olivier Kah

1. Introduction 76
2. Sex Steroids 77
3. Corticoid Receptors 91
4. Metabolic Hormones 95
5. Concluding Remarks 99

References 100

Section II. Functional Neuroendocrinology **113**

3. The GnRH System and the Neuroendocrine Regulation of Reproduction
Glen Van Der Kraak

1. Introduction 116
2. Gonadotropin-releasing Hormone (GnRH) 117
3. Steroid Feedback Regulation of GnRH, FSH and LH 121
4. Monoamine and Amino Acid Neurotransmitters 124
5. Neuropeptides 129
6. Protein Hormones 132
7. Integrated Neuroendocrine Control of Gonadal Development 134
8. Perspectives 137
References 140

4. Growth Hormone Regulation in Fish: A Multifactorial Model with Hypothalamic, Peripheral and Local Autocrine/Paracrine Signals
John P. Chang and Anderson O. L. Wong

1. Introduction 152
2. Growth Hormone and Growth Hormone Receptors 152
3. Biological Actions of Growth Hormone in Fish 154
4. Regulation of Growth Hormone Secretion and Synthesis 155
5. Functional Interactions of GH Regulators at the Hypothalamic and Pituitary Levels 179
6. Summary Remarks 183
References 184

5. The Neuroendocrine Regulation of Prolactin and Somatolactin Secretion in Fish
Hiroshi Kawauchi, Stacia A. Sower, and Shunsuke Moriyama

1. Introduction 198
2. The Neuroendocrine Regulation of Prolactin Secretion 199
3. The Neuroendocrine Regulation of Somatolactin Secretion 214
4. Concluding Remarks 220
References 221

6. Regulation and Contribution of the Corticotropic, Melanotropic and Thyrotropic Axes to the Stress Response in Fishes
Nicholas J. Bernier, Gert Flik, and Peter H. M. Klaren

1. Introduction 236
2. Hypothalamic Regulation of Pituitary Cells 237
3. Targets and Functions of the Corticotrope, Melanotrope and Thyrotrope Secretions 264
4. Targets, Functions and Feedback Effects of Cortisol and Thyroid Hormones 272
5. Contribution of the Corticotropic, Melanotropic, and Thyrotropic Axes to the Stress Response 282
6. Perspectives 288
References 289

7. Neuroendocrine–Immune Interactions in Teleost Fish
B. M. Lidy Verburg-Van Kemenade, Ellen H. Stolte, Juriaan R. Metz, and Magdalena Chadzinska

1. Neuroendocrine–Immune Interaction 313
2. Immune Modulation by Neuroendocrine Factors 324
3. Neuroendocrine Modulation by Cytokines 342
4. Conclusions and Future Perspectives 346
References 349

8. The Neuroendocrine Regulation of Fluid Intake and Fluid Balance
Yoshio Takei and Richard J. Balment

1. Introduction 366
2. Regulation of Fluid Intake 372
3. Regulation of Fluid Balance: Renal and Extrarenal Mechanisms 389
4. Perspectives 403
References 405

9. The Endocrine Regulation of Food Intake
Hélène Volkoff, Suraj Unniappan, and Scott P. Kelly

1. Introduction 422
2. Endocrine Regulation 430
3. Influence of Intrinsic Factors 444
4. Influence of Extrinsic Factors 447

5. Proposed Model of Endocrine Circuitries Involved in Feeding in Fishes and Concluding Remarks 449
References 450

10. The Neuronal and Endocrine Regulation of Gut Function
Susanne Holmgren and Catharina Olsson

1. Introduction 468
2. Anatomy of the Gut Neuronal and Endocrine Systems 468
3. Neurotransmitters and Hormones of the Gut 471
4. Development of Gut Innervation and Neuroendocrine System 479
5. Control of Gut Motility 481
6. Control of Secretion and Digestion 490
7. Control of Nutrient Absorption 494
8. Control of Water and Ion Transport 494
9. Control of Splanchnic Circulation 495
10. Concluding Remarks 499
References 499

INDEX 513

OTHER VOLUMES IN THE SERIES 527

CONTRIBUTORS

The numbers in parentheses indicate the pages on which the authors' contributions begin.

RICHARD. J. BALMENT *(365), Faculty of Life Sciences, The University of Manchester, Manchester*

NICHOLAS J. BERNIER *(235), Department of Integrative Biology, University of Guelph, Ontario, Canada*

LUIS FABIÁN CANOSA *(3), Laboratorio de Neuroendocrinología Comparada, Instituto de Investigaciones Biotecnológicas-Instituto Tecnológico de Chascomús (IIB-INTECH), Camino de Circunvalación Laguna, Chascomús, Provincia de Buenos Aires, Argentina*

JOSE MIGUEL CERDÁ-REVERTER *(3), Instituto de Acuicultura de Torre de la Sal (IATS), Consejo Superior de Investigaciones Cientificas (CSIC), Torre de la Sal s/n, Ribera de Cabanes, Castellon, Spain*

MAGDALENA CHADZINSKA *(313), Cell Biology & Immunology Group, Wageningen University, Marijkeweg, Wageningen, The Netherlands*

JOHN P. CHANG *(151), Department of Biological Sciences, University of Alberta, Biological Sciences Centre Edmonton, Alberta, Canada*

GERT FLIK *(235), Department of Organismal Animal Physiology, Radboud University Nijmegen, Toernooiveld, Nijmegen, The Netherlands*

SUSANNE HOLMGREN *(467), Göteborg University, Department of Zoophysiology Göteburg, Sweden*

OLIVIER KAH *(75), Endocrinologie Moléculaire de la Reproduction, Université de Rennes 1, France*

HIROSHI KAWAUCHI *(197), Showa, Aoba, Sendai, Miyagi, Japan*

SCOTT P. KELLY *(421), Department of Biology, York University, Toronto, Ontario, Canada*

B. M. Lidy Verburg-van Kemenade *(313), Cell Biology and Immunology Group, Wageningen University, Marijkeweg, Wageningen, The Netherlands*

Peter H. M. Klaren *(235), Department of Organismal Animal Physiology, Radboud University Nijmegen, Toernooiveld, Nijmegen, The Netherlands*

Glen Van Der Kraak *(115), Department of Integrative Biology, University of Guelph, Ontario, Canada*

Juriaan R. Metz *(313), Department of Organismal Animal Physiology, Radboud University Nijmegen, Toernooiveld, Nijmegen, The Netherlands*

Shunsuke Moriyama *(197), Laboratory of Molecular Endocrinology, School of Fisheries Sciences, Kitasato University, Sanriku, Iwate, Japan*

Catharina Olsson *(467), Göteborg University, Department of Zoophysiology Göteburg, Sweden*

Stacia A. Sower *(197), Center for Molecular and Comparative Endocrinology University of New Hampshire, Durham, USA*

Ellen H. Stolte *(313), Cell Biology & Immunology Group, Wageningen University, Marijkeweg, Wageningen, The Netherlands*

Yoshio Takei *(365), Ocean Research Institute, The University of Tokyo, Nakano, Tokyo, Japan*

Suraj Unniappan *(421), Laboratory of Integrative Neuroendocrinology, Department of Biology, York University, Toronto, Ontario, Canada*

Hélène Volkoff *(421), Departments of Biology and Biochemistry, Memorial University of Newfoundland, Canada*

Anderson O. L. Wong *(151), School of Biological Sciences, The University of Hong Kong, Pokfulam Road, Hong Kong SAR, China*

PREFACE

The field of fish neuroendocrinology continues to have significant impacts on our general understanding of the functional roles and evolution of a variety of neurochemical messengers and systems. Not only do fish possess certain unique neuroendocrine features, they have and continue to be an important and tractable vertebrate model for the discovery of new neuropeptides. During the last fifty years, neuroendocrinologists have documented a complex, and sometimes seemingly infinite, number of interactions between hormones and nerve structures. Emerging from this knowledge is an understanding of the specific neurohormonal pathways and the messengers responsible for maintaining homeostasis in an aquatic environment and for regulating the functional systems that allow for the highly diverse life histories and reproductive tactics of fish.

We now have a vast and rapidly expanding knowledge of the neuroendocrine regulatory mechanisms in fish. But until now there was no single text that covered the discipline of fish neuroendocrinology. In fact, other than a few mammalian neuroendocrinology textbooks, there is a void of major reference texts in comparative neuroendocrinology.

Consequently, this volume of *Fish Physiology* provides the first comprehensive reference in the discipline, with sections dedicated to both anatomy and function. Section I includes chapters that describe the anatomy of the neuroendocrine systems of the fish brain and the endocrine targets of the hypothalamus and pituitary. The chapters in Section II focus on the mechanisms of action of neurohormones at the molecular, cellular and systems levels. In doing so, they review the functional roles of the major neurohormones in the regulation of pituitary hormones and the control of key body processes such as fluid balance, food intake, and gut function. The volume also includes a chapter dealing with the neuroendocrine control of the immune system.

We wish to thank all the authors for their commitment and dedication to this volume. We also gratefully acknowledge all the referees for their insightful and constructive comments. Finally, we thank the staff at Elsevier for their encouragement in seeing this project to fruition.

We dedicate this book to the memory of Dr. Richard E. (Dick) Peter. In many respects, Dick pioneered the discipline of fish neuroendocrinology. His introduction of several key techniques to the field and contribution of fish brain atlases allowed generations of neuroendocrinologists to gain access and study the neuroendocrine domains of the fish brain. Dick's major scientific achievements to various aspects of the neuroendocrine regulation of growth, reproduction and feed intake in fish resonate throughout this volume. Dick was also a great mentor and contributed immensely to the establishment and advancement of fish neuroendocrinology.

Nicholas J. Bernier
Glen Van Der Kraak
Anthony P. Farrell
Colin J. Brauner

ABBREVIATIONS

Abbreviations used throughout this volume are listed below. Standard abbreviations are not defined in the text. Nonstandard abbreviations are defined in each chapter and listed below for reference purposes.

STANDARD ABBREVIATIONS

AMP, ADP, ATP	-adenosine 5′-mono-, di-, and triphosphates
cDNA	-complementary DNA
cAMP	-3′,5′-cyclic AMP
DNA	-deoxyribonucleic acid
GDP, GMP, GTP	-guanosine 5′-mono-, di-, and triphosphates
EC_{50}	-median effective concentration
ED_{50}	-median effective dose
IC_{50}	-median inhibitory concentration
LD_{50}	-median lethal dose
mRNA	-messenger RNA
NMR	-nuclear magnetic resonance
PCR	-polymerase chain reaction
RIA	-radioimmunoassay
RT-PCR	-reverse transcription-PCR
RNA	-ribonucleic acid

NONSTANDARD ABBREVIATIONS

A

A	-anterior thalamic nucleus
AA	-arachidonic acid
AC	-adenylate cyclase

ACE -angiotensin-converting enzyme
ACh -acetylcholine
ACTH -adrenocorticotropic hormone
AD -adrenaline
AD -epinephrine à adrenaline
AgRP -agouti-related protein
AM -adrenomedullin
ANG II -angiotensin II
ANP -atrial natriuretic peptide
AP -area postrema
AR -androgen receptors
ARC -arcuate nucleus
ASP -agouti-signaling peptide
AVP -arginine-vasopressin
AVT -arginine-vasotocin

B

BBB -blood-brain barrier
BBS -bombesin
BNP -B-type natriuretic peptide

C

CAM -calmodulin
CART -cocaine- and amphetamine-regulated transcript
CB1 & CB2 -cannabinoid receptor subtypes
CCK -cholecystokinin
CE -cerebellum
CFTR -cystic fibrosis transmembrane regulator
CGRP -calcitonin gene-related peptide
ChAT -choline acetyl transferase
CLIP -corticotropin-like intermediate lobe peptide
CLR -calcitonin receptor-like receptor
CNP -C-type natriuretic peptide
CNS -central nervous system
CNS -central nervous system
CNSS -caudal neurosecretory system
COX -cyclooxygenase
CP -central posterior thalamic nucleus
CR -corticosteroid receptor

CRF-BP -CRF binding protein
CRF -corticotropin-releasing factor
CRF_1 and CRF_2 -CRF receptor subtypes
CSF -cerebrospinal fluid
CVO -circumventricular organ

D

DA -dopamine
Dc -dorsal telencephalon pars centralis
Dd -dorsal telencephalon pars dorsalis
DHT -dihydrotestosterone
Dl -dorsal telencephalon pars lateralis
DL -dorsolateral thalamic nucleus
Dm -dorsal telencephalon pars medialis
DOC -deoxycorticosterone
DOPAC -3,4 dihydroxyphenylacetic acid
Dp -dorsal telencephalon pars posterior
DP -dorsal posterior thalamic nucleus
DYN -dynorphin
17, 20βP -17, 20β dihydroxy-4-pregnen-3-one

E

E2 -17β-estradiol
E -ventral telencephalon entopeduncular nuclei
EC -endocannabinoids
EDC -endocrine disrupting compound
EE2 -ethinylestradiol
END -endorphin
ENS -enteric nervous system
EPO -erythropoietin
ER -estrogen receptor
ERK1/2 -extracellular signal-regulated kinase 1/2

F

FR -fasciculus retroflexus
FSH -follicle-stimulating hormone
FSH -gonadotropin I

G

GABA -γ-amino butyric acid
GAD -glutamic acid decarboxylase
GAL -galanin
GALP -galanin-like peptide
GEP -gastroenteropancreatic
GFR -glomerular filtration rate
GH -growth hormone (somatotropin)
GHR -growth hormone receptor
GHRH -growth hormone-releasing hormone
GHS-R -growth hormone secretagogue receptor
GI -gastrointestinal
GIP -gastric inhibitory peptide
GLP -glucagon-like peptide
GnIH -gonadotropin inhibitor hormone
GnRH-R -gonadotropin-releasing hormone receptor
GnRH -gonadotropin-releasing hormone
GPCR -G protein-coupled receptor
GR -glucocorticoid receptor
GRE -glucocorticoid responsive element
GRK -G-protein-coupled receptor kinase
GRL-R -ghrelin receptor
GRL -ghrelin
GRP -gastrin-releasing peptide
GTH -gonadotropins
GVC -glossopharyngeal-vagal motor complex

H

Ha -anterior hypothalamus
Hc -caudal hypothalamus
HCo -horizontal commissure
Hd -dorsal hypothalamus
HPA -hypothalamic-pituitary adrenal
HPG -hypothalamic-pituitary gonadal
HPI -hypothalamic-pituitary interrenal
HPT -hypothalamic-pituitary thyroid
HRP -horseradish peroxidase
Hsp -heat shock protein
Hv -ventral hypothalamus

HYP -hypothalamus
5-HIAA -5-hydroxy indole acetic acid
5-HT -5-hydroxytryptamine (serotonin)

I

ICCs -interstitial cells of Cajal
ICL -internal cell layer of the olfactory bulb
icv -intracerebroventricular
IFN -interferon
Ig -immunoglobulin
IGF-I / IGF-II -insulin-like growth factor I / II
IL (e.g., IL-1, IL-6) -interleukin
IL -inferior lobe of the hypothalamus
im -intramuscular(-ly)
IP3 -inositol 1,4,5-trisphosphate
ip -intraperitoneal(-ly)
ir -immunoreactive (-ity)
IST -isotocin
iv -intravenous(-ly)

J

JAK -janus kinase

K

KKS -kallikrein-kinin system
11KA -11-ketoandrostenedione
11KT -11-ketotestosterone

L

LE -leu-enkephalin
LFB -lateral forebrain bundle
LH -gonadotropin II
LH -luteinizing hormone
LH -lateral hypothalamic nucleus
LIF -leukemia inhibitory factor
LPH -lipotropin

LPS -lipopolysaccharide
LR -lateral recess
LT -lymphotoxin

M

α2M -macroglobulin alpha-2
MAPK -mitogen-activated protein kinase
MC -melanocortin
MCH -melanin-concentrating hormone
MCR (MC1R – MC5R) -melanocortin receptor
ME -met-enkephalin
MHC -major histocompatibility complex
MPO -median preoptic nucleus
MR -mineralocorticoid receptor
MRCs -mitochondria-rich cells
MSH -melanocyte-stimulating hormone

N

NA -noradrenaline
NA -nucleus ambiguous
NAPv -anterior periventricular nucleus
NAT -anterior tuberal nucleus
NCC -non specific cytotoxic cells
NCC -commissural nucleus of Cajal
NCLI -central nucleus of the inferior lobe (m: mediall: lateral)
NDLI -diffuse nucleus of the inferior lobe
NE -entopeduncular nucleus
NEO -neoendorphin
NH -neurohypophysial
NK -natural killer
NKA -neurokinin A
NLT -lateral tuberal nucleus (a: anterior; d: dorsal; i: inferior; l: lateral; m: medial; p: posterior ventral)
NMDA -N-methyl-D-aspartate
NMLI -medial nucleus of the inferior lobe
NO -nitric oxide
NOR -nucleus olfactoretinalis
NOS -nitric oxide synthase
NP -natriuretic peptide

NP -paracommissural nucleus
NpCp -copeptine
NPO -nucleus preopticus (av: anteroventral; gc: gigantocellular; mc: magnocellular; p: paraventricular division; pc: parvocellular; r: rostral; s: supraoptic division)
NPP -N-terminal peptide of pro-opiomelanocortin
NPP -preoptic parvocellular nucleus
NPPv -posterior periventricular nucleus
NPR -natriuretic peptide receptor
NPT -posterior tuberal nucleus
NPY -neuropeptide Y
NRL -lateral recess nucleus (d: dorsal; i: inferior; l: lateral; v: ventral)
NRP -posterior recess nucleus
NSC -suprachiasmatic nucleus
NSV -nucleus of the saccus vasculosus
NTe -nucleus of the thalamic eminentia
NTS -nucleus tractus solitarius

O

OB -olfactory bulb
OpN -optic nerve
OT -optic tectum
OVLT -organon vasculosum of the lamina terminalis
OX_1 and OX_2 -orexin (hypocretin) receptors

P

$P450_{c11}$ -11β-hydroxylase enzyme
$P450_{scc}$ -cytochrome P450 side chain cleavage enzyme
P -pituitary
PACAP -pituitary adenylate cyclase-activating polypeptide
PAMPs -pathogen-associated molecular patterns
PC -prohormone convertase
PD -pituitary pars distalis
PDYN -pro-dynorphin
PENK -pro-enkephalin
PI3K -phosphatidylinositol 3-kinase
PI -pituitary pars intermedia
PKA -protein kinase A
PKC -protein kinase C

PLC -phospholipase C
PM -magnocellular preoptic nucleus (gc: gigantocellular; mc: magnocellular; pc: parvocellular)
PN -pituitary pars nervosa
POA -preoptic area
POMC -pro-opiomelanocortin
PP -pancreatic polypeptide
PP -parvocellular preoptic nucleus
PPa -parvocellular preoptic nucleus anterior pars
PPd -dorsal periventricular pretectal nucleus
PPD -pituitary proximal pars distalis
PPp -parvocellular preoptic nucleus posterior pars
PPv -ventral periventricular pretectal nucleus
PR -progesterone receptor
PRL -prolactin
PRLR -prolactin receptor
PRP -PACAP-related peptide
PRR -pathogen recognition receptors
PrRP -prolactin-releasing peptide
PSS -preprosomatostatin
PTN -posterior tuberal nucleus
PVN -paraventricular nucleus
PVO -paraventricular organ
PY -peptide Y
PYY -peptide tyrosine-tyrosine / peptide YY

R

RAMP -receptor activity-modifying protein
RAS -renin angiotensin system
RFa -arginyl-phenylalanyl-amide peptides
RFRP -RFamide-related peptide
ROS -reactive oxygen species
RPD -pituitary rostral pars distalis
rT_3 -3,3′,5′-triiodothyronine

S

SAA -serum amyloid A
SCO -subcommissural organ
SFO -subfornical organ
SL -somatolactin
SLR -somatolactin receptor
SOCS -suppressor of cytokine signaling
SON -supraoptic nucleus
SS -somatostatin (SRIF, somatotropin release inhibiting factor)
SST_{1-5} -somatostatin receptors 1-5
StAR -steroidogenic acute regulatory protein
STAT -signal transducer and activator of transcription

T

T -testosterone
T_3 -3,5,3′-triiodothyronine
T_4 -thyroxine
TCR -T-cell receptor
TEL -telencephalon
TGF -transforming growth factor
TH -tyrosine hydroxylase
TK -tachykinin
TLR -toll-like receptor
TN -tuberal nucleus
TNF -tumor necrosis factor
TR -thyroid hormone receptor
TRH -thyrotropin-releasing hormone
TSH -thyrotropin (thyroid-stimulating hormone)
TTX -tetrodotoxin
TV -ventral tuberal nucleus

U

UCN -urocortin
UI -urotensin I
UII -urotensin II
URP -urotensin II-related peptide

V

Vc -ventral telencephalon central nuclei
Vd -ventral telencephalon dorsal nuclei
Vi -ventral telencephalon intermediate nuclei
VIP -vasoactive intestinal polypeptide
Vl -ventral telencephalon lateral nuclei
VM -ventromedial thalamic nucleus
VNP -ventricular natriuretic peptide
Vp -ventral telencephalon postcommissural nuclei
Vs -ventral telencephalon supracommissural nuclei
VSCC -voltage-sensitive calcium channel
Vv -ventral telencephalon ventral nuclei

Z

ZL -zona limitans diencephali

SECTION I

ANATOMICAL NEUROENDOCRINOLOGY

1

NEUROENDOCRINE SYSTEMS OF THE FISH BRAIN

JOSE MIGUEL CERDÁ-REVERTER
LUIS FABIÁN CANOSA

1. Introduction
2. Cytoarchitecture of the Main Hypophysiotropic Territories
   2.1. Telencephalon
   2.2. Preoptic Area
   2.3. Hypothalamus
3. Hypophysiotropic Territories in the Fish Brain: Tract-Tracing Studies
4. Central Neurohormones
   4.1. Arginine-Vasotocin (AVT)
   4.2. Isotocin (IST)
   4.3. Melanin-concentrating Hormone (MCH)
5. Hypophysiotropic Peptides
   5.1. Cholecystokinin (CCK)
   5.2. Cocaine- and Amphetamine-regulated Transcript (CART)
   5.3. Corticotropin-releasing Factor (CRF) and Related Peptides
   5.4. Galanin
   5.5. Gastrin-releasing Peptide (GRP)
   5.6. Gonadotropin-releasing Hormone (GnRH)
   5.7. Growth Hormone-releasing Hormone (GHRH)/Pituitary Adenyl Cyclase Activating Polypeptide (PACAP)
   5.8. Melanocortin System
   5.9. Neuropeptide Y (NPY) Family of Peptides
   5.10. Orexins
   5.11. RF-amide Peptides
   5.12. Somatostatin
   5.13. Thyrotropin-releasing Hormone (TRH)
6. Hypothalamic Neurotransmitters
   6.1. Amino Acid Neurotransmitters: Glutamate and Gamma-amino Butyric Acid (GABA)
   6.2. Dopamine
   6.3. Serotonin
7. Concluding Remarks

Fish Neuroendocrinology: Volume 28
FISH PHYSIOLOGY

DOI: 10.1016/S1546-5098(09)28001-0

The study of the neuronal systems innervating the pituitary is a key point for understanding the regulation of a wide array of vital processes. This chapter offers a study of the neuroendocrine territories in fish. We first describe the anatomy of the main neuroendocrine territories of the teleost brains, making correspondences between the differently proposed nomenclatures. Teleost fish lack a canonical median eminence, and the hypothalamic neurons terminate very close to the adenohypophysial cells or make synaptoid contact upon them. This anatomical characteristic allows the study of the hypothalamo-hypophysial system by retrograde tracing experiments. Tract-tracing techniques have corroborated early studies showing the preoptic area and tuberal hypothalamus as loci for the neuronal cell bodies whose axons reach the neuro- and adenohypophysis along well-defined fiber tracts. We review the different neuronal systems that produce hypothalamic releasing or inhibitory peptides and neurotransmitters, and innervate the pituitary. Finally, the chapter describes the peptidergic innervation of the different pituitary domains and the association with the different secretory cells.

1. INTRODUCTION

The pioneering works of Ernest Scharrer suggested the presence of secretory cells in the brain of teleost fish (Scharrer, 1928). These studies suggested that the glandular nerve cells were responsible for the secretion of hormones in the pars nervosa of the pituitary. Scharrer's ideas, exclusively based on morphological criteria, met with powerful resistance and were not immediately accepted but they did establish a new discipline in neurosciences, neuroendocrinology. Subsequently staining studies identified two areas in the ventral forebrain, the preoptic area and hypothalamus, as loci for the neuronal cell bodies whose axons reach the neurohypophysis along well-defined fiber tracts, the hypophysial and the preoptico-hypophysial tracts (Palay, 1945). These pioneering studies revealed the existence of a hypothalamo-hypophysial system linking the central nervous (CNS) and endocrine systems to regulate an array of vital processes.

The hypothalamo-hypophysial system in fish is divided into three main areas: the hypothalamus, which is part of the diencephalon; the neurohypophysis, which derives from the ventral diencephalon and represents the neural compartment of the pituitary; and the adenohypophysis, which is the non-neuronal part of the gland (Figure 1.1; Pogoda and Hammerschmidt, 2007). The neurohypophysis is made up of the nerve terminals whose cell bodies are found mainly in the preoptic area and cells exhibiting glial-like

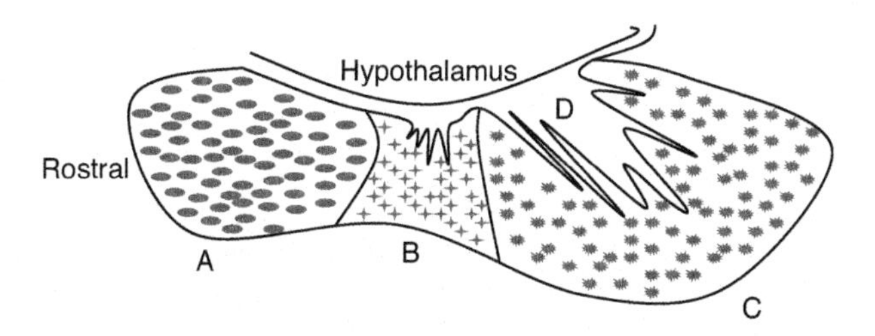

Fig. 1.1. Structure of the fish pituitary. A, Rostral pars distalis or pro-adenohypophysis; B, proximal pars distalis or meso-adenohypophysis; C, pars intermedia or meta-adenohypophysis. Regions A, B, and C are parts of the adenohypophysis. D, Pars nervosa of the neurohypophysis.

properties or pituicytes, which have a supportive function. In some fish species, like plainfin midshipman (*Porichthys notatus*), the neurohypophysis draws away from the brain and forms a stalk of nerve tissue (Sathyanesan, 1965), while in other fish there is no infundibular stalk or the gland is placed at the end of a short infundibular stalk (see Gorbman *et al.*, 1983). In elasmobranchs (van de Kamer and Zandbergen, 1981) and non-teleost bony ray-finned fishes (Lagios, 1968), the neurohypophysis is divided into the median eminence and the pars nervosa. The former includes the portal system, a blood capillary network where the hypothalamic neurons release their secretory products into the vascular system for subsequent delivery to the adenohypophysis. Agnathan species including lampreys and hagfishes have no canonical median eminence but exhibit an anterior neurohemal region through which materials can rapidly diffuse from the brain into the adenohypophysis (neuro-diffuse system, Nozaki *et al.*, 1994). On the contrary, there is no median eminence and no portal system in teleost fish, in which hypothalamic neurons terminate very close to adenohypophysial cells, reducing the diffusional distance, or make synaptoid contact with adenohypophysial cells. This means that the hypothalamic control over the adenohypophysis can be exerted through direct action upon the secretory cells. Figure 1.2 shows the possible adenohypophysial innervation patterns observed in fish. There is some controversy concerning whether or not central neurosecretions indirectly reach the adenohypophysis through the vascular plexus in the neurohypophysis, which would mean the possible existence of a dual mechanism operating by direct (neural) and indirect (vascular) pathways in teleost fish (Hill and Henderson, 1968). However, it is clear that teleost fish do not have a vascular pattern like that of tetrapod species in which a median eminence-like primary plexus leads to a portal vein which subsequently leads to a secondary capillary plexus (Gorbman *et al.*, 1983).

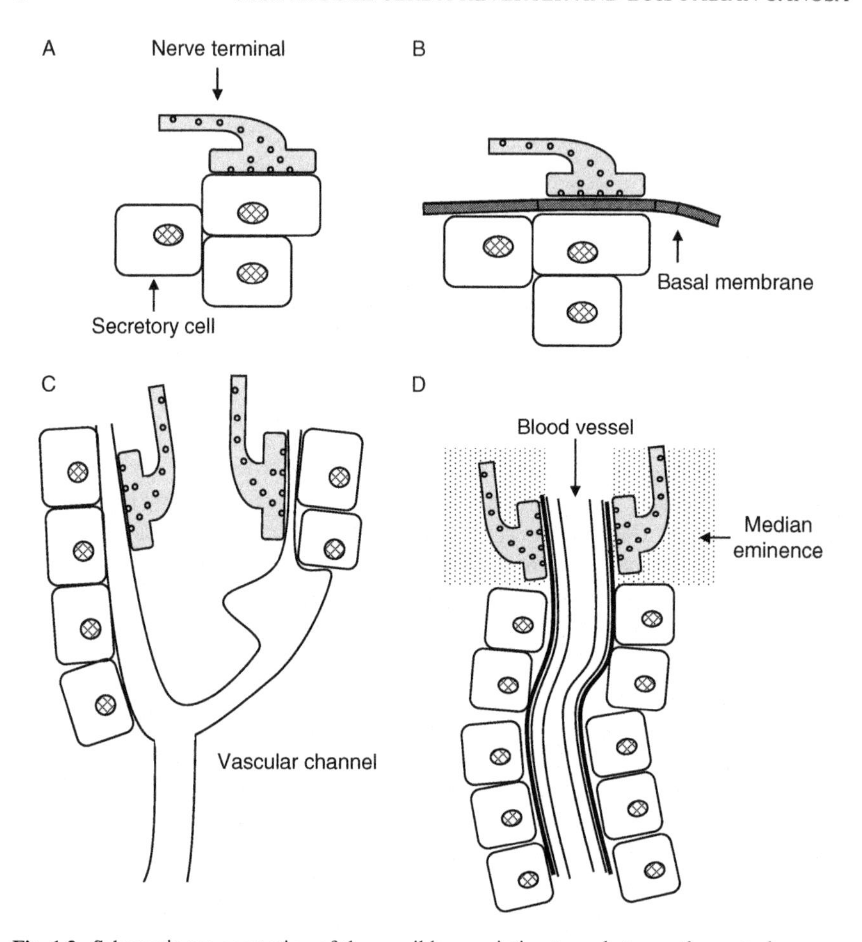

Fig. 1.2. Schematic representation of the possible association types between the central nervous system and the adenohypophysial endocrine cells in teleost fish (A–C) and non-teleost fish (D). (A) Direct synaptic contact between the neurosecretory terminals and endocrine cell in the seahorse (*Hippocampus* sp.). (B) The basement membrane is interposed between the neurosecretory ends and the endocrine cell in the tench (*Tinca* sp.). (C) In the eel, (*Anguilla anguilla*), the neurosecretory terminals lie over the vascular channels among the endocrine cells. (D) Physical separation between the neurosecretory terminals and the adenohypophysial cells. Hypothalamic neurons release their secretory products into the hypothalamic vascular system (median eminence) for subsequent delivery to the adenohypophysis. This later type of innervation is also observed in tetrapod vertebrates (Modified from Vollrath, 1967).

Adenohypophysial cells synthesize their own hormones and remain organized in discrete domains (Pogoda and Hammerschmidt, 2007). In contrast to tetrapod species, fish adenohypophysis lacks a pars tuberis. Two different nomenclatures have been proposed for the regionalization of fish adenohypophysis (Green, 1951; Pickford and Atz, 1957). Both terminologies recognize three distinct zones: the most rostral part, termed the rostral pars distalis (RPD) (Green, 1951) or pro-adenohypophysis (Pickford and Azt, 1957); the remaining pars distalis (PD), termed the proximal pars distalis (PPD) or meso-adenohypophysis; and the pars intermedia (PI) or meta-adenohyopophysis. In contrast to mammalian species, there is no strict morphological separation between the PD and the PI. The pars nervosa (PN) or the posterior part of the neurohypophysis is fully interdigitated with the PI of the adenohypophysis to form the neurointermediate lobe (NIL). The PD also is invaded by tongues of the neurohypophysis. The PN contains nerve terminals from neurons in the preoptic area which color with neurosecretory stains. The anterior part of the neurohypophysis that is interdigitated with the PD is not stained by the above techniques and contains endings of nerve fibers from the lateral tuberal nucleus and some other ventral areas of the forebrain that may be considered as hypophysiotropic areas (Gorbman *et al.*, 1983).

Fish adenohypophysis synthesizes at least eight different hormones (Figure 1.3), which can be divided into three different categories (Pogoda and Hammerschmidt, 2007). Category 1 includes hormones belonging to the growth hormone (GH)/prolactin (PRL)/somatolactin family including the growth hormone (GH) made in the somatotropes, prolactin (PRL) generated by lactotropes, and somatolactin (SL) produced in the somatolactotropes (Kawauchi and Sower, 2006). Category 2 consists of thyroid-stimulating hormone or thyrotropin (TSH), and two gonadotropins (GTH): follicle-stimulating hormone (FSH/GTH-I) and luteinizing hormone (LH/GTHII). All three hormones are heterodimeric glycoproteins, formed by a common α subunit and a hormone-specific β subunit encoded by different genes. While the α subunit gene is commonly expressed in the gonadotropes and thyrotropes, β-subunit gene expression is cell-specific (Querat *et al.*, 2000). Category 3 comprises peptides derived from a common precursor, proopiomelanocortin (POMC). It is comprised of three main domains: N-terminal pro-γ-melanocyte-stimulating hormone (MSH), central adrenocorticotropic hormone (ACTH) and C-terminal β-lipotropin. Each domain contains one MSH peptide, i.e., γ-MSH in pro-γ-MSH, α-MSH as the N-terminal sequence of ACTH and β-MSH in the β-lipotropin domain. The last domain further includes C-terminal β-endorphin peptide. Teleost fish lack γ-MSH domain but retain both α- and β-MSH regions and a β-endorphin deduced peptide. Elasmobranchs have a fourth MSH domain termed δ-MSH (Takahashi *et al.*, 2001). Following a general pattern in teleost fish (Laiz-Carrión *et al.*, 2003), PRL-producing cells (lactotropes) and ACTH-producing cells (corticotropes) are localized in the RPD, with lactotropes ventrally positioned to the corticotropes. Somatotropes (GH-producing cells) and

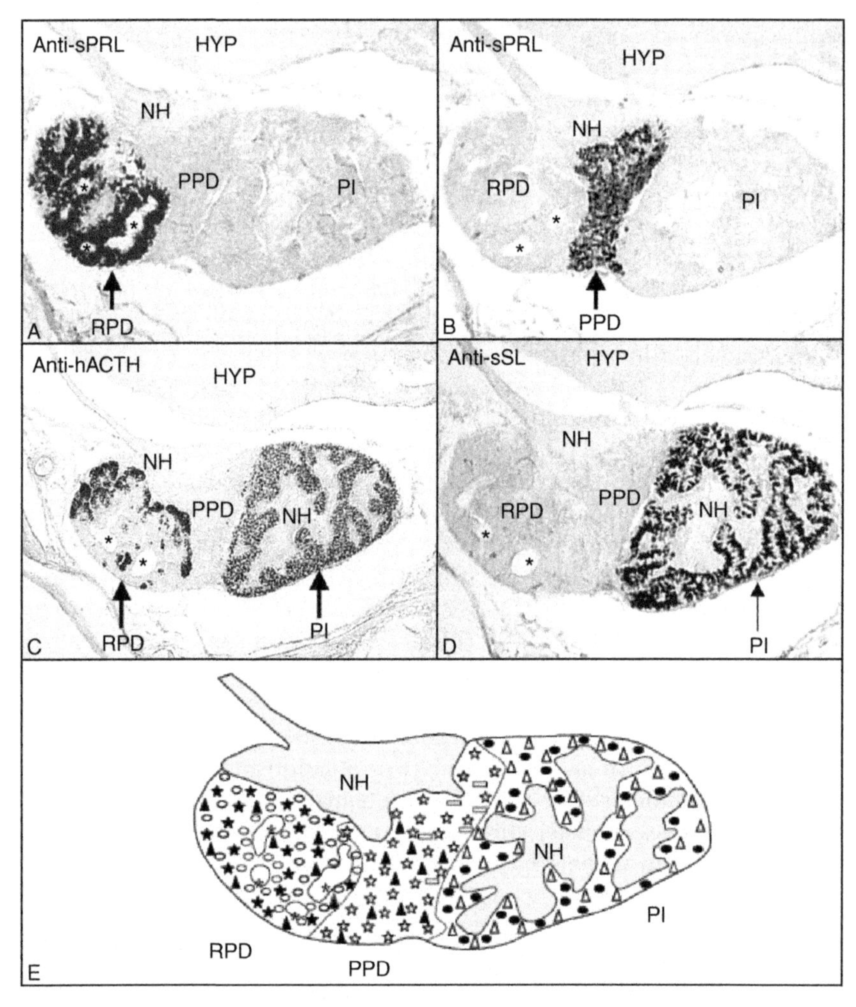

Fig. 1.3. Immunoreactive demonstration of lactotropes. (A) Anti-chum salmon prolactin (sPRL)-immunoreactive (ir) cells), somatotropes; (B) anti-seabream growth hormone (GH)-ir cells, corticotropes; (C) anti-human adrenocorticotropin (ACTH)-ir cells and somatolactotropes; (D) anti-chum salmon somatolactin (SL)-ir cells distribution in the adenohypophysis of the American shad (*Alosa sapidissima*). (E) Schematic sagittal representation of the pituitary showing the distribution of adenohypophysial cells. PRL (★), ACTH (○), GH (☆), GTH (▲), TSH (□), SL (•) and MSH (Δ). Asterisk indicates cavities; HYP, hypothalamus; NH, neurohypophysis; PI, pars intermedia; PPD proximal pars distalis; RPD, rostral pars distalis. Modified from Laiz-Carrión *et al.* (2003).

thyrotropes (TSH-producing cells) are concentrated within the following domain or PPD whereas somatolactotropes (SL producing cells) occupy the caudal-most domain, the PI, where they are intermingled with the melanotropes (α-MSH-cells). Finally, both FSHβ and LHβ are produced in different gonadotropes in the PPD, with some LHβ cells which line the external border of the PI. In the region posterior to the NIL and on the midline of the inferior hypothalamic lobules, most bony and cartilaginous fishes have the saccus vasculosus, a circumventricular organ consisting of highly vascularized neuroepithelium, which is formed by an exclusive cell type termed coronet cells, as well as by neurons contacting the cerebrospinal fluid and supporting cells (Sueiro *et al.*, 2007).

2. CYTOARCHITECTURE OF THE MAIN HYPOPHYSIOTROPIC TERRITORIES

The organization of the CNS can be studied by specific tissue-staining techniques. Nissl's staining, which uses basic aniline dyes, remains the routine technique for the investigation of the cytoarchitecture of the CNS. This technique is an indispensable first step in the analysis of the CNS and one of the most powerful tools in neuroanatomy. This method stains RNA blue, revealing the localization of the so-called "Nissl's bodies," which are granular basophilic bodies found in the cytoplasm of neurons and composed of rough endoplasmic reticulum and free polyribosomes. The staining shows exclusively neuronal perikarya while dendrites, axons and terminals are not stained. It allows clusters of neuronal somata called "nuclei" to be delimited. The nuclei can be identified according to several morphological criteria including size, shape and staining intensity of the perikarya, packing density and distribution pattern of the cell bodies, neuropil surrounding cell groups and the consistency of cell groups. In this section, we will describe the Nissl cytoarchitecture of the main hypophysiotropic territories in the fish brain. The description will focus briefly on the morphology of the teleost forebrain that includes the telencephalon and diencephalon (Figure 1.4). We use the nomenclature of Braford and Northcutt (1983) with additional elaborations by Cerdá-Reverter (2001a,b). For a more extensive comparative study, readers should consult Braford and Northcutt (1983), Nieuwenhuys *et al.* (1998) and Butler and Hodos (2005).

2.1. Telencephalon

In teleost fish, the topology of the telencephalon is highly distorted. During the initial steps of development in most vertebrates (lampreys, cartilaginous fish, amphibians and amniotes) the telencephalon develops through a process called evagination. During this process, the lumen of the neural tube enlarges to form the telencephalic ventricles as the telencephalon

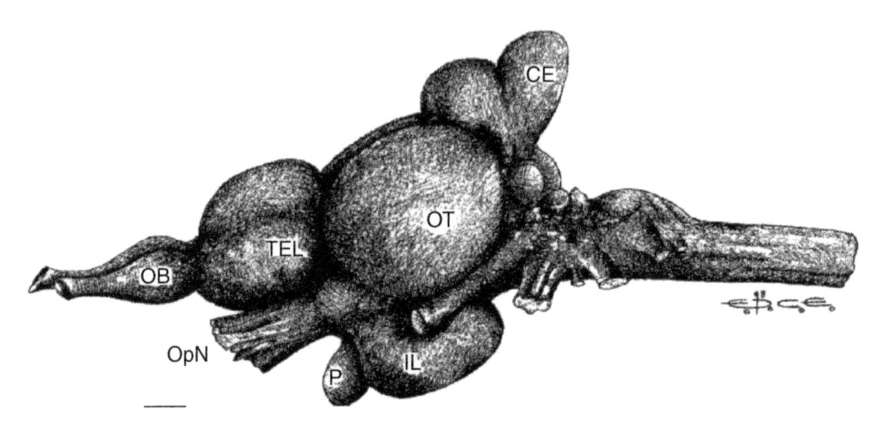

Fig. 1.4. Lateral view of the brain of the sea bass (*Dicentrarchus labrax*). CE, cerebellum; IL, inferior lobe of the hypothalamus; OB, olfactory bulb; OpN, optic nerve; OT, optic tectum; P, pituitary; TEL, telencephalon. Scale bar = 1 mm. (Modified from Cerdá-Reverter *et al.*, 2000a).

protrudes and expands. In contrast, the telencephalon of ray-finned fish undergoes a process of eversion in which the roof of the neural tube extends laterally so that the paired dorsal parts roll out lateroventrally. As a result of eversion, the telencephalon is covered by thin tela (Figure 1.5; Northcutt, 1995). This developmental difference makes it difficult to infer the homology between the structures of the telencephala of ray-finned fish and other vertebrate species, and the independent nomenclature has been proposed (Northcutt and Davis, 1983; Meek and Nieuwenhuys, 1998). Recently, however, some potential homologies have been established using neuronal circuitry patterns and distribution of chemical transmitters and peptides (Wullimann and Mueller, 2004).

The telencephalon comprises telencephalic hemispheres and olfactory bulbs (Figure 1.4). The hemispheres can be divided into two main regions: an area dorsalis telencephali (pallium) and an area ventralis telencephali (subpallium). The area dorsalis is the everted part of the telencephalon and exhibits a number of large histologically distinct zones where most of the cells lie far from the ventricle surface. In contrast, the area ventralis telencephali is organized into nuclei, most of which are in proximity to the ventricular area (Wullimann and Mueller, 2004).

The dorsal area or pallium represents the everted part and largest region of the teleost telencephalon. Using cytoarchitectonic criteria it can be divided into three periventricular, externally placed, zones – the pars medialis (Dm), pars dorsalis (Dd) and pars lateralis (Dl) – one central area, the pars centralis (Dc), and a posterior area, the pars posterior (Dp). Neurons in the dorsal telencephalon do not project to the pituitary. The ventral area is the

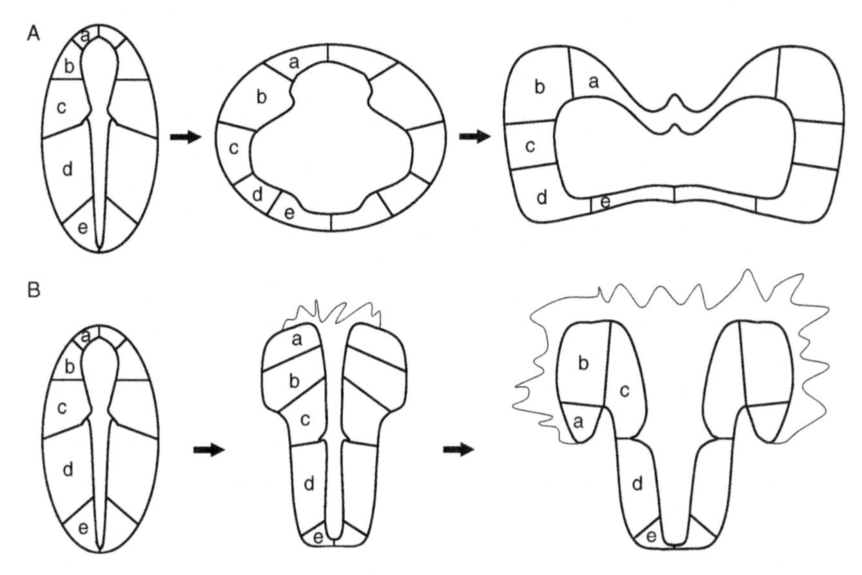

Fig. 1.5. (A) Process of evagination in the forebrain of most vertebrates during early development. (B) Process of eversion in the forebrain of ray-finned fish. Comparison of the position of lower case letters makes for easy comparisons between evagination and eversion processes. Modified from Butler and Hodos, 2005.

non-everted part of the teleost telencephalon placed rostral to the preoptic area. This area is split into precommissural and postcommissural nuclei, where the anterior commissure is the inflexion point. Dorsal (Vd), ventral (Vv) and lateral (Vl) nuclei are the main precommissural nuclei, whereas central (Vc), supracommissural (Vs), postcommissural (Vp), intermediate (Vi) and entopeduncular nuclei (E) are found in postcommissural position (Figure 1.6). From the neuroendocrine perspective, both the Vv and Vc nuclei have been reported to project to the pituitary in several teleost fish (see later). In some species such as sea bass (*Dicentrarchus labrax*; Cerdá-Reverter *et al.*, 2001a), the Vv is characterized by the presence of an ependymal column of small, densely stained packed cells lining the ventricle and associated with the dorsal-most aspect of the nucleus. The central nucleus (Vc) contains large and intensely stained perikarya intermingled with the fibers of the lateral forebrain bundle (LFB). Neuronal circuitry patterns and distribution of chemical transmitters and peptides suggest that the Vd and Vc nuclei of the area ventralis telencephali represent the mammalian striatal formation, whereas the Vl and Vv nuclei correspond to the septal formation. One of the arguments used to defend the homology between the Vl-Vv and the septal formation is the massive descending output to the midline hypothalamus, which is also a characteristic of the septal formation in amniotes (Wullimann and Mueller, 2004).

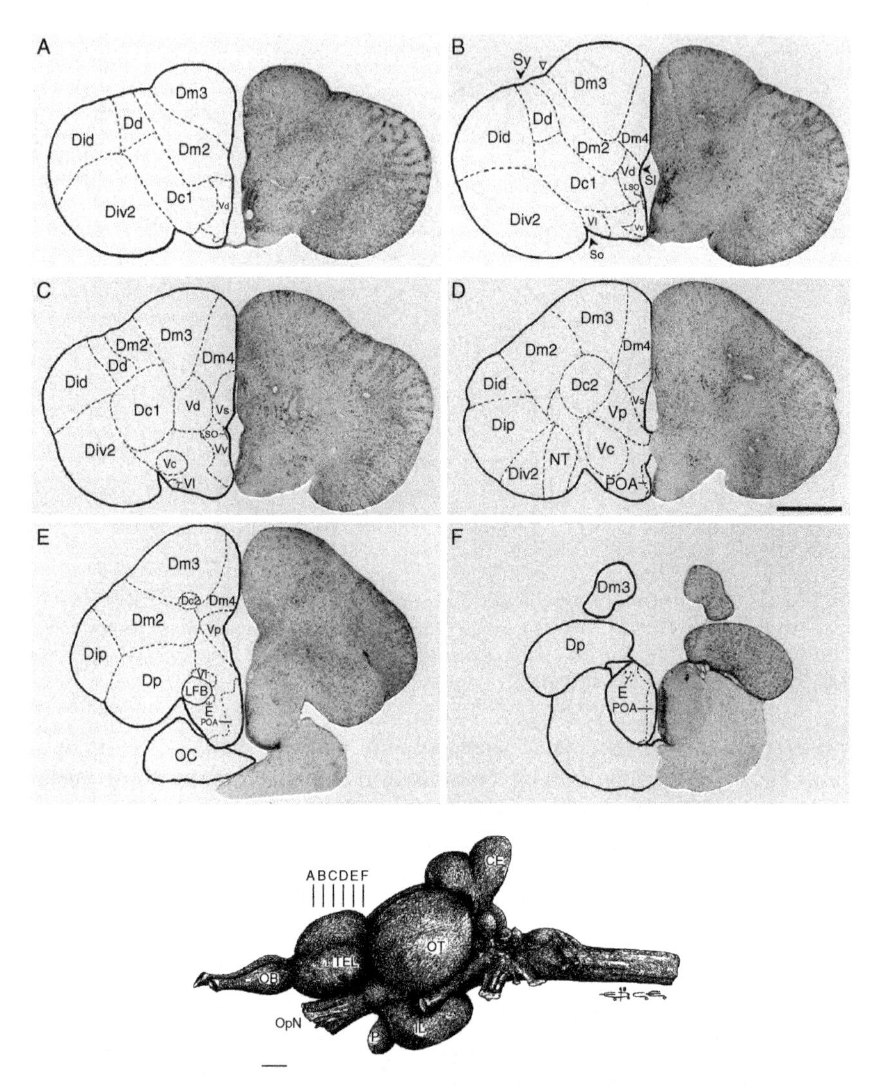

Fig. 1.6. Photomicrographs of cross-sections of sea bass (*Dicentrarchus labrax*) brain stained with cresyl violet showing different tencephalic cell masses. Dc1, dorsal telencephalic area, central part, subdivision 1; Dc2, dorsal telencephalic area, central part, subdivision 2; Dd, dorsal telencephalic area, dorsal part,; Dld, dorsal telencephalic area, lateral part,; Dlp, dorsal telencephalic area, lateral posterior part; Dlv1, dorsal telencephalic area, lateral ventral, subdivision 1; Dlv2, dorsal telencephalic area, lateral ventral part, subdivision 2; Dm2, dorsal telencephalic area, medial part, subdivision 2; Dm3, dorsal telencephalic area, medial part, subdivision 3; Dm4, dorsal telencephalic area, medial part, subdivision 4; Dp, dorsal telencephalic area, posterior part; E, entopeduncular nucleus; LSO, lateral septal organ; NT, nucleus taenia; OB, olfactory bulb; OC, optic chiasma; OpN, optic nerve; POA, preoptic area; Sy, sulcus ypsiliformis; Vc, ventral telencephalic

2.2. Preoptic Area

The preoptic area forms a structural and functional continuum with the hypothalamus. It is convenient to treat these two entities as forming a single complex and to allocate this preoptic–hypothalamic continuous to the diencephalon (Nieuwenhuys *et al.*, 1998). Endocrinologists often include the preoptic area within the hypothalamus when discussing the hypophysiotropic control of pituitary activity.

The preoptic region surrounds the preoptic periventricular recess and is located between the anterior commissure and the optic chiasm. Following Braford and Northcutt (1983) and Meek and Nieuwenhuys (1998), the preoptic area can be divided into a magnocellular (PM) and parvocellular preoptic (PP) nucleus (Figure 1.7). The magnocellular preoptic nucleus contains large neurosecretory cells and it is subdivided into the parvocellular (PMpc), magnocellular (PMmc) and gigantocellular (PMgc) parts. This nucleus is shaped, in longitudinal plane, like an inverted L with the vertical rod located rostrally and the horizontal rod extending dorsocaudally. The smaller rostroventral cells constitute the parvocellular parts of the PM and the larger dorsocaudal cells constitute the magno- and gigantocellular parts. In turn, the parvocellular preoptic nucleus can be further subdivided into anterior (PPa) and posterior pars (PPp), which are located anterior and caudoventral to the magnocellular preoptic nucleus. According to Peter's nomenclature for goldfish (*Carassius auratus*), killifish (*Fundulus heteroclitus*) and two salmonid species (Peter *et al.*, 1975; Peter and Gill, 1975; Billard and Peter, 1982; Peter *et al.*, 1991), the PPa is called the preoptic parvocellular nucleus (NPP), the magnocellular division is termed the preoptic nucleus (NPO) and the PPp is termed the periventricular nucleus and is subdivided into the anterior (NAPv) and posterior (NPPv) periventricular nucleus. However, the caudal region of the NPPv corresponds to the zona limitans diencephali (ZL) of Braford and Northcutt (1983). According to Cerdá-Reverter's nomenclature for the sea bass, the NPP (Peter and Gill, 1975) or PPa (Braford and Northcutt, 1983) is termed the parvocellular preoptic nucleus and it is subdivided into the anteroventral (NPOav) and parvocellular pars (NPOpc). The NPOav is termed the supraoptic division of the preoptic nucleus (NPOs) in the catfish (*Clarias batrachus*; Prasada Rao *et al.*, 1993). According to Maler's nomenclature for brown ghost knifefish (*Apteronotus leptorhynchus*; Maler *et al.*, 1991) the PM (Braford and

area, central part; Vd, ventral telencephalic area, dorsal part; Vi, ventral telencephalic area, intermedial part; Vl, ventral telencephalic area, lateral part; Vp, ventral telencephalic area, postcommissural part; Vs, ventral telencephalic area, supracommissuralis part; Vv, ventral telencephalic area, ventral part. Scale bar = 100 μm. From Cerdá-Reverter *et al.* (2001b).

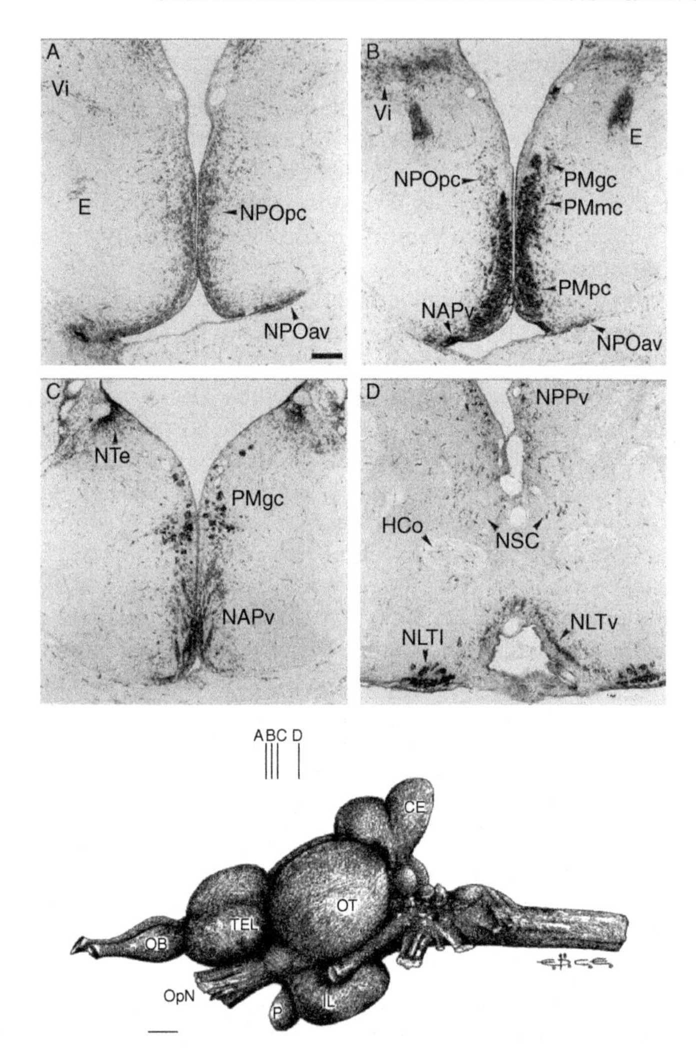

Fig. 1.7. Photomicrographs of cross-sections of sea bass (*Dicentrarchus labrax*) brain stained with cresyl violet showing different preoptic cell masses. E, entopeduncular nucleus; HCo, horizontal commissure; NAPv, anterior periventricular nucleus; NLTl, lateral part of the lateral tuberal nucleus; NLTv, ventral part of the lateral tuberal nucleus; NPOav, anteroventral part of the parvocellular preoptic nucleus; NPOpc, parvocellular part of the parvocellular preoptic nucleus; NPPv, posterior periventricular nucleus; NSC, suprachiasmatic nucleus; NTe, nucleus of the thalamic eminentia; PMgc, gigantocellular part of the magnocellular preoptic nucleus; PMmc, magnocellular part of the magnocellular preoptic nucleus; PMpc, parvocellular part of the magnocellular preoptic nucleus; Vi, intermediate nucleus of the ventral telencephalon. Scale bar = 100 μm. From Cerdá-Reverter *et al.* (2001b).

Northcutt, 1983) has been referred to as anterior hypothalamic nucleus (Ha) and the PPp (Braford and Northcutt, 1983) is subdivided into the proper PPp and the anterior periventricular nucleus (nAPv). Finally, all the nomenclatures include the suprachiasmatic nucleus (NSC) in the most caudoventral region of the preoptic area although in some species it is not recognized as a separate nucleus (Gómez-Segade and Anadón, 1988).

2.3. Hypothalamus

As the name indicates, the hypothalamus is located below the thalamus, caudal to the preoptic area. It appears abruptly caudal to the chiasmatic ridge and is easily distinguished from the preoptic area by its densely packed and stained cells surrounding the nascent infundibular recess (Figure 1.8). The hypothalamus is the largest diencephalic area and is connected to the pituitary via the pituitary stalk, which contains hypothalamic and preoptic neuroendocrine fibers (see Section 1). In several teleost species, the infundibular recess or hypothalamic portion of the third ventricle is laterally expanded to form caudally directed paired diverticula termed the lateral recess. In the caudal area, the third ventricle gives rise to the posterior recess, which also has small dorsolaterally directed diverticula (Figure 1.8; Braford and Northcutt, 1983). Cytoarchitectonic studies of the teleostean hypothalamus point to considerable variation (Peter *et al.*, 1975; Braford and Northcutt, 1983; Gómez-Segade and Anadón, 1988; Striedter, 1990; Maler *et al.*, 1991; Wullimann *et al.*, 1996; Cerdá-Reverter *et al.*, 2001b), however it is uncertain whether the same subdivisions in different teleost species are comparable or whether subdivisions merely reflect different criteria for subdivision (Meek and Nieuwenhuys, 1998).

According to Meek and Nieuwenhuys (1998), the teleostean hypothalamus can be divided into three main regions, i.e., the periventricular region, the medially located tuberal region and the paired inferior lobes, which are separated from the tuberal hypothalamus by a deep ventral sulcus. However, other authors subdivide the hypothalamus into dorsal (Hd), ventral (Hv) and caudal (Hc) regions (Braford and Northcutt, 1983; Striedter, 1990; Wullimann *et al.*, 1996). Ventral and caudal zones constitute most of the median tuberal portion of the hypothalamus. All three subdivisions display periventricular cell populations bordered by laterally migrated nuclei. According to Braford and Northcutt (1983), the rostral area of the hypothalamus is occupied by two periventricular cell populations, the dorsal (Hd) and ventral (Hv) periventricular hypothalamus, and two laterally migrated populations, the anterior tuberal nucleus (NAT) and the lateral hypothalamic nucleus (LH). The border between the two periventricular subdivisions is very diffuse. The ventral aspect of the third ventricle is bound by cells of the

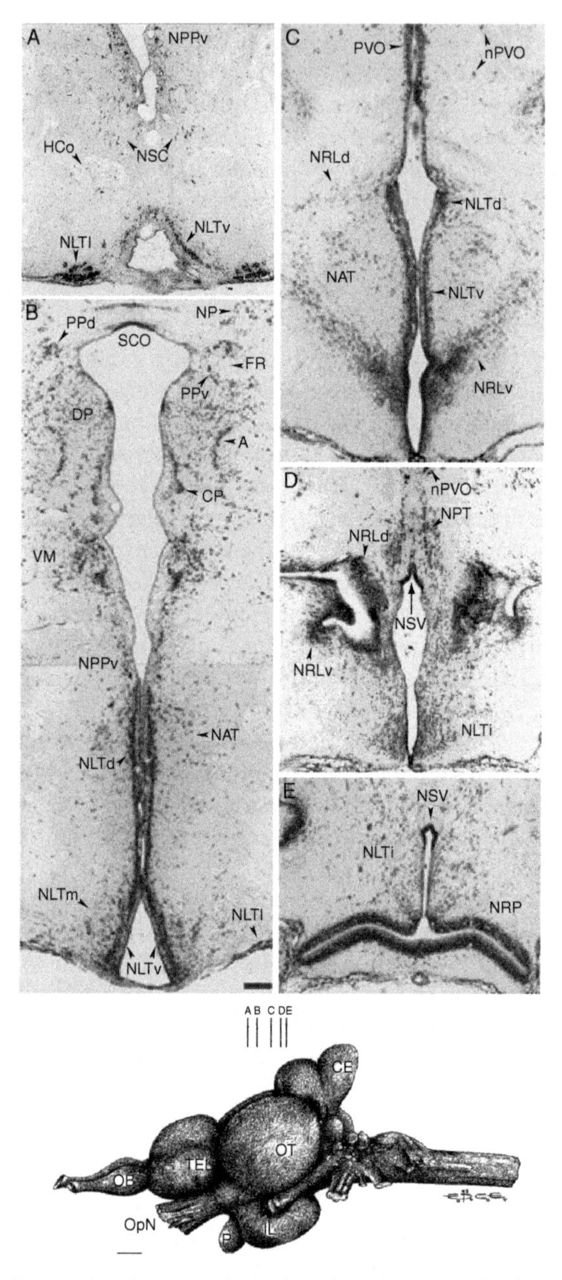

Fig. 1.8. Photomicrographs of cross-sections of sea bass (*Dicentrarchus labrax*) brain stained with cresyl violet showing different diencephalic cell masses. A, Anterior thalamic nucleus; CP, central posterior thalamic nucleus; DP, dorsal posterior thalamic nucleus; FR, fasciculus

Hv, which are densely packed along the ventricular wall, forming a 1–10 cell-thick lamina. Cells of the Hd are slightly bigger, less densely packed and form a 1–4 cell-thick layer (Cerdá-Reverter *et al.*, 2001b). The anterior tuberal nucleus (NAT) is formed by scattered and large fusiform cells located lateral to the Hd. All nomenclatures are coincident with the designation of a NAT but its location is variable (Peter and Gill, 1975; Braford and Northcutt, 1983; Gómez-Segade and Anadón, 1988; Striedter, 1990; Maler *et al.*, 1991; Wullimann *et al.*, 1996). Some authors consider that the NAT extends to the periventricular wall (Peter and Gill, 1975; Striedter, 1990; Maler *et al.*, 1991), while others further subdivide the nucleus into rostral and ventral regions (Striedter, 1990). In the zebrafish (*Danio rerio*) brain atlas, Wullimann *et al.* (1996) considered that the Hv extends farther rostrally than the Hd and therefore the rostral part of the Hd is designated Hv. Consequently, the NAT lies laterally to the Hv.

The lateral aperture of the third ventricle delimits the rostrocaudal level where the Hd emerges. Just before the lateral opening of the hypothalamic ventricle, the small cells of the LH extend dorsolaterally. This nucleus is located lateral to the caudal region of the Hv and its cells seem to coat the ventral aspect of the lateral recess (LR). In fact, this lateral aperture of the third ventricle could be considered as a transitional zone leading to the Hc. At this level, the cells of the Hd coat the LR, whereas the dorsal area of the third ventricle is massively occupied by the cells of the posterior tuberal nucleus (NPT). Slightly caudal and approximately coinciding with the complete separation of the LR from the medial hypothalamic ventricle, the Hc emerges.This is a small area bordering the caudal aspect of the posterior recess dorsally limited by the NPT. The Hc arises as a thickening of the ependymal cell layer and more caudally the nucleus expands laterally to adopt a columnar disposition along the laterally directed diverticula of the posterior recess. The Hc contains a small cell population termed the nucleus of the saccus vasculosus (NSV) as well as the posterior recess nucleus (NRP) of Peter and Gill (1975).

---

retroflexus; NAT, anterior tuberal nucleus; NLTd, dorsal part of the lateral tuberal nucleus; NLTi, inferior part of the lateral tuberal nucleus; NLTm, medial part of the lateral tuberal nucleus; NLTl, lateral part of the lateral tuberal nucleus; NLTv, ventral part of the lateral tuberal nucleus; NP, paracommissural nucleus; NPPv, posterior periventricular nucleus; NPT, posterior tuberal nucleus; nPVO, nucleus of the paraventricular organ; NRLd, dorsal part of the nucleus of the lateral recess; NRLv, ventral part of the nucleus of the lateral recess; NRP, nucleus of the posterior recess; NSC, suprachiasmatic nucleus; NSV, nucleus of the saccus vasculosus; PPd, dorsal periventricular pretectal nucleus; PPv, ventral periventricular pretectal nucleus; PVO, paraventricular organ; SCO, subcommissural organ; VM, ventromedial thalamic nucleus. Scale bar = 100 μm. From Cerdá-Reverter *et al.* (2001b).

Following the nomenclature of Cerdá-Reverter *et al.* (2001b), which is partially based on Peter and Gill (1975), the rostral periventricular region is collectively termed the lateral tuberal nucleus (NLT). In the transversal plane, this nucleus presents two differentiable subdivisions, i.e., dorsal (NLTd) and ventral (NLTv), which correspond to the rostral region of the Hd and Hv of Braford and Northcutt (1983), respectively. Within this rostral hypothalamic area, two additional laterally migrated populations, the lateral (NLTl) and medial (NLTm) part of the NLT, have been described in the sea bass (Cerdá-Reverter *et al.*, 2001b). Cells of the NLTl are the first cells of the hypothalamus to appear in a cross-section. In the sea bass, the NLTl consists of very large and darkly stained cells lying ventrolateral to the horizontal commissure and lateral to the incipient infundibular recess (Cerdá-Reverter *et al.*, 2001b). Caudally, these neurons are smaller and located on the ventrolateral surface of the brain. The large neurons of the rostral portion were also recognized by Braford and Northcutt (1983) and designated ventral tuberal nucleus (TV), whereas the smaller caudal neurons were named LH. Neurons of the NLTl are chemically differentiable because they synthesize melanin-concentrating hormone (MCH), a neurohormone of the teleost brain (Baker and Bird, 2002). The NLTm exhibits scattered cells located between the ventral and lateral parts of the NLT. This subdivision was previously reported in the African cichlid fish (*Haplochromis burtoni*; Fernald and Shelton, 1985). The NLTm of the sea bass corresponds partially to the anterior and posterior parts of the NLT of Peter and Gill (1975) and to the ventral hypothalamus of Braford and Northcutt (1983).

Two cell masses that will later surround the LR, i.e., the dorsal and ventral parts of the lateral recess nucleus (NRLd and NRLv, respectively), are visible immediately before the lateral opening of the ventricle. The NRLv is located ventral to the NLTv and both nuclei coexist as far as the ventral periventricular zone is caudally occupied by both the nucleus of the posterior recess and the laterally placed inferior part of the lateral tuberal nucleus (NLTi). The NRLv corresponds to the LH of Braford and Northcutt (1983), Striedter (1990) and Wullimann *et al.* (1996) and also contains the ventricular cells of the caudal Hv. Cells of the NRLd and the paraventricular organ (PVO) presumably migrate laterally, coating the dorsal aspect of the nascent lateral recess whereas the ventral aspect is coated by cells of the NRLv. According to Braford and Northcutt (1983), the NRLd matches the caudal region of Hd.

The inferior lobes are by far the largest part of the teleost hypothalamus and are clearly visible on the ventral surface of the brain. Cells, presumably migrated from the dorsal periventricular populations, "fill up" the inferior lobes and thus these could be considered as a part of the Hd which expands laterally (Wullimann *et al.*, 1996). According to Braford and Northcutt (1983) and Wullimann *et al.* (1996), three populations can clearly be distinguished: the Hd, the diffuse nucleus of the inferior lobe (NDLI) and the central nucleus of the inferior lobe (NCLI). The Hd coats the LR which occupies a central position within the inferior lobe. The NDLI is comprised of very small, lightly stained scattered cells, whereas the NCLI arises more caudally in a ventromedial position to the lateral torus which is made up of large fusiform and ovoid cells. In the sea bass, the NDLI is further subdivided into medial (NDLIm) and lateral (NDLIl) parts (Cerdá-Reverter *et al.*, 2001b), whereas the population coating the LR is called lateral part of the lateral recess nucleus (NRLl). In addition, based on cytoarchitectonic criteria, we described the medial nucleus of the inferior lobe (NMLI), located in the vicinity of the NRLl, ventromedial to the nucleus glomerulosus and lateral to the corpus mamillare. This nucleus was later confirmed by study of neuronal circuitry in other teleosts (Ahrens and Wullimann, 2002).

3. HYPOPHYSIOTROPIC TERRITORIES IN THE FISH BRAIN: TRACT-TRACING STUDIES

The lack of median eminence and portal system in teleost fish has facilitated the study of hypophysiotropic neurons in this vertebrate group using tract-tracing techniques. These studies are based on the retrograde transport of marker substances, mainly lectins or dextran amines. Carbocyanine fluorescent dyes such as DiI (1-1′-dioctadecyl-3,3,3′,3′-tetramethylindocarbocyanin) offer several advantages over other techniques such as autoradiography, horseradish peroxidase (HRP) or cobalt tracing. These advantages include the possibility of using the carbocyanine tracers in paraformaldehyde-fixed tissue, and their compatibility with other fluorescent tracers as well as with immunohistochemical techniques (Holmqvist *et al.*, 1992). The technique is based on the implantation of small crystals of dye into the structure by using insect pins or glass electrodes and subsequent sectioning after incubation. Using tracing techniques, the hypophysiotropic neurons of the catfish (*Clarias batrachus*; Rama Krishna and Subhedar, 1989; Prasada Rao *et al.*, 1993), goldfish (Fryer and Maler, 1981; Anglade *et al.*, 1993), brown ghost knifefish (Johnston and Maler, 1992; Zupanc *et al.*, 1999; Corrêa, 2004) and Atlantic salmon (*Salmo salar*; Holmqvist *et al.*, 1992; Holmqvist and Ekström, 1995) have been studied. As mentioned above (see Section 1), tract-tracing studies have revealed two main areas where the neuronal bodies project toward the pituitary: the preoptic area and the tuberal hypothalamus. However, the alternative nomenclatures used to describe the different nuclei of fish brain sometimes make it difficult to compare species. Table 1.1 summarizes the distribution of the hypophysiotropic structures revealed by tract-tracing techniques in teleost fish.

Table 1.1
Main hypophysiotropic territories of the teleost brain

	Apteronotus leptorhynchus	*Salmo salar*	*Carassius auratus*	*Clarias batrachus*	Bradford and Northcutt (1983)
Technique	DiI Dextran Biocytin	DiI	HRP DiI	HRP Cobaltous-lysine	
Telencephalon					
Olfactory bulb (OB)			X		OB
Ventral part of the ventral telencephalon (Vv)	X	X	X		Vv
Supracommissural part of the ventral telencephalon (Vs)			X		Vs
Central part of the ventral telencephalon (Vc)		X	X		Vc
Preoptic area					
Periventricular preoptic nucleus, anterior part (PPa)	X	X			PPa
Periventricular preoptic nucleus, posterior part (PPp)	X	X			PPp
Suprachiasmatic nucleus (NSC)	X	X	X		NSC
Periventricular preoptic nucleus (NPP)			X		PPa
Preoptic nucleus, supraoptic division (NPOs or SO)				X	PPa
Preoptic nucleus paraventricular division (NPOp or PV)				X	PM
Preoptic nucleus parvocellular part (NPOpc)		X			PMpc
			X		PPa
Preoptic nucleus magnocellular part (NPOmc)		X			PMmc
			X		PMmc-PMgc
Preoptic nucleus gigantocellular part (NPOgc)		X			PMgc
Anterior periventricular nucleus (NAPv)		X	X		PPp
Posterior periventricular nucleus (NPPv)			X		PPp
Anterior hypothalamus (Ha)	X				PM
Tuberal hypothalamus					
Dorsal hypothalamus (Hd)	X				Hd
Ventral hypothalamus (Hv)	X				Hv
Caudal hypothalamus (Hc)	X				Hc

Lateral hypothalamus (Hl)	X				LH, partially
Anterior tuberal nucleus (NAT)	X		X		NAT
Lateral tuberal nucleus (NLT)		X		X	Hd/Hv/LH
Lateral tuberal nucleus, anterior part (NLTa)	X				Hv and Hc
Lateral tuberal nucleus, posterior part (NLTp)	X				Hc
Lateral tuberal nucleus, lateral part (NLTl)			X		LH
Lateral tuberal nucleus, anterior part (NLTa)			X		Hv
Lateral tuberal nucleus, posterior part (NLTp)			X		Hd/Hv
Lateral tuberal nucleus, inferior part (NLTi)			X		Hv
Lateral recess nucleus (NRL)		X	X	X	Hd
Posterior recess nucleus (NRP)		X	X	X	Hc
Nucleus of the saccus vasculosus (NSV)		X			PTN
Central nucleus of the inferior lobe (NCLI)	X				NCLI
Posterior tubercle					
Paraventricular organ (PVO)		X		X	PVO
Thalamus					
Central posterior nucleus (CP)	X				CP
Dorsolateral thalamic nucleus (DL)			X		Vm/DP/I/A
Reticular formation	X				R

In the telencephalon several studies have demonstrated the presence of retrograde labeled cells and fibers have been consistently found within the ventral (Vv) and central areas (Vc) of several teleost species. In addition, some hypophysiotropic cells were also placed within the olfactory bulbs and supracommissural part of the ventral telencephalon (Vs) of the goldfish.

The preoptic area is the major field innervating the teleost pituitary. Studies in several species have demonstrated that all three anterior (PPa), posterior (PPp) periventricular and suprachiasmatic nuclei (NSC) project towards the pituitary (see Table 1.1). In addition, retrograde cell bodies were found in the all three subdivisions of the preoptic nucleus, i.e., NPOpc, NPOgc and NPOmc. Studies in Atlantic salmon also reported DiI-labeled cells in the most rostral preoptic region, which were similar in size and morphology to those of the NPOgc but clearly separated from the main cell masses of the NPO. They suggested naming this cell group the rostral preoptic area (NPOr).

The tuberal hypothalamus is the second major area innervating the teleost pituitary. Studies in several species have reported consistently the presence of labeled cells of the tuberal hypothalamus within the dorsal, ventral lateral and caudal hypothalamus. The presence of hypophysiotropic cells within the anterior tuberal nucleus (NAT) seems to be species dependent (see Table 1.1) and only one study has reported the presence of retrogradely labeled cells in the inferior lobe (Corrêa, 2004). In addition, studies in goldfish using DiI showed labeled cells within the ventral tuberal nucleus (TV).

4. CENTRAL NEUROHORMONES

4.1. Arginine-Vasotocin (AVT)

Arginine-vasotocin (AVT) peptide belongs to a family of neuro-nonapeptides that includes isotocin (IST) as well as the mammalian homologues, arginine-vasopressin (AVP) and oxytocin (Acher, 1996). AVT has a key role in the endocrine control of the salt/water homeostasis and vascular function (McCormick and Bradshaw, 2006) as well as in a wide range of physiological processes (Balment *et al.*, 2006) and social behavior (Goodson and Bass, 2001). AVT is one of the main neurohypophysial hormones in lower vertebrates; it is synthesized by preoptic neurons, travels through the preoptico-hypophysial tract and is released into the vascular system via neurohypophysial axon terminals. Studies in several teleost fish have consistently demonstrated that AVT-immunoreactive (ir) neurons are exclusively localized in the parvocellular and magnocellular neurons (see Table 1.2) of

Table 1.2
Main hypophysiotropic territories containing neuropeptides

Peptide	Telencephalon	Preoptic area	Hypothalamus
AgRP			Hv
AVT	–	PPa, PPp PMmc [1]	–
CART	Ventral telencephalon	PMmc	Hv, Hc, NCLI
CCK	Ventral telencephalon	PPa, PMpc PMmc	Hv, Hc, Hd, PVO
CRF	OB, ventral telencephalon	PPa, PPp, PM, NSC	Hv, Hc
Galanin	OB, Vs	PPa, PM	Hv, LH, Hc
GHRH	–	PPa, PPp, PMpc, PMmc	Hv, LH, DF(NDLI)
GnRH	OB, Vv	PPa	LH
GRP	Ventral TEL	PPa, PM, PPp	Hv, NAT, Hd, PVO
IST	–	PPa, PMmc	–
MCH	–	–	LH, PVO, Hv
NPY	OB, Vd, Vv, Vs, Vl, Vc	PPa, PPp PMgc,	Hv, LH, NAT, Hc
Orexin	–	PMmc, PMgc, PPp, NSC	PVO, Hc
PACAP	–	PPa, PPp, PMgc	Hv,
RF-amide	OB	-	Hv (PrRP)
SS	OB, Vl, NE	PPa, PMpc, PMmc, PPp, NSC	Hv, Hd, Hc PVO
TRH	OB, Ventral TEL,	PPa, PMmc, PPp, NSC	Hv, NAT, NPT, Hc
UI	Vv, Vd, Vp,	PPa, PPp	Hv, Hc
α–MSH	–	Hv, LH	

Nomenclature of nuclei was adapted following Bradford and Northcutt (1983).
See Table 1.1 for abbreviations of hypophysiotropic territories.
AgRP, agouti-related peptide; AVT, arginine-vasotocin; CART, cocaine- and amphetamine-regulated transcript; CCK, cholecystokinin; CRF, corticotropin-releasing factor; GHRH, growth hormone releasing hormone; GnRH, gonadotropin-releasing hormone; GRP, gastrin-releasing peptide; IST, isotocin; MCH, melanin-concentrating hormone; MSH, melanin-stimulating hormone; NPY, neuropeptide Y; PACAP, pituitary adenylate cyclase activating polypeptide; SS, somatostatin; TRH, thyrotropin-releasing hormone; UI, urotensin I.

the preoptic nucleus (Goossens *et al.*, 1977; Batten *et al.*, 1990a; Duarte *et al.*, 2001; Lema and Nevitt, 2004; Saito *et al.*, 2004; Mukuda *et al.*, 2005; Bond *et al.*, 2007; Maruska *et al.*, 2007). However, immunostaining studies in plainfin midshipman showed an additional AVT population in the ventral hypothalamus. This second neuronal group is composed of a few small and round AVT-ir neurons often embedded in the preoptico-hypophysial tract (Goodson and Bass, 2001). In this species, most of the AVT neurons are placed in the PPa and only some weakly labeled cells are localized in the PPp. Within the PM, most AVT cells are placed in the magnocellular subdivision and these neurons are more intensely labeled than those in the PPa and PPp. Such a distribution of AVT-ir cells in the preoptic area is conserved in most teleost species studied although some divergences have been reported. Studies in valley pupfish (*Cyprinodon nevadensis*) reported a similar pattern but no AVT-ir cells were found in the PPp whereas peptide-ir in the PM was found in all three subdivisions, i.e., PMpc, PMmc and PMgc (Lema and Nevitt, 2004). Similar results were reported in rainbow trout (*Oncorhynchus mykiss*) where AVT neurons occupy almost 50% of neurosecretory cells in the gigantocellular part of the PM but only 15% in the PPa. Neuronal tract tracing combined with AVT immunohistochemical studies in Atlantic salmon showed that most DiI-labelled neurons in the PM are AVT-ir. However, not all AVT-ir neurons are hypophysiotropic since some cell bodies, especially in the PMgc, were not DiI labeled. The AVT neurons in the PPa did not show retrograde labeling and therefore were suggested to be AVT-ir non-hypophysiotropic preoptic neurons. This study reported hypothalamic and extrahypothalamic projections of the preoptic neurosecretory AVT-ir cells (Holmqvist and Ekström, 1995) as also reported in the plainfin midshipman (Goodson and Bass, 2001). Detailed work in rainbow trout using single-cell staining techniques supported these studies and further showed that in all studied cases, AVT-ir neurons simultaneously project toward the pituitary and the extrahypothalamic regions, including the ventral telencephalon, thalamus and mesencephalon. Multiple projections of single neurosecretory cells are not known in other vertebrates. These projections are suitable for the coordinated control of peripheral and central outputs through the synchronization of the electrical activity under different physiological challenges (Saito *et al.*, 2004). In the PM of goldfish, rainbow trout and eel (*Anguilla* sp.), AVT neurons frequently cluster together by means of soma–somatic apposition near the ventricular wall (Cumming *et al.*, 1982; Saito *et al.*, 2004). The function of this anatomical apposition is unknown but it could be involved in communication by changing local field potentials and/or the somatodendritic release of peptides (Saito *et al.*, 2004). A significant feature of AVT neurons in the PM is the frequent contacts between their proximal processes, originated in the same or different clusters, which may also be involved in the communication among AVT neurons (Saito *et al.*, 2004). Ultrastructural studies in the goldfish demonstrated axo-dendritic synapses with AVT-ir axon terminals in the PM. It suggests the existence of synaptic contacts between AVT cells and the presence of a local neuronal circuitry in the PM (Cumming *et al.*, 1982), probably acting in the synchronization of the neurosecretory cells to fit the secretory activity to physiological requirements (Saito and Urano, 2001; Saito *et al.*, 2004). The AVT neurons in the PM have been further demonstrated to contact directly with the systemic circulation in the eel (Mukuda *et al.*, 2005), receiving chemical information from the periphery. In fact, AVT neurons in all three subdivisions of the PM express glucocorticoid receptor in the flounder

(*Platichthys flesus*) and are modulated by confinement stress (Bond *et al.*, 2007). AVT neurons have also been shown to colocalize with CRF and galanin in the PM and PPa, respectively, in several species (see later). A second notable characteristic of the AVT neuronal system is sexual dimorphism observed in several species (Grober and Sunobe, 1996; Maruska *et al.*, 2007), as well as the differences observed between males exhibiting different reproductive tactics (Foran and Bass, 1998; Goodson and Bass, 2001). In addition, sex reversal in the marine goby (*Trimma okinawae*) is correlated with a variation in the AVT-ir cell size, but no differences in the number of cells were observed (Grober and Sunobe, 1996).

As stated above, preoptic AVT neurons mainly project to the neurohypophysis. However, in the deep interdigitations of the neurohypophysis into the RPD a close association between AVT terminal and corticotropic cells was observed in the sea bass pituitary (see Table 1.3). Within the proximal pars distalis, AVT fibers were intermingled with TSH and GH cells and occasionally with gonadotropic cells. In the PI, numerous AVT fibers were in contact with both MSH and somatolactin cells (Moon *et al.*, 1989). Ultrastructural studies in green molly (*Poecilia latipina)* demonstrated that AVT fibers were mainly restricted to the pericapillary areas of the neurohypophysis but some immunoreactive profiles abutted the basement lamina and occasionally discontinuities in the lamina allowed direct contact between AVT fibers and ACTH cells and rarely between AVT fibers and prolactin cells. In the central pituitary, discontinuities of the basal lamina allowed intermingling of AVT fibers and GH and TSH cells and rarely with gonadotropic cells. Similar to the sea bass, the AVT fibers in the PI contact MSH and somatolactin cells (Batten, 1986).

4.2. Isotocin (IST)

Isotocin (IST) is the peptide homologue of mammalian oxytocin in bony fish (Acher, 1996). As in the AVT neuronal system, the distribution of IST-ir in the teleost brain seems to be quite similar in several species examined, i.e., goldfish, European plaice (*Pleuronectes platessa*; Goossens *et al.*, 1977), green molly (Batten *et al.*, 1990a), plainfin midshipman, gulf toadfish (*Opsanus beta;* Goodson *et al.*, 2003), rainbow trout (Saito *et al.*, 2004) and white sea bream (*Diplodus sargus*; Duarte *et al.*, 2001). In all fish species, IST is produced in the PPa and PM (Table 1.2). However, studies in plainfin midshipman (Goodson *et al.*, 2003) also reported a few IST-ir perikarya in the PPp. IST projections are widely spread through the teleost brain suggesting that IST exerts very broad modulatory action on physiological,

Table 1.3
Peptidergic innervation of fish pituitary and phenotype of the target cells

Peptide	Target cell type	Observations	References
AVT	MSH, ACTH, SL, GH, TSH	Occasionally GTH and PRL. Most AVT-ir terminals surround capillaries	Batten (1986); Moons *et al.* (1989); Batten *et al.* (1999).
CART	GH, TSH	Only some fibers in the PI	Singru *et al.* (2007)
CCK	GH, non-specified RPD cells	CCK innervation was rarely observed in goldfish	Moons *et al.* (1989); Himick *et al.* (1993); Batten *et al.* (1999)
CRF	MSH, TSH, ACTH	No contacts with GH and GTH cells	Moons *et al.* (1989); Batten *et al.* (1999); Matz and Hofeldt (1999); Duarte *et al.* (2001); Pepels *et al.* (2002)
Galanin	SL, GH, TSH, GTH, ACTH, PRL	Innervation of SL was shown in platyfish and trout, but not observed in four-eyed fish	Moons *et al.* (1989); Magliulo-Cepriano *et al.* (1993); Anglade *et al.* (1994); Power *et al.* (1996); Jadhao and Pinelli (2001)
GHRH	MSH, LS, GH, less extensive GTH, ACTH (?)	Innervation to PI was shown in goldfish, sea bass, sculpin, and pejerrey, but was not observed in cod, eel, and trout. Innervation to RPD was shown in the sea bass, grey mullet, sculpin, but not in eel, carp, goldfish, cod and salmonids	Marivoet *et al.* (1988); Moons *et al.* (1988, 1989); Pan *et al.* (1985)>; Olivereau *et al.* (1990); Miranda *et al.* (2002)
GnRH	SL, GTH, GH, PRL (?)	Innervation to RPD was shown in pejerrey and Nile perch but not in other teleost	Parhar and Iwata (1994); Parhar *et al.* (1995); Parhar (1997); Batten *et al.* (1999); Vissio *et al.* (1999); Mousa and Mousa (2003); Pandolfi *et al.* (2005)
GRP	GH (?), non-specified PI cells	Not direct innervation to PPD; it may diffuse from PI.	Himick and Peter (1995)

(*continued*)

Table 1.3 (*continued*)

Peptide	Target cell type	Observations	References
IST	MSH, ACTH, SL, GH, TSH,	Occasionally GTH and PRL, practically overlapping AVT innervation. Most IST-ir terminals surround capillaries	Batten (1986); Moons *et al.* (1989); Batten *et al.* (1999).
MCH	Blood capillaries, MSH, SL, ACTH	MCH-ir terminals surround capillaries	Batten and Baker (1988); Powell and Baker (1988); Batten *et al.* (1999); Amano *et al.* (2003); Pandolfi *et al.* (2003)
NPY	MSH, GH, GTH	Innervation species dependent, see Section 5.9 for details	Moons *et al.* (1989); Pontet *et al.* (1989); Zandbergen *et al.* (1994); Marchetti *et al.* (2000); Gaidwad *et al.* (2004)
PACAP	MSH, SL, β-endorphins, GH, PRL, ACTH		Montero *et al.* (1998); Wong *et al.* (1998); Matsuda *et al.* (2005a,b)
RF-amide	Non-specified PPD cells, PRL (PrRP)	Local production in RPD (potential paracrine effect). PrRP innervation	Rama Krishna *et al.* (1992); Magliuo-Cepriano *et al.* (1993); Amano *et al.* (2007)
SS	MSH, SL, GH, less extensive GTH, PRL, ACTH	Innervation of RPD was shown in tilapia, sea bass, catfish but not in killifish, mudsucker and sea bream	Olivereau *et al.* (1984a); Batten *et al.* (1985); Grau et al, (1985); Moons *et al.* (1989); Power *et al.* (1996); Batten *et al.* (1999)
TRH	MSH, GH, TSH (?)	TSH cells were not positively identified	Batten *et al.* (1999); Díaz *et al.* (2001, 2002)

ACTH, adrenocorticotropin-releasing hormone; AVT, arginine-vasotocin; CART, cocaine- and amphetamine-regulated transcript; CCK, cholecystokinin; CRF, corticotropin-releasing factor; GH, growth hormone; GHRH, growth hormone releasing hormone; GnRH, gonadotropin-releasing hormone; GRP, gastrin-releasing peptide; GTH, gonadotrophin; IST, isotocin; MCH, melanin-concentrating hormone; MSH, melanin-stimulating hormone; NPY, neuropeptide Y; PACAP, pituitary adenylate cyclase activating polypeptide; PI, pars intermedia; PPD, proximal pars distalis; PRL, prolactin; PrRP, prolactin-releasing peptide; RPD, rostral pars distalis; SL, somatolactin; SS, somatostatin; TRH, thyrotropin-releasing hormone; TSH, thyroid-stimulating hormone.

behavioral and sensorimotor processes (Goodson *et al.*, 2003). In the rainbow trout IST-ir neurons are placed more laterally in the PM, whereas AVT neurons occupy a medial position (Saito *et al.*, 2004). AVT neurons extended more caudally than IST neurons in the white sea bream (Duarte *et al.*, 2001) but neither of the AVT-/IST-ir perikarya showed a preferential position in goldfish, European plaice or green molly (Goossens *et al.*, 1977; Batten *et al.*, 1990a). Similar to AVT neurons, IST neurons cluster together and display contact between their processes in the PM regions of the rainbow trout. However, intracellular staining demonstrated the incidence of dye coupling among the IST neurons, but not AVT, of the same or different clusters in the PM of rainbow trout. This suggests the involvement of the electrical coupling in the communication among IST neurons. IST neurons can use electrical and/or chemical synapses for local neuronal circuitry. Taken together, the evidence suggests that AVT and IST neurons in the PM form cell-type specific neurons, from which neurohypophysial and extrahypothalamic projections arise (Saito *et al.*, 2004).

Ultrastructural studies in green molly demonstrated that IST fibers outnumber AVT fibers (2:1) in the posterior neurohypophysis but both types occurred in similar number in the rostral area (Batten, 1986). Few IST fibers contact the basement lamina of the RPD with similar relationship to ACTH and PRL cells as AVT fibers but most of them were centrally located surrounding the capillaries (Table 1.3). At the interface between neuro- and adenohypophysis IST fibers are placed around all cell types in the PPD and PI, particularly TSH, GH and MSH cells. A similar fiber distribution was also observed in the sea bass pituitary (Moons *et al.*, 1989).

4.3. Melanin-concentrating Hormone (MCH)

From an evolutionary perspective, MCH is probably one of the most fascinating peptides in the vertebrate CNS. It was originally purified from chum salmon (*Oncorhynchus keta*) pituitaries and characterized as a circulating-cyclic heptadecapeptide mediating color change in teleost fish (Kawauchi *et al.*, 1983), where it induces reversible centripetal aggregation of melanosomes through the dendritic processes of the melanophores. This central accumulation of the pigmentary organelles results in a change in the refractive index, which makes the scales appear paler, thus allowing cryptic camouflage (reviewed by Kawauchi and Baker, 2004). In teleost fish, MCH works as a neurohypophysial hormone that is released to the blood stream when fish move onto a pale-colored background. Most of the MCH axons entering the teleost pituitary terminate in the posterior area of the neural lobe. In addition, positive fibers are commonly distributed in the neurohypophysial digitations of the PI, where MSH and somatolactin cells are localized

(Table 1.3). Some fibers also penetrate the pars distalis, thus suggesting its involvement in the control of synthesis/release of adenohypophysial hormones (Batten and Baker, 1988; Powell and Baker, 1988; Gröneveld *et al.*, 1995; Baker and Bird, 2002; Amano *et al.*, 2003; Pandolfi *et al.*, 2003). Electron microscopy studies in eel (Powell and Baker, 1988) and molly (Batten and Baker, 1988) showed that MCH-like terminals are placed close to the thick basement membrane which separates the PI from the neural tissue. MCH terminals are located mainly opposite the α-MSH cells and less commonly terminate near the blood capillaries. On the contrary, contact between the MCH fibers and the basement membrane surrounding the capillaries is prominent within the rostral and central region of the neurohypophysis. Therefore, MCH terminals are well sited for the released hormone to diffuse along this membrane into the vascular system and also to diffuse into the PI (Powell and Baker, 1988). The ultrastructural studies in green molly further showed that some MCH fibers can come into direct contact with the endocrine cells of the PI, most often MSH cells, through the discontinuities in the basement membrane. Both studies also reported synaptoid contacts between the MCH fibers and pituicytes of the neurohypophysis but the functional significance of these contacts remains unknown. Suggestively, Gröneveld *et al.* (1995) reported the expression of MCH by *in situ* hybridization in the neurohypophysis of tilapia (*Oreochromis mossambicus*) and suggest that probably subpopulations of pituicytes could produce MCH mRNA.

The neuronal cell bodies producing MCH are commonly localized within the LH or NLTl of fish (Table 1.2). However, the number of MCH-producing cell populations, the cellular morphology and the position with respect to the ventricular surface is variable (Baker and Bird, 2002). In polypteriform and chondrostean species, the MCH neurons are placed mainly on the hypothalamic periventricular surface. A first group of neurons is associated to the PVO over the dorsal surface of the third ventricle, whereas a second group of neurons occurs around the lateral surface of the third ventricle. A migrated MCH-producing group placed in the LH was first seen in holostean species which share the periventricular populations observed in polypteriformes and chondrostean species. In teleosts, most MCH neurons migrate away from the periventricular areas and it is within euteleosts that hypothalamic hypophysial magnocellular neurons of the LH represent the most prominent MCH-producing area (Baker and Bird, 2002). These neurons in the lateral hypothalamus are the main neurons responsible for the pituitary innervation and are present in all the euteleost species studied. More caudally, coinciding with the lateral aperture of the hypothalamic region of the third ventricle, a second group of small perikarya is associated dorsally to the PVO. At the same level, some scattered cell bodies producing

MCH are also placed in the Hv, close to the ventral surface of the lateral aperture of the third ventricle. The function of these caudal hypothalamic MCH neurons is unknown, but a specific increase of MCH mRNA levels has been observed in repeatedly disturbed tilapia. MCH gene expression levels in these caudal hypothalamic cells were also modified by salt water challenge (Gröneveld *et al.*, 1995).

5. HYPOPHYSIOTROPIC PEPTIDES

5.1. Cholecystokinin (CCK)

Cholecystokinin (CCK) and gastrin constitute a family of peptides characterized by the common C-terminus of Trp-Met-Asp-Phe-NH2 (for review see Chandra and Liddle, 2007; Rehfeld *et al.*, 2007). The structure of the C-terminus octapeptide of CCK has been well conserved during evolution and is identical in mammals, chicken, turtle and frog, with only one amino acid substitution in fish (Johnsen, 1998; Peyon *et al.*, 1998). This octapeptide is the main CCK peptide produced in the nervous system, while longer peptides such as CCK 58, 33 and 22 are also found in peripheral tissues and circulation (Rehfeld *et al.*, 2007).

CCK/gastrin-ir perikarya and fibers are widely distributed in the forebrain, midbrain and hindbrain of goldfish (Himick and Peter, 1994). A highly concentrated and extensive CCK/gastrin-ir perikarya and fiber system is prominent in the posterior ventromedial and ventrolateral hypothalamus, and the inferior hypothalamic lobes of goldfish (Himick and Peter, 1994). Thus, CCK/gastrin-ir cell bodies were found scattered in the preoptic area, periventricular and lateral Hv and Hc and the dorsal and ventral thalamus (Table 1.2). A similar brain distribution of CCK-ir has been described in the green molly and rainbow trout (Notenboom *et al.*, 1981; Batten *et al.*, 1990a). In rainbow trout, CCK/gastrin-like cell bodies were located in the ventromedial hypothalamus in the caudal region of the Hv, from where the pituitary innervation originates. In the guppy, CCK-ir was found throughout the brain. Notably, CCK-ir cell bodies were detected in the ventral telencephalon, the preoptic area and tuberal and periventricular hypothalamus (Batten *et al.*, 1990a). Furthermore, studies in goldfish have revealed that CCK mRNA, similar to CCK-ir, is expressed widely in the brain of goldfish, the hypothalamus being the main brain area expressing CCK (Peyon *et al.*, 1999). *In situ* hybridization analysis has shown CCK expression in the ventroposterior hypothalamus (Peyon *et al.*, 1999). In rainbow trout, three different mRNA for CCK (CCK-N, CCK-L CCK-T) have been isolated, which differ from each other in one amino acid substitution (Jensen *et al.*, 2001). Each of these variants shows a specific expression pattern in the brain, as well as in the peripheral tissue. CCK-N and CCK-L were found in the preoptic area and ventroposterior hypothalamus, respectively (Jensen *et al.*, 2001).

CCK/gastin-ir fibers, originating in the ventral hypothalamus, enter the hypophysis as a large bundle towards the neurohypophysis (Table 1.3). This bundle branches into smaller bundles and even into single fibers at the level of the PPD. These immunoreactive fibers appear to terminate on the basal lamina separating the neuro- from the adenohypophysis (Notenboom *et al.*, 1981; Batten *et al.*, 1999). Studies have shown CCK-ir fibers in close relationship with GH cells (Moons *et al.*, 1989; Himick *et al.*, 1993; Batten *et al.*, 1999), strongly supporting a role for CCK in the neuroendocrine regulation of GH release (Canosa *et al.*, 2007). CCK-ir fibers were not observed to innervate the teleost PI (Batten *et al.*, 1999; Himick *et al.*, 1993), neither were they found in the RPD of the green molly, African catfish or sea bass (Batten *et al.*, 1999), but were only rarely seen in goldfish RPD (Himick *et al.*, 1993).

5.2. Cocaine- and Amphetamine-regulated Transcript (CART)

Cocaine- and amphetamine-regulated transcript (CART) peptide was originally isolated from ovine hypothalamus and suggested to be an SS-like peptide (Spiess *et al.*, 1981). Subsequently, CART mRNA expression was found to increase in the rat striatum after the administration of psychostimulant drugs such as cocaine and amphetamine (Douglas *et al.*, 1995). In rat, CART gene is alternatively spliced to produce a peptide of either 129 or 116 amino acids containing a 27 amino acid signal peptide thus resulting in a pro-peptide of either 102 (long form) or 89 (short form) residues (Douglas *et al.*, 1995). These pro-peptides are further processed to release the same main product, referred to as rat long CART [55–102] or rat short CART [42–89] depending on the length of the precursor (Dylag *et al.*, 2006).

In fish, CART mRNA has been characterized in goldfish (Volkoff and Peter, 2001) and cod (*Gadus morhua*), but DNA sequences are also available for zebrafish and pufferfish (*Takifugu rubripes*; Kehoe and Volkoff, 2006). Immunohistochemical localization of the CART peptide in fish brain has only been studied in the catfish by using antibodies against rat CART [55–102] (Singru *et al.*, 2007). The results demonstrated that CART immunoreactivity is widely distributed within the catfish brain, where it is localized within the main neuroendocrine territories (Table 1.2). CART cells were reported within the ventral telencephalon including Vv V, Vc and Vs nuclei. Intense immunoreactivity was detected in the PMgc. CART-ir terminals were seen in the neurohypophysis and several long fibers were detected throughout the PPD, where somatotropes and thyrotropes are concentrated

(Table 1.3). Only a few fibers were seen in the PI of the catfish pituitary. These hypophysial CART terminals could also originate in the tuberal hypothalamus, where few neurons were immunoreactive to CART peptide. The authors termed this area the arcuate nucleus, which corresponds to the Hv. Additionally, some authors have reported CART-ir neurons in the Hc or NRP and NCLI, both shown to be hypophysiotropic areas in catfish and brown ghost knifefish brain, respectively.

5.3. Corticotropin-releasing Factor (CRF) and Related Peptides

In fish, the corticotropin-releasing factor (CRF) system comprises four neuropeptides, i.e., CRF, urotensin I (UI) and the orthologues of the mammalian urocortin 2 and 3, two G-protein-coupled receptors (CRF_1 and CRF_2) and a binding protein (CRF-BP, Bernier, 2006). The CRF system is pivotal in the coordination of the stress response. This system regulates the pituitary–adrenal axis by modulating the activity of the adenohypophysial ACTH and MSH cells. However, it has also been involved in the regulation of additional physiological processes (Flik *et al.*, 2006; see Chapter 6, this volume).

The distribution of the CRF-expressing neurons has been studied in the white sucker (*Catostomus commersoni*; Okawara *et al.*, 1992) and recently in the zebrafish (Alderman and Bernier, 2007) by *in situ* hybridization. In the latter species, CRF is widely expressed in several hypophysiotropic areas of the brain including the ventral olfactory bulb, telencephalon, preoptic area and tuberal hypothalamus (Table 1.2). In addition, some expression levels were also found within the dorsal telencephalon, dorsal and ventral thalamus, optic tectum (OT) and tegmentum. Within the ventral telencephalon, CRF is expressed in theVv and Vc (Table 1.2) both projecting towards the pituitary gland in salmon and goldfish (Table 1.1). Immunostaining studies in tilapia (Pepels *et al.*, 2002) localized CRF-ir in the Vl and Vc whereas only the CRF population in the lateral region was reported in green molly (Batten *et al.*, 1990a). However, no immunoreactive cell bodies were reported in the telencephalon of some species (Olivereau *et al.*, 1984b; Yulis and Lederis, 1987; Olivereau and Olivereau, 1988; Mancera and Fernández-Llébrez, 1995; Matz and Hofeldt, 1999; Zupanc *et al.*, 1999; Duarte *et al.*, 2001).

The preoptic area is a major CRF-producing region. There are four populations of CRF-expressing neurons in the preoptic area of the zebrafish localized in the PPa, PPp, PM and NSC (Alderman and Bernier, 2007). Expression in the PPa and PM was previously reported in white sucker (Yulis and Lederis, 1987; Okawara *et al.*, 1992). In addition, the confinement-induced stress was shown to stimulate CRF synthesis in the preoptic area of the rainbow trout (Ando *et al.*, 1999). CRF-like-ir in the preoptic area has been demonstrated in almost all studies reporting the CRF distribution in the fish brain although the neuronal CRF circuits differ depending on the species. Only a few cells were seen in the PM of green molly (Batten *et al.*, 1990a). Both populations (PM and PPa) exhibited CRF-ir in salmonids, eel, brown ghost knifefish and cyprinid species (Olivereau *et al.*, 1984b; Olivereau and Olivereau, 1988; Zupanc *et al.*, 1999). Studies carried out in sparid species, white (Duarte *et al.*, 2001) and gilthead sea bream (*Sparus aurata*; Mancera and Fernández-Llébrez, 1995), found no CRF neurons in the entire rostral preoptic area and only some scattered perikarya in the PPp, just rostral to the lateral aperture of the tuberal recess (ZL). Some CRF neurons in the PM have been shown to contain AVT in the eel (Olivereau *et al.*, 1988), green molly (Batten *et al.*, 1990a), carp (Huising *et al.*, 2004) and tilapia (Pepels *et al.*, 2002), whereas all CRF neurons in the PM of white sucker displayed AVT-ir (Yulis and Lederis, 1988). This colocalization was also found in fibers forming the preoptico-hypophysial tract and in the neurohypophysial interdigitations projecting to the neurointermediate lobe. In fact, CRF cells in the PM seem to innervate the neurointermediate lobe whereas cells in the ventral hypothalamus project to the RPD of the adenohypophysis in the white sucker (Table 1.3). Peptide colocalization was also suggested in the preoptic area of rainbow trout (Ando *et al.*, 1999). Neuronal tract tracing combined with CRF immunohistochemical studies in brown ghost knifefish have demonstrated that approximately 6% of the retrogradely traced cells in the PPa and PM exhibited CRF-like-ir.

Within the tuberal hypothalamus, CRF expression has been localized in the Hv and Hc of the zebrafish. In addition, some CRF-mRNA expressing neurons were also detected in the PTN (Alderman and Bernier, 2007). However, no hybridization signal was reported in the tuberal hypothalamus of the white sucker (Okawara *et al.*, 1992) but peptide-ir was detected in the Hv (Yulis and Lederis, 1987). Similarly, CRF-ir in the Hv has been reported in sparid species (Mancera and Fernández-Llébrez, 1995; Duarte *et al.*, 2001), tilapia (Pepels *et al.*, 2002) and green molly (Batten *et al.*, 1990a). In the latter species, CRF neuronal bodies are located close to the pituitary stalk suggesting the hypophysial innervation. In several teleost fish including brown ghost knifefish, salmonid species and eel, CRF-ir in the tuberal hypothalamus was absent (Olivereau and Olivereau, 1988; Zupanc *et al.*, 1999) but was the only CRF system projecting to the pituitary in the gilthead and white sea bream (Mancera and Fernández-Llébrez, 1995; Duarte *et al.*, 2001).

The distribution of CRF fibers in the pituitary has also been studied in several species (Table 1.3). Early studies in the sea bass demonstrated that the great mass of CRF fibers run into the posterior neurohypophysis and the PI of the sea bass (Moons *et al.*, 1989). However, a portion of CRF-positive fibers reached RPD and PPD. A similar scenario has been shown in some teleost species (Yulis and Lederis, 1987; Matz and Hofeldt, 1999), but no CRF-ir was reported in the PPD of the eel, salmon (Olivereau and Olivereau, 1988), white sea bream (Duarte *et al.*, 2001), tilapia (Pepels *et al.*, 2002) and carp (Huising *et al.*, 2004). The RPD, where ACTH cells are placed, seems to be the main destination of the CRF fibers in some species (Olivereau *et al.*, 1984b) but in others this area receives no CRF-ir fibers (Mancera and Fernández-Llébrez, 1995). In some species, it has been suggested that CRF may reach ACTH cells via blood vessels or in a paracrine fashion. In the RPD of the sea bass, CFR fibers are placed in the vicinity of the ACTH cells, in line with the well-defined action of ACTH-releasing factor, whereas in the PPD they are in close apposition to the TSH cells but no contacts with somatotropes and gonadotropes are seen. Similar results were reported in Chinook salmon (*Oncorhynchus tshawytscha*) in which immunostaining of adjacent sections with CRF and TSH antibodies demonstrated partial overlapping (Matz and Hofeldt, 1999). *In vitro* studies previously demonstrated the thyrotropin-releasing activity of CRF and UI in the latter species (Larsen *et al.*, 1998). Double immunostaining methods showed CRF fibers running close to the ACTH cells whereas immunoreactive fibers in the neurohypophysis ended close to MSH cells in the salmonids, eel (Olivereau and Olivereau, 1988) and cyprinid (Huising *et al.*, 2004) adenohypophysis. A similar situation has also been reported in white sea bream, thus suggesting CRF control of the pituitary MSH system (Duarte *et al.*, 2001). In fact, CRF is known to stimulate MSH secretion in teleost fish (see Chapter 6, this volume).

Early studies on the fish brain demonstrated the presence of UI-like-ir in the diencephalon of white sucker (Yulis *et al.*, 1986). Immunoreactivity was located within the Vp, Hv and Hc. Additional studies in white sucker further reported UI-ir in the Vv and pretectal area (Yulis and Lederis, 1986). UI-like fibers in the pituitary were restricted to the PDP in close proximity to the adenohypophysial cells (Table 1.3). Similarly, studies in eel reported the presence of UI transcript in the dorsal and ventral subdivisions of the PVO (Kawauchi *et al.*, 2003). The latter areas probably correspond to the caudal region of the dorsal and ventral hypothalamus, respectively, according to Braford and Northcutt (1983). Recent expression studies have shown a wider expression of UI mRNA in the brain of zebrafish (Alderman and Bernier, 2007). UI transcripts were localized within the dorsal and ventral telencephalon, preoptic area and tuberal hypothalamus, mesencephalic tectal and tegmental areas. In the telencephalon, UI-mRNA-expressing cell bodies were located within the Vd as well as in the caudal region, probably the Vp. The expression in the preoptic region is restricted to the parvocellular region with stronger labeling in the PPa than in the PPp. Within the tuberal hypothalamus, some expression was detected in the Hv and Hc.

To our knowledge, the distribution of urocortin 2 and 3, and the participation of these peptides in the neuroendocrine system of fish, are unknown.

5.4. Galanin

Galanin is a 29-amino acid N-terminal peptide originally isolated from the porcine intestine. It is proteolytically processed from the prepropeptide along with a peptide known as galanin message-associated peptide. This neuropeptide binds to three different G-protein coupled receptors which exhibit substantial differences in their functional coupling and subsequent signaling activities. Galanin has been shown to have a wide distribution in the central and peripheral nervous system of many mammalian species and multiple biological effects, including feeding and metabolism, osmotic regulation and water intake, learning and memory consolidation, anxiety and related behaviors, arousal and sleep regulation, reproduction and nociception (reviewed by Lang *et al.*, 2007). Galaninergic fibers have been reported in the pituitary of several vertebrate groups including fish, suggesting that this peptide could be involved in the regulation of the secretion of this gland (Moons *et al.*, 1989; Batten *et al.*, 1990a; Olivereau and Olivereau, 1991; Magliulo-Cepriano *et al.*, 1993; Anglade *et al.*, 1994; Rodríguez-Gómez *et al.*, 2000a; Jadhao and Pinelli, 2001). Ultrastructural studies in fish have further demonstrated the presence of galanin-immunoreactive terminals in the central region of the neurohypophysis where some fibers follow the interdigitations of neural tissue into the proximal par distalis ending in the basal lamina opposite to PRL, ACTH, GH, TSH and GTH cells (Table 3) but did not show particular association with any one endocrine cell type of the PPD (Batten *et al.*, 1990c, 1999). Immunocytochemical studies in the sea bass suggested that galanin fibers directly innervate ACTH and PRL adenohypophysial cells. Similarly, studies in platyfish (*Xhiphophorus maculates*) demonstrated that galanin immunoreactive fibers colocalize with SL, PRL and GH cells. Accordingly, binding studies in the latter species showed galanin receptors to be confined within the area occupied by the prolactin cells in the rostral part of the adenohypophysis (Moons *et al.*, 1991).

Galanin-ir cell populations are mainly located within the preoptic area and tuberal hypothalamus (Table 1.2) but sexual dimorphism has been reported in several species (Prasada Rao *et al.*, 1996; Jadhao and Meyer, 2000; Rodríguez *et al.*, 2003). However, inter-sex dimorphism of the central galaninergic system was not found in "four eyed" fish (*Anableps anableps*; Jadhao and Pinelli, 2001). In addition, galanin-ir neuronal populations have also been reported within the olfactory bulb of the brown trout (*Salmo trutta fario*; Rodríguez *et al.*, 2003) and in the Vs in the goldfish (Prasada Rao *et al.*, 1996) and rainbow trout brain (Anglade *et al.*, 1994). In most teleost species, galanin-ir and/or gene expressing cells have been commonly localized within the PPa. The number of immunoreactive cells in the PPa was consistently greater in goldfish (Prasada Rao *et al.*, 1996) and brown trout males than in females (Rodríguez *et al.*, 2003), but no sex differences were reported in the PPa of the sockeye salmon (*Oncorhynchus nerka*; Jadhao and Meyer, 2000) or green molly (Cornbrooks and Parsons, 1991). A second population of galanin-ir neurons has been reported in the PM of several teleost species (Batten *et al.*, 1990c; Magliulo-Cepriano *et al.*, 1993; Prasada Rao *et al.*, 1996; Jadhao and Meyer, 2000; Jadhao and Pinelli, 2001; Rodríguez *et al.*, 2003; Adrio *et al.*, 2005). This neuronal population was absent in goldfish and molly females but conspicuous in males (Cornbrooks and Parsons, 1991; Prasada Rao *et al.*, 1996). Opposite results were reported for sockeye salmon (Jadhao and Meyer, 2000). Our expression studies in the goldfish reported the presence of galanin mRNA in the PPa but not in the PM (Unniappan *et al.*, 2004). Double labeling studies suggested the coexistence of galanin- and CRH-ir in some neurons of the PM of molly and killfish (Batten *et al.*, 1990c). Galanin-ir cells were not found in the PM of rainbow trout (Anglade *et al.*, 1994) and Senegalese sole (*Solea senegalensis*; Rodríguez-Gómez *et al.*, 2000a).

The third population of galanin-producing cells is located in the tuberal hypothalamus, within the posterior and caudal areas. *In situ* hybridization studies in the goldfish localized galanin mRNA within the LH as well as in the Hv (Unniappan *et al.*, 2004). The cell population of the Hv has also been reported in several species (Power *et al.*, 1996; Rodríguez-Gómez *et al.*, 2000a) but this locus is sometimes called the posterior part of the lateral tuberal nucleus (Batten *et al.*, 1990c; Magliulo-Cepriano *et al.*, 1993; Anglade *et al.*, 1994; Prasada Rao *et al.*, 1996; Jadhao and Meyer, 2000; Jadhao and Pinelli, 2001) or NAT (Rodríguez *et al.*, 2003). This tuberal population seems to be continuous with that of the Hc and PTN described in some species (Anglade *et al.*, 1994; Power *et al.*, 1996; Prasada Rao *et al.*, 1996; Jadhao and Pinelli, 2001; Rodríguez *et al.*, 2003). Sexual dimorphism within the tuberal hypothalamus has only been reported in the brown trout, in which the number of galanin-ir cells was greater in males than in females.

5.5. Gastrin-releasing Peptide (GRP)

Gastrin-releasing peptide (GRP), a 27-amino acid peptide, is part of the bombesin (BBS) and neuromedin B family of peptides (Martínez and Taché, 2000). This peptide has been isolated from several mammalian species and is also found in other vertebrates including fish (McDonald *et al.*, 1979; Holmgren and Jensen, 1994). This group of peptides is characterized by a highly conserved C-terminus, which is important for biological actions

(Martínez and Taché, 2000). Due to the structural similarity between BBS and GRP, exogenous BBS effects most likely reflect the function of endogenous GRP. BBS/GRP peptides are widely distributed in the gastrointestinal tract and CNS (McCoy and Avery, 1990), and have been shown to be potent anorexigenic substances when administered intraperitoneally or centrally in mammals and fish (Gibbs *et al.*, 1979; Flynn, 1991; Himick and Peter, 1995; Volkoff *et al.*, 2005). In addition, a BBS-like peptide has also been shown to be involved in the regulation of gut motility and visceral activity in fish species (Holmgren and Jonsson, 1988; see Chapter 10, this volume). In goldfish, BBS-ir fibers were present in the ventral telencephalon, the preoptic area, the tuberal hypothalamus, the posterior hypothalamus, areas associated with the putative feeding center in fish and some thalamic nuclei (Himick and Peter, 1995). In addition, BBS-ir perikarya were found in the preoptic area and the periventricular regions including NAT, Hv (cell population coating the lateral recess) and Hc (Himick and Peter, 1995). In rainbow trout, BBS-ir cell bodies were found in the PPp, Hv (coating the lateral recess) and Hc (Cuadrado *et al.*, 1994). BBS-ir fibers were widely distributed in the diencephalon, midbrain and hindbrain, the hypothalamus being the most densely labeled (Cuadrado *et al.*, 1994). The mRNA encoding for GRP has been identified in goldfish (Volkoff *et al.*, 2000). RT-PCR analysis shows that GRP mRNA is widely expressed in the brain of goldfish, as well as in the ovary, gill, skin, gut and pituitary (Volkoff *et al.*, 2000). Within the pituitary gland, BBS/GRP fibers were observed mainly in the PI.

5.6. Gonadotropin-releasing Hormone (GnRH)

In vertebrates, gonadotropin-releasing hormone (GnRH) is a decapeptide that has a cyclic structure with a pyro-glutamate modification in the amino terminus and an amide function at the carboxyl terminus. As many as 24 molecular isoforms of GnRH have been characterized so far and eight variants have been found in the teleost brain, six of them being exclusive to teleost species (Kah *et al.*, 2007). White and colleagues (1998) have proposed grouping the GnRH diversity into three types. As such, GnRH type 1 represents hypophysiotropic GnRH variants such as mGnRH, cfGnRH, pjGnRH and sbGnRH. Type 2 includes the cGnRH-II from all species studied which are localized in the midbrain, whereas type 3 includes sGnRH from several fish species (Lethimonier *et al.*, 2004). While it seems clear that type 1 has hypophysiotropic functions, the role for type 2 and 3 has not been established. There is some evidence for type 2 being involved in sexual behavior (Volkoff and Peter, 1999; Canosa *et al.*, 2008); however, the case of type 3 is more elusive. In several teleost species in which both type 1 and 3 coexist in the ventral forebrain, fibers of type 3 GnRH also innervate the pituitary (Kah *et al.*, 2007; González-Martínez *et al.*, 2002; Pandolfi *et al.*, 2005) suggesting that the type 3 isoform retains some hypophysiotropic activity.

The forebrain distribution of GnRH peptides expands from the olfactory bulb, ganglion cells of the terminal nerve (TNgc), along the ventral telencephalon, the preoptic area, and in some cases, the anteroventral hypothalamus (Lepretre *et al.*, 1993; Montero *et al.*, 1994; González-Martínez *et al.*, 2002; Mohamed *et al.*, 2005; Pandolfi *et al.*, 2005). However, the GnRH hypophysiotropic areas in teleost fish are mainly confined to the parvocellular preoptic area with a few cells in the mediobasal hypothalamus (Lepretre *et al.*, 1993; Montero *et al.*, 1994; Yamamoto *et al.*, 1998; González-Martínez *et al.*, 2002; Mousa and Mousa, 2003). In species with only two GnRH systems (type 1 or 3 in the forebrain and type 2 in the midbrain) the same peptide is found all along the forebrain distribution area. On the other hand, for those species which carry three GnRH systems, types 1 and 3 greatly overlap in the forebrain distribution area and pituitary innervation (González-Martínez *et al.*, 2002; Pandolfi *et al.*, 2005). In general, though, type 1 (sGnRH) is principally found in the olfactory region and TNgc, while type 3 (hgGnRH, sbGnRH, pjGnRH) is mainly expressed in preoptic area neuronal cells that innervate the pituitary (González-Martínez *et al.*, 2002; Pandolfi *et al.*, 2005).

In terms of GnRH brain distribution, goldfish are unique in being the only species that expresses cGnRH-II in the forebrain (Yu *et al.*, 1988). An immunohistochemical study has shown sGnRH-ir cell bodies localized in the TNgc, the ventral telencephalon, preoptic area and hypothalamus (Kim *et al.*, 1995). Within the forebrain, cGnRH-II-expressing neurons were found in lower numbers but in the same areas as sGnRH. Moreover, sGnRH-ir and cGnRH-II-ir fibers were distributed in the hypothalamus as well as in the pituitary gland (Kim *et al.*, 1995). Recent studies correlating GnRH mRNA levels with serum LH levels during ovulation and spawning in goldfish suggest that forebrain cGnRH-II in this species does not initiate the ovulatory LH surge (Canosa *et al.*, 2008).

The presence of GnRH fibers in the PI has been documented in salmonid and Nile perch (*Lates niloticus*) in close proximity with SL cells (Parhar and Iwata, 1994; Parhar *et al.*, 1995) and SL and MSH cells (Mousa and Mousa, 2003), respectively. In addition, South American cichlid (*Cichlasoma dimerus*) shows a strong presence of sbGnRH fibers in the PI running along the border and probably interacting with PI endocrine cells (Pandolfi *et al.*, 2005). In pejerrey fish, double immunostaining for GnRH and SL showed a close apposition betweeen pjGnRH fibers and SL cells in the PI (Vissio *et al.*, 1999). Furthermore, GnRH binding sites have been detected on dispersed SL cells from pejerrey fish pituitaries (Stefano *et al.*, 1999).

GnRH-ir was found in tissue penetrating the GTH zone in the PPD of green molly, sea bass and Indian catfish (Moons *et al.*, 1989; Batten *et al.*, 1999). In tilapia, GnRH fibers were also found in the PPD innervating GTH and GH cells (Melamed *et al.*, 1995; Parhar, 1997). In European eel, an extensive mGnRH innervation of the pituitary was observed, while only a few cGnRH-II-ir fibers were detected (Montero *et al.*, 1994). Thus, the hypophysiotropic form of GnRH for different species has been shown to innervate the PPD in close proximity with GH- and GTHs-producing cells (Vissio *et al.*, 1999; Mousa and Mousa, 2003; Pandolfi *et al.*, 2005).

The presence of GnRH fibers in RPD and its function on RPD hormonal secretion is scarce and limited to a few teleost species. In tilapia, GnRH stimulated the release of PRL *in vitro* (Weber *et al.*, 1997). Furthermore, GnRH binding sites have been shown on dispersed PRL cells from pejerrey fish (Stefano *et al.*, 1999). In this species, pjGnRH-ir fibers were found to penetrate the RPD (Vissio *et al.*, 1999); however, no GnRH fibers were detected in RPD of a South American cichlid (Pandolfi *et al.*, 2005).

5.7. Growth Hormone-releasing Hormone (GHRH)/Pituitary Adenyl Cyclase Activating Polypeptide (PACAP)

Growth hormone-releasing hormone (GHRH) and pituitary adenyl cyclase activating polypeptide (PACAP) belong to the glucagon/vasoactive intestinal peptide (VIP)/secretin superfamily of peptides and have a number of structural and functional similarities (for review, see Sherwood *et al.*, 2000; Vaudry *et al.*, 2000). Apparently, PACAP is the most conserved member of the family and represents the ancestral molecule from which other members have arisen by in-tandem exon duplication and subsequent gene duplication (Sherwood *et al.*, 2000). PACAP shows that amino acid sequence identity varies by between 88 and 97% across vertebrates, whereas GHRH sequences have only 32–45% identity between human and non-mammalian vertebrates (Sherwood *et al.*, 2000; Vaudry *et al.*, 2000). In mammals, separate genes encode for each peptide. In addition, a C-peptide, with no known function, is encoded along with GHRH whereas PACAP-related peptide (PRP) is present in the same transcript as PACAP (Mayo *et al.*, 1985; Hosoya *et al.*, 1992). It was thought that in tunicates, teleost fish, amphibians and birds, the PACAP gene included an upstream exon that encoded for GHRH (see for review Sherwood *et al.*, 2000; Vaudry *et al.*, 2000); however, the existence of a different set of genes coding for GHRH and its specific receptors was recently proved in fish (Lee *et al.*, 2007). This GHRH gene would be homologous with the mammalian counterpart since the derived peptide is more potent in eliciting GH release than the previous GHRH peptides described in fish (Lee *et al.*, 2007). Therefore, the previous GHRH peptides would represent the homologue of mammalian PACAP-related peptide (PRP) rather than GHRH. This finding would also explain why these previous peptides have shown low potency in GH stimulation compared to PACAP (see Canosa *et al.*, 2007, and Chapter 4, this volume, for further discussion). Nevertheless, PACAP seems to have clear hypophysiotropic effects in fish (Montero *et al.*, 1998; Wong *et al.*, 1998; Kong *et al.*, 2007; Canosa *et al.*, 2008; Mitchell *et al.*, 2008).

The immunohistological distribution of GHRH has been studied in the brains of several teleost species (Pan *et al.*, 1985; Marivoet *et al.*, 1988; Luo and McKeown, 1989; Batten *et al.*, 1990a; Olivereau *et al.*, 1990; Rao *et al.*, 1996; Miranda *et al.*, 2002). Early studies in cod, using antiserum against the C-terminus and middle portions of human GHRH forms, identified two main groups of immunoreactive cell bodies in the PM, LH and Hv. In the PM, both parvocellular and magnocellular neuronal perikarya were GHRH positive while in the Hv the cells were of the magnocellular type (Pan *et al.*, 1985). In the sea bass, positively stained neurons were located in the PMpc and PMmc and Hv (Marivoet *et al.*, 1988). In rainbow trout, GHRH-ir perikarya were located mainly in the LH and Hv. A small group of GHRH-ir cell bodies were also found in the caudal region of the PM (Luo and McKeown, 1989). Several species of fish including goldfish, carp, eel, salmonids and sculpin (*Myoxocephalus octodecimspinosus*) present GHRH-ir perikarya in PMpc and PMmc and occasionally in the Hv (Olivereau *et al.*, 1990). In the green molly, GHRH-ir cell bodies were only found in the PMgc and PMpc, and these cells also expressed AVT (Batten *et al.*, 1990a). More recently, in goldfish, GHRH-ir perikarya were identified in the preoptic region, Hv, the pineal nucleus and the lateral lemniscus in the midbrain tegmentum (Rao *et al.*, 1996). GHRH-ir fibers were found in the ventral telencephalon, preoptic region, pituitary, mesencephalic tegmentum and the hypothalamic inferior lobes of goldfish. The ontogeny of GHRH-ir has been examined in pejerrey (Miranda *et al.*, 2002).

The brain distribution of PACAP has been studied in European eel (*Anguilla anguilla*; Montero *et al.*, 1998) and the stargazer, (*Uranoscopus japonicus*; Matsuda *et al.*, 2005a). PACAP has a similar distribution to GHRH, being found in the preoptic and hypothalamic regions, and is also present in some neurons in the hindbrain (Matsuda *et al.*, 2005a). In the European eel, PACAP-ir perikarya were observed in the preoptic area, in the PPa and PPp and in the PM. In addition, groups of immunoreactive perikarya were also found in the thalamus within the ventral and dorsal thalamic nuclei (Montero *et al.*, 1998). No other brain regions were positive for PACAP. Immunoreactive fibers were widespread in the brain of the European eel (Montero *et al.*, 1998). In the stargazer, PACAP-ir cell bodies were found in the PMpc and PMgc. In addition, some PACAP-ir cell bodies were also observed in the Hv, corpus cerebelli and in the ventral horn of the spinal cord (Matsuda *et al.*, 2005a).

Both GHRH and PACAP fibers innervate the adenohypophysis and intermingle among the endocrine cells in all three PD subdivisions (Table 1.3). GHRH-ir fibers were shown running along blood vessels in the interdigitations of PN and penetrating into the PI in sea bass, goldfish, carp, sculpin and pejerrey (Marivoet *et al.*, 1988; Moons *et al.*, 1989; Olivereau *et al.*, 1990; Miranda *et al.*, 2002). Within this pituitary zone GHRH fibers contact both the MSH and the SL (or PAS-positive) cells (Marivoet *et al.*, 1988; Moons *et al.*, 1989). Conversely, no GHRH-ir fibers were found in cod, eel and trout (Pan *et al.*, 1985; Olivereau *et al.*, 1990). GHRH-ir fibers originating in the hypothalamus project to the PPD in a variable number depending on the species (Pan *et al.*, 1985; Olivereau *et al.*, 1990). These fibers run among GH cells showing terminals in close relationship with GH cells and, less obviously, with GTH cells (Marivoet *et al.*, 1988; Moons *et al.*, 1988, 1989). In addition, a close relationship between GHRH fibers and TSH cells was documented in sea bass (Moons *et al.*, 1989). Recently, an ontogenetic study demonstrated the appearance of GHRH immunohistological perikarya and fibers in the preoptic region and PPD of the pituitary about 1 week after hatching of the pejerrey (Miranda *et al.*, 2002), indicating a possible early involvement of GHRH in the regulation of GH release. Whereas in the sea bass, sculpin and grey mullet (*Mugil cephalus*) GHRH-ir nerve fibers, originating in the hypothalamus, projected to the RPD (Marivoet *et al.*, 1988; Moons *et al.*, 1989; Olivereau *et al.*, 1990), in eel, carp, goldfish, salmonids and cod, GHRH-ir fibers did not enter the RPD (Pan *et al.*, 1985; Olivereau *et al.*, 1990). Double staining studies showed a close relationship between GHRH-ir nerve fibers and ACTH cells (Marivoet *et al.*, 1988; Moons *et al.*, 1989).

PACAP-ir fibers were seen going through the PN and in close proximity to SL-, POMC (MSH)- and endorphin-ir endocrine cells in the PI of the stargazer (Matsuda *et al.*, 2005a,b) and European eel (Montero *et al.*, 1998). In addition, PACAP-ir fibers were found in the PPD of European eel (Montero *et al.*, 1998), goldfish (Wong *et al.*, 1998) and stargazer (Matsuda *et al.*, 2005b). PACAP-ir fibers were observed in the proximity of somatotrophs in goldfish (Wong *et al.*, 1998) as well as young stargazer but not in adults (Matsuda *et al.*, 2005b). In contrast, no PACAP-ir fibers were found innervating GTH cells. In the RPD, PACAP-ir fibers were found ending in close proximity to PRL and ACTH cells (Montero *et al.*, 1998; Matsuda *et al.*, 2005a,b).

5.8. Melanocortin System

The central melanocortin system is composed of POMC-expressing neurons, neurons expressing the endogenous melanocortin antagonist, agouti-related protein (AgRP), and downstream targets of these neurons expressing central melanocortin receptors. The POMC gene encodes a complex precursor exhibiting three main domains, each containing an MSH peptide (see Section 1). α-MSH is found within the second domain as the N-terminal sequence of ACTH (Cerdá-Reverter *et al.*, 2003a). POMC is mainly produced in the pituitary and its post-transcriptional processing occurs in a tissue-specific manner. The proteolytic cleavage by proconvertase 1 (PC1) generates ACTH and β-lipotropin in the corticotropes in the RPD, whereas cleavage by PC1 and PC2 produces α-MSH and β-endorphin in the melanotropes in the PI. POMC is also expressed in some neuronal populations of the CNS, where it is mainly processed to α-MSH and β-endorphin (reviewed by Castro and Morrison, 1997).

The melanocortins ACTH and MSH are involved in a wide range of physiological functions and exert their physiological role by binding to a family of G-protein coupled receptors. In tetrapods, five melanocortin receptor subtypes (MC1R–MC5R) have been cloned (reviewed by Schiöth *et al.*, 2005). The zebrafish genome contains six MCR subtypes, since MC5R is duplicated, whereas only four subtypes were found in the genome of *Fugu rubripes* (MC1R, MC2R, MC4R and MC5Rl; Logan *et al.*, 2003).

Expression studies using *in situ* hybridization techniques have reported that POMC is exclusively produced in the lateral tuberal nucleus or Hv in salmon and goldfish brains (Salbert *et al.*, 1992; Cerdá-Reverter *et al.*, 2003a). Similarly, immunostaining studies using antibodies against either α-MSH or ACTH (1-39) commonly reported that melanocortin-ir is exclusively localized in the tuberal hypothalamus of several species (Kishida *et al.*, 1988; Bird *et al.*, 1989; Olivereau and Olivereau, 1990; Amano *et al.*, 2005; Forlano and Cone, 2007; Table 2). Generally, α-MSH-ir perikarya are located immediately adjacent to the hypothalamic floor in the posterior part of the ventral hypothalamus. In carp, MSH-ir has been observed also in the LH coinciding with the MCH neurons (Kishida *et al.*, 1988). Only one study reported the presence of ACTH-ir in the fish brain using an antibody directed against the α-MSH-corticotropin-like intermediate peptide (CLIP) transition of carp ACTH (10Gly–23Tyr) that did not cross-react against α-MSH or CLIP. ACTH-ir cell bodies were localized in the PMgc. ACTH-containing cell bodies were negative for AVT and no ACTH fibers were observed in the pituitary. The function of ACTH in the fish brain is unknown but POMC expression in the PM, as determined by PCR, is up-regulated during stress (Metz *et al.*, 2004). However, two previous studies in carp could not find melanocortin immunoreactivity in the PM after immunostaining with antiserum against synthetic α-MSH, synthetic ACTH (11-24) or salmon N-terminal POMC peptide (Kishida *et al.*, 1988; Bird *et al.*, 1989). α-MSH fibers are widespread in the teleost brain but pituitary innervation by melanocortinic axon terminals does not seem prominent (Table 1.3). Studies in goldfish reported that MSH fibers do not penetrate deep into the neurohypophysis. No MSH projections from hypothalamic neurons to the pituitary were seen in the barfin flounder (Amano *et al.*, 2005) but a dense tract was observed coursing ventrally through the pituitary stalk and terminating in the neurohypophysis of the rainbow trout. Some fibers were also intermingled in the pars distalis, suggesting a neuroendocrine role of α-MSH (Vallarino *et al.*, 1989). In addition, MSH cells in the LH in carp seem to project to the pituitary gland (Kishida *et al.*, 1988).

Atypically, melanocortin signaling is not exclusively regulated by the binding of endogenous agonists, since naturally occurring antagonists, agouti and agouti-related protein (AgRP), compete with melanocortin peptides by binding to MCRs. In mammalian species, AgRP is mainly produced within the hypothalamic arcuate nucleus, in the same neurons where NPY is expressed, and it is potent in inhibiting melanocortin signaling at MC3R and MC4R (Cerdá-Reverter and Peter, 2003). Binding studies with zebrafish receptors demonstrated that AgRP acts as a competitive antagonist at MC3R, MC4R and MC5R, all three of which are expressed in the brain (Song and Cone, 2007). Only one study to date has precisely localized AgRP mRNA in brain regions in teleost fish (Cerdá-Reverter and Peter, 2003) whereas one other has reported AgRP-ir in the fish brain (Forlano and Cone, 2007). Both studies identified AgRP production within the posterior region of the ventral hypothalamus of the sea bass and zebrafish, respectively. However, there is no information about the pituitary innervation by AgRP fibers. In the zebrafish, AgRP and α-MSH projections strikingly match the nuclei that express MC4R (Cerdá-Reverter *et al.*, 2003b) and MC5R mRNA (Cerdá-Reverter *et al.*, 2003c) in the goldfish.

5.9. Neuropeptide Y (NPY) Family of Peptides

The neuropeptide tyrosine (NPY) family of peptides consists of 36-amino acid peptides exhibiting carboxy terminal (C-terminal) amidation (Cerdá-Reverter and Larhammar, 2000). The family comprises three different peptides: the NPY, tyrosine–tyrosine peptide (PYY) and the pancreatic polypeptide (PP). Tetrapod species produce all three peptides, whereas non-tetrapod vertebrates have only two types, i.e., NPY and PYY. Teleost fish, which have undergone a third round of genome duplication, synthesize two different versions of each peptide, i.e., NPY and PYY but no PP (Sundström *et al.*, 2008). Several studies in fish have reported the involvement of NPY in the control of the endocrine secretion of adenohypophysial cells (see Chapters 4, 5 and 6, this volume) including GH and LH (Kah *et al.*, 1989; Peng *et al.*, 1993; Cerdá-Reverter *et al.*, 1999).

Several studies in fish have reported the involvement of NPY in the control of the endocrine secretion of adenohypophysial cells (see Chapters 4, 5 and 6, this volume) including GH and LH (Kah *et al.*, 1989; Peng *et al.*, 1993; Cerdá-Reverter *et al.*, 1999). Neuroanatomical studies have further characterized the presence of NPY terminals in the neurohypophysis of several teleosts (Pontet *et al.*, 1989; Batten *et al.*, 1990a; Danger *et al.*, 1991; Cepriano and Schreibman, 1993; Zandbergen *et al.*, 1994; Chiba *et al.*, 1996; Subhedar *et al.*, 1996; Rodríguez-Gómez *et al.*, 2001; Gaikwad *et al.*, 2004; Chiba, 2005). In many species, NPY fibers reach the adenohypophysis but the innervation of the different parts is species dependent (Table 1.3). Studies in catfish (Gaikwad *et al.*, 2004), platyfish (Cepriano and Schreibman, 1993) and zebrafish (Mathieu *et al.*, 2002) demonstrated the presence of NPY fibers in all three adenohypophysial parts. Electron microscopy revealed NPY-immunoreactive particles bound to the cytoplasmic vesicles in the cells of all three adenohypophysial divisions in the catfish (Gaikwad *et al.*, 2004). Similar results were previously reported in goldfish in which ultrastructural studies detected NPY-immunoreactive neurosecretory vesicles in direct contact with most of the cell types including gonadotropes (LF–FSH), somatotropes (GH) and melanotropes (Pontet *et al.*, 1989). Immunoreactive studies also reported NPY terminals in the PPD and PI of the Senegalese sole (Rodríguez-Gómez *et al.*, 2001) whereas studies in tilapia showed a massive NPY-ir innervation of the PI and relatively few fibers in the pars distalis (Sakharkar *et al.*, 2005). However, in the catfish (*Clarias gariepinus*, Zandbergen *et al.*, 1994) and carp (Marchetti *et al.*, 2000) NPY immunoreactivity was confined to the PI.

Studies in ayu (*Plecoglossus altivelis*) demonstrated that neurohypophysial NPY innervation exhibits seasonal variations, supporting an involvement in the regulation of the reproductive axis. Double immunostaining has occasionally demonstrated close apposition of NPY terminals to the GnRH cells in the preoptic area. However, NPY and GnRH terminals were distinct though intermingled in the neurohypophysis (Chiba *et al.*, 1996). Similar apposition was found in the median eminence of the gar (*Lepisosteus oculatus*, Chiba, 2005) and catfish (Gaikwad *et al.*, 2005). Ultrathin sectioning in the latter species demonstrated that neurosecretory axons carrying NPY-ir gold particles were occasionally seen in close association with GnRH-containing cells in the PPD, suggesting the involvement of NPY in the regulation of GnRH secretion. In fact, the central administration of NPY induced increases of the GnRH contents in the olfactory bulb, medial olfactory tract, telencephalon/preoptic area and pituitary (Gaikwad *et al.*, 2005).

Immunochemical and *in situ* hybridization studies have localized NPY neuronal cell bodies in several hypophysiotropic areas in the teleost brain. The rostral-most NPY immunoreactivity in the teleost brain is localized in the olfactory bulb. In the sea bass, there is a conspicuous NPY-mRNA expressing population within the internal cell layer (ICL; Cerdá-Reverter *et al.*, 2000). Some controversy exists about NPY production in the nucleus olfactoretinalis or TNgc. Immunostaining studies reported the presence on NPY-ir within the TNgc of catfish (Gaikwad *et al.*, 2004), ayu (Chiba *et al.*, 1996), gar (Chiba, 2005), tilapia (Sakharkar *et al.*, 2005), zebrafish (Mathieu *et al.*, 2002), paltyfish (Cepriano and Schreibman, 1993) and killifish (Subhedar *et al.*, 1996) but no immunoreactivity was detected in the same structure of rainbow trout (Danger *et al.*, 1991), carp (Marchetti *et al.*, 2000), Senegalese sole (Rodríguez-Gómez *et al.*, 2001), goldfish (Pontet *et al.*, 1989) or catfish (*Clarias garepimnus*; Zandbergen *et al.*, 1994). Expression studies in the sea bass (Cerdá-Reverter *et al.*, 2000), goldfish (Peng *et al.*, 1994; Vecino *et al.*, 1994) and salmon (Silverstein *et al.*, 1998) have never reported expression of NPY gene in the TNgc. The ventral telencephalon is considered as a major component of the NPYergic system in teleost and non-teleost fish (Cerdá-Reverter and Larhammar, 2000). Gene expression or immunoreactivity of NPY has been reported in the Vd, Vv, Vl, Vs, Vc (or NE) nuclei of the ventral telencephalon. The Vc probably exhibits the highest levels of NPY production in the teleost brain and has been reported as the NPYergic nucleus in all studied fish species (see references above). Neurons in the Vs, Vv and Vc may participate in the control of pituitary since they project to the pituitary in the goldfish (Table 1.1). Studies in male tilapia and goldfish suggest that NPY neurons in the Vc may act as an important center for processing the sex-steroid signaling (Peng *et al.*, 1994; Sakharkar *et al.*, 2005). Actually, castration dramatically increases NPY-ir in the Vc whereas testosterone reverses the effects (Sakharkar *et al.*, 2005).

Studies have also reported the presence of NPY-ir or mRNA transcripts in the preoptic area of teleost fish but cell numbers are much lower than those observed in the telencephalon. However, studies in catfish (Zandbergen *et al.*, 1994) and carp (Marchetti *et al.*, 2000) failed to demonstrate NPY-ir in the entire preoptic area, whereas studies in goldfish (Pontet *et al.*, 1989) and Senegalese sole (Rodríguez-Gómez *et al.*, 2001) found NPY-ir to be confined to the most caudal area of the PPp. The precise location of the NPY cell bodies within the preoptic area differs among species. The most rostral NPY population has been reported in the PPa of the sea bass (Cerdá-Reverter *et al.*, 2000), ayu (Chiba *et al.*, 1996), killifish (Subhedar *et al.*, 1996), male tilapia (Sakharkar *et al.*, 2005), zebrafish (Mathieu *et al.*, 2002) and catfish (Gaikwad *et al.*, 2004). In the two last species, the PPa is the only NPY-producing nucleus in the preoptic area. NPY-ir or gene expression has also been reported in the PM of the goldfish (Peng *et al.*, 1994), killifish (Subhedar *et al.*, 1996) tilapia (Sakharkar *et al.*, 2005) and salmonids (Danger *et al.*, 1991; Silverstein *et al.*, 1998). Expression in the PM is up-regulated by treatment with sex steroids in goldfish (Peng *et al.*, 1994) but not in male tilapia (Sakharkar *et al.*, 2005) and no seasonality of NPY-ir was reported in the preoptic area of the ayu (Chiba *et al.*, 1996). Finally, NPY-ir and/or expression has also been reported in the tuberal hypothalamus of several species, including ayu (Chiba *et al.*, 1996), sea bass (Cerdá-Reverter *et al.*, 2000), Senegalese sole (Rodríguez-Gómez *et al.*, 2001), zebrafish (Mathieu *et al.*, 2002) and male tilapia (Sakharhar *et al.*, 2005). In the sea bass, NPY-expressing cells are located in the Hv, NAT and Hc. Immunoreactive levels are not affected by castration and subsequent testosterone treatment in male tilapia (Sakharkar *et al.*, 2005) but undergo a seasonal variation in ayu (Chiba *et al.*, 1996). This seasonality of NPY immunoreactivity is concomitant with that observed in the NPYergic neurohypophysial fibers, thus suggesting its participation in the control of pituitary secretion.

5.10. Orexins

Orexins, also called hypocretins, are excitatory neuromodulatory peptides produced from a common precursor of the incretin family (de Lecea *et al.*, 1998). Hypocretin 1 and 2 exert their biological function by binding to two G-protein coupled receptors which exhibit different affinity for the hypocretins as well as different distribution in the CNS (Sutcliffe and de Lecea, 2000).

In fish, hypocretins were initially characterized in pufferfish, by searching in the public DNA sequence database (Alvarez and Sutcliffe, 2002). Unlike mammals, only one receptor has been characterized in zebrafish, pufferfish, medaka and three-spined stickleback (Yokogawa *et al.*, 2007). Studies in several species have demonstrated that hypocretins stimulate food intake (Nakamachi *et al.*, 2006), whereas fasting raises orexin mRNA levels in the hypothalamus (Novak *et al.*, 2005; Nakamachi *et al.*, 2006; see Chapter 9, this volume). In addition, orexin has been shown to increase locomotor activity in goldfish (Nakamachi *et al.*, 2006). Orexin overexpression induces an insomnia-like phenotype in zebrafish (Prober *et al.*, 2006). As in mammals, zebrafish lacking a functional orexin receptor exhibit disruptions in the consolidation of sleep/wake behavior but the phenotype of zebrafish orexin receptor mutant is only sleep fragmentation and decreased sleep in darkness (Yokogawa *et al.*, 2007; see Chapter 9, this volume). Studies in goldfish demonstrated that orexin-containing neurons are found in the same areas where the MCH hormone neurons are localized (Huesa *et al.*, 2005; Nakamachi *et al.*, 2006), i.e., LH, ZL and PTN. Neurons in the ZL correspond to the PVO-associated neuronal population expressing MCH (see above). Only the latter orexin neuronal population was described in medaka (Amiya *et al.*, 2007). Studies in zebrafish have identified two prominent orexin-ir neuronal groups in the preoptic area and tuberal hypothalamus (Table 1.2). Within the preoptic area, orexin neurons were localized in the PMmc and PMgc, PPp and NSC. The second cluster was found dorsally above the lateral recess, coinciding with the cell population described in goldfish and medaka. Only the tuberal hypothalamic population could be considered to be synthesizing hypocretin neurons since both orexin-mRNA and processed peptide were found. On the contrary, the preoptic neuronal population lacks preprohypocretin mRNA (Kaslin *et al.*, 2004). The vast majority of the orexin fibers from the preoptic area were directed to the pituitary through the preoptico-hypophysial tract, turning ventrally behind the optic chiasm and running the ventral surface of the hypothalamus to the pituitary. No orexin-immunoreactive fibers were found in the pituitary of medaka (Amiya *et al.*, 2007) but fibrous staining profiles were found in the sea perch (*Lateolabrax japonicus*) neurohypophysis (Suzuki *et al.*, 2007). Orexin-ir cells were reported in the pituitary of medaka (Amiya *et al.*, 2007) and sea perch (Suzuki *et al.*, 2007). In sea perch, orexin-ir was located in the GH cells, both peptides coexisting in the secretory granules.

5.11. RF-amide Peptides

The first member of this class of peptides was discovered in molluscs by Price and Greenberg (1977). Since then, increasing immunohistochemical evidence has shown that peptides similar to molluscan FMRF-amide exist in the brain of vertebrates. For instance, FMRF-amide-ir was found in the TNgc in the goldfish co-expressed with GnRH (Stell *et al.*, 1984). Similarly, in swordtail (*Xyphophorus helleii*), FMRF-amide-ir was localized in cell bodies of the TNgc (Magliulo-Cepriano *et al.*, 1993), sometimes colocalizing with GnRH-ir (Table 1.2). Projections of these cells could be traced to the rostral area of the tuberal hypothalamus through the preoptic area. However, fibers containing FMRF-amide-ir had a broader distribution than GnRH fibers, being also found in the dorsal forebrain, OT and Hc (Magliulo-Cepriano *et al.*, 1993). Of particular note was the fact that fiber tracts that reached the tuberal hypothalamus had completed their development by the time sexual maturation was achieved. Additionally, in both catfish and swordtail, FMRF-amide-ir fibers that originate in the TNgc innervate all subdivisions of the pituitary gland (Rama Krishna *et al.*, 1992; Magliulo-Cepriano *et al.*, 1993). However, the endogenous peptide responsible for this immunoreaction still remains unknown (Table 1.3). In this context, several neuropeptides with an RF-amide motif have been identified in fish such as prolactin-releasing peptide (PrRP; Moriyama *et al.*, 2002; Seale *et al.*, 2002; Sakamoto *et al.*, 2003b; Montefusco-Siegmund *et al.*, 2006; Amano *et al.*, 2007), RF-amide-related peptide (RFRP; Hinuma *et al.*, 2000; Fukusumi *et al.*, 2001; Ukena *et al.*, 2002), other LPXRF-amides (Tsutsui and Ukena, 2006) and kisspeptin (van Aerle *et al.*, 2008).

PrRP was discovered from bovine hypothalamic extracts as a ligand for the anterior pituitary orphan G-protein coupled receptor (hGR3; Hinuma *et al.*, 1998). This peptide showed a capacity to release PRL (Hinuma *et al.*, 1998; Fujimoto *et al.*, 2006). PrRP homologues have been isolated from brain extracts of Japanese crucian carp (*Carassius cuvieri*; Fujimoto *et al.*, 1998), goldfish (Kelly and Peter, 2006), tilapia (Seale *et al.*, 2002), chum salmon (Moriyama *et al.*, 2002), and Atlantic salmon (Montefusco-Siegmund *et al.*, 2006). In teleost, PrRP facilitates *in vitro* and *in vivo* PRL secretion (Moriyama *et al.*, 2002; Seale *et al.*, 2002; Sakamoto *et al.*, 2003a,b), elevates PRL gene expression (Sakamoto *et al.*, 2003a) and regulates osmotic balance and food intake in goldfish (Fujimoto *et al.*, 2006; Kelly and Peter, 2006). Immunocytochemical staining in rainbow trout revealed that PrRP-ir cell bodies were located in the posterior part of hypothalamus (Moriyama *et al.*, 2002). In addition, in adult guppies, PrRP-ir cell bodies were detected in the Hv (Amano *et al.*, 2007). However, in the Atlantic salmon, immunoreactivity to the synthetic putative PrRP peptide was found mainly in the cerebellum and few weakly stained cell bodies in the Hc, and almost no positive fibers were seen in the pituitary stalk (Montefusco-Siegmund *et al.*, 2006). In the lamprey, two RF-amide peptides (A and B) homologous to teleost PrRP were identified (Moriyama *et al.*, 2007). In this study, using an anti-salmon PrRP antiserum, PrRP-ir cell bodies were observed in the ventral part of the periventricular arcuate nucleus of the hypothalamus (most probably corresponding to Hv). Within the pituitary, a small number of PrRP-ir fibers were observed adjacent to the PRL cell in the guppy (Amano *et al.*, 2007).

Kisspeptins, endogenous peptides displaying agonist activity on the orphan G-protein coupled receptor 54 (GPCR54), were isolated from human placenta (Ohtaki *et al.*, 2001). These 54-, 14-, 13-, and 10-amino acid peptides, with a common RF-amide C-terminus, are derived from the product of KISS-1, a metastasis suppressor gene for melanoma cells. It has recently been demonstrated in mammals that kisspeptins are critical factors for the onset of puberty (Tena-Sempere, 2006). GPCR54 has been studied in very few teleost species but was found to be expressed in GnRH neurons, while the number of neurons expressing GPCR54, as well as the level of GPCR54 expression, increased with gonadal maturation (Parhar *et al.*, 2004). A kisspeptin counterpart has been recently identified in fish by computational genome analysis (van Aerle *et al.*, 2008), but no brain and/or pituitary mapping specific for kisspeptins has been carried out yet. Of note, the FMRF-ir material studied in platyfish and swordtail (Magliulo-Cipriano *et al.*, 1993) may represent kisspeptin since, as was described above, it was shown to reach its maximum development with adulthood and sexual maturation innervating nucleus in the basal hypothalamus.

5.12. Somatostatin

Somatostatin (SS) is a potent inhibitor of basal and stimulated GH secretion in teleosts (Canosa *et al.*, 2007). This peptide belongs to a multifunctional family of peptides which includes up to three different SS precursors encoded by different genes in teleosts (for reviews, see Nelson and Sheridan, 2005; Canosa *et al.*, 2007). The most conserved form, PSS-I, encodes for SS-14 at its C-terminus and has an identical amino acid sequence in all vertebrate species studied. A second cDNA only isolated from teleosts, named PSS-II, encodes for SS peptides of variable length between 22 and 28 amino acids. These peptides are characterized by [Tyr7,Gly10]SS-14 at their C-terminus. PSS-III has been isolated from several species of fish and other vertebrates (Nelson and Sheridan, 2005; Canosa *et al.*, 2007). The peptides encoded by PSS-III characteristically bear proline in the second position of the 14-amino acid C-terminus peptide. The biology of these alternative peptides is not fully understood and most of the information on brain localization and pituitary innervation is based on immunostaining to SS-14. Although anti-SS-14 antibodies may cross-react against other SS peptides (Canosa *et al.*, 2004), the data obtained by immunostaining is considered as SS-14.

Early work in goldfish showed SS-ir in the PPa, PM and PPp in the preoptic area, the Hv in the tuberal hypothalamus and the ventral thalamus (Kah *et al.*, 1982) (Table 2). The distribution of SS in the brains of several teleosts was studied by Olivereau *et al.* (1984b), who found the SS-ir present in the Vc, PPa, PPp, Hv and Hc. In tilapia, SS-ir neurons were located in the anterior part of the Dl and Vc in the telencephalon, the preoptic area and the tuberal hypothalamus (Grau *et al.*, 1985). Brain localization of SS perikarya in the green molly showed SS cells in the Vl and Vc in the telencephalon, PPa, PM and rostral area of the PPp of the preoptic area, Hv and NAT and Hc of the tuberal hypothalamus (Batten *et al.*, 1985). A similar distribution pattern of SS-ir cell bodies was shown in the European minnow and the European eel (Vigh-Teichmann *et al.*, 1983). Many of these SS-ir cell bodies are CSF-contacting neurons, particularly those in the anterior, lateral and posterior recesses such as Hv and Hc. Furthermore, the brain distribution of SS-ir was studied in brown ghost knifefish, goldfish and sea bream (Sas and Maler, 1991; Pickavance *et al.*, 1992; Power *et al.*, 1996) and pointed to a similar distribution pattern to that mentioned in previous reports. In the hypothalamus of sea bream, SS-ir neurons were found in the PM, Hv and Hc and a network of varicose fibers projecting toward the pituitary through the preoptico-hypophysial tract (Power *et al.*, 1996).

Goldfish is the only teleost species in which three SS genes have been described (Lin *et al.*, 1999). *In situ* hybridization studies show that, although there is an overlapping brain distribution of these three mRNAs, co-expression is unlikely to occur in the same cell (Canosa *et al.*, 2004). The overall distribution of SS-positive cell bodies is in agreement with previous reports, as mentioned above, in other species of fish as well as in other vertebrates. PSS-I and PSS-III overlap in the ventral telencephalon, tuberal hypothalamus, ventral and dorsal thalamus and most areas of the ventroposterior hypothalamus, the mesencephalon and hindbrain. On the other hand, PSS-II was restricted to some nuclei in the hypothalamus and ventral thalamus: Hv, NCLI, VM and NTP (Canosa *et al.*, 2004). One of the most striking differences in the expression pattern of PSS-I and PSS-III is found in the preoptic area, where the former is expressed in the PM but not in the anterior part of the PPp while the latter shows the opposite distribution (Canosa *et al.*, 2004).

All three PD zones are innervated by SS-ir fibers (Table 1.3). In sculpin, the green molly and the catfish, SS-ir fibers in the caudal PN were found to approach the PI and to end among the MSH and SL cells (Olivereau *et al.*, 1984a; Batten *et al.*, 1999). Double immunostaining for both SS and MSH in the molly suggested that SS fibers innervate groups of MSH cells in the PI (Batten *et al.*, 1985). SS-ir fibers were also found to penetrate into the PPD in several species of fish (Kah *et al.*, 1982; Olivereau *et al.*, 1984a,b; Batten *et al.*, 1985; Grau *et al.*, 1985; Moons *et al.*, 1989; Sas and Maler, 1991; Power *et al.*, 1996). SS-ir fibers were found in contact with GH cells (Moons *et al.*, 1989; Power *et al.*, 1996) or in contact to the basal lamina GH and GTH islets where the GH cells are located peripherally (Olivereau *et al.*, 1984a,b; Batten *et al.*, 1999). In tilapia, SS-ir fibers were clearly visualized within the RPD where they innervate the PRL cells (Grau *et al.*, 1985). This pattern of SS distribution seems to be specific to tilapia since neither killifish nor mudsucker (*Gillichthys mirabilis*) showed SS-ir in the RPD (Grau *et al.*, 1985). Additionally, SS fibers were scarce or absent and did not innervate PRL or ACTH cells in the sea bream (Power *et al.*, 1996) and were never seen to penetrate the RPD of salmonid species (Olivereau *et al.*, 1984a). However, in sea bass, double staining showed SS fibers in close contact with ACTH cells (Moons *et al.*, 1989). Furthermore, in the sea bass and, to some extent, the Indian catfish, SS-ir terminals were seen to make synaptoid contact with the basal membrane in the RPD (Batten *et al.*, 1999).

5.13. Thyrotropin-releasing Hormone (TRH)

Thyrotropin-releasing hormone (TRH) was the first hypothalamic releasing factor to be chemically characterized from pig and sheep hypothalami (Schally *et al.*, 1969; Guillemin, 1970). In tetrapods but not amphibians, its role in controlling pituitary TSH release is well established (Fliers *et al.*, 2006). In teleost fish, TRH stimulates several pituitary hormones such as PRL (Batten and Wigham, 1984; Wigham and Batten, 1984; Barry and Grau, 1986), GH (Canosa *et al.*, 2007) and MSH (Tran *et al.*, 1989; Lamers *et al.*, 1991), but its actions on the pituitary–thyroid gland axis are more controversial (see Chapter 6, this volume) since contradictory results have been published, showing either stimulatory (Tsuneki and Fernholm, 1975; Eales and Himick, 1988), no (Peter and McKeown, 1975) or even inhibitory effects (Bromage, 1975). There is only one report showing TRH innervation in the PPD of catfish (Batten *et al.*, 1999) (Table 3). In sea bass, small TRH-ir cell bodies were only found in NAT and the region of the Hv coating the lateral recess (Batten *et al.*, 1990b) and in the PPa in the green molly (Batten *et al.*, 1990a) (Table 2). Similar results have been observed in carp (Hamano *et al.*, 1990). In salmonid fish, on the other hand, the distribution of TRH cell bodies seems much wider. In chinook salmon, TRH-positive cell bodies are observed in the internal cellular layer of the olfactory bulb, the supracommissural nucleus of the ventral telencephalon (Vs) and the preoptic region. In the latter area, the TRH-ir cell bodies were observed in the PMmc (Matz and Takahashi, 1994). Similar results have been shown in sockeye salmon using *in situ* hybridization (Ando *et al.*, 1998). More recently, in trout and zebrafish, TRH-ir perikarya were observed in the OB and several areas of the ventral telencephalon, in the preoptic and suprachiasmatic region, the ventromedial hypothalamus, and some thalamic nuclei (Díaz *et al.*, 2001, 2002) (Table 2).

6. HYPOTHALAMIC NEUROTRANSMITTERS

6.1. Amino Acid Neurotransmitters: Glutamate and Gamma-amino Butyric Acid (GABA)

Glutamate is considered the major excitatory neurotransmitter in the vertebrate CNS (Nakanishi, 1992; Trudeau *et al.*, 2000b). Glutamate is also an important hypophysiotropic regulator and is involved in the control of LH, GH, PRL (Trudeau *et al.*, 1996; Holloway and Leatherland, 1997; Trudeau *et al.*, 2000b; Bellinger *et al.*, 2006) and probably MSH or SL since glutamate-ir fibers were found in the NIL (Trudeau *et al.*, 1996). Moreover, glutamatergic fibers have been revealed by immunohistochemistry in goldfish pituitary (Trudeau *et al.*, 1996). However, in teleost fish, there is no report showing the central origin of this glutamatergic innervation (Table 1.3).

Gamma-amino butyric acid (GABA) represents the major inhibitory neurotransmitter in the CNS, although excitatory responses to GABA have been reported (Wagner *et al.*, 1997). In addition, GABA has an important role in the control of pituitary hormone secretion (Khan and Thomas, 1999; Mañanos *et al.*, 1999; Trudeau *et al.*, 2000a,b; Martyniuk *et al.*, 2007b). GABA is synthesized principally from glutamate in a single enzymatic step catabolized by the enzyme glutamic acid decarboxylase (GAD). This enzyme is therefore used as a marker for GABAergic cell bodies and fibers (Kah *et al.*, 1987; Martinoli *et al.*, 1990; Medina *et al.*, 1994; Anglade *et al.*, 1999). Thus, in goldfish, GAD-ir cell bodies were detected in the olfactory bulbs and telencephalon. In the diencephalon, GABA-containing cell bodies were found in the hypothalamus, particularly in the preoptic and tuberal regions. Furthermore, the ventroposterior hypothalamus, dorsal and ventromedial thalamus and pretectal area exhibited numerous GABA-positive perikarya (Martinoli *et al.*, 1990). A similar distribution of GAD-ir cell bodies was revealed in the forebrain of the silver eel (Medina *et al.*, 1994) and rainbow trout (Anglade *et al.*, 1999). Employing non-radioactive *in situ* hybridization, the distribution of GABAergic cell bodies was analyzed in goldfish brain (Martyniuk *et al.*, 2007a). In this report, transcripts for GAD65, GAD67, along with the GABA-metabolizing enzyme GABA-T, were mainly detected in the medial and ventral regions of the telencephalon, the PM in the preoptic area and the lateral aspect of the Hv, Hc. Several reports have shown that GABA plays a role in regulating LH secretion (Kah *et al.*, 1992; Sloley *et al.*, 1992; Trudeau *et al.*, 1993; Anglade *et al.*, 1998; Trudeau *et al.*, 2000b). Moreover, GABAergic fibers have been found innervating all pituitary lobes in goldfish (Kah *et al.*, 1987).

6.2. Dopamine

Dopamine (DA) is an important neurotransmitter in the CNS of vertebrates and possesses key hypophysiotropic functions. Using antibodies or riboprobes to detect tyrosine hydroxylase, an enzyme involved in DA synthesis, the dopaminergic system in the fish nervous system has been widely studied (Ma, 1997; Smeets and González, 2000; Rink and Wullimann, 2001). In addition, studies using antisera toward DA itself have also been performed (Meek *et al.*, 1989; Roberts *et al.*, 1989; Pierre *et al.*, 1997).

In fish, the highest concentration of dopaminergic neurons is localized in the posterior tuberculum and adjacent hypothalamic regions (see Ma, 2003; Ma and Lopez, 2003, and references therein). In particular, large numbers of dopamine-containing neurons are found in nuclei closely associated with the ventricle and its recesses (Table 1.4). Besides, dopaminergic neurons are located in the olfactory bulbs, the ventral regions of the preoptic area and

Table 1.4
Main hypophysiotropic territories containing neurotransmitters

Neurotransmitters	Telencephalon	Preoptic area	Hypothalamus
DA	OB	PPa, PMmc, PMgc, PPp, NSC	Hv, Hc, PVO
GABA	OB, ventral TEL	PPa, PMpc, PMmc, PPp NSC	Hv, LH, Hc, NAT, PVO, NDLI
Serotonin	OB, Vv	PMpc, PPp, NSC	Hv,Hd, Hc, PVO

Nomenclature of nuclei was adapted following Bradford and Northcutt (1983).
See Table 1.1 for abbreviations of hypophysiotropic territories.

tuberal hypothalamus (Meek *et al.*, 1989; Roberts *et al.*, 1989; Pierre *et al.*, 1997). Early studies have shown that the preoptic area and the posterior hypothalamus in goldfish contain dopaminergic neurones that project toward the pituitary gland innervating all three zones of the PD (Kah *et al.*, 1984; Fryer *et al.*, 1985; Kah *et al.*, 1986). In Atlantic salmon, major dopaminergic innervation of the pituitary originates from a subset of the anterior region of the PPp and the suprachiasmatic nucleus (Holmqvist and Ekström, 1995). In zebrafish, dopaminergic neurons have mainly been found in the preoptic area in the PPa, PM, NSC, the posterior tuberculum, and the ventroposterior hypothalamus (Ma, 2003). More recently, in the European eel, it has been shown that dopaminergic neurons innervating the gonadotropes originate in the rostral region of the PPa and these cells are responsive to steroid hormones (Weltzien *et al.*, 2006).

6.3. Serotonin

Serotonin or 5-hydroxytryptamine (5-HT) is an indoleamine neurotransmitter that constitutes the monoamine group along with the catecholamine neurotransmitters DA and NA. It has been shown that 5-HT has both behavioral (De Pedro *et al.*, 1998; Lin *et al.*, 2000; Johansson *et al.*, 2004) and neuroendocrine functions (Khan and Thomas, 1992; Trudeau, 1997; Winberg *et al.*, 1997; Canosa *et al.*, 2007). The distribution of serotoninergic cells and fibers has been studied in several vertebrate species, including fish (Kah and Chambolle, 1983; Margolis-Kazan *et al.*, 1985; Ekström and Ebbesson, 1989; Meek and Joosten, 1989; Johnston *et al.*, 1990; Khan and Thomas, 1993; Rodríguez-Gómez *et al.*, 2000b). In summary, there are two main localizations of the 5-HT system in the teleost brain: one anterior, which covers the nuclei associated with the ventricle and its recess in the caudal hypothalamus, and one posterior in the brainstem (Table 1.4).

The rostral 5-HT system mainly consists of CSF-contacting cell bodies in the PVO, where it extends to the ventromedial hypothalamus and the preoptic area (Johnston *et al.*, 1990; Khan and Thomas, 1993). Although a direct effect of 5-HT on pituitary LH secretion has been demonstrated (Somoza and Peter, 1991; Khan and Thomas, 1992), several studies in teleosts have shown very low or even undetectable levels of 5-HT or its major metabolite, 5- hydroxyindoleacetic acid (5-HIAA), in the pituitary (Sloley *et al.*, 1992; Hernández-Rauda *et al.*, 1996). Hernández-Rauda and Aldegunde (2002) analyzed the role of neurotransmitters during gonadal development in yellow snapper (*Lutjanus argentiventris*) and found that neither 5-HT nor 5-HIAA was detectable in the pituitary at any gonadal stage studied. Similarly, serotoninergic innervation of the pituitary in sea bass is scant (Batten *et al.*, 1993). In the Senegalese sole, serotoninergic-ir fibers were absent in the pituitary (Rodríguez-Gómez *et al.*, 2000b). However, 5-HT-ir has been located in the PPD of goldfish (Kah and Chambolle, 1983), and serotoninergic fibers were observed in the pituitary stalk and in all regions of the pituitary gland of the catfish (Corio *et al.*, 1991). Furthermore, serotoninergic fibers were found in PPD and PI of the platyfish (Margolis-Kazan *et al.*, 1985) and the Atlantic croaker (Khan and Thomas, 1993). In the pituitary of the Altantic croaker (*Micropogonias undulatus*), the strongest 5-HT-ir was observed in the PPD while a few scattered cells were bordering the PI and their fibers were contacting the endocrine cells in the PPD (Khan and Thomas, 1993) (Table 3).

7. CONCLUDING REMARKS

The hypothalamus–pituitary complex in teleost species shows a particular specialization in which the median eminence is greatly reduced or absent. As a consequence, the hypothalamic control of the pituitary is exerted by an important pituitary innervation that penetrates to the adenohypophysis as interdigitation of neuronal tissue. This neuronal circuitry integrates incoming information from both external and internal environments by expressing the appropriate set of hormonal receptors (Chapter 2, this volume), as well as through interneuronal communication. There is some information about how these neuronal circuits are regulated by internal factors and how sensory information is conveyed to the "regulatory brain." However, little is known about how these hypothalamic regulatory systems integrate both internal and external information and how they elaborate a coordinated final response. How do these neuronal regulatory systems interplay to coordinate different physiological functions, i.e., nutrition and reproduction?

Recent retrograde tract-tracing studies have corroborated earlier studies showing the preoptic area and tuberal hypothalamus as loci for the neuronal cell bodies whose axons reach the neuro- and adenohypophysis along well-defined fiber tracts (see Section 3). These neurons release hormones to the vascular system (see Section 4) or use different releasing peptides (see Section 5) and neurotransmitters (see Section 6) to control a wide array of physiological functions through their anatomical and functional links with the pituitary gland. Studies using double-labeling, tract-tracing, and pituitary cell culture have led to the localization of the neurons that regulate the different secretory cells of the pituitary gland, as well as the identification of the releasing peptides and/or neurotransmitters used by these neuronal systems. Direct innervation within the RPD, in proximity of either ACTH or PRL cells, is exerted by neurons that contain AVT, IST, MCH, GHRH, PACAP, SS, CRF, galanin, CCK and PrRP. The endocrine cells of the PPD (GH, LH, FSH and TSH) are innervated by fibers containing AVT, IST, NPY, GHRH, PACAP, SS, CRF, galanin, GnRH and CCK, and to a lesser extent by TRH-containing neurons. Finally, the PI is interdigitated by neuronal fibers containing AVT, IST, MCH, NPY, GHRH, PACAP, SS, CRF, galanin, GnRH, CCK, BBS and TRH, in close relationship with either MSH or SL cells. Neurosecretion within the PN mainly involves AVT, IST and MCH neurons.

ACKNOWLEDGEMENTS

We want to express our gratitude to Dick Peter, who recently passed away, for his invaluable support and friendship. We would like to thank Maria M. Corvi for her help with the preparation of this manuscript. Authors have been financially supported by the Spanish Ministry of Education and Science (AGL2004-08137-C04-04, AGL2007-65744-C03-02, CSD2007-00002), the Killam Trust Association to JMC-R, Consejo Nacional de Investigaciones Científicas y Técnicas (Argentina) and research grants from Agencia Nacional de Promoción Científica y Tecnológica (PICT06-074 and PICT06-519) and Comisión de Investigaciones Científicas (673/07), Province of Buenos Aires, Argentina, to LFC.

REFERENCES

Acher, R. (1996). Molecular evolution of fish neurohypophysial hormones: Neutral and selective evolutionary mechanisms. *Gen. Comp. Endocrinol.* **102,** 157–172.

Adrio, F., Rodríguez, M. Á., and Rodríguez-Moldes, I. (2005). Distribution of galanin like immunoreactivity in the brain of the Siberian sturgeon (*Acipenser baeri*). *J. Comp. Neurol.* **487,** 54–74.

Ahrens, K., and Wullimann, M. F. (2002). Hypothalamic inferior lobe and lateral torus connections in a percomorph teleost, the red cichlid (*Hemichromis lifalili*). *J. Comp. Neurol.* **449,** 43–64.

Alderman, S. L., and Bernier, N. J. (2007). Localization of corticotropin-releasing factor, urotensin I, and CRF-binding protein gene expression in the brain of the zebrafish, *Danio rerio*. *J. Comp. Neurol.* **502,** 783–793.

Alvarez, C. E., and Sutcliffe, J. G. (2002). Hypocretin is an early member of the incretin gene family. *Neurosci. Lett.* **324,** 169–172.

Amano, M., Takahashi, A., Oka, Y., Yamanome, T., Kawauchi, H., and Yamamori, K. (2003). Immunocytochemical localization and ontogenic development of melaninconcentrating hormone in the brain of a pleuronectiform fish, the barfin flounder. *Cell Tissue Res.* **311,** 71–77.

Amano, M., Takahashi, A., Yamanome, T., Oka, Y., Amiya, N., Kawauchi, H., and Yamamori, K. (2005). Immunocytochemical localization and ontogenic development of α-melanocyte-stimulating hormone (α-MSH) in the brain of a pleuronectiform fish, barfin flounder. *Cell Tissue Res.* **320,** 127–134.

Amano, M., Oka, Y., Amiya, N., and Yamamori, K. (2007). Immunohistochemical localization and ontogenic development of prolactin-releasing peptide in the brain of the ovoviviparous fish species *Poecilia reticulata* (guppy). *Neurosci. Lett.* **413,** 206–209.

Amiya, N., Amano, M., Oka, Y., Iigo, M., Takahashi, A., and Yamamori, K. (2007). Immunohistochemical localization of orexin/hypocretin-like immunoreactive peptides and melanin-concentrating hormone in the brain and pituitary of medaka. *Neurosci. Lett.* **427,** 16–21.

Ando, H., Ando, J., and Urano, A. (1998). Localization of mRNA encoding thyrotropin-releasing hormone precursor in the brain of sockeye salmon. *Zool. Sci.* **15,** 945–953.

Ando, H., Hasegawa, M., Ando, J., and Urano, A. (1999). Expression of salmon corticotropin-releasing hormone precursor gene in the preoptic nucleus in stressed rainbow trout. *Gen. Comp. Endocrinol.* **113,** 87–95.

Anglade, I., Zanbergen, T., and Kah, O. (1993). Origin of the pituitary innervation in the goldfish. *Cell Tissue Res.* **273,** 345–355.

Anglade, I., Wang, Y., Jensen, J., Tramu, G., Kah, O., and Conlon, J. M. (1994). Characterization of trout galanin and its distribution in trout brain and pituitary. *J. Comp. Neurol.* **350,** 63–74.

Anglade, I., Douard, V., Le Jossic-Corcos, C., Mañanos, E. L., Mazurais, D., Michel, D., and Kah, O. (1998). The GABAergic system: A possible component of estrogenic feedback on gonadotropin secretion in rainbow trout (*Oncorhynchus mykiss*). *BFPP – Bulletin Francais de la Peche et de la Protection des Milieux Aquatiques* **71,** 647–654.

Anglade, I., Mazurais, D., Douard, V., Le Jossic-Corcos, C., Mañanos, E. L., Michel, D., and Kah, O. (1999). Distribution of glutamic acid decarboxylase mRNA in the forebrain of the rainbow trout as studied by in situ hybridization. *J. Comp. Neurol.* **410,** 277–289.

Baker, B. I., and Bird, D. J. (2002). Neuronal organization of melanin concentrating hormone system in primitive actinopterygians: Evolutionary changes leading to teleost. *J. Comp. Neurol.* **442,** 99–114.

Balment, R. J., Lu, W., Weybourne, E., and Warne, J. M. (2006). Arginine vasotocin a key hormone in fish physiology and behaviour: A review with insights from mammalian models. *Gen. Comp. Endocrinol.* **147,** 9–16.

Barry, T. P., and Grau, E. G. (1986). Estradiol-17b and thyrotropin-releasing hormone stimulate prolactin release from the pituitary gland of a teleost fish *in vitro*. *Gen. Comp. Endocrinol.* **62,** 306–314.

Batten, T. F. C. (1986). Ultrastructural characterization of neurosecretory fibres immunoreactive for vasotocin, isotocin, somatostatin, LHRH and CRF in the pituitary of a teleost fish, *Poecilia latipinna*. *Cell Tissue Res.* **244,** 661–672.

Batten, T. F. C., and Baker, B. I. (1988). Melanin-concentrating hormone (MCH) immunoreactive hypophysial neurosecretory system in the teleost *Poecilia latipinna*: Light and electron microscopic study. *Gen. Comp. Endocrinol.* **70,** 193–205.

Batten, T. F. C., and Wigham, T. (1984). Effects of TRH and somatostatin on releases of prolactin and growth hormone in vitro by the pituitary of *Poecilia latipinna*. II. Electron-microscopic morphometry using automatic image analysis. *Cell Tissue Res.* **237,** 595–603.

Batten, T. F. C., Groves, D. J., and Ball, J. N. (1985). Immunocytochemical investigation of forebrain control by somatostatin of the pituitary in the teleost *Poecilia latipinna. Cell Tissue Res.* **242,** 115–125.

Batten, T. F. C., Cambre, M. L., Moons, L., and Vandesande, F. (1990a). Comparative distribution of neuropeptide-immunoreactive systems in the brain of the green molly, *Poecilia latipinna. J. Comp. Neurol.* **302,** 893–919.

Batten, T. F. C., Moons, L., Cambre, M. L., Vandesande, F., Seki, T., and Suzuki, M. (1990b). Thyrotropin-releasing hormone-immunoreactive system in the brain and pituitary gland of the sea bass (*Dicentrarchus labrax*, teleostei). *Gen. Comp. Endocrinol.* **79,** 385–392.

Batten, T. F. C., Moons, L., Cambre, M., and Vandesande, F. (1990c). Anatomical distribution of galanin-like immunoreactivity in the brain and pituitary of teleost fishes. *Neurosci. Lett.* **111,** 12–17.

Batten, T. F. C., Berry, P. A., Maqbool, A., Moons, L., and Vandesande, F. (1993). Immunolocalization of catecholamine enzymes, serotonin, dopamine and L-dopa in the brain of *Dicentrarchus labrax* (Teleostei). *Brain Res. Bull.* **31,** 233–252.

Batten, T. F. C., Moons, L., and Vandesande, F. (1999). Innervation and control of the adenohypophysis by hypothalamic peptidergic neurons in teleost fishes: EM immunohistochemical evidence. *Microsc. Res. Techniq.* **44,** 19–35.

Bellinger, F. P., Fox, B. K., Wing, Y. C., Davis, L. K., Andres, M. A., Hirano, T., Grau, E. G., and Cooke, I. M. (2006). Ionotropic glutamate receptor activation increases intracellular calcium in prolactin-releasing cells of the adenohypophysis. *Am. J. Physiol.* **291,** E1188–E1196.

Bernier, N. J. (2006). The corticotropin-releasing factor system as a mediator of the appetite-suppressing effects of stress in fish. *Gen. Comp. Endocrinol.* **146,** 45–55.

Billard, R., and Peter, R. E. (1982). A stereotaxic atlas and technique for the nuclei of the diencephalon of rainbow trout (*Salmo gairdneri*). *Reprod. Nutr. Develop.* **22,** 1–25.

Bird, D. J., Baker, B. I., and Kawauchi, H. (1989). Immunocytochemical demonstration of melanin-concentrating hormone and proopiomelanocortin-like products in the brain of the trout and carp. *Gen. Comp. Endocrinol.* **74,** 442–450.

Bond, H., Warne, J. M., and Balment, R. J. (2007). Effect of acute restraint on hypothalamic pro-vasotocin mRNA expression in flounder, *Platichthys flesus. Gen. Comp. Endocrinol.* **153,** 221–227.

Braford, M. R., Jr., and Northcutt, R. G. (1983). Organization of the diencephalon and pretectum of the ray-finned fishes. *In* "Fish Neurobiology" (R. G. Northcutt, and R. E. Davis, Eds.), Vol. 2, pp. 117–163. University of Michigan Press, Ann Arbor.

Bromage, N. R. (1975). The effects of mammalian thyrotropin releasing hormone on the pituitary thyroid axis of teleost fish. *Gen. Comp. Endocrinol.* **25,** 292–297.

Butler, A. B., and Hodos, W. (2005). "Comparative Vertebrate Neuroanatomy". Wiley- Interscience, New Jersey.

Canosa, L. F., Cerdá-Reverter, J. M., and Peter, R. E. (2004). Brain mapping of three somatostatin encoding genes in the goldfish. *J. Comp. Neurol.* **474,** 43–57.

Canosa, L. F., Chang, J. P., and Peter, R. E. (2007). Neuroendocrine control of growth hormone in fish. *Gen. Comp. Endocrinol.* **151,** 1–26.

Canosa, L. F., Stacey, N., and Peter, R. E. (2008). Changes in brain mRNA levels of gonadotropin-releasing hormone, pituitary adenylate cyclase activating polypeptide, and somatostatin during ovulatory luteinizing hormone and growth hormone surges in goldfish. *Am. J. Physiol.* **295,** R1815–R1821.

Castro, M. G., and Morrison, E. (1997). Post-translational processing of proopiomelanocortin in the pituitary and the brain. *Crit. Rev. Neurobiol.* **11,** 35–57.
Cepriano, K. M., and Schreibman, M. P. (1993). The distribution of neuropeptide Y and dynorphin immunoreactivity in the brain and pituitary gland of the platyfish, *Xiphophorus maculatus*, from birth to sexual maturity. *Cell Tissue Res.* **271,** 87–92.
Cerdá-Reverter, J. M., and Larhammar, D. (2000). Neuropeptide Y family of peptides: structure, anatomical expression, function and molecular evolution. *Biochem. Cell Biol.* **78,** 371–392.
Cerdá-Reverter, J. M., and Peter, R. E. (2003). Endogenous melanocortin antagonist in fish. Structure, brain mapping and regulation by fasting of the goldfish agouti-related protein gene. *Endocrinology* **144,** 4552–4561.
Cerdá-Reverter, J. M., Sorbera, L., Carrillo, M., and Zanuy, S. (1999). Energetic dependence of NPY-induced LH secretion in a teleost fish (*Dicentrarchus labrax*). *Am. J. Physiol.* **46,** R1627–R1634.
Cerdá-Reverter, J. M., Anglade, I., Martínez-Rodríguez, G., Mazurais, D., Muñoz-Cueto, J. A., Carrillo, M., Kah, O., and Zanuy, S. (2000). Characterization of neuropeptide Y expression in the brain of a perciform fish, the sea bass (*Dicentrarchus labrax*). *J. Chem. Neuroanat.* **19,** 197–210.
Cerdá-Reverter, J. M., Zanuy, S., and Munoz-Cueto, J. A. (2001a). Cytoarchitectonic study of the brain of a perciform species, the sea bass (*Dicentrarchus labrax*). I. The telencephalon. *J. Morphol.* **247,** 217–228.
Cerdá-Reverter, J. M., Zanuy, S., and Munoz-Cueto, J. A. (2001b). Cytoarchitectonic study of the brain of a perciform species, the sea bass (*Dicentrarchus labrax*). II. The diencephalon. *J. Morphol.* **247,** 229–251.
Cerdá-Reverter, J. M., Schiöth, H. B., and Peter, R. E. (2003a). The central melanocortin system regulates food intake in goldfish. *Regul. Pept.* **115,** 101–113.
Cerdá-Reverter, J. M., Ringholm, A., Schiöth, H. B., and Peter, R. E. (2003b). Molecular cloning, pharmacological characterization and brain mapping of the melanocortin 4 receptor in the goldfish: Involvement in the control of food intake. *Endocrinology* **144,** 2336–2349.
Cerdá-Reverter, J. M., Ling, M., Schiöth, H. B., and Peter, R. E. (2003c). Molecular cloning, pharmacological characterization and brain mapping of the melanocortin 5 receptor in the goldfish. *J. Neurochem.* **87,** 1354–1367.
Chandra, R., and Liddle, R. A. (2007). Cholecystokinin. *Curr. Opin. Endocrinol. Diabetes Obes.* **14,** 63–67.
Chiba, A. (2005). Neuropeptide Y-immunoreactive (NPY-ir) structures in the brain of the gar *Lepisosteus oculatus* (Lepisosteiformes, Osteichthyes) with special regard to their anatomical relations to gonadotropin-releasing hormone (GnRH)-ir structures in the hypothalamus and the terminal nerve. *Gen. Comp. Endocrinol.* **142,** 336–346.
Chiba, A., Chang, Y., and Honma, Y. (1996). Distribution of neuropeptide Y and gonadotropin-releasing hormone immunoreactivities in the brain and hypophysis of the ayu, *Plecoglossus altivelis* (Teleostei). *Arch. Histol. Cytol.* **59,** 137–148.
Corio, M., Peute, J., and Steinbusch, H. W. M. (1991). Distribution of serotonin- and dopamine-immunoreactivity in the brain of the teleost *Clarias gariepinus*. *J. Chem. Neuroanat.* **4,** 79–95.
Cornbrooks, E. B., and Parsons, R. L. (1991). Sexually dimorphic distribution of a galanin-like peptide in the central nervous system of the teleost fish *Poecilia latipinna*. *J. Comp. Neurol.* **304,** 639–657.
Corrêa, S. (2004). Re-evaluation of the afferent connections of the pituitary in the weakly electric fish *Apteronutus leptorhynchus*: An *in vitro* tract-tracing study. *J. Comp. Neurol.* **470,** 39–49.
Cuadrado, M. I., Coveñas, R., and Tramu, G. (1994). Distribution of gastrin-releasing peptide/bombesin-like immunoreactivity in the rainbow trout brain. *Peptides* **15,** 1027–1032.

Cumming, R., Reaves, T. A., Jr., and Hayward, J. N. (1982). Ultrastructural immunocytochemical characterization of isotocin, vasotocin and neurophysin neurons in the magnocellular preoptic nucleus of the goldfish. *Cell Tissue Res.* **223,** 685–694.

Danger, J.-M., Breton, B., Vallarino, M., Fournier, A., Pelletier, G., and Vaudry, H. (1991). Neuropeptide-Y in the trout brain and pituitary: Localization, characterization, and action on gonadotropin release. *Endocrinology* **128,** 2360–2368.

De Lecea, L., Kilduff, T. S., Peyron, C., Gao, X.-B., Foye, P. E., Danielson, P. E., Fukuhara, C., Battenberg, E. L. F., Gautvik, V. T., Bartlett, F. S., II, Frankel, W. N., Van Den Pol, A. N., Bloom, F. E., Gautvik, K. M., and Sutcliffe, J. G. (1998). The hypocretins: Hypothalamus-specific peptides with neuroexcitatory activity. *Proc. Natl. Acad. Sci. USA* **95,** 322–327.

De Pedro, N., Pinillos, M. L., Valenciano, A. I., Alonso-Bedate, M., and Delgado, M. J. (1998). Inhibitory effect of serotonin on feeding behavior in goldfish: Involvement of CRF. *Peptides* **19,** 505–511.

Díaz, M. L., Becerra, M., Manso, M. J., and Anadón, R. (2001). Development of thyrotropin-releasing hormone immunoreactivity in the brain of the brown trout *Salmo trutta fario*. *J. Comp. Neurol.* **429,** 299–320.

Díaz, M. L., Becerra, M., Manso, M. J., and Anadón, R. (2002). Distribution of thyrotropin-releasing hormone (TRH) immunoreactivity in the brain of the zebrafish (*Danio rerio*). *J. Comp. Neurol.* **450,** 45–60.

Douglas, J., McKinzie, A. A., and Couceyro, P. (1995). PCR differential display identifies a rat brain mRNA that is transcriptionally regulated by cocaine and amphetamine. *J. Neurosci.* **15,** 2471–2481.

Duarte, G., Segura-Noguera, M. M., Martín del Río, M. P., and Mancera, J. M. (2001). The hypothalamo-hypophyseal system of the white seabream *Diplodus sargus*: Immunocytochemical identification of arginine-vasotocin, isotocin, melanin-concentrating hormone and corticotropin-releasing factor. *Histochem. J.* **33,** 569–578.

Dylag, T., Kotlinska, J., Rafalski, P., Pachuta, A., and Siberring, J. (2006). The activity of CART peptide fragments. *Peptides* **27,** 1926–1933.

Eales, J. G., and Himick, B. A. (1988). The effect of TRH on plasma thyroid hormone levels of rainbow trout (*Salmo gairdneri*) and arctic charr (*Salvelinus alpinus*). *Gen. Comp. Endocrinol.* **72,** 333–339.

Ekström, P., and Ebbesson, S. O. (1989). Distribution of serotonin-immunoreactive neurons in the brain of sockeye salmon fry. *J. Chem. Neuroanat.* **2,** 201–213.

Fernald, R. D., and Shelton, L. C. (1985). The organization of the diencephalon and pretectum in the cichlid fish, *Haplochromis burtoni*. *J. Comp. Neurol.* **238,** 202–217.

Fliers, E., Unmehopa, U. A., and Alkemade, A. (2006). Functional neuroanatomy of thyroid hormone feedback in the human hypothalamus and pituitary gland. *Mol. Cell. Endocrinol.* **251,** 1–8.

Flik, G., Klaren, P. H. M., Van Den Burg, E. H., Metz, J. R., and Huising, M. O. (2006). CRF and stress in fish. *Gen. Comp. Endocrinol.* **146,** 36–44.

Flynn, F. W. (1991). Effects of fourth ventricle bombesin injection on meal-related parameters and grooming behavior. *Peptides* **12,** 761–765.

Foran, C. M., and Bass, A. H. (1998). Preoptic AVT immunoreactive neurons of a teleost fish with alternative reproductive tactics. *Gen. Comp. Endocrinol.* **111,** 271–282.

Forlano, P. M., and Cone, R. D. (2007). Conserved neurochemical pathways involved in hypothalamic control of energy homeostasis. *J. Comp. Neurol.* **505,** 235–248.

Fryer, J. N., and Maler, L. (1981). Hypophysiotrophic neurons in the goldfish hypothalamus demonstrated by retrograde transport of horseradish peroxidase. *Cell Tissue Res.* **218,** 93–102.

Fryer, J. N., Boudreault-Chateauvert, C., and Kirby, R. P. (1985). Pituitary afferents originating in the paraventricular organ (PVO) of the goldfish hypothalamus. *J. Comp. Neurol.* **242,** 475–484.

Fujimoto, M., Takeshita, K. I., Wang, X., Takabatake, I., Fujisawa, Y., Teranishi, H., Ohtani, M., Muneoka, Y., and Ohta, S. (1998). Isolation and characterization of a novel bioactive peptide, *Carassius* RFamide (C-RFa), from the brain of the Japanese crucian carp. *Biochem. Bioph. Res. Co.* **242,** 436–440.

Fujimoto, M., Sakamoto, T., Kanetoh, T., Osaka, M., and Moriyama, S. (2006). Prolactin-releasing peptide is essential to maintain the prolactin level and osmotic balance in freshwater teleost fish. *Peptides* **27,** 1104–1109.

Fukusumi, S., Habata, Y., Yoshida, H., Iijima, N., Kawamata, Y., Hosoya, M., Fujii, R., Hinuma, S., Kitada, C., Shintani, Y., Suenaga, M., Onda, H., *et al.* (2001). Characteristics and distribution of endogenous RFamide-related peptide-1. *Biochim. Biophys. Acta Mol. Cell Res.* **1540,** 221–232.

Gaikwad, A., Biju, K. C., Saha, S. G., and Subhedar, N. (2004). Neuropeptide Y in the olfactory system, forebrain and pituitary of the teleost, *Clarias batrachus*. *J. Chem. Neuroanat.* **27,** 55–70.

Gaikwad, A., Biju, K. C., Muthal, P. L., Saha, S., and Subhedar, N. (2005). Role of neuropeptide Y in the regulation of gonadotropin releasing hormone system in the forebrain of *Clarias batrachus* (Linn.): Immunocytochemistry and high performance liquid chromatography-electrospray ionization-mass spectrometric analysis. *Neuroscience* **133,** 267–279.

Gibbs, J., Fauser, D. J., and Rowe, E. A. (1979). Bombesin suppresses feeding in rats. *Nature* **282,** 208–210.

Gómez-Segade, P., and Anadón, R. (1988). Specialization in the diencephalon of advanced teleost. *J. Morphol.* **197,** 71–103.

González-Martínez, D., Zmora, N., Mañanos, E., Saligaut, D., Zanuy, S., Zohar, Y., Elizur, A., Kah, O., and Muñoz-Cueto, J. A. (2002). Immunohistochemical localization of three different prepro-GnRHs in the brain and pituitary of the European sea bass (*Dicentrarchus labrax*) using antibodies to the corresponding GnRH-associated peptides. *J. Comp. Neurol.* **446,** 95–113.

Goodson, J. L., and Bass, A. H. (2001). Social behavior functions and related anatomical characteristics of vasotocin/vasopressin systems in vertebrates. *Brain Res. Rev.* **35,** 246–265.

Goodson, J. L., Evans, A. K., and Bass, A. H. (2003). Putative isotocin distributions in sonic fish: Relation to vasotocin and vocal-acoustic circuitry. *J. Comp. Neurol.* **462,** 1–14.

Goossens, N., Dierickx, K., and Vandesande, F. (1977). Immunocytochemical localization of vasotocin and isotocin in the preopticohypophysial neurosecretory system of teleosts. *Gen. Comp. Endocrinol.* **32,** 371–375.

Gorbman, A., Dickhoff, W.W, Vigna, S. R., Clark, N. B., and Ralph, C. L. (1983). "Comparative Endocrinology". Wiley-Interscience, New York.

Grau, E. G., Nishioka, R. S., Young, G., and Bern, H. A. (1985). Somatostatin-like immunoreactivity in the pituitary and brain of three teleost fish species – somatostatin as a potential regulator of prolactin cell-function. *Gen. Comp. Endocrinol.* **59,** 350–357.

Green, J. D. (1951). The comparative anatomy of the hypophysis with special reference to its blood supply and innervation. *Am. J. Anat.* **88,** 225–311.

Grober, M. S., and Sunobe, T. (1996). Serial adult sex change involves rapid and reversible changes in forebrain neurochemistry. *Neuroreport* **7,** 2945–2949.

Gröneveld, D., Balm, P. H. M., Martens, G. J. M., and Wenderlaar Bonga, S. E. (1995). Differential melanin-concentrating hormone expression in two hypothalamic nuclei of the teleost tilapia in response to environmental changes. *J. Neuroendocrinol.* **7,** 527–533.

Guillemin, R. (1970). Hormones secreted by the brain. Isolation, molecular structure and synthesis of the first hypophysiotropic hypothalamic hormone (to be discovered), TRF (thyrotropin-releasing factor). *Science* **68,** 64–67.

Hamano, K., Inoue, K., and Yanagisawa, T. (1990). Immunohistochemical localization of thyrotropin-releasing hormone in the brain of carp, *Cyprinus carpio*. *Gen. Comp. Endocrinol.* **80,** 85–94.

Hernández-Rauda, R., and Aldegunde, M. (2002). Changes in dopamine, norepinephrine and serotonin levels in the pituitary, telencephalon and hypothalamus during gonadal development of male *Lutjanus argentiventris* (Teleostei). *Mar. Biol.* **141,** 209–216.

Hernández-Rauda, R., Otero, J., Rey, P., Rozas, G., and Aldegunde, M. (1996). Dopamine and serotonin in the trout (*Oncorhynchus mykiss*) pituitary: Main metabolites and changes during gonadal recrudescence. *Gen. Comp. Endocrinol.* **103,** 13–23.

Hill, J. J., and Henderson, N. E. (1968). The vascularization of the hypothalamichypophysial region of the eastern brook trout, *Salvelinus fontinalis*. *Am. J. Anat.* **122,** 301–316.

Himick, B. A., and Peter, R. E. (1994). CCK/gastrin-like immunoreactivity in brain and gut, and CCK suppression of feeding in goldfish. *Am. J. Physiol.* **267,** R841–R851.

Himick, B. A., and Peter, R. E. (1995). Bombesin-like immunoreactivity in the forebrain and pituitary and regulation of anterior pituitary hormone-release by bombesin in goldfish. *Neuroendocrinology* **61,** 365–376.

Himick, B. A., Golosinski, A. A., Jonsson, A. C., and Peter, R. E. (1993). CCK/gastrinlike immunoreactivity in the goldfish pituitary – Regulation of pituitary hormone secretion by CCK-like peptides *in vitro*. *Gen. Comp. Endocrinol.* **92,** 88–103.

Hinuma, S., Habata, Y., Fujii, R., Kawamata, Y., Hosoya, M., Fukusumi, S., Kitada, C., Masuo, Y., Asano, T., Matsumoto, H., Sekiguchi, M., Kurokawa, T., *et al.* (1998). A prolactin-releasing peptide in the brain. *Nature* **393,** 272–276.

Hinuma, S., Shintani, Y., Fukusumi, S., Iijima, N., Matsumoto, Y., Hosoya, M., Fujii, R., Watanabe, T., Kikuchi, K., Terao, Y., Yano, T., Yamamoto, T., *et al.* (2000). New neuropeptides containing carboxyterminal RFamide and their receptor in mammals. *Nat. Cell Biol.* **2,** 703–708.

Holloway, A. C., and Leatherland, J. F. (1997). The effects of N-methyl-D,L-aspartate and gonadotropin-releasing hormone on *in vitro* growth hormone release in steroidprimed immature rainbow trout, *Oncorhynchus mykiss*. *Gen. Comp. Endocrinol.* **107,** 32–43.

Holmgren, S., and Jensen, J. (1994). Comparative aspects on the biochemical identity of neurotransmitters of autonomic neurons. *In* "Comparative physiology and evolution of the autonomic nervous system" (S. Nilsson, and S. Holmgren, Eds.), pp. 69–95. Harwood Academic, Chur, Switzerland.

Holmgren, S., and Jonsson, A. C. (1988). Occurrence and effects on motility of bombesin related peptides in the gastrointestinal tract of the Atlantic cod, *Gadus morhua*. *Comp. Biochem. Physiol.* **89C,** 249–256.

Holmqvist, B. I., and Ekström, P. (1995). Hypophysiotrophic systems in the brain of the Atlantic salmon: Neuronal innervation of the pituitary and the origin of pituitary dopamine and nonapeptides identified by means of combined carbocyanine tract tracing and immunocytochemistry. *J. Chem. Neuroanat.* **8,** 125–145.

Holmqvist, B. I., Östholm, T., and Ekström, P. (1992). DiI tracing in combination with the immunocytochemistry for analysis of connectivities and chemoarchitectonics of specific neuronal systems in a teleost, the Atlantic salmon. *J. Neurosci. Meth.* **42,** 45–63.

Hosoya, M., Kimura, C., Ogi, K., Ohkubo, S., Miyamoto, Y., Kugoh, H., Shimizu, M., Onda, H., Oshimura, M., Arimura, A., and Fujino, M. (1992). Structure of the human pituitary adenylate cyclase activating polypeptide (PACAP) gene. *Biochim. Biophys. Acta – Gene Struct. Expression* **1129,** 199–206.

Huesa, G., van den Pol, A. N., and Finger, T. E. (2005). Differential distribution of hypocretin (orexin) and melanin-concentrating hormone in the goldfish brain. *J. Comp. Neurol.* **488,** 476–491.

Huising, M. O., Metz, J. R., van Schooten, C., Taverne-Thiele, A. J., Hermsen, T., Verburg-van Kemenade, B. M. L., and Flik, G. (2004). Structural characterisation of a cyprinid (*Cyprinus carpio* L.) CRH, CRH-BP and CRH-R1, and the role of these proteins in the acute stress response. *J. Mol. Endocrinol.* **32,** 627–648.

Jadhao, A. G., and Meyer, D. L. (2000). Sexually dimorphic distribution of galanin in the preoptic area of red salmon, *Oncorhynchus nerka*. *Cell Tissue Res.* **302,** 199–203.

Jadhao, A., and Pinelli, C. (2001). Galanin-like immunoreactivity in the brain and pituitary of the "four-eyed" fish, *Anableps anableps*. *Cell Tissue Res.* **306,** 309–318.

Jensen, H., Rourke, I. J., Moller, M., Jonson, L., and Johnsen, A. H. (2001). Identification and distribution of CCK-related peptides and mRNAs in the rainbow trout, *Oncorhynchus mykiss*. *Biochim. Biophys. Acta – Gene Struct. Expression* **1517,** 190–201.

Johansson, V., Winberg, S., Jonsson, E., Hall, D., and Bjornsson, B. T. (2004). Peripherally administered growth hormone increases brain dopaminergic activity and swimming in rainbow trout. *Horm. Behav.* **46,** 436–443.

Johnsen, A. H. (1998). Phylogeny of the cholecystokinin/gastrin family. *Front. Neuroendocrinol.* **19,** 73–99.

Johnston, S. A., and Maler, L. (1992). Anatomical organization of the hypophysiotrophic systems in the electric fish, *Apteronotus leptorhynchus*. *J. Comp. Neurol.* **317,** 421–437.

Johnston, S. A., Maler, L., and Tinner, B. (1990). The distribution of serotonin in the brain of *Apteronotus leptorhynchus*: An immunohistochemical study. *J. Chem. Neuroanat.* **3,** 429–465.

Kah, O., and Chambolle, P. (1983). Serotonin in the brain of the goldfish, *Carassius auratus*. An immunocytochemical study. *Cell Tissue Res.* **234,** 319–333.

Kah, O., Chambolle, P., Dubourg, P., and Dubois, M. P. (1982). Localisation immunocytochimique de la somatostatine dans le cerveau anterieur et l'hypophyse de deux teleosteens, le cyprin (*Carassius auratus*) et *Gambusia sp*. *Comptes Rendus des Seances de l'Academie des Sciences* **294,** 519–524.

Kah, O., Chambolle, P., Thibault, J., and Geffard, M. (1984). Existence of dopaminergic neurons in the preoptic region of the goldfish. *Neurosci. Lett.* **48,** 293–298.

Kah, O., Dubourg, P., and Onteniente, B. (1986). The dopaminergic innervation of the goldfish pituitary. An immunocytochemical study at the electron-microscope level using antibodies against dopamine. *Cell Tissue Res.* **244,** 577–582.

Kah, O., Dubourg, P., Martinoli, M. G., Rabhi, M., Gonnet, F., Geffard, M., and Calas, A. (1987). Central GABAergic innervation of the pituitary in goldfish: A radioautographic and immunocytochemical study at the electron microscope level. *Gen. Comp. Endocrinol.* **67,** 324–332.

Kah, O., Pontet, A., Danger, J.-M., Dubourg, P., Pelletier, G., Vaudry, H., and Calas, A. (1989). Characterization, cerebral distribution and gonadotropin release activity of neuropeptide Y (NPY) in the goldfish. *Fish Physiol. Biochem.* **7,** 69–76.

Kah, O., Trudeau, V. L., Sloley, B. D., Chang, J. P., Dubourg, P., Yu, K. L., and Peter, R. E. (1992). Influence of GABA on gonadotrophin release in the goldfish. *Neuroendocrinology* **55,** 396–404.

Kah, O., Lethimonier, C., Somoza, G., Guilgur, L. G., Vaillant, C., and Lareyre, J. J. (2007). GnRH and GnRH receptors in metazoa: A historical, comparative, and evolutive perspective. *Gen. Comp. Endocrinol.* **153,** 346–364.

Kaslin, J., Nystedt, J. M., Östergård, M., Peitsaro, N., and Panula, P. (2004). The orexin/hypocretin system in zebrafish is connected to the aminergic and cholinergic systems. *J. Neurosci.* **24,** 2678–2689.

Kawauchi, H., and Baker, B. I. (2004). Melanin-concentrating hormone signaling systems in fish. *Peptides* **25,** 1577–1584.

Kawauchi, H., and Sower, S. A. (2006). The dawn and evolution of hormones in the adenohypophysis. *Gen. Comp. Endocrinol.* **148,** 3–14.

Kawauchi, H., Kawazoe, I., Tsubokawa, M., Kishida, M., and Baker, B. I. (1983). Characterization of melanin-concentrating hormone in chum salmon pituitaries. *Nature* **305,** 321–323.

Kawauchi, N., Okubo, K., and Aida, K. (2003). The expression and localization of corticotropin-releasing hormone and urotensin I transcripts in the Japanese eel, *Anguilla japonica. Fish Physiol. Biochem.* **28,** 43–44.

Kehoe, A. S., and Volkoff, H. (2006). Cloning and characterization of neuropeptide Y (NPY) and cocaine and amphetamine regulated transcript (CART) in Atlantic cod (*Gadus morhua*). *Comp. Biochem. Physiol.* **146A,** 451–461.

Kelly, S. P., and Peter, R. E. (2006). Prolactin-releasing peptide, food intake, and hydromineral balance in goldfish. *Am. J. Physiol.* **291,** R1474–R1481.

Khan, I. A., and Thomas, P. (1992). Stimulatory effects of serotonin on maturational gonadotropin release in the Atlantic croaker, *Micropogonias undulatus. Gen. Comp. Endocrinol.* **88,** 388–396.

Khan, I. A., and Thomas, P. (1993). Immunocytochemical localization of serotonin and gonadotropin-releasing hormone in the brain and pituitary gland of the Atlantic croaker *Micropogonias undulatus. Gen. Comp. Endocrinol.* **91,** 167–180.

Khan, I. A., and Thomas, P. (1999). GABA exerts stimulatory and inhibitory influences on gonadotropin II secretion in the Atlantic croaker (*Micropogonias undulatus*). *Neuroendocrinology* **69,** 261–268.

Kim, M. H., Oka, Y., Amano, M., Kobayashi, M., Okuzawa, K., Hasegawa, Y., Kawashima, S., Suzuki, Y., and Aida, K. (1995). Immunocytochemical localization of sGnRH and cGnRH-II in the brain of goldfish, *Carassius auratus. J. Comp. Neurol.* **356,** 72–82.

Kishida, M., Baker, B. I., and Bird, D. J. (1988). Localisation and identification of melanocyte-stimulating hormones in the fish brain. *Gen. Comp. Endocrinol.* **71,** 229–242.

Kong, H. S., Zhou, H., Yang, Y., He, M., Jiang, Y., and Wong, A. O. L. (2007). Pituitary Adenylate Cyclase-Activating Polypeptide (PACAP) as a Growth Hormone (GH)-releasing factor in grass carp: II. Solution structure of a brainspecific PACAP by nuclear magnetic resonance spectroscopy and functional studies on GH release and gene expression. *Endocrinology* **148,** 5042–5059.

Lagios, M. D. (1968). Tetrapod-like organization of the pituitary gland of the polypteriformid fishes, *Calamoichthys calabaricus* and *Polypterus palmas. Gen. Comp. Endocrinol.* **11,** 300–315.

Laiz-Carrión, R., Segura-Noguera, M. M., Martín del Río, M. P., and Mancera, J. M. (2003). Ontogeny of adenohypophyseal cells in the pituitary of American shad (*Alosa sapidissima*). *Gen. Comp. Endocrinol.* **132,** 454–464.

Lamers, A. E., Balm, P. H. M., Haenen, H. E. M. G., Jenks, B. G., and Wendelaar Bonga, S. E. (1991). Regulation of differential release of α-melanocyte-stimulating hormone forms from the pituitary of a teleost fish, *Oreochromis mossambicus. J. Endocrinol.* **129,** 179–187.

Lang, R., Gundlach, A. L., and Kofler, B. (2007). The galanin peptide family: Receptor pharmacology, pleiotropic biological actions, and implications in health and disease. *Pharmacol. Therapeut.* **115,** 177–207.

Larsen, D. A., Swanson, P., Dickey, J. T., Rivier, J., and Dickhoff, W. W. (1998). *In vitro* thyrotropin-releasing activity of corticotropin-releasing hormone-family peptides in coho salmon, *Oncorhynchus kisutch. Gen. Comp. Endocrinol.* **109,** 276–285.

Lee, L. T. O., Siu, F. K. Y., Tam, J. K. V., Lau, I. T. Y., Wong, A. O. L., Lin, M. C. M., Vaudry, H., and Chow, B. K. C. (2007). Discovery of growth hormone releasing hormones and receptors in nonmammalian vertebrates. *Proc. Natl. Acad. Sci. USA* **104,** 2133.

Lema, S. C., and Nevitt, G. A. (2004). Variation in vasotocin immunoreactivity in the brain of recently isolated populations of a death valley pupfish, *Cyprinodon nevadensis*. *Gen. Comp. Endocrinol.* **135,** 300–309.

Lepretre, E., Anglade, I., Williot, P., Vandesande, F., Tramu, G., and Kah, O. (1993). Comparative distribution of mammalian GnRH (gonadotrophin-releasing hormone) and chicken GnRH-II in the brain of the immature Siberian sturgeon (*Acipenser baeri*). *J. Comp. Neurol.* **337,** 568–583.

Lethimonier, C., Madigou, T., Munoz-Cueto, J. A., Lareyre, J. J., and Kah, O. (2004). Evolutionary aspects of GnRHs, GnRH neuronal systems and GnRH receptors in teleost fish. *Gen. Comp. Endocrinol.* **135,** 1–16.

Lin, X. W., Otto, C. J., and Peter, R. E. (1999). Expression of three distinct somatostatin messenger ribonucleic acids (mRNAs) in goldfish brain: Characterization of the complementary deoxyribonucleic acids, distribution and seasonal variation of the mRNAs, and action of a somatostatin-14 variant. *Endocrinology* **140,** 2089–2099.

Lin, X. W., Volkoff, H., Narnaware, Y., Bernier, N. J., Peyon, P., and Peter, R. E. (2000). Brain regulation of feeding behavior and food intake in fish. *Comp. Biochem. Physiol.* **126A,** 415–434.

Logan, D. W., Bryson-Richardson, R. J., Pagán, K. E., Taylor, M. S., Currie, P. D., and Jackson, I. J. (2003). The structure and evolution of the melanocortin and MCH receptors in fish and mammals. *Genomics* **81,** 184–191.

Luo, D., and McKeown, B. A. (1989). Immunohistochemical detection of a substance resembling growth hormone-releasing factor in the brain of the rainbow trout (*Salmo gairdneri*). *Experientia* **45,** 577–580.

Ma, P. M. (1997). Catecholaminergic systems in the zebrafish. III. Organization and projection pattern of medullary dopaminergic and noradrenergic neurons. *J. Comp. Neurol.* **381,** 411.

Ma, P. M. (2003). Catecholaminergic systems in the zebrafish. IV. Organization and projection pattern of dopaminergic neurons in the diencephalon. *J. Comp. Neurol.* **460,** 13–37.

Ma, P. M., and Lopez, M. (2003). Consistency in the number of dopaminergic paraventricular organ-accompanying neurons in the posterior tuberculum of the zebrafish brain. *Brain Res.* **967,** 267–272.

Magliulo-Cepriano, L., Schreibman, M. P., and Blum, V. (1993). The distribution of immunoreactive FMRF-amide, neurotensin, and galanin in the brain and pituitary gland of three species of *Xiphophorus* from birth to sexual maturity. *Gen. Comp. Endocrinol.* **92,** 269–280.

Maler, L., Sas, E., Johnston, S., and Ellis, W. (1991). An atlas of the brain of the electric fish *Apteronotus leptorhynchus*. *J. Chem. Neuroanat.* **4,** 1–38.

Mañanos, E. L., Anglade, I., Chyb, J., Saligaut, C., Breton, B., and Kah, O. (1999). Involvement of γ-aminobutyric acid in the control of GTH-1 and GTH-2 secretion in male and female rainbow trout. *Neuroendocrinology* **69,** 269–280.

Mancera, J. M., and Fernández-Llébrez, P. (1995). Localization of corticotropin-releasing factor immunoreactivity in the brain of the teleost *Sparus aurata*. *Cell Tissue Res.* **281,** 569–572.

Marchetti, G., Cozzi, B., Tavanti, M., Russo, V., Pellegrini, S., and Fabiani, O. (2000). The distribution of Neuropeptide Y-immunoreactive neurons and nerve fibers in the forebrain of the carp *Cyprinus carpio* L. *J. Chem. Neuroanat.* **20,** 129–139.

Margolis-Kazan, H., Halpern-Sebold, L. R., and Schreibman, M. P. (1985). Immunocytochemical localization of serotonin in the brain and pituitary gland of the platyfish, *Xiphophorus maculatus*. *Cell Tissue Res.* **240,** 311–314.

Marivoet, S., Moons, L., and Vandesande, F. (1988). Localization of growth hormone releasing factor-like immunoreactivity in the hypothalamo-hypophyseal system of the frog (*Rana temporaria*) and the sea bass (*Dicentrarchus labrax*). *Gen. Comp. Endocrinol.* **72,** 72–79.

Martínez, V., and Tache, Y. (2000). Bombesin and the brain–gut axis. *Peptides* **21,** 1617–1625.

Martinoli, M. G., Dubourg, P., Geffard, M., Calas, A., and Kah, O. (1990). Distribution of GABA-immunoreactive neurons in the forebrain of the goldfish, *Carassius auratus*. *Cell Tissue Res.* **260,** 77–84.

Martyniuk, C. J., Awad, R., Hurley, R., Finger, T. E., and Trudeau, V. L. (2007a). Glutamic acid decarboxylase 65, 67, and GABA-transaminase mRNA expression and total enzyme activity in the goldfish (*Carassius auratus*) brain. *Brain Res.* **1147,** 154–166.

Martyniuk, C. J., Chang, J. P., and Trudeau, V. L. (2007b). The effects of GABA agonists on glutamic acid decarboxylase, GABA-transaminase, activin, salmon gonadotrophin-releasing hormone and tyrosine hydroxylase mRNA in the goldfish (*Carassius auratus*) neuroendocrine brain. *J. Neuroendocrinol.* **19,** 390–396.

Maruska, K. P., Mizobe, M. H., and Tricas, T. C. (2007). Sex and seasonal co-variation of arginine vasotocin (AVT) and gonadotropin-releasing hormone (GnRH) neurons in the brain of the halfspotted goby. *Comp. Biochem. Physiol.* **147A,** 129–144.

Mathieu, M., Tagliafierro, G., Bruzzone, F., and Vallarino, M. (2002). Neuropeptide tyrosine-like immunoreactive system in the brain, olfactory organ and retina of the zebrafish, *Danio rerio*, during development. *Dev. Brain Res.* **139,** 255–265.

Matsuda, K., Nagano, Y., Uchiyama, M., Onoue, S., Takahashi, A., Kawauchi, H., and Shioda, S. (2005a). Pituitary adenylate cyclase-activating polypeptide (PACAP)-like immunoreactivity in the brain of a teleost, *Uranoscopus japonicus*: Immunohistochemical relationship between PACAP and adenohypophysial hormones. *Regul. Pept.* **126,** 129–136.

Matsuda, K., Nagano, Y., Uchiyama, M., Takahashi, A., and Kawauchi, H. (2005b). Immunohistochemical observation of pituitary adenylate cyclase-activating polypeptide (PACAP) and adenohypophysial hormones in the pituitary of a teleost, *Uranoscopus japonicus*. *Zool. Sci.* **22,** 71–76.

Matz, S. P., and Hofeldt, G. T. (1999). Immunohistochemical localization of corticotropin-releasing factor in the brain and corticotropin-releasing factor and thyrotropin-stimulating hormone in the pituitary of Chinook salmon (*Oncorhynchus tshawytscha*). *Gen. Comp. Endocrinol.* **114,** 151–160.

Matz, S. P., and Takahashi, T. T. (1994). Immunohistochemical localization of thyrotropin-releasing hormone in the brain of chinook salmon (*Oncorhynchus tshawytscha*). *J. Comp. Neurol.* **345,** 214–223.

Mayo, K. E., Cerelli, G. M., and Lebo, R. V. (1985). Gene encoding human growth hormone-releasing factor precursor: Structure, sequence, and chromosomal assignment. *P. Natl. Acad. Sci. USA* **82,** 63–67.

McCormick, S. D., and Bradshaw, D. (2006). Hormonal control of salt and water balance in vertebrates. *Gen. Comp. Endocrinol.* **147,** 3–8.

McCoy, J. G., and Avery, D. D. (1990). Bombesin: Potential integrative peptide for feeding and satiety. *Peptides* **11,** 595–607.

McDonald, T. J., Jornvall, H., and Nilsson, G. (1979). Characterization of a gastrin releasing peptide from porcine non-antral gastric tissue. *Biochem. Biophys. Res. Commun.* **90,** 227–233.

Medina, M., Reperant, J., Dufour, S., Ward, R., Le Belle, N., and Miceli, D. (1994). The distribution of GABA-immunoreactive neurons in the brain of the silver eel (*Anguilla anguilla* L.). *Anat. Embryol.* **189,** 25–39.

Meek, J., and Joosten, H. W. J. (1989). Distribution of serotonin in the brain of the mormyrid teleost *Gnathonemus petersii*. *J. Comp. Neurol.* **281,** 206–224.

Meek, J., and Nieuwenhuys, R. D. (1998). Holosteans and teleost. *In* "The Central Nervous System of Vertebrates". (R. Nieuwenhuys, H. J. Ten Donkelaar, and C. Nicholson, Eds.), Vol. 1, pp. 759–937. Springer-Verlag, Heidelberg.

Meek, J., Joosten, H. W. J., and Steinbusch, H. W. M. (1989). Distribution of dopamine immunoreactivity in the brain of the mormyrid teleost *Gnathonemus petersii*. *J. Comp. Neurol.* **2**(81), 362–383.

Melamed, P., Eliahu, N., Levavi-Sivan, B., Ofir, M., Farchi-Pisanty, O., Rentier-Delrue, F., Smal, J., Yaron, Z., and Naor, Z. (1995). Hypothalamic and thyroidal regulation of growth hormone in tilapia. *Gen. Comp. Endocrinol.* **97,** 13–30.

Metz, J. R., Huising, M. O., Meek, J., Taverne-Thiele, A. J., Bonga, S. E. W., and Flik, G. (2004). Localization, expression and control of adrenocorticotropic hormone in the nucleus preopticus and pituitary gland of common carp (*Cyprinus carpio* L.). *J. Endocrinol.* **182,** 23–31.

Miranda, L. A., Strobl-Mazzulla, P. H., and Somoza, G. M. (2002). Ontogenetic development and neuroanatomical localization of growth hormone-releasing hormone (GHRH) in the brain and pituitary gland of pejerrey fish *Odontesthes bonariensis*. *Int. J. Dev. Neurosci.* **20,** 503–510.

Mitchell, G., Sawisky, G. R., Grey, C. L., Wong, C. J., Uretsky, A. D., and Chang, J. P. (2008). Differential involvement of nitric oxide signaling in dopamine and PACAP stimulation of growth hormone release in goldfish. *Gen.Comp. Endocrinol.* **155,** 318–327.

Mohamed, J. S., Thomas, P., and Khan, I. A. (2005). Isolation, cloning, and expression of three prepro-GnRH mRNAs in Atlantic croaker brain and pituitary. *J. Comp. Neurol.* **488,** 384.

Montefusco-Siegmund, R. A., Romero, A., Kausel, G., Muller, M., Fujimoto, M., and Figueroa, J. (2006). Cloning of the prepro C-RFa gene and brain localization of the active peptide in *Salmo salar*. *Cell Tissue Res.* **325,** 277.

Montero, M., Vidal, B., King, J. A., Tramu, G., Vandesande, F., Dufour, S., and Kah, O. (1994). Immunocytochemical localization of mammalian GnRH (gonadotropin-releasing hormone) and chicken GnRH-II in the brain of the European silver eel (*Anguilla anguilla* L.). *J. Chem. Neuroanat.* **7,** 227–241.

Montero, M., Yon, L., Rousseau, K., Arimura, A., Fournier, A., Dufour, S., and Vaudry, H. (1998). Distribution, characterization, and growth hormone-releasing activity of pituitary adenylate cyclase-activating polypeptide in the European eel, *Anguilla anguilla*. *Endocrinology* **139,** 4300–4310.

Moons, L., Cambre, M., Marivoet, S., Batten, T. F., Vanderhaeghen, J. J., Ollevier, F., and Vandesande, F. (1988). Peptidergic innervation of the adrenocorticotropic hormone (ACTH)- and growth hormone (GH)-producing cells in the pars distalis of the sea bass (*Dicentrarchus labrax*). *Gen. Comp. Endocrinol.* **72,** 171–180.

Moons, L., Cambré, M., Ollevier, F., and Vandesande, F. (1989). Immunocytochemical demonstration of close relationships between neuropeptidergic nerve fibers and hormone-producing cell types in the adenohypophysis of the sea bass (*Dicentrarchus labrax*). *Gen. Comp. Endocrinol.* **73,** 270–283.

Moons, L., Batten, T. F. C., and Vandesande, F. (1991). Autoradiographic distribution of galanin binding sites in the brain and pituitary of the sea bass (*Dicentrarchus labrax*). *Neurosci. Lett.* **123,** 49–52.

Moriyama, S., Toshihiro, I. T. O., Takahashi, A., Amano, M., Sower, S. A., Hirano, T., Yamamori, K., and Kawauchi, H. (2002). A homolog of mammalian PRL-releasing peptide (fish arginyl-phenylalanyl-amide peptide) is a major hypothalamic peptide of prl release in teleost fish. *Endocrinology* **143,** 2071–2079.

Moriyama, S., Kasahara, M., Amiya, N., Takahashi, A., Amano, M., Sower, S. A., Yamamori, K., and Kawauchi, H. (2007). RFamide peptides inhibit the expression of melanotropin and growth hormone genes in the pituitary of an agnathan, the sea lamprey, *Petromyzon marinus*. *Endocrinology* **148,** 3740–3749.

Mousa, M. A., and Mousa, S. A. (2003). Immunohistochemical localization of gonadotropin releasing hormones in the brain and pituitary gland of the Nile perch, *Lates niloticus* (Teleostei, Centropomidae). *Gen. Comp. Endocrinol.* **130,** 245–255.

Mukuda, T., Matsunaga, Y., Kawamoto, K., Yamaguchi, K.-I., and Ando, M. (2005). "Blood-contacting neurons" in the brain of the Japanese eel *Anguilla japonica. J. Exp. Zool.* **303A,** 366–376.

Nakamachi, T., Matsuda, K., Maruyama, K., Miura, T., Uchiyama, M., Funahashi, H., Sakurai, T., and Shioda, S. (2006). Regulation by orexin of feeding behaviour and locomotor activity in the goldfish. *J. Neuroendocrinol.* **18,** 290–297.

Nakanishi, S. (1992). Molecular diversity of glutamate receptors and implications for brain function. *Science* **258,** 597–603.

Nelson, L. E., and Sheridan, M. A. (2005). Regulation of somatostatins and their receptors in fish. *Gen. Comp. Endocrinol.* **142,** 117–133.

Nieuwenhuys, R., Ten Donkelaar, H. J., and Nicholson, C. (1998). "The Central Nervous System of Vertebrates". Springer-Verlag, Heidelberg.

Northcutt, R. G. (1995). The forebrain of gnathostomes: in search of a morphotype. *Brain Behav. Evol.* **46,** 275–318.

Northcutt, R. G., and Davis, R. E. (1983). Telencephalic organization in ray-finned fishes. *In* "Fish Neurobiology" (R. G. Northcutt, and R. E. Davis, Eds.), Vol. 2, pp. 203–236. University of Michigan Press, Ann Arbor.

Notenboom, C. D., Garaud, J. C., Doerr-Schott, J., and Terlou, M. (1981). Localization by immunofluorescence of a gastrin-like substance in the brain of the rainbow trout, *Salmo gairdneri. Cell Tissue Res.* **214,** 247–255.

Novak, C. M., Jiang, X., Wang, C., Teske, J. A., Kotz, C. M., and Levine, J. A. (2005). Caloric restriction and physical activity in zebrafish (*Danio rerio*). *Neurosci. Lett.* **383,** 99–104.

Nozaki, M., Gorbman, A., and Sower, S. A. (1994). Diffusion between the neurohypophysis and the adenohypophysis of lampreys, *Petromyzon marinus. Gen. Comp. Endocrinol.* **96,** 385–391.

Ohtaki, T., Shintani, Y., Honda, S., Matsumoto, H., Hori, A., Kanehashi, K., Terao, Y., Kumano, S., Takatsu, Y., Masuda, Y., Ishibashi, Y., Watanabe, T., *et al.* (2001). Metastasis suppressor gene KiSS-1 encodes peptide ligand of a G-protein-coupled receptor. *Nature* **411,** 613–617.

Okawara, Y., Ko, D., Morley, S. D., Richter, D., and Lederis, K. P. (1992). *In situ* hybridization of corticotropin-releasing factor-encoding messenger RNA in the hypothalamus of the white sucker, *Catostomus commersoni. Cell Tissue Res.* **267,** 545–549.

Olivereau, M., and Olivereau, J. (1988). Localization of CRF-like immunoreactivity in the brain and pituitary of teleost fish. *Peptides* **9,** 13–21.

Olivereau, M., and Olivereau, J. M. (1990). Corticotropin-like immunoreactivity in the brain and pituitary of three teleost species (goldfish, trout and eel). *Cell Tissue Res.* **262,** 115–123.

Olivereau, M., and Olivereau, J. M. (1991). Immunocytochemical localization of a galanin-like peptidergic system in the brain and pituitary of some teleost fish. *Histochemistry* **96,** 343–354.

Olivereau, M., Ollevier, F., Vandesande, F., and Olivereau, J. (1984a). Somatostatin in the brain and the pituitary of some teleosts. Immunocytochemical identification and the effect of starvation. *Cell Tissue Res.* **238,** 289–296.

Olivereau, M., Ollevier, F., Vandesande, F., and Verdonck, W. (1984b). Immunocytochemical identification of CRF-like and SRIF-like peptides in the brain and the pituitary of cyprinid fish. *Cell Tissue Res.* **237,** 379–382.

Olivereau, M., Moons, L., Olivereau, J., and Vandesande, F. (1988). Coexistence of corticotropin-releasing factor-like immunoreactivity and vasotocin in perikarya of the preoptic nucleus in the eel. *Gen. Comp. Endocrinol.* **70,** 41–48.

Olivereau, M., Olivereau, J., and Vandesande, F. (1990). Localization of growth hormone-releasing factor-like immunoreactivity in the hypothalamo-hypophysial system of some teleost species. *Cell Tissue Res.* **259,** 73–80.

Palay, S. L. (1945). Neurosecretion. VII. The preoptico-hypophyseal pathway in fishes. *J. Comp. Neurol.* **82,** 129–143.

Pan, J. X., Lechan, R. M., Lin, H. D., and Jackson, I. M. D. (1985). Immunoreactive neuronal pathways of growth hormone-releasing hormone (GRH) in the brain and pituitary of the teleost *Gadus morhua. Cell Tissue Res.* **241,** 487–493.

Pandolfi, M., Cánepa, M. M., Ravaglia, M. A., Maggese, M. C., Paz, D. A., and Vissio, P. G. (2003). Melanin-concentrating hormone system in the brain and skin of the cichlid fish *Cichlasoma dimerus*: Anatomical localization ontogeny and distribution in comparison to α-melanocyte-stimulating hormone-expressing cells. *Cell Tissue Res.* **311,** 61.

Pandolfi, M., Cueto, J. A. M., Lo Nostro, F. L., Downs, J. L., Paz, D. A., Maggese, M. C., and Urbanski, H. F. (2005). GnRH systems of *Cichlasoma dimerus* (Perciformes, Cichlidae) revisited: A localization study with antibodies and riboprobes to GnRH-associated peptides. *Cell Tissue Res.* **321,** 219–232.

Parhar, I. S. (1997). GnRH in tilapia: three genes, three origins and their roles. *In* "GnRH Neurons, Gene to behavior" (I. S. Parhar, and Y. Sakuma, Eds.), pp. 99–122. Brain Shuppan, Tokyo.

Parhar, I. S., and Iwata, M. (1994). Gonadotropin releasing hormone (GnRH) neurons project to growth hormone and somatolactin cells in the Steelhead trout. *Histochemistry* **102,** 195–203.

Parhar, I. S., Iwata, M., Pfaff, D. W., and Schwanzel-Fukuda, M. (1995). Embryonic development of gonadotropin-releasing hormone neurons in the Sockeye salmon. *J. Comp. Neurol.* **362,** 256–270.

Parhar, I. S., Ogawa, S., and Sakuma, Y. (2004). Laser-captured single digoxigenin-labeled neurons of gonadotropin-releasing hormone types reveal a novel G protein-coupled receptor (GPR54) during maturation in cichlid fish. *Endocrinology* **145,** 3613–3618.

Peng, C., Chang, J. P., Yu, K. L., Wong, A. O.-L., Van Goor, F., Peter, R. E., and Rivier, J. E. (1993). Neuropeptide-Y stimulates growth hormone and gonadotropin-II secretion in the goldfish pituitary: Involvement of both presynaptic and pituitary cell actions. *Endocrinology* **132,** 1820–1829.

Peng, C., Gallin, W., Peter, R. E., Blomqvist, A. G., and Larhammar, D. (1994). Neuropeptide-Y gene expression in the goldfish brain: distribution and regulation by ovarian steroids. *Endocrinology* **134,** 1095–1103.

Pepels, P. P. L. M., Meek, J., Wendelaar Bonga, S. E., and Balm, P. H. M. (2002). Distribution and quantification of corticotropin-releasing hormone (CRH) in the brain of the teleost fish *Oreochromis mossambicus* (tilapia). *J. Comp. Neurol.* **453,** 247–268.

Peter, R. E., and Gill, V. E. (1975). A stereotaxical atlas and technique for forebrain nuclei of the goldfish, *Carassius auratus. J. Comp. Neurol.* **159,** 69–101.

Peter, R. E., and McKeown, B. A. (1975). Hypothalamic control of prolactin and thyrotropin secretion in teleosts, with special reference to recent studies on the goldfish. *Gen. Comp. Endocrinol.* **25,** 153–165.

Peter, R. E., Macey, M. J., and Gill, V. E. (1975). A stereotaxic atlas and technique for forebrain nuclei of the killifish *Fundulus heteroclitus. J. Comp. Neurol.* **159,** 103–127.

Peter, R. E., Crim, L. W., and Billard, R. (1991). A stereotaxical atlas and implantation technique for the nuclei of the diencephalon of Atlantic salmon (*Salmo salar*) parr. *Reprod. Nutr. Dev.* **31,** 167–186.

Peyon, P., Lin, X. W., Himick, B. A., and Peter, R. E. (1998). Molecular cloning and expression of cDNA encoding brain preprocholecystokinin in goldfish. *Peptides* **19,** 199–210.

Peyon, P., Saied, H., Lin, X., and Peter, R. E. (1999). Postprandial, seasonal and sexual variations in cholecystokinin gene expression in goldfish brain. *Mol. Brain Res.* **74,** 190–196.

Pickavance, L. C., Staines, W. A., and Fryer, J. N. (1992). Distributions and colocalization of neuropeptide-Y and somatostatin in the goldfish brain. *J. Chem. Neuroanat.* **5,** 221–233.

Pickford, G. E., and Atz, J. W. (1957). "The Physiology of the Pituitary Gland of Fish". New York Zoological Society, New York.

Pierre, J., Mahouche, M., Suderevskaya, E. I., Repérant, J., and Ward, R. (1997). Immunocytochemical localization of dopamine and its synthetic enzymes in the central nervous system of the lamprey *Lampetra fluviatilis*. *J. Comp. Neurol.* **380,** 119–135.

Pogoda, H.-M., and Hammerschmidt, M. (2007). Molecular genetics of the pituitary development in zebrafish. *Semin. Cell Dev. Biol.* **18,** 543–558.

Pontet, A., Danger, J. M., Dubourg, P., Pelletier, G., Vaudry, H., Calas, A., and Kah, O. (1989). Distribution and characterization of neuropeptide Y-like immunoreactivity in the brain and pituitary of the goldfish. *Cell Tissue Res.* **255,** 529–538.

Powell, K. A., and Baker, B. I. (1988). Structural studies of nerve terminals containing melanin-concentrating hormone in the eel, *Anguilla anguilla*. *Cell Tissue Res.* **251,** 433–439.

Power, D. M., Canario, A. V. M., and Ingleton, P. M. (1996). Somatotropin release-inhibiting factor and galanin innervation in the hypothalamus and pituitary of seabream (*Sparus aurata*). *Gen. Comp. Endocrinol.* **101,** 264–274.

Prasada Rao, P. D., Job, T. C., and Screibman, M. P. (1993). Hypophysiotrophic neurons in the hypothalamus of the catfish *Clarias batrachus*: A cobaltous lysine and HRP study. *Brain Behav. Evol.* **42,** 24–38.

Prasada Rao, P. D., Murthy, C. K., Cook, H., and Peter, R. E. (1996). Sexual dimorphism of galanin-like immunoreactivity in the brain and pituitary of goldfish, *Carassius auratus*. *J. Chem. Neuroanat.* **10,** 119–135.

Price, D. A., and Greenberg, M. J. (1977). Structure of a molluscan cardioexcitatory neuropeptide. *Science* **197,** 670–671.

Prober, D. A., Rihel, J., Onah, A. A., Sung, R.-J., and Schier, A. F. (2006). Hypocretin/orexin overexpression induces an insomnia-like phenotype in zebrafish. *J. Neurosci.* **26,** 13400–13410.

Querat, B., Sellouk, A., and Salmon, C. (2000). Phylogenetic analysis of the vertebrate glycoprotein hormone family including new sequences of sturgeon (*Acipenserbaeri*) beta subunits of the two gonadotropins and the thyroid-stimulating hormone. *Biol. Reprod.* **63,** 222–228.

Rama Krishna, N. S., and Subhedar, N. (1989). Hypothalamic innervation of the pituitary in the catfish, *Clarias batrachus* (L.): a retrograde horseradish peroxidase study. *Neurosci. Lett.* **107,** 39–44.

Rama Krishna, N. S., Subhedar, N., and Schreibman, M. P. (1992). FMRFamide-like immunoreactive nervus terminalis innervation to the pituitary in the catfish, *Clarias batrachus* (Linn.): Demonstration by lesion and immunocytochemical techniques. *Gen. Comp. Endocrinol.* **85,** 111–117.

Rao, S. D., Prasada Rao, P. D., and Peter, R. E. (1996). Growth hormone-releasing hormone immunoreactivity in the brain, pituitary, and pineal of the goldfish, *Carassius auratus*. *Gen. Comp. Endocrinol.* **102,** 210–220.

Rehfeld, J. F., Lennart, F.-H., Goetze, J. P., and Hansen, T. V. O. (2007). The biology of cholecystokinin and gastrin peptides. *Curr. Top. Med. Chem.* **7,** 1154–1165.

Rink, E., and Wullimann, M. F. (2001). The teleostean (zebrafish) dopaminergic system ascending to the subpallium (striatum) is located in the basal diencephalon (posterior tuberculum). *Brain Res.* **889,** 316–330.

Roberts, B. L., Meredith, G. E., and Maslam, S. (1989). Immunocytochemical analysis of the dopamine system in the brain and spinal cord of the European eel, *Anguilla anguilla. Anat. Embryol.* **180,** 401–412.

Rodríguez-Gómez, F. J., Rendón-Unceta, M. C., Sarasquete, C., and Muñoz-Cueto, J. A. (2000a). Localization of galanin-like immunoreactive structures in the brain of the Senegalese sole, *Solea senegalensis. Histochem. J.* **32,** 123–131.

Rodríguez-Gómez, F. J., Rendón-Unceta, M. C., Sarasquete, C., and Muñoz-Cueto, J. A. (2000b). Distribution of serotonin in the brain of the Senegalese sole, *Solea senegalensis*: An immunohistochemical study. *J. Chem. Neuroanat.* **18,** 103–118.

Rodríguez-Gómez, F. J., Rendón-Unceta, C., Sarasquete, C., and Muñoz-Cueto, J. A. (2001). Distribution of neuropeptide Y-like immunoreactivity in the brain of the Senegalese sole (*Solea senegalensis*). *Anat. Rec.* **262,** 227–237.

Rodríguez, M. A., Anadón, R., and Rodríguez-Moldes, I. (2003). Development of galanin-like immunoreactivity in the brain of the brown trout (*Salmo trutta fario*), with some observations on sexual dimorphism. *J. Com. Neurol.* **465,** 263–285.

Saito, D., and Urano, A. (2001). Synchronized periodic Ca^{2+} pulses define neurosecretory activities in magnocellular vasotocin and isotocin neurons. *J. Neurosci.* **21,** RC178.

Saito, D., Komatsuda, M., and Urano, A. (2004). Functional organization of preoptic vasotocin and isotocin neurons in the brain of rainbow trout: Central and neurohypophysial projections of single neurons. *Neuroscience* **124,** 973–984.

Sakamoto, T., Agustsson, T., Moriyama, S., Itoh, T., Takahashi, A., Kawauchi, H., Björnsson, B. T., and Ando, M. (2003a). Intra-arterial injection of prolactin-releasing peptide elevates prolactin gene expression and plasma prolactin levels in rainbow trout. *J. Comp. Physiol. B* **173B,** 333–337.

Sakamoto, T., Fujimoto, M., and Ando, M. (2003b). Fishy tales of prolactin-releasing peptide. *Int. Rev. Cytol.* **225,** 91.

Sakharkar, A. J., Singru, P. S., Sarkar, K., and Subhedar, N. K. (2005). Neuropeptide Y in the forebrain of the adult male cichlid fish *Oreochromis mossambicus*: Distribution, effects of castration and testosterone replacement. *J. Comp. Neurol.* **489,** 148–165.

Salbert, G., Chauveau, I., Bonnec, G., Valotaire, Y., and Jego, P. (1992). One of the two trout proopiomelanocortin messenger RNAs potentially encodes new peptides. *Mol. Endocrinol.* **6,** 1605–1613.

Sas, E., and Maler, L. (1991). Somatostatin-like immunoreactivity in the brain of an electric fish (*Apteronotus leptorhynchus*) identified with monoclonal antibodies. *J. Chem. Neuroanat.* **4,** 155–186.

Sathyanesan, A. G. (1965). Hypothalamo-neurohypohyseal system in the normal and hypophysectomized teleost *Porichthys notatus* Girard and its response to continuous light. *J. Morphol.* **117,** 25–48.

Schally, A. V., Redding, T. W., Bowers, C. Y., and Barrett, J. F. (1969). Isolation and properties of porcine thyrotropin-releasing hormone. *J. Biol. Chem.* **244,** 4077–4088.

Scharrer, E. (1928). Untersuchungen über das Zwischenhirn der fische. I. *Z. Vergleich. Physiol.* **7,** 1–38.

Schiöth, H. B., Haitina, T., Ling, M. K., Ringholm, A., Fredriksson, R., Cerdá-Reverter, J. M., and Klovins, J. (2005). Evolutionary conservation of the structural, pharmacological and genomic characteristics of the melanocortin receptors subtypes. *Peptides* **26,** 1886–1900.

Seale, A. P., Itoh, T., Moriyama, S., Takahashi, A., Kawauchi, H., Sakamoto, T., Fujimoto, M., Riley, L. G., Hirano, T., and Grau, E. G. (2002). Isolation and characterization of a homologue of mammalian prolactin releasing peptide from the tilapia brain and its effect on prolactin release from the tilapia pituitary. *Gen. Comp. Endocrinol.* **125,** 328–339.

Sherwood, N. M., Krueckl, S. L., and McRory, J. E. (2000). The origin and function of the pituitary adenylate cyclase-activating polypeptide (PACAP)/glucagon superfamily. *Endocr. Rev.* **21,** 619–670.

Silverstein, J. T., Breininger, J., Baskin, D. G., and Plisetskaya, E. M. (1998). Neuropeptide Y-like gene expression in the salmon brain increases with fasting. *Gen. Comp. Endocrinol.* **110,** 157–165.

Singru, P. S., Mazumdar, M., Sakharkar, A. J., Lechan, R. M., Thim, L., Clausen, J. T., and Subhedar, N. K. (2007). Immunohistochemical localization of cocaine- and amphetamine-regulated transcript peptide in the brain of the catfish, *Clarias batrachus* (Linn.). *J. Comp. Neurol.* **502,** 215–235.

Sloley, B. D., Kah, O., Trudeau, V. L., Dulka, J. G., and Peter, R. E. (1992). Amino acid neurotransmitters and dopamine in brain and pituitary of the goldfish: involvement in the regulation of gonadotropin secretion. *J. Neurochem.* **58,** 2254–2262.

Smeets, W. J. A. J., and González, A. (2000). Catecholamine systems in the brain of vertebrates: new perspectives through a comparative approach. *Brain Res. Rev.* **33,** 308–379.

Somoza, G. M., and Peter, R. E. (1991). Effects of serotonin on gonadotropin and growth hormone release from in vitro perfused goldfish pituitary fragments. *Gen. Comp. Endocrinol.* **82,** 103–110.

Song, Y., and Cone, R. D. (2007). Creation of a genetic model of obesity in a teleost. *FASEB J.* **21,** 2042–2049.

Spiess, J., Villarreal, J., and Vale, W. (1981). Isolation and sequence analysis of a somatostatin-like polypeptide from ovine hypothalamus. *Biochemistry* **20,** 1982–1988.

Stefano, A. V., Vissio, P. G., Paz, D. A., Somoza, G. M., Maggese, M. C., and Barrantes, G. E. (1999). Colocalization of GnRH binding sites with gonadotropin-, somatotropin-, somatolactin-, and prolactin-expressing pituitary cells of the pejerrey, *Odontesthes bonariensis, in vitro*. *Gen. Comp. Endocrinol.* **116,** 133–139.

Stell, W. K., Walker, S. E., Chohan, K. S., and Ball, A. K. (1984). The goldfish nervus terminalis: a luteinizing hormone-releasing hormone and molluscan cardioexcitatory peptide immunoreactive olfactoretinal pathway. *Proc. Natl. Acad. Sci. USA* **81,** 940–944.

Striedter, G. F. (1990). The diencephalon of the channel catfish, *Ictalurus punctatus*. I Nuclear organization. *Brain Behav. Evol.* **36,** 329–354.

Subhedar, N., Cerdá, J., and Wallace, R. A. (1996). Neuropeptide Y in the forebrain and retina of the killifish, *Fundulus heteroclitus*. *Cell Tissue Res.* **283,** 313–323.

Sueiro, C., Carrera, I., Ferreiro, S., Molist, P., Adrio, F., Anadón, R., and Rodríguez-Moldes, I. (2007). New insights on saccus vasculosus evolution: A developmental and immunohistochemical study in elasmabranchs. *Brain Behav. Evol.* **70,** 187–204.

Sundström, G., Larsson, T. A., Brenner, S., Venkatesh, B., and Larhammar, D. (2008). Evolution of neuropeptide y family: New genes by chromosome duplications in early vertebrates and in teleost fishes. *Gen. Comp. Endocrinol.* **155,** 705–716.

Sutcliffe, J. G., and de Lecea, L. (2000). The hypocretins: excitatory neuromodulatory peptides for multiple homeostatic systems, including sleep and feeding. *J. Neurosci. Res.* **62,** 161–168.

Suzuki, H., Miyoshi, Y., and Yamamoto, T. (2007). Orexin-A (hypocretin 1)-like immunoreactivity in growth hormone-containing cells of the Japanese seaperch (*Lateolabrax japonicus*) pituitary. *Gen. Comp. Endocrinol.* **150,** 205–211.

Takahashi, A., Amemiya, Y., Nozaki, M., Sower, S. A., and Kawauchi, H. (2001). Evolutionary significance of proopiomelanocortin in agnatha and chondrichthyes. *Comp. Biochem. Physiol.* **29B,** 283–289.

Tena-Sempere, M. (2006). The roles of kisspeptins and G protein-coupled receptor-54 in pubertal development. *Curr. Opin. Pediatr.* **18,** 442–447.

Tran, T. N., Fryer, J. N., Bennett, H. P. J., Tonon, M. C., and Vaudry, H. (1989). TRH stimulates the release of POMC-derived peptides from goldfish melanotropes. *Peptides* **10,** 835–841.

Trudeau, V. L. (1997). Neuroendocrine regulation of gonadotrophin II release and gonadal growth in the goldfish, *Carassius auratus*. *Rev. Reprod.* **2,** 55–68.

Trudeau, V. L., Sloley, B. D., and Peter, R. E. (1993). GABA stimulation of gonadotropin-II release in goldfish: involvement of GABAA receptors, dopamine, and sex steroids. *Am. J. Physiol.* **265,** R348–55.

Trudeau, V. L., Sloley, B. D., Kah, O., Mons, N., Dulka, J. G., and Peter, R. E. (1996). Regulation of growth hormone secretion by amino acid neurotransmitters in the goldfish (I): Inhibition by N-methyl-D, L-aspartic acid. *Gen. Comp. Endocrinol.* **103,** 129–137.

Trudeau, V. L., Kah, O., Chang, J. P., Sloley, B. D., Dubourg, P., Fraser, E. J., and Peter, R. E. (2000a). The inhibitory effects of (gamma)-aminobutyric acid (GABA) on growth hormone secretion in the goldfish are modulated by sex steroids. *J. Exp. Biol.* **203,** 1477–1485.

Trudeau, V. L., Spanswick, D., Fraser, E. J., Lariviere, K., Crump, D., Chiu, S., MacMillan, M., and Schulz, R. W. (2000b). The role of amino acid neurotransmitters in the regulation of pituitary gonadotropin release in fish. *Biochem. Cell Biol.* **78,** 241–259.

Tsuneki, K., and Fernholm, B. (1975). Effect of thyrotropin-releasing hormone on the thyroid of a teleost, *Chasmichthys dolicognathus*, and a hagfish, *Eptatretus burgeri*. *Acta Zool.* **56,** 61–65.

Tsutsui, K., and Ukena, K. (2006). Hypothalamic LPXRF-amide peptides in vertebrates: Identification, localization and hypophysiotropic activity. *Peptides* **27,** 1121–1129.

Ukena, K., Iwakoshi, E., Minakata, H., and Tsutsui, K. (2002). A novel rat hypothalamic RFamide-related peptide identified by immunoaffinity chromatography and mass spectrometry. *FEBS Lett.* **512,** 255–258.

Unniappan, S., Cerdá-Reverter, J. M., and Peter, R. E. (2004). In situ localization of preprogalanin mRNA in the goldfish brain and changes in its expression during feeding and starvation. *Gen. Comp. Endocrinol.* **136,** 200–207.

Vallarino, M., Delbende, C., Ottonello, I., Tranchand-Bunel, D., Jegou, S., and Vaudry, H. (1989). Immunocytochemical localization and biochemical characterization of α-melanocyte-stimulating hormone in the brain of the rainbow trout, *Salmo gairdneri*. *J. Neuroendocrinol.* **1,** 53–60.

van Aerle, R., Kille, P., Lange, A., and Tyler, C. R. (2008). Evidence for the existence of a functional Kiss1/Kiss1 receptor pathway in fish. *Peptides* **29,** 57–64.

van de Kamer, J. C., and Zandbergen, M. A. (1981). The hypothalamic–hypophyseal system and its evolutionary aspects in *Scyliorhinus caniculus*. *Cell Tissue Res.* **214,** 575–582.

Vaudry, D., Gonzalez, B. J., Basille, M., Yon, L., Fournier, A., and Vaudry, H. (2000). Pituitary adenylate cyclase-activating polypeptide and its receptors: from structure to functions. *Pharmacol. Rev.* **52,** 269–324.

Vecino, E., Perez, M.-T. R., and Ekstrom, P. (1994). *In situ* hybridization of neuropeptide Y (NPY) mRNA in the goldfish brain. *NeuroReport* **6,** 127–131.

Vigh-Teichmann, I., Vigh, B., Korf, H. W., and Oksche, A. (1983). CSF-contacting and other somatostatin-immunoreactive neurons in the brains of *Anguilla anguilla, Phoxinus phoxinus*, and *Salmo gairdneri* (Teleostei). *Cell Tissue Res.* **233,** 319–334.

Vissio, P. G., Stefano, A. V., Somoza, G. M., Maggese, M. C., and Paz, D. A. (1999). Close association of gonadotropin-releasing hormone fibers and gonadotropin, growth hormone, somatolactin and prolactin expressing cells in pejerrey, *Odontesthes bonariensis*. *Fish Physiol. Biochem.* **21,** 121–127.

Volkoff, H., and Peter, R. E. (1999). Actions of two forms of gonadotropin releasing hormone and a GnRH antagonist on spawning behavior of the goldfish *Carassius auratus*. *Gen. Comp. Endocrinol.* **116,** 347–355.

Volkoff, H., and Peter, R. E. (2001). Characterization of two forms of cocaine- and amphetamine-regulated transcript (CART) peptide precursors in goldfish: molecular cloning and distribution, modulation of expression by nutritional status, and interactions with leptin. *Endocrinology* **142,** 5076–5088.

Volkoff, H., Peyon, P., Lin, X., and Peter, R. E. (2000). Molecular cloning and expression of cDNA encoding a brain bombesin/gastrin-releasing peptide-like peptide in goldfish. *Peptides* **21,** 639–648.

Volkoff, H., Canosa, L. F., Unniappan, S., Cerda-Reverter, J. M., Bernier, N. J., Kelly, S. P., and Peter, R. E. (2005). Neuropeptides and the control of food intake in fish. *Gen. Comp. Endocrinol.* **142,** 3–19.

Vollrath, L. (1967). On neurosecretory innervation of the adenohypophysis in teleost fishes, especially in the Hippocampus cuda and Tinca tinca. *Z. Zellforsch. Mikrosk. Anat.* **78,** 234–260.

Wagner, S., Castel, M., Gainer, H., and Yarom, Y. (1997). GABA in the mammalian suprachiasmatic nucleus and its role in diurnal rhythmicity. *Nature* **387,** 598–603.

Weber, G. M., Powell, J. F. F., Park, M., Fischer, W. H., Craig, A. G., Rivier, J. E., Nanakorn, U., Parhar, I. S., Ngamvongchon, S., Grau, E. G., and Sherwood, N. M. (1997). Evidence that gonadotropin-releasing hormone (GnRH) functions as a prolactin-releasing factor in a teleost fish (*Oreochromis mossambicus*) and primary structures for three native GnRH molecules. *J. Endocrinol.* **155,** 121–132.

Weltzien, F. A., Pasqualini, C., Sébert, M. E., Vidal, B., Le Belle, N., Kah, O., Vernier, P., and Dufour, S. (2006). Androgen-dependent stimulation of brain dopaminergic systems in the female European eel (*Anguilla anguilla*). *Endocrinology* **147,** 2964–2973.

White, R. B., Eisen, J. A., Kasten, T. L., and Fernald, R. D. (1998). Second gene for gonadotropin-releasing hormone in humans. *Proc. Natl. Acad. Sci. USA* **95,** 305–309.

Wigham, T., and Batten, T. F. C. (1984). In vitro effects of thyrotropin-releasing hormone and somatostatin on prolactin and growth hormone release by the pituitary of *Poecilia latipinna*. I. An electrophoretic study. *Gen. Comp. Endocrinol.* **55,** 444–449.

Winberg, S., Nilsson, A., Hylland, P., Söderstöm, V., and Nilsson, G. E. (1997). Serotonin as a regulator of hypothalamic–pituitary–interrenal activity in teleost fish. *Neurosci. Lett.* **230,** 113–116.

Wong, A. O. L., Leung, M. Y., Shea, W. L. C., Tse, L. Y., Chang, J. P., and Chow, B. K. C. (1998). Hypophysiotropic action of pituitary adenylate cyclase-activating polypeptide (PACAP) in the goldfish: Immunohistochemical demonstration of PACAP in the pituitary, PACAP stimulation of growth hormone release from pituitary cells, and molecular cloning of pituitary type I PACAP receptor. *Endocrinology* **139,** 3465–3479.

Wullimann, M., and Mueller, T. (2004). Teleostean and mammalian forebrains contrasted: Evidence from genes to behaviour. *J. Comp Neurol.* **475,** 143–162.

Wullimann, M. F., Rupp, B., and Reichert, H. (1996). "Neuroanatomy of Zebrafish Brain: A Topological Atlas". Birkhaeuser Verlag, Switzerland.

Yamamoto, N., Parhar, I. S., Sawai, N., Oka, Y., and Ito, H. (1998). Preoptic gonadotropin-releasing hormone (GnRH) neurons innervate the pituitary in teleosts. *Neurosci. Res.* **31,** 31–38.

Yokogawa, T., Marin, W., Faraco, J., Pézeron, G., Appelbaum, L., Zhang, J., Rosa, F., Mourrain, P., and Mignot, E. (2007). Characterization of sleep in zebrafish and insomnia in hypocretin receptor mutants. *PLoS Biol.* **5,** 2379–2397.

Yu, K. L., Sherwood, N. M., and Peter, R. E. (1988). Differential distribution of two molecular forms of gonadotropin-releasing hormone in discrete brain areas of goldfish (*Carassius auratus*). *Peptides* **9,** 625–630.

Yulis, C. R., and Lederis, K. (1986). The distribution of 'extraurophyseal' urotensin I-immunoreactivity in the central nervous system of *Catostomus commersoni* after urophysectomy. *Neurosci. Lett.* **70,** 75–80.

Yulis, C. R., and Lederis, K. (1987). Co-localization of the immunoreactivities of corticotropin-releasing factor and arginine vasotocin in the brain and pituitary system of the teleost *Catostomus commersoni*. *Cell Tissue Res.* **247,** 267–273.

Yulis, C. R., and Lederis, K. (1988). Occurrence of an anterior spinal, cerebrospinal fluid-contacting, urotensin II neuronal system in various fish species. *Gen. Comp. Endocrinol.* **70,** 301–311.

Yulis, C. R., Lederis, K., Wong, K.-L., and Fisher, A. W. F. (1986). Localization of urotensin I- and corticotropin-releasing factor-like immunoreactivity in the central nervous system of *Catostomus commersoni*. *Peptides* **7,** 79–86.

Zandbergen, M. A., Voormolen, A. H. T., Kah, O., and Goos, H. J. T. (1994). Immunohistochemical localizatin of neuropeptide Y positive cell bodies and fibres in forebrain and pituitary of the African catfish, *Clarias gariepinus*. *Neth. J. Zool.* **44,** 43–54.

Zupanc, G. K., Horschke, I., and Lovejoy, D. A. (1999). Corticotropin releasing factor in the brain of the gymnotiform fish, *Apteronotus leptorhynchus*: immunohistochemical studies combined with neuronal tract tracing. *Gen. Comp. Endocrinol.* **114,** 349–364.

2

ENDOCRINE TARGETS OF THE HYPOTHALAMUS AND PITUITARY

OLIVIER KAH

1. Introduction
2. Sex Steroids
   2.1. Steroid Production in the Brain of Fish
   2.2. Steroid Receptors
3. Corticoid Receptors
   3.1. Mineralocorticoid Receptors
   3.2. Glucocorticoid Receptors
4. Metabolic Hormones
   4.1. Leptin and Leptin Receptors
   4.2. Insulin and Insulin-like Growth Factor
   4.3. Thyroid Hormone Receptors
5. Concluding Remarks

Maintenance of the internal milieu, also called homeostasis, allows organisms to effectively adapt to a broad range of environmental conditions. In vertebrates, including fishes, this is achieved by a fine tuning of the endocrine responses through multiple feedback mechanisms to maintain hormonal changes within a certain range. The best documented examples of these feedback mechanisms are those exerted by a number of peripheral hormones on the hypothalamo–pituitary complex, for example sex steroids onto the reproductive axis, cortisol onto the corticotropin-releasing factor (CRF)/adrenocorticotropin (ACTH) axis or insulin-like growth factor I (IGF-I) onto growth hormone (GH) production. This is achieved by expression in the brain and the pituitary of receptors to those peripheral hormones. Such mechanisms are particularly well understood in the context of the reproductive cycle and include well-documented feedback effects of sex steroids on the neuroendocrine circuits controlling the reproductive axis.

Fish Neuroendocrinology: Volume 28
FISH PHYSIOLOGY

DOI: 10.1016/S1546-5098(09)28002-2

However, of equally crucial importance are the effects exerted by a number of metabolic hormones, such as insulin, IGF-I, leptin and cortisol, onto the higher regulatory centers. These complex regulatory mechanisms and their cross-talks are essential to ensure energy allocation between basal metabolism, reproductive events and stress responses, all of which are crucial for individual and species survival. This review aims at synthesizing the currently available information on the sites of expression of the receptors to sex steroids, corticoids, leptin, insulin, IGF-I and thyroid hormone receptors in the brain of fish. It also aims at providing information, when available, on the relationships between those receptors and the neuroendocrine circuits controlling the pituitary.

1. INTRODUCTION

In 1865, Claude Bernard introduced the concept of homeostasis or fixity of the internal milieu and claimed that the "constance of the internal environment is the condition for a free and independent life." This notably implies that any significant tendency to a change of the internal milieu will be counteracted by mechanisms resisting this change. This applies to many physiological mechanisms, notably to the maintenance of energy homeostasis that relies on a permanent and exquisitely regulated interplay between the central nervous system, namely the hypothalamus, and the peripheral endocrine glands. The former controls the activity of the latter, which in turn feed back to the former allowing the organism to constantly adapt the response, within certain limits. This is what neuroendocrinology is all about. As stated by another Frenchman, Claude Kordon, a pioneer in the field, "Neuroendocrinology is the science looking at the reciprocal interactions between the brain and the hormones."

Many, but by no means all, of these interactions take place in the hypothalamus, a brain center that, on the one hand, integrates metabolic, environmental and hormonal signals and, on the other hand, activates specific neuronal pathways coordinating behavioral and reproductive responses. This process involves an exquisite interplay between specialized neuronal populations that are able to monitor incoming peripheral signals to then tune up the energy balance, metabolism, growth and reproductive activity. Throughout evolution, mechanisms were selected that allow strict maintenance of homeostasis, "the condition for free and independent life" while allowing allocation of energy to sustain successful reproduction, the condition for species survival.

These interactions occur in all vertebrate species and are well documented in mammals in which they have been dealt with in excellent recent reviews to which the reader is referred (Gamba and Pralong, 2006; Tena-Sempere, 2006; Navarro *et al.*, 2007; Gao and Horvath, 2008; Goulis and Tarlatzis, 2008; Popa *et al.*, 2008). Much less is known in fish. However, there is now no doubt that those brain/hormones relationships appeared in early vertebrates together with the emergence of the pituitary and the control of pituitary hormone synthesis and release. Given the conserved functions of these pituitary hormones, there is no reason to believe that the mechanisms controlling their secretion will be much different in fish as compared to mammals. As a result, many concepts established in mammals also apply to fish as shown by comparative endocrinology studies. The present chapter intends to review what is known on the hormonal targets in the brain and the pituitary of fish with special emphasis on sex hormones, glucocorticoids and metabolic hormones (Figure 2.1).

2. SEX STEROIDS

The gonadal steroid hormones are well known for playing key roles in the regulation of reproduction, but they also contribute to the regulation of energy balance (Gomez, 2007; Lovejoy and Sainsbury, 2009; O'Sullivan, 2009).

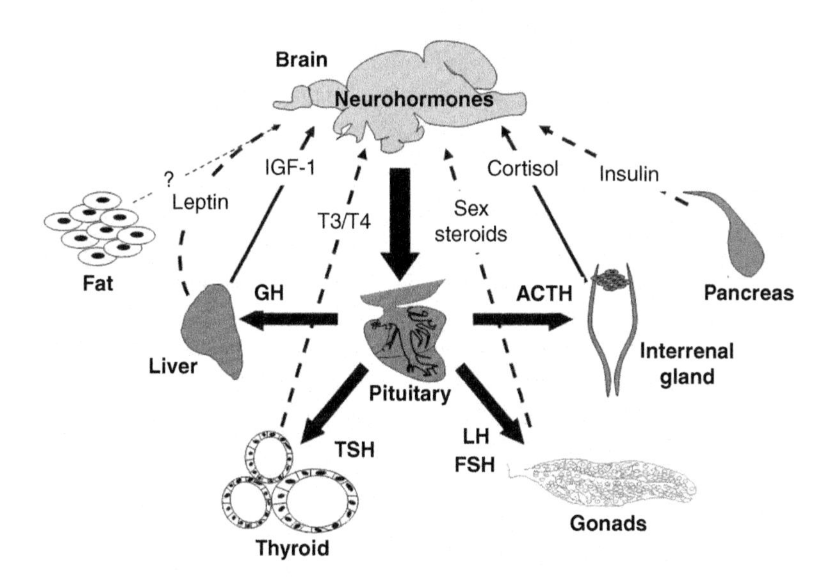

Fig. 2.1. Schematic representation of the main brain/hormone relationships that ensure maintenance of the hormonal milieu in fishes. Through synthesis of neurohormones, the brain controls the liberation of the trophic pituitary hormones, which stimulate secretion of peripheral hormones. These hormones in turn feedback on the brain/pituitary complex to regulate activity of the neuroendocrine systems.

Originally, both negative and positive feedback mechanisms were reported following castration/hormone replacement experiments (Donaldson and McBride, 1967; Olivereau and Olivereau, 1979; Crim and Evans, 1983; Trudeau, 1997). Such positive and negative effects of sex steroids were often contradictory and shown to depend on the steroid considered, the species and the physiological status, the target tissue (brain or pituitary) and the studied parameter (synthesis or release). One of the difficulties in getting a clear picture arises from the fact that reproductive development in many teleost species is asynchronous resulting in complex steroid profiles and thus confusing interpretation.

2.1. Steroid Production in the Brain of Fish

The classical view regarding steroid effects on the hypothalamo–pituitary complex is that steroids are produced by peripheral glands, mainly the gonads but not solely, in response to stimulation by pituitary tropic hormones, gonadotropins (GTH). In turn, these steroids will then feedback onto the neuroendocrine systems controlling the secretion of these tropic hormones to permanently adapt their activity to the ongoing physiological situation. However, an emerging concept is that the brain is itself a steroidogenic organ that expresses several steroidogenic enzymes and thus produces a number of steroids whose functional significance is poorly understood. The neurosteroid hypothesis was proposed by E.E. Baulieu's laboratory after the finding that pregnenolone and dehydroepiandrosterone concentrations in the mammalian brain are greater than those in plasma. In addition, brain contents in pregnenolone and dehydroepiandrosterone did not decrease after both adrenalectomy and castration (Corpechot *et al.*, 1981; Robel *et al.*, 1995). In mammals, it is now clear that at least some brain regions, notably the hippocampus, are able to produce estrogens and androgens *de novo* from cholesterol (Hojo *et al.*, 2004, 2008).

In fish, evidence for brain expression of steroidogenic enzymes essentially concerns P450 aromatase B (AroB), the product of the cyp19a1b gene and 5α-reductase, two enzymes that convert testosterone into estradiol and dihydrotestosterone, respectively (Callard *et al.*, 1978, 1981; Pasmanik and Callard, 1985; Pasmanik *et al.*, 1988). These pioneering studies indicated that testosterone could be efficiently metabolized in the brain of fish. However, until now, there has been a deficit of information regarding other steroidogenic enzymes necessary for production of steroids. The cleavage of the lateral chain of cholesterol requires the presence of P450scc yielding pregnenolone. Further metabolization of pregnenolone is performed by two key steroidogenic enzymes, namely, 3β-hydroxysteroid dehydrogenase/D4–D5 isomerase (3βHSD) and cytochrome P450c17 (CYP17). While 3βHSD causes dehydrogenation and isomerization of pregnenolone into progesterone, CYP17 causes hydroxylation of the C21 steroids (17α-hydroxylase activity), followed by cleavage of the two-carbon side chain (C17,20 lyase activity). This will generate the C19 steroids androstenedione or dehydroepiandrosterone, respectively. Expression of all these enzymes is documented in the brain of fish but in different species and there is need for more accurate work using complementary techniques to further document the capacity of the brain of fish to produce a variety of active steroids *de novo*.

By *in toto* hybridization, P450scc was shown to be highly expressed in the brain of adult zebrafish (*Danio rerio*), notably in the forebrain (Hsu *et al.*, 2002). Expression of CYP17 was evidenced in the brain of fish only by PCR (Halm *et al.*, 2003; Wang and Ge, 2004; Tomy *et al.*, 2007), although preliminary *in situ* hybridization data indicate that CYP17 messengers are widely present in the forebrain, together with 3β-HSD transcripts (Tong S.K., Kah O., Diotel N., and Chung B.C., unpublished data). Expression and activity of 3βHSD was also documented in the brain of adult zebrafish (Sakamoto *et al.*, 2001). Using an antibody against purified bovine adrenal 3βHSD, clusters of immunoreactive (ir) cell bodies were localized in the dorsal telencephalic areas, central posterior thalamic nucleus, preoptic nuclei, posterior tuberal nucleus, paraventricular organ, and nucleus of medial longitudinal fascicle (Sakamoto *et al.*, 2001). Immunoreactivity for 3βHSD-like was also observed in the soma of cerebellar Purkinje neurons. These data strongly suggest that the brain of fish is likely able to produce progesterone or dehydroepiandrosterone. Regarding 17βHSD, available information is limited to PCR data showing that three candidate homologues for both HSD17B1 and HSD17B3 exist in zebrafish and are expressed in the brain (Mindnich *et al.*, 2004).

It is worth pointing out that, recently, in the brain of the protandrous black porgy (*Acanthopagrus schlegeli*), the presence of key steroidogenic enzymes was detected as early as 60 days after hatching, before gonadal sex differentiation. This suggests that the steroid biosynthetic capacity in brain precedes the histological differentiation of the gonads (Tomy *et al.*, 2007). In the black porgy, mRNAs of these genes showed a synchronous peak at 120 days after hatching, indicating that estradiol may be locally formed at this time in both the forebrain and the midbrain (Tomy *et al.*, 2007). Similar data were obtained in whole male populations of rainbow trout (*Oncorhynchus mykiss*; Vizziano *et al.*, unpublished results) again suggesting that steroid production in the brain of developing fish is independent from that of the gonad. Recent data in zebrafish suggest that estrogen production in the brain of fish might also be linked to neurogenesis (Pellegrini *et al.*, 2007; Mouriec *et al.*, 2008).

In contrast with other enzymes, there is substantial information on brain aromatase in fish. Several studies have now documented the fact that aromatase B expression in developing and adult fish is strictly confined to radial glial cells (Figure 2.2). Such cells are characterized by a small nucleus adjacent to the ventricle and long radial processes terminating by end feet at the brain surface (Rakic, 1978; Bentivoglio and Mazzarello, 1999). Radial cells are involved in embryonic neurogenesis and, in contrast to mammals where they disappear at the end of neurogenesis, they largely persist in the adult brain of non-mammalian species, notably in fish. Strong expression of aromatase B in radial glial cells of fish was first shown in the plainfin midshipman (*Porichthys notatus*) (Forlano *et al.*, 2005) and then in rainbow trout, zebrafish, pejerrey (*Odontesthes bonariensis*) and bluehead wrasse (*Thalassoma bifasciatum*) (Menuet *et al.*, 2003, 2005; Pellegrini *et al.*, 2005; Strobl-Mazzulla *et al.*, 2005; Marsh *et al.*, 2006; Kallivretaki *et al.*, 2007; Pellegrini *et al.*, 2007; Strobl-Mazzulla *et al.*, 2008). These aromatase-expressing cells are most abundant in the forebrain, notably in the olfactory bulbs, the telencephalon, the preoptic area and the mediobasal hypothalamus, in particular along the lateral and posterior recesses. However, consistent with the distribution of the messengers, aromatase B positive cells were also observed in the periventricular layers bordering the optic tectum, the torus semicircularis and along the fourth ventricle (Menuet *et al.*, 2002, 2005; Pellegrini *et al.*, 2007). In conclusion, it seems now clear that aromatase expression in the brain of fish is restricted to radial glial cells. The reason for the high expression of aromatase in the brain of fish is that the *cyp19a1b* gene is strongly regulated by estrogens and aromatizable androgens (Menuet *et al.*, 2005). Whether the other steroidogenic enzymes are also expressed in radial glial cells is unknown.

Aromatase-expressing cells and their processes are often in close association with neurons. This is notably the case of gonadotropin-releasing hormone (GnRH) neurons, which in some regions are sometimes totally surrounded by the radial processes (Figure 2.2C). Although GnRH neurons do not express estrogen receptors (ER; see below), one cannot exclude that estrogens produced by such cells influence activity of GnRH neurons through membrane receptors. In the pituitary, aromatase B is strongly expressed in cells of the proximal pars distalis, most likely in gonadotrophs, and in small cells within the neurohypophysis (Figure 2.2D).

As for the function of aromatase expression in radial cells, recent data suggest a link with the neurogenic activity (Mouriec *et al.*, 2008). Indeed, using BrdU immunohistochemistry and aromatase B as a marker of radial glial cells it was found that, at short survival times (12 and 24 hours), a large majority of cells exhibiting BrdU labeling corresponded to AroB-positive radial cells (Adolf *et al.*, 2006; Pellegrini *et al.*, 2007). The radial nature of proliferative cells in the telencephalon and diencephalon was also indicated using antibodies to brain lipid binding protein (Figure 2.2E and 2.2F).

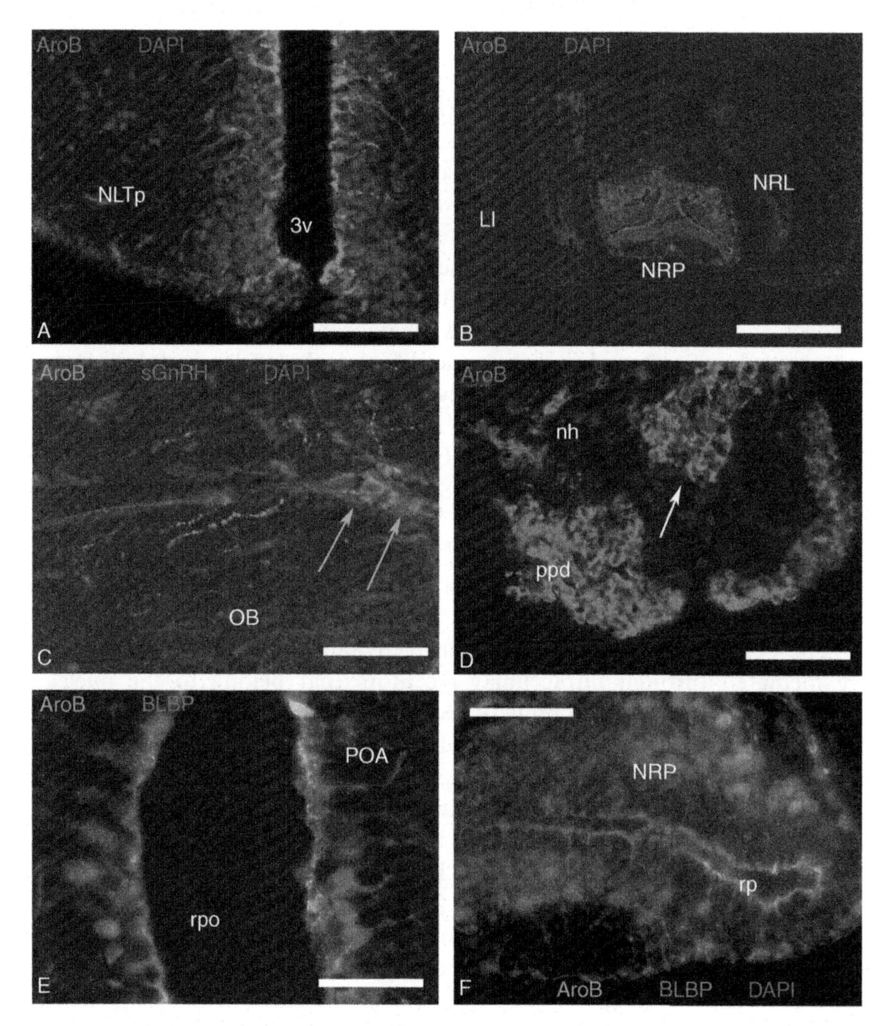

Fig. 2.2. (A) Immunohistochemical demonstration of aromatase B (AroB) at the level of the posterior nucleus lateralis tuberis (NLTp) of zebrafish. The red signal is only detected in cells bordering the third ventricle (3v) and sending long radial processes towards the ventral surface of the brain. Bar = 50 μm. (B) Immunohistochemical demonstration of AroB at the level of the nucleus of the posterior recess (NRP) of zebrafish. The red signal is only detected in cells bordering the third ventricle (3v), sending short processes towards the posterior recess and long radial processes towards the ventral surface of the brain. Note that a much weaker, but still significant, signal is observed in the caudal part nucleus of the lateral recess (NRL) at the level of the inferior lobe (LI). Bar = 500 μm. (C) Sagittal section at the level of the olfactory bulbs (OB) of zebrafish showing the presence of AroB (in green) in radial processes around a cluster of salmon GnRH (sGnRH) cell bodies in red. Whether estrogens produced by such AroB-expressing cells can influence activity of GnRH neurons is unknown at the moment. Bar = 50 μm.

In addition, it was shown that, over time, newborn cells clearly move away from the periventricular proliferative zones, as indicated by double BrdU/PCNA staining (Adolf *et al.*, 2006; Pellegrini *et al.*, 2007). In the zebrafish, many of the newborn cells differentiate into neurons, as shown by combining BrdU and the use of several neuronal markers such as Hu or acetylated tubulin (Zupanc *et al.*, 2005; Adolf *et al.*, 2006; Pellegrini *et al.*, 2007).

Estrogens can be metabolized into catechol-estrogens through the action of the enzyme estrogen 2-hydroxylase. These oxidized estrogens, such as 2-hydroxyestrogen, are potentially bifunctional molecules that can mediate estrogen effects on neuroendocrine and behavioral functions. Catechol-estrogens have been reported in fish to modify catecholamine metabolism by inhibiting tyrosine hydroxylase activity in a non-competitive manner. Catechol-estrogens also are competitive substrate for catechol-O-methyltransferase, one of the degrading enzymes of catecholamine (de Leeuw *et al.*, 1985; Goos *et al.*, 1985; Timmers *et al.*, 1988; Joy and Senthilkumaran, 1998; Joy *et al.*, 1998; Chaube and Joy, 2003).

2.2. Steroid Receptors

The best-documented effects of estrogens, androgens and progesterone are mediated through intracellular receptors that belong to the nuclear receptor family. This superfamily of receptors regulates essential physiological functions from development to the fine tuning of homeostasis. Traditionally, the nuclear receptors are best known for acting as ligand-activated transcription factors and they all share the same modular organization in different functional domains. Among the distinct structural and functional domains are a conserved zinc finger DNA-binding domain and a ligand-binding domain. The nuclear receptor family contains 48 members including the glucocorticoid receptor, progesterone receptor, androgen receptor, vitamin D receptor, thyroid hormone receptor, retinoid acid, and also a large number of orphan receptors for which there is no identified corresponding ligand (Robinson-Rechavi *et al.*, 2003; Germain *et al.*, 2006). For this reason, it is assumed that the nuclear receptor family appeared early during evolution

(D) Immunohistochemical demonstration of AroB in the proximal pars distalis (ppd) of the zebrafish. A very strong signal is observed in secretory cells, most likely gonadotrophs, while a much weaker signal is detected in smaller cells in the neurohypophysis (nh). Bar = 50 μm. (E) and (F) show co-expression of AroB and brain lipid binding protein (BLBP) in cells of the preoptic area (POA) and cells of the nucleus (NRP) of the posterior recess (rp) of zebrafish. Note that BLBP in red is present in both the cytoplasm and the nucleus of the cells while AroB is only present in the cytoplasm. As BLBP is a marker of radial glial cells in mammals, this further confirms the radial nature of AroB-expressing cells. (A) Bar = 25 μm; (F) Bar = 50 μm. (See Color Insert.)

and that the ancestral nuclear receptor emerged from an orphan receptor that gained ligand-binding ability during evolution of metazoans. Nuclear receptors classically regulate the expression of target genes by direct interaction with specific DNA sequences (Wuertz *et al.*, 2007). Analysis of the three-dimensional (3D) structure of several nuclear receptors demonstrated that upon ligand binding, the ligand-binding domain (LBD) undergoes a conformational change allowing the recruitment of coactivator proteins. These cofactors that are recruited in an ordered series of sequential receptor–coactivator interactions possess diverse functional domains with notably acetyltransferase, methyltransferase or ubiquitin ligase activity. By relaxing chromatin structure, the cyclic recruitment of these cofactors on the promoters allows easier access of the transcription machinery to the DNA (Metivier *et al.*, 2003, 2008).

More recently, a family of steroid membrane receptors has emerged that makes it necessary to re-examine the current state of the art under the light of this new concept (Thomas *et al.*, 2006; Pang *et al.*, 2008; Pang and Thomas, 2009). Such receptors will not be dealt with in this chapter as they are reviewed in another section.

2.2.1. Estrogen Receptors (ERs)

Following the discovery of ERβ in mammals, it became rapidly obvious that all vertebrates have two ERs, ERα (esr1) and ERβ (esr2), generated by two distinct genes that have partially distinct expression patterns and binding activities towards some ligands (Kuiper *et al.*, 1998; Katzenellenbogen *et al.*, 2000). In fish, it is assumed that an additional genome duplication (Steinke *et al.*, 2006) gave rise to two ERβ, called ERβ1 (esr2b) and ERβ2 (esr2a). At the C-terminus of ER, the A/B domain contains a ligand-independent activation function (AF1) and is poorly conserved between the two ERs. In contrast, domain C, the DNA-binding domain, is highly conserved between both ERs (more than 95% of identity), and also between the members of the NR superfamily (40–50% of identity). This region contains the two zinc fingers permitting the recognition of specific DNA hormone-responsive sequences. In the case of ERs, this element, known as estrogen-responsive element (ERE), is a palindromic sequence with the consensus sequence (AGGTCAnnnTGACCT) defined from the *Xenopus* vitellogenin gene (Klein-Hitpass *et al.*, 1986). The poorly conserved D domain is involved in the three-dimensional structure of the ERs and permits the stabilization of the DNA binding. The E/F-domain is multifunctional and includes the LBD together with a hormone-dependent transactivation function (AF2). It presents a conserved arrangement of 12 helices (H1–H12) forming an hydrophobic ligand pocket closed by an antiparallel β-sheet and by helix H12. The LBD of ERs is highly promiscuous and binds a variety of compounds, exhibiting a great diversity in their size and in their chemical properties (Thomas, 2000; Singleton and Khan, 2003). This property is responsible for the fact that many synthetic chemicals act as xeno-estrogens. Upon agonist binding, there is a repositioning of H12 known to be involved in the transactivation function AF2 and able to interact with cofactors.

We have little information in fish as to the subcellular localization of the unliganded receptor. In mammals, although they are found principally in the nucleus, a small proportion can also be found in the cytosol. In the nucleus, ERs modulate gene transcription, and translated proteins determine the physiological effects of estrogens. Although not documented in fish, there is increasing evidence that ER can bind to DNA at regulatory sequences other than ERE (Carroll *et al.*, 2006). In particular, ERα was able to modulate their transcription by establishing protein–protein interactions with other transcription factors such as AP-1 or Sp1 complexes (Nilsson *et al.*, 2001; Safe and Kim, 2008).

In addition to the classic "dogma" of molecular mechanisms of ER interactions with the transcriptional machinery, co-activators/co-repressors and other transcription factors, multiple levels of cross-talks between ERs and other intracellular signalling pathways have rapidly emerged. Notably, membrane-associated ERs can activate several signal transduction cascades such as mitogen-activated protein kinase (MAPK), protein kinase C (PKC) and phosphatidylinositol 3-kinase (PI3K), and could be responsible for the so-called "non genomic" effects of estrogens. Moreover, AF1, the ligand-independent activation of the ER can be activated through phosphorylation by growth factors causing activation of downstream cellular kinases (MAPKs, PI3K, PKC) and other signaling pathways. Although similar non-classical mechanisms of ER actions are not documented in fish, their existence cannot be excluded at the present stage.

Nuclear estrogen receptors are the best characterized steroid receptors in fish. This is largely due to their key roles in regulating vitellogenin expression in the liver. This process is strictly estrogen dependent, providing an outstanding model to study the underlying molecular mechanisms. The first ER was cloned from a rainbow trout liver library (Pakdel *et al.*, 1990, 1991). Later, it was found that vertebrates have two ER genes, ERα and ERβ, and thus this rainbow trout ER was renamed ERα (esr1). However, due to the teleost-specific genome duplication (3R) (Siegel *et al.*, 2007), two ERβ forms have been cloned in a number of species (Hawkins *et al.*, 2000; Menuet *et al.*, 2002; Nagler *et al.*, 2007; Muriach *et al.*, 2008a,b). Originally termed ERβ and ERγ in the Atlantic croaker (*Micropogonias undulates*) (Hawkins *et al.*, 2000), these receptors have been renamed ERβ1 (esr2b) and ERβ2 (esr2a) when it became clear that the two ERβ forms arose from the 3R duplication. Estrogen receptors have now been cloned and at least partly characterized in a number of teleost species (Pakdel *et al.*, 1991; Munoz-Cueto *et al.*, 1999; Xia *et al.*, 1999; Ma *et al.*, 2000; Pakdel *et al.*, 2000; Socorro *et al.*, 2000; Menuet *et al.*, 2002; Andreassen *et al.*, 2003; Teves *et al.*, 2003; Urushitani *et al.*, 2003; Halm *et al.*, 2004; Filby and Tyler, 2005; Caviola *et al.*, 2007; Greytak and Callard, 2007; Fu *et al.*, 2008). However, the number of species in which the three receptor sequences are available is still limited.

Brain distribution of estrogen receptors. The distribution of these receptors has been studied in a number of species, mainly at the messenger levels, and certainly much remains to be learnt regarding the expression of the three receptor subtypes. However, long before the molecular characterization of fish ER, the distribution of estrogen-concentrating cells had been studied following injection of tritiated hormones and autoradiography (Davis *et al.*, 1977; Kim *et al.*, 1978, 1979a,b). For example in the goldfish (*Carassius auratus*) and platyfish (*Xiphophorus maculatus*), estrogen-concentrating cells were found mainly in the periventricular midline regions of the ventral precommissural and dorsal supracommissural telencephalic areas, and in the diencephalon the preoptic, central hypothalamic and thalamic areas (Kim *et al.*, 1978, 1979b). These authors also demonstrated that 80% of the gonadotrophs in the pituitary picked up tritiated estrogens (Kim *et al.*, 1979a). These pioneering results were next confirmed using *in situ* hybridization. The best-documented expression is that of the ERα of the rainbow trout, the only one that was studied at both the protein and messenger levels in the same species (Anglade *et al.*, 1994; Pakdel *et al.*, 1994; Navas *et al.*, 1995; Linard *et al.*, 1996; Menuet *et al.*, 2001; Salbert *et al.*, 1991, 1993). These results pointed out first that the data obtained by pioneer authors using autoradiography were perfectly valid (Kim *et al.*, 1978, 1979b), and also showed that the data obtained by immunohistochemistry and *in situ* hybridization (Figure 2.3A and 2.3B) were very consistent with each other (Salbert *et al.*, 1991; Anglade *et al.*, 1994; Pakdel *et al.*, 1994).

Since then data have been obtained in different species, mostly by *in situ* hybridization (Salbert *et al.*, 1991, 1993; Anglade *et al.*, 1994; Pakdel *et al.*, 1994; Navas *et al.*, 1995; Menuet *et al.*, 2001, 2002; Andreassen *et al.*, 2003; Menuet *et al.*, 2003). Most studies point to the ventral telencephalon, the preoptic region (including the magnocellular preoptic nucleus) and the mediobasal hypothalamus as the regions with strongest ERα expression. This is also the case for the expression of ERβ which is documented in only a few species. In general, it was found that ERα and ERβ have a largely overlapping, but differential distribution (Figures 2.2A and 2.3C) as shown in zebrafish (Hawkins *et al.*, 2000) and rainbow trout (Lethimonier C., and Kah O., unpublished data). In zebrafish, ERβ1 and ERβ2 have a larger distribution than ERα. They are notably present above the periventricular areas of the telencephalon and diencephalon, where they could be involved in

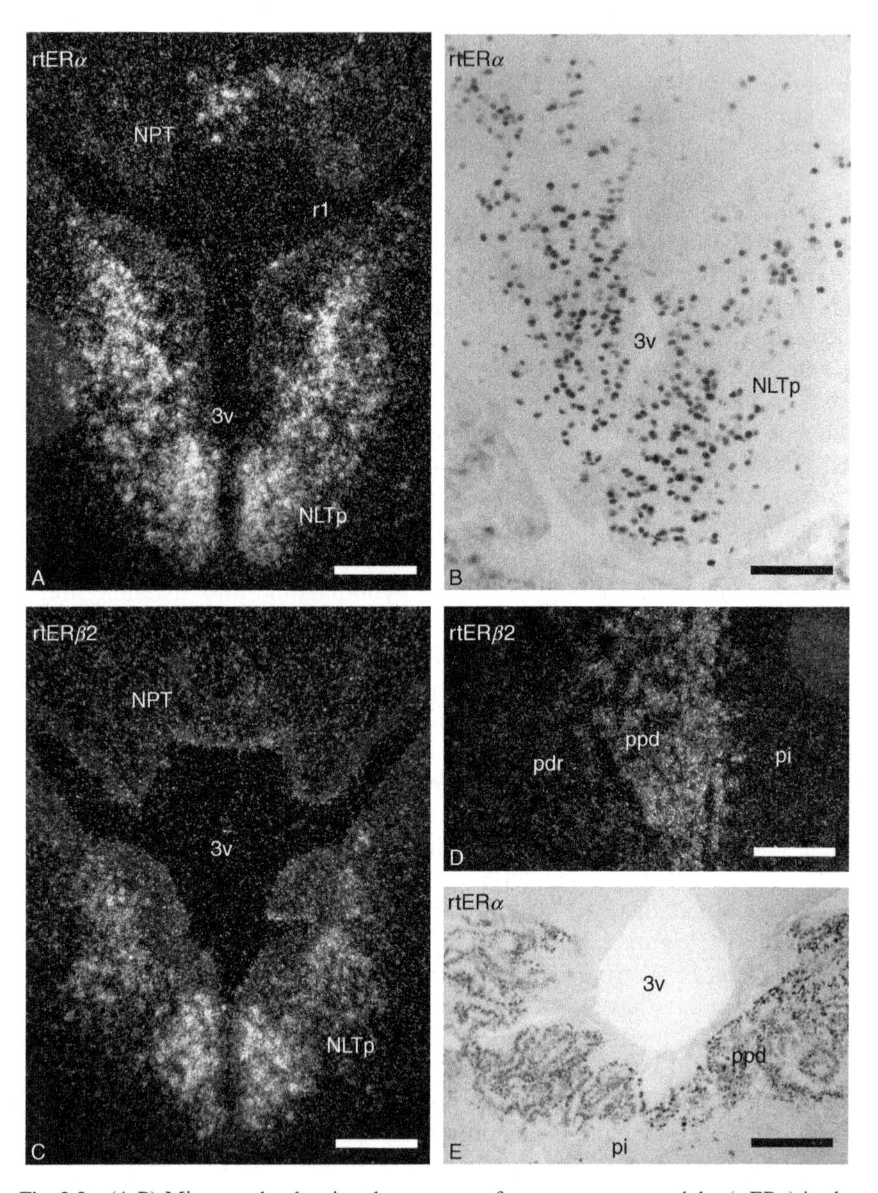

Fig. 2.3. (A,B) Micrographs showing the presence of estrogen receptor alpha (rtERα) in the brain of the rainbow trout as demonstrated by *in situ* hybridization (A) or immunohistochemistry (B). Bar = 60 μm. (C) shows a section adjacent to that in (A) evidencing expression of rtERβ2 in the posterior nucleus lateralis tuberis (NLTp). Note that expression of rtERβ2 is weaker in the NLT and is absent from the nucleus posterioris tuberis (NPT). Bar = 60 μm. (D) Regional distribution of rtERβ2 on a sagittal section of the pituitary of a male rainbow trout showing expression only in the proximal pars distalis (ppd), while the rostral pars distalis (pdr) and the pars intermedia (pi) are devoid of signal. Bar = 300 μm. (E) Immunohistochemical demonstration of rtERα in the ppd of the pituitary of the rainbow trout. The pi exhibits much weaker immunoreactivity. Bar = 250 μm. rl, lateral recess; 3v, third ventricle.

the estrogenic regulation of AroB expression. As observed in mammals, ERα and ERβ are found often in the same cells (Adrio F., and Kah O., unpublished data). In the European sea bass (*Dicentrachus labrax*), the largest distribution of ERα was reported in the mesencephalic tectum and tegmentum and rhombencephalon (Muriach *et al.*, 2008b). However, as these results are not substantiated by a technique cross-validation, these new results await confirmation. Estrogen receptor alpha mRNAs were also detected by Northern blotting in the retina and pineal of the rainbow trout where estradiol is able to modulate melatonin secretion (Begay *et al.*, 1994).

Nature of the estrogen receptor expressing cells in the brain of fish. The phenotype of the ERα-expressing cells is poorly documented and has been investigated only in rainbow trout in relation with the neuroendocrine control of reproduction.

KISS neurons. Kisspeptins (KISS) in mammals have emerged as key actors in the control of GnRH neuron activity (Popa *et al.*, 2008; Roa *et al.*, 2008). KISS1 neurons seem to integrate a large variety of external and internal signals such as gonadal steroids, metabolic factors, photoperiod and season. According to recent data, KISS1 neurons in the arcuate nucleus would regulate the negative feedback effect of gonadal steroids on GnRH and GTH secretion in both sexes. In contrast, at least in rodents, the expression of KISS1 in the anteroventral periventricular nucleus (AVPV) is sexually dimorphic, and KISS1 neurons in the AVPV would mediate the positive feedback of E2 to trigger the pre-ovulatory GnRH/luteinizing hormone (LH) in females (Popa *et al.*, 2008; Roa *et al.*, 2008). There is also accumulating evidence to state that the KISS1 neurons play a role in activating the GnRH system at puberty. In the sheep, it was shown that 100% of the KISS1 neurons in the medial preoptic area express ERα (Franceschini *et al.*, 2006).

Kisspeptins also exist in fish where there seem to be two KISS genes, KISS1 and KISS2, at least in medaka (*Oryzias melastigma*), zebrafish and sea bass (Felip *et al.*, 2009; Kitahashi *et al.*, 2009). There are limited data as to the significance of these two genes in terms of puberty control. However, a recent study in medaka indicates that KISS-expressing neurons are sexually dimorphic and sensitive to steroids. Indeed, the number of KISS1 neurons differs between fish under breeding and non-breeding conditions (Kanda *et al.*, 2008).

GnRH neurons. In both mammals and fish, activity of GnRH neurons is strongly influenced by estrogens and aromatizable androgens. Double immunostaining studies failed to evidence ERα expression onto 550 GnRH neurons observed in male and female adult rainbow trout (Navas *et al.*, 1995). This is consistent with observations in the brain of mammals where several studies also reported that ERα is not expressed by GnRH neurons (Wintermantel *et al.*, 2006). However, the potential expression of membrane

ERβ receptors capable of mediating rapid responses has been considered (Abraham *et al.*, 2003, 2004). Recent data in mice using transgenic techniques demonstrated that neurons expressing ERα are critical but established that ovulation is driven indirectly by estrogen actions upon ERα-expressing neuronal afferents to GnRH neurons (Wintermantel *et al.*, 2006). Therefore, although effects of steroids on GnRH neurons have been reported in fish (Trudeau *et al.*, 1992; Montero *et al.*, 1995; Breton and Sambroni, 1996; Dubois *et al.*, 1998; Parhar *et al.*, 2001; Vetillard *et al.*, 2006), they are likely to be mediated by interneurons, probably KISS neurons, similar to the situation in rodents.

Dopaminergic neurons. It is well documented that dopamine (DA) inhibits gonadotrophin release in a number of teleosts, such as the goldfish (Chang and Peter, 1983; Chang *et al.*, 1990), eel (*Anguilla anguilla*) (Weltzien *et al.*, 2006) or rainbow trout (Linard *et al.*, 1995; Saligaut *et al.*, 1998). The DA neurons implicated in this mechanism are located in the anterior ventral preoptic region (Peter and Paulencu, 1980; Kah *et al.*, 1987b; Vetillard *et al.*, 2002; Weltzien *et al.*, 2006) as shown by a combination of immunohistochemistry, *in situ* hybridization and lesioning studies. In the rainbow trout, using double immunohistochemistry, it was found that all dopamine neurons of the anterior ventral preoptic region strongly express ERα (Linard *et al.*, 1996).

GABA neurons. The last chemically identified cells bearing ERα in fish are GABA neurons (Anglade I., and Kah O., unpublished results). GABA (γ-aminobutyric acid) is strongly expressed in the brain and pituitary of fish and there is clear evidence for its implication in the control of anterior pituitary function, in particular GTH release (Kah *et al.*, 1987a, 1992; Trudeau, 1997; Anglade *et al.*, 1999; Khan and Thomas, 1999; Mananos *et al.*, 1999; Trudeau *et al.*, 2000; Fraser *et al.*, 2002). Double immunohistochemical studies in rainbow trout showed that a large proportion of ERα-expressing cells in the preoptic area and mediobasal hypothalamus do express ERα (Anglade I., and Kah O., unpublished data). These data well support the fact that sex steroids deeply affect the response of the gonadotropic axis to GABA agonists or antagonists.

Melanin-concentrating hormone. Both testosterone and estradiol stimulate melanin-concentrating hormone (MCH) mRNA expression in the hypothalamus in a sex-dependent manner, with females showing the greatest responsiveness. In addition, *in vitro* experiments demonstrated that graded doses of salmon MCH stimulate LH, but not GH, secretion from dispersed pituitary cells. Results suggest that hypothalamic MCH may participate in the steroid positive feedback loop on pituitary LH secretion (Cerda-Reverter *et al.*, 2006).

Estrogen receptor in the pituitary. The first report on the presence of estrogen-concentrating cells was obtained using a histochemical procedure: fluorescent steroid–hormone conjugates were localized in the cytoplasm and nucleus of the gonadotropes of the caudal pars distalis (CPD) and in cells of the pars intermedia (PI) previously demonstrated to contain immunoreactive GTH (Schreibman *et al.*, 1982). Following the cloning of ERs, the presence of ER mRNAs in the pituitary was demonstrated in a number of species including the rainbow trout (Figure 2.3D and 2.3E), and the eelpout (*Zoarces viviparous*) (Andreassen *et al.*, 2003).

2.2.2. Androgen Receptors (AR)

In teleosts, there is a dramatic deficit in detailed information regarding AR receptors and their expression in the brain pituitary complex. The first data concerning the presence of ARs in the brain of fish came from pioneering works using tritiated testosterone (T) or dihydrotestosterone (DHT) (Davis *et al.*, 1977; Fine *et al.*, 1982; Bass *et al.*, 1986; Fine *et al.*, 1996). However, it must be kept in mind that both of these androgens can be converted in estrogenic compounds, T in E2 and DHT in 5 alpha-androstan-3 beta, 17 beta-diol, a known metabolite of DHT with estrogenic activity (Kuiper *et al.*, 1998; Mouriec *et al.*, 2009). Thus, tritiated androgens can be converted in substances binding to estrogen receptors and caution must be taken when interpreting those studies that demonstrated an abundance of cells taking up the tracer in the telencephalon and diencephalon. In the oyster toadfish (*Opsanus tau*) for instance, cells were found in the dorsal telencephalon, the supracommissural nucleus of the ventral telencephalon, the nucleus preopticus parvocellularis anterior and other preoptic nuclei, the ventral, dorsal and caudal hypothalamus (Fine *et al.*, 1996). Additionally, labeled cells were located in the optic tectum, torus semicircularis, nucleus lateralis valvulae and inferior reticular formation (Fine *et al.*, 1996). In the mormyrid fish of Africa, cells concentrating ^{3}H-DHT were detected within the reticular formation that lies adjacent to the medullary relay nucleus that innervates the spinal electromotorneurons (Bass *et al.*, 1986). In the goldfish, a high density of ARs was reported with a peak in mature animals as compared to immature (Pasmanik and Callard, 1988). Despite the fact that AR have been cloned in a number of species (Takeo and Yamashita, 1999, 2000; Todo *et al.*, 1999; Touhata *et al.*, 1999; Blazquez and Piferrer, 2005; Olsson *et al.*, 2005; Hossain *et al.*, 2008; Liu *et al.*, 2009), data gained by *in situ* hybridization are limited to one or two species (Burmeister *et al.*, 2007; Harbott *et al.*, 2007).

In the cichlid *Astatotilapia burtoni*, two ARs, ARα and ARβ, were cloned and their distribution studied by *in situ* hybridization. The main brain areas showing overlapping expression of ARα and ARβ were the preoptic area and the ventral hypothalamus. This is not surprising given the well-established roles of these regions in the control of GTH release, sexual differentiation and reproductive behavior in fish, all events that rely on androgen signaling.

In the pituitary, ARα and ARβ are expressed with similar patterns, but ARα mRNA was present at a higher level (Harbott *et al.*, 2007). These results are in total agreement with those obtained in rainbow trout (Menuet *et al.*, 1999). In this latter species, both ARα and ARβ (Takeo and Yamashita, 1999) were detected in the pallial region, the preoptic region, mediobasal hypothalamus, optic tectum and pituitary where ARα messengers most likely overlapped with the gonadotropes (Menuet *et al.*, 1999). Expression of ARα in the gonadotropes is consistent with the fact that in trout testosterone, but not estradiol, strongly increases LH release, indicating a clear androgenic effect at the gonadotrope level (Breton and Sambroni, 1996). In the zebrafish, androgen receptors were found in the brain, in the pineal organ primordium and the retina at 3–5 days post fertilization. In the adult brain, AR was expressed in discrete regions of the telencephalon, in the preoptic area, and in the periventricular hypothalamus (Gorelick *et al.*, 2008), similar to the rainbow trout and tilapia (*Astatotilapia burtoni*).

In the Atlantic croaker, two distinct nuclear AR, termed AR1 and AR2, have been biochemically characterized. Interestingly, AR1 was found only in brain tissue, whereas AR2 was found in both gonadal and brain tissues (Sperry and Thomas, 1999a, 2000). In addition, the two ARs have different steroid-binding specificities. While AR1 has a high affinity for testosterone, AR2 exhibits a broader affinity for androgens and a greater affinity for structurally diverse androgens, including 5α-reduced steroids (Sperry and Thomas, 1999a,b, 2000). According to Thomas and colleagues, the presence of two ARs could explain the differences in physiological activity of T and 11KT in male teleosts. Testosterone could act through both receptors, while testosterone, the major androgen in female fish, would preferentially bind to AR1. These authors also assume that 5α-reductase, abundantly expressed in fish brain (Pasmanik and Callard, 1985), could be important in determining which AR is activated. However, the physiological importance of DHT in teleosts has not been thoroughly examined. In the zebrafish, it was recently shown that DHT can activate the cyp19a1b (aromatase B) gene after conversion into 5 alpha-androstan-3 beta, 17 beta-diol, a known metabolite of DHT with estrogenic activity (Mouriec *et al.*, 2009). It would be highly interesting to possess detailed information regarding the brain expression of these two receptor subtypes, as compared to that of the steroidogenic enzymes, notably 5α-reductase and 3β-hydroxysteroid dehydrogenase.

Little is known on the regulation of AR expression in the brain. In Astatotilapia, a positive relationship was found between the dominance status, the size of the testis, the levels of sex steroids and the expression of ER and AR. This possibly suggests that expression of these receptors is up-regulated by sex steroids and thus that dominant male brains are more sensitive to sex steroids (Burmeister *et al.*, 2007). In the cyprinid *Spinibarbus denticulatus*, abundance of AR mRNA in male fish significantly increased in pituitary at fully recrudesced stage and brain at late recrudescing phase, respectively (Liu *et al.*, 2009).

2.2.3. Progesterone Receptors (PRs)

In addition to classical nuclear progesterone receptors (PR), there is strong evidence for the existence of membrane PRs (mPR) in fish and other vertebrates. These mPRs mediate the non-classical action of progestins to induce oocyte maturation in fish (Zhu *et al.*, 2003; Thomas *et al.*, 2004; Hanna *et al.*, 2006; Thomas *et al.*, 2007). Accordingly, in spotted sea trout (*Cynoscion nebulosus*) (Zhu *et al.*, 2003) and other fish (Hanna *et al.*, 2006; Mourot *et al.*, 2006), mPRα is expressed in the ovary but also in a wide range of other reproductive and non-reproductive tissues, including the brain and the pituitary (Thomas, 2008). There is so far no detailed study on the expression sites of these new receptors in fish brain and pituitary. However, potential effects of progestins in the brain–pituitary complex through these mPRs must now be taken into account (Thomas, 2008).

As for nuclear PRs, information is also very limited. They are expressed in the brain with a very large distribution as shown in zebrafish (Zhu *et al.*, unpublished data).

3. CORTICOID RECEPTORS

Corticosteroids are steroid hormones produced in the adrenal cortex of tetrapods and the interrenal gland in fish. Corticosteroids are involved in the regulation of many physiological systems such as the stress response, the immune response, the carbohydrate metabolism, the blood electrolyte levels and also behavior. Glucocorticoids have been named after their role on glucose mobilization as a response to restore homeostasis upon stress, but cortisol also plays important regulatory roles in metabolism, development and immune function. Mineralocorticoids such as aldosterone (in mammals) control electrolyte and water levels, mainly by promoting sodium retention in the kidney. While it is clear that cortisol is the most abundant circulating glucocorticoid in fish, there has also been a long debate regarding the exact nature of corticosteroids secreted by the interrenal. The current view is that the adrenal in fish lacks the necessary enzyme that accomplishes the last step of aldosterone biosynthesis (Jiang *et al.*, 1998). However, following the cloning of a mineralocorticoid-like receptor in the rainbow trout (Colombe *et al.*, 2000), it has been recently suggested that 11-deoxycorticosterone (DOC) is the endogenous ligand for this fish mineralocorticoid receptor (MR) (Sturm *et al.*, 2005; Prunet *et al.*, 2006). Similarly an MR has also been cloned in a cichlid fish (Greenwood *et al.*, 2003), together with two glucocorticoid receptors (GR) as also found in rainbow trout (Ducouret *et al.*, 1995; Bury *et al.*, 2003) and more recently in common carp, *Cyprinus carpio* (Stolte *et al.*, 2008).

3.1. Mineralocorticoid Receptors

Until now, precise information regarding expression of these receptors in the brain and pituitary of fish has been scarce. Nevertheless, there is indication that the brain is the major site of MR expression in the rainbow trout and carp (Sturm *et al.*, 2005; Stolte *et al.*, 2008).

3.2. Glucocorticoid Receptors

Brain cortisol receptors were first characterized in salmonids through binding studies that demonstrated high-affinity, low-capacity, binding sites for the synthetic glucocorticoid triamcinolone acetonide. Binding was displaced by cortisol, dexamethasone and to a lesser extent RU38486 (Lee *et al.*, 1992; Knoebl *et al.*, 1996). Following these studies, the first glucocorticoid receptor was cloned in rainbow trout (Ducouret *et al.*, 1995). This receptor, now named rtGR1, has the typical structure of a steroid nuclear receptor with the exception of a 9-amino acid insertion between the two zinc fingers. This additional sequence in the rtGR confers a better binding affinity of the receptor to a single glucocorticoid-responsive element (GRE), as shown by gel shift experiments with GST–DBDrtGR fusion proteins, deleted or not of the 9 additional amino acids. This higher affinity is correlated with a higher constitutive transcriptional activity of the receptor on a reporter gene driven by a single GRE, but not with the ligand-induced transcriptional activity (Ducouret *et al.*, 1995; Lethimonier *et al.*, 2002b).

3.2.1. Distribution of Glucocorticoid Receptors

Pioneer studies by Teitsma and colleagues (1997) on the brain distribution of this receptor using *in situ* hybridization showed extensive signal in the forebrain, namely in the pallial and subpallial regions, in all subdivisions of the preoptic nucleus (Figure 2.4A and 2.4C) and the nucleus lateralis tuberis, which are the main hypophysiotrophic regions in fish (Anglade *et al.*, 1993). In particular, a strong hybridization signal was observed on the magnocellular neurons of the nucleus preopticus, known for producing vasotocin, isotocin and CRF (Teitsma *et al.*, 1997). A weaker signal was detected in the nucleus anterioris periventricularis, nucleus suprachiasmaticus and thalamic region.

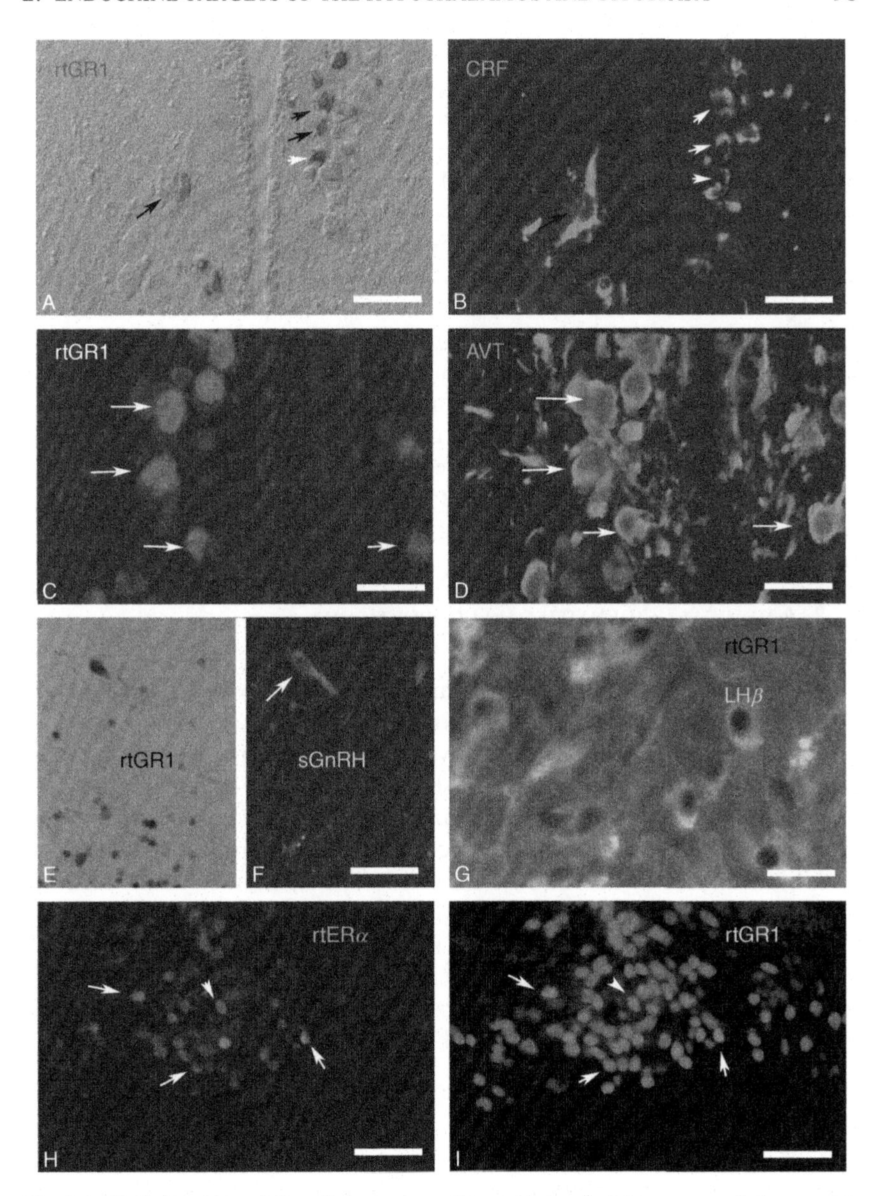

Fig. 2.4. (A,B) Double staining of the same section at the level of the magnocellular preoptic nucleus of the rainbow trout with rtGR1 (A) and corticotrophin-releasing factor (CRF) (B) antibodies. Note that rtGR1 staining (arrows) in (A) often corresponds to the nucleus of CRF-expressing neurons in (B). Bar = 50 μm. (C,D) Double staining of the same section at the level of the magnocellular preoptic nucleus of the rainbow trout with rtGR1 (C) and vasotocin (AVT) (D) antibodies. Note that rtGR1 staining (arrows) in (C) often corresponds to the nucleus of

Until now antibodies to fish GR have only been obtained against the first 165 amino acids of the NH2 terminal A/B domain of the rtGR1 receptor (Tujague *et al.*, 1998). Given the similarity between the rtGR1 and rtGR2 sequences (Bury *et al.*, 2003), it is likely that this antibody also cross-reacts with rtGR2. Nevertheless, immunohistochemistry largely validated the *in situ* hybridization results (Teitsma *et al.*, 1997, 1998, 1999), which is in agreement with the fact that in the common carp GR1 and GR2 messengers exhibit the same distribution patterns (Stolte *et al.*, 2008) highly similar to that reported for the rtGR1 (Teitsma *et al.*, 1997). The immunohistochemical staining extended from the telencephalon to the spinal cord, with the highest density in the neuroendocrine component of the brain, the preoptic region and the mediobasal hypothalamus, and in the periventricular zone of the optic tectum. That same antibody was also used in tilapia (*Oreochromis mossambicus*) with slightly different results (Pepels *et al.*, 2004). However, given the low conservation of the A/B domains between the tilapia and the rainbow trout GRs, this result requires confirmation and cross-validation by other techniques.

3.2.2. Chemically Identified Glucocorticoid Receptor-Expressing Cells

The high concentration of rtGR-immunoreactive cells in the most important neuroendocrine areas and the pituitary points toward a major role of cortisol in modulating both adaptive and stress responses in fish. As expected, colocalization experiments showed that 100% of the CRF-immunoreactive neurons in the parvocellular and magnocellular preoptic nucleus strongly express glucocorticoid receptors (Figure 2.4A and 2.4B), indicating a feedback loop of cortisol on neurons controlling its own secretion (Teitsma *et al.*, 1998). Data in the common carp and tilapia also point to the magnocellular preoptic nucleus as one of the main sites of expression for GR1 and GR2 (Pepels *et al.*, 2004; Stolte *et al.*, 2008). In the rainbow trout, using double immunostaining, rtGR1 was observed in most magnocellular neurons expressing vasotocin (Figure 2.4C and 2.4D) (Teitsma *et al.*, 1998).

AVT-expressing neurons (D). Bar = 30 μm. (E,F) Double staining of the same section at the level of the ventral telencephalon of the rainbow trout with rtGR1 (E) and salmon GnRH (sGnRH) (F) antibodies. Note that rtGR1 staining (arrow) in (E) corresponds to the nucleus of sGnRH neurons (F). Bar = 50 μm. (G) Confocal microscope image of the proximal pars distalis of the rainbow trout showing that most LHβ-positive cells (in green) express the rtGR1 receptor (in black). Bar = 30 μm. (H,I) Double staining of the same section at the level of the ventral preoptic region of the rainbow trout with rtERα (H) and rtGR1 (I) antibodies. Note that the two stainings correspond to cell nuclei and often overlap (arrows). Bar = 100 μm. (See Color Insert.)

In the pituitary, the rtGR immunoreactivity was consistently found in the rostral pars distalis, most likely on the corticotropes (Teitsma *et al.*, 1998; Stolte *et al.*, 2008), with the exception of the prolactin cells, and in the proximal pars distalis that in trout contains thyrotropes, gonadotropes (Figure 2.4G), and somatotropes (Teitsma *et al.*, 1998, 1999).

Stress and cortisol are well known for impacting reproduction at various levels including the liver, the brain and the pituitary. The GnRH system is notably a potential target for cortisol as shown in territorial tilapia (*Astatotilapia burtoni*) (Fox *et al.*, 1997; Greenwood and Fernald, 2004). In vitellogenic rainbow trout, it was found that a large majority of GnRH neurons of the caudal ventral telencephalon and anterior preoptic region expressed the rtGR1 receptor (Figure 2.4E and 2.4F) (Teitsma *et al.*, 1999). Interestingly, rtGR1 is also strongly expressed in the dopaminergic neurons of the anterior ventral preoptic region known for inhibiting gonadotropin release in a number of species (Teistma *et al.*, 1999). In such neurons, potential interactions between rtGR1 and ERα are likely. In the rainbow trout liver, it was shown that cortisol decreases vitellogenin by disturbing the molecular mechanisms involved in the autoregulation of ERα by estradiol (Lethimonier *et al.*, 2002a). This indicates that GRs, when co-expressed with ERα, have the potential to interfere with ERα-mediated events. In the brain of rainbow trout, extensive co-expression of ERα and rtGR1 was observed (Figure 2.4H and 2.4I).

4. METABOLIC HORMONES

In fish, as in any animals, there is a conflict between energy targeted to growth and mobilization of resources for reproduction. This conflict is even more pronounced in oviparous species, given the lower survival rate of progeny and the high energetic cost of gametogenesis. Thus, any individual suffering metabolic stress or reduced energy stores will exhibit retarded puberty and/or reduced fertility. The cross-talk between energy status and the reproductive axis is ensured through sensing of metabolic signals by neuroendocrine circuits involving neuropeptides [such as neuropeptide Y (NPY), cocaine- and amphetamine-regulated transcript (CART), KISS or GnRH], and through a number of neuropeptide hormones and metabolic cues, whose nature and mechanisms of action have begun to be deciphered only in recent years. In this context, the emergence of kisspeptins, encoded by the KiSS-1 gene, and their receptor, GPR54, as mandatory signals for normal pubertal maturation and gonadal function, has raised the possibility that the KiSS-1/GRP54 system might also participate in coupling body energy status and reproduction. In rodents, there is accumulating evidence that the KiSS-1 system is integral in mediating metabolic information onto the centers governing reproductive function, through a putative leptin–kisspeptin–GnRH pathway (Roa *et al.*, 2008).

The KISS/GPR54 system is just starting to be deciphered in fish and so far detailed morphological bases are lacking. There is now good evidence for the presence of two KISS genes in fish, encoding KISS1 and KISS2 (Felip *et al.*, 2009; Kitahashi *et al.*, 2009; Lee *et al.*, 2009), and thus there is need for more data on their respective expression sites. Information is for the moment limited to KISS1 in medaka (Kanda *et al.*, 2008) and preliminary data on KISS2 in zebrafish and medaka (Kitahashi *et al.*, 2009). However, one may speculate that negative inputs on the GnRH systems via the KISS neurons may explain the well-documented alterations of fertility linked to conditions of disturbed energy balance in fish, similar to mammals.

4.1. Leptin and Leptin Receptors

The hormone leptin, discovered in 1994, is produced by the adipose tissue of mammalian species and leptin levels are roughly proportional to fat stores (Robertson *et al.*, 2008). This index of energy stores is important to inform the brain on the level of reserves. Leptin regulates food intake by acting on the appetite center in the ventromedial nucleus of the hypothalamus. Leptin works by inhibiting NPY and agouti-related peptide (AgRP), and by stimulating the activity of α-melanocyte-stimulating hormone (α-MSH). In terms of reproduction, leptin is one of the parameters used by the brain to assess energy levels and give the "green light" to the highly demanding reproductive function. In mammals, there are six types of receptors (LepRa–LepRf), LepRb being the only isoform that contains active intracellular signaling domains. Accordingly, the hypothalamus contains high levels of LepRb mRNA expression. In mammals, leptin strongly impacts reproductive functions as shown by the fact that leptin restores reproduction in infertile ob/ob mice. It is believed that leptin activates GnRH neurons but GnRH neurons do not express LepRb (Finn *et al.*, 1998; Gamba and Pralong, 2006). More recently, it has been suggested that leptin regulates expression of hypothalamic KISS1 and there is some evidence for the presence of LepRb on KISS1 neurons (Smith *et al.*, 2006).

In mammals, LepRb is expressed in neurons expressing the orexigenic neuropeptides AgRP and NPY and the anorexigenic pro-hormone proopiomelanocortin (POMC) (Robertson *et al.*, 2008).

In fish, the primary source of leptin is the liver (Kurokawa *et al.*, 2005; Huising *et al.*, 2006) and there is so far little evidence regarding the action of leptin on feeding and reproduction. This is mainly due to the fact that leptin proved to be very difficult to characterize in non-mammalian species.

Early studies in European sea bass and rainbow trout showed direct action of recombinant mammalian leptin on FSH and/or LH release (Peyon *et al.*, 2001; Weil *et al.*, 2003). In rainbow trout, this effect was observed only after the onset of gametogenesis, suggesting that leptin is not the unique trigger for the activation of the gonadotropic axis (Weil *et al.*, 2003). Although fish leptin has now been characterized and produced, there is still a great need for detailed information on its effects upon the neuroendocrine circuits controlling reproduction and feeding or the expression of leptin receptors (Huising *et al.*, 2006; Nagasaka *et al.*, 2006; Murashita *et al.*, 2008; Yacobovitz *et al.*, 2008). It is possible that, as in mammals, leptin will emerge as one of the actors ensuring the dialogue between the growth/nutrition axis and the reproductive axis.

4.2. Insulin and Insulin-like Growth Factor

Insulin is a key peptide well known as an essential anabolic hormone responsible for maintaining glucose homeostasis in vertebrates. Insulin is produced by the pancreas from the proinsulin precursor molecule under the action of prohormone convertases. However, proinsulin is also expressed in the developing brain where it acts as a survival factor involved in early morphogenesis. At this stage, unprocessed proinsulin stimulates proliferation and cell survival in the developing brain and retina (de la Rosa *et al.*, 1994; Hernandez-Sanchez *et al.*, 2006; Papasani *et al.*, 2006). In teleost fish, insulin is secreted from Brockmann bodies, but appears to be also produced at a low level in the brain and pituitary gland as shown in tilapia (*Oreochromis niloticus*) (Hrytsenko *et al.*, 2007). Although production of insulin in the brain of fish was discussed for some time (Plisetskaya *et al.*, 1993), recent data showed mRNA and protein expression in several brain regions, notably the hypothalamus, the optic tectum, the cerebellum and the medulla oblongata (Hrytsenko *et al.*, 2007).

IGF-I is produced primarily by the liver as an endocrine hormone as well as in target tissues in a paracrine/autocrine fashion. Its production is stimulated by GH and its main action is mediated by binding to specific IGF receptors present on many cell types in virtually all tissues, especially in muscle, cartilage, bone, liver, kidney, brain, skin and lungs. In addition, IGF-I is also expressed in the brain, notably in the olfactory bulb, the hippocampus and the cerebellum (Werther *et al.*, 1990).

Insulin and IGF-I receptor-binding and receptor intrinsic tyrosine kinase activity were detected in the brain of common carp and brown trout (*Salmo trutta fario*). Both ligands stimulated phosphorylation of exogenous substrates in a dose-dependent manner (Leibush *et al.*, 1996). The distribution of IGF-I receptors was studied in detail in juvenile and adult brown trout.

A large distribution of [^{125}I]IGF-I binding was observed in many forebrain regions (olfactory bulbs, hypothalamus, thalamus) and was highest in cerebellum and optic tectum (Smith *et al.*, 2005). It is believed that this large distribution of IGF-I receptor is linked to the growth of the fish brain during early development and adulthood. In zebrafish, IGF-I signaling regulates embryonic growth and development by promoting cell survival and cell cycle progression (Schlueter *et al.*, 2007).

Many studies point to close interactions between estrogen receptor and IGF-I receptors in the brain of mammals (Quesada *et al.*, 2007). For the moment, such information in fish is limited to studies showing that E2 modulates IGF-I and IGF-I receptor in multiple tissues, notably the pituitary and the brain, with potential consequences for the functioning of many physiological processes, not just reproduction (Filby *et al.*, 2006).

Evidence for IGF-I production in the pituitary was obtained in Nile tilapia. Both IGF-I mRNA and peptide were present in the majority of ACTH cells in all individuals and in the GH cells with a high individual variation (Eppler *et al.*, 2007). Only IGF-I mRNA but not IGF-I peptide was detected in α-MSH cells, suggesting immediate release of IGF-I upon synthesis. IGF-I released from the GH cells may serve as an auto/paracrine mediator of a negative feedback mechanism in addition to liver-derived endocrine IGF-I. Local IGF-I may regulate synthesis and release of pituitary hormones in an autocrine and/or paracrine manner as well as prevent apoptosis and stimulate proliferation of endocrine cells (Eppler *et al.*, 2007). In masu salmon (*Oncorhynchus masu*), IGF-I was shown to modulate directly expression of GTH subunit genes during sexual maturation. In particular, IGF-I differently affects sGnRH-induced GTH subunit gene expression, depending on reproductive stages (Ando *et al.*, 2006; Furukuma *et al.*, 2008).

4.3. Thyroid Hormone Receptors

Through thyroid hormone receptors (TRs), thyroid hormones (THs) exert pleiotropic effects on growth, differentiation, metamorphosis and reproduction. The THs are iodinated derivatives of tyrosine and are present as two major forms: thyroxin (T_4) containing four iodine residues and triiodothyronine (T_3), containing only three iodine residues, which is the more potent and the major biological TH. Some of these effects are due to feedback of TH onto the brain pituitary complex according to highly complex mechanisms (Bernal, 2007; Alkemade *et al.*, 2008). Similar to steroid receptors (see above), TRs belong to the superfamily of nuclear receptors. In mammals, there are two TR genes, TRα and TRβ, with splice variant subtypes produced by TRβ (TRβ1and TRβ2). In fish, THs are also involved in several aspects of development, growth and regulation of metabolism.

THs also play crucial functions in metamorphosis in fish and amphibians and in the parr–smolt transformation in salmon (Power *et al.*, 2001; Eales, 2006). TRα1 and TRβ forms, also generating spliced variants, have now been evidenced in a number of teleost fishes (Essner *et al.*, 1999; Marchand *et al.*, 2004; Kawakami *et al.*, 2007; Nelson and Habibi, 2009).

Thyroid hormones are strongly involved in development of the fish brain as shown in zebrafish (Essner *et al.*, 1999) in which over-expression of TRα1 during embryogenesis disrupts hindbrain patterning. This effect implicates repression of retinoic acid receptors in the control of hox gene expression (Essner *et al.*, 1999). In the Japanese conger eel (*Conger myriaster*), two TRαs and two TRβs were identified. In adult fish, TRs are widely expressed in many tissues and, notably, expression is very strong in the brain and the pituitary (Kawakami *et al.*, 2003, 2007), but there is no detailed information on the precise sites of expression. In the pituitary, recent data in the Japanese eel (*Anguilla japonica*) have documented the strong expression of TRβB mRNAs in the whole of the pars distalis, most likely corresponding to TSH, ACTH and GH and/or gonadotropic hormone cells (Kawakami *et al.*, 2007). Clearly, there is a need for more information regarding TH actions in the developing and adult brain.

5. CONCLUDING REMARKS

Over the last 30 years, our understanding of physiological processes has made tremendous progress, especially in the field of sex steroid feedback. There is of course a great need for more studies before we can get a clear picture. Areas for future research concern a better understanding of the actions of sex steroids in the brain of fish. What are the respective parts of true estrogenic and true androgenic effects? On what targets, when and where? What is the role of brain aromatase? How is aromatase activity regulated? All these questions are awaiting answers.

There are also many black boxes regarding the roles of the membrane steroid receptors, their localization and the regulation of their expression. One may expect fascinating findings in this field. Hopefully, tools will be developed that permit us to address these issues and many others such as the cross-talks between the growth/metabolism axis, the stress axis and the reproductive axis. These cross-talks are integral for the concept of homeostasis by preventing deleterious responses when the physiological status or the environmental conditions are inappropriate.

Finally, many achievements have been made since the democratization of molecular biology techniques that boosted the availability of molecular and immunological tools. Without these tools, no accurate information can be gained within the brain. The complexity of the organization of the central nervous system and its heterogeneity require investigating the morphological relationships at the cellular and subcellular scales and at both the messenger and post-protein levels. The accessibility and sensitivity of the PCR techniques are certainly of great help in many circumstances, but it should not be forgotten that the important functional molecule is the protein.

ACKNOWLEDGMENTS

Original data presented in this chapter were supported by the CNRS, the French Ministry of Research and Technology and the European Union. Thanks to Farzad Pakdel, Christine Teitsma, Christhèle Lethimonier, Fatima Adrio, Bernadette Ducouret, Isabelle Anglade, Elisabeth Pellegrini, Nicolas Diotel, Karen Mouriec and Arnaud Menuet for their input in this research.

REFERENCES

Abraham, I. M., Han, S. K., Todman, M. G., Korach, K. S., and Herbison, A. E. (2003). Estrogen receptor beta mediates rapid estrogen actions on gonadotropin-releasing hormone neurons *in vivo*. *J. Neurosci.* **23,** 5771–5777.

Abraham, I. M., Todman, M. G., Korach, K. S., and Herbison, A. E. (2004). Critical *in vivo* roles for classical estrogen receptors in rapid estrogen actions on intracellular signaling in mouse brain. *Endocrinology* **145,** 3055–3061.

Adolf, B., Chapouton, P., Lam, C. S., Topp, S., Tannhauser, B., Strahle, U., Gotz, M., and Bally-Cuif, L. (2006). Conserved and acquired features of adult neurogenesis in the zebrafish telencephalon. *Dev. Biol.* **295,** 278–293.

Alkemade, A., Visser, T. J., and Fliers, E. (2008). Thyroid hormone signaling in the hypothalamus. *Curr. Opin. Endocrinol. Diabetes Obes.* **15,** 453–458.

Ando, H., Luo, Q., Koide, N., Okada, H., and Urano, A. (2006). Effects of insulin-like growth factor I on GnRH-induced gonadotropin subunit gene expressions in masu salmon pituitary cells at different stages of sexual maturation. *Gen. Comp. Endocrinol.* **149,** 21–29.

Andreassen, T. K., Skjoedt, K., Anglade, I., Kah, O., and Korsgaard, B. (2003). Molecular cloning, characterisation, and tissue distribution of oestrogen receptor alpha in eelpout (*Zoarces viviparus*). *Gen. Comp. Endocrinol.* **132,** 356–368.

Anglade, I., Zandbergen, A. M., and Kah, O. (1993). Origin of the pituitary innervation in the goldfish. *Cell Tissue Res.* **273,** 345–355.

Anglade, I., Pakdel, F., Bailhache, T., Petit, F., Salbert, G., Jego, P., Valotaire, Y., and Kah, O. (1994). Distribution of estrogen receptor-immunoreactive cells in the brain of the rainbow trout (*Oncorhynchus mykiss*). *J. Neuroendocrinol.* **6,** 573–583.

Anglade, I., Mazurais, D., Douard, V., Le Jossic-Corcos, C., Mananos, E. L., Michel, D., and Kah, O. (1999). Distribution of glutamic acid decarboxylase mRNA in the forebrain of the rainbow trout as studied by *in situ* hybridization. *J. Comp. Neurol.* **410,** 277–289.

Bass, A. H., Segil, N., and Kelley, D. B. (1986). Androgen binding in the brain and electric organ of a mormyrid fish. *J. Comp. Physiol. [A]* **159,** 535–544.

Begay, V., Valotaire, Y., Ravault, J. P., Collin, J. P., and Falcon, J. (1994). Detection of estrogen receptor mRNA in trout pineal and retina: Estradiol-17 beta modulates melatonin production by cultured pineal photoreceptor cells. *Gen. Comp. Endocrinol.* **93,** 61–69.

Bentivoglio, M., and Mazzarello, P. (1999). The history of radial glia. *Brain Res. Bull.* **49,** 305–315.

Bernal, J. (2007). Thyroid hormone receptors in brain development and function. *Nat. Clin. Pract. Endocrinol. Metab.* **3,** 249–259.

Blazquez, M., and Piferrer, F. (2005). Sea bass (*Dicentrarchus labrax*) androgen receptor: cDNA cloning, tissue-specific expression, and mRNA levels during early development and sex differentiation. *Mol. Cell Endocrinol.* **237,** 37–48.

Breton, B., and Sambroni, E. (1996). Steroid activation of the brain-pituitary complex gonadotropic function in the triploid rainbow trout *Oncorhynchus mykiss. Gen. Comp. Endocrinol.* **101,** 155–164.

Burmeister, S. S., Kailasanath, V., and Fernald, R. D. (2007). Social dominance regulates androgen and estrogen receptor gene expression. *Horm. Behav.* **51,** 164–170.

Bury, N. R., Sturm, A., Le Rouzic, P., Lethimonier, C., Ducouret, B., Guiguen, Y., Robinson-Rechavi, M., Laudet, V., Rafestin-Oblin, M. E., and Prunet, P. (2003). Evidence for two distinct functional glucocorticoid receptors in teleost fish. *J. Mol. Endocrinol.* **31,** 141–156.

Callard, G. V., Petro, Z., and Ryan, K. J. (1978). Phylogenetic distribution of aromatase and other androgen-converting enzymes in the central nervous system. *Endocrinology* **103,** 2283–2290.

Callard, G. V., Petro, Z., and Ryan, K. J. (1981). Estrogen synthesis *in vitro* and *in vivo* in the brain of a marine teleost (Myoxocephalus). *Gen. Comp. Endocrinol.* **43,** 243–255.

Carroll, J. S., Meyer, C. A., Song, J., Li, W., Geistlinger, T. R., Eeckhoute, J., Brodsky, A. S., Keeton, E. K., Fertuck, K. C., Hall, G. F., Wang, Q., Bekiranov, S., *et al.* (2006). Genome-wide analysis of estrogen receptor binding sites. *Nat. Genet.* **38,** 1289–1297.

Caviola, E., Dalla Valle, L., Belvedere, P., and Colombo, L. (2007). Characterisation of three variants of estrogen receptor beta mRNA in the common sole, *Solea solea* L. (Teleostei). *Gen. Comp. Endocrinol.* **153,** 31–39.

Cerda-Reverter, J. M., Canosa, L. F., and Peter, R. E. (2006). Regulation of the hypothalamic melanin-concentrating hormone neurons by sex steroids in the goldfish: Possible role in the modulation of luteinizing hormone secretion. *Neuroendocrinology* **84,** 364–377.

Chang, J. P., and Peter, R. E. (1983). Effects of dopamine on gonadotropin release in female goldfish, *Carassius auratus. Neuroendocrinology* **36,** 351–357.

Chang, J. P., Yu, K. L., Wong, A. O., and Peter, R. E. (1990). Differential actions of dopamine receptor subtypes on gonadotropin and growth hormone release *in vitro* in goldfish. *Neuroendocrinology* **51,** 664–674.

Chaube, R., and Joy, K. P. (2003). *In vitro* effects of catecholamines and catecholestrogens on brain tyrosine hydroxylase activity and kinetics in the female catfish *Heteropneustes fossilis. J. Neuroendocrinol.* **15,** 273–279.

Colombe, L., Fostier, A., Bury, N., Pakdel, F., and Guiguen, Y. (2000). A mineralocorticoid-like receptor in the rainbow trout, *Oncorhynchus mykiss*: Cloning and characterization of its steroid binding domain. *Steroids* **65,** 319–328.

Corpechot, C., Robel, P., Axelson, M., Sjovall, J., and Baulieu, E. E. (1981). Characterization and measurement of dehydroepiandrosterone sulfate in rat brain. *Proc. Natl. Acad. Sci. USA* **78,** 4704–4707.

Crim, L. W., and Evans, D. M. (1983). Influence of testosterone and/or luteinizing hormone releasing hormone analogue on precocious sexual development in the juvenile rainbow trout. *Biol. Reprod.* **29,** 137–142.

Davis, R. E., Morrell, J. I., and Pfaff, D. W. (1977). Autoradiographic localization of sex steroid-concentrating cells in the brain of the teleost *Macropodus opercularis* (Osteichthyes: Belontiidae). *Gen. Comp. Endocrinol.* **33,** 496–505.

de la Rosa, E. J., Bondy, C. A., Hernandez-Sanchez, C., Wu, X., Zhou, J., Lopez-Carranza, A., Scavo, L. M., and de Pablo, F. (1994). Insulin and insulin-like growth factor system components gene expression in the chicken retina from early neurogenesis until late development and their effect on neuroepithelial cells. *Eur. J. Neurosci.* **6,** 1801–1810.

de Leeuw, R., Smit-van Dijk, W., Zigterman, J. W., van der Loo, J. C., Lambert, J. G., and Goos, H. J. (1985). Aromatase, estrogen 2-hydroxylase, and catechol-O-methyltransferase activity in isolated, cultured gonadotropic cells of mature African catfish, *Clarias gariepinus* (Burchell). *Gen. Comp. Endocrinol.* **60,** 171–177.

Donaldson, E. M., and McBride, J. R. (1967). The effects of hypophysectomy in the rainbow trout *Salmo gairdnerii* (Rich.) with special reference to the pituitary-interrenal axis. *Gen. Comp. Endocrinol.* **9,** 93–101.

Dubois, E. A., Florijn, M. A., Zandbergen, M. A., Peute, J., and Goos, H. J. (1998). Testosterone accelerates the development of the catfish GnRH system in the brain of immature African catfish (*Clarias gariepinus*). *Gen. Comp. Endocrinol.* **112,** 383–393.

Ducouret, B., Tujague, M., Ashraf, J., Mouchel, N., Servel, N., Valotaire, Y., and Thompson, E. B. (1995). Cloning of a teleost fish glucocorticoid receptor shows that it contains a deoxyribonucleic acid-binding domain different from that of mammals. *Endocrinology* **136,** 3774–3783.

Eales, J. G. (2006). Modes of action and physiological effects of thyroid hormones in fish. *In* "Fish Endocrinology" (Reinecke, M., *et al.*, Eds.), pp. 767–808. Science Publishers, Enfield, NH.

Eppler, E., Shved, N., Moret, O., and Reinecke, M. (2007). IGF-I is distinctly located in the bony fish pituitary as revealed for *Oreochromis niloticus*, the Nile tilapia, using real-time RT-PCR, *in situ* hybridisation and immunohistochemistry. *Gen. Comp. Endocrinol.* **150,** 87–95.

Essner, J. J., Johnson, R. G., and Hackett, P. B., Jr. (1999). Overexpression of thyroid hormone receptor alpha 1 during zebrafish embryogenesis disrupts hindbrain patterning and implicates retinoic acid receptors in the control of hox gene expression. *Differentiation* **65,** 1–11.

Felip, A., Zanuy, S., Pineda, R., Pinilla, L., Carrillo, M., Tena-Sempere, M., and Gómez, A. (2009). Evidence for two distinct KiSS genes in non-placental vertebrates that encode kisspeptins with different gonadotropin-releasing activities in fish and mammals. *Mol. Cell Endocrinol.* in press.

Filby, A. L., and Tyler, C. R. (2005). Molecular characterization of estrogen receptors 1, 2a, and 2b and their tissue and ontogenic expression profiles in fathead minnow (*Pimephales promelas*). *Biol. Reprod.* **73,** 648–662.

Filby, A. L., Thorpe, K. L., and Tyler, C. R. (2006). Multiple molecular effect pathways of an environmental oestrogen in fish. *J. Mol. Endocrinol.* **37,** 121–134.

Fine, M. L., Keefer, D. A., and Leichnetz, G. R. (1982). Testosterone uptake in the brainstem of a sound-producing fish. *Science* **215,** 1265–1267.

Fine, M. L., Chen, F. A., and Keefer, D. A. (1996). Autoradiographic localization of dihydrotestosterone and testosterone concentrating neurons in the brain of the oyster toadfish. *Brain Res.* **709,** 65–80.

Finn, P. D., Cunningham, M. J., Pau, K. Y., Spies, H. G., Clifton, D. K., and Steiner, R. A. (1998). The stimulatory effect of leptin on the neuroendocrine reproductive axis of the monkey. *Endocrinology* **139,** 4652–4662.

Forlano, P. M., Deitcher, D. L., and Bass, A. H. (2005). Distribution of estrogen receptor alpha mRNA in the brain and inner ear of a vocal fish with comparisons to sites of aromatase expression. *J. Comp. Neurol.* **483,** 91–113.

Fox, H. E., White, S. A., Kao, M. H., and Fernald, R. D. (1997). Stress and dominance in a social fish. *J. Neurosci.* **17,** 6463–6469.

Franceschini, I., Lomet, D., Cateau, M., Delsol, G., Tillet, Y., and Caraty, A. (2006). Kisspeptin immunoreactive cells of the ovine preoptic area and arcuate nucleus co-express estrogen receptor alpha. *Neurosci. Lett.* **401,** 225–230.

Fraser, E. J., Bosma, P. T., Trudeau, V. L., and Docherty, K. (2002). The effect of water temperature on the GABAergic and reproductive systems in female and male goldfish (*Carassius auratus*). *Gen. Comp. Endocrinol.* **125,** 163–175.

Fu, K. Y., Chen, C. Y., Lin, C. T., and Chang, W. M. (2008). Molecular cloning and tissue distribution of three estrogen receptors from the cyprinid fish *Varicorhinus barbatulus*. *J. Comp. Physiol. [B]* **178,** 189–197.

Furukuma, S., Onuma, T., Swanson, P., Luo, Q., Koide, N., Okada, H., Urano, A., and Ando, H. (2008). Stimulatory effects of insulin-like growth factor 1 on expression of gonadotropin subunit genes and release of follicle-stimulating hormone and luteinizing hormone in masu salmon pituitary cells early in gametogenesis. *Zoolog. Sci.* **25,** 88–98.

Gamba, M., and Pralong, F. P. (2006). Control of GnRH neuronal activity by metabolic factors: The role of leptin and insulin. *Mol. Cell Endocrinol.* **254–255,** 133–139.

Gao, Q., and Horvath, T. L. (2008). Cross-talk between estrogen and leptin signaling in the hypothalamus. *Am. J. Physiol. Endocrinol. Metab.* **294,** E817–E826.

Germain, P., Staels, B., Dacquet, C., Spedding, M., and Laudet, V. (2006). Overview of nomenclature of nuclear receptors. *Pharmacol. Rev.* **58,** 685–704.

Gomez, J. M. (2007). Serum leptin, insulin-like growth factor-I components and sex-hormone binding globulin. Relationship with sex, age and body composition in healthy population. *Protein Pept. Lett.* **14,** 708–711.

Goos, H. J., van der Loo, J. C., Smit-van Dijk, W., and de Leeuw, R. (1985). Steroid aromatase, 2-hydroxylase and COMT activity in gonadotropic cells of the African catfish, *Clarias gariepinus*. *Cell Biol. Int. Rep.* **9,** 529.

Gorelick, D. A., Watson, W., and Halpern, M. E. (2008). Androgen receptor gene expression in the developing and adult zebrafish brain. *Dev. Dyn.* **237,** 2987–2995.

Goulis, D. G., and Tarlatzis, B. C. (2008). Metabolic syndrome and reproduction: I. testicular function. *Gynecol. Endocrinol.* **24,** 33–39.

Greenwood, A. K., and Fernald, R. D. (2004). Social regulation of the electrical properties of gonadotropin-releasing hormone neurons in a cichlid fish (*Astatotilapia burtoni*). *Biol. Reprod.* **71,** 909–918.

Greenwood, A. K., Butler, P. C., White, R. B., DeMarco, U., Pearce, D., and Fernald, R. D. (2003). Multiple corticosteroid receptors in a teleost fish: Distinct sequences, expression patterns, and transcriptional activities. *Endocrinology* **144,** 4226–4236.

Greytak, S. R., and Callard, G. V. (2007). Cloning of three estrogen receptors (ER) from killifish (*Fundulus heteroclitus*): Differences in populations from polluted and reference environments. *Gen. Comp. Endocrinol.* **150,** 174–188.

Halm, S., Kwon, J. Y., Rand-Weaver, M., Sumpter, J. P., Pounds, N., Hutchinson, T. H., and Tyler, C. R. (2003). Cloning and gene expression of P450 17alpha-hydroxylase,17,20-lyase cDNA in the gonads and brain of the fathead minnow *Pimephales promelas*. *Gen. Comp. Endocrinol.* **130,** 256–266.

Halm, S., Martinez-Rodriguez, G., Rodriguez, L., Prat, F., Mylonas, C. C., Carrillo, M., and Zanuy, S. (2004). Cloning, characterisation, and expression of three oestrogen receptors (ERalpha, ERbeta1 and ERbeta2) in the European sea bass, *Dicentrarchus labrax*. *Mol. Cell Endocrinol.* **223,** 63–75.

Hanna, R., Pang, Y., Thomas, P., and Zhu, Y. (2006). Cell-surface expression, progestin binding, and rapid nongenomic signaling of zebrafish membrane progestin receptors alpha and beta in transfected cells. *J. Endocrinol.* **190,** 247–260.

Harbott, L. K., Burmeister, S. S., White, R. B., Vagell, M., and Fernald, R. D. (2007). Androgen receptors in a cichlid fish, *Astatotilapia burtoni*: Structure, localization, and expression levels. *J. Comp. Neurol.* **504,** 57–73.

Hawkins, M. B., Thornton, J. W., Crews, D., Skipper, J. K., Dotte, A., and Thomas, P. (2000). Identification of a third distinct estrogen receptor and reclassification of estrogen receptors in teleosts. *Proc. Natl. Acad. Sci. USA* **97,** 10751–10756.

Hernandez-Sanchez, C., Mansilla, A., de la Rosa, E. J., and de Pablo, F. (2006). Proinsulin in development: New roles for an ancient prohormone. *Diabetologia* **49,** 1142–1150.

Hojo, Y., Hattori, T. A., Enami, T., Furukawa, A., Suzuki, K., Ishii, H. T., Mukai, H., Morrison, J. H., Janssen, W. G., Kominami, S., Harada, N., Kimoto, T., *et al.* (2004). Adult male rat hippocampus synthesizes estradiol from pregnenolone by cytochromes P45017alpha and P450 aromatase localized in neurons. *Proc. Natl. Acad. Sci. USA* **101,** 865–870.

Hojo, Y., Murakami, G., Mukai, H., Higo, S., Hatanaka, Y., Ogiue-Ikeda, M., Ishii, H., Kimoto, T., and Kawato, S. (2008). Estrogen synthesis in the brain – role in synaptic plasticity and memory. *Mol. Cell Endocrinol.* **290,** 31–43.

Hossain, M. S., Larsson, A., Scherbak, N., Olsson, P. E., and Orban, L. (2008). Zebrafish androgen receptor: Isolation, molecular, and biochemical characterization. *Biol. Reprod.* **78,** 361–369.

Hrytsenko, O., Wright, J. R., Jr., Morrison, C. M., and Pohajdak, B. (2007). Insulin expression in the brain and pituitary cells of tilapia (*Oreochromis niloticus*). *Brain Res.* **1135,** 31–40.

Hsu, H. J., Hsiao, P., Kuo, M. W., and Chung, B. C. (2002). Expression of zebrafish cyp11a1 as a maternal transcript and in yolk syncytial layer. *Gene Expr. Patterns* **2,** 219–222.

Huising, M. O., Geven, E. J., Kruiswijk, C. P., Nabuurs, S. B., Stolte, E. H., Spanings, F. A., Verburg-van Kemenade, B. M., and Flik, G. (2006). Increased leptin expression in common Carp (*Cyprinus carpio*) after food intake but not after fasting or feeding to satiation. *Endocrinology* **147,** 5786–5797.

Jiang, J. Q., Young, G., Kobayashi, T., and Nagahama, Y. (1998). Eel (*Anguilla japonica*) testis 11beta-hydroxylase gene is expressed in interrenal tissue and its product lacks aldosterone synthesizing activity. *Mol. Cell Endocrinol.* **146,** 207–211.

Joy, K. P., and Senthilkumaran, B. (1998). Annual and diurnal variations in, and effects of altered photoperiod and temperature, ovariectomy, and estradiol-17 beta replacement on catechol-O-methyltransferase level in brain regions of the catfish, *Heteropneustes fossilis*. *Comp. Biochem. Physiol. C Pharmacol. Toxicol. Endocrinol.* **119,** 37–44.

Joy, K. P., Senthilkumaran, B., and Sudhakumari, C. C. (1998). Periovulatory changes in hypothalamic and pituitary monoamines following GnRH analogue treatment in the catfish *Heteropneustes fossilis*: A study correlating changes in plasma hormone profiles. *J. Endocrinol.* **156,** 365–372.

Kah, O., Dubourg, P., Martinoli, M. G., Rabhi, M., Gonnet, F., Geffard, M., and Calas, A. (1987a). Central GABAergic innervation of the pituitary in goldfish: A radioautographic and immunocytochemical study at the electron microscope level. *Gen. Comp. Endocrinol.* **67,** 324–332.

Kah, O., Dulka, J. G., Dubourg, P., Thibault, J., and Peter, R. E. (1987b). Neuroanatomical substrate for the inhibition of gonadotrophin secretion in goldfish: Existence of a dopaminergic preoptico-hypophyseal pathway. *Neuroendocrinology* **45,** 451–458.

Kah, O., Trudeau, V. L., Sloley, B. D., Chang, J. P., Dubourg, P., Yu, K. L., and Peter, R. E. (1992). Influence of GABA on gonadotrophin release in the goldfish. *Neuroendocrinology* **55,** 396–404.

Kallivretaki, E., Eggen, R. I., Neuhauss, S. C., Kah, O., and Segner, H. (2007). The zebrafish, brain-specific, aromatase cyp19a2 is neither expressed nor distributed in a sexually dimorphic manner during sexual differentiation. *Dev. Dyn.* **236,** 3155–3166.

Kanda, S., Akazome, Y., Matsunaga, T., Yamamoto, N., Yamada, S., Tsukamura, H., Maeda, K., and Oka, Y. (2008). Identification of KiSS-1 product kisspeptin and steroid-sensitive sexually dimorphic kisspeptin neurons in medaka (*Oryzias latipes*). *Endocrinology* **149,** 2467–2476.

Katzenellenbogen, B. S., Choi, I., Delage-Mourroux, R., Ediger, T. R., Martini, P. G., Montano, M., Sun, J., Weis, K., and Katzenellenbogen, J. A. (2000). Molecular mechanisms of estrogen action: Selective ligands and receptor pharmacology. *J. Steroid Biochem. Mol. Biol.* **74,** 279–285.

Kawakami, Y., Tanda, M., Adachi, S., and Yamauchi, K. (2003). cDNA cloning of thyroid hormone receptor betas from the conger eel, *Conger myriaster*. *Gen. Comp. Endocrinol.* **131,** 232–240.

Kawakami, Y., Adachi, S., Yamauchi, K., and Ohta, H. (2007). Thyroid hormone receptor beta is widely expressed in the brain and pituitary of the Japanese eel, *Anguilla japonica*. *Gen. Comp. Endocrinol.* **150,** 386–394.

Khan, I. A., and Thomas, P. (1999). GABA exerts stimulatory and inhibitory influences on gonadotropin II secretion in the Atlantic croaker (*Micropogonias undulatus*). *Neuroendocrinology* **69,** 261–268.

Kim, Y. S., Stumpf, W. E., and Sar, M. (1978). Topography of estrogen target cells in the forebrain of goldfish, *Carassius auratus*. *J. Comp. Neurol.* **182,** 611–620.

Kim, Y. S., Sar, M., and Stumpf, W. E. (1979a). Estrogen target cells in the pituitary of platyfish, *Xiphophorus maculatus*. *Cell Tissue Res.* **198,** 435–440.

Kim, Y. S., Stumpf, W. E., and Sar, M. (1979b). Topographical distribution of estrogen target cells in the forebrain of platyfish, *Xiphophorus maculatus*, studied by autoradiography. *Brain Res.* **170,** 43–59.

Kitahashi, T., Ogawa, S., and Parhar, I. S. (2009). Cloning and expression of kiss2 in the zebrafish and medaka. *Endocrinology* **150,** 821–831.

Klein-Hitpass, L., Schorpp, M., Wagner, U., and Ryffel, G. U. (1986). An estrogen-responsive element derived from the 5′ flanking region of the Xenopus vitellogenin A2 gene functions in transfected human cells. *Cell* **46,** 1053–1061.

Knoebl, I., Fitzpatrick, M. S., and Schreck, C. B. (1996). Characterization of a glucocorticoid receptor in the brains of chinook salmon, *Oncorhynchus tshawytscha*. *Gen. Comp. Endocrinol.* **101,** 195–204.

Kuiper, G. G., Shughrue, P. J., Merchenthaler, I., and Gustafsson, J. A. (1998). The estrogen receptor beta subtype: A novel mediator of estrogen action in neuroendocrine systems. *Neuroendocrinol.* **19,** 253–286.

Kurokawa, T., Uji, S., and Suzuki, T. (2005). Identification of cDNA coding for a homologue to mammalian leptin from pufferfish, *Takifugu rubripes*. *Peptides* **26,** 745–750.

Lee, P. C., Goodrich, M., Struve, M., Yoon, H. I., and Weber, D. (1992). Liver and brain glucocorticoid receptor in rainbow trout, *Oncorhynchus mykiss*: Down-regulation by dexamethasone. *Gen. Comp. Endocrinol.* **87,** 222–231.

Lee, Y. R., Tsunekawa, K., Moon, M. J., Um, H. N., Hwang, J. I., Osugi, T., Otaki, N., Sunakawa, Y., Kim, K., Vaudry, H., Kwon, H. B., Seong, J. Y., *et al.* (2009). Molecular evolution of multiple forms of kisspeptins and GPR54 receptors in vertebrates. *Endocrinology* Jan 22. [Epub ahead of print].

Leibush, B., Parrizas, M., Navarro, I., Lappova, Y., Maestro, M. A., Encinas, M., Plisetskaya, E. M., and Gutierrez, J. (1996). Insulin and insulin-like growth factor-I receptors in fish brain. *Regul. Pept.* **61,** 155–161.

Lethimonier, C., Flouriot, G., Kah, O., and Ducouret, B. (2002a). The glucocorticoid receptor represses the positive autoregulation of the trout estrogen receptor gene by preventing the enhancer effect of a C/EBPbeta-like protein. *Endocrinology* **143,** 2961–2974.

Lethimonier, C., Tujague, M., Kern, L., and Ducouret, B. (2002b). Peptide insertion in the DNA-binding domain of fish glucocorticoid receptor is encoded by an additional exon and confers particular functional properties. *Mol. Cell Endocrinol.* **194,** 107–116.

Linard, B., Bennani, S., and Saligaut, C. (1995). Involvement of estradiol in a catecholamine inhibitory tone of gonadotropin release in the rainbow trout (*Oncorhynchus mykiss*). *Gen. Comp. Endocrinol.* **99,** 192–196.

Linard, B., Anglade, I., Corio, M., Navas, J. M., Pakdel, F., Saligaut, C., and Kah, O. (1996). Estrogen receptors are expressed in a subset of tyrosine hydroxylase-positive neurons of the anterior preoptic region in the rainbow trout. *Neuroendocrinology* **63,** 156–165.

Liu, X., Su, H., Zhu, P., Zhang, Y., Huang, J., and Lin, H. (2009). Molecular cloning, characterization and expression pattern of androgen receptor in *Spinibarbus denticulatus*. *Gen. Comp. Endocrinol.* **160,** 93–101.

Lovejoy, J. C., and Sainsbury, A. (2009). Sex differences in obesity and the regulation of energy homeostasis. *Obes. Rev.* **10,** 154–167.

Ma, C. H., Dong, K. W., and Yu, K. L. (2000). cDNA cloning and expression of a novel estrogen receptor beta-subtype in goldfish (*Carassius auratus*). *Biochim. Biophys. Acta* **1490,** 145–152.

Mananos, E. L., Anglade, I., Chyb, J., Saligaut, C., Breton, B., and Kah, O. (1999). Involvement of gamma-aminobutyric acid in the control of GTH-1 and GTH-2 secretion in male and female rainbow trout. *Neuroendocrinology* **69,** 269–280.

Marchand, O., Duffraisse, M., Triqueneaux, G., Safi, R., and Laudet, V. (2004). Molecular cloning and developmental expression patterns of thyroid hormone receptors and T3 target genes in the turbot (*Scophtalmus maximus*) during post-embryonic development. *Gen. Comp. Endocrinol.* **135,** 345–357.

Marsh, K. E., Creutz, L. M., Hawkins, M. B., and Godwin, J. (2006). Aromatase immunoreactivity in the bluehead wrasse brain, *Thalassoma bifasciatum*: Immunolocalization and co-regionalization with arginine vasotocin and tyrosine hydroxylase. *Brain Res.* **1126,** 91–101.

Menuet, A., Guigen, Y., Teitsma, C., Pakdel, F., Mañanos, E., Mazurais, D., Kah, O., and Anglade, I. (1999). Localization of nuclear androgen receptor messengers in the brain and pituitary of rainbow trout. *In* "Reproductive Physiology of Fish," (Norberg, B., Kjesby, O. S., Taranger, G. L., Andersson, E., and Stefansson, S. O., Eds.). Bergen, John Grieg AS.

Menuet, A., Anglade, I., Flouriot, G., Pakdel, F., and Kah, O. (2001). Tissue-specific expression of two structurally different estrogen receptor alpha isoforms along the female reproductive axis of an oviparous species, the rainbow trout. *Biol. Reprod.* **65,** 1548–1557.

Menuet, A., Pellegrini, E., Anglade, I., Blaise, O., Laudet, V., Kah, O., and Pakdel, F. (2002). Molecular characterization of three estrogen receptor forms in zebrafish: Binding characteristics, transactivation properties, and tissue distributions. *Biol. Reprod.* **66,** 1881–1892.

Menuet, A., Anglade, I., Le Guevel, R., Pellegrini, E., Pakdel, F., and Kah, O. (2003). Distribution of aromatase mRNA and protein in the brain and pituitary of female rainbow trout: Comparison with estrogen receptor alpha. *J. Comp. Neurol.* **462,** 180–193.

Menuet, A., Pellegrini, E., Brion, F., Gueguen, M. M., Anglade, I., Pakdel, F., and Kah, O. (2005). Expression and estrogen-dependent regulation of the zebrafish brain aromatase gene. *J. Comp. Neurol.* **485,** 304–320.

Metivier, R., Penot, G., Hubner, M. R., Reid, G., Brand, H., Kos, M., and Gannon, F. (2003). Estrogen receptor-alpha directs ordered, cyclical, and combinatorial recruitment of cofactors on a natural target promoter. *Cell* **115,** 751–763.

Metivier, R., Gallais, R., Tiffoche, C., Le Peron, C., Jurkowska, R. Z., Carmouche, R. P., Ibberson, D., Barath, P., Demay, F., Reid, G., Benes, V., Jeltsch, A., *et al.* (2008). Cyclical DNA methylation of a transcriptionally active promoter. *Nature* **452,** 45–50.

Mindnich, R., Deluca, D., and Adamski, J. (2004). Identification and characterization of 17 beta-hydroxysteroid dehydrogenases in the zebrafish, *Danio rerio*. *Mol. Cell Endocrinol.* **215,** 19–30.

Montero, M., Le Belle, N., King, J. A., Millar, R. P., and Dufour, S. (1995). Differential regulation of the two forms of gonadotropin-releasing hormone (mGnRH and cGnRH-II) by sex steroids in the European female silver eel (*Anguilla anguilla*). *Neuroendocrinology* **61,** 525–535.

Mouriec, K., Pellegrini, E., Anglade, I., Menuet, A., Adrio, F., Thieulant, M. L., Pakdel, F., and Kah, O. (2008). Synthesis of estrogens in progenitor cells of adult fish brain: Evolutive novelty or exaggeration of a more general mechanism implicating estrogens in neurogenesis? *Brain Res. Bull.* **75,** 274–280.

Mouriec, K., Gueguen, M. L., Manuel, C., Percevault, F., Thieulant, M. L., Pakdel, F., and Kah, O. (2009). Androgens upregulate cyp19a1b (aromatase B) gene expression in the brain of zebrafish (*Danio rerio*) through estrogen receptors. *Biol. Reprod.* in press.

Mourot, B., Nguyen, T., Fostier, A., and Bobe, J. (2006). Two unrelated putative membrane-bound progestin receptors, progesterone membrane receptor component 1 (PGMRC1) and membrane progestin receptor (mPR) beta, are expressed in the rainbow trout oocyte and exhibit similar ovarian expression patterns. *Reprod. Biol. Endocrinol.* **4,** 6.

Munoz-Cueto, J. A., Burzawa-Gerard, E., Kah, O., Valotaire, Y., and Pakdel, F. (1999). Cloning and sequencing of the gilthead sea bream estrogen receptor cDNA. *DNA Seq.* **10,** 75–84.

Murashita, K., Uji, S., Yamamoto, T., Ronnestad, I., and Kurokawa, T. (2008). Production of recombinant leptin and its effects on food intake in rainbow trout (*Oncorhynchus mykiss*). *Comp. Biochem. Physiol. B Biochem. Mol. Biol.* **150,** 377–384.

Muriach, B., Carrillo, M., Zanuy, S., and Cerda-Reverter, J. M. (2008a). Distribution of estrogen receptor 2 mRNAs (Esr2a and Esr2b) in the brain and pituitary of the sea bass (*Dicentrarchus labrax*). *Brain Res.* **1210,** 126–141.

Muriach, B., Cerda-Reverter, J. M., Gomez, A., Zanuy, S., and Carrillo, M. (2008b). Molecular characterization and central distribution of the estradiol receptor alpha (ERalpha) in the sea bass (*Dicentrarchus labrax*). *J. Chem. Neuroanat.* **35,** 33–48.

Nagasaka, R., Okamoto, N., and Ushio, H. (2006). Increased leptin may be involved in the short life span of ayu (*Plecoglossus altivelis*). *J. Exp. Zoolog. A Comp. Exp. Biol.* **305,** 507–512.

Nagler, J. J., Cavileer, T., Sullivan, J., Cyr, D. G., and Rexroad, C., 3rd. (2007). The complete nuclear estrogen receptor family in the rainbow trout: Discovery of the novel ERalpha2 and both ERbeta isoforms. *Gene* **392,** 164–173.

Navarro, V. M., Castellano, J. M., Garcia-Galiano, D., and Tena-Sempere, M. (2007). Neuro-endocrine factors in the initiation of puberty: The emergent role of kisspeptin. *Rev. Endocr. Metab. Disord.* **8,** 11–20.

Navas, J. M., Anglade, I., Bailhache, T., Pakdel, F., Breton, B., Jego, P., and Kah, O. (1995). Do gonadotrophin-releasing hormone neurons express estrogen receptors in the rainbow trout? A double immunohistochemical study. *J. Comp. Neurol.* **363,** 461–474.

Nelson, E. R., and Habibi, H. R. (2009). Thyroid receptor subtypes: Structure and function in fish. *Gen. Comp. Endocrinol.* **161,** 90–96.

Nilsson, S., Makela, S., Treuter, E., Tujague, M., Thomsen, J., Andersson, G., Enmark, E., Pettersson, K., Warner, M., and Gustafsson, J. A. (2001). Mechanisms of estrogen action. *Physiol. Rev.* **81,** 1535–1565.

Olivereau, M., and Olivereau, J. (1979). Effect of estradiol-17 beta on the cytology of the liver, gonads and pituitary, and on plasma electrolytes in the female freshwater eel. *Cell Tissue Res.* **199,** 431–454.

Olsson, P. E., Berg, A. H., von Hofsten, J., Grahn, B., Hellqvist, A., Larsson, A., Karlsson, J., Modig, C., Borg, B., and Thomas, P. (2005). Molecular cloning and characterization of a nuclear androgen receptor activated by 11-ketotestosterone. *Reprod. Biol. Endocrinol.* **3,** 37.

O'Sullivan, A. J. (2009). Does oestrogen allow women to store fat more efficiently? A biological advantage for fertility and gestation. *Obes. Rev.* **10,** 168–177.

Pakdel, F., Le Gac, F., Le Goff, P., and Valotaire, Y. (1990). Full-length sequence and *in vitro* expression of rainbow trout estrogen receptor cDNA. *Mol. Cell Endocrinol.* **71,** 195–204.

Pakdel, F., Feon, S., Le Gac, F., Le Menn, F., and Valotaire, Y. (1991). *In vivo* estrogen induction of hepatic estrogen receptor mRNA and correlation with vitellogenin mRNA in rainbow trout. *Mol. Cell Endocrinol.* **75,** 205–212.

Pakdel, F., Petit, F., Anglade, I., Kah, O., Delaunay, F., Bailhache, T., and Valotaire, Y. (1994). Overexpression of rainbow trout estrogen receptor domains in *Escherichia coli*: Characterization and utilization in the production of antibodies for immunoblotting and immunocytochemistry. *Mol. Cell Endocrinol.* **104,** 81–93.

Pakdel, F., Metivier, R., Flouriot, G., and Valotaire, Y. (2000). Two estrogen receptor (ER) isoforms with different estrogen dependencies are generated from the trout ER gene. *Endocrinology* **141,** 571–580.

Pang, Y., and Thomas, P. (2009). Involvement of estradiol-17beta and its membrane receptor, G protein coupled receptor 30 (GPR30) in regulation of oocyte maturation in zebrafish, *Danio rario*. *Gen. Comp. Endocrinol.* **161,** 58–61.

Pang, Y., Dong, J., and Thomas, P. (2008). Estrogen signaling characteristics of Atlantic croaker G protein-coupled receptor 30 (GPR30) and evidence it is involved in maintenance of oocyte meiotic arrest. *Endocrinology* **149,** 3410–3426.

Papasani, M. R., Robison, B. D., Hardy, R. W., and Hill, R. A. (2006). Early developmental expression of two insulins in zebrafish (*Danio rerio*). *Physiol. Genomics* **27,** 79–85.

Parhar, I. S., Tosaki, H., Sakuma, Y., and Kobayashi, M. (2001). Sex differences in the brain of goldfish: gonadotropin-releasing hormone and vasotocinergic neurons. *Neuroscience* **104,** 1099–1110.

Pasmanik, M., and Callard, G. V. (1985). Aromatase and 5 alpha-reductase in the teleost brain, spinal cord, and pituitary gland. *Gen. Comp. Endocrinol.* **60,** 244–251.

Pasmanik, M., and Callard, G. V. (1988). A high abundance androgen receptor in goldfish brain: Characteristics and seasonal changes. *Endocrinology* **123,** 1162–1171.

Pasmanik, M., Schlinger, B. A., and Callard, G. V. (1988). *In vivo* steroid regulation of aromatase and 5 alpha-reductase in goldfish brain and pituitary. *Gen. Comp. Endocrinol.* **71,** 175–182.

Pellegrini, E., Menuet, A., Lethimonier, C., Adrio, F., Gueguen, M. M., Tascon, C., Anglade, I., Pakdel, F., and Kah, O. (2005). Relationships between aromatase and estrogen receptors in the brain of teleost fish. *Gen. Comp. Endocrinol.* **142,** 60–66.

Pellegrini, E., Mouriec, K., Anglade, I., Menuet, A., Le Page, Y., Gueguen, M. M., Marmignon, M. H., Brion, F., Pakdel, F., and Kah, O. (2007). Identification of aromatase-positive radial glial cells as progenitor cells in the ventricular layer of the forebrain in zebrafish. *J. Comp. Neurol.* **501,** 150–167.

Pepels, P. P., Van Helvoort, H., Wendelaar Bonga, S. E., and Balm, P. H. (2004). Corticotropin-releasing hormone in the teleost stress response: Rapid appearance of the peptide in plasma of tilapia (*Oreochromis mossambicus*). *J. Endocrinol.* **180,** 425–438.

Peter, R. E., and Paulencu, C. R. (1980). Involvement of the preoptic region in gonadotropin release-inhibition in goldfish, *Carassius auratus*. *Neuroendocrinology* **31,** 133–141.

Peyon, P., Zanuy, S., and Carrillo, M. (2001). Action of leptin on *in vitro* luteinizing hormone release in the European sea bass (*Dicentrarchus labrax*). *Biol. Reprod.* **65,** 1573–1578.

Plisetskaya, E. M., Bondareva, V. M., Duan, C., and Duguay, S. J. (1993). Does salmon brain produce insulin? *Gen. Comp. Endocrinol.* **91,** 74–80.

Popa, S. M., Clifton, D. K., and Steiner, R. A. (2008). The role of kisspeptins and GPR54 in the neuroendocrine regulation of reproduction. *Annu. Rev. Physiol.* **70,** 213–238.

Power, D. M., Llewellyn, L., Faustino, M., Nowell, M. A., Bjornsson, B. T., Einarsdottir, I. E., Canario, A. V., and Sweeney, G. E. (2001). Thyroid hormones in growth and development of fish. *Comp. Biochem. Physiol. C Toxicol. Pharmacol.* **130,** 447–459.

Prunet, P., Sturm, A., and Milla, S. (2006). Multiple corticosteroid receptors in fish: From old ideas to new concepts. *Gen. Comp. Endocrinol.* **147,** 17–23.

Quesada, A., Romeo, H. E., and Micevych, P. (2007). Distribution and localization patterns of estrogen receptor-beta and insulin-like growth factor-1 receptors in neurons and glial cells of the female rat substantia nigra: Localization of ERbeta and IGF-1R in substantia nigra. *J. Comp. Neurol.* **503,** 198–208.

Rakic, P. (1978). Neuronal migration and contact guidance in the primate telencephalon. *Postgrad. Med. J.* **54**(Suppl 1), 25–40.

Roa, J., Aguilar, E., Dieguez, C., Pinilla, L., and Tena-Sempere, M. (2008). New frontiers in kisspeptin/GPR54 physiology as fundamental gatekeepers of reproductive function. *Front. Neuroendocrinol.* **29,** 48–69.

Robel, P., Young, J., Corpechot, C., Mayo, W., Perche, F., Haug, M., Simon, H., and Baulieu, E. E. (1995). Biosynthesis and assay of neurosteroids in rats and mice: Functional correlates. *J. Steroid Biochem. Mol. Biol.* **53,** 355–360.

Robertson, S. A., Leinninger, G. M., and Myers, M. G., Jr. (2008). Molecular and neural mediators of leptin action. *Physiol. Behav.* **94,** 637–642.

Robinson-Rechavi, M., Escriva Garcia, H., and Laudet, V. (2003). The nuclear receptor superfamily. *J. Cell Sci.* **116,** 585–586.

Safe, S., and Kim, K. (2008). Non-classical genomic estrogen receptor (ER)/specificity protein and ER/activating protein-1 signaling pathways. *J. Mol. Endocrinol.* **41,** 263–275.

Sakamoto, H., Ukena, K., and Tsutsui, K. (2001). Activity and localization of 3beta-hydroxysteroid dehydrogenase/Delta5-Delta4-isomerase in the zebrafish central nervous system. *J. Comp. Neurol.* **439,** 291–305.

Salbert, G., Bonnec, G., Le Goff, P., Boujard, D., Valotaire, Y., and Jego, P. (1991). Localization of the estradiol receptor mRNA in the forebrain of the rainbow trout. *Mol. Cell Endocrinol.* **76,** 173–180.

Salbert, G., Atteke, C., Bonnec, G., and Jego, P. (1993). Differential regulation of the estrogen receptor mRNA by estradiol in the trout hypothalamus and pituitary. *Mol. Cell Endocrinol.* **96,** 177–182.

Saligaut, C., Linard, B., Mananos, E. L., Kah, O., Breton, B., and Govoroun, M. (1998). Release of pituitary gonadotrophins GtH I and GtH II in the rainbow trout (*Oncorhynchus mykiss*): Modulation by estradiol and catecholamines. *Gen. Comp. Endocrinol.* **109,** 302–309.

Schlueter, P. J., Peng, G., Westerfield, M., and Duan, C. (2007). Insulin-like growth factor signaling regulates zebrafish embryonic growth and development by promoting cell survival and cell cycle progression. *Cell Death Differ.* **14,** 1095–1105.

Schreibman, M. P., Pertschuk, L. P., Rainford, E. A., Margolis-Kazan, H., and Gelber, S. J. (1982). The histochemical localization of steroid binding sites in the pituitary gland of a teleost (the platyfish). *Cell Tissue Res.* **226,** 523–530.

Siegel, N., Hoegg, S., Salzburger, W., Braasch, I., and Meyer, A. (2007). Comparative genomics of ParaHox clusters of teleost fishes: Gene cluster breakup and the retention of gene sets following whole genome duplications. *BMC Genomics* **8,** 312.

Singleton, D. W., and Khan, S. A. (2003). Xenoestrogen exposure and mechanisms of endocrine disruption. *Front. Biosci.* **8,** s110–s118.

Smith, A., Chan, S. J., and Gutierrez, J. (2005). Autoradiographic and immunohistochemical localization of insulin-like growth factor-I receptor binding sites in brain of the brown trout, *Salmo trutta*. *Gen. Comp. Endocrinol.* **141,** 203–213.

Smith, J. T., Acohido, B. V., Clifton, D. K., and Steiner, R. A. (2006). KiSS-1 neurones are direct targets for leptin in the ob/ob mouse. *J. Neuroendocrinol.* **18,** 298–303.

Socorro, S., Power, D. M., Olsson, P. E., and Canario, A. V. (2000). Two estrogen receptors expressed in the teleost fish, *Sparus aurata*: CDNA cloning, characterization and tissue distribution. *J. Endocrinol.* **166,** 293–306.

Sperry, T. S., and Thomas, P. (1999a). Characterization of two nuclear androgen receptors in Atlantic croaker: Comparison of their biochemical properties and binding specificities. *Endocrinology* **140,** 1602–1611.

Sperry, T. S., and Thomas, P. (1999b). Identification of two nuclear androgen receptors in kelp bass (*Paralabrax clathratus*) and their binding affinities for xenobiotics: Comparison with Atlantic croaker (*Micropogonias undulatus*) androgen receptors. *Biol. Reprod.* **61,** 1152–1161.

Sperry, T. S., and Thomas, P. (2000). Androgen binding profiles of two distinct nuclear androgen receptors in Atlantic croaker (*Micropogonias undulatus*). *J. Steroid Biochem. Mol. Biol.* **73,** 93–103.

Steinke, D., Hoegg, S., Brinkmann, H., and Meyer, A. (2006). Three rounds (1R/2R/3R) of genome duplications and the evolution of the glycolytic pathway in vertebrates. *BMC Biol.* **4,** 16.

Stolte, E. H., de Mazon, A. F., Leon-Koosterziel, K. M., Jesiak, M., Bury, N. R., Sturm, A., Savelkoul, H. F. J., Verburg van Kemenade, B. M. L., and Flik, G. (2008). Corticosteroid receptors involved in stress regulation in common carp, *Cyprinus carpio*. *J. Endocrinol.* **198,** 403–417.

Strobl-Mazzulla, P. H., Moncaut, N. P., Lopez, G. C., Miranda, L. A., Canario, A. V., and Somoza, G. M. (2005). Brain aromatase from pejerrey fish (*Odontesthes bonariensis*): CDNA cloning, tissue expression, and immunohistochemical localization. *Gen. Comp. Endocrinol.* **143,** 21–32.

Strobl-Mazzulla, P. H., Lethimonier, C., Gueguen, M. M., Karube, M., Fernandino, J. I., Yoshizaki, G., Patino, R., Strussmann, C. A., Kah, O., and Somoza, G. M. (2008). Brain aromatase (Cyp19A2) and estrogen receptors, in larvae and adult pejerrey fish *Odontesthes bonariensis*: Neuroanatomical and functional relations. *Gen. Comp. Endocrinol.* **158,** 191–201.

Sturm, A., Bury, N., Dengreville, L., Fagart, J., Flouriot, G., Rafestin-Oblin, M. E., and Prunet, P. (2005). 11-deoxycorticosterone is a potent agonist of the rainbow trout (*Oncorhynchus mykiss*) mineralocorticoid receptor. *Endocrinology* **146,** 47–55.

Takeo, J., and Yamashita, S. (1999). Two distinct isoforms of cDNA encoding rainbow trout androgen receptors. *J. Biol. Chem.* **274,** 5674–5680.

Takeo, J., and Yamashita, S. (2000). Rainbow trout androgen receptor-alpha fails to distinguish between any of the natural androgens tested in transactivation assay, not just 11-ketotestosterone and testosterone. *Gen. Comp. Endocrinol.* **117,** 200–206.

Teitsma, C. A., Bailhache, T., Tujague, M., Balment, R. J., Ducouret, B., and Kah, O. (1997). Distribution and expression of glucocorticoid receptor mRNA in the forebrain of the rainbow trout. *Neuroendocrinology* **66,** 294–304.

Teitsma, C. A., Anglade, I., Toutirais, G., Munoz-Cueto, J. A., Saligaut, D., Ducouret, B., and Kah, O. (1998). Immunohistochemical localization of glucocorticoid receptors in the forebrain of the rainbow trout (*Oncorhynchus mykiss*). *J. Comp. Neurol.* **401,** 395–410.

Teitsma, C. A., Anglade, I., Lethimonier, C., Le Drean, G., Saligaut, D., Ducouret, B., and Kah, O. (1999). Glucocorticoid receptor immunoreactivity in neurons and pituitary cells implicated in reproductive functions in rainbow trout: A double immunohistochemical study. *Biol. Reprod.* **60,** 642–650.

Tena-Sempere, M. (2006). KiSS-1 and reproduction: Focus on its role in the metabolic regulation of fertility. *Neuroendocrinology* **83,** 275–281.

Teves, A. C., Granneman, J. C., van Dijk, W., and Bogerd, J. (2003). Cloning and expression of a functional estrogen receptor-alpha from African catfish (*Clarias gariepinus*) pituitary. *J. Mol. Endocrinol.* **30,** 173–185.

Thomas, P. (2000). Chemical interference with genomic and nongenomic actions of steroids in fishes: Role of receptor binding. *Mar. Environ. Res.* **50,** 127–134.

Thomas, P. (2008). Characteristics of membrane progestin receptor alpha (mPRalpha) and progesterone membrane receptor component 1 (PGMRC1) and their roles in mediating rapid progestin actions. *Front. Neuroendocrinol.* **29,** 292–312.

Thomas, P., Pang, Y., Zhu, Y., Detweiler, C., and Doughty, K. (2004). Multiple rapid progestin actions and progestin membrane receptor subtypes in fish. *Steroids* **69,** 567–573.

Thomas, P., Dressing, G., Pang, Y., Berg, H., Tubbs, C., Benninghoff, A., and Doughty, K. (2006). Progestin, estrogen and androgen G-protein coupled receptors in fish gonads. *Steroids* **71,** 310–316.

Thomas, P., Pang, Y., Dong, J., Groenen, P., Kelder, J., de Vlieg, J., Zhu, Y., and Tubbs, C. (2007). Steroid and G protein binding characteristics of the seatrout and human progestin membrane receptor alpha subtypes and their evolutionary origins. *Endocrinology* **148,** 705–718.

Timmers, R. J., Granneman, J. C., Lambert, J. G., and van Oordt, P. G. (1988). Estrogen-2-hydroxylase in the brain of the male African catfish, *Clarias gariepinus*. *Gen. Comp. Endocrinol.* **72,** 190–203.

Todo, T., Ikeuchi, T., Kobayashi, T., and Nagahama, Y. (1999). Fish androgen receptor: CDNA cloning, steroid activation of transcription in transfected mammalian cells, and tissue mRNA levels. *Biochem. Biophys. Res. Commun.* **254,** 378–383.

Tomy, S., Wu, G. C., Huang, H. R., Dufour, S., and Chang, C. F. (2007). Developmental expression of key steroidogenic enzymes in the brain of protandrous black porgy fish, *Acanthopagrus schlegeli*. *J. Neuroendocrinol.* **19,** 643–655.

Touhata, K., Kinoshita, M., Tokuda, Y., Toyohara, H., Sakaguchi, M., Yokoyama, Y., and Yamashita, S. (1999). Sequence and expression of a cDNA encoding the red sea bream androgen receptor. *Biochim. Biophys. Acta* **1449,** 199–202.

Trudeau, V. L. (1997). Neuroendocrine regulation of gonadotrophin II release and gonadal growth in the goldfish, *Carassius auratus*. *Rev. Reprod.* **2,** 55–68.

Trudeau, V. L., Somoza, G. M., Nahorniak, C. S., and Peter, R. E. (1992). Interactions of estradiol with gonadotropin-releasing hormone and thyrotropin-releasing hormone in the control of growth hormone secretion in the goldfish. *Neuroendocrinology* **56,** 483–490.

Trudeau, V. L., Spanswick, D., Fraser, E. J., Lariviere, K., Crump, D., Chiu, S., MacMillan, M., and Schulz, R. W. (2000). The role of amino acid neurotransmitters in the regulation of pituitary gonadotropin release in fish. *Biochem. Cell Biol.* **78,** 241–259.

Tujague, M., Saligaut, D., Teitsma, C., Kah, O., Valotaire, Y., and Ducouret, B. (1998). Rainbow trout glucocorticoid receptor overexpression in *Escherichia coli*: Production of antibodies for western blotting and immunohistochemistry. *Gen. Comp. Endocrinol.* **110,** 201–211.

Urushitani, H., Nakai, M., Inanaga, H., Shimohigashi, Y., Shimizu, A., Katsu, Y., and Iguchi, T. (2003). Cloning and characterization of estrogen receptor alpha in mummichog, *Fundulus heteroclitus*. *Mol. Cell Endocrinol.* **203,** 41–50.

Vetillard, A., Benanni, S., Saligaut, C., Jego, P., and Bailhache, T. (2002). Localization of tyrosine hydroxylase and its messenger RNA in the brain of rainbow trout by immunocytochemistry and *in situ* hybridization. *J. Comp. Neurol.* **449,** 374–389.

Vetillard, A., Ferriere, F., Jego, P., and Bailhache, T. (2006). Regulation of salmon gonadotrophin-releasing hormone gene expression by sex steroids in rainbow trout brain. *J. Neuroendocrinol.* **18,** 445–453.

Vizziano, D., Pellegrini, E., Mouriec, K., Gueguen, M. M., Anglade, I., Guiguen, Y., and Kah, O. Sexual dimorphism in the expression of steroidogenic enzymes in the brain of rainbow trout around the period of gonadal sex differentiation (Unpublished).

Wang, Y., and Ge, W. (2004). Cloning of zebrafish ovarian P450c17 (CYP17, 17alpha-hydroxylase/17, 20-lyase) and characterization of its expression in gonadal and extra-gonadal tissues. *Gen. Comp. Endocrinol.* **135,** 241–249.

Weil, C., Le Bail, P. Y., Sabin, N., and Le Gac, F. (2003). *In vitro* action of leptin on FSH and LH production in rainbow trout (*Onchorynchus mykiss*) at different stages of the sexual cycle. *Gen. Comp. Endocrinol.* **130,** 2–12.

Weltzien, F. A., Pasqualini, C., Sebert, M. E., Vidal, B., Le Belle, N., Kah, O., Vernier, P., and Dufour, S. (2006). Androgen-dependent stimulation of brain dopaminergic systems in the female European eel (*Anguilla anguilla*). *Endocrinology* **147,** 2964–2973.

Werther, G. A., Abate, M., Hogg, A., Cheesman, H., Oldfield, B., Hards, D., Hudson, P., Power, B., Freed, K., and Herington, A. C. (1990). Localization of insulin-like growth factor-I mRNA in rat brain by *in situ* hybridization – relationship to IGF-I receptors. *Mol. Endocrinol.* **4,** 773–778.

Wintermantel, T. M., Campbell, R. E., Porteous, R., Bock, D., Grone, H. J., Todman, M. G., Korach, K. S., Greiner, E., Perez, C. A., Schutz, G., and Herbison, A. E. (2006). Definition of estrogen receptor pathway critical for estrogen positive feedback to gonadotropin-releasing hormone neurons and fertility. *Neuron* **52,** 271–280.

Wuertz, S., Nitsche, A., Jastroch, M., Gessner, J., Klingenspor, M., Kirschbaum, F., and Kloas, W. (2007). The role of the IGF-I system for vitellogenesis in maturing female sterlet, *Acipenser ruthenus* Linnaeus, 1758. *Gen. Comp. Endocrinol.* **150,** 140–150.

Xia, Z., Patino, R., Gale, W. L., Maule, A. G., and Densmore, L. D. (1999). Cloning, *in vitro* expression, and novel phylogenetic classification of a channel catfish estrogen receptor. *Gen. Comp. Endocrinol.* **113,** 360–368.

Yacobovitz, M., Solomon, G., Gusakovsky, E. E., Levavi-Sivan, B., and Gertler, A. (2008). Purification and characterization of recombinant pufferfish (*Takifugu rubripes*) leptin. *Gen. Comp. Endocrinol.* **156,** 83–90.

Zhu, Y., Rice, C. D., Pang, Y., Pace, M., and Thomas, P. (2003). Cloning, expression, and characterization of a membrane progestin receptor and evidence it is an intermediary in meiotic maturation of fish oocytes. *Proc. Natl. Acad. Sci. USA* **100,** 2231–2236.

Zupanc, G. K., Hinsch, K., and Gage, F. H. (2005). Proliferation, migration, neuronal differentiation, and long-term survival of new cells in the adult zebrafish brain. *J. Comp. Neurol.* **488,** 290–319.

SECTION II

FUNCTIONAL NEUROENDOCRINOLOGY

3

THE GnRH SYSTEM AND THE NEUROENDOCRINE REGULATION OF REPRODUCTION

GLEN VAN DER KRAAK

1. Introduction
2. Gonadotropin-releasing Hormone (GnRH)
   2.1. Multiplicity of GnRHs
   2.2. GnRH Receptors
   2.3. Actions of GnRH
3. Steroid Feedback Regulation of GnRH, FSH and LH
   3.1. Negative Feedback
   3.2. Positive Feedback
4. Monoamine and Amino Acid Neurotransmitters
   4.1. Dopamine
   4.2. Serotonin
   4.3. Noradrenaline
   4.4. Melatonin
   4.5. γ-Aminobutyric Acid
5. Neuropeptides
   5.1. Kisspeptins
   5.2. Pituitary Adenylate Cyclase-activating Polypeptide
   5.3. Neuropeptide Y
   5.4. Other Neuropeptides
6. Protein Hormones
   6.1. Insulin-like Growth Factor I
   6.2. Activin and Follistatin
7. Integrated Neuroendocrine Control of Gonadal Development
8. Perspectives

This chapter reviews the current knowledge of the neuroendocrine pathways that regulate the synthesis and release of the pituitary gonadotropins, follicle-stimulating hormone (FSH) and luteinizing hormone (LH), in teleost fish. A primary focus is on describing the multiple forms of both gonadotropin-releasing hormone (GnRH) and its receptors in relation to their actions within

Fish Neuroendocrinology: Volume 28
FISH PHYSIOLOGY

DOI: 10.1016/S1546-5098(09)28003-4

the brain and pituitary. Recent studies have demonstrated that a vast number of neurotransmitters, neuropeptides and other factors exert inhibitory and/or stimulatory control over the GnRH–FSH/LH system. The identity and roles of these factors at different stages of gonadal development are discussed. Finally, the chapter examines the manner in which the actions of sex steroids and the inhibitory neurotransmitter dopamine coordinate the activation of the GnRH–FSH/LH system during puberty and in relation to sexual maturation and spawning.

1. INTRODUCTION

Reproduction in fish is controlled by the actions of follicle-stimulating hormone (FSH) and luteinizing hormone (LH). These belong to the glycoprotein hormone family and are heterodimeric glycoproteins composed of a common alpha subunit (α-glycoprotein hormone; αGP) and a hormone-specific β-subunit. LH has been characterized in a number of teleost species whereas we know much less about FSH as it has only been characterized relatively recently. Although the precise roles of LH and FSH are not known, they are produced in separate cells in the pituitary, exhibit distinct patterns of expression at different stages of the reproductive cycle and have been shown to act on the gonads to produce sex steroids and other gonadal factors which play important roles in the development and maturation of gametes (Devlin and Nagahama, 2002; Yaron *et al.*, 2003; Kusakabe *et al.*, 2006; Zmora *et al.*, 2007). Based in large part on knowledge of their actions in salmonids, FSH has been implicated in the control of gametogenesis because it stimulates the production of 17β-estradiol and the incorporation of vitellogenin into the developing oocyte. In males, FSH stimulates Sertoli cell proliferation and maintenance of quantitatively normal spermatogenesis. LH is very low or undetectable during early stages of reproductive development but high before maturation where it is known to stimulate gonadal steroidogenesis and is involved in oocyte maturation, ovulation and spermiation.

Much progress has been made in understanding the neuroendocrine regulation of LH and FSH synthesis and release in fish (Peter *et al.*, 1986; Yaron 1995; Yaron *et al.*, 2003). While some of this progress has been driven by the desire to control reproduction in farmed species, there are unique aspects of the anatomy of the hypothalamus and pituitary and a diversity of reproductive strategies that make teleost fish an interesting model group for study. For example, the adenohypophysial cells of fish are directly innervated by hypothalamic nerve fibers and in contrast to the situation in mammals, teleosts do not possess a hypothalamic–hypophysial portal system (see Chapter 1, this volume). The neuroendocrine control of gonadotropin secretion in fish parallels what has been described for other vertebrates in terms of regulation by gonadotropin-releasing hormone (GnRH) and the feedback effects of steroid hormones. However there are other unique features. Compared to mammals, teleosts exhibit a greater number and diversity of GnRH peptides and GnRH receptor subtypes (Lethimonier *et al.*, 2004). Dopamine (DA) functions as a gonadotropin release inhibitory factor in fish, and unlike mammals appears to be a key factor controlling both puberty and timing of sexual maturity and ovulation (Peter *et al.*, 1986; Vidal *et al.*, 2004). As has occurred in mammals there has been the identification of a wide array of neurotransmitters, neurohormones and hormones in fish that affect LH and FSH secretion. These add new levels of complexity to the regulation of reproduction and in turn have prompted much interest in defining how these regulators act.

The purpose of this chapter is to review recent developments that have contributed to understanding of the neuroendocrine regulation of LH and FSH by hypothalamic factors, peripheral hormones and local regulators produced within the pituitary. This includes a description of the different forms of GnRH and GnRH receptors in relation to their actions within the brain and pituitary. The actions of neurotransmitters, neuropeptides and other factors on the GnRH–FSH/LH system are discussed. Finally, the chapter examines the manner in which the actions of sex steroids and the inhibitory neurotransmitter DA coordinate the activation of the GnRH–FSH/LH system during puberty and in relation to sexual maturation and spawning. Recent studies have identified the extrapituitary expression of gonadotropins and GnRHs, and these are not included as part of this review (Wong and Zohar, 2004; Andreu-Vieyra *et al.*, 2005; So *et al.*, 2005).

2. GONADOTROPIN-RELEASING HORMONE (GnRH)

2.1. Multiplicity of GnRHs

GnRHs are closely related decapeptides produced by distinct but phylogenetically related genes. To date, 14 different variants of the GnRH peptide have been identified in vertebrates, eight of which are found in teleosts (Table 3.1; Lethimonier *et al.*, 2004; Kah *et al.*, 2007). Phylogenetic analysis of the preproGnRH has shown that GnRHs can be classified into four clades: GnRH1, GnRH2, GnRH3 and GnRH4 (Lethimonier *et al.*, 2004; Silver *et al.*, 2004; Tello *et al.*, 2008). Teleost fish contain members of the first three clades while lampreys are represented in the fourth clade. The amino acid sequences of GnRH2 and GnRH3 are conserved, whereas the structure of GnRH1 varies considerably across vertebrate species. All vertebrates have

Table 3.1
Summary of the eight variants of GnRH identified in teleosts

Position	1	2	3	4	5	6	7	8	9	10
GnRH 1 clade										
Mammalian GnRH	pGlu	His	Trp	Ser	Tyr	Gly	Leu	Arg	Pro	Gly-H2
Pejerrey GnRH	—	—	—	—	Phe	—	—	Ser	—	—
Sea bream GnRH	—	—	—	—	—	—	—	Ser	—	—
Catfish GnRH	—	—	—	—	His	—	—	Asn	—	—
Herring GnRH	—	—	—	—	His	—	—	Ser	—	—
Whitefish GnRH	—	—	—	—	—	—	Met	Asn	—	—
GnRH 2 clade										
Chicken GnRH-II	—	—	—	—	His	—	Trp	Tyr	—	—
GnRH 3 clade										
Salmon GnRH	—	—	—	—	—	—	Trp	Leu	—	—

two or three distinct forms of GnRH. Perhaps not surprisingly the different forms of GnRH have distinct distributions within the brain and pituitary of fishes (Lethimonier *et al.*, 2004; Chapter 2, this volume). The GnRH produced in the preoptic area (POA) of the hypothalamus is considered the hypophysiotropic hormone since in teleosts virtually all POA neurons project to the pituitary gland. While it is well established that GnRH is necessary for reproduction there is increasing evidence that GnRHs have additional neuroregulatory and neuromodulatory roles (Soga *et al.*, 2005; Kah *et al.*, 2007).

The GnRH1 clade is regarded as the species-specific form and in teleosts this includes mammalian GnRH (mGnRH), sea bream GnRH (sbGnRH), pejerrey GnRH (pjGnRH), catfish GnRH (catGnRH), herring GnRH (hGnRH) and whitefish GnRH (wfGnRH) (Table 3.1; Lethimonier *et al.*, 2004; Millar *et al.*, 2004; Kah *et al.*, 2007). GnRH1 is found primarily in the POA, caudal hypothalamus and nerve fibers innervating the pituitary. Based on the relative abundance and seasonal fluctuation of mRNA and peptide levels in the brain and pituitary it has become clear that the GnRH1 subtype originating from the POA is the form that induces gonadotropin release from the pituitary and gonadal development (Kah *et al.*, 2007; Nocillado *et al.*, 2007).

The GnRH2 clade is represented by a single peptide [His5 Trp7 Tyr8] GnRH or chicken II GnRH (cGnRH-II). This peptide is found in the midbrain tegmentum of all jawed vertebrates and is believed to play a role in reproductive behavior and/or in control of appetite and metabolism (Volkoff and Peter, 1999; Kah *et al.*, 2007). It has been shown that cGnRH-II fibers innervate the pituitary in several species of fish including the goldfish (*Carassius auratus*), European eel (*Anguilla anguilla*), African catfish (*Clarias gariepinus*), tilapia (*Oreochromis mossambicus*), hybrid striped bass (*Morone saxatilus*), European sea bass (*Dicentrarchus labrax*) and herring (*Clupea harengus*) (see Peter *et al.*, 1986; Canosa *et al.*, 2008). Several studies have shown that chicken II GnRH is capable of inducing LH release from the pituitary (Chang *et al.*, 2009).

The GnRH3 system is unique to fishes and encodes for a single peptide, salmon GnRH (sGnRH). This peptide is distributed in the terminal nerve (TN) and in the hypothalamus of some species. The function of GnRH3 in the TN is not clear although a role in reproductive behavior has been suggested (Ogawa *et al.*, 2006). The GnRH1 and GnRH3 neuronal elements have been shown to overlap in certain anterior forebrain regions, including the POA of several species (Gonzalez-Martinez *et al.*, 2002; Kah *et al.*, 2007). Cyprinids, including the goldfish, roach (*Rutilus rutilus*), zebrafish (*Danio rerio*) and fathead minnow (*Pimephales promelas*), have only the GnRH2 and GnRH3 subtypes (Lin and Peter, 1996; Penlington *et al.*, 1997; Steven *et al.*, 2003; Filby *et al.*, 2008). In these species GnRH1 appears to have been lost during evolution with the result that many of the functions of this peptide have been taken over by GnRH3. In cyprinids, GnRH3 is located in the TN, telencephalon and POA where it acts as the hypophysiotropic form.

2.2. GnRH Receptors

Much work has been done on the characterization of GnRH receptors in teleost fish (Millar *et al.*, 2004; Lethimonier *et al.*, 2004; Kah *et al.*, 2007; Tello *et al.*, 2008). GnRH receptors are members of the G-protein coupled receptor family and can be grouped into distinct classes (types 1, 2 and 3) based on their nucleotide sequences (Millar *et al.*, 2004; Tello *et al.*, 2008). Type 1 GnRH receptors (GnRHr1) are found in mammals and fish whereas the type 2 GnRH receptor (GnRHr2) includes forms found primarily in amphibians and humans. The third branch includes type 3 GnRH receptors (GnRHr3) of evolved fish, mainly perciform species (Levavi-Sivan and Avitan, 2005; Flanagan *et al.*, 2007; Tello *et al.*, 2008). The situation for GnRH receptors in teleosts is further complicated owing to the genome duplication event that occurred since they diverged from tetrapods. The result is that typically more than one receptor of each GnRH receptor subtype has been identified in the fish species studied to date (Lethimonier *et al.*, 2004; Flanagan *et al.*, 2007; Tello *et al.*, 2008). There have been five GnRH receptors reported for two species of pufferfish (*Fugu ruprides* and *Tetraodon nigrovidis*), cherry salmon (*Oncorhynchus masou*), and the European sea bass and four distinct GnRH receptors in the zebrafish (Kah *et al.*, 2007; Tello *et al.*, 2008).

Given the multiplicity of GnRH receptors there is some uncertainty as to which receptor subtypes are involved in LH and FSH release. This was illustrated in the Nile tilapia *Oreochromis niloticus* where three forms of GnRH receptor were identified in a single gonadotrope (Parhar *et al.*, 2005). Early studies with the goldfish showed that two forms of GnRH receptor are expressed in the pituitary, although one receptor form is better correlated to GnRH action on LH release and the other receptor form with growth hormone release (Illing *et al.*, 1999). Levavi-Sivan *et al.* (2006) identified two types of GnRH receptor in Nile tilapia and these had different distributions. GnRHr3 was found mainly in brain areas related to reproductive function and was highly expressed in the posterior part of the pituitary that contains LH and FSH cells, suggesting that this receptor type may be important for regulating reproduction. Conversely, GnRHr1 was detected widely throughout the brain, from the olfactory bulb to the medulla, as well as in the dorsal anterior and posterior parts of the pituitary, suggesting that this receptor type may be involved in the modulation of sensory input and growth (Levavi-Sivan *et al.*, 2006). In the cichlid *Astatotilapia burtoni*, the GnRH receptor subtype GnRH-R1SHS colocalizes with gonadotropes, whereas a second form GnRH-R2PEY colocalizes with somatotropes, suggesting that these receptors have different roles (Flanagan *et al.*, 2007). A third form of GnRH receptor in this species, GnRH-R2PEY, colocalized with neurons that produce GnRH, and may play a role in the feedback control of GnRH production. There is much work still to be done to characterize the responses mediated by the different GnRH receptor subtypes in the brain and pituitary of teleosts.

2.3. Actions of GnRH

Studies involving *in vivo* injections, *in vitro* incubations of pituitaries, or isolated gonadotropes have been used in demonstrating the stimulatory effect of GnRH on LH synthesis and release (Yaron *et al.*, 2003; Ando and Urano, 2005; Chang *et al.*, 2009). Indeed we know much about the signaling pathways which mediate these effects (Chang *et al.*, 2009). By comparison, we know less about the effects of GnRH on FSH release. This is due in large part to the limited number of immunoassay methods available to quantity FSH in teleosts. Increasingly the measurement of FSHβ mRNA expression has been used to fill this gap. Now studies have shown that GnRH also stimulates FSH release in coho salmon (*Oncorhynchus kisutch*; Dickey and Swanson, 2000), rainbow trout (*Oncorhynchus mykiss*; Vacher *et al.*, 2000), masu salmon (Ando and Urano, 2005), and Nile tilapia (Levavi-Sivan *et al.*, 2006; Aizen *et al.*, 2007). Interestingly, GnRH had no effect on *fshb* expression in the European sea bass (Mateos *et al.*, 2002) or prepubertal red sea bream *Pagrus major* (Kumakura *et al.*, 2003). It has been suggested that there may be constitutive expression of *fshb* during the early stages of development in some species (Swanson *et al.*, 2003).

It is apparent that the appearance of GnRH and GnRH receptors plays an integral role in controlling the onset of gonadal growth and the seasonality of reproduction (Schulz and Goos, 1999). For example, increases in the GnRH content and the size of the GnRH cell bodies of the POA have been reported in several fish species in association with sexual maturation (Amano *et al.*, 1994; Van Der Kraak *et al.*, 1998; Yaron *et al.*, 2003). More recently several studies have estimated GnRH and GnRH receptor mRNAs by real-time quantitative RT-PCR at their site(s) of expression in both male and female fish with immature and recrudescing gonads in order to understand their potential role during gonadal recrudescence (e.g., Amano *et al.*, 2006; Mohammed and Khan, 2006; Moles *et al.*, 2007; Nocillado *et al.*, 2007; Filby *et al.*, 2008; Martinez-Chavez *et al.*, 2008). This approach has consistently shown that there are increases in GnRH in the POA that correlate with the onset of reproductive development. For example in the grey mullet (*Mugil cephalus*), there was elevated expression of *gnrh1, 2* and *3* at the onset of pubertal development (Nocillado *et al.*, 2007). While the increase in *gnrh1* was consistent with its presence in neurons that innervate the pituitary and its role in releasing gonadotropins, the importance of changes in the other forms of GnRH is largely unknown. Similarly in the fathead minnow there was a marked increase in the expression of *gnrh3* which is the hypophysiotropic form in this species (Filby *et al.*, 2008). There was also an increase in the expression of *gnrh2* although its role in early development is poorly understood.

Recent studies have shown a differential regulation of *gnrh2* and *gnrh3* expression during ovulation in goldfish (Canosa *et al.*, 2008). Based on their expression patterns these studies suggest that GnRH3 controls LH secretion whereas GnRH2 correlates with spawning behavior. This conclusion was further supported by the finding that non-ovulated fish induced to perform spawning behavior by prostaglandin F2aα treatment exhibited increased *gnrh2* expression in both forebrain and midbrain, but decreased *gnrh3* expression in the forebrain.

3. STEROID FEEDBACK REGULATION OF GnRH, FSH AND LH

Gonadal steroids play major roles in controlling the synthesis and release of LH and FSH in teleosts (Goos, 1987; Trudeau, 1997; Van Der Kraak *et al.*, 1998; Yaron *et al.*, 2003; Levavi-Sivan *et al.*, 2006). Both positive and negative feedback effects of testosterone and 17β-estradiol have been demonstrated. These actions are complex in that sex steroid feedback regulation of

LH and FSH release involves actions at the level of the hypothalamus and pituitary. In addition to effects on LH and FSH synthesis within the pituitary, testosterone and 17β-estradiol also affect the GnRH system and other neuroendocrine factors which control LH synthesis and release. Owing to high levels of aromatase in the brain and pituitary of teleosts, it can be difficult to distinguish whether the effects of testosterone are mediated through the androgen receptor or result from metabolism to estrogens and actions via estrogen receptors. Consequently studies often involve comparing the actions of non-aromatizable androgens such as 11-ketoandrostenedione (11KA) and 11-ketotestosterone (11-KT) to that of testosterone and 17β-estradiol in order to differentiate the effects through these two receptors. Other steroids including the maturation-inducing steroid 17, 20β-dihydroxy-4-pregnen-3-one (17, 20βP) and the corticosteroid cortisol have been implicated in the regulation of LH and FSH release (Levavi-Sivan *et al.*, 2006; Chapter 2) but these effects are considered to be secondary to the dominant actions of testosterone and 17β-estradiol.

3.1. Negative Feedback

Gonadectomy and steroid hormone replacement have been common approaches used to evaluate the effects of gonadal steroids on LH and FSH secretion in teleosts. These studies have shown that the effects of gonadal steroids are highly dependent on the stage of gonadal development. Negative feedback effects of steroids are evident during the latter stages of vitellogenesis and spermatogenesis as gonad removal increases LH secretion in goldfish (Kobayashi and Stacey, 1990), African catfish (de Leeuw *et al.*, 1986), Atlantic croaker (*Micropogonias undulates*; Khan *et al.*, 1999), Indian catfish (*Heteropneustes fossilis*; Senthilkumaran and Joy, 1996), hybrid striped bass (Klenke and Zohar, 2003), European sea bass (Mateos *et al.*, 2002) and some salmonids (Bommelaer *et al.*, 1981; Larson and Swanson, 1997). The stimulatory effect of gonadectomy on LH can be reversed by treatment with testosterone or 17β-estradiol. Other studies have shown that FSH is also controlled by a steroid-dependent negative feedback loop in rainbow trout (Saligaut *et al.*, 1998; Chyb *et al.*, 1999; Vetillard *et al.*, 2003), goldfish (Kobayashi *et al.*, 2000), coho salmon (Larsen and Swanson, 1997), Atlantic croaker (Banerjee and Khan, 2008) and European sea bass (Mateos *et al.*, 2002). In these cases, administration of either testosterone or 17β-estradiol reduced the effects of gonadectomy on FSH secretion.

There are several mechanisms by which steroids may mediate negative feedback effects on LH and FSH secretion. Several studies suggest that the negative feedback effects of steroids are not primarily mediated at the level of LH and FSH synthesis in the pituitary, as *in vitro* studies have shown that the expression of LH and FSH is often unchanged or increases in response to testosterone or 17β-estradiol (Dickey and Swanson, 1998; Huggard-Nelson *et al.*, 2002; Levavi-Sivan *et al.*, 2006). There are a number of studies showing that testosterone or 17β-estradiol administration decreases the levels of GnRH mRNA, suggesting that the primary targets for negative feedback are the brain GnRH neurons (Vacher *et al.*, 2002; Levavi-Sivan *et al.*, 2006).

Additionally, the negative feedback effects of steroids may be mediated through effects on other regulators of LH and FSH release. In part, steroid-negative feedback may be mediated by reducing the stimulatory influence of γ-aminobutyric acid (GABA) (Section 4.5) or by increasing the inhibitory dopaminergic tone on GnRH release (Section 3.1). In support of these hypotheses, there are steroid receptors in dopaminergic and GABAergic neurons (Linard *et al.*, 1995, 1996; Vetillard *et al.*, 2003). The inhibition of LH synthesis could be associated with steroid hormone effects on stimulatory gonadal peptides such as activin (Section 6.2).

3.2. Positive Feedback

Many studies have shown that testosterone and 17β-estradiol exert a positive feedback control of LH and FSH synthesis (Trudeau *et al.*, 1991; Mateos *et al.*, 2002; Banerjee and Khan, 2008). This effect is seen commonly in immature fish although it can also be seen in adults (Schulz *et al.*, 1995). The positive feedback effects of steroids are mediated directly at the level of the pituitary and through effects on the GnRH system.

Many studies have shown that the LH content of the pituitary in juvenile fish increases in response to estrogens and aromatizable androgens whereas the non-aromatizable androgens are far less effective (Crim *et al.*, 1981; Borg *et al.*, 1998; Dickey and Swanson, 1998). This suggests that the response may be mediated by 17β-estradiol derived by the aromatization of testosterone in the gonadotropes. Other studies with goldfish and eel pituitaries showed that testosterone increased LH mRNA levels (Huggard *et al.*, 1996; Vidal *et al.*, 2004; Aroua *et al.*, 2007). In goldfish, 17β-estradiol increased the mRNA levels for all gonadotropin subunits, whereas testosterone increases *lhb* but decreases *fshb* mRNA levels (Huggard *et al.*, 1996; Sohn *et al.*, 2001; Huggard-Nelson *et al.*, 2002). There is some variation in response as in studies with European eel pituitaries *in vitro* the levels of LHβ mRNA were increased by non-aromatizable androgens but not 17β-estradiol (Huang *et al.*, 1997; Aroua *et al.*, 2007). Collectively these results suggest that there is widespread evidence for the positive (and negative) feedback effects of testosterone and 17β-estradiol in the regulation of LH and FSH yet there is considerable diversity in the mechanisms and sites of action across species (Mateos *et al.*, 2002; Aroua *et al.*, 2007).

Sex steroids also have stimulatory effects on the GnRH system which in turn stimulate LH and FSH release. Testosterone and 11-ketotestosterone increased the GnRH content in the brain of a variety of species (Amano *et al.*, 1994; Breton and Sambroni, 1996; Borg *et al.*, 1998; Dubois *et al.*, 1998). 17β-estradiol also acts at the level of the GnRH receptors as in the Nile tilapia 17β-estradiol increases the expression of gnrhr1 and gnrhr3 (Levavi-Sivan *et al.*, 2006).

4. MONOAMINE AND AMINO ACID NEUROTRANSMITTERS

4.1. Dopamine

Studies with goldfish have shown that lesioning of the anterior nucleus preopticus periventricularis (NPP), destruction of the pituitary stalk or transplantation of the pituitary to a recipient leads to a marked and prolonged increase in LH secretion (Peter *et al.*, 1986; Trudeau, 1997). These responses supported the concept that LH secretion in the goldfish was controlled by the tonic actions of a gonadotropin release inhibitory factor. Subsequent studies showed that DA acts as a gonadotropin release inhibitory factor in goldfish where it directly inhibits basal as well as GnRH-stimulated LH release (Peter *et al.*, 1986; Trudeau 1997). It soon became apparent that there are marked species differences in the actions of DA. Whereas DA was demonstrated to produce powerful inhibitory tone in cyprinids (Peter *et al.*, 1986; Blazquez *et al.*, 1998a), its effect was much less pronounced in salmonids (Van Der Kraak *et al.*, 1986; Breton *et al.*, 1998; Saligaut *et al.*, 1998) and absent in the Atlantic croaker (Copeland and Thomas, 1989).

It is now well established that DA inhibits basal and GnRH-stimulated LH secretion and does so by several mechanisms (Van Der Kraak *et al.*, 1998; Popesku *et al.*, 2008; Chang *et al.*, 2009). The primary action of DA involves interfering with the intracellular GnRH signaling pathways that mediate LH release (see Yaron *et al.*, 2003). *In vitro* experiments with pituitaries from goldfish (Chang *et al.*, 1990), rainbow trout (Saligaut *et al.*, 1999; Vacher *et al.*, 2000), Nile tilapia (Levavi-Sivan *et al.*, 1995), African catfish (Van Asselt *et al.*, 1990), eel (Vidal *et al.*, 2004) and mullet (Aizen *et al.*, 2005) have provided evidence for the involvement of DA D2-like but not D1-like receptors in the inhibition of LH secretion. While several studies failed to detect a direct effect of DA on LH synthesis (Melamed *et al.*, 1998; Yaron *et al.*, 2003), recent studies have demonstrated DA inhibitory control of GnRH-induced LH synthesis in the European eel (Vidal *et al.*, 2004). DA was also shown in the goldfish to exert inhibitory effects on GnRH neurons by blocking the synthesis of the peptide or inhibiting its release from the pituitary nerve terminals (Yu and

Peter, 1990). In the POA, DA inhibits GnRH by a D1-like receptor mechanism (Yu and Peter, 1990, 1992). DA also acts at the level of the gonadotropes to inhibit GnRH activity by down-regulating the synthesis of GnRH receptors (Kumakura *et al.*, 2003; Levavi-Sivan *et al.*, 2004) and GnRH binding capacity (de Leeuw *et al.*, 1989). These multiple actions account for the potent inhibitory effects of DA on LH release in most teleost species.

Far fewer studies have looked at the involvement of DA in FSH release. Studies with rainbow trout have shown that DA modulates FSH release and that the effects are most prominent at the end of the reproductive cycle when gonadotropic cells have a high sensitivity to GnRH (Vacher *et al.*, 2000, 2002).The catecholamine synthesis inhibitor α-methyl-p-tyrosine increased FSH concentrations in mature but not vitellogenic trout (Saligaut *et al.*, 1998; Vacher *et al.*, 2000, 2002). The DA D2 agonist bromocriptine significantly inhibited basal FSH release from cultured gonadotropes of mature but not vitellogenic trout. Bromocriptine blocked sGnRH-induced FSH release from cultured gonadotropes of both vitellogenic and mature trout (Vacher *et al.*, 2002).

DA also interacts with other hormones and neurotransmitters to modulate gonadotropin release. Perhaps the most important of these interactions relates to the negative feedback effects of 17β-estradiol on LH secretion. 17β-estradiol is essential for maintenance of the dopaminergic inhibition of LH release in trout and goldfish (Trudeau *et al.*, 1993a; Linard *et al.*, 1995; Saligaut *et al.*, 1999; Vacher *et al.*, 2002). For example, 17β-estradiol activates catecholamine turnover and magnifies the DA inhibition of LH release as vitellogenesis proceeds (Saligaut *et al.*, 1992). Studies with Nile tilapia have shown that increasing concentrations of 17β-estradiol resulted in increased DA receptor 2 mRNA levels both *in vivo* and *in vitro*, inhibition of LH and FSH release, and inhibition of *lhb* mRNA levels *in vivo* (Levavi-Sivan *et al.*, 2006). In rainbow trout, dopaminergic neurons of the preoptic area express the estrogen receptor (Linard *et al.*, 1996) providing a neuroanatomical substrate for DA–17β-estradiol interactions. The recent finding that the promoter of grey mullet drd2 possesses a functional 17β-estradiol-responsive element (Nocillado *et al.*, 2005) adds support to the hypothesis that 17β-estradiol negatively modulates LH secretion through effects on dopaminergic neurons. Overall, the negative feedback effect of 17β-estradiol on LH and FSH in vitellogenic and prespawning fish appears to be mediated via effects on dopaminergic inhibitory tone (Levavi-Sivan *et al.*, 2006).

DA also affects the actions of the factors including GABA and neuropeptide Y (NPY) which are stimulators of LH release (Sections 4.5 and 5.3, respectively). Depletion of DA leads to increased synthesis of GABA and the GABA synthetic enzyme glutamic acid decarboxylase 67 (GAD67) and contributes to a potentiation of GABA-mediated LH release (Trudeau,

1997; Trudeau *et al.*, 2000; Popesku *et al.*, 2008). DA has also been shown to block NPY-induced LH secretion in the goldfish pituitaries incubated *in vitro* (Peng *et al.*, 1993a).

There is considerable evidence that removal of dopaminergic inhibition, together with increased GnRH stimulation, constitute important neuroendocrine mechanisms leading to the preovulatory LH surge and ovulation in many species (Peter *et al.*, 1988). This knowledge has had considerable application in aquaculture. In many species the strong endogenous inhibitory impact of DA compromises the ability of externally applied GnRH to increase LH release, and consequently the fish fail to ovulate (Yaron, 1995). The widely adopted method of spawning induction in fish with strong dopaminergic inhibition relies on the simultaneous application of a potent DA receptor antagonist and a GnRH superactive analogue (Peter *et al.*, 1988; Yaron, 1995). There is evidence that a reduced DA inhibitory tone is also associated with sex pheromone-induced LH secretion in male goldfish (Dulka *et al.*, 1992).

4.2. Serotonin

Serotonin (5-hydroxytryptamine, 5-HT) has been shown to increase LH secretion in several teleost species (reviewed in Kreke and Dietrich, 2007; Popesku *et al.*, 2008). In goldfish, 5-HT injected systemically results in a significant increase in serum LH levels but had no effect when administered into the brain third ventricle, suggesting a stimulatory action on LH secretion at the level of the pituitary gland (Somoza *et al.*, 1988). In the Atlantic croaker, 5-HT alone did not elevate LH levels but potentiated the effects of GnRH on LH levels (Khan and Thomas, 1992). Other studies with perfused pituitary fragments from goldfish (Somoza and Peter, 1991; Wong *et al.*, 1998) and Atlantic croaker (Khan and Thomas, 1992) showed that 5-HT stimulated LH release. In both goldfish and Atlantic croaker, the actions of 5HT were blocked by ketanserin which antagonizes 5-HT type T2 receptors. 5-HT also stimulated GnRH release from the POA and pituitary of sexually mature female goldfish (Yu *et al.*, 1991). In the red sea bream, 5-HT induced sbGnRH secretion from the POA of juvenile and adult fish but had no effect at the level of the pituitary (Senthilkumaran *et al.*, 2001). This action also seemed to be mediated by 5-HT2 receptors, since the effect was blocked by ketanserin. Collectively, these studies demonstrate that LH release is stimulated by 5-HT but that this may involve actions at the level of the hypothalamus or pituitary.

Recent studies have shown that fish exposed to hypoxic conditions exhibit marked reductions in gonadal development. Follow-up laboratory studies with the Atlantic croaker have shown that hypoxia was associated with decreased

GnRH mRNA expression in the POA, reduced release of LH in response to GnRH, and decreases in hypothalamic 5-HT content and the activity of the 5-HT biosynthetic enzyme, tryptophan hydroxylase (Thomas *et al.*, 2007). Pharmacological replacement of hypothalamic 5-HT levels restored neuroendocrine function, suggesting that the stimulatory serotonergic neuroendocrine pathway is a major site of hypoxia-induced inhibition of reproduction in the Atlantic croaker. Other studies have shown that polychlorinated biphenyls negatively influence reproduction in fish and do so through effects on the serotoninergic system (see Thomas, 2008), reinforcing the importance of this pathway in the neuroendocrine control of LH secretion in fish.

4.3. Noradrenaline

A number of studies have demonstrated that noradrenaline acting in the pituitary via α1-like receptors stimulates LH release in goldfish, rainbow trout and Indian catfish (Chang *et al.*, 1991; Linard *et al.*, 1995; Senthilkumaran and Joy, 1996; Wong *et al.*, 1998). Noradrenaline also stimulates the production of GnRH in the hypothalamus and its release in the pituitary (Yu and Peter, 1992).

4.4. Melatonin

Melatonin is released from the pineal gland at night and is known to play a fundamental role in the regulation of reproduction in mammals. Perhaps surprisingly, there have been very few studies that have examined the role of melatonin in the neuroendocrine regulation of LH or FSH secretion in fish. Studies involving pinealectomy and melatonin administration have demonstrated the involvement of the pineal in reproduction in fish but the responses have been highly variable (see Khan and Thomas, 1996; Sebert *et al.*, 2008). Administration of melatonin to Indian catfish led to a decrease in the number and size of gonadotropes in the pituitary (Sundararaj and Keshavanath, 1976). In the Atlantic croaker with fully developed gonads, both injection studies and *in vitro* incubations of pituitary fragments showed that melatonin increased LH secretion (Khan and Thomas, 1996). However, in Indian catfish intraperitoneal injections of melatonin decreased plasma LH levels (Senthilkumaran and Joy, 1995). Some of the most comprehensive studies investigating the effects of melatonin on reproduction in fish come from studies with silver eels where long-term melatonin treatment was shown to increase brain tyrosine hydroxylase (TH, the rate-limiting enzyme of DA synthesis) mRNA expression (Sebert *et al.*, 2008). Melatonin stimulated the dopaminergic system of the preoptic area thereby enhancing the inhibitory control of LH and FSH synthesis and release. Melatonin had no effects on the two eel native forms of GnRH (mGnRH and cGnRH-II) mRNA expression but decreased both LHβ and FSHβ mRNA expression. There is still much work to be done to understand the role that melatonin may play in controlling reproduction in fish.

4.5. γ-Aminobutyric Acid

The amino acid neurotransmitter GABA stimulates LH release in teleosts (see Trudeau *et al.*, 2000; Popesku *et al.*, 2008). *In vivo* studies showed that administration of GABA or GABA agonists increased plasma LH in goldfish (Kah *et al.*, 1992; Martyniuk *et al.*, 2007), Atlantic croaker (Khan and Thomas, 1999) and rainbow trout (Mañanos *et al.*, 1999). There is also evidence that GABA stimulates FSH release in rainbow trout (Mañanos *et al.*, 1999).

In studies using dispersed pituitary cells from male rainbow trout, GABA stimulated basal LH and FSH secretion, and potentiated the actions of sGnRH on both hormones (Mañanos *et al.*, 1999). However, for females, GABA did not affect basal or sGnRH-induced LH release from dispersed pituitary cells. Other experiments using dispersed pituitary cells from goldfish and Atlantic croaker failed to demonstrate direct effects of GABA on LH release (Khan and Thomas, 1999). Subsequently it was shown that in female goldfish the stimulatory effect of GABA on LH release was associated with a stimulation of GnRH release from nerve terminals in pituitary fragments and the hypothalamus (Kah *et al.*, 1992). Other studies have shown that GABA stimulates *in vitro* release of sbGnRH from POA slices but not from pituitary slices in the red sea bream (Senthilkumaran *et al.*, 2001). GABA and muscimol, a GABA agonist, stimulated GnRH production and release in adult female lamprey during the final stages of reproduction (Root *et al.*, 2004). Apart from effects on GnRH neurons, another possible mechanism whereby GABA may regulate LH release in fish involves modulation of dopaminergic neurons. There is evidence that GABA effects dopaminergic neurons in the goldfish but not the African catfish or Atlantic croaker (reviewed in Trudeau *et al.*, 2000), suggesting that these effects are likely of secondary importance to actions at the level of the GnRH neurons.

Studies with goldfish suggest that the effects of GABA depend on the steroid environment and the reproductive stage of the fish (Kah *et al.*, 1992; Trudeau *et al.*, 1993a,b). GABA stimulated LH release at early stages of gonadal development but was ineffective in mature or sexually regressed goldfish. 17β-estradiol decreased GABA-stimulated LH release (Trudeau *et al.*, 1993a). Testosterone, but not 17β-estradiol, can restore GABA actions on LH release in these fish (Kah *et al.*, 1992; Trudeau *et al.*, 1993a). In contrast, GABA had a stimulatory effect on LH release in mature female rainbow trout, which is interesting in that these fish are characterized by high testosterone levels and decreasing 17β-estradiol levels (Mañanos *et al.*, 1999). Moreover, both testosterone and 17β-estradiol have been shown to modulate GABA synthesis in telencephalon and hypothalamus of sexually regressed fish (Trudeau *et al.*, 1993a).

Recent studies in goldfish have shown that GABA induces a rapid and substantial increase in activin βa mRNA, and a concomitant increase in LH release (Martyniuk *et al.*, 2007). Given the stimulatory effects of activin on LH release, this may represent a novel mechanism by which GABA stimulates LH release in fish (see Section 6.2).

5. NEUROPEPTIDES

5.1. Kisspeptins

Kisspeptins are a family of neuropeptides encoded by the *kiss1* gene. They are the endogenous ligands for the kisspeptin receptor *(kiss 1r)*, which is also known as the G protein-coupled receptor GPR54. Kisspeptins play a major role in the maintenance of the reproductive axis in mammals and may be critical for the onset of puberty (Seminara and Kaiser, 2005; Kauffman *et al.*, 2007). There is growing evidence to suggest the importance of kisspeptins in fish. Using the genomic database information, *kiss* orthologous sequences have been identified in zebrafish (van Aerle *et al.*, 2008; Biran *et al.*, 2008; Kitahashi *et al.*, 2009), pufferfish (van Aerle *et al.*, 2008) and medaka (*Oryzias latipes*; Kanda *et al.*, 2008; Kitahashi *et al.*, 2009). In addition, a single type *gpr54* cDNA has been cloned from Nile tilapia (Parhar *et al.*, 2004), grey mullet (Nocillado *et al.*, 2007), cobia (*Rachycentron canadun*; Mohamed *et al.*, 2007) and fathead minnow (Filby *et al.*, 2008). Two types of *gpr54* have been identified in the zebrafish (Biran *et al.*, 2008).

Several studies point to the importance of kisspeptins to reproductive function in fish. Early studies identified mRNA for *gpr54* in GnRH1, GnRH2 and GnRH3 neurons in the Nile tilapia (Parhar *et al.*, 2004). In the brain of zebrafish and medaka, *in situ* hybridization and laser capture microdissection coupled with real-time PCR showed *kiss1* expression in the ventro-medial habenula and the periventricular hypothalamic nucleus (Kitahashi *et al.*, 2009). Expression of *kiss2* was observed in the posterior tuberal nucleus and the periventricular hypothalamic nucleus. In the medaka, two populations of neurons expressing the *kiss1* gene were identified; both are located in the hypothalamic nuclei, at the nucleus posterioris periventricularis (NPPv) and the nucleus ventral tuberis (NVT) which are important regulators of LH and FSH release.

Elizur (2009) summarized the results of studies that evaluated the expression of kiss and GPR54 in relation to the onset of puberty. Collectively the expression studies of GPR54 in fish support its involvement in puberty. For example, GPR54 levels are significantly higher in brains of mature Nile tilapia males than in immature males (Parhar *et al.*, 2004). In the female grey mullet, GPR54 expression in the brain was the highest in the early stages of gonadal development, and dropped in the intermediate and advanced stages of gonadal development (Nocillado *et al.*, 2007). The profile of GPR54 expression in the grey mullet brain was comparable to that of the three GnRHs. Other studies with fathead minnow, Nile tilapia and zebrafish showed that GPR54 expression peaked at the time of onset of gonadal development and declined with reproductive maturity (Biran *et al.*, 2008; Filby *et al.*, 2008; Martinez-Chavez *et al.*, 2008). In studies with fathead minnow, the expression of GPR54 closely aligned with that of the GnRH genes and in particular *gnrh3* which is the hypophysiotropic form (Filby *et al.*, 2008). Quantitative real-time PCR analysis showed a significant increase in zebrafish *kiss1, kiss2, gnrh2* and *gnrh3* mRNA levels during zebrafish development, starting at the pubertal phase, and these remained high in adulthood (Kitahashi *et al.*, 2009).

Some of the most compelling evidence for the physiological importance of the kisspeptin/GPR54 system comes from work with sexually mature female zebrafish, where Kiss2 but not Kiss1 administration significantly increased *fshb* and *lhb* mRNAs in the pituitary (Kitahashi *et al.*, 2009). Whether the activation of gonadotropins by Kiss2 is via preoptic GnRH neurons or through GPR54 at the level of gonadotrophs needs investigation, because it is possible that peripheral kisspeptin could directly stimulate gonadotropin secretion through *gpr54-b* in the pituitary of fish. In other studies, pubertal fathead minnow receiving an intraperitoneal injection of mammalian kisspeptin-10 showed enhanced expression of *gnrh3* in the brain (Filby *et al.*, 2008). Collectively these studies provide evidence that kisspeptin may be involved in stimulating the onset of puberty and sexual maturation in fish.

5.2. Pituitary Adenylate Cyclase-activating Polypeptide

In goldfish, pituitary adenylate cyclase-activating polypeptide (PACAP)-immunoreactive hypothalamic fibers directly innervate the pituitary, terminating in the vicinity of gonadotropes. Both *in vitro* and *in vivo* studies using PACAP and the PACAP antagonist PACAP6-38 indicate that PACAP stimulates LH release by directly acting on gonadotropes (Wong *et al.*, 2000; Chang *et al.*, 2001; Sawisky and Chang, 2005). In Nile tilapia, PACAP increases mRNA levels for GPα, LHβ and FSHβ (Yaron *et al.*, 2003). PACAP involvement in FSH release in fish has not been examined.

5.3. Neuropeptide Y

Neuropeptide Y (NPY) stimulates LH release in goldfish, common carp, rainbow trout and European sea bass (Peng *et al.*, 1990; Breton *et al.*, 1991; Cerdá-Reverter *et al.*, 1999). In goldfish, NPY, acting through Y2 receptors, stimulates LH release by direct actions on pituitary cells and indirectly by increasing GnRH release from terminals in the pituitary (Peng *et al.*, 1993a). In other studies, NPY stimulated the *in vitro* release of sbGnRH from slices of the POA of the hypothalamus but not from the pituitary of immature red sea bream (Senthilkumaran *et al.*, 2001). In this case, NPYY1 and NPYY2 specific agonists were equally effective. Immunohistochemical studies suggest that NPY affects the GnRH–LH system via Y1 receptors in catfish (*Clarias batrachus*; Mazumdar *et al.*, 2006). NPY elevates the expression of *lhb* and *gpa*, but not *fshb*, in Nile tilapia (Yaron *et al.*, 2003). NPY was also shown to induce gonad reversal in the protogynous bluehead wrasse (*Thalassoma bifasciatum*; Kramer and Imbriano, 1997).

There is evidence that the actions of NPY on LH secretion are modulated by the reproductive and energetic status of the fish. In European sea bass, under negative energetic status imposed by chronic fasting and characterized by a significant decrease in insulin and glucose, plasma levels of LH were increased in a dose-dependent manner by NPY. In contrast, positive energetic status suppressed the ability of NPY to stimulate LH secretion (Cerdá-Reverter *et al.*, 1999). Similarly, dispersed pituitary cells from fasted animals incubated in restricted media lacking essential nutrients displayed higher responsiveness to NPY than those of fasted animals incubated in L-15. Sex steroids are a likely candidate responsible for the differential regulation of NPY. However, both plasma testosterone and 17β-estradiol from fasted animals were not significantly different from those of the fed animals. Studies with both goldfish and rainbow trout have shown that the response to NPY was dependent on reproductive status and that the NPY effects on LH were modulated by 17β-estradiol and testosterone (Peng *et al.*, 1993b, 1994). There may be a complex interplay between nutritional status and seasonal effects on NPY-induced LH secretion.

5.4. Other Neuropeptides

A number of other neuropeptides have been implicated in the regulation of LH and FSH release. Ghrelin acts directly on goldfish pituitary cells to release LH (Unniappan and Peter, 2004). Central and peripheral injections of ghrelin stimulate LH release and *lhb* expression, indicating that ghrelin is also involved in the synthesis of LH. Given the stimulatory effects of ghrelin on food intake (see Chapter 9, this volume), these results suggest that ghrelin maybe one of the signals linking the physiology of food intake, growth, and reproduction (Unniappan and Peter, 2004, 2005).

Neuropeptides containing a C-terminal Arg–Phe–NH2 sequence (RFamide peptides) have been identified in the brains of several vertebrates including fish. This family of peptides acts as gonadotropin release-inhibitory hormones in tetrapods, yet they stimulate FSH and LH release from the pituitary cells of sockeye salmon (*Oncorhynchus nerka*; Amano *et al.*, 2006). Interestingly, RFamide peptides decrease LH release in goldfish (Chang *et al.*, 2009).

Secretogranin-II (Sg-II) and its proteolytic product secretoneurin (SN) are potential paracrine regulators of goldfish LH release and production. Goldfish pituitary cells express Sg-II mRNA, and SN increases LH release from pituitary fragments (Blazquez *et al.*, 1998b; Zhao *et al.*, 2006).

Melanin-concentrating hormone (MCH), a cyclic heptadecapeptide produced within the hypothalamus, may regulate LH release in the goldfish (Cerdá-Reverter *et al.*, 2006). Salmon MCH stimulated LH secretion from dispersed pituitary cells, suggesting a direct action on gonadotrope cells. MCH expression was increased in response to 17β-estradiol and testosterone, suggesting that MCH may participate in the steroid positive feedback loop on LH secretion (Cerdá-Reverter *et al.*, 2006).

The neurohypophysial hormones vasotocin and isotocin, which are members of the vasopressin and oxytocin family of peptides, play a role in fish reproduction and in particular in aspects of reproductive behavior (Balment *et al.*, 2006; Popesku *et al.*, 2008). However, little is known of whether these hormones affect LH and FSH release.

6. PROTEIN HORMONES

6.1. Insulin-like Growth Factor I

Insulin-like growth factor I (IGF-I) is an important regulator of gonadotropin secretion in teleosts (Table 3.2). IGF-I increased the release and cell content of LH in pituitary cells of juvenile European eels (Huang *et al.*, 1998, 1999). In rainbow trout pituitary cells, co-administration of IGF-I and sGnRH elevated sensitivity to sGnRH as measured by the release of LH and FSH (Weil *et al.*, 1999). IGF-I increased GnRH-stimulated FSH release and cell content of FSH and LH in coho salmon pituitary cells (Baker *et al.*, 2000). All of these observations support the idea that IGF-I has a role in

Table 3.2
Effects of IGF-I on LH and FSH secretion

Species	Test conditions	Response to IGF-I	Reference
Juvenile (yellow stage) female eels (*Anguilla anguilla* L.)	*In vitro* cultures of pituitary cells	Increased LH release and pituitary cell content	Huang *et al.* (1998, 1999)
Rainbow trout (*Oncorhynchus mykiss*)	*In vitro* cultures of pituitary cells from both sexes	No effect on LH or FSH release but potentiated GnRH particularly in early stages of gametogenesis	Weil *et al.* (1999)
Coho salmon (*Oncorhynchus kisutch*)	*In vitro* cultures of pituitary cells	No effect on FSH release potentiated actions of GnRH	Baker *et al.* (2000)
Masu salmon (*Oncorhynchus masou*)	*In vitro* cultures of primary pituitary cells	IGF-I differently modulates sGnRH-induced GTH subunit gene expression, in sexually maturing depending on reproductive stages	Ando *et al.* (2006); Furukama *et al.* (2008)

controlling LH and FSH production, particularly through its interactions with GnRH.

Some insights into the role of IGF-I come from studies with fish at different stages of sexual maturity. In studies with primary pituitary cell cultures from male cherry salmon IGF-I increased GPα, FSHβ and LHβ mRNA expression early in gametogenesis but not in pituitary cultures from maturing or spawning fish (Ando *et al.*, 2006; Furukuma *et al.*, 2008). In females, IGF-I stimulated release of FSH and LH early in gametogenesis, whereas no stimulatory effects on the subunit mRNA levels were observed. IGF-I plus sGnRH stimulated release of FSH and LH at all stages in both sexes, but had different effects on the subunit mRNA levels depending on subunit and stage. These results were consistent with earlier studies with rainbow trout pituitaries, where the potentiating effects of IGF-I on LH and FSH release were more pronounced in fish at earlier stages of gametogenesis than in sexually mature individuals (Weil *et al.*, 1999). Collectively these studies suggest that IGF-I may act as a signaling molecule that transmits information on growth and nutritional status to the gonadotropic axis at the onset of puberty.

6.2. Activin and Follistatin

Several studies have evaluated the actions of activin and its binding protein follistatin on the regulation of LH and FSH secretion in the goldfish. Activin βB is the main form of activin subunit expressed in the goldfish pituitary (Lau and Ge, 2005) and studies have shown that recombinant goldfish activin stimulates *fshb* expression while suppressing *lhb* expression (Yuen and Ge, 2004; Yam *et al.*, 1999; Cheng *et al.*, 2007). Follistatin is also expressed in the goldfish pituitary and it has been shown to reverse the effects of activin. Specifically, follistatin decreases basal *fshb* expression, increases *lhb* expression, and counteracts activin-induced changes in LH and FSH subunit mRNA levels (Yuen and Ge, 2004). Recent studies using primary cultures of zebrafish pituitary cells confirmed the responses seen in goldfish as activin stimulated zebrafish *fshb* expression but suppressed the expression of *lhb*, whereas follistatin caused the reverse responses of *fshb* and *lhb* (Lin and Ge, 2009).

Given the co-expression of activin and follistatin in the pituitary, it has been suggested that the relative expression levels of FSH and LH may be influenced by the balance between these two proteins. Since the expression level of activin βB in the pituitary is rather stable both *in vitro* and *in vivo*, it is conceivable that follistatin may play a pivotal regulatory role in the intra-pituitary activin system. To test this hypothesis, Cheng *et al.* (2007) examined the effects of testosterone and 17β-estradiol on the expression of follistatin, activin βB and FSHβ. Both testosterone and 17β-estradiol significantly increased follistatin and LHβ expression in cultured pituitary cells within the physiological range of concentration. However, neither testosterone nor 17β-estradiol had any effect on the expression of activin βB and FSHβ. The expression of follistatin but not activin was regulated by sex steroids, suggesting that it is follistatin that serves as the point of the system that is subject to the endocrine or neuroendocrine regulation (Fig. 3.1; Cheng *et al.*, 2007).

7. INTEGRATED NEUROENDOCRINE CONTROL OF GONADAL DEVELOPMENT

There is much interest in understanding the neuroendocrine pathways controlling key reproductive events such as puberty and the onset of sexual maturity and spawning in teleosts. It is apparent that these events involve the coordinated actions of multiple neuroendocrine pathways controlling LH and FSH release and that sex steroids and DA are major players. In controlling reproductive development teleosts also integrate information from environmental cues such as temperature and photoperiod thereby

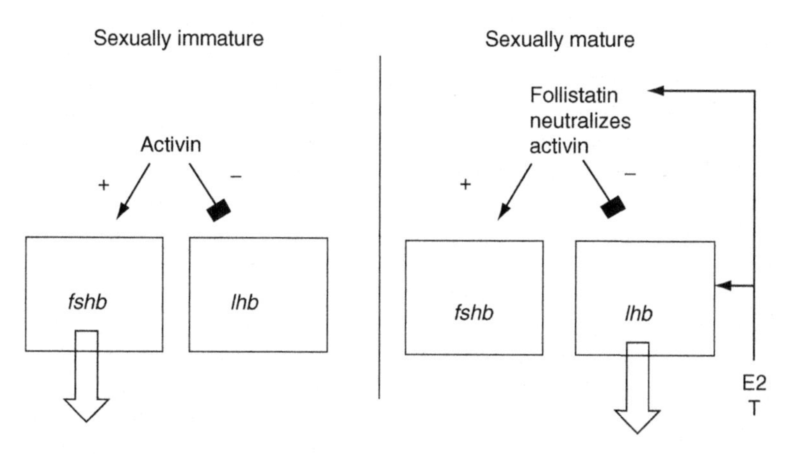

Fig. 3.1. A generalized model of the interactive effects of activin, the activin-binding protein follistatin and sex steroids on the expression of *fshb* and *lhb* in the pituitary of sexually immature and sexually mature goldfish (modified from Cheng *et al.*, 2007). In sexually immature fish, activin stimulates the expression of *fshb* and inhibits *lhb*. In sexually mature fish, follistatin expression is induced by 17β-estradiol, testosterone and activin. Follistatin neutralizes the effects of activin and the expression of *fshb* is reduced. 17β-estradiol and testosterone also act on the LH cells to increase *lhb* expression and in the presence of follistatin the inhibitory effects of activin on *lhb* are diminished. E2, 17β-estradiol; T, testosterone.

adding a further level of complexity. There is growing evidence of considerable diversity in the neuroendocrine control of reproduction in teleosts and perhaps this is not surprising given that fishes represent almost 50% of all vertebrates and exhibit some of the most diverse reproductive strategies in the animal kingdom.

Several studies have implicated DA as playing a role in controlling the timing of puberty. For example, removal of DA inhibition is important for GnRH stimulation of LH gene expression in pubertal European eel (Vidal *et al.*, 2004). In the juvenile spadefish *Chaetodipterus faber*, a decrease in dopaminergic activity was observed in the hypothalamus at the time of puberty (Marcano *et al.*, 1995). Studies with the grey mullet showed that the abundance of DA D2 receptors in the pituitary decreased as puberty progressed (Nocillado *et al.*, 2007). DA levels are also linked to energy status given that starvation has been shown to increase telencephalic and diencephalic DA levels in the lungfish (*Protopterus annectens*) and this may lead to reduced reproductive development. The importance of DA to reproductive development is reinforced through studies showing that blocking DA receptors with haloperidol initiated sex reversal in the saddleback wrasse *Thalassoma duperrey* (Larson *et al.*, 2003). Conversely in this species, the DA agonist apomorphine prevented animals from completing sex reversal under permissive social conditions. However, the importance of DA to puberty is not universal. In juvenile hybrid striped bass and red sea bream juvenile females there is no evidence that DA plays a role controlling LH secretion at the time of puberty (Holland *et al.*, 1998; Kumakura *et al.*, 2003).

Despite the positive feedback effects of steroids on the pituitary content of LH and FSH in juvenile teleosts there is often only limited release of these hormones. In the case of eel pretreated with 17β-estradiol, GnRH treatment alone was unsuccessful in inducing LH release (Dufour *et al.*, 1988). Both in the eel (Vidal *et al.*, 2004) and hybrid striped bass (Holland *et al.*, 1998), testosterone was necessary to observe a stimulatory effect of GnRH on LH release. Testosterone appears to play a facilitatory role by increasing the sensitivity of the pituitary to GnRH (Vidal *et al.*, 2004). Although the mechanisms mediating the stimulatory actions of testosterone are poorly understood these may involve an increase in GnRH-receptor gene expression and/or a modulation of GnRH-receptor coupling and signaling pathway (Trudeau *et al.*, 1991; Lo and Chang, 1998). There is evidence from studies with African catfish that aromatizable androgens of testicular origin lead to the puberty-associated increase in immunoreactive gonadotropes and increased LH mRNA and protein content in the pituitary (Covaco *et al.*, 2001). Collectively, these observations are consistent with the hypothesis that androgens play a key role in the activation of the brain–pituitary–gonadal axis at puberty in some juvenile teleosts.

At this time we do not have a cohesive picture of the steroid feedback regulation of FSH. Negative feedback effects of steroids on FSH release dominate the literature. It seems that rising levels of 17β-estradiol are coincident with reductions in FSHβ mRNA levels and declining plasma FSH levels (Larsen and Swanson, 1997; Dickey and Swanson, 1998; Sohn *et al.*, 1998; Mateos *et al.*, 2002; Banerjee and Khan, 2008). The situation is far more variable for immature fish. 17β-estradiol had no effect on plasma FSH or FSHβ mRNA levels in immature and early recrudescent coho salmon (Dickey and Swanson, 1995) whereas positive feedback effects of steroids on pituitary and plasma FSH levels have been described in rainbow trout (Breton *et al.*, 1997) and Atlantic salmon (*Salmo salar*; Borg *et al.*,1998). Other studies with female silver eels have shown that *in vivo* treatment steroids have no effect on FSH mRNA expression although there was a modest increase seen in cultured pituitary cells treated with 17β-estradiol *in vitro*. Similar results were found for goldfish pituitary cells (Huggard-Nelson *et al.*, 2002). Collectively these studies emphasize that there are marked differences in response at the level of the pituitary and hypothalamus and that there may be differential regulation of FSH between species and developmental stages.

Gonadal steroids and changes in steroid feedback regulation appear to play a role in mediating the seasonal/maturational differences in the availability of LH and FSH for release. There are gonadal stage dependent effects of 17β-estradiol on LH mRNA expression in that 17β-estradiol has positive feedback effects on LH mRNA in juvenile goldfish and Atlantic croaker but not in gonadally mature fish (Kobayashi *et al.*, 2000; Banerjee and Kahn, 2008). In the case of mature fish, gonadectomy does not influence LH secretion and 17β-estradiol had no effect on LH mRNA expression. In both species, the primary action of estrogen feedback in maturing fish appears to involve blocking of GnRH-induced LH release. In the final stages of sexual maturation in these species, plasma estrogen levels decline as progestins levels increase to promote oocyte maturation and ovulation. This decline has the effect of reducing the negative feedback effects on GnRH release, thereby enabling the LH surge that promotes ovulation.

Photoperiod can modulate the feedback regulation of steroids on LH and FSH expression in the pituitary. In studies with the three-spine stickleback, *Gasterosteus aculeatus*, under both long-day and short photoperiods, LHβ mRNA levels were lower in castrated males compared to sham-operated males, and treatment with 11KA and T increased LHβ mRNA expression, indicating a positive feedback mechanism (Hellqvist *et al.*, 2008). A positive feedback was also found on *fshb* expression under long photoperiod, where castration decreased, and androgen replacement restored, *fshb* expression. However, the response was different under short photoperiod. Castration under short photoperiod instead increased *fshb* levels, whereas treatment with 11KA and T decreased *fshb* expression, indicating that under these conditions a negative feedback mechanism controlled *fshb* (Hellqvist *et al.*, 2008). The shift between negative and positive feedback under different photoperiods is not found in mammals, where instead there are changes in effectiveness of negative feedbacks.

8. PERSPECTIVES

Access to the tools in the molecular biology toolbox has done much to advance the field of fish neuroendocrinology. These have been instrumental in demonstrating the multiplicity of GnRH and its receptors and in making rapid advances in the identification of newly emerging regulators of hypothalamic and pituitary function such as the kisspeptins. Identification of the components of the regulatory pathways is only just a beginning (Fig. 3.2) and now the task of unraveling the actions of these regulators is needed. Much of the recent emphasis has been on identification of the gene products and there is need to move to studies examining both the presence and functionality of

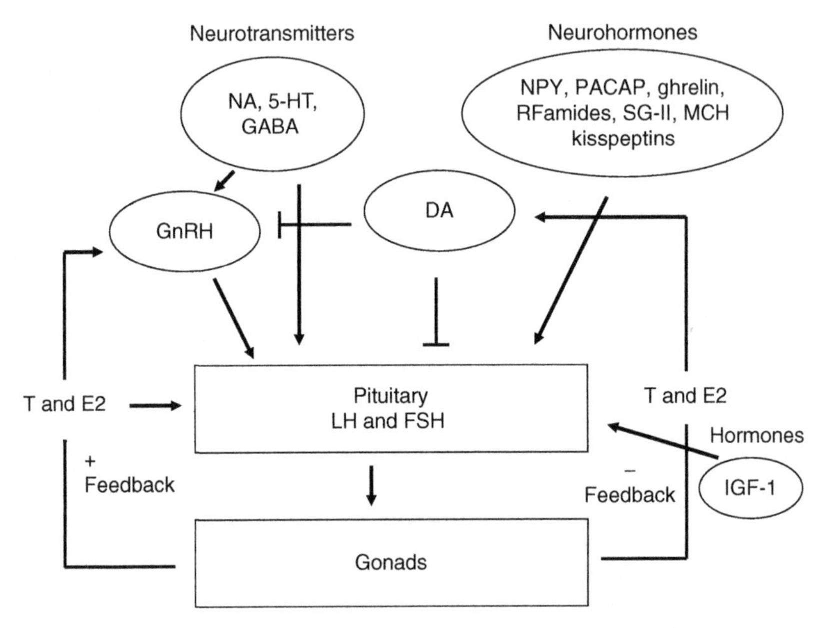

Fig. 3.2. Summary of the major neuroendocrine factors controlling LH and FSH release in teleosts. There are a large number of stimulatory factors including GnRH, neurotransmitters, neurohormones and peripheral protein hormones. The dominant inhibitory factor is dopamine. The sex steroids testosterone and 17β-estradiol may exert positive or negative effects on LH and FSH release depending on the developmental stage. Activin and its binding protein exert differential effects on LH and FSH release and are not included in this diagram. The relative importance of the various factors changes throughout the course of sexual development. 5-HT, serotonin; DA, dopamine; E2, 17β-estradiol; GABA, γ-aminobutyric acid; GnRH, gonadotropin-releasing hormone; IGF-I, insulin-like growth factor I; MCH, melanin-concentrating hormone; NA, noradrenaline; NPY, neuropeptide Y; PACAP, pituitary adenylate cyclase-activating polypeptide; RFamides, RF amide peptides; SG-II, secretogranin-II; T, testosterone.

these proteins. As an example, the study of FSH (and to some extent LH) in teleosts has been hampered by the availability of purified proteins for biological testing. Now with the development of methods to express LH and FSH (e.g., Aizen *et al.*, 2007; Zmora *et al.*, 2007; Kazeto *et al.*, 2008) and other hormones (Yuen and Ge, 2004) there will be new opportunities for the testing of activities of homologous proteins, thereby alleviating the concerns associated with the testing of heterologous (mammalian) hormones. The availability of recombinant FSH may also provide the opportunity to develop new immunoassay methods for the measurement of this hormone, which will fill a very large knowledge gap. Access to methods for the expression of fish genes in mammalian cells as was done for GnRH receptors

(Tello *et al.*, 2008) is opening new opportunities for studies on the functionality of the receptors and the signaling pathways they employ.

It is increasingly recognized that the GnRH–FSH/LH system is regulated by multiple inputs from both hypothalamic and peripheral sources. It will be a significant challenge to identify those inputs that are of primary importance from those that are redundant and may be of lesser significance. Here again the molecular biology toolbox may provide some important insights. To date there are few studies that have taken advantage of techniques that are available to silence or over-express genes and proteins of interest and to investigate loss or gain of function. Valuable insight into the factors controlling gene transcription may also be gained from sequence analysis of the promoter region and the identification of consensus sequences of transcriptional regulators of gene expression. The recent demonstration of different steroid hormone receptor subtypes and identification of non-classical membrane receptors is causing scientists to re-evaluate the actions of steroid hormones. Here is an area where the availability of different steroid hormone selective agonists and antagonists may add significantly to our understanding of how steroids affect the GnRH–FSH/LH system.

The diversity of reproductive strategies used by teleosts makes them an interesting model group for studies of the neuroendocrine control of reproduction. To date studies have been restricted to a relatively small number of species and we will benefit from studies in the future that examine species that utilize a diversity of reproductive strategies. Many of the current studies have focused on seasonal breeders and often those with group synchronous ovarian development. There may be some opportunities for new insights based on studies of fish that are continuous breeders with asynchronous ovarian development such as the medaka and zebrafish, for which genome sequences exist. The latter two species are also ones that are used extensively in biomedical research for developmental biology studies and they may be valuable in studies examining the ontogeny of the GnRH–FSH/LH system. Finally it is important not to forget the major roles that environmental factors such as temperature and photoperiod play in the control of reproductive function in teleosts. Certainly the new techniques that are available for studies of the GnRH–FSH/LH system can open new doors into how these factors affect reproduction (Hellqvist *et al.*, 2008). The same can be said for studies of anthropogenic chemicals as many of these have the potential to disrupt endocrine homeostasis. Often these compounds function as steroid hormone mimics and, given the involvement of steroids in positive and negative feedback regulation of the GnRH-FSH/LH system, it would not be surprising to see that these affect the neuroendocrine control of reproduction (Van Der Kraak *et al.*, 1992; Thomas, 2008). Studies showing that anthropogenic chemicals and hypoxia affect the serotoninergic system (Thomas *et al.*, 2007) suggest that there may be multiple targets affected. Clearly this is an exciting time where there are new opportunities to get answers to fundamental and applied questions concerning the neuroendocrine control of reproduction in teleosts.

REFERENCES

Aizen, J., Meiri, I., Tzchori, I., Levavi-Sivan, B., and Rosenfeld, H. (2005). Enhancing spawning in the grey mullet (*Mugil cephalus*) by removal of dopaminergic inhibition. *Gen. Comp. Endocrinol.* **142,** 212–222.

Aizen, J., Kasuto, H., Golan, M., Zakay, H., and Levavi-Sivan, B. (2007). Tilapia follicle-stimulating hormone (FSH): Immunochemistry, stimulation by gonadotropin-releasing hormone, and effect of biologically active recombinant FSH on steroid secretion. *Biol. Reprod.* **76,** 692–700.

Amano, M., Hyodo, S., Urano, A., Okumoto, N., Kitamura, S., Ikuta, K., Suzuki, Y., and Aida, K. (1994). Activation of salmon gonadotropin-releasing hormone synthesis by 17-methyl-testosterone administration in yearling masu salmon, *Oncorhynchus masou. Gen. Comp. Endocrinol.* **95,** 374–380.

Amano, M., Moriyama, S., Iigo, M., Kitamura, S., Amiya, N., Yamamori, K., Ukena, K., and Tsutsui, K. (2006). Novel fish hypothalamic neuropeptides stimulate the release of gonadotrophins and growth hormone from the pituitary of sockeye salmon. *J. Endocrinol.* **188,** 417–423.

Ando, H., and Urano, A. (2005). Molecular regulation of gonadotropin secretion by gonadotropin-releasing hormone in salmonid fishes. *Zool. Sci.* **22,** 379–389.

Ando, H., Luo, Q., Koide, N., Okada, H., and Urano, A. (2006). Effects of insulin-like growth factor I on GnRH-induced gonadotropin subunit gene expressions in masu salmon pituitary cells at different stages of sexual maturation. *Gen. Comp. Endocrinol.* **149,** 21–29.

Andreu-Vieyra, C. V., Buret, A. G., and Habibi, H. R. (2005). Gonadotropin-releasing hormone induction of apoptosis in the testes of goldfish (*Carassius auratus*). *Endocrinology* **146,** 1588–1596.

Aroua, S., Weltzien, F. A., Belle, N. L., and Dufour, S. (2007). Development of real-time RT-PCR assays for eel gonadotropins and their application to the comparison of *in vivo* and *in vitro* effects of sex steroids. *Gen. Comp. Endocrinol.* **153,** 333–343.

Baker, D. M., Davies, B., Dickhoff, W. W., and Swanson, P. (2000). Insulin-like growth factor I increases follicle-stimulating hormone (FSH) content and gonadotropin-releasing hormone-stimulated FSH release from coho salmon pituitary cells *in vitro*. *Biol. Reprod.* **63,** 865–871.

Balment, R. J., Lu, W., Weybourne, E., and Warne, J. K. (2006). Arginine vasotocin a key hormone in fish physiology and behaviour: A review with insights from mammalian models. *Gen. Comp. Endocrinol.* **147,** 9–16.

Banerjee, A., and Khan, I. A. (2008). Molecular cloning of FSH and LH b subunits and their regulation by estrogen in Atlantic croaker. *Gen. Comp. Endocrinol.* **155,** 827–837.

Biran, J., Ben-Dor, S., and Levavi-Sivan, B. (2008). Molecular identification and functional characterization of the kisspeptin/kisspeptin receptor system in lower vertebrates. *Biol. Reprod.* **79,** 776–786.

Blazquez, M., Bosma, P. T., Fraser, E. J., Van Look, K. J. W., and Trudeau, V. L. (1998a). Fish as models for the neuroendocrine regulation of reproduction and growth. *Comp. Biochem. Physiol. Part C* **119,** 345–364.

Blazquez, M., Bosma, P. T., Chang, J. P., Docherty, K., and Trudeau, V. L. (1998b). Gamma aminobutyric acid up-regulates the expression of a novel secretogranin-II messenger ribonucleic acid in the goldfish pituitary. *Endocrinology* **139,** 4870–4880.

Bommelaer, M. C., Billard, R., and Breton, B. (1981). Changes in plasma gonadotropin after ovariectomy and estradiol supplementation at different stages at the end of the reproductive cycle in the rainbow trout (*Salmo gairdneri* R.). *Reprod. Nutr. Dev.* **21,** 989–997.

Borg, B., Antonopoulu, E., Mayer, I., Andersson, E., Berglund, I., and Swanson, P. (1998). Effects of gonadectomy and androgen treatments on pituitary and plasma levels of gonadotropins in mature male Atlantic salmon, *Salmo salar*, parr-positive feedback control of both gonadotropins. *Biol. Reprod.* **58,** 814–820.

Breton, B., and Sambroni, E. (1996). Steroid activation of the brain-pituitary complex gonadotropic function in the triploid rainbow trout *Oncorhynchus mykiss*. *Gen. Comp Endocrinol.* **101,** 155–164.

Breton, B., Mikolajczyk, T. W., Popek, T. W., Bieniarz, K., and Epler, P. (1991). Neuropeptide Y stimulates *in vivo* gonadotropin secretion in teleost fish. *Gen. Comp. Endocrinol.* **84,** 277–283.

Breton, B., Sambroni, É., Govoroun, M., and Weil, C. (1997). Effects of steroids on GTH I and GTH II secretion and pituitary concentration in the immature rainbow trout *Oncorhynchus mykiss*. *C. R. Acad. Sci. Paris, Life Sci.* **320,** 783–789.

Breton, B., Govoroun, M., and Mikolajczyk, T. (1998). GTH I and GTH II secretion profiles during the reproductive cycle in female rainbow trout: Relationship with pituitary responsiveness to GnRH-A stimulation. *Gen. Comp. Endocrinol.* **111,** 38–50.

Canosa, L. F., Stacey, N., and Peter, R. E. (2008). Changes in brain mRNA levels of gonadotropin-releasing hormone, pituitary adenylate cyclase activating polypeptide, and somatostatin during ovulatory luteinizing hormone and growth hormone surges in goldfish. *Am. J. Physiol. Regul. Integr. Comp. Physiol.* **295,** R1815–R1821.

Cerdá-Reverter, J. M., Sorbera, L. A., Carrillo, M., and Zanuy, S. (1999). Energetic dependence of NPY-induced LH secretion in a teleost fish (*Dicentrarchus labrax*). *Am. J. Physiol. Reg. Int. Comp. Physiol.* **277,** R1627–R1634.

Cerdá -Reverter, J. M., Canosa, L. F., and Peter, R. E. (2006). Regulation of the hypothalamic melanin-concentrating hormone neurons by sex steroids in the goldfish: Possible role in the modulation of luteinizing hormone secretion. *Neuroendocrinology* **84,** 364–377.

Chang, J. P., Yu, K. L., Wong, A. O. L., and Peter, R. E. (1990). Differential actions of dopamine receptor subtypes on gonadotropin and growth-hormone release *in vitro* in goldfish. *Neuroendocrinology* **51,** 664–674.

Chang, J. P., Van Goor, F., and Acharya, S. (1991). Influences of norepinephrine, and adrenergic agonists and antagonists on gonadotropin secretion from dispersed pituitary cells of goldfish, *Carassius auratus*. *Neuroendocrinology* **54,** 202–210.

Chang, J. P., Wirachowsky, N. R., Kwong, P., and Johnson, J. D. (2001). PACAP stimulation of gonadotropin-II secretion in goldfish pituitary cells: Mechanism of actions and interaction with gonadotropin releasing hormone signaling. *J. Neuroendocrinol.* **13,** 540–550.

Chang, J. P., Johnson, J. D., Sawisky, G. R., Grey, C. L., Mitchell, G., Booth, M., Volk, M. M., Parks, S. K., Thompson, E., Goss, G. G., Klausen, C., and Habibi, H. R. (2009). Signal transduction in multifactorial neuroendocrine control of gonadotropin secretion and synthesis in teleosts – studies on the goldfish model. *Gen .Comp. Endocrinol.* **161,** 42–52.

Cheng, G. F., Yuen, C. W., and Ge, W. (2007). Evidence for the existence of a local activin follistatin negative feedback loop in the goldfish pituitary and its regulation by activin and gonadal steroids. *J. Endocrinol.* **195,** 373–384.

Chyb, J., Mikolajczyk, T., and Breton, B. (1999). Post-ovulatory secretion of pituitary gonadotropins GtH I and GtH II in the rainbow trout (*Oncorhynchus mykiss*): Regulation by steroids and possible role of non-steroidal gonadal factors. *J. Endocrinol.* **163,** 87–97.

Copeland, P. A., and Thomas, P. (1989). Control of gonadotropin release in Atlantic croaker: Evidence for a lack of dopaminergic inhibition. *Gen. Comp. Endocrinol.* **74,** 474–483.

Covaco, J. E. B., van Baal, J., van Dijk, W., Hassing, G. A. M., Goos, H. J.Th., and Schulz, R. W. (2001). Steroid hormones stimulate gonadotrophs in juvenile male African catfish (*Clarias gariepinus*). *Biol. Reprod.* **64,** 1358–1365.

Crim, L. W., Peter, R. E., and Billard, R. (1981). Onset of gonadotropic hormone accumulation in the immature trout pituitary gland in response to estrogen or aromatizable androgen steroid hormones. *Gen. Comp. Endocrinol.* **44,** 374–381.

de Leeuw, R., Wurth, Y. A., Zandbergen, M. A., Peute, J., and Goos, H. J. (1986). The effects of aromatizable androgens, non-aromatizable androgens, and estrogens on gonadotropin release in castrated African catfish, *Clarias gariepinus* (Burchell): A physiological and ultrastructural study. *Cell Tissue Res.* **243,** 587–594.

de Leeuw, R., Habibi, H. R., Nahorniak, C. S., and Peter, R. E. (1989). Dopaminergic regulation of pituitary gonadotrophin-releasing hormone receptor activity in the goldfish (*Carassius auratus*). *J. Endocrinol.* **121,** 239–247.

Devlin, R., and Nagahama, Y. (2002). Sex determination and sex differentiation on fish: An overview of genetic, physiological and environmental influences. *Aquaculture* **208,** 191–364.

Dickey, J. T., and Swanson, P. (1998). Effects of sex steroids on gonadotropin (FSH and LH) regulation in coho salmon (*Oncorhynchus kisutch*). *J. Mol. Endocrinol.* **21,** 291–306.

Dickey, J. T., and Swanson, P. (2000). Effects of salmon gonadotropin-releasing hormone on follicle stimulating hormone secretion and subunit gene expression in coho salmon (*Oncorhynchus kisutch*). *Gen. Comp. Endocrinol.* **118,** 436–449.

Dubois, E. A., Florijn, M. A., Zandbergen, M. A., Peute, J., and Goos, H. J. T. (1998). Testosterone accelerates the development of the catfish GnRH system in the brain of immature African catfish (*Clarias gariepinus*). *Gen. Comp. Endocrinol.* **112,** 383–393.

Dufour, S., Lopez, E., Le Menn, F., Le Belle, N., Baloche, S., and Fontaine, Y. A. (1988). Stimulation of gonadotropin release and of ovarian development, by the administration of a gonadoliberin agonist and of dopamine antagonists, in female silver eel pretreated with estradiol. *Gen. Comp. Endocrinol.* **70,** 20–30.

Dulka, J. G., Sloley, B. D., Stacey, N. E., and Peter, R. E. (1992). A reduction in pituitary dopamine turnover is associated with sex pheromone-induced gonadotropin secretion in male goldfish. *Gen. Comp. Endocrinol.* **86,** 496–505.

Elizur, A. (2009). The KiSS1/GPR54 system in fish. *Peptides* **30,** 164–170.

Filby, A. L., Aerle, R. V., Duitman, J., and Tyler, C. R. (2008). The kisspeptin/gonadotropin-releasing hormone pathway and molecular signaling of puberty in fish. *Biol. Reprod.* **78,** 278–289.

Flanagan, C. A., Chen, C. C., Coetsee, M., Mamputha, S., Whitlock, K. E., Bredenkamp, N., Grosenick, L., Fernald, R. D., and Illing, N. (2007). Expression, structure, function, and evolution of gonadotropin-releasing hormone (GnRH) receptors GnRH-R1SHS and GnRH-R2PEY in the teleost, *Astatotilapia burtoni*. *Endocrinology* **148,** 5060–5071.

Furukuma, S., Onuma, T., Swanson, P., Luo, Q., Koide, N., Okada, H., Urano, A., and Ando, H. (2008). Stimulatory effects of insulin-like growth factor 1 on expression of gonadotropin subunit genes and release of follicle-stimulating hormone and luteinizing hormone in masu salmon pituitary cells early in gametogenesis. *Zoolog. Sci.* **25,** 88–98.

Gonzalez-Martinez, D., Zmora, N., Zanuy, S., Sarasquete, C., Elizur, A., Kah, O., and Munoz-Cueto, J. A. (2002). Developmental expression of three different prepro-GnRH (gonadotrophin-releasing hormone) messengers in the brain of the European sea bass (*Dicentrarchus labrax*). *J. Chem. Neuroanat.* **23,** 255–267.

Goos, H. J. (1987). Steroid feedback on pituitary gonadotropin secretion. *In* "Proceedings of the Third International Symposium on Reproductive Physiology of Fish" (Idler, D. R., Crim, L. W., and Walsh, J., Eds.), pp. 16–20. St. John's University, St. John, NS, Canada.

Hellqvist, A., Schmitz, M., and Borg, B. (2008). Effects of castration and androgen treatment on the expression of FSH-beta and LH-beta in the three-spine stickleback, *Gasterosteus aculeatus* – Feedback differences mediating the photoperiodic maturation response? *Gen. Comp. Endocrinol.* **158,** 178–182.

Holland, M. C., Hassin, S., and Zohar, Y. (1998). Effects of long-term testosterone, gonadotropin-releasing hormone agonist, and pimozide treatments on gonadotropin II levels and ovarian development in juvenile female striped bass (*Morone saxatilis*). *Biol. Reprod.* **59,** 1153–1162.

Huang, Y. S., Schmitz, M., Le Belle, N., Chang, C. F., Quérat, B., and Dufour, S. (1997). Androgens stimulate gonadotropin-II β subunit in eel pituitary cells *in vitro*. *Mol. Cell. Endocrinol.* **131,** 157–166.

Huang, Y. S., Rousseau, K., Le Belle, N., Vidal, B., Burzawa-Gérard, E., Marchelidon, J., and Dufour, S. (1998). Insulin-like growth factor-I stimulates gonadotrophin production from eel pituitary cells: A possible metabolic signal for induction of puberty. *J. Endocrinol.* **159,** 43–52.

Huang, Y. S., Rousseau, K., Le Belle, N., Vidal, B., Burzawa-Gérard, E., Marchelidon, J., and Dufour, S. (1999). Opposite effects of insulin-like growth factors (IGF-Is) on gonadotropin (GtH-II) and growth hormone (GH) production by primary culture of European eel (*Anguilla anguilla*) pituitary cells. *Aquaculture* **177,** 73–83.

Huggard, D., Khakoo, Z., Kassam, G., Mahmoud, S. S., and Habibi, H. R. (1996). Effect of testosterone on maturational gonadotropin subunit messenger ribonucleic acid levels in the goldfish pituitary. *Biol. Reprod.* **54,** 1184–1191.

Huggard-Nelson, D. L., Nathwani, P. S., Kermouni, A., and Habibi, H. R. (2002). Molecular characterization of LH-beta and FSH-beta subunits and their regulation by estrogen in the goldfish pituitary. *Mol. Cell. Endocrinol.* **188,** 171–193.

Illing, N., Troskie, B. E., Nahorniak, C. S., Hapgood, J. P., Peter, R. E., and Millar, R. P. (1999). Two gonadotropin-releasing hormone receptor subtypes with distinct ligand selectivity and differential distribution in brain and pituitary in the goldfish (*Carassius auratus*). *Proc. Natl. Acad. Sci. USA* **96,** 2526–2531.

Kah, O., Trudeau, V. L., Sloley, B. D., Chang, J. P., Dubourg, P., Yu, K. L., and Peter, R. E. (1992). Influence of GABA on gonadotrophin release in the goldfish. *Neuroendocrinology* **55,** 396–404.

Kah, O., Lethimonier, C., Somoza, G., Guilgur, L. G., Vaillant, C., and Lareyre, J. J. (2007). GnRH and GnRH receptors in metazoa: A historical, comparative, and evolutive perspective. *Gen. Comp. Endocrinol.* **153,** 346–364.

Kanda, S., Akazome, Y., Matsunaga, T., Yamamoto, N., Yamada, S., Tsukamura, H., Maeda, K., and Oka, Y. (2008). Identification of KiSS-1 product kisspeptin and steroid-sensitive sexually dimorphic kisspeptin neurons in medaka (*Oryzias latipes*). *Endocrinology* **149,** 2467–2476.

Kauffman, A. S., Clifton, D. K., and Steiner, R. A. (2007). Emerging ideas about kisspeptin-GPR54 signaling in the neuroendocrine regulation of reproduction. *Trends Neurosci.* **30,** 504–511.

Kazeto, Y., Kohara, M., Miura, T., Miura, C., Yamaguchi, S., Trant, J. M., Adachi, S., and Yamauchi, K. (2008). Japanese eel follicle-stimulating hormone (FSH) and luteinizing hormone (LH): Production of biologically active recombinant FSH and LH by Drosophila S2 cells and their differential actions on the reproductive biology. *Biol. Reprod.* **79,** 938–946.

Khan, I. A., and Thomas, P. (1992). Stimulatory effects of serotonin on maturational gonadotropin release in the Atlantic croaker, *Micropogonias undulatus*. *Gen. Comp. Endocrinol.* **88,** 388–396.

Khan, I. A., and Thomas, P. (1996). Melatonin influences gonadotropin II secretion in the Atlantic croaker (*Micropogonias undulatus*). *Gen. Comp. Endocrinol.* **104,** 231–242.

Khan, I. A., and Thomas, P. (1999). GABA exerts stimulatory and inhibitory influences on gonadotropin II secretion in the Atlantic croaker (*Micropogonias undulatus*). *Neuroendocrinology* **69,** 261–268.

Khan, I. A., Hawkins, M. B., and Thomas, P. (1999). Gonadal stage-dependent effects of gonadal steroids on gonadotropin II secretion in the Atlantic croaker (*Micropogonias undulatus*). *Biol. Reprod.* **61,** 834–841.

Kitahashi, T., Ogawa, S., and Parhar, I. S. (2009). Cloning and expression of kiss2 in the zebrafish and medaka. *Endocrinology* **150,** 821–823.

Klenke, U., and Zohar, Y. (2003). Gonadal regulation of gonadotropin subunit expression and pituitary LH protein content in female hybrid striped bass. *Fish Physiol. Biochem.* **28,** 25–27.

Kobayashi, M., and Stacey, N. E. (1990). Effects of ovariectomy and steroid hormone implantation on serum gonadotropin levels in female goldfish. *Zool. Sci.* **7,** 715–721.

Kobayashi, M. A., Sohn, Y. C., Yoshiura, Y. A., and Aida, K. A. (2000). Effects of sex steroids on the mRNA levels of gonadotropin subunits in juvenile and ovariectomized goldfish *Carassius auratus*. *Fisher. Sci.* **66,** 223–231.

Kramer, C. R., and Imbriano, M. A. (1997). Neuropeptide Y (NPY) induced gonad reversal in the protogynous bluehead wrasses, *Thalassoma bifasciatum* (Teleostei: Labridea). *J. Exp. Zool.* **279,** 133–144.

Kreke, N., and Dietrich, D. R. (2007). Physiological endpoints for potential SSRI interactions in fish. *Crit. Rev. Toxicol.* **38,** 215–247.

Kumakura, N., Okuzawa, K., Gen, K., and Kagawa, H. (2003). Effects of gonadotropin-releasing hormone agonist and dopamine antagonist on hypothalamus–pituitary–gonadal axis of prepubertal female red seabream (*Pagrus major*). *Gen. Comp. Endocrinol.* **131,** 264–273.

Kusakabe, M., Nakamura, I., Evans, J., Swanson, P., and Young, G. (2006). Changes in mRNAs encoding steroidogenic acute regulatory protein, steroidogenic enzymes and receptors for gonadotropins during spermatogenesis in rainbow trout testes. *J. Endocrinol.* **189,** 541–554.

Larsen, D. A., and Swanson, P. (1997). Effects of gonadectomy on plasma gonadotropins I and II in coho salmon, *Oncorhynchus kisutch*. *Gen. Comp. Endocrinol.* **108,** 152–160.

Larson, E. T., Norris, D. O., Grau, E. G., and Summers, C. H. (2003). Monoamines stimulate sex reversal in the saddleback wrasse. *Gen. Comp. Endocrinol.* **130,** 289–298.

Lau, M. T., and Ge, W. (2005). Cloning of Smad2, Smad3, Smad4, and Smad7 from the goldfish pituitary and evidence for their involvement in activin regulation of goldfish FSHβ promoter activity. *Gen. Comp. Endocrinol.* **141,** 22–38.

Lethimonier, C., Madigou, T., Munoz-Cueto, J. A., Lareyre, J. J., and Kah, O. (2004). Evolutionary aspects of GnRHs, GnRH neuronal systems and GnRH receptors in teleost fish. *Gen. Comp. Endocrinol.* **135,** 1–16.

Levavi-Sivan, B., and Avitan, A. (2005). Sequence analysis, endocrine regulation, and signal transduction of GnRH receptors in teleost fish. *Gen. Comp. Endocrinol.* **142,** 67–73.

Levavi-Sivan, B., Biran, J., and Fireman, E. (2006). Sex steroids are involved in the regulation of gonadotropin-releasing hormone and dopamine D2 receptors in female tilapia pituitary. *Biol. Reprod.* **75,** 642–650.

Levavi-Sivan, B., Ofir, M., and Yaron, Z. (1995). Possible sites of dopaminergic inhibition of gonadotropin release from the pituitary of a teleost fish, tilapia. *Mol. Cell Endocrinol.* **109,** 87–95.

Levavi-Sivan, B., Safarian, H., Rosenfeld, H., Elizur, A., and Avitan, A. (2004). Regulation of gonadotropin-releasing hormone (GnRH)-receptor gene expression in tilapia: Effect of GnRH and dopamine. *Biol. Reprod.* **70,** 1545–1551.

Lin, S. W., and Ge, W. (2009). Differential regulation of gonadotropins (FSH and LH) and growth hormone (GH) by neuroendocrine, endocrine, and paracrine factors in the zebrafish – An *in vitro* approach. *Gen. Comp. Endocrinol.* **160,** 183–193.

Lin, X. W., and Peter, R. E. (1996). Expression of salmon gonadotropin-releasing hormone (GnRH) and chicken GnRH-II precursor messenger ribonucleic acids in the brain and ovary of goldfish. *Gen. Comp. Endocrinol.* **101,** 282–296.

Linard, B., Bennani, S., and Saligaut, C. (1995). Involvement of estradiol in a catecholamine inhibitory tone of gonadotropin release in rainbow trout (*Oncorhyncus mykiss*). *Gen. Comp. Endocrinol.* **99,** 192–196.

Linard, B., Anglade, I., Corio, M., Navas, J. M., Pakdel, F., Saligaut, C., and Kah, O. (1996). Estrogen receptors are expressed in a subset of tyrosine hydroxylase-positive neurons of the anterior preoptic region in the rainbow trout. *Neuroendocrinology* **63,** 156–165.

Lo, A., and Chang, J. P. (1998). *In vitro* application of testosterone potentiates gonadotropin-releasing hormone-stimulated gonadotropin-II secretion from cultured goldfish pituitary cells. *Gen. Comp. Endocrinol.* **111,** 334–346.

Mañanos, E. L., Anglade, I., Chyb, J., Saligaut, C., Breton, B., and Kah, O. (1999). Involvement of gamma-aminobutyric acid in the control of GTH-1 and GTH-2 secretion in male and female rainbow trout. *Neuroendocrinology* **69,** 269–280.

Marcano, D., Guerrero, H. Y., Gago, N., Cardillo, E., Requena, M., and Ruiz, L. (1995). Monoamine metabolism in the hypothalamus of the juvenile teleost fish, *Chaetodipterus faber*. *In* "Proceedings of the Fifth International Symposium on Reproductive Physiology of Fish" (Goetz, F. W., and Thomas, P., Eds.), pp. 64–66. University of Texas, Austin, TX.

Martinez-Chavez, C. C., Minghetti, M., and Migaud, H. (2008). GPR54 and rGnRH I gene expression during the onset of puberty in Nile tilapia. *Gen. Comp. Endocrinol.* **156,** 224–233.

Martyniuk, C. J., Chang, J. P., and Trudeau, V. L. (2007). The effects of GABA agonists on glutamic acid decarboxylase, GABA-transaminase, activin, salmon gonadotrophin-releasing hormone and tyrosine hydroxylase mRNA in the goldfish (*Carassius auratus*) neuroendocrine brain. *J. Neuroendocrinol.* **19,** 390–396.

Mateos, J., Mananos, E., Carrillo, M., and Zanuy, S. (2002). Regulation of follicle stimulating hormone (FSH) and luteinizing hormone (LH) gene expression by gonadotropin-releasing hormone (GnRH) and sexual steroids in the Mediterranean Sea bass. *Comp. Biochem. Physiol. B Biochem. Mol. Biol.* **132,** 75–86.

Mazumdar, M., Lal, B., Sakharkar, A. J., Deshmukh, M., Singru, P. S., and Subhedar, N. (2006). Involvement of neuropeptide Y Y1 receptors in the regulation of LH and GH cells in the pituitary of the catfish, *Clarias batrachus*: An immunocytochemical study. *Gen. Comp. Endocrinol.* **149,** 190–196.

Melamed, P., Rosenfeld, H., Elizur, A., and Yaron, Z. (1998). Endocrine regulation of gonadotropin and growth hormone gene transcription in fish. *Comp. Biochem. Physiol. C Pharmacol. Toxicol. Endocrinol.* **119,** 325–338.

Millar, R. P., Lu, Z. L., Pawson, A. J., Flanagan, C. A., Morgan, K., and Maudsley, S. R. (2004). Gonadotropin-releasing hormone receptors. *Endocr. Rev.* **25,** 235–275.

Mohamed, J. S., and Khan, I. A. (2006). Molecular cloning and differential expression of three GnRH mRNAs in discrete brain areas and lymphocytes in red drum. *J. Endocrinol.* **188,** 407–416.

Mohamed, J. S., Benninghoff, A. D., Holt, G. J., and Khan, I. A. (2007). Developmental expression of the G protein-coupled receptor 54 and three GnRH mRNAs in the teleost fish cobia. *J. Mol. Endocrinol.* **38,** 235–244.

Moles, G., Carrillo, M., Mañanósa, E., Mylonas, C. C., and Zanuy, S. (2007). Temporal profile of brain and pituitary GnRHs, GnRH-R and gonadotropin mRNA expression and content during early development in European sea bass (*Dicentrarchus labrax* L.). *Gen. Comp. Endocrinol.* **150,** 75–86.

Nocillado, J. N., Levavi-Sivan, B., Avitan, A., Carrick, F., and Elizur, A. (2005). Isolation of dopamine D2 receptor (D2R) promoters in *Mugil cephalus. Fish Physiol. Biochem.* **31,** 149–152.

Nocillado, J. N., Levavi-Sivan, B., Carrick, F., and Elizur, A. (2007). Temporal expression of G protein-coupled receptor 54 (GPR54), gonadotropin-releasing hormones (GnRH), and dopamine receptor D2 (drd2) in pubertal female grey mullet, *Mugil cephalus. Gen. Comp. Endocrinol.* **150,** 278–287.

Ogawa, S., Akiyama, G., Kato, S., Soga, T., Sakuma, Y., and Parhar, I. S. (2006). Immunoneutralization of gonadotropin-releasing hormone type-III suppresses male reproductive behavior of cichlids. *Neurosci. Lett.* **403,** 201–205.

Parhar, I. S., Ogawa, S., and Sakuma, Y. (2004). Laser-captured single digoxigenin-labeled neurons of gonadotropin-releasing hormone types reveal a novel G protein-coupled receptor (Gpr54) during maturation in cichlid fish. *Endocrinology* **145,** 3613–3618.

Parhar, I. S., Ogawa, S., and Sakuma, Y. (2005). Three GnRH receptor types in laser-captured single cells of the cichlid pituitary display cellular and functional heterogeneity. *Proc. Natl. Acad. Sci. USA* **102,** 2204–2209.

Peng, C., Huang, Y.-P., and Peter, R. E. (1990). Neuropeptide Y stimulates growth hormone and gonadotropin release from the goldfish pituitary *in vitro. Neuroendocrinology* **52,** 28–34.

Peng, C., Humphries, S., Peter, R. E., Rivier, J. E., Blomqvist, A. G., and Larhammar, D. (1993a). Actions of goldfish neuropeptide Y on secretion of growth hormone and gonadotropin-II in female goldfish. *Gen. Comp. Endocrinol.* **90,** 306–317.

Peng, C., Trudeau, V., and Peter, R. E. (1993b). Seasonal variation of neuropeptide Y actions on growth hormone and gonadotropin-II secretion in the goldfish: Effects of sex steroids. *J. Neuroendocrinol.* **5,** 273–280.

Peng, C., Gallin, W., Peter, R. E., Blomqvist, A. G., and Larhammar, D. (1994). Neuropeptide-Y gene expression in the goldfish brain: Distribution and regulation by ovarian steroids. *Endocrinology* **134,** 1095–1103.

Penlington, M. C., Williams, M. A., Sumpter, J. P., Rand-Weaver, M., Hoole, D., and Arme, C. (1997). Isolation and characterisation of mRNA encoding the salmon and chicken-II type gonadotrophin-releasing hormones in the teleost fish *Rutilus rutilus* (Cyprinidae). *J. Mol. Endocrinol.* **19,** 337–346.

Peter, R. E., Chang, J. P., Nahorniak, C. S., Omeljaniuk, R. J., Sokolowska, M., Shih, S. H., and Billard, R. (1986). Interactions of catecholamines and GnRH in regulation of gonadotropin secretion in teleost fish. *Recent Progress Hormone Res.* **42,** 513–548.

Peter, R. E., Lin, H. R., and Van Der Kraak, G. (1988). Induced ovulation and spawning of cultured freshwater fish in China: Advances in application of GnRH analogues and dopamine antagonists. *Aquaculture* **74,** 1–10.

Popesku, J. T., Martyniuk, C. J., Mennigen, J., Xiong, H., Zhang, D., Xia, X., Cossins, A. R., and Trudeau, V. L. (2008). The goldfish (*Carassius auratus*) as a model for neuroendocrine signalling. *Mol. Cell. Endocrinol.* **293,** 43–56.

Root, A. R., Sanford, J. D., Kavanaugh, S. I., and Sower, S. A. (2004). *In vitro* and *in vivo* effects of GABA, muscimol, and bicuculline on lamprey GnRH concentration in the brain of the sea lamprey (*Petromyzon marinus*). *Comp. Biochem. Physiol. A Mol. Integr. Physiol.* **138,** 493–501.

Saligaut, C., Garnier, D. H., Bennani, S., Salbert, G., Bailhache, T., and Jego, P. (1992). Effects of estradiol on brain aminergic turnover of the female rainbow trout (*Oncorhynchus mykiss*) at the beginning of the vitellogenesis. *Gen. Comp. Endocrinol.* **88,** 209–216.

Saligaut, C., Linard, B., Mañanós, E. L., Kah, O., Breton, B., and Govoroun, M. (1998). Release of pituitary gonadotrophins GtH I and GtH II in the rainbow trout (*Oncorhynchus mykiss*): Modulation by estradiol and catecholamines. *Gen. Comp. Endocrinol.* **109,** 302–309.

Saligaut, C., Linard, B., Breton, B., Anglade, I., Bailhache, T., Kah, O., and Jego, P. (1999). Brain aminergic systems in salmonids and other teleosts in relation to steroid feedback and gonadotropin release. *Aquaculture* **177,** 13–20.

Sawisky, G. R., and Chang, J. P. (2005). Intracellular calcium involvement in pituitary adenylate cyclase-activating polypeptide stimulation of growth hormone and gonadotrophin secretion in goldfish pituitary cells. *J. Neuroendocrinol.* **17,** 353–371.

Schulz, R., and Goos, H. J.Th. (1999). Puberty in male fish: Concepts and recent developments with special reference to the African catfish (*Clarias gariepinus*). *Aquaculture* **177,** 5–12.

Schulz, R. W., Bogerd, J., Bosma, P. T., Peute, J., Rebers, F. E. M., Zandbergen, M.A, and Goos, H. J. Th. (1995).*In* "Proceedings of the Fifth International Symposium on Reproductive Physiology of Fish" (Goetz, F. W., and Thomas, P., Eds.), pp. 2–6. University of Texas, Austin, TX.

Sebert, M.-E., Legros, C., Weltzien, F.-A., Malpaux, B., Chemineau, P., and Dufour, S. (2008). Melatonin activates brain dopaminergic systems in the eel with an inhibitory impact on reproductive function. *J. Neuroendocrinol.* **20,** 917–929.

Seminara, S. B., and Kaiser, U. B. (2005). New gatekeepers of reproduction: GPR54 and its cognate ligand, KiSS-1. *Endocrinology* **146,** 1686–1688.

Senthilkumaran, B., and Joy, K. P. (1995). Effects of melatonin, p-chlorophenylalanine, and α-methyltyrosine on plasma gonadotropin level and ovarian activity in the catfish, *Heteropneustes fossilis*: A study correlating changes in hypothalamic monoamines. *Fish Physiol. Biochem.* **14,** 471–480.

Senthilkumaran, B., and Joy, K. P. (1996). Effects of administration of some monoamine-synthesis blockers and precursors on ovariectomy-induced rise in plasma gonadotropin II in the catfish, *Heteropneustes fossilis*. *Gen. Comp. Endocrinol.* **101,** 220–226.

Senthilkumaran, B., Okuzawa, K., Gen, K., and Kagawa, H. (2001). Effects of serotonin, GABA and neuropeptide Y on seabream gonadotropin releasing hormone release *in vitro* from preoptic-anterior hypothalamus and pituitary of red seabream, *Pagrus major*. *J. Neuroendocrinol.* **13,** 395–400.

Silver, M. R., Kawauchi, H., Nozaki, M., and Sower, S. A. (2004). Cloning and analysis of the lamprey GnRH-III cDNA from eight species of lamprey representing the three families of Petromyzoniformes. *Gen. Comp. Endocrinol.* **139,** 85–94.

So, W.- K., Kwok, H.-F., and Ge, Y. (2005). Zebrafish gonadotropins and their receptors: II. Cloning and characterization of zebrafish follicle-stimulating hormone and luteinizing hormone subunits – their spatial-temporal expression patterns and receptor specificity. *Biol. Reprod.* **72,** 1382–1396.

Soga, T., Ogawa, S., Millar, R. P., Sakuma, Y., and Parhar, I. S. (2005). Localization of the three GnRH types and GnRH receptors in the brain of a cichlid fish: Insights into their neuroendocrine and neuromodulator functions. *J. Comp. Neurology* **487,** 28–41.

Sohn, Y. C., Yoshiura, Y., Kobayashi, M., and Aida, K. (1998). Effect of sex steroids on the mRNA levels of gonadotropin I and II subunits in the goldfish *Carassius auratus*. *Fish. Sci.* **64,** 715–721.

Sohn, Y. C., Kobayashi, M., and Aida, K. (2001). Regulation of gonadotropin beta subunit gene expression by testosterone and gonadotropin-releasing hormones in the goldfish, *Carassius auratus*. *Comp. Biochem. Physiol. B Biochem. Mol. Biol.* **129,** 419–426.

Somoza, G. M., and Peter, R. E. (1991). Effects of serotonin on gonadotropin and growth hormone release from *in vitro* perfused goldfish pituitary fragments. *Gen. Comp. Endocrinol.* **82,** 103–110.

Somoza, G. M., Yu, K. L., and Peter, R. E. (1988). Serotonin stimulates gonadotropin release in female and male goldfish, *Carassius auratus*, L. *Gen. Comp. Endocrinol.* **72,** 364–382.

Steven, C., Lehnen, N., Kight, K., Ijiri, S., Klenke, U., Harris, W. A., and Zohar, Y. (2003). Molecular characterization of the GnRH system in zebrafish (*Danio rerio*): Cloning of chicken GnRH-II, adult brain expression patterns and pituitary content of salmon GnRH and chicken GnRH-II. *Gen. Comp. Endocrinol.* **133,** 27–37.

Sundararaj, B. I., and Keshavanath, P. (1976). Effects of melatonin and prolactin treatment in hypophyseal-ovarian system in *Heteropneustes fossilis* (Bl.). *Gen. Comp. Endocrinol.* **29,** 84–96.

Swanson, P., Dickey, J. T., and Campbell, B. (2003). Biochemistry and physiology of fish gonadotropins. *Fish Physiol. Biochem.* **28,** 53–59.

Tello, J. A., Wu, S., Rivier, J. E., and Sherwood, N. M. (2008). Four functional GnRH receptors in zebrafish: Analysis of structure, signalling, synteny and phylogeny. *Integ. Comp. Biology* **48,** 570–587.

Thomas, P. (2008). The endocrine system. *In* "The Toxicology of Fishes" (Di Giulio, R. T., and Hinton, D. E., Eds.), pp. 457–488. CRC Press, Boca Raton, FL.

Thomas, P., Rahman, M. S., Khan, I. A., and Kummer, J. A. (2007). Widespread endocrine disruption and reproductive impairment in an estuarine fish population exposed to seasonal hypoxia. *Proc Biol. Sci.* **274,** 2693–2701.

Trudeau, V. L. (1997). Neuroendocrine regulation of gonadotrophin II release and gonadal growth in the goldfish, *Carassius auratus*. *Rev. Reprod.* **2,** 55–68.

Trudeau, V. L., Peter, R. E., and Sloley, B. D. (1991). Testosterone and estradiol potentiate the serum gonadotropin response to gonadotropin-releasing hormone in goldfish. *Biol. Reprod.* **44,** 951–960.

Trudeau, V.L, Sloley, B. D., and Peter, R. E. (1993a). GABA stimulation of gonadotropin secretion in the goldfish: Involvement of GABA receptors, dopamine, and sex steroids. *Am. J. Physiol.* **265,** R348–R355.

Trudeau, V. L., Sloley, B. D., and Peter, R. E. (1993b). Testosterone enhances GABA and taurine but not N-methyl-D-L-aspartate stimulation of gonadotropin secretion in the goldfish: Possible sex steroid feedback mechanisms. *J. Neuroendocrinol.* **5,** 129–136.

Trudeau, V. L., Spanswick, D., Fraser, E. J., Lariviere, K., Crump, D., Chiu, S., MacMillan, M., and Schulz, R. W. (2000). The role of amino acid neurotransmitters in the regulation of pituitary gonadotropin release in fish. *Biochem. Cell Biol.* **78,** 241–259.

Unniappan, S., and Peter, R. E. (2004). *In vitro* and *in vivo* effects of ghrelin on luteinizing hormone and growth hormone release in goldfish. *Am. J. Physiol. Regul. Integr. Comp. Physiol.* **286,** R1093–R1101.

Unniappan, S., and Peter, R. E. (2005). Structure, distribution and physiological functions of ghrelin in fish. *Comp. Biochem. Physiol. A Mol. Integr. Physiol.* **140,** 396–408.

Vacher, C., Mananos, E. L., Breton, B., Marmignon, M.H, and Saligaut, C. (2000). Modulation of pituitary dopamine D1 or D2 receptor and secretion of follicle stimulating hormone and luteinizing hormone during the annual reproductive cycle of female rainbow trout. *J. Neuroendocrinol.* **12,** 1219–1226.

Vacher, C., Ferrière, F., Marmignon, M. H., Pellegrini, E., and Saligaut, C. (2002). Dopamine D2 receptors and secretion of FSH and LH: Role of sexual steroids on the pituitary of the female rainbow trout. *Gen. Comp. Endocrinol.* **127,** 198–206.

van Aerle, R., Kille, P., Lange, A., and Tyler, C. R. (2008). Evidence for the existence of a functional Kiss1/Kiss1 receptor pathway in fish. *Peptides* **29,** 57–64.

Van Asselt, L. A., Goos, H. J., de Leeuw, R., Peter, R. E., Hol, E. M., Wassenberg, F. P., and Van Oordt, P. G. (1990). Characterization of dopamine D2 receptors in the pituitary of the African catfish, *Clarias gariepinus*. *Gen. Comp. Endocrinol.* **80,** 107–115.

Van Der Kraak, G., Donaldson, E. M., and Chang, J. P. (1986). Dopamine involvement in the regulation of gonadotropin secretion in coho salmon. *Can. J. Zool.* **64,** 1245–1248.

Van Der Kraak, G., Munkittrick, K. R., McMaster, M. E., Portt, C. B., and Chang, J. P. (1992). Exposure to bleached kraft pulp mill effluent disrupts the pituitary–gonadal axis of white sucker at multiple sites. *Tox. Appl. Pharmacol.* **115,** 224–233.

Van Der Kraak, G. J., Chang, J. P., and Janz, D. M. (1998). Reproduction. *In* "The Physiology of Fishes, Second Edition" (Evans, D. H., Ed.), pp. 465–488. CRC Press, Boca Raton, FL.

Vetillard, A., Atteke, C., Saligaut, C., Jego, P., and Bailhache, T. (2003). Differential regulation of tyrosine hydroxylase and estradiol receptor expression in the rainbow trout brain. *Mol. Cell. Endocrinol.* **199,** 37–47.

Vidal, B., Pasqualini, C., Le Belle, N., Holland, M. C., Sbaihi, M., Vernier, P., Zohar, Y., and Dufour, S. (2004). Dopamine inhibits luteinizing hormone synthesis and release in the juvenile European eel: A neuroendocrine lock for the onset of puberty. *Biol. Reprod.* **71,** 1491–1500.

Volkoff, H., and Peter, R. E. (1999). Actions of two forms of gonadotropin releasing hormone and GnRH antagonists on spawning behavior of the goldfish *Carassius auratus*. *Gen. Comp. Endocrinol.* **116,** 347–355.

Weil, C., Carré, F., Blaise, O., Breton, B., and Le Bail, P-Y. (1999). Differential effect of insulin-like growth factor I on *in vitro* gonadotropin (I and II) and growth hormone secretions in rainbow trout (*Oncorhynchus mykiss*) at different stages of the reproductive cycle. *Endocrinology* **140,** 2054–2062.

Wong, A. O., Li, W. S., Lee, E. K., Leung, M. Y., Tse, L. Y., Chow, B. K., Lin, H. R., and Chang, J. P. (2000). Pituitary adenylate cyclase activating polypeptide as a novel hypophysiotropic factor in fish. *Biochem. Cell Biol.* **78,** 329–343.

Wong, A. O. L., Murphy, C. K., Chang, J. P., Neumann, C. M., Lo, A., and Peter, R. E. (1998). Direct actions of serotonin on gonadotropin-II and growth hormone release from goldfish pituitary cells: Interactions with gonadotropin-releasing hormone and dopamine and further evaluation of serotonin receptor specificity. *Fish Physiol. Biochem.* **19,** 23–34.

Wong, T. T., and Zohar, Y. (2004). Novel expression of gonadotropin subunit genes in oocytes of the gilthead seabream (*Sparus aurata*). *Endocrinology* **145,** 5210–5220.

Yam, K. M., Yoshiura, Y., Kobayashi, M., and Ge, W. (1999). Recombinant goldfish activin B stimulates gonadotropin-Ib but inhibits gonadotropin-IIb expression in the goldfish, *Carassius auratus*. *Gen. Comp. Endocrinol.* **116,** 81–89.

Yaron, Z. (1995). Endocrine control of gametogenesis and spawning induction in the carp. *Aquaculture* **129,** 49–73.

Yaron, Z., Gur, G., Melamed, P., Rosenfeld, H., Elizur, A., and Levavi-Sivan, B. (2003). Regulation of fish gonadotropins. *Intl. Rev. Cytol.* **225,** 131–185.

Yu, K. L., and Peter, R. E. (1990). Dopaminergic regulation of brain gonadotropin-releasing hormone in male goldfish during spawning behaviour. *Neuroendocrinology* **52,** 276–283.

Yu, K. L., and Peter, R. E. (1992). Adrenergic and dopaminergic regulation of gonadotropin-releasing hormone release from goldfish preoptic-anterior hypothalamus and pituitary *in vitro*. *Gen. Comp. Endocrinol.* **85,** 138–146.

Yu, K. L., Rosenblum, P. M., and Peter, R. E. (1991). *In vitro* release of gonadotropin releasing hormone from the brain preoptic-anterior hypothalamic region and pituitary of female goldfish. *Gen. Comp. Endocrinol.* **81,** 256–267.

Yuen, C. W., and Ge, W. (2004). Follistatin suppresses FSHb but increases LHb expression in the goldfish – evidence for an activin-mediated autocrine/paracrine system in fish pituitary. *Gen. Comp. Endocrinol.* **135,** 108–115.

Zhao, E., Basak, A., and Trudeau, V. L. (2006). Secretoneurin stimulates goldfish pituitary luteinizing hormone production. *Neuropeptides* **40,** 275–282.

Zmora, N., Kazeto, Y., Kumar, R. S., Schulz, R. W., and Trant, J. M. (2007). Production of recombinant channel catfish (*Ictalurus punctatus*) FSH and LH in S2 Drosophila cell line and an indication of their different actions. *J. Endocrinol.* **194,** 407–416.

4

GROWTH HORMONE REGULATION IN FISH: A MULTIFACTORIAL MODEL WITH HYPOTHALAMIC, PERIPHERAL AND LOCAL AUTOCRINE/PARACRINE SIGNALS

JOHN P. CHANG
ANDERSON O. L. WONG

1. Introduction
2. Growth Hormone and Growth Hormone Receptors
3. Biological Actions of Growth Hormone in Fish
4. Regulation of Growth Hormone Secretion and Synthesis
   4.1. General Considerations and Patterns
   4.2. Hypothalamic Signals from CNS
   4.3. Signals from Peripheral Organs/Tissues
   4.4. Autocrine/Paracrine Signals within the Pituitary
5. Functional Interactions of GH Regulators at the Hypothalamic and Pituitary Levels
   5.1. Integration of Neuroendocrine Signals at the Level of Signal Transduction
   5.2. Influence of Sex Steroids on Hypothalamic GH Regulators
   5.3. Interactions between GH Regulators in Mediating Changes in Somatotrope Activity Following Feeding and Food Deprivation
   5.4. Other Interactions
6. Summary Remarks

In fish, growth hormone (GH) affects many functions, including somatic growth, energy metabolism, reproduction, feeding, osmoregulation and immune functions. GH release and synthesis are controlled by neuroendocrine factors from the brain and peripheral tissues. Hypothalamic regulators influence the expression of one another, forming an interacting network in GH regulation. GH release is inhibited tonically by somatostatin with insulin-like growth factor as a major feedback regulator; however, the actual amount of GH released reflects the balance of total inhibitory and stimulatory

Fish Neuroendocrinology: Volume 28
FISH PHYSIOLOGY

DOI: 10.1016/S1546-5098(09)28004-6

influences. Sex steroids and nutritional status also modulate the expression and pituitary actions of hypothalamic factors. Intrapituitary regulators, including GH, gonadotropin and inhibin/activin, provide autocrine/paracrine control over GH synthesis and secretion. At the somatotrope level, receptor expression for a multitude of neuroendocrine factors can integrate the regulatory signals from various regulators. The distinct and yet overlapping signaling cascades utilized by different regulators allow for ligand- and function-specificity for GH regulation.

1. INTRODUCTION

Growth hormone (GH) was first identified by its ability to stimulate elongation of long bones. As discussed in this and other chapters of this book, fish GH is a pleiotropic hormone with diverse functions. GH effects are generally thought to be indirect through GH stimulation of somatomedin (insulin-like growth factors, IGFs) release; however, GH also has direct effects independent of IGF (Nordgarden *et al.*, 2006; Wong *et al.*, 2006). The adenohypophysis is the main source of GH; however, extrapituitary GH also exists (e.g., in immune cells, gonads and brain). This chapter describes the neuroendocrine control of GH secretion and synthesis in the fish pituitary by hypothalamic factors, peripheral hormones and local intrapituitary signals. Given that the control of extrapituitary GH release and synthesis in fish is still unknown, this topic will not be discussed in this chapter.

2. GROWTH HORMONE AND GROWTH HORMONE RECEPTORS

GH, prolactin (PRL) and somatolactin (SL) are members of the cytokine family. It is commonly accepted that SL and PRL were derived from the ancestral GH during early phases of gnathostome evolution. SL is unique in fish models as the gene has been lost in tetrapods during the period of land invasion (Fukamachi and Meyer, 2007). Teleostean GH is a single-chain polypeptide protein of 21–23 kDa, and like other tetrapod GHs, contains two highly conserved intramolecular disulfide bonds important for biological activity, as well as a site for N-linked glycosylation. However, GH of ostariophysan fish contains an additional unpaired cysteine residue with unknown function. The GH gene structure is more variable in fish than in tetrapods. The typical 5 exon/4 intron structure of higher vertebrates is found in lampreys, cypriniforms and siluriforms while a 6 exon/5 intron structure occurs in salmoniforms, perciforms and tetradontiforms. The latter configuration is thought to have resulted from the insertion of a fifth intron, dividing the fifth exon in two (5 and 6). The 5′-flanking region of fish GH genes has been reported to contain multiple transcription regulatory elements, including the binding sites for growth hormone factor-1 (GHF-1/Pit-1), activator protein 1 (AP-1) and cAMP response element (CRE). Interestingly, while cDNA cloning reveals the presence of one GH transcript in some fish species [e.g., orange-spotted grouper (*Epinephelus coioides*)], other studies demonstrate the presence of two GH transcripts [e.g., rainbow trout (*Oncorhynchus mykiss*) and goldfish (*Carassius auratus*)]. By comparing the genome databases for model fish [e.g., zebrafish (*Danio rerio*), medaka (*Oryzias melastigma*) and fugu (*Takifugu rubripes*)] with those of other vertebrates, it is commonly accepted that two genome duplications have occurred during the evolution of modern-day bony fish. During the process, tetraploidization of the fish genome may give rise to multiple copies of GH transcripts in some teleost species (e.g., salmonids). (For review, see Kawauchi and Sower, 2006).

As in mammals, fish GH receptors (GHRs) belong to the type I cytokine receptor family and contain single extracellular, transmembrane and intracellular domains. The extracellular domain has conserved cysteine residues and a sandwich of two antiparallel β-sheets containing the GH binding site while the intracellular domain has two conserved proline-rich sequences, Box 1 and 2. Box 1 typically is a Janus kinase 2 (JAK 2) binding site essential for GHR signaling and Box 2 is involved in receptor internalization (for review, see Lichanska and Waters, 2008). Two GHR subtypes, GHR1 and GHR2, have been reported in teleost species as a result of gene duplication. GHR1 is structurally closer to tetrapod GHRs and contains 6–7 extracellular cysteine residues while GHR2 is unique for teleosts and contains only 4–5 extracellular cysteines. Both GHR1 and GHR2 are genuine GH receptors (Fukamachi and Meyer, 2007). In sea bream (*Sparus aurata*), GHR1 but not GHR2 triggers c-fos promoter activity in expression systems, suggesting that the two GHR subtypes may have dissimilar signal transduction systems. Where examined, tissue distribution of GHR1 and GHR2 differs; similarly, the intensity of GHR1 and GHR2 mRNA expression within a tissue can also be different (Jiao *et al.*, 2006). Furthermore, gonadal steroids and corticosteroids, as well as osmotic and salinity changes, have been reported to regulate the levels of GHR1 and GHR2 mRNA differentially (Pierce *et al.*, 2007). The presence of two GHR subtypes may be a part of the mechanisms by which the multitude of tissue-specific GH effects are mediated and differentially modulated by other regulatory signals.

3. BIOLOGICAL ACTIONS OF GROWTH HORMONE IN FISH

As reported in other vertebrates, GH released from the anterior pituitary serves as a major regulator of body growth and metabolism in fish models. Treatment with fish and mammalian GH are effective in stimulating both somatic and linear growth in fish species, including rainbow trout, salmon and carps (Peng and Peter, 1997). These growth-promoting effects have been further confirmed by transgenesis studies using mammalian GH [e.g., common carp (*Cyprinus carpio*)] and fish GH transgenes (e.g., salmon and tilapia species) (Zbikowska, 2003). In these previous studies, a drop in condition factor (defined as body weight $\times$ 100/length^3) is commonly observed during the growth enhancement induced by GH, indicating that the fish become relatively "leaner" with a concurrent gain in body weight. In representative species, e.g., rainbow trout, GH treatment can increase *de novo* protein synthesis in various tissues. During the process, lipolysis is also activated as reflected by rapid rises in free fatty acid and glycerol levels in circulation and this catabolic effect probably is caused by the differential actions of GH on hepatic triacylglycerol lipase and acetyl-coenzyme A carboxylase activities (Bjornsson, 1997). In fish models, GH is also known to play a role in modulating the behavior pattern during foraging. In salmonids (e.g., trout), GH treatment can increase the appetite and dominant feeding behavior with a drop in avoidance responses to predators. Similar behaviors are also observed in transgenic fish with GH over-expression, suggesting that the metabolic demand for growth enhancement may increase the risk-taking behavior during foraging (Sundstrom *et al.*, 2004). These behavioral changes are suspected to be the results of GH modulation of dopaminergic activities/ neuronal circuitry within the central nervous system (Bjornsson *et al.*, 2002).

Similar to mammals, in which GH has been proposed to be a "co-gonadotropin" (co-GTH), GH also interacts with the gonadotropic axis and contributes to sexual maturation, gametogenesis and gonadal steroidogenesis in fish models. The reproductive functions of GH are mediated mainly by GH receptors expressed in the gonad [e.g., in rainbow trout and tilapia (*Oreochromis mossambicus*)]. At the gonadal level, GH is also produced locally and plays a role in inducing steroidogenesis via direct actions on ovarian tissues. In some species, GH can also act indirectly by potentiating GTH stimulation on steroid production (e.g., goldfish). These stimulatory effects probably are the results of GH induction of ovarian aromatase activity via activation of cAMP-dependent cascades (Kajimura *et al.*, 2004; Li *et al.*, 2005). In addition to reproductive functions, GH is also essential for seawater adaptation and a rise in serum GH level is commonly observed during the parr–smolt transformation of anadromous salmons.

GH enhances the tolerance/survival of fish species to hyperosmotic stress, mainly by increasing gill chloride cell proliferation, stimulating branchial Na^+/K^+-ATPase activity, and activation of Na^+, K^+, $2Cl^-$-cotransporter and ion channels [e.g., cystic fibrosis transmembrane conductance regulator (CFTR) channels] involved in osmoregulation (Sakamoto and McCormick, 2006; Makino *et al.*, 2007). These stimulatory actions by GH can be further enhanced by cortisol, a major signal from the hypothalamic–pituitary–adrenal (HPA) axis in fish during osmotic stress, and this synergism is partly caused by GH-induced cortisol receptor expression in the gills (Pelis and McCormick, 2001). Given that (i) IGF-I can mimic GH induction of chloride cell proliferation and Na^+/K^+-ATPase activity, (ii) a rise in IGF-I level in circulation is observed in fish during hyperosmotic stress, and (iii) both IGF-I mRNA and IGF-I binding sites can be detected in the gill epithelium, it is commonly accepted that both endocrine and autocrine/paracrine components of IGF-I are involved in the osmoregulatory functions of GH in fish models (Sakamoto and McCormick, 2006).

In euryhaline fish, the elevation in GH release during hyperosmotic stress also occurs with concurrent activation of immune functions, consistent with the immunomodulatory effects of GH reported in mammals. Hypophysectomy in fish models [e.g., channel catfish (*Ictalurus punctatus*) and rainbow trout] can suppress immune responses (e.g., by reducing Ig-secreting leukocytes), which can be partially reversed by GH replacement (Yada, 2007). Furthermore, GH administration can enhance the survival of fish species against bacterial infection and artificial vibriosis (Sakai *et al.*, 1997). The immunoprotective effects of GH can be attributed to (i) its stimulation on antibody production and immune cell proliferation, (ii) activation of phagocytic and non-specific cytotoxic activities in leukocytes, (iii) induction of superoxide production and lysozyme activity, and (iv) anti-inflammatory actions via production of ceruloplasmin (Yada, 2007). (For review on neuroendocrine modulation of immune system in fish, see Chapter 7.)

4. REGULATION OF GROWTH HORMONE SECRETION AND SYNTHESIS

4.1. General Considerations and Patterns

In mammals, basal GH release decreases after disruption of hypothalamic connection with the pituitary, suggesting the dominance of stimulatory influence in neuroendocrine control of GH secretion. It is believed that GH-releasing hormone (GHRH) and somatostatin (SS) are the main stimulatory and inhibitory factors, respectively. Release of GHRH and SS from the median eminence into the portal blood are 180° out of phase of one another, forming a "dual regulator" control system. Inhibition by SS is responsible for the low GH levels during the trough while increased GHRH release induces the surge in GH secretion during each GH release episode. GHRH and SS neurons also form a reciprocal control network. SS neurons suppress GHRH release while GHRH increases SS secretion. This, together with GH feedback action, forms the basis of pulsatile GH secretion. This "dual regulator" system may be overly simplistic as other neuroendocrine regulators, such as pituitary adenylate cyclase activating polypeptide (PACAP) and ghrelin, have been shown to be important in the control of GH secretion (Lengyel, 2006; Goldenberg and Barkan, 2007).

In contrast to mammals, neuroendocrine control of GH release in fish, especially in terms of the involvement of stimulatory hypothalamic factors, is multifactorial in nature (Rousseau and Dufour, 2004). In fish models, GHRH may not be the dominant GH-releasing factor (Montero *et al.*, 2000) and GH secretion at the pituitary level is under tonic inhibitory, rather than stimulatory control (Rousseau and Dufour, 2004). Pulsatile GH release has been demonstrated in grass carp (*Ctenopharyngodon idellus*; Zhang *et al.*, 1994) but the universality of episodic GH release in fish has not been established. This is mainly due to the difficulties in obtaining repeated samples in sufficiently close time intervals to properly evaluate the kinetics of GH release *in vivo*. On the other hand, a circadian pattern of GH release has been clearly demonstrated in several salmonid (Bjornsson *et al.*, 2002) and cyprinid species (Marchant and Peter, 1986; Zhang *et al.*, 1994). In general, the pattern is characterized by the presence of one or more irregular peaks in blood GH levels during the photophase and with blood GH generally higher in the scotophase. The pattern of the peaks in blood GH level over a 24 h period can also be modified by temperature, feeding, photoperiod regimes and/or development. For example, a peak in serum GH can be noted at the beginning of the scotophase in goldfish acclimated to short photoperiod and low temperature but this is not observed in fish acclimated to long photoperiod and higher temperature (Marchant and Peter, 1986). GH secretion is often affected by food deprivation and can be entrained by feeding (see Chapter 9). Recently, a diurnal pattern of GH mRNA expression in the pituitary has been reported in juvenile rabbitfish (*Siganus guttatus*), with GH mRNA levels being lower in the photophase than in the scotophase (Ayson and Takemura, 2006). Based on similarities in the overall diurnal pattern of GH mRNA expression in rabbitfish and the general diurnal pattern of serum GH levels in other teleosts, it is tempting to speculate that both events are coordinated by a common factor. In trout, melatonin has been reported to stimulate GH release from pituitary tissues *in vitro* (Falcon *et al.*, 2003). Whether melatonin actually participates in the generation of the diurnal rhythms of GH mRNA levels and GH release has not been examined. Furthermore, dissociation between GH release and GH gene transcription has been reported in fish models, e.g., in goldfish (Johnson *et al.*, 2002) and sea bream (Chan *et al.*, 2004a), which indicates the complexity of the mechanisms involved in GH regulation.

Annual/seasonal variations in blood GH levels also exist in fish; however, the annual period of maximal somatic growth in fish is often not well correlated with the period of highest circulating GH levels partly because of the involvement of IGF in mediating GH effects on somatic growth and a role of GH in reproduction. For example, growth rate in gilthead sea bream is correlated with blood IGF-I levels but is delayed in terms of blood GH levels (Mingarro *et al.*, 2002). Similarly, in goldfish, although maximal growth occurs in July (summer) while serum GH is still high relative to values in late fall, the highest serum GH level occurs during the spring spawning season (Marchant and Peter, 1986). Increases in serum GH levels are often observed at times of gonadal growth, vitellogenesis and preovulatory LH surge in fish species, e.g., in white sucker (*Catostomus commersoni*), carp, gilthead sea bream, Atlantic halibut (*Hippoglossus hippoglossus*), Atlantic salmon (*Salmo salar*), chum salmon (*Oncorhynchus keta*), and rainbow trout (for review, see Canosa *et al.*, 2008). These observations on season changes of serum GH levels correlate well with the proposed role of GH as a co-GTH over the seasonal reproductive cycle of fishes (see Section 3 above). GH mRNA and protein levels in the pituitary of carp species are relatively higher in summer-adapted fish than in winter-adapted fish (Figueroa *et al.*, 2005), indicating that seasonal variation in pituitary GH synthesis also occurs in fish.

4.2. Hypothalamic Signals from CNS

4.2.1. Inhibitors

Somatostatin (SS). In teleosts, blood level of GH usually goes up following pituitary ectopic autotransplant and basal GH secretion from pituitary tissue or cell culture is high, suggesting that pituitary GH release is under tonic inhibitory control. Evidence indicating that SS represents this major tonic inhibitory influence comes from the initial studies in goldfish, in which electroradiofrequency lesions of the periventricular area with the cell bodies of SS neurons projecting into the pituitary significantly elevate serum GH levels. In the same study, the seasonal patterns of hypothalamic and pituitary SS contents also approximate the mirror image of that of serum GH levels (Marchant *et al.*, 1989; Figure 4.1). Basal GH release from primary cultures of goldfish pituitary cells is generally higher in late fall and early winter

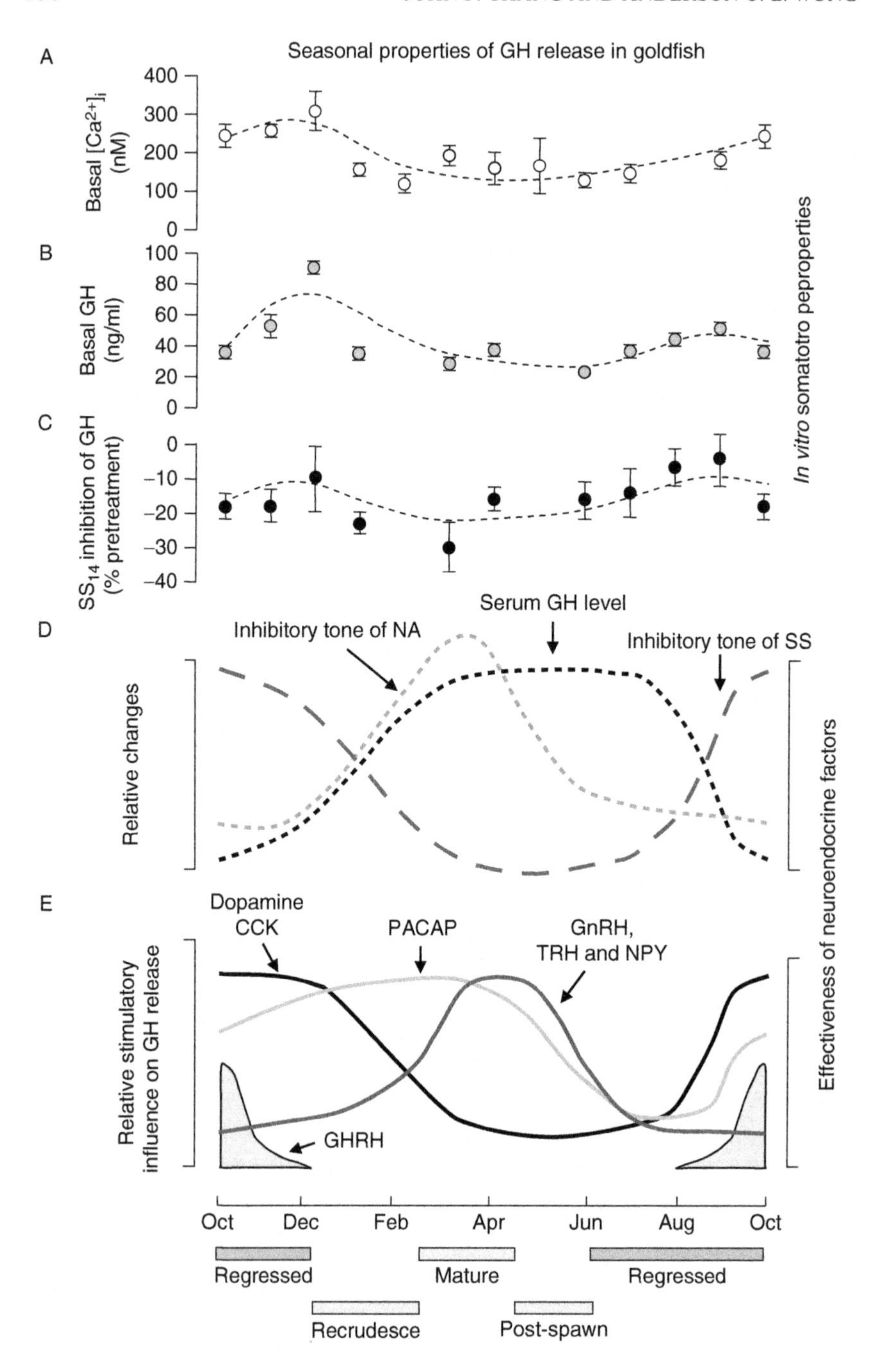
A
Seasonal properties of GH release in goldfish
Basal [Ca2+]i (nM)
B
Basal GH (ng/ml)
C
SS14 inhibition of GH (% pretreatment)
In vitro somatotro peproperties
D
Serum GH level
Inhibitory tone of NA
Inhibitory tone of SS
Relative changes
E
Dopamine
CCK
PACAP
GnRH,
TRH and NPY
GHRH
Relative stimulatory influence on GH release
Effectiveness of neuroendocrine factors
Oct
Dec
Feb
Apr
Jun
Aug
Oct
Regressed
Mature
Regressed
Recrudesce
Post-spawn

(October to December) than in spring (March to May), perhaps reflecting the release from endogenous SS inhibitory tone; in contrast, no significant seasonal differences can be detected in the relative ability of SS to decrease basal GH secretion *in vitro* (WK Yunker and JP Chang, unpublished; Figure 4.1). The ability of cysteamine, an SS-inhibiting agent, to increase serum GH levels in grass carp also supports the hypothesis that SS provides a tonic inhibitory control over GH release (Xiao and Lin, 2003). In the case of the turbot (*Psetta maxima*), tonic inhibition of GH release *in vivo* is so intense such that pituitary basal GH release *in vitro* is extremely high, approaching maximal release capacity (Rousseau *et al.*, 2001).

SS is a multifunctional and multimember family of peptides. Its tissue distribution, possible phylogeny and extrapituitary functions have been extensively reviewed elsewhere and peripheral sources of SS are not believed to play a physiological role in the regulation of pituitary GH secretion (Nelson and Sheridan, 2005; Klein and Sheridan, 2008). In terms of hypothalamic hypophysiotropic SS, the classical molecule is SS-14, a 14-amino acid long peptide first discovered in mammals. SS-14 is highly conserved among all vertebrates including fish. In fish, up to three separate genes encoding prepro-SS peptides (PSSs) have been identified and their molecular evolution has been reviewed recently (Tostivint *et al.*, 2008). SS-14 is encoded by preprosomatostatin peptide-1 (PSS-1). In mammals, PSS-I is processed to form SS-14 and the N-terminal extended SS-28. However, it is believed that PSS-I in fish only yields SS-14. PSS-II appears to be unique in fish and encodes for SS peptides of variable lengths (generally 22–28 amino acids) all having [Tyr^7, Gly^{10}]SS-14 at the C-terminus. In goldfish, PSS-II yields goldfish brain (gb)SS-28 (Lin *et al.*, 2000). However, there are two PSS-II peptides in rainbow trout, namely PSS-II′ and PSS-II″, yielding salmon SS-28 and SS-25, respectively (Holloway *et al.*, 2000). PSS-III encodes for [Pro^2]SS-14 and its

Fig. 4.1. Seasonal changes in neuroendocrine regulation of GH release *in vivo* versus *in vitro* GH responses at the somatotrope level. Seasonal changes of (A) intracellular Ca^{2+} ($[Ca^{2+}]_i$), (B) basal GH release *in vitro*, and (C) magnitude of GH inhibition caused by somatostatin-14 (SS_{14}) treatment in primary cultures of goldfish pituitary cells are presented together with GH secretion *in vivo* with respect to the reproductive cycle of the goldfish (D). The seasonality of plasma GH levels can be correlated with the seasonal changes in neuroendocrine input by (D) somatostatin (SS) and noradrenaline (NA), and (E) dopamine, cholecystokinin (CCK), pituitary adenylate cyclase activating polypeptide (PACAP), neuropeptide Y (NPY), thyrotropin-releasing hormone (TRH), GH-releasing hormone (GHRH) and gonadotropin-releasing hormone (GnRH). Seasonal changes of basal GH release from goldfish pituitary cell cultures correlate well with $[Ca^{2+}]_i$ but not with serum GH levels, suggesting that neuroendocrine regulators play an important role in the seasonal changes in circulating GH levels. The magnitude of GH release *in vivo* at any point in time is the result of the balance of multiple inhibitory and stimulatory influences acting at the pituitary cell level. Modified and updated from Chang and Habibi, 2002. (See Color Insert.)

variants and is believed to be the mammalian cortistatin orthologue. In goldfish, cDNA encoding for all three PSS forms has been found in the hypophysiotropic hypothalamus, and PSS-I and PSS-II have also been located in the pituitary (Lin and Peter, 2001; Cerdà-Reverter and Canosa, Chapter 1 of this volume). Likewise, messages for all three PSS forms are present in the hypothalamus and pituitary of the orange-spotted grouper (Xing *et al.*, 2005). These observations suggest that products from all three PSSs play a role in the neuroendocrine control of pituitary functions, at least in some fish species.

Products of PSS-I, PSS-II and PSS-III are active at the level of the pituitary. SS-14 is a potent inhibitor of basal GH secretion in many fish species, including goldfish, rainbow trout, tilapia, salmon, eel (*Anguilla anguilla*) and turbot (Rousseau *et al.*, 1998, 2001; Lin *et al.*, 2000). Similarly, SS-14 effectively reduces/blocks stimulated GH secretion, including the GH responses to dopamine (DA), gonadotropin-releasing hormone (GnRH), PACAP, neuropeptide Y (NPY), ghrelin and/or corticotropin-releasing factor (CRF) in a number of fish models (for review, see Lin and Peter, 2001). In goldfish, SS-14 and [Pro^2]SS-14 are equipotent in suppressing basal GH secretion while gbSS-28 is the most potent of the three SS forms. On the other hand, gbSS-28 is less effective than SS-14 and [Pro^2]SS-14 in inhibiting the stimulated GH release response to DA and PACAP. In addition, salmon SS-25 and catfish SS-22 have no effects on goldfish GH secretion (Peter and Chang, 1999; Lin *et al.*, 2000). These results indicate that goldfish somatotropes selectively respond to the different PSS products native to the species, as well as discriminate between SS forms from other species. When taken together, these and other results indicate that the different SS isoforms may differentially modulate GH responses to various neuroendocrine stimulators, as well as play selective neuroendocrine roles in the integrated neuroendocrine control of GH release (see Section 5).

In mammals, five SS receptor (SST) subtypes (SST_{1-5}) exist and with each SST subtype coupled to distinct suites of intracellular second message systems. To date, cloned fish SSTs fall into SST_{1-3} and SST_5 groups and with multiple messages for a single receptor subtype in a species (Lin *et al.*, 2000; Nelson and Sheridan, 2005). For example, two variants of SST_1 (1A and 1B) are known in rainbow trout (Nelson and Sheridan, 2006). In goldfish, eight SSTs ($gfSST_{1A,\ 1B,\ 2,\ 3A,\ 3B,\ 5A,\ 5B,}$ and $_{5C}$) have been identified and multiple SSTs are expressed in the pituitary, with $gfSST_2$ and $gfSST_5$ mRNA being the most abundant. When expressed in COS-7 cells, $gfSST_2$ can be activated by SS-14 and [Pro^2]SS-14, but not gbSS-28, whereas $gfSST_5$ has a higher affinity for gbSS-28 than for SS-14 and [Pro^2]SS-14 (Lin *et al.* 2000; Lin and Peter, 2001). Thus, the complement of pituitary SST subtypes in goldfish processes the required properties to mediate the differential actions of the three endogenous SS forms.

Information on how different fish SS isoforms act at the level of intracellular signaling leading to inhibition of GH release is largely provided by the studies in goldfish pituitary cells (for review, see Chang *et al.*, 2000; Table 4.1). Basal GH release is sensitive to intracellular Ca^{2+} levels ($[Ca^{2+}]_i$) and the level of cAMP. Although SS-14, [Pro^2]SS-14 and gbSS-28 all lower cAMP, their relative potency in altering cAMP production is not correlated with that of suppression of basal GH release (Yunker *et al.*, 2003). Furthermore, SS-14 has no effects on basal $[Ca^{2+}]_i$ in identified goldfish somatotropes. These results imply that actions on other signaling components mediating basal GH release may also be involved (Yunker and Chang, 2004). SS-14 effectively inhibits the GH responses triggered by activation of cAMP, Ca^{2+}, protein kinase C (PKC), nitric oxide (NO) and arachidonic acid (AA) signaling cascades using pharmacological agents, suggesting that SS-14 may target intracellular events downstream of the respective pathways to inhibit stimulated GH release (Kwong and Chang, 1997; Yunker and Chang, 2001, 2004). Interestingly, while the spectrum of activity of [Pro^2] SS-14 is largely similar to that of SS-14, gbSS-28 is much less effective in

Table 4.1

Summary of the actions for GH-release inhibitors on intracellular signal transduction in fish somatotropes leading to GH secretion and GH gene expression

Signaling components	IGF	SS_{14}	[Pro^2]SS_{14}	$gbSS_{28}$	NA	5HT
PI3K	↑					
MAPK	↑					
CAM	↑*					
Actions Downstream of Pathway						
Ca^{2+}		↓	↓	↓/↔	↓*	↔ ?
NO		↓	↓	↔		
AA		↓			↓	
cAMP		↓	↓	↔	↓*	
PKC		↓	↓	↓/↔	↓	↓ ?
Induced Signal Generation						
cAMP		↓			↓	
Ca^{2+}		↓/↑			↓	

Findings were based on the reports in goldfish and grass carp pituitary cells. Asterisk (*) denotes the involvement of the respective signaling pathways in GH gene expression. (Keys: ↑, increase/ activate; ↓, decrease/inhibit; ↔, little or no effect; ?, findings based on preliminary data.) [Abbreviations: IGF, Insulin-like Growth Factor; SS14, Somatostatin-14; [Pro^2]SS_{14}, [Pro^2] variant of somatostatin-14; $gbSS_{28}$, Goldfish Brain Somatostatin-28; NA, Noradrenaline; 5HT, Serotonin; PI3K, Phosphatidylinositol-3-Kinase; MAPK, Mitogen-activated Protein Kinase; CAM, Calmodulin; NO, Nitric Oxide; AA, Arachidonic Acid; cAMP, cyclic AMP; PKC, Protein Kinase C] inhibiting the GH response to cAMP/PKC activation (Yunker *et al.*, 2003). The differential action on cAMP-dependent GH release may explain the different efficiency of gfSS-28 and SS-14 in suppressing the GH responses to DA and PACAP described above. The three SS isoforms may also act at the level of second messenger generation to affect stimulated GH release. SS-14 inhibition of the GH responses to salmon (s)GnRH, DA and PACAP is associated with decreases in the maximal amplitude of the $[Ca^{2+}]_i$ responses to these ligands. In contrast, the Ca^{2+} signal induced by chicken (c)GnRH-II is unaffected by SS-14 and SS-14 suppression of the GH response to direct PKC activation is even associated with increases in $[Ca^{2+}]_i$ (Yunker and Chang, 2004). Thus, regardless of the effects of SS at the level of second messenger generation, SS actions at sites downstream of these messengers are of great importance.

Despite its potent inhibitory actions on GH release, SS-14 does not alter steady-state GH mRNA levels in tilapia and rainbow trout (Melamed *et al.*, 1998). Whether this is true in other fish species and whether all SS isoforms behave similarly is unknown.

Serotonin (5-HT). In goldfish, 5-HT reduces basal GH release from pituitary fragments and cells *in vitro*; in addition, 5-HT inhibits sGnRH- and DA-elicited GH release (Somoza and Peter, 1991; Wong *et al.*, 1998). These actions of 5-HT are mediated by 5-HT2-like receptors. Whether 5-HT similarly affects GH secretion in other fish species is not known. In teleosts such as goldfish, rainbow trout, torpedo fish (*Torpedo marmorata*), African catfish (*Clarias gariepinus*) and Atlantic croaker (*Micropogonias undulatus*), 5-HT immunoreactive fibers have been detected in the pars distalis of the pituitary gland (Kah and Chambolle, 1983; Frankenhuis-van den Heuvel and Nieuwenhuys, 1984; Bonn and König, 1990; Corio *et al.*, 1991; Khan and Thomas, 1993). In addition, 5-HT has been measured in the pituitary of Indian catfish (*Heteropneustes fossilis*; Joy *et al.*, 1998) and goldfish (Sloley *et al.*, 1991) indicating that pituitary cells are exposed to 5-HT. Apparently, 5-HT from hypothalamic origin can act directly at the level of pituitary cells in bony fishes to inhibit GH secretion. However, 5-HT immunoreactivity has also been detected in gonadotropes and pars intermedia cells in platyfish (Margolis-Nunno *et al.*, 1986), implying that a local pituitary source of 5-HT is also possible in some species. At present, not much is known about the intracellular signaling system mediating 5-HT-induced inhibition of GH secretion but preliminary evidence suggests that 5-HT acts downstream of PKC but not distal to increases in $[Ca^{2+}]_i$ (Yu *et al.*, 2008; Table 4.1).

γ-Aminobutyric acid (GABA). In goldfish, GABA inhibits GH release when applied *in vivo* but had no effects on pituitary GH release *in vitro* (Trudeau *et al.*, 2000a). Since GABA also inhibits DA turnover in the brain of fish species, it raises the possibility that GABA may act indirectly via suppression of hypothalamic dopaminergic input to goldfish somatotropes (Trudeau *et al.*, 2000b; see section on dopamine below), which can reduce D1 receptor-mediated DA stimulation of GH release (Wong *et al.*, 1993).

4.2.2. Stimulators

Growth hormone-releasing hormone (GHRH). GHRH is a member of the glucagon superfamily of peptides that also includes PACAP. Immunohistochemical studies suggest that GHRH fibers originating from the preoptic area innervate the pars distalis of fish (see Chapter 1). Synthetic common carp GHRH stimulates GH release *in vivo* and *in vitro* in rainbow trout, tilapia and goldfish (Peng and Peter, 1997). The ability of GHRH to increase GH release from pituitary fragment and cell preparations indicates that GHRH directly acts on pituitary somatotropes. However, the efficacy of GHRH in stimulating GH release is relatively low and its ability to enhance GH release is not consistently demonstrated. In goldfish, GHRH is only effective at very restricted periods of the seasonal reproductive cycle (Chang and Habibi, 2002). Thus it is believed that, unlike mammals, GHRH may not be the dominant neuroendocrine stimulator of GH release in fish (for review, see Montero *et al.*, 2000).

It had been thought that GHRH and PACAP were encoded on the same gene in fish; however, more recent analyses indicate that GHRH and PACAP are each encoded by a separate gene although there is a GHRH-like peptide (also called PACAP-related peptide) on the PACAP gene (Tam *et al.*, 2007). Goldfish GHRH activates cAMP production in Chinese hamster ovarian cells expressing goldfish GHRH receptors and stimulates GH release from goldfish pituitary cells in a dose-dependent manner (Lee *et al.*, 2007, Table 4.2). These results are consistent with the mammalian model where GHRH stimulates GH release via a cAMP-dependent pathway. These recent findings raise the question whether previous observations with GHRH actually represent effects with GHRH-like peptides rather than authentic GHRH.

Gonadotropin-releasing hormone (GnRH). Multiple GnRHs have been identified in fish, and fish GnRH forms are represented in all four clades of the classification scheme proposed by Silver *et al.* (2004). In general, either two or three forms are present in each fish species and the one form of preoptic GnRH (either GnRH1 or GnRH3) is delivered to the pituitary (Kah *et al.*, 2007; see also Cerdà-Reverter and Canosa, Chapter 1 of this volume). GnRH2 (cGnRH-II) is thought to be exclusively a mid-brain GnRH form derived from circumventricular ependymal cells posterior to the neural plate; however, an anterior olfactory area derived GnRH2 neuronal population is present in the preoptic area of zebrafish (Palevitch *et al.*, 2007).

Table 4.2

Summary of the actions for GH-releasing factors on intracellular signal transduction in fish somatotropes leading to GH release and GH gene expression

Signaling components	sGnRH	cGnRH-II	DA	PACAP	GHRH	ghrelin	GH	LH	E2
Major Pathways									
AC/cAMP/PKA	x	x	↑	↑*	↑ ?			↑*	
PLC				↑		↑ ?			
VSCC	↑	↑	↑	↑*		↑ ?			
PKC	↑, x*	↑, x*	x	↑/ x					↑
CAM	↑	↑	↑	↑*					
AA	x	x	↑						
NO/cGMP	↑	↑	↑	x					
PI3K							↑*		
JAK_2							↑*		
ERK/MAPK	↑*	↑*				↑ ?	↑*		
Na^+/K^+ antiport	↑	↑	x						
Intracellular Ca^{2+} Components									
IP3-sensitive	↑	x		↲					
TMB8-sensitive	↑	↑	↑	↑					
Thapsigargin-sensitive	x	x	↲	↲					
BHQ-sensitive	x	x	↑	↑					
CPA-sensitive	x	x							
Caffeine-sensitive	↑	↑	x	↑					
Ryanodine-sensitive	x	↑	x	x					
Dantrolene-sensitive	x	x		x					
cADP ribose				x					
mitochondria	↲	x		x					

Findings were based on the reports in goldfish and grass carp pituitary cells. Asterisk (*) denotes the involvement of the respective signaling pathways in GH gene expression. (Keys: ↑, increase/ activate; ↓, decrease/inhibit; **x**, not involved; ↲, indirect modulation; **?**, findings based on preliminary data.) [Abbreviations: sGnRH, Salmon Gonadotropin-releasing Hormone; cGnRH-II, Chicken Gonadotropin-releasing hormone II; DA, Dopamine; PACAP, Pituitary Adenylate Cyclase-Activating Polypeptide; GHRH, Growth Hormone-Releasing Hormone; GH, Growth Hormone; LH, Luteinizing Hormone; E_2, Estradiol; AC, Adenyate Cyclase; cAMP, Cyclic AMP; PKA, Protein Kinase A; PLC, Phospholipase C; VSCC, Voltage-Sensitive Ca^{2+} Channels; PKC, Protein Kinase C; CAM, Calmodulin; AA, Arachidonic Acid; NO, Nitric Oxide; cGMP, Cyclic GMP; PI3K, Phosphatidylinositol-3-Kinase; JAK_2, Janus Kinase 2; MAPK, Mitogen-activated Protein Kinase; IP3, Inositol 1,4,5-Triphospate; TMB8, 8-(N,N-diethylamino)-octyl-3,4,5-Trimethoxybennzoate ; CPA, Cycloplanzonic Acid; cADP Ribose, Cyclic ADP Ribose]

This raises the possibility that GnRH2 also has a hypophysiotropic function in some fish species. GnRH stimulation of GH release has been demonstrated in goldfish, common carp, tilapia, rainbow trout and grass carp, but not in African catfish and European eel (Peng and Peter, 1997; Bosma *et al.*, 1997;

Rousseau *et al.*, 1999). The ability of GnRH to increase GH release from pituitary cell preparations and the presence of GnRH receptors on somatotropes in goldfish and pejerrey (*Odontesthes bonariensis*) demonstrate that GnRH acts directly on somatotropes (Peter and Chang, 1999; Stefano *et al.*, 1999). Seasonal differences in the ability of GnRH to increase GH release exist, with the GH responses to GnRH being greatest in sexually mature and lowest in sexually regressed goldfish and rainbow trout (Chang and Habibi, 2002). Part of the seasonal/reproductive stage-dependent variations in responsiveness to GnRH may be due to changes in GnRH receptor (GnRHR) expression (Jodo *et al.*, 2005). In goldfish, a GH surge occurs in conjunction with the preovulatory increase in GTH (Peter and Chang, 1999). As described in Section 3, GH plays an enhancing role in gonadal steroidogenesis. These observations, taken together with the ability of a potent GnRH antagonist to suppress serum GH levels in sexually mature goldfish (Peter and Chang, 1999), not only indicate that GnRH is a physiological regulator of GH secretion but also suggest that GnRH provides the neuroendocrine link between GH and GTH secretion when increased release of both hormones is essential (e.g., during oocyte maturation). In the case of the goldfish, two GnRH forms, sGnRH (GnRH3) and cGnRH-II (GnRH2) are delivered to and released at pituitary although the origin of cGnRH-II has not been firmly established. In goldfish, both sGnRH and cGnRH-II are effective and equipotent in stimulating GH secretion.

Numerous GnRHRs, all belonging to the class A family of G-protein coupled receptors, have been cloned in fish. Up to five GnRHRs have been discovered in some species (Guilgur *et al.*, 2006) with up to four GnRHR forms being present in the pituitary [e.g., masu salmon (*Oncorhynchus masu*) Jodo *et al.*, 2003; European sea bass (*Dicentrarchus labrax*) Moncaut *et al.*, 2005]. In goldfish where two GnRHR forms belonging to the same GnRHR subtype (type I) have been cloned, one form is best correlated to the control of GH based on *in situ* hybridization and pharmacological properties (Illing *et al.*, 1999), but the possibility that fish somatotropes possess more than one GnRHR type cannot be discounted. Information on the intracellular signaling cascades mediating GnRH stimulation of GH release is derived from studies on goldfish, common carp and tilapia, with the most detailed information from the goldfish. Elevation in $[Ca^{2+}]_i$ produced by mobilization of Ca^{2+} from intracellular stores and entry of extracellular Ca^{2+} through voltage-sensitive Ca^{2+} channels (VSCC), activation of calmodulin (CAM) kinase, and activation of PKC are the common signaling pathways shown to be involved (Figure 4.2). In addition, GnRH induction of GH release in goldfish involves nitric oxide synthase/NO/cGMP and Na^+/H^+ antiport downstream of PKC. Whether these elements participate in GnRH signaling in somatotropes of other fish species is unknown. In addition, sGnRH and

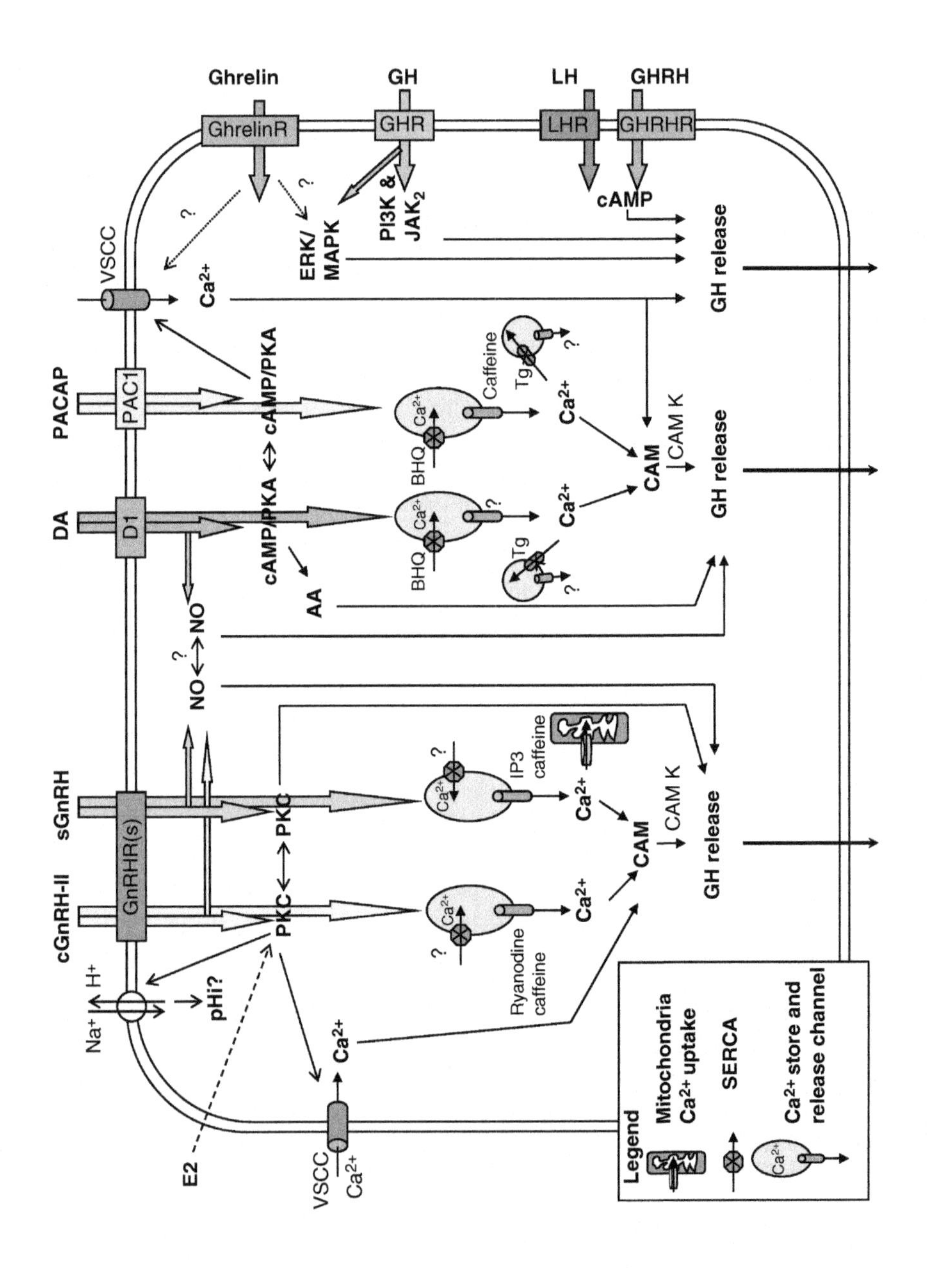

Ghrelin
GH
LH
GHRH
GhrelinR
GHR
LHR
GHRHR
cAMP
PI3K &
JAK2
ERK/
MAPK
VSCC
Ca2+
GH release
PACAP
PAC1
DA
D1
cAMP/PKA
cAMP/PKA ⟺ cAMP/PKA
BHQ
Caffeine
Tg
CAM
CAM K
AA
NO ⟷ NO
sGnRH
cGnRH-II
GnRHR(s)
PKC ⟷ PKC
IP3
caffeine
Ryanodine
caffeine
Na+
H+
pHi?
E2
Legend
Mitochondria
Ca2+ uptake
SERCA
Ca2+ store and
release channel

cGnRH-II differ in their use of pharmacologically distinct intracellular Ca^{2+} stores. Despite sharing some common Ca^{2+} stores, sGnRH alone utilizes an IP3-sensitive pool and a mitochondrial Ca^{2+} pool, whereas only cGnRH-II uses a ryanodine-sensitive but caffeine-insensitive Ca^{2+} store (for review, see Chang *et al.*, 2000; Table 4.2).

GnRH also elevates pituitary GH mRNA levels in grouper, masu salmon, goldfish, grass carp and sockeye salmon (*Oncorhynchus nerka*), but not in tilapia (Melamed *et al.*, 1998; Taniyama *et al.*, 2000; Li *et al.*, 2002; Bhandari *et al.*, 2003; Ran *et al.*, 2004; Klausen *et al.*, 2005). In masu salmon, the ability of GnRH to increase pituitary GH mRNA is seasonal, being only effective in March, a time prior to initiation of gonadal maturation, and GnRH stimulation of Pit-1 gene expression leads to increases in GH gene expression (Onuma *et al.*, 2005). Based on the studies in goldfish, activation of extracellular signal-regulated kinase (ERK), rather than PKC mediates GnRH effects on GH gene expression (Klausen *et al.*, 2005). Both increases and decreases in GH mRNA levels have been observed in response to treatments that increase $[Ca^{2+}]_i$; perhaps the source of Ca^{2+} and the mechanism by which Ca^{2+} is mobilized is critical (Johnson *et al.*, 2002).

Dopamine (DA). The pars distalis of fish, including goldfish and rainbow trout, is also innervated by dopaminergic fibers (see Chapter 1). DA stimulates GH release *in vivo* and/or *in vitro* in goldfish, tilapia, grass carp, rainbow trout, African catfish and grouper (for review, see Canosa *et al.*, 2007). Results from pharmacological, radioligand binding and fluorescence ligand

---

Fig. 4.2. Summary of post-receptor signaling cascades mediating the direct actions of salmon gonadotropin-releasing hormone (sGnRH), chicken gonadotropin-releasing hormone II (cGnRH- II), dopamine (DA), pituitary adenylate cyclase activating polypeptide (PACAP), ghrelin, growth hormone (GH), luteinizing hormone (LH) and GH-releasing hormone (GHRH) on fish somatotropes. The model presented was mainly based on the findings reported in goldfish and grass carp. At the somatotrope level, sGnRH and cGnRH-II stimulate GH release through PKC cascades coupled to Ca^{2+} entry via voltage-sensitive Ca^{2+} channels (VSCC) and mobilization of ryanodine- and IP3-sensitive intracellular Ca^{2+} ($[Ca^{2+}]_i$) stores. The subsequent rise in $[Ca^{2+}]_i$ can trigger GH exocytosis via activation of calmodulin (CAM) and CaM kinases (CAM K). The stimulatory effects of the two GnRH via PKC are under the modulation of estradiol (E2) and may also involve Na^+/H^+ exchange at the membrane level as well as activation of nitric oxide (NO)-dependent mechanisms. Unlike GnRH, the stimulatory effects of DA and PACAP are mediated by cAMP/PKA cascades coupling to Ca^{2+} entry via VSCC and $[Ca^{2+}]_i$ mobilization from stores that are sensitive to butylhydroquinone [BHQ, a sarcoplasmic endoplasmic reticulum Ca^{2+} ATPase (SERCA) inhibitor] and caffeine manipulation. In addition, DA and PACAP actions are also modulation by thapsigargin (Tg)-sensitive SERCA. Although cAMP-dependent mechanisms are involved in GH release triggered by LH and GHRH stimulation, the ERK/MAPK and JAK_2/ PI3K cascades represent the major pathways mediating GH autoregulation via activation of GH receptors (GHR). At present, the signaling mechanism responsible for ghrelin induction of GH release is unclear but may involve Ca^{2+} entry via VSCC and parallel activation of ERK/ MAPK pathways. Modified and updated from Canosa *et al.*, 2007. (See Color Insert.)

binding imaging studies indicate that DA acts via DA D1 receptors on somatotropes to stimulate GH secretion (Chang *et al.*, 2000). In support for a physiological role for DA in the regulation of GH release, injection of a DA D1 antagonist reduced serum GH levels in goldfish (Wong *et al.*, 1993); furthermore, DA has been detected electrochemically in pituitary extracts of goldfish and Indian catfish (Sloley *et al.*, 1991; Joy *et al.*, 1998). Interestingly, rainbow trout receiving GH implants for 7 days have increased levels of the DA metabolite dihydroxyphenylacetic acid (DOPAC) in the hypothalamus (Johansson *et al.*, 2004). Given that a rise in DOPAC in general can be taken as an index for activation of dopaminergic neurons, these results raise the possibility that the dopaminergic neuronal system may serve as a target for GH feedback regulation.

The ability of DA to stimulate GH secretion in goldfish can be detected at all times of the annual seasonal reproductive cycle; however, the efficacy is highest at times of year when the gonads are regressed and lowest at times of pre-spawning (Chang and Habibi, 2002). Based almost entirely on information on cyprinids, DA stimulation of GH release is mediated mainly through action of the adenylate cyclase (AC)/cAMP/PKA signaling pathway and subsequent stimulation of extracellular Ca^{2+} entry through VSCC and activation of CAM kinase (Figure 4.2). Activation of AA mobilization and metabolism through the lipoxygenase pathway is another component of DA signaling downstream of cAMP. Recent results in goldfish indicate that DA also utilizes NOS/NO signaling and intracellular Ca^{2+} stores to stimulate GH release (Wong *et al.*, 2001; Chang *et al.*, 2003; Mitchell *et al.*, 2008); however, DA- and GnRH-sensitive Ca^{2+} stores have distinct pharmacological properties suggesting that these two groups of neuroendocrine regulators utilize dissimilar intracellular Ca^{2+} stores (Chang *et al.*, 2003; Mitchell *et al.*, 2008; Table 4.2).

DA also increases GH mRNA levels in pituitaries of grouper, rainbow trout and tilapia (Melamed *et al.*, 1998; Ran *et al.*, 2004). The involvement of cAMP/PKA in mediating DA effects on GH synthesis has been directly observed in tilapia (Melamed *et al.*, 1998). Whether this is true for other fish species is not currently known but cAMP-response elements are involved in the regulation of chinook salmon (*Oncorhynchus tshawytscha*) GH gene promoter activity (Wong *et al.*, 1996).

Pituitary adenylate cyclase activating polypeptide (PACAP). PACAP fibers innervate the pars distalis of goldfish, grass carp, European eel and stargazer (*Uranoscopus japonicus*) (for review, see Wong *et al.*, 2000 and 2005; Cerdà-Reverter and Canosa, Chapter 1 of this volume). PACAP has been isolated and/or cloned in a number of teleosts, including grass carp, zebrafish, stargazer, grouper, lake whitefish (*Coregonus clupeaformis*), Arctic grayling (*Thymallus arcticus*), yellowtail flounder (*Pleuronectes ferrugineus*), Atlantic halibut, channel catfish, stingray (*Dasyatis akajei*) and the African catfish, as well as in a chondrostean, the white sturgeon (*Ascipenser transmontanus*) (McRory *et al.*, 1995; Matsuda *et al.*, 1998; Wei *et al.*, 1998; Small and Nonneman, 2001; Jiang *et al.*, 2003; Wang *et al.*, 2003; Matsuda *et al.*, 2005; Sze *et al.*, 2007). The three-dimensional molecular structure of grass carp PACAP has been recently established by nuclear magnetic resonance (NMR) techniques and found to be highly comparable to that of mammals (Sze *et al.*, 2007). While a single copy gene encodes for PACAP in some fish species (e.g., grass carp), two copies are present in others (e.g., zebrafish). Often, two PACAPs (~38 and 27 amino acid in length) are found with high homologies to the equivalent peptides in tetrapods. PACAP stimulates GH release in goldfish, salmon, European eel, turbot, common carp and grass carp, and fish PACAP and mammalian PACAP38 are usually effective where tested. Where compared, PACAP more consistently and effectively stimulates GH release than GHRH, suggesting that PACAP is more important than GHRH in the regulation of GH secretion in fish. In goldfish, PACAP stimulation of GH release is noted at all times of the seasonal reproductive cycle but seasonal variation in the efficacy exists. Maximal effectiveness in PACAP action occurs at times when the gonads are matured during the pre-spawning period (Chang, J.P., unpublished data; Figure 4.1). The above findings suggest that PACAP is a physiological regulator of GH release in fish, but loss of function experiments would help to strengthen this conclusion.

Information on the signal transduction pathway mediating PACAP stimulation of GH release is largely derived from grass carp (Wong *et al.*, 2005) and goldfish (Wong *et al.*, 2000; Mitchell *et al.*, 2008; Sawisky and Chang, 2005; Table 4.2). PACAP binding to type I PACAP (PAC-I) receptors activates AC/cAMP/PKA, increases $[Ca^{2+}]_i$ by enhancing Ca^{2+} entry through VSCC and Ca^{2+} release from intracellular stores, and activates Ca^{2+}/CAM kinase (Figure 4.2). The Ca^{2+} stores involved have different pharmacological properties from those utilized by DA and GnRH. In addition, unlike DA, PACAP does not use NOS/NO signaling. Although the involvement of phospholipase C (PLC) in PACAP's action has been demonstrated, whether this leads to activation of the traditional IP3/Ca^{2+} and PKC signaling or novel pathways is controversial. In one study in goldfish, the involvement of PKC in GH mRNA expression is refuted (Klausen *et al.*, 2005) whereas this component is shown to be involved in another study (Wong *et al.*, 2007). Splice variants of PAC1 receptors exist in zebrafish and goldfish (Fradinger *et al.*, 2005; Kwok *et al.*, 2006); this may explain the observed differences in PACAP signaling. PACAP also increases GH mRNA level in grass carp pituitary via activation of cAMP/PKA, VSCC and CAM kinase (Wong *et al.*, 2005).

Neuropeptide Y (NPY). That NPY is a GH-releasing factor was first demonstrated in goldfish. The presence of NPY in the preoptic hypothalamus and NPY-immunoreactive fibers in the vicinity of somatotropes suggest a role in regulating GH release. Results with human and goldfish NPY on pituitary cells and fragments preparations show that NPY acts through Y2 receptors to stimulate GH release via direct actions on pituitary cells and/or indirect actions through GnRH release from nerve terminals in the pituitary. The ability of NPY to stimulate GH release is seasonal, being most effective on pituitaries obtained from sexually mature/pre-spawning goldfish (Chang and Habibi, 2002; Figure 4.1). In Indian catfish (*Clarias batrachus*), intracranial injection of NPY decreases pituitary GH immunoreactivity via activation of Y1 receptors, and these findings were interpreted as a result of cellular depletion of GH stores in the pituitary caused by increased GH exocytosis (Mazumdar *et al.*, 2006). Whether this Y1 effect is exerted directly at the level of the pituitary cells is not known. The receptor signaling mechanisms mediating the direct and indirect actions of NPY on GH release and whether NPY affects GH synthesis need further investigation.

Thyroid-releasing hormone (TRH). The presence of TRH transcript and/or immunoreactivity in the hypophysiotropic hypothalamus of fishes suggests that this peptide also regulates pituitary cell functions in fish (see Chapter 1). TRH stimulates GH release in goldfish *in vivo* and from goldfish pituitary fragments *in vitro* (Trudeau *et al.*, 1992). The magnitude of the GH release response *in vitro* is greater from pituitaries from pre-spawning than from regressed goldfish, suggesting the presence of seasonal reproductive influences on TRH action (Chang and Habibi, 2002). However, in tilapia, TRH elevates serum GH levels but has no effects on GH release from pituitary cells, suggesting that the GH-releasing effects of TRH are indirect (Melamed *et al.*, 1998). That the GH-releasing effect of TRH is not through direct actions on somatotropes may explain the negative findings in European eel and turbot pituitary cells (Rousseau *et al.*, 2001). On the other hand, application of TRH stimulates *de novo* synthesis of GH protein in pituitary cells of common carp suggesting that TRH acts at the level of pituitary cells to stimulate GH production (Kagabu *et al.*, 1998).

Corticotropin-releasing factor (CRF). CRF is another well-known hypophysiotropic factor in fish (see Chapter 1). CRF stimulates GH release from European eel pituitary cell cultures, suggesting a direct action at the level of pituitary cells (Rosseau *et al.*, 1999). Whether CRF has similar GH-releasing action in other fish species is still unknown.

Cholecystokinin (CCK). CCK is an anorexigenic factor (Chapter 9) and CCK fibers are found in the forebrain and pars distalis of fish species. Besides having GH-releasing activity when applied intraperitoneally or intracranially, sulfated CCK-8 stimulates GH release from goldfish pituitary fragments, suggesting a direct action at the level of the pituitary (Himick *et al.*, 1993). Whether CCK acts directly on somatotropes is not known, but CCK decreases PSS-I mRNA in the forebrain of goldfish (Canosa and Peter, 2004), indicating that CCK can increase GH release indirectly by altering SS neuronal activity. The GH-releasing effect of CCK changes with the reproductive cycle of the fish, being most effective on pituitaries obtained from sexually regressed fish (Figure 4.1). Because of the high expression level in the gut, peripheral CCK may be an important source of the peptide in the neuroendocrine regulation of GH release.

Gastrin-releasing peptide (GRP). GRP is a member of the bombesin (BBS) family of peptides and has been isolated from a number of fish species (Volkoff *et al.*, 2005). GRP is an effective anorexigenic peptide (see Chapter 9). In goldfish, BBS stimulates GH release from pituitary fragments *in vitro* (Himick *et al.*, 1993). In addition, BBS/GRP binding sites have been detected in the pars distalis and GRP-immunoreactive fibers have been detected at the interface of the pars intermedia and pars distalis, as well as in the forebrain. GRP mRNA is also found in the pituitary (Himick *et al.*, 1995; Himick and Peter, 1995). These observations suggest that GRP acts as a hypothalamic factor to regulate GH release by direct action at the level of pituitary cells in goldfish. In addition, BBS treatment decreases forebrain PSS-I and PSS-II mRNA levels indicating that down-regulation of SS expression and neuronal activity is part of the mechanism by which BSS stimulates GH release (Canosa and Peter, 2004). However, GRP mRNA is also widely expressed in peripheral tissues including the skin and gut; as a consequence, the possibility that this neuroendocrine factor also serves as peripheral signal carried by the blood cannot be excluded. Whether GRP has similar GH-releasing activity in other fish species as in goldfish is unknown.

Galanin. Galanin has been identified in cyclostomes (e.g., lamprey, *Lampetra fluviatilis*), chondrychthians (e.g., dogfish, *Scyliorhinus canicula*) and teleosts [e.g., rainbow trout, bowfin (*Amia calva*) and goldfish]; furthermore, galanin fibers and/or mRNA are present in the hypothalamus and pituitary of fish species including salmon and goldfish (see Chapters 1 and 9). *In vivo* galanin administration stimulates GH release in coho salmon (*Oncorhynchus kisutch*; Diez *et al.*, 1992). These observations suggest that hypothalamic galanin in fish may act as a GH-releasing factor through direct stimulation of pituitary GH release. However, direct action of galanin at the level of fish somatotropes has yet to be confirmed.

4.2.3. Others

Other "non-classical" neuroendocrine regulators of GH release have been identified but their roles are controversial and not well understood.

Glutamate. Injection of the glutamate agonist N-methyl-D-aspartate (NMDA) reduces serum GH in goldfish and this response is enhanced by estradiol (E2) treatment (Trudeau *et al.*, 1996). In contrast, NMDA stimulates GH secretion in steroid-primed immature rainbow trout

(Holloway and Leatherland, 1997). The role of glutamate in regulating GH release in fish is still controversial.

RFamides. Recently, RFamide peptides having the general LPXRFamide motif were identified in the brains of tetrapods and fishes, including the cyclostomes (e.g., sea lamprey, *Petromyzon marinus*) and teleosts [e.g., salmon, goldfish and mudskipper (*Periophthalmus modestus*)]. Results from immunohistochemistry and mRNA studies indicate that hypothalamic RFamide peptides innervate the pars distalis and pars intermedia, suggesting that these peptides may have hypophysiotropic functions in teleosts (Osugi *et al.*, 2006; Cerdà-Reverter and Canosa, Chapter 1 of this volume). Three goldfish RFamides stimulated GH release from sockeye salmon pituitary cell cultures (Amano *et al.*, 2006), indicating that these peptides can act as GH-releasing peptides. On the other hand, two synthetic sea lamprey RFamides inhibited GH mRNA expression in lamprey *in vitro* (Moriyama *et al.*, 2007). These results suggest that RFamides have direct effects on somatotropes but their roles in regulating somatotrope functions still need to be further clarified.

Natriuretic peptides. The gene for C-type natriuretic peptide (CNP) is expressed in the tilapia brain and pituitary. Tilapia CNP, atrial natriuretic peptide (ANP) and B-type natriuretic peptide (BNP) stimulate GH mRNA expression, while ANP and BNP increase GH release in studies with tilapia pituitary cell culture. Interestingly, ANP and BNP, but not CNP, increase intracellular cGMP levels. These results suggest at least ANP and BNP regulate somatotrope functions in teleosts via cGMP. Whether brain and/or peripheral tissues are the source of natriuretic peptides that affect somatotrope functions is unknown since the level of brain ANP and BNP is very low in tilapia (Fox *et al.*, 2007a).

4.3. Signals from Peripheral Organs/Tissues

Regulators released from peripheral organs/tissues control GH release in fish. While some of these fit into classical "feedback" regulation modes, others may not.

4.3.1. Inhibitors

Insulin-like growth factor (IGF). IGF participates in a classical "long-loop negative feedback" regulation of GH secretion (Wong *et al.*, 2006). The molecular structures of IGFs and the IGF-I receptors in fish are highly conserved. Both IGF-I and IGF-II are expressed in teleosts and elasmobranches, whereas only IGF-I appears to be expressed in agnathans. GH increases liver IGF-1 gene expression in several fish species including common carp, sea bream, salmon, goldfish, rainbow trout and tilapia

(Wong *et al.*, 2006; Chen *et al.*, 2007). IGF-I receptors are present in pituitaries of fishes (Fruchtman *et al.*, 2002; Filby and Tyler, 2007) and IGF-I and/or IGF-II decrease GH release and GH mRNA levels in pituitary cells of rainbow trout (Weil *et al.*, 1999) and grass carp (Huo *et al.*, 2005). These data indicate that circulating IGFs exert negative feedback regulation on fish somatotrope functions as in mammals. IGF-I effects on GH release are associated with activation of phosphatidylinositol 3-kinase (PI3K) and mitogen-activated protein kinases (MAPK) (Fruchtman *et al.*, 2001), whereas IGF-I and IGF-II inhibition of GH synthesis is mediated by CAM/ calcineurin mechanisms and up-regulation of CAM gene expression (Huo *et al.*, 2005; Table 4.1). IGF-I receptors are also expressed in the brain of fish including brown trout (*Salmo trutta*), fathead minnow (*Pimephales promelas*) and grouper (Smith *et al.*, 2005; Kuang *et al.*, 2005; Filby and Tyler, 2007), in addition, IGF-I-expressing neurons are found in the brain of juvenile tilapia (Shved *et al.*, 2007). Whether IGF also acts at the level of the release of hypothalamic regulators to affect somatotrope functions in fish is unknown. Interestingly, IGF-I mRNA is expressed in the pituitary of fathead minnow (Filby and Tyler, 2007) raising the possibility that IGF participates in local autocrine/paracrine regulation of somatotrope functions in fish (see Section 4.4 below).

Noradrenaline (NA). NA decreases serum GH levels in goldfish and directly reduces basal GH secretion from primary cultures of goldfish and grass carp pituitary cells. NA also inhibits the GH responses to GnRH, DA and PACAP at the pituitary level through alpha 2 adrenergic receptors. NA inhibitory action is exerted at levels downstream of PKC activation, mobilization of Ca^{2+} and AA, and cAMP generation. In addition, NA blocks the ability of PACAP to elevate cAMP and $[Ca^{2+}]_i$, suggesting that NA also acts at the level of second messenger generation (Table 4.1). In grass carp, NA alpha 2 inhibition can reduce GH mRNA, GH primary transcripts and GH promoter activity, suggesting that NA also blocks GH synthesis by down-regulation of GH gene transcription. Interestingly, although NA suppression of basal GH release is not associated with a decrease in $[Ca^{2+}]_i$, the rebound increase in GH release following removal of NA is correlated with an increase in $[Ca^{2+}]_i$. A greater rebound increase in GH release is observed following NA inhibition of GnRH and PACAP stimulation of GH secretion and is associated with a higher $[Ca^{2+}]_i$ relative to basal situation (Wang *et al.*, 2007; Wong *et al.*, 2007). These results raise the possibility that NA blocks the ability of Ca^{2+} mobilizing intracellular signaling molecules to increase $[Ca^{2+}]_i$ such that when NA inhibition is removed, a sudden rapid "overshoot" in $[Ca^{2+}]_i$ occurs, the magnitude of which is dependent on the stimulatory intensity provided by the neuroendocrine regulator. Since the amount of NA detected in pituitary of fish species (e.g., goldfish and rainbow trout) is relatively low, the importance of NA as a hypothalamic factor acting on pituitary somatotropes has been questioned. On the other hand, the amount of circulating NA can be significantly increased in teleosts under stress and NA from systemic circulation may reach the pituitary gland and affect GH secretion under these conditions. The relative importance of brain and peripheral NA in the regulation of GH release needs to be examined.

4.3.2. Stimulators

Ghrelin. The structure and function of ghrelin in fishes and other vertebrates have been extensively reviewed recently (Unniappan and Peter, 2005; Kaiya *et al.*, 2008). Briefly, ghrelin has been cloned from several actinopterygian fish species, and two species of sharks. Fish ghrelin, like mammalian ghrelin, is n-acylated. Although n-octanylation (C-8) is believed to be the prevalent modification, both a C-8 and C-10 (decanylated) form have been purified from tilapia (Fox *et al.*, 2007b). In addition, two ghrelin receptor (GH secretagogue receptor, GHS-R) splice variants are present in some fish species [e.g., black sea bream (*Acanthopagrus schlegeli*), Chan and Cheng, 2004]. Ghrelin is an orexigenic factor and is produced by the digestive tract and released into the blood, forming a link between food intake and GH release. The increase in circulating ghrelin and GH during long-term food deprivation in rainbow trout is one of the observations supporting such a hypothesis (Jönsson *et al.*, 2007; Volkoff *et al.*, Chapter 9 of this volume). Stomach is believed to be the major source of ghrelin, and ghrelin has been measured in the blood of fish. Whether the hypothalamus is another source of ghrelin that affects the pituitary is controversial as ghrelin mRNA is present in the hypothalamus of rainbow trout (Kaiya *et al.*, 2008) and goldfish (Unniappan and Peter, 2005) but not in other species (e.g., sea bream, Yeung *et al.*, 2006). However, the interaction of ghrelin with other hypothalamic factors such as GRP/BBS is likely a part of the hypothalamic network regulating pre- and post-prandial changes in serum GH levels (see Section 5 below).

Ghrelin stimulates GH release from cultured pituitary cells of several fish species and co-application of GHS-R antagonists blocks such responses (Unniappan and Peter, 2005; Kaiya *et al.*, 2008). These data indicate that ghrelin acts directly on somatotropes to stimulate GH release via activation of GHS-Rs. Ghrelin also increases GH mRNA levels in most, but not all, fish species examined (for review, see Unniappan and Peter, 2005). Stimulation of sea bream GHS-R1a expressed in HEK293 cells by GRP6 and L163,540, artificial agonists for GHS-R, demonstrates that stimulation of PLC, increases in $[Ca^{2+}]_i$, and activation of VSCC and ERK1/2 are possible mechanisms linked to ghrelin action in fish (Chan *et al.*, 2004b). Consistent with the involvement of Ca^{2+} signaling pathway in ghrelin action on GH release in fish, goldfish ghrelin increases $[Ca^{2+}]_i$ in identified goldfish somatotropes (Grey and Chang, 2009; Table 4.2).

Thyroid hormones. As discussed in the section on thyrotropin-releasing hormone in Section 4.2.2, TRH induction of GH release *in vivo* cannot always be attributed to direct actions of TRH at the level of the pituitary. Thyroid hormone receptors are present in the pituitary of fish, e.g., common carp and fathead minnow, suggesting that thyroid hormones directly affect somatotrope functions. In support of this idea, T_3/T_4 treatments induce GH release *in vitro* in rainbow trout and GH gene expression in tilapia and rainbow trout pituitary. In addition, GH increases extrathyroidal T_3 production by activating 5′-monodeiodinase in peripheral tissues. In contrast, T_4 and T_3 treatments *in vitro* and *in vivo* inhibit GH release and synthesis in the eel pituitary, suggesting that species differences exist in terms of how thyroid hormone affects somatotropes. Taken together, these results suggest that thyroid hormone participates in a regulatory feedback loop (negative or positive depending on the species) on pituitary GH secretion and production. The existence of a complex TRH/GH/thyroid axis interaction is further supported by the observations that TSH can also directly stimulate GH release in tilapia and that T_3 increases GHRH-induced GH release in rainbow trout (Melamed *et al.*, 1998; Rousseau *et al.*, 2002; Wong *et al.*, 2006; Filby and Tyler, 2007).

Cortisol. Like GH, cortisol is involved in the regulation of energy metabolism, gill ion transport activity, osmoregulation, immune function and stress. Perhaps reflecting such a close relationship between the two hormones, cortisol also affects GH secretion and *vice versa*. Cortisol has direct action at the level of the pituitary cells, and glucocorticoid receptors are also expressed in the pituitary of fish models, e.g., fathead minnow (Filby and Tyler, 2007). Cortisol increases GH release from tilapia pituitary organ cultures (Uchida *et al.*, 2004). In rainbow trout pituitary cells, dexamethasone increases basal as well as GHRH-induced GH secretion (Luo and McKeown, 1991). In addition, cortisol can also affect GH release indirectly by limiting IGF-I feedback activity on somatotropes. Dexamethasone treatment decreases GH-induced increase in IGF-1 mRNA in salmon hepatocyte cultures (Pierce *et al.*, 2005) and cortisol injection decreases hepatic IGF-1 mRNA but elevates circulating IGF-binding protein levels in tilapia (Kajimura *et al.*, 2003). Feeding with cortisol supplemented diet also increases GH mRNA levels in the pituitary of channel catfish (Peterson and Small, 2005) although the site of action for cortisol in GH gene expression is still unclear. Conversely, GH treatment increases circulating cortisol levels in rainbow trout (Biga *et al.*, 2004) and glucocorticoid receptor affinity and capacity in the gill of Atlantic salmon (Shrimpton and McCormick, 1998).

These data suggest that while GH increases cortisol release and activity, cortisol can exert positive feedback on GH release and synthesis via direct actions at the pituitary level or indirect actions by reducing the intensity of IGF-I negative feedback.

Sex steroids. Changes in serum GH during gonadal maturation and seasonal reproduction suggest that sex steroids modulate GH secretion (Makino *et al.*, 2007). Supporting this idea, *in vivo* estradiol (E2) treatments elevate serum GH in tilapia, rainbow trout and goldfish. *In vivo* treatment with testosterone (T) also increases serum GH level in goldfish and this effect requires aromatization of T to E2 (for review, see Canosa *et al.*, 2007). Interestingly, pituitary aromatase activity has been associated with somatotropes in the sculpin (*Myoxocephalus*; Olivereau and Callard, 1985) and gonadotropes in tilapia (Melamed *et al.*, 1999), indicating that local conversion of T to E2 is a distinct possibility. However, similar treatments with T and 11-ketotestosterone (11KT) have no effect on serum GH levels in coho salmon, suggesting that species differences in the response to androgens exist (Larsen *et al.*, 2004). Although *in vitro* exposure to E2 did not alter GH release in goldfish pituitary cells, E2 and T treatments elevated basal GH secretion from carp pituitary cells (Melamed *et al.*, 1998), indicating that the actions of sex steroids on serum GH levels can be exerted directly on basal GH release, at least in some species). Sex steroids may also act indirectly through modulation of the production of other peripheral factors and/or their effectiveness on GH secretion. *In vitro* treatment with E2 decreases liver IGF-I and IGF-II mRNA levels in sea bream (Carnevali *et al.*, 2005) and exposure to ethinylestradiol (EE2) reduces hepatic GHR and IGF-I expression in fathead minnow (Filby and Tyler, 2007). On the other hand, T and 11KT increase serum IGF-binding protein and IGF-I levels in coho salmon (Larsen *et al.*, 2004). In addition, E2 lowers plasma SS-14 levels and pituitary responsiveness to SS-14 in rainbow trout (Holloway *et al.*, 1997) and downregulates SST_2 mRNA levels in goldfish pituitary (Cardenas *et al.*, 2003). Since GH is known to increase gonadal steroidogenesis and serum E and/or T levels in a number of fish species (see Section 3 above), a positive feedback regulatory loop involving sex steroids can be proposed. In this model, GH-elicited elevation in circulating sex steroid levels decreases GH-induced liver IGF production in some species and elevates IGF-binding protein in others. This decreases IGF feedback inhibition on GH release. In addition, sex steroids lower pituitary SST expression, reducing SS inhibition on GH secretion. The cumulative result is an increase in GH secretion.

Sex steroids affect GH synthesis in some fish but the effects are controversial even within a single species. Sex steroids (E2, androgens or synthetic estrogenic agonists) have no effects on GH mRNA levels in European eel, adult tilapia, rainbow trout, common carp, juvenile Atlantic salmon and masu salmon (Yadetie and Male, 2002; Onuma *et al.*, 2005; Wong *et al.*, 2006). In contrast, *in vivo* treatments with E2 increase pituitary GH protein content in goldfish (Zou *et al.*, 1997), and *in vitro* exposure to E2 and a xenoestrogen elevate GH mRNA levels in rainbow trout pituitary culture (Elango *et al.*, 2006). These results suggest that sex steroids can stimulate GH synthesis. On the other hand, *in vivo* administration of E2 and T decreases GH mRNA levels in tilapia pituitaries in other studies (Melamed *et al.*, 1998; Shved *et al.*, 2007), suggesting that the effects of sex steroids on GH synthesis can vary among fish species. Life stage, gender, species differences, the neuroendocrine milieu and direct versus indirect effects probably contribute to the diversity in the observed results. Further to these peripheral actions, sex steroids can also act centrally to affect GH secretion in fish (see Section 5 below).

Leptin. Leptin has been cloned in fish models, including pufferfish (Kurokawa *et al.*, 2005; Yacobovitz *et al.*, 2008) and common carp (Huising *et al.*, 2006). Similar to mammals, leptin inhibits food intake in fish via suppression of NPY expression in the hypothalamus, e.g., in goldfish (Volkoff *et al.*, 2003) and trout (Murashita *et al.*, 2008). Since (i) icv injection of leptin can also elevate cholecystokinin (CCK) expression in the brain of goldfish (Volkoff *et al.*, 2003) and (ii) both CCK and NPY are known to stimulate GH release in the fish pituitary (Peter and Chang, 1999), it is conceivable that leptin may play a role in GH regulation via indirect actions within the CNS. In fish models, direct action of leptin at the pituitary level in regulating GH release and synthesis has not been reported. However, leptin is known to trigger prolactin (e.g., tilapia) and somatolactin secretion (e.g., sea bass) in fish pituitary cells (Peyon *et al.*, 2003; Tipsmark *et al.*, 2008), indicating that leptin can also act on cell targets other than somatotrophs. (For reviews on feeding control in fish, see Volkoff, 2006, and Chapter 9.)

4.4. Autocrine/Paracrine Signals within the Pituitary

4.4.1. Intrapituitary Feedback by Local Actions of LH and GH

A zonal distribution of endocrine cells in the pars distalis is commonly observed in teleosts and the close proximity between gonadotropes and somatotropes suggests that they interact via autocrine/paracrine mechanisms. The presence of an intrapituitary feedback loop for GH regulation has been recently demonstrated in the carp pituitary (Wong *et al.*, 2006). In this model, LH released from gonadotropes triggers GH secretion in somatotropes through cAMP-dependent mechanisms. This paracrine action of LH can be further enhanced by GH-induced GH release and GH gene expression via direct actions on somatotropes through JAK_2/MAPK and JAK_2/PI3K cascades (Figure 4.2, Table 4.2). Meanwhile, local elevation in

GH release exerts a negative feedback to suppress LH secretion in neighboring gonadotropes. The autoregulation by GH-induced GH release in the carp pituitary is different from the "ultra-short feedback" in mammals, in which GH suppresses GH secretion at the pituitary level. Of note, a similar inhibition on GH release by GH treatment has been reported in whole pituitary culture in rainbow trout (Agustsson and Bjornsson, 2000). The discrepancy observed suggests that species-specific variations may exist regarding the local actions of GH at the pituitary level. In carp pituitary cells, GH gene expression can be induced by GH-releasing factors, including GnRH, PACAP and dopamine, and these stimulatory effects can be blocked by removing endogenous GH secreted into culture medium by immunoneutralization (Wong *et al.*, 2006). Apparently, local release of GH not only can regulate GH secretion and synthesis at the pituitary level but also play a role in maintaining the sensitivity of somatotropes to stimulation by hypothalamic factors.

4.4.2. Activin/Follistatin System in the Pituitary

Similar to mammals, the activin/follistatin system participates in the regulation of reproductive functions in fish. Besides its involvement in steroidogenesis and oocyte maturation, activin also acts at the pituitary level in goldfish to modulate GTH secretion and gene expression (Ge, 2000). Activin subunits, activin receptors and the activin-binding protein follistatin are expressed in fish pituitary (e.g., goldfish) and this intrapituitary activin/follistatin system is functionally coupled to Smad2 and Smad3 activation (Yuen and Ge, 2004; Lau and Ge, 2005). In mammals, activin is primarily expressed in gonadotropes and activin consistently inhibits GH release; however, in fish models, activin is expressed in somatotropes, but not gonadotropes, and activin stimulates GH secretion by direct actions at the pituitary level (Ge and Peter, 1994). These findings indicate that activin acts as a local stimulator of fish somatotropes via autocrine/paracrine actions. In goldfish, activin modulates LH secretion and gene expression and these regulatory actions are blocked by follistatin (Yuen and Ge, 2004). Given that LH can act as a local stimulator for GH secretion and synthesis in the carp species (Section 4.4.1), it is possible that activin affects GH release indirectly through regulation of local LH production.

4.4.3. IGF-I Produced Locally in the Pituitary

In mammals, the autocrine/paracrine effects of IGF-I produced locally in target tissues are well recognized and form the basis of the revised model of somatomedin hypothesis (Kaplan and Cohen, 2007). In fish models, IGF-I transcript and immunoreactivity can be identified in the pituitary with signals consistently detected in corticotropes. In some species (e.g., tilapia), IGF-I transcripts are also detected in somatotropes and gonadotropes (Melamed *et al.*, 1999; Eppler *et al.*, 2007). Furthermore, IGF-I binding sites are ubiquitously expressed in the pituitary of fish species (Fruchtman *et al.*, 2002) and IGF-I produced locally at the pituitary level can exert anti-apoptotic effect to maintain the population size of somatotropes (Melamed *et al.*, 1999). These findings, as a whole, suggest that pituitary IGF-I may be involved in GH regulation in addition to the long-loop feedback by liver-derived IGF-I. In mammals, functional cross-talk between IGF-I and insulin via their respective receptors is well documented. Given that (i) insulin can stimulate GH gene expression in carp pituitary cells (Huo *et al.*, 2005) and (ii) insulin expression has been recently demonstrated in the pituitary of tilapia (Hrytsenko *et al.*, 2007), it raises the possibility that insulin may also serve as a local regulator for GH expression. The functional interactions between IGF-I and insulin may represent a new facet of GH regulation by autocrine/paracrine mechanisms and clearly warrant future investigations.

5. FUNCTIONAL INTERACTIONS OF GH REGULATORS AT THE HYPOTHALAMIC AND PITUITARY LEVELS

Many of the neuroendocrine regulators of GH release described above also interact at the level of the brain and pituitary. These interactions are complex because GH is involved in multiple physiological systems. For example, integration of metabolic regulation with feeding and satiation would provide one system whereby complex neuroendocrine regulation of GH release would be necessary (Chapter 9), while the integration of reproduction, growth, and energy metabolism and partition, likely involving GnRH and sex steroids, represents another. Information on the physical interactions between many of these neuroendocrine systems is reviewed in Chapters 1 and 2. This section will focus on selected topics, including (i) the ability of somatotropes to differentiate and integrate the diverse neuroendocrine signals, (ii) the central sites of sex steroids action in relationship to the control of somatotrope functions, and (iii) the relationship between ghrelin and other GH regulators in mediating the GH response to food deprivation and feeding.

5.1. Integration of Neuroendocrine Signals at the Level of Signal Transduction

Since multiple neuroendocrine factors directly affect hormone release and synthesis in fish somatotropes, a question that needs to be addressed is how these influences are integrated and differentiated at the level of the somatotropes. It is possible that receptors for the different regulatory factors are differentially expressed among somatotropes. However, Ca^{2+} imaging and fluorescent-receptor ligand studies in identified goldfish somatotropes indicate that receptors for DA, PACAP and GnRH exist on the same cell, indicating that receptors for multiple regulators are expressed simultaneously in a single cell at least in some somatotropes. In goldfish and common carp, the maximal GH responses to GnRH and DA, and PACAP and GnRH are additive, whereas those for GnRH and cGnRH-II, and PACAP and DA are not. Thus, in addition to receptor specificity, ligand- and function-selectivity must reside at the level of post-receptor intracellular signaling (Chang *et al.*, 2000; Wong *et al.*, 2006).

As described in Section 4.1, the suite of signaling pathways mediating the stimulatory actions of sGnRH, cGnRH-II, DA, PACAP and ghrelin are distinct, though partly overlapping (Table 4.2). Interestingly, the maximal GH responses to direct activation of PKC and PKA by pharmacological manipulations are additive in goldfish and carp models (Chang *et al.*, 2000) suggesting that two major cascades lead to GH release: (i) PKC and its downstream component, Na^{+}/H^{+} exchange, and (ii) PKA and subsequent activation of AA. This hypothesis explains why GH responses to DA or PACAP are additive to GnRH, but neither the responses between the two GnRHs nor between DA and PACAP are additive. Although both PKC and PKA signaling are associated with increases in $[Ca^{2+}]_i$ in somatotropes, the presence of multiple pharmacologically distinct Ca^{2+} stores provides another level of complexity in which mobilization of Ca^{2+} and Ca^{2+} effects can be differentiated. How the abilities of GnRH, DA and PACAP to increase GH synthesis interact has not been examined. In addition, how NOS/NO/cGMP, PI3K, ERK and MAPK signaling interacts with the PKC and cAMP/PKA pathways to regulate GH release and synthesis is a question that remains to be addressed.

Knowledge of how the inhibitory neuroendocrine regulators of GH release interact at the level of somatotropes is more limited than the stimulatory regulators. As in the case of the stimulatory factors, the known intracellular signaling targets affected by the inhibitory regulators such as SS-14, [Pro^{2}]SS-14, gbSS-28, NE, and 5HT are partly dissimilar (Section 4.2, Table 4.1). Differences in intracellular signaling likely form the basis to explain the differential abilities of the three SSs, NA and 5-HT to modulate the GH release response to neuroendocrine stimulators. In addition, these differences may also explain the ability of NA to further decrease basal GH release in the presence of SS-14 in preliminary experiments with goldfish pituitary cells (Yunker W.K., and Chang J.P., unpublished observation). Recently, some information on how NA inhibits PACAP-induced GH synthesis has been reported (see section on noradrenaline in Section 4.3.1) but much needs to be done in order to understand how inhibition of GH synthesis is manifested at the level of intracellular signaling.

5.2. Influence of Sex Steroids on Hypothalamic GH Regulators

Part of the ability of sex steroids to alter GH release and GH gene expression resides at the level of the hypothalamus where sex steroids alter the activity of neurons involved in neuroendocrine regulation of somatotropes. In goldfish, E2 increases forebrain expression of NPY and PACAP (Canosa *et al.*, 2007) and elevates DA turnover in preoptic areas of the hypothalamus (Trudeau *et al.*, 1993a). In addition, E2 increases GnRH content and release in cultured brain cells from the black porgy (*Acanthopagrus schlegeli*; Lee *et al.*, 2004). Thus the stimulatory effect of E2 on GH release in this species can be related to increased neuronal activities of these four GH-releasing neuroendocrine factors. Likewise, E2 also increases DA turnover in the brain of rainbow trout (Saligaut *et al.*, 1992), consistent with the abilities of E2 and DA to increase GH mRNA levels in this species (see dopamine in Section 4.2.2 and sex steroids in Section 4.3.2).

In addition to DA turnover, E2 also affects other neurotransmitter systems. E2 decreases brain 5-HT content in tilapia (Tsai *et al.*, 2000) and reduces pituitary 5-HT turnover in rainbow trout (Hernandez-Rauda and Aldegunde, 2002). Similarly, E2 increases NA turnover in the telencephalon/preoptic area and hypothalamus of goldfish (Trudeau *et al.*, 1993b). These results suggest that E2 stimulatory effects on GH release are mediated by reduction of the activities of GH-release inhibitory neuroendocrine systems. In contrast, E2 treatment increases PPS-I and PSS-III expression in the forebrain of goldfish (Canosa *et al.*, 2002). This latter observation can be reconciled with the other findings if it represents the result of feedback regulation secondary to the increase in GH secretion.

Besides affecting GABA, SS, and glutamate actions (see GHRH in Section 4.2.2, GABA in Section 4.2.1 and glutamate in Section 4.2.3), sex steroids also modulate the ability of other neuroendocrine regulators to alter GH secretion. T enhances the GH response to NPY in goldfish. Likewise, E2 increases the GH-releasing ability of GnRH and TRH, but decreases the response to DA *in vivo*. How sex steroids affect the responsiveness of somatotropes to NPY and TRH has not been directly examined but some information is available in the case of GnRH. *In vitro* treatment of goldfish pituitary cells with E2 has no effect on total cellular GH content but increases the GH-releasing capabilities of sGnRH, cGnRH-II and a PKC activator, but not that of a Ca^{2+} ionophore. These results indicate that E2 enhances the GH response to GnRH by direct action at the pituitary level to selectively modulate the effectiveness of specific signaling cascades for GH release (i.e., PKC) independent of GH synthesis Reviewed in Canosa *et al.*, 2007; Table 4.2. Thus, sex steroids regulation of somatotrope functions can be manifested at multiple levels including: (i) GH synthesis, (ii) pituitary responsiveness to neuroendocrine regulators, (iii) hypothalamic network activity within the CNS, and (iv) at various sites along the GH–IGF feedback loop.

5.3. Interactions between GH Regulators in Mediating Changes in Somatotrope Activity Following Feeding and Food Deprivation

The interactions between the neuroendocrine factors for GH release during feeding and starvation are complex since some of these hypothalamic regulators also affect feeding behavior and satiety. In many fish species, long-term food deprivation increases serum GH and pituitary GH mRNA levels (e.g., tilapia, Weber and Grau, 1999; goldfish, Unniappan and Peter, 2005; grouper, Pedroso *et al.*, 2006; rabbitfish, Ayson *et al.*, 2007; Atlantic salmon, Wilkinson *et al.*, 2006). In goldfish, fasting increases hypothalamic ghrelin and NPY mRNA but decreases CCK mRNA. The increases in ghrelin and NPY mRNA may be related to the increase in GH secretion and synthesis during starvation. The down-regulation in CCK mRNA, an anorexigenic factor, in conjunction with the increase in ghrelin, an orexigenic factor, may reflect a "hungry period" in anticipation of eventual food availability (see Chapter 9). A transient post-prandial increase in serum GH has been reported in goldfish, which peaks at 30 min and returns to pre-prandial levels by 1.5 h (Himick and Peter, 1994). Insights into the interactions between GH-regulating neuroendocrine factors mediating these events can be deduced from experiments in goldfish. In this species, there is a short-term post-prandial increase in forebrain CCK mRNA level and injection of CCK decreases forebrain PSS-I levels. These data suggest that the post-prandial increase in GH release may in part be mediated by CCK-induced decrease in SS neuronal activity. At 3 hours post-feeding, brain ghrelin and PSS-II mRNA levels are reduced. BBS application decreases PSS-I and PSS-II mRNA in the brain. Injection of ghrelin suppresses brain PSS-II mRNA levels, increases PSS-I mRNA and blocks the BBS induced decrease in PSS-I. As a result, the combined treatment of BBS and ghrelin counteract each other's activity on PSS-I and ghrelin and BBS decreases only brain PSS-II mRNA (Canosa *et al.*, 2005). These data suggest that increased ghrelin and BSS neuronal activity mediates the transient rise in serum GH associated with feeding. Stimulation of somatotrope activity is likely due to a combination of direct effects of ghrelin and BBS on GH release, as well as indirect effects through suppression of PSS-II neuronal activity.

5.4. Other Interactions

PACAP and SS neuronal systems also interact and cross-regulate one another. In goldfish, injection of gbSRIF and [Pro^2]SS-14 increase, whereas SS-14 decreases, forebrain PACAP mRNA levels. Conversely, PACAP reduces forebrain SS-14 and [Pro^2]SS-14 mRNA levels (Canosa *et al.*, 2007). Although the physiological condition(s) whereby these changes would be relevant is not known, these results clearly demonstrate that PACAP and SS neuronal activity not only act directly at the level of pituitary to control somatotrope activities but they can also act indirectly by altering the activity of other neuroendocrine regulatory neurons.

Recently, cGnRH-II has also been identified as an anorexigenic factor in goldfish (Matsuda *et al.*, 2008). sGnRH and cGnRH-II also induce female spawning behavior in this species (Volkoff and Peter, 1999). Under natural conditions, only recently ovulated female goldfish undergo spawning behavior, and feeding usually stops during spawning acts. Thus it seems likely that changes in GnRH neuronal activities in different part of the brain not only integrate the periovulatory GH and gonadotropin surges, but also the increase in spawning behavior and suppression of feeding (see Chapter 3 for a review of neuroendocrine regulation of reproduction). As discussed in Section 5.2, E2 increases GnRH content and release from cultured black porgy brain neurons, suggesting that positive feedback effects of E2 on the GnRH network are also part of this neuroendocrine circuit integrating GH and gonadotropin secretion. How the brain GnRH circuits, sex steroids and other neuroendocrine factors regulating somatotrope functions interact during these processes would be a fruitful model for understanding the complex neuroendocrine control of GH release in the future.

6. SUMMARY REMARKS

GH release is controlled by multiple neuroendocrine factors from both hypothalamic and peripheral sources. Due to the nature of GH biological action, the integrated neuroendocrine control of GH release and synthesis is necessarily complex, with interactions occurring at both the brain and pituitary levels. Seasonal variation in the effectiveness of stimulatory and inhibitory neuroendocrine regulators at the level of the pituitary, at least in part resulting from the effects of sex steroids, leads to seasonal fluctuations in circulating GH levels; this is further modified by feeding and food availability on a shorter term basis. Local factors released at the level of the pituitary provide further paracrine/autocrine control over GH release and synthesis. At the level of the somatotropes, distinct suites of signal transduction mechanisms utilized by the different neuroendocrine factors form the basis to integrate ligand- and function-specificity. Understanding the complex neuroendocrine regulation of somatotropes also has important environmental and aquacultural significance. Many of the environmental disruptors act on nuclear receptors and/or affect steroid metabolism. Thus, endocrine disruptors can influence the neuroendocrine regulation of GH release and synthesis through actions on sex steroids and thyroid hormone receptors at multiple levels of the brain–pituitary axis and exert effects on the numerous physiological functions that GH directly or indirectly mediates, including growth, energy metabolism and reproduction. Likewise, increasing our knowledge on the neuroendocrine regulation of GH by immune cytokines and stress hormones, an area that is not well studied in fish, will be essential to develop aquacultural practices that maximize growth and health of cultured fish species.

ACKNOWLEDGMENTS

Supported by a NSERC DG to J.P.C. and Hong Kong RGC grants to A.O.L.W. This chapter is dedicated to the memory of the late Dr. Richard E. Peter whose work has contributed greatly to the understanding of neuroendocrine regulation of GH release in fish.

REFERENCES

Agustsson, T., and Bjornsson, B. T. (2000). Growth hormone inhibits growth hormone secretion from the rainbow trout pituitary *in vitro*. *Comp. Biochem. Physiol. C Toxicol. Pharmacol.* **126,** 299–303.

Amano, M., Moriyama, S., Iigo, M., Kitamura, S., Amiya, N., Yamamori, K., Ukena, K., and Tsutsui, K. (2006). Novel fish hypothalamic neuropeptides stimulate the release of gonadotrophins and growth hormone from the pituitary of sockeye salmon. *J. Endocrinol.* **188,** 417–423.

Ayson, F. G., and Takemura, A. (2006). Daily expression patterns for mRNAs of GH, PRL, SL, IGF-I and IGF-II in juvenile rabbitfish, *Siganus guttatus*, during 24-h light and dark cycles. *Gen. Comp. Endocrinol.* **149,** 261–268.

Ayson, F. G., de Jesus-Ayson, E. G., and Takemura, A. (2007). mRNA expression patterns for GH, PRL, SL, IGF-I and IGF-II during altered feeding status in rabbitfish, *Siganus guttatus*. *Gen. Comp. Endocrinol.* **150,** 196–204.

Bhandari, R. K., Taniyama, S., Kitahashi, T., Ando, H., Yamauchi, K., Zohar, Y., Ueda, H., and Urano, A. (2003). Seasonal changes of responses to gonadotropin-releasing hormone analog in expression of growth hormone/prolactin/somatolactin genes in the pituitary of masu salmon. *Gen. Comp. Endocrinol.* **130,** 55–63.

Biga, P. R., Cain, K. D., Hardy, R. W., Schelling, G. T., Overturf, K., Roberts, S. B., Goetz, F. W., and Ott., T. L. (2004). Growth hormone differentially regulates muscle myostatin1 and -2 and increases circulating cortisol in rainbow trout (*Oncorhynchus mykiss*). *Gen. Comp. Endocrinol.* **138,** 32–41.

Bjornsson, B. (1997). The biology of salmon growth hormone: From day light to dominance. *Fish Physiol. Biochem.* **17,** 9–24.

Bjornsson, B., Johnsson, J., Benedet, S., Einarsdottir, I., Hildahl, J., Agustsson, T., and Johnsson, E. (2002). Growth hormone endocrinology of salmonids: Regulatory mechanisms and mode of action. *Fish Physiol. Biochem.* **27,** 227–242.

Bonn, U., and König, B. (1990). Serotonin-immunoreactive neurons in the brain of *Eigenmannia lineata* (Gymnotiformes, Teleostei). *J. Hirnforsch.* **31,** 297–306.

Bosma, P. T., Kolk, S. M., Rebers, F. E., Lescroart, O., Roelants, I., Willems, P. H., and Schulz, R. W. (1997). Gonadotrophs but not somatotrophs carry gonadotrophin-releasing hormone receptors: Receptor localisation, intracellular calcium, and gonadotrophin and GH release. *J. Endocrinol.* **152,** 437–446.

Canosa, L. F., and Peter, R. E. (2004). Effects of cholecystokinin and bombesin on the expression of preprosomatostatin-encoding genes in goldfish forebrain. *Regul. Pept.* **121,** 99–105.

Canosa, L. F., Lin, X., and Peter, R. E. (2002). Regulation of expression of somatostatin genes by sex steroid hormones in goldfish forebrain. *Neuroendocrinology* **76,** 8–17.

Canosa, L. F., Unniappan, S., and Peter, R. E. (2005). Periprandial changes in growth hormone release in goldfish: Role of somatostatin, ghrelin, and gastrin-releasing peptide. *Am. J. Physiol. Regul. Integr. Comp. Physiol.* **289,** R125–133.

Canosa, L. F., Chang, J. P., and Peter, R. E. (2007). Neuroendocrine control of growth hormone in fish. *Gen. Comp. Endocrinol.* **151,** 1–26.

Canosa, L. F., Stacey, N., and Peter, R. E. (2008). Changes in brain mRNA levels of gonadotropin-releasing hormone, pituitary adenylate cyclase activating polypeptide, and somatostatin during ovulatory luteinizing hormone and growth hormone surges in goldfish. *Am. J. Physiol. Regul. Integr. Comp. Physiol.* **295,** R1815–1821.

Cardenas, R., Lin, X., Canosa, L. F., Luna, M., Aramburo, C., and Peter, R. E. (2003). Estradiol reduces pituitary responsiveness to somatostatin (SRIF-14) and down-regulates the expression of somatostatin SST2 receptors in goldfish pituitary. *Gen. Comp. Endocrinol.* **132,** 119–124.

Carnevali, O., Cardinali, M., Maradonna, F., Parisi, M., Olivotto, I., Polzonetti-Magni, A. M., Mosconi, G., and Funkenstein, B. (2005). Hormonal regulation of hepatic IGF-I and IGF-II gene expression in the marine teleost *Sparus aurata. Mol. Reprod. Dev.* **71,** 12–18.

Chan, C. B., and Cheng, C. H. (2004). Identification and functional characterization of two alternatively spliced growth hormone secretagogue receptor transcripts from the pituitary of black seabream *Acanthopagrus schlegeli. Mol. Cell. Endocrinol.* **214,** 81–95.

Chan, C. B., Fung, C. K., Fung, W., Tse, M. C., and Cheng, C. H. (2004a). Stimulation of growth hormone secretion from seabream pituitary cells in primary culture by growth hormone secretagogues is independent of growth hormone transcription. *Comp. Biochem. Physiol. C Toxicol. Pharmacol.* **139,** 77–85.

Chan, C. B., Leung, P. K., Wise, H., and Cheng, C. H. (2004b). Signal transduction mechanism of the seabream growth hormone secretagogue receptor. *FEBS Lett.* **577,** 147–153.

Chang, J. P., and Habibi, H. R. (2002). Intracellular integration of multifactorial neuroendocrine regulation of goldfish somatotrophe functions. *In* "Developments in understanding fish growth" (B. Small, and D. MacKinlay, Eds.), pp. 5–14. University of British Columbia, VancouverSymposium Proceedings of the International Congress on the Biology of Fish.

Chang, J. P., Johnson, J. D., Van Goor, F., Wong, C. J. H., Yunker, W. K., Uretsky, A. D., Taylor, D., Jobin, R. M., Wong, A. O. L., and Goldberg, J. I. (2000). Signal transduction mechanisms mediating secretion in goldfish gonadotropes and somatotropes. *Biochem. Cell Biol.* **78,** 139–153.

Chang, J. P., Wong, C. J., Davis, P. J., Soetaert, B., Fedorow, C., Sawisky, G. (2003). Role of Ca^{2+} stores in dopamine- and PACAP-evoked growth hormone release in goldfish. *Mol. Cell. Endocrinol.* **206,** 63–74.

Chen, M. H., Li, Y. H., Chang, Y., Hu, S. Y., Gong, H. Y., Lin, G. H., Chen, T. T., and Wu, JL. (2007). Co-induction of hepatic IGF-I and progranulin mRNA by growth hormone in tilapia, *Oreochromis mossambiccus*. *Gen. Comp. Endocrinol.* **150,** 212–218.

Corio, M., Peute, J., and Steinbusch, H. W. (1991). Distribution of serotonin- and dopamine-immunoreactivity in the brain of the teleost *Clarias gariepinus*. *J. Chem. Neuroanat.* **4,** 79–95.

Diez, J. M., Giannico, G., McLean, E., and Donaldson, E. M. (1992). The effect of somatostatin (SRIF-14, 25, 28), galanin and anti-SRIF on plasma growth hormone levels in coho salmon (*Oncorhynchus kituch*, Walbaum). *J. Fish Biol.* **40,** 877–893.

Elango, A., Shepherd, B., and Chen, T. T. (2006). Effects of endocrine disrupters on the expression of growth hormone and prolactin mRNA in the rainbow trout pituitary. *Gen. Comp. Endocrinol.* **145,** 116–127.

Eppler, E., Shved, N., Moret, O., and Reinecke, M. (2007). IGF-I is distinctly located in the bony fish pituitary as revealed for *Oreochromis niloticus*, the Nile tilapia, using real-time RT-PCR, *in situ* hybridisation and immunohistochemistry. *Gen. Comp. Endocrinol.* **150,** 87–95.

Falcon, J., Besseau, L., Fazzari, D., Attia, J., Gaildrat, P., Beauchaud, M., and Boeuf, G. (2003). Melatonin modulates secretion of growth hormone and prolactin by trout pituitary glands and cells in culture. *Endocrinology* **144,** 4648–4658.

Figueroa, J., Martín, R. S., Flores, C., Grothusen, H., and Kausel, G. (2005). Seasonal modulation of growth hormone mRNA and protein levels in carp pituitary: Evidence for two expressed genes. *J. Comp. Physiol. [B]* **175,** 185–192.

Filby, A. L., and Tyler, C. R. (2007). Cloning and characterization of cDNAs for hormones and/or receptors of growth hormone, insulin-like growth factor-I, thyroid hormone, and corticosteroid and the gender-, tissue-, and developmental-specific expression of their mRNA transcripts in fathead minnow (*Pimephales promelas*). *Gen. Comp. Endocrinol.* **150,** 151–163.

Fox, B. K., Naka, T., Inoue, K., Takei, Y., Hirano, T., and Grau, E. G. (2007a). *In vitro* effects of homologous natriuretic peptides on growth hormone and prolactin release in the tilapia, *Oreochromis mossambicus*. *Gen. Comp. Endocrinol.* **150,** 270–277.

Fox, B. K., Riley, L. G., Dorough, C., Kaiya, H., Hirano, T., and Grau, E. G. (2007b). Effects of homologous ghrelins on the growth hormone/insulin-like growth factor-I axis in the tilapia, *Oreochromis mossambicus*. *Zoolog. Sci.* **24,** 391–400.

Fradinger, E. A., Tello, J. A., Rivier, J. E., and Sherwood, N. M. (2005). Characterization of four receptor cDNAs: PAC1, VPAC1, a novel PAC1 and a partial GHRH in zebrafish. *Mol. Cell. Endocrinol.* **231,** 49–63.

Frankenhuis-van den Heuvel, T. H., and Nieuwenhuys, R. (1984). Distribution of serotonin-immunoreactivity in the diencephalon and mesencephalon of the trout, *Salmo gairdneri*. Cellbodies, fibres and terminals. *Anat. Embryol. (Berl.)* **169,** 193–204.

Fruchtman, S., McVey, D. C., and Borski, R. J. (2002). Characterization of pituitary IGF-I receptors: Modulation of prolactin and growth hormone. *Am. J. Physiol. Regul. Integr. Comp. Physiol.* **283,** R468–476.

Fruchtman, S., Gift, B., Howes, B., and Borski, R. (2001). Insulin-like growth factor-I augments prolactin and inhibits growth hormone release through distinct as well as overlapping cellular signaling pathways. *Comp. Biochem. Physiol. B Biochem. Mol. Biol.* **129,** 237–42.

Fukamachi, S., and Meyer, A. (2007). Evolution of receptors for growth hormone and somatolactin in fish and land vertebrates: Lessons from the lungfish and sturgeon orthologues. *J. Mol. Evol.* **65,** 359–372.

Ge, W. (2000). Roles of the activin regulatory system in fish reproduction. *Can. J. Physiol. Pharmacol.* **78,** 1077–1085.

Ge, W., and Peter, R. E. (1994). Activin-like peptides in somatotrophs and activin stimulation of growth hormone release in goldfish. *Gen. Comp. Endocrinol.* **95,** 213–221.

Goldenberg, N., and Barkan, A. (2007). Factors regulating growth hormone secretion in humans. *Endocrinol. Metab. Clin. North. Am.* **36,** 37–55.

Grey, C. L., and Chang, J. P. (2009). Ghrelin-induced growth hormone release from goldfish pituitary cells involves voltage-gated calcium channels. *Gen. Comp. Endocrinol.* **160,** 148–157.

Guilgur, L. G., Moncaut, N.P, Canário, A. V., and Somoza, G. M. (2006). Evolution of GnRH ligands and receptors in gnathostomata. *Comp. Biochem. Physiol. A Mol. Integr. Physiol.* **144,** 272–283.

Hernandez-Rauda, R., and Aldegunde, M. (2002). Effects of acute 17alpha-methyltestosterone, acute 17beta-estradiol, and chronic 17alpha-methyltestosterone on dopamine, norepinephrine and serotonin levels in the pituitary, hypothalamus and telencephalon of rainbow trout (*Oncorhynchus mykiss*). *J. Comp. Physiol. [B]* **172,** 659–667.

Himick, B. A., and Peter, R. E. (1994). Bombesin acts to suppress feeding behavior and alter serum growth hormone in goldfish. *Physiol. Behav.* **55,** 65–72.

Himick, B. A., and Peter, R. E. (1995). Bombesin-like immunoreactivity in the forebrain and pituitary and regulation of anterior pituitary hormone release by bombesin in goldfish. *Neuroendocrinology* **61,** 365–376.

Himick, B. A., Golosinski, A. A., Jonsson, A. C., and Peter, R. E. (1993). CCK/gastrin-like immunoreactivity in the goldfish pituitary: Regulation of pituitary hormone secretion by CCK-like peptides *in vitro*. *Gen. Comp. Endocrinol.* **92,** 88–103.

Himick, B. A., Vigna, S. R., and Peter, R. E. (1995). Characterization and distribution of bombesin binding sites in the goldfish hypothalamic feeding center and pituitary. *Regul. Pept.* **60,** 167–176.

Holloway, A. C., and Leatherland, J. F. (1997). The effects of N-methyl-D,L-aspartate and gonadotropin-releasing hormone on *in vitro* growth hormone release in steroid-primed immature rainbow trout, *Oncorhynchus mykiss*. *Gen. Comp. Endocrinol.* **107,** 32–43.

Holloway, A. C., Sheridan, M. A., and Leatherland, J. F. (1997). Estradiol inhibits plasma somatostatin 14 (SRIF-14) levels and inhibits the response of somatotrophic cells to SRIF-14 challenge *in vitro* in rainbow trout, *Oncorhynchus mykiss*. *Gen. Comp. Endocrinol.* **106,** 407–414.

Holloway, A. C., Melroe, G. T., Ehrman, M. M., Reddy, P. K., Leatherland, J. F., and Sheridan, M. A. (2000). Effect of 17beta-estradiol on the expression of somatostatin genes in rainbow trout (*Oncorhynchus mykiss*). *Am. J. Physiol. Regul. Integr. Comp. Physiol.* **279,** R389–393.

Hrytsenko, O., Wright, J. R., Jr., Morrison, C. M., and Pohajdak, B. (2007). Insulin expression in the brain and pituitary cells of tilapia (*Oreochromis niloticus*). *Brain Res.* **1135,** 31–40.

Huising, M. O., Geven, E. J., Kruiswijk, C. P., Nabuurs, S. B., Stolte, E. H., Spanings, F. A., Verburg-van Kemenade, B. M., and Flik, G. (2006). Increased leptin expression in common Carp (*Cyprinus carpio*) after food intake but not after fasting or feeding to satiation. *Endocrinology* **147,** 5786–5797.

Huo, L., Fu, G., Wang, X., Ko, W. K., and Wong, A. O. (2005). Modulation of calmodulin gene expression as a novel mechanism for growth hormone feedback control by insulin-like growth factor in grass carp pituitary cells. *Endocrinology* **146,** 3821–3835.

Illing, N., Troskie, B. E., Nahorniak, C. S., Hapgood, J. P., Peter, R. E., and Millar, R. P. (1999). Two gonadotropin-releasing hormone receptor subtypes with distinct ligand selectivity and differential distribution in brain and pituitary in the goldfish (*Carassius auratus*). *Proc. Natl. Acad. Sci. USA* **96,** 2526–2531.

Jiang, Y., Li, W. S., Xie, J., and Lin, H. R. (2003). Sequence and expression of a cDNA encoding both pituitary adenylate cyclase activating polypeptide and growth hormone-releasing hormone in grouper (*Epinephelus coioides*). *Sheng Wu Hua Xue Yu Sheng Wu Wu Li Xue Bao (Shanghai)* **35,** 864–872.

Jiao, B., Huang, X., Chan, C. B., Zhang, L., Wang, D., and Cheng, C. H. (2006). The co-existence of two growth hormone receptors in teleost fish and their differential signal transduction, tissue distribution and hormonal regulation of expression in seabream. *J. Mol. Endocrinol.* **36,** 23–40.

Jodo, A., Ando, H., and Urano, A. (2003). Five different types of putative GnRH receptor gene are expressed in the brain of masu salmon (*Oncorhynchus masou*). *Zoolog. Sci.* **20,** 1117–1125.

Jodo, A., Kitahashi, T., Taniyama, S., Bhandari, R. K., Ueda, H., Urano, A., and Ando, H. (2005). Seasonal variation in the expression of five subtypes of gonadotropin-releasing hormone receptor genes in the brain of masu salmon from immaturity to spawning. *Zoolog. Sci.* **22,** 1331–1338.

Johansson, V., Winberg, S., Jönsson, E., Hall, D., and Björnsson, B. T. (2004). Peripherally administered growth hormone increases brain dopaminergic activity and swimming in rainbow trout. *Horm. Behav.* **46,** 436–443.

Johnson, J. D., Klausen, C., Habibi, H. R., and Chang, J. P. (2002). Function-specific calcium stores selectively regulate growth hormone secretion, storage, and mRNA level. *Am. J. Physiol. Endocrinol. Metab.* **282,** E810–819.

Jönsson, E., Forsman, A., Einarsdottir, I. E., Kaiya, H., Ruohonen, K, and Björnsson, B. T. (2007). Plasma ghrelin levels in rainbow trout in response to fasting, feeding and food composition, and effects of ghrelin on voluntary food intake. *Comp. Biochem. Physiol. A Mol. Integr. Physiol.* **147,** 1116–1124.

Joy, K. P., Senthilkumaran, B., and Sudhakumari, C. C. (1998). Periovulatory changes in hypothalamic and pituitary monoamines following GnRH analogue treatment in the catfish *Heteropneustes fossilis*: A study correlating changes in plasma hormone profiles. *J. Endocrinol.* **156,** 365–372.

Kagabu, Y., Mishiba, T., Okino, T., and Yanagisawa, T. (1998). Effects of thyrotropin-releasing hormone and its metabolites, Cyclo(His-Pro) and TRH-OH, on growth hormone and prolactin synthesis in primary cultured pituitary cells of the common carp, *Cyprinus carpio*. *Gen. Comp. Endocrinol.* **111,** 395–403.

Kah, O., and Chambolle, P. (1983). Serotonin in the brain of the goldfish, *Carassius auratus*. An immunocytochemical study. *Cell Tissue Res.* **234,** 319–333.

Kah, O., Lethimonier, C., Somoza, G., Guilgur, L. G., Vaillant, C., and Lareyre, J. J. (2007). GnRH and GnRH receptors in metazoa: A historical, comparative, and evolutive perspective. *Gen. Comp. Endocrinol.* **153,** 346–364.

Kaiya, H., Miyazato, M., Kangawa, K., Peter, R. E., and Unniappan, S. (2008). Ghrelin: A multifunctional hormone in non-mammalian vertebrates. *Comp. Biochem. Physiol. A Mol. Integr. Physiol.* **149,** 109–128.

Kajimura, S., Hirano, T., Visitacion, N., Moriyama, S., Aida, K., and Grau, E. G. (2003). Dual mode of cortisol action on GH/IGF-I/IGF binding proteins in the tilapia, *Oreochromis mossambicus*. *J. Endocrinol.* **178,** 91–99.

Kajimura, S., Kawaguchi, N., Kaneko, T., Kawazoe, I., Hirano, T., Visitacion, N., Grau, E. G., and Aida, K. (2004). Identification of the growth hormone receptor in an advanced teleost, the tilapia (*Oreochromis mossambicus*) with special reference to its distinct expression pattern in the ovary. *J. Endocrinol.* **181,** 65–76.

Kaplan, S. A., and Cohen, P. (2007). The somatomedin hypothesis 2007: 50 years later. *J. Clin. Endocrinol. Metab.* **92,** 4529–4535.

Kawauchi, H., and Sower, S. A. (2006). The dawn and evolution of hormones in the adenohypophysis. *Gen. Comp. Endocrinol.* **148,** 3–14.

Khan, I. A., and Thomas, P. (1993). Immunocytochemical localization of serotonin and gonadotropin-releasing hormone in the brain and pituitary gland of the Atlantic croaker *Micropogonias undulatus*. *Gen. Comp. Endocrinol.* **91,** 167–180.

Klausen, C., Tsuchiya, T., Chang, J. P., and Habibi, H. R. (2005). PKC and ERK are differentially involved in gonadotropin-releasing hormone-induced growth hormone gene expression in the goldfish pituitary. *Am. J. Physiol. Regul. Integr. Comp. Physiol.* **289,** R1625–1633.

Klein, S. E., and Sheridan, M. A. (2008). Somatostatin signaling and the regulation of growth and metabolism in fish. *Mol. Cell. Endocrinol.* **286,** 148–154.

Kuang, Y. M., Li, W. S., and Lin, H. R. (2005). Molecular cloning and mRNA profile of insulin-like growth factor type 1 receptor in orange-spotted grouper, *Epinephelus coioides. Acta Biochim. Biophys. Sin. (Shanghai)* **37,** 327–334.

Kurokawa, T., Uji, S., Suzuki, T. (2005). Identification of cDNA coding for a homologue to mammalian leptin from pufferfish, *Takifugu rubripes. Peptides.* **26,** 745–750.

Kwok, Y. Y., Chu, J. Y., Vaudry, H., Yon, L., Anouar, Y., and Chow, B. K. (2006). Cloning and characterization of a PAC1 receptor hop-1 splice variant in goldfish (*Carassius auratus*). *Gen. Comp. Endocrinol.* **145,** 188–196.

Kwong, P., and Chang, J. P. (1997). Somatostatin inhibition of growth hormone release in goldfish: Possible targets of intracellular mechanisms of action. *Gen. Comp. Endocrinol.* **108,** 446–456.

Larsen, D. A., Shimizu, M., Cooper, K. A., Swanson, P., and Dickhoff, W. W. (2004). Androgen effects on plasma GH, IGF-I, and 41-kDa IGFBP in coho salmon (*Oncorhynchus kisutch*). *Gen. Comp. Endocrinol.* **139,** 29–37.

Lau, M. T., and Ge, W. (2005). Cloning of Smad2, Smad3, Smad4, and Smad7 from the goldfish pituitary and evidence for their involvement in activin regulation of goldfish FSHbeta promoter activity. *Gen. Comp. Endocrinol.* **141,** 22–38.

Lee, L. T., Siu, F. K., Tam, J. K., Lau, I. T., Wong, A. O., Lin, M. C., Vaudry, H., and Chow, B. K. (2007). Discovery of growth hormone-releasing hormones and receptors in nonmammalian vertebrates. *Proc. Natl. Acad. Sci. USA* **104,** 2133–2138.

Lee, Y. H., Du, J. L., Shih, Y. S., Jeng, S. R., Sun, L. T., and Chang, C. F. (2004). *In vivo* and *in vitro* sex steroids stimulate seabream gonadotropin-releasing hormone content and release in the protandrous black porgy, *Acanthopagrus schlegeli. Gen. Comp. Endocrinol.* **139,** 12–19.

Lengyel, A. M. (2006). Novel mechanisms of growth hormone regulation: Growth hormone-releasing peptides and ghrelin. *Braz. J. Med. Biol. Res.* **39,** 1003–1011.

Li, W. S., Lin, H. R., and Wong, A. O. (2002). Effects of gonadotropin-releasing hormone on growth hormone secretion and gene expression in common carp pituitary. *Comp. Biochem. Physiol. B Biochem. Mol. Biol.* **132,** 335–341.

Li, W. S., Chen, D., Wong, A. O., and Lin, H. R. (2005). Molecular cloning, tissue distribution, and ontogeny of mRNA expression of growth hormone in orange-spotted grouper (*Epinephelus coioides*). *Gen. Comp. Endocrinol.* **144,** 78–89.

Lichanska, A. M., and Waters, M. J. (2008). New insights into growth hormone receptor function and clinical implications. *Horm. Res.* **69,** 138–145.

Lin, X., and Peter, R. E. (2001). Somatostatins and their receptors in fish. *Comp. Biochem. Physiol. B Biochem. Mol. Biol.* **129,** 543–550.

Lin, X., Otto, C. J., Cardenas, R., and Peter, R. E. (2000). Somatostatin family of peptides and its receptors in fish. *Can. J. Physiol. Pharmacol.* **78,** 1053–1066.

Luo, D., and McKeown, B. A. (1991). The effect of thyroid hormone and glucocorticoids on carp growth hormone-releasing factor (GRF)-induced growth hormone (GH) release in rainbow trout (*Oncorhynchus mykiss*). *Comp. Biochem. Physiol. A* **99,** 621–626.

Makino, K., Onuma, T. A., Kitahashi, T., Ando, H., Ban, M., and Urano, A. (2007). Expression of hormone genes and osmoregulation in homing chum salmon: A minireview. *Gen. Comp. Endocrinol.* **152,** 304–309.

Marchant, T. A., and Peter, R. E. (1986). Seasonal variations in body growth rates and circulating levels of growth hormone in the goldfish, *Carassius auratus*. *J. Exp. Zool.* **237,** 231–239.

Marchant, T. A., Dulka, J. G., and Peter, R. E. (1989). Relationship between serum growth hormone levels and the brain and pituitary content of immunoreactive somatostatin in the goldfish, *Carassius auratus L. Gen. Comp. Endocrinol.* **73,** 458–468.

Margolis-Nunno, H., Halpern-Sebold, L., and Schreibman, M. P. (1986). Immunocytochemical changes in serotonin in the forebrain and pituitary of aging fish. *Neurobiol. Aging* **7,** 17–21.

Matsuda, K., Yoshida, T., Nagano, Y., Kashimoto, K., Yatohgo, T., Shimomura, H., Shioda, S., Arimura, A., and Uchiyama, M. (1998). Purification and primary structure of pituitary adenylate cyclase activating polypeptide (PACAP) from the brain of an elasmobranch, stingray, *Dasyatis akajei*. *Peptides* **19,** 1489–1495.

Matsuda, K., Nagano, Y., Uchiyama, M., Onoue, S., Takahashi, A., Kawauchi, H., and Shioda, S. (2005). Pituitary adenylate cyclase-activating polypeptide (PACAP)-like immunoreactivity in the brain of a teleost, *Uranoscopus japonicus*: Immunohistochemical relationship between PACAP and adenohypophysial hormones. *Regul. Pept.* **126,** 129–136.

Matsuda, K., Nakamura, K., Shimakura, S. I., Miura, T., Kageyama, H., Uchiyama, M., Shioda, S., and Ando, H. (2008). Inhibitory effect of chicken gonadotropin-releasing hormone II on food intake in the goldfish, *Carassius auratus*. *Horm. Behav.* **54,** 83–89.

Mazumdar, M., Lal, B., Sakharkar, A. J., Deshmukh, M., Singru, P. S., and Subhedar, N. (2006). Involvement of neuropeptide Y Y1 receptors in the regulation of LH and GH cells in the pituitary of the catfish, *Clarias batrachus*: An immunocytochemical study. *Gen. Comp. Endocrinol.* **149,** 190–196.

McRory, J. E., Parker, D. B., Ngamvongchon, S., and Sherwood, N. M. (1995). Sequence and expression of cDNA for pituitary adenylate cyclase activating polypeptide (PACAP) and growth hormone-releasing hormone (GHRH)-like peptide in catfish. *Mol. Cell. Endocrinol.* **108,** 169–177.

Melamed, P., Rosenfeld, H., Elizur, A., and Yaron, Z. (1998). Endocrine regulation of gonadotropin and growth hormone gene transcription in fish. *Comp. Biochem. Physiol. C Pharmacol. Toxicol. Endocrinol.* **119,** 325–338.

Melamed, P., Gur, G., Rosenfeld, H., Elizur, A., and Yaron, Z. (1999). Possible interactions between gonadotrophs and somatotrophs in the pituitary of tilapia: Apparent roles for insulin-like growth factor I and estradiol. *Endocrinology* **140,** 1183–1191.

Mingarro, M., Vega-Rubin de Celis, S., Astola, A., Pendon, C., and Perez-Sanchez, J. (2002). Endocrine mediators of seasonal growth in gilthead sea bream (*Sparus aurata*): The growth hormone and somatolactin paradigm. *Gen. Comp. Endocrinol.* **128,** 102–111.

Mitchell, G., Sawisky, G. R., Grey, C. L., Wong, C.J, Uretsky, A. D., and Chang, J. P. (2008). Differential involvement of nitric oxide signaling in dopamine and PACAP stimulation of growth hormone release in goldfish. *Gen. Comp. Endocrinol.* **155,** 318–327.

Moncaut, N., Somoza, G., Power, D. M., and Canário, A. V. (2005). Five gonadotrophin-releasing hormone receptors in a teleost fish: Isolation, tissue distribution and phylogenetic relationships. *J. Mol. Endocrinol.* **34,** 767–779.

Montero, M., Yon, L., Kikuyama, S., Dufour, S., and Vaudry, H. (2000). Molecular evolution of the growth hormone-releasing hormone/pituitary adenylate cyclase-activating polypeptide gene family. Functional implication in the regulation of growth hormone secretion. *J. Mol. Endocrinol.* **25,** 157–168.

Moriyama, S., Kasahara, M., Amiya, N., Takahashi, A., Amano, M., Sower, S. A., Yamamori, K., and Kawauchi, H. (2007). RFamide peptides inhibit the expression of melanotropin and growth hormone genes in the pituitary of an Agnathan, the sea lamprey, *Petromyzon marinus*. *Endocrinology* **148,** 3740–3749.

Murashita, K., Uji, S., Yamamoto, T., Ronnestad, I., and Kurokawa, T. (2008). Production of recombinant leptin and its effects of food intake in rainbow trout (*Oncorhynchus mykiss*). *Comp. Biochem. Physiol. B Biochem. Mol. Biol.* **150,** 377–384.

Nelson, L. E., and Sheridan, M. A. (2005). Regulation of somatostatins and their receptors in fish. *Gen. Comp. Endocrinol.* **142,** 117–133.

Nelson, L. E., and Sheridan, M. A. (2006). Insulin and growth hormone stimulate somatostatin receptor (SSTR) expression by inducing transcription of SSTR mRNAs and by upregulating cell surface SSTRs. *Am. J. Physiol. Regul. Integr. Comp. Physiol.* **291,** R163–169.

Nordgarden, U., Fjelldal, P. G., Hansen, T., Björnsson, B. T., and Wargelius, A. (2006). Growth hormone and insulin-like growth factor-I act together and independently when regulating growth in vertebral and muscle tissue of Atlantic salmon postsmolts. *Gen. Comp. Endocrinol.* **149,** 253–260.

Olivereau, M., and Callard, G. (1985). Distribution of cell types and aromatase activity in the sculpin (*Myoxocephalus*) pituitary. *Gen. Comp. Endocrinol.* **58,** 280–290.

Onuma, T., Ando, H., Koide, N., Okada, H., and Urano, A (2005). Effects of salmon GnRH and sex steroid hormones on expression of genes encoding growth hormone/prolactin/somatolactin family hormones and a pituitary-specific transcription factor in masu salmon pituitary cells *in vitro*. *Gen. Comp. Endocrinol.* **143,** 129–141.

Osugi, T., Ukena, K., Sower, S. A., Kawauchi, H., and Tsutsui, K. (2006). Evolutionary origin and divergence of PQRFamide peptides and LPXRFamide peptides in the RFamide peptide family. Insights from novel lamprey RFamide peptides. *FEBS J.* **273,** 1731–1743.

Palevitch, O., Kight, K., Abraham, E., Wray, S., Zohar, Y., and Gothilf, Y. (2007). Ontogeny of the GnRH systems in zebrafish brain: *In situ* hybridization and promoter-reporter expression analyses in intact animals. *Cell Tissue Res.* **327,** 313–322.

Pedroso, F. L., de Jesus-Ayson, E. G., Cortado, H. H., Hyodo, S., and Ayson, F. G. (2006). Changes in mRNA expression of grouper (*Epinephelus coioides*) growth hormone and insulin-like growth factor I in response to nutritional status. *Gen. Comp. Endocrinol.* **145,** 237–246.

Pelis, R. M., and McCormick, S. D. (2001). Effects of growth hormone and cortisol on Na^{+}-K^{+}-$2Cl^{-}$ cotransporter localization and abundance in the gills of Atlantic salmon. *Gen. Comp. Endocrinol.* **124,** 134–143.

Peng, C., and Peter, R. E. (1997). Neuroendocrine regulation of growth hormone secretion and growth in fish. *Zool. Stud.* **36,** 79–89.

Peter, R. E., and Chang, J. P. (1999). Brain regulation of growth hormone secretion and food intake in fish. *In* "Regulation of the Vertebrate Endocrine System" (P. D. P. Rao, and R. E. Peter, Eds.), pp. 55–67. Plenum, New York.

Peterson, B. C., and Small, B. C. (2005). Effects of exogenous cortisol on the GH/IGF-I/IGFBP network in channel catfish. *Domest. Anim. Endocrinol.* **28,** 391–404.

Peyon, P., Vega-Rubín de Celis, S., Gómez-Requeni, P., Zanuy, S., Pérez-Sánchez, J., and Carrillo, M. (2003). *In vitro* effect of leptin on somatolactin release in the European sea bass (*Dicentrarchus labrax*): Dependence on the reproductive status and interaction with NPY and GnRH. *Gen. Comp. Endocrinol.* **132,** 284–292.

Pierce, A. L., Fukada, H., and Dickhoff, W. W. (2005). Metabolic hormones modulate the effect of growth hormone (GH) on insulin-like growth factor-I (IGF-I) mRNA level in primary culture of salmon hepatocytes. *J. Endocrinol.* **184,** 341–349.

Pierce, A. L., Fox, B. K., Davis, L. K., Visitacion, N., Kitahashi, T., Hirano, T., and Grau, E. G. (2007). Prolactin receptor, growth hormone receptor, and putative somatolactin receptor in Mozambique tilapia: Tissue specific expression and differential regulation by salinity and fasting. *Gen. Comp. Endocrinol.* **154,** 31–40.

Ran, X. Q., Li, W. S., and Lin, H. R. (2004). Stimulatory effects of gonadotropin-releasing hormone and dopamine on growth hormone release and growth hormone mRNA expression in *Epinephelus coioides*. *Sheng Li Xue Bao* **56,** 644–650.

Rousseau, K., and Dufour, S. (2004). Phylogenetic evolution of the neuroendocrine control of growth hormone: Contribution from teleosts. *Cybium* **28,** 181–198.

Rousseau, K., Huang, Y. S., Le Belle, N., Vidal, B., Marchelidon, J., Epelbaum, J., and Dufour, S. (1998). Long-term inhibitory effects of somatostatin and insulin-like growth factor 1 on growth hormone release by serum-free primary culture of pituitary cells from European eel (*Anguilla anguilla*). *Neuroendocrinology* **67,** 301–309.

Rousseau, K., Le Belle, N., Marchelidon, J., and Dufour, S. (1999). Evidence that corticotropin-releasing hormone acts as a growth hormone-releasing factor in a primitive teleost, the European eel (*Anguilla anguilla*). *J. Neuroendocrinol.* **11,** 385–392.

Rousseau, K., Le Belle, N., Pichavant, K., Marchelidon, J., Chow, B. K., Boeuf, G., and Dufour, S. (2001). Pituitary growth hormone secretion in the turbot, a phylogenetically recent teleost, is regulated by a species-specific pattern of neuropeptides. *Neuroendocrinology* **74,** 375–385.

Rousseau, K., Le Belle, N., Sbaihi, M., Marchelidon, J., Schmitz, M., and Dufour, S. (2002). Evidence for a negative feedback in the control of eel growth hormone by thyroid hormones. *J. Endocrinol.* **175,** 605–613.

Sakai, M., Kajita, Y., Kobayashi, M., and Kawauchi, H. (1997). Immunostimulating effect of growth hormone: *In-vivo* administration of growth hormone in rainbow trout enhances resistance to *Vibrio anguillarum* infection. *Vet. Immunol. Immunopathol.* **57,** 147–152.

Sakamoto, T., and McCormick, S. D. (2006). Prolactin and growth hormone in fish osmoregulation. *Gen. Comp. Endocrinol.* **147,** 24–30.

Saligaut, C., Garnier, D. H., Bennani, S., Salbert, G., Bailhache, T., and Jego, P. (1992). Effects of estradiol on brain aminergic turnover of the female rainbow trout (*Oncorhynchus mykiss*) at the beginning of vitellogenesis. *Gen. Comp. Endocrinol.* **88,** 209–216.

Sawisky, G. R., and Chang, J. P. (2005). Intracellular calcium involvement in pituitary adenylate cyclase-activating polypeptide stimulation of growth hormone and gonadotropin secretion in goldfish pituitary cells. *J. Neuroendocrinol.* **17,** 353-371.

Shrimpton, J. M., and McCormick, S. D. (1998). Regulation of gill cytosolic corticosteroid receptors in juvenile Atlantic salmon: Interaction effects of growth hormone with prolactin and triiodothyronine. *Gen. Comp. Endocrinol.* **112,** 262–274.

Shved, N., Berishvili, G., D'Cotta, H., Baroiller, J. F., Segner, H., Eppler, E., and Reinecke, M. (2007). Ethinylestradiol differentially interferes with IGF-I in liver and extrahepatic sites during development of male and female bony fish. *J. Endocrinol.* **195,** 513–523.

Silver, M. R., Kawauchi, H., Nozaki, M., and Sower, S. A. (2004). Cloning and analysis of the lamprey GnRH-III cDNA from eight species of lamprey representing the three families of Petromyzoniformes. *Gen. Comp. Endocrinol.* **139,** 85–94.

Sloley, B. D., Trudeau, V. L., Dulka, J. G., and Peter, R. E. (1991). Selective depletion of dopamine in the goldfish pituitary caused by domperidone. *Can. J. Physiol. Pharmacol.* **69,** 776–781.

Small, B. C., and Nonneman, D. (2001). Sequence and expression of a cDNA encoding both pituitary adenylate cyclase activating polypeptide and growth hormone-releasing hormone-like peptide in channel catfish (*Ictalurus punctatus*). *Gen. Comp. Endocrinol.* **122,** 354–363.

Smith, A., Chan, S. J., and Gutiérrez, J. (2005). Autoradiographic and immunohistochemical localization of insulin-like growth factor-I receptor binding sites in brain of the brown trout, *Salmo trutta*. *Gen. Comp. Endocrinol.* **141,** 203–213.

Somoza, G. M., and Peter, R. E. (1991). Effects of serotonin on gonadotropin and growth hormone release from *in vitro* perifused goldfish pituitary fragments. *Gen. Comp. Endocrinol.* **82,** 103–110.

Stefano, A. V., Vissio, P. G., Paz, D. A., Somoza, G. M., Maggese, M. C., and Barrantes, G. E. (1999). Colocalization of GnRH binding sites with gonadotropin-, somatotropin-, somatolactin-, and prolactin-expressing pituitary cells of the pejerrey, *Odontesthes bonariensis, in vitro*. *Gen. Comp. Endocrinol.* **116,** 133–139.

Sundstrom, L. F., Lohmus, M., Johnsson, J. I., and Devlin, R. H. (2004). Growth hormone transgenic salmon pay for growth potential with increased predation mortality. *Proc. Biol. Sci.* **271**(Suppl 5), S350–352.

Sze, K. H., Zhou, H., Yang, Y., He, M., Jiang, Y., and Wong, A. O. (2007). Pituitary adenylate cyclase-activating polypeptide (PACAP) as a growth hormone (GH)-releasing factor in grass carp: II. Solution structure of a brain-specific PACAP by nuclear magnetic resonance spectroscopy and functional studies on GH release and gene expression. *Endocrinology* **148,** 5042–5059.

Tam, J. K., Lee, L. T., and Chow, B. K. (2007). PACAP-related peptide (PRP) – molecular evolution and potential functions. *Peptides* **28,** 1920–1929.

Taniyama, S., Kitahashi, T., Ando, H., Kaeriyama, M., Zohar, Y., Ueda, H., and Urano, A. (2000). Effects of gonadotropin-releasing hormone analog on expression of genes encoding the growth hormone/prolactin/somatolactin family and a pituitary-specific transcription factor in the pituitaries of prespawning sockeye salmon. *Gen. Comp. Endocrinol.* **118,** 418–424.

Tipsmark, C. K., Strom, C. N., Bailey, S. T., and Borski, R. J. (2008). Leptin stimulates pituitary prolactin release through an extracellular signal-regulated kinase-dependent pathway. *J. Endocrinol.* **196,** 275–281.

Tostivint, H., Lihrmann, I., and Vaudry, H. (2008). New insight into the molecular evolution of the somatostatin family. *Mol. Cell. Endocrinol.* **286,** 5–17.

Trudeau, V. L., Somoza, G. M., Nahorniak, C., and Peter, R. E. (1992). Interactions of estradiol with gonadotropin-releasing hormone and thyrotropin-releasing hormone in the control of growth hormone secretion in the goldfish. *Neuroendocrinology* **56,** 483–490.

Trudeau, V. L., Sloley, B. D., Wong, A. O., and Peter, R. E. (1993a). Interactions of gonadal steroids with brain dopamine and gonadotropin-releasing hormone in the control of gonadotropin-II secretion in the goldfish. *Gen. Comp. Endocrinol.* **89,** 39–50.

Trudeau, V. L., Sloley, B. D., and Peter, R. E. (1993b). Norepinephrine turnover in the goldfish brain is modulated by sex steroids and GABA. *Brain Res.* **624,** 29–34.

Trudeau, V. L., Sloley, B. D., Kah, O., Mons, N., Dulka, J. G., and Peter, R. E. (1996). Regulation of growth hormone secretion by amino acid neurotransmitters in the goldfish (I): Inhibition by N-methyl-D, L-aspartic acid. *Gen. Comp. Endocrinol.* **103,** 129–137.

Trudeau, V. L., Kah, O., Chang, J. P., Sloley, B. D., Dubourg, P., Fraser, E. J., and Peter, R. E. (2000a). The inhibitory effects of γ-aminobutyric acid (GABA) on growth hormone secretion in the goldfish are modulated by sex steroids. *J. Exp. Biol.* **203,** 1477–1485.

Trudeau, V. L., Spanswick, D., Fraser, E. J., Lariviere, K., Crump, D., Chiu, S., MacMillan, M., and Schulz, R. W. (2000b). The role of amino acid neurotransmitters in the regulation of pituitary gonadotropin release in fish. *Biochem. Cell Biol.* **78,** 241–259.

Tsai, C. L., Wang, L. H., Chang, C. F., and Kao, C. C. (2000). Effects of gonadal steroids on brain serotonergic and aromatase activity during the critical period of sexual differentiation in tilapia, *Oreochromis mossambicus*. *J. Neuroendocrinol.* **12,** 894–898.

Uchida, K., Yoshikawa-Ebesu, J. S., Kajimura, S., Yada, T., Hirano, T., and Gordon Grau, E. (2004). *In vitro* effects of cortisol on the release and gene expression of prolactin and growth hormone in the tilapia, *Oreochromis mossambicus*. *Gen. Comp. Endocrinol.* **135,** 116–125.

Unniappan, S., and Peter, R. E. (2005). Structure, distribution and physiological functions of ghrelin in fish. *Comp. Biochem. Physiol. A Mol. Integr. Physiol.* **140,** 396–408.

Volkoff, H. (2006). The role of neuropeptide Y, orexins, cocaine and amphetamine-related transcript, cholecystokinin, amylin and leptin in the regulation of feeding in fish. *Comp. Biochem. Physiol. A Mol. Integr. Physiol.* **144,** 325–331.

Volkoff, H., and Peter, R. E. (1999). Actions of two forms of gonadotropin releasing hormone and a GnRH antagonist on spawning behavior of the goldfish *Carassius auratus*. *Gen. Comp. Endocrinol.* **116,** 347–355.

Volkoff, H., Eykelbosh, A. J., and Peter, R. E. (2003). Role of leptin in the control of feeding of goldfish: Interactions with cholecystokinin, neuropeptide Y and orexin-A, and modulation by fasting. *Brain Res.* **972,** 90–109.

Volkoff, H., Canosa, L. F., Unniappan, S., Cerdá-Reverter, J. M., Bernier, N. J., Kelly, S. P., and Peter, R. E. (2005). Neuropeptides and the control of food intake in fish. *Gen. Comp. Endocrinol.* **142,** 3–19.

Wang, X., Chu, M. M., and Wong, A. O. (2007). Signaling mechanisms for alpha2-adrenergic inhibition of PACAP-induced growth hormone secretion and gene expression grass carp pituitary cells. *Am. J. Physiol. Endocrinol. Metab.* **292,** E1750–1762.

Wang, Y., Wong, A.O, and Ge, W. (2003). Cloning, regulation of messenger ribonucleic acid expression, and function of a new isoform of pituitary adenylate cyclase-activating polypeptide in the zebrafish ovary. *Endocrinology* **144,** 4799–4810.

Weber, G. M., and Grau, E. G. (1999). Changes in serum concentrations and pituitary content of the two prolactins and growth hormone during the reproductive cycle in female tilapia, *Oreochromis mossambicus*, compared with changes during fasting. *Comp. Biochem. Physiol. C Pharmacol. Toxicol. Endocrinol.* **124,** 323–335.

Wei, Y., Martin, S. C., Heinrich, G., and Mojsov, S. (1998). Cloning and functional characterization of PACAP-specific receptors in zebrafish. *Ann. N. Y. Acad. Sci.* **865,** 45–48.

Weil, C., Carre, F., Blaise, O., Breton, B., and Le Bail, P. Y. (1999). Differential effect of insulin-like growth factor I on *in vitro* gonadotropin and growth hormone secretions in rainbow trout (*Oncorhynchus mykiss*) at different stages of the reproductive cycle. *Endocrinology* **140,** 2054–2062.

Wilkinson, R. J., Porter, M., Woolcott, H., Longland, R., and Carragher, J. F. (2006). Effects of aquaculture related stressors and nutritional restriction on circulating growth factors (GH, IGF-I and IGF-II) in Atlantic salmon and rainbow trout. *Comp. Biochem. Physiol. A Mol. Integr. Physiol.* **145,** 214–224.

Wong, A. O., Le Drean, Y., Liu, D., Hu, Z. Z., Du, S. J., and Hew, C. L. (1996). Induction of chinook salmon growth hormone promoter activity by the adenosine 3′,5′-monophosphate (cAMP)-dependent pathway involves two cAMP-response elements with the CGTCA motif and the pituitary-specific transcription factor Pit-1. *Endocrinology* **137,** 1775–1784.

Wong, A. O., Li, W., Leung, C. Y., Huo, L., and Zhou, H. (2005). Pituitary adenylate cyclase-activating polypeptide (PACAP) as a growth hormone (GH)-releasing factor in grass carp. I. Functional coupling of cyclic adenosine 3′,5′-monophosphate and Ca^{2+}/calmodulin-dependent signaling pathways in PACAP-induced GH secretion and GH gene expression in grass carp pituitary cells. *Endocrinology* **146,** 5407–5424.

Wong, A. O., Zhou, H., Jiang, Y., and Ko, W. K. (2006). Feedback regulation of growth hormone synthesis and secretion in fish and the emerging concept of intrapituitary feedback loop. *Comp. Biochem. Physiol. A Mol. Integr. Physiol.* **144,** 284–305.

Wong, A. O., Chuk, M. C., Chan, H. C., and Lee, E. K. (2007). Mechanisms for gonadotropin-releasing hormone potentiation of growth hormone rebound following norepinephrine inhibition in goldfish pituitary cells. *Am. J. Physiol. Endocrinol Metab.* **292,** E203–214.

Wong, A. O. L., Chang, J. P., and Peter, R. E. (1993). *In vitro* and *in vivo* evidence that dopamine exerts growth hormone releasing activity in the goldfish, *Carassius auratus*. *Am. J. Physiol.* **264,** E925–932.

Wong, A. O. L., Murthy, C. K., Chang, J. P., Neumann, C. M., Lo, A., and Peter, R. E. (1998). Direct actions of serotonin on gonadotropin-II and growth hormone release from goldfish pituitary cells: Interactions with gonadotropin-releasing hormone and dopamine and further evaluation of serotonin receptor specificity. *Fish Physiol. Biochem.* **19,** 22–34.

Wong, A. O. L., Li, W. S., Lee, E. K. Y., Leung, M. Y., Tse, L. Y., Chow, B. K. C., Lin, H. R., and Chang, J. P. (2000). Pituitary adenylate cyclase-activating polypeptide as a novel hypophysiotropic factor in fish. *Biochem. Cell Biol.* **78,** 329–343.

Wong, C. J., Johnson, J. D., Yunker, W. K., and Chang, J. P. (2001). Caffeine stores and dopamine differentially require Ca^{2+} channels in goldfish somatotropes. *Am. J. Physiol. Regul. Integr. Comp. Physiol.* **280,** R494–R503.

Xiao, D., and Lin, H. R. (2003). Effects of cysteamine – a somatostatin-inhibiting agent – on serum growth hormone levels and growth in juvenile grass carp (*Ctenopharyngodon idellus*). *Comp. Biochem. Physiol. A Mol. Integr. Physiol.* **134,** 93–99.

Xing, Y., Wensheng, L., and Haoran, L. (2005). Polygenic expression of somatostatin in orange-spotted grouper (*Epinephelus coioides*): Molecular cloning and distribution of the mRNAs encoding three somatostatin precursors. *Mol. Cell. Endocrinol.* **241,** 62–72.

Yacobovitz, M., Solomon, G., Gusakovsky, E. E., Levavi-Sivan, B., and Gertler, A. (2008). Purification and characterization of recombinant pufferfish (*Takifugu rubripes*) leptin. *Gen. Comp. Endocrinol.* **156,** 83–90.

Yada, T. (2007). Growth hormone and fish immune system. *Gen. Comp. Endocrinol.* **152,** 353–358.

Yadetie, F., and Male, R. (2002). Effects of 4-nonylphenol on gene expression of pituitary hormones in juvenile Atlantic salmon (*Salmo salar*). *Aquat. Toxicol.* **58,** 113–129.

Yeung, C. M., Chan, C. B., Woo, N. Y., and Cheng, C. H. (2006). Seabream ghrelin: CDNA cloning, genomic organization and promoter studies. *J. Endocrinol.* **189,** 365–379.

Yu, Y., Wong, A. O., and Chang, J. P. (2008). Serotonin interferes with Ca^{2+} and PKC signaling to reduce gonadotropin-releasing hormone-stimulated GH secretion in goldfish pituitary cells. *Gen. Comp. Endocrinol.* **159,** 58–66.

Yuen, C. W., and Ge, W. (2004). Follistatin suppresses FSHbeta but increases LHbeta expression in the goldfish – evidence for an activin-mediated autocrine/paracrine system in fish pituitary. *Gen. Comp. Endocrinol.* **135,** 108–115.

Yunker, W. K., and Chang, J. P. (2001). Somatostatin actions on a protein kinase C-dependent growth hormone secretagogue cascade. *Mol. Cell. Endocrinol.* **175,** 193–204.

Yunker, W. K., and Chang, J. P. (2004). Somatostatin-14 actions on dopamine- and pituitary adenylate cyclase-activating polypeptide-evoked Ca^{2+} signals and growth hormone secretion. *J. Neuroendocrinol.* **16,** 684–694.

Yunker, W. K., Smith, S., Graves, C., Unniappan, S., Rivier, J. E., Peter, R. E., and Chang, J. P. (2003). Endogenous hypothalamic somatostatins differentially regulate growth hormone secretion from goldfish pituitary somatotropes *in vitro*. *Endocrinology* **144,** 4031–4041.

Zbikowska, H. M. (2003). Fish can be first – advances in fish transgenesis for commercial applications. *Transgenic Res.* **12,** 379–389.

Zhang, W. M., Lin, H. R., and Peter, R. E. (1994). Episodic growth hormone secretion in the grass carp, *Ctenopharyngodon idellus*. *Gen. Comp. Endocrinol.* **95,** 337–341.

Zou, J. J., Trudeau, V. L., Cui, Z., Brechin, J., Mackenzie, K., Zhu, Z., Houlihan, D. F., and Peter, R. E. (1997). Estradiol stimulates growth hormone production in female goldfish. *Gen. Comp. Endocrinol.* **106,** 102–112.

5

THE NEUROENDOCRINE REGULATION OF PROLACTIN AND SOMATOLACTIN SECRETION IN FISH

HIROSHI KAWAUCHI
STACIA A. SOWER
SHUNSUKE MORIYAMA

1. Introduction
2. The Neuroendocrine Regulation of Prolactin Secretion
   2.1. Prolactin and its Receptor
   2.2. Functions of Prolactin
   2.3. Prolactin Secretion by Hypothalamic Peptides
3. The Neuroendocrine Regulation of Somatolactin Secretion
   3.1. Somatolactin and its Receptor
   3.2. Functions of Somatolactin
   3.3. Hypothalamic Control of Somatolactin Secretion
4. Concluding Remarks

This chapter describes the neuroendocrine control of synthesis and secretion of two homologous pituitary hormones, prolactin (PRL) and somatolactin (SL), in teleost fish. PRL and SL are produced in the rostral pars distalis and the pars intermedia of Osteichthyes, respectively. The synthesis and secretion of PRL and SL are regulated by two antagonistic hypothalamic neurohormones. Specific action of each hypothalamic neurohormone is determined by the presence of specific G-protein coupled receptors. Both PRL and SL have been implicated in many physiological processes through specific Janus kinase (JAK) associated receptors. The molecular characteristics and biological functions of these hormones and their receptors are summarized briefly and mode of actions of hypothalamic stimulating and

Fish Neuroendocrinology: Volume 28
FISH PHYSIOLOGY

DOI: 10.1016/S1546-5098(09)28005-8

inhibiting factors, including some of our recent findings on PRL-releasing peptide (PrRP), are discussed.

1. INTRODUCTION

Prolactin (PRL) and somatolactin (SL) along with growth hormone (GH) form a family of pituitary hormones that belong to the class I helical cytokines superfamily. All members of this superfamily share a similar three-dimensional fold into a bundle of four α-helices, signal via related receptors and activate a similar intracellular signaling cascade (Huising *et al.*, 2006). PRLs and GHs have been identified in many representative species from all classes of vertebrates except for PRL in the Chondrichthyes and Agnathans, while SLs have been found only within the Osteichthyes (Kawauchi *et al.*, 2002). In teleosts, PRL, GH and SL cells are localized in three distinct areas of the pituitary gland: the rostral pars distalis (RPD), proximal pars distalis (PPD) and pars intermedia (PI), respectively.

The first recognized action for PRL was the lactogenic effects on mammary gland of female rabbits for which the hormone was named. Since then, an increasing number of biological activities for PRL have been reported in various vertebrates. Following the reviews for PRL functions by Nicoll (1993), Bole-Feysot *et al.* (1998) rearranged over 300 functions into seven categories: 1) water and electrolyte balance, 2) growth and development, 3) endocrinology and metabolism, 4) brain and behavior, 5) reproduction, 6) immunoregulation and protection, and 7) actions associated with pathological disease states. The osmoregulatory actions of PRL are particularly important for fish that reside in water with different salinities. GHs almost exclusively stimulate somatic growth of the vertebrates, primarily through induction of insulin-like growth factor (IGF), and promote sexual maturation and reproductive function, and seawater adaptation in some teleost fish. SL, a fish-specific pituitary hormone, has been implicated in many physiological processes like PRL, including background adaptation, energy homeostasis, fat metabolism, osmoregulation, stress response, some aspects of reproduction, and acid-base regulation in teleosts. PRL, SL and GH exert their biological effects by interacting with and dimerizing specific single transmembrane-domain receptors. They belong to the cytokine class-1 receptor super family with no intrinsic tyrosine kinase, which use the JAK (Janus kinase)–STAT (signal transducers and activators of transcription) cascade as a major signaling pathway (Bole-Feysot *et al.*, 1998). Thus, these hormones and their receptors are thought to have co-evolved as a result of gene duplication and subsequent divergence.

The synthesis and secretion of the pituitary hormones are regulated by hypothalamic neurohormones, which are generally classified into two categories: releasing or inhibiting hormones. Specific action of each hypothalamic neurohormone is determined by the presence of a highly specific receptor on a target cell in the pituitary. These receptors belong to the G-protein coupled receptors (GPCR), which are seven trans-membrane-segments and bind to and activate or inhibit G proteins.

In this chapter, we will briefly summarize the molecular characteristics and biological functions of PRL and SL and review the hypothalamic neurohormones in controlling the secretion of PRL and SL in teleost fish.

2. THE NEUROENDOCRINE REGULATION OF PROLACTIN SECRETION

2.1. Prolactin and its Receptor

Prolactin (PRL). PRL is a single chain polypeptide produced in the pars distalis of the pituitary gland in gnathostomes but not in agnathans (Kawauchi *et al.*, 2002). In most teleost fish, PRL cells are segregated into a nearly homologous mass within the most anterior portion of the pituitary, RPD, whereas only a small number of another group of endocrine cells, adrenocorticotropic hormone (ACTH) cells, are on the border along the neurohypophysis. Therefore, this region can be isolated to obtain a group of nearly homologous PRL cells and be used for incubation and perfusion to investigate the effects of neurohormones and other factors by *in vitro* incubations.

Teleost PRL genes have been characterized for the common carp (*Cyprinus carpio*) (X52881), chinook salmon (*Oncorhynchus tshawytscha)* (S66606), Mozambique tilapia (*Oreochromis mossambicus*) (X92380) and gilthead sea bream (*Sparus aurata*) (AJ509807). All known PRL genes consist of five exons and four introns, as in SLs and the majority of GHs. Only advanced teleost fish GHs consists of six exons, in which the fifth intron is inserted in the fifth exon of other GH genes during diversification of teleosts. The TATA box of teleost PRL genes is about 20 bp upstream from the transcriptional initiation site. Several Pit-1/GHF-1 binding consensus sequences have been identified in the 5′-flanking region of PRL genes in rainbow trout (*Oncorhynchus mykiss*) (X95907), tilapia (X92380) and gilthead sea bream (AJ509807).

Protein and cDNA sequences for fish PRLs are available from over 30 teleost species. Two structural variants of PRL have been identified in some teleosts. They differ by only a few amino acid substitutions in salmonids and cyprinoids, suggesting there is no functional difference between the two isoforms. In contrast, two PRLs of tilapia, PRL_{177} (P09318) and PRL_{188} (P09319), are different in chain length and share only 69% sequence identity. PRL_{188} is more similar than PRL_{177} to the PRLs of other fish. Therefore some functional differences become evident. In amino acid sequence comparison, the most noticeable feature of teleost PRL is the lack of an N-terminal disulfide bond in tetrapods, lungfish (*Protopterus aethiopicus*) and sturgeon (*Acipenser gueldenstaedtii*) PRL, suggesting the loss of this segment during the divergence of the ray-finned fish.

The expression of the PRL gene has been demonstrated by using reverse transcriptase polymerase chain reaction (RT-PCR) in extrapituitary tissues, such as liver, intestine and gonads, but not in the brain, of the sea bream (Santos *et al.*, 1999) and in the liver, kidney, spleen, gill, muscle, gonads and brain, but not in the intestine, of goldfish (*Carassius auratus*) (Imaoka *et al.*, 2000) and rainbow trout (Sakamoto *et al.*, 2003a). PRL may act in an autocrine or paracrine manner in these extrapituitary tissues and poses an interesting area for future research.

PRL receptor. PRL binds to a cell surface receptor, a member of the cytokine/hematopoietin receptor superfamily. In mammals, there are long, intermediate and short PRLR forms that are generated by alternative splicing and promoter usage from a single PRLR gene (Freeman *et al.,* 2000). These isoforms differ in the length and composition of their intracellular domains and their signal transduction mechanisms. The ligand binding triggers the formation of a homodimer consisting of one molecule of PRL and two molecules of receptor. In the extracellular domain of PRLR, two pairs of disulfide bonds and a Trp-Ser-Xaa-Trp-Ser motif (WS motif) are required for the proper folding and trafficking of the receptor, although they are not responsible for binding itself (Bole-Feysot *et al.*, 1998; Freeman *et al.*, 2000). In the intracellular domains there are two relatively conserved regions termed box 1 and box 2. Box 1 is a membrane-proximal, Pro-rich motif necessary for the consensus folding of the molecule recognized by the transducing molecules. Box 2 is less conserved and is missing in the short isoform of the prolactin receptor. Site 1 on PRL binds to one receptor molecule, after a second receptor molecule binds to site 2 on the hormone. At these sites, the PRLR molecule associates with a cytoplasmic tyrosine kinase, JAK2, and another kinase such as STAT. Dimerization of the receptor induces activation of the Jak kinase followed by phosphorylation of the receptor and STAT. The phosphorylated STAT dimerizes and translocates to the nucleus and binds to specific promoter elements on PRL-responsive genes. In addition, the Ras/Raf/MAP kinase pathway is also activated by PRL and may be involved in the proliferative effects of the hormone (Figure 5.1).

In teleosts, PRLRs have been cloned in tilapia (Sandra *et al.*, 1995; Prunet *et al.*, 2000), goldfish (Tse *et al.*, 2000), rainbow trout (Prunet *et al.*, 2000), sea

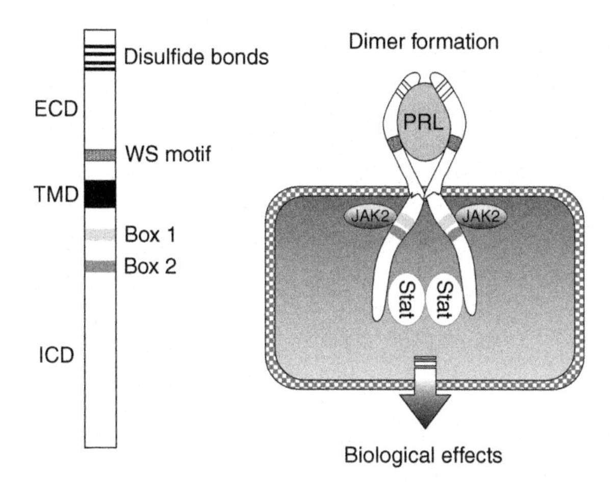

Fig. 5.1. Schematic representation of PRLR (left side) and PRL and PRLR mechanism of action (right side). PRL induces a dimerization of two PRL receptors. ECD, extracellular domain; TMD, transmembrane domain; ICD, intracellular domain. Binding of PRL to the PRL dimerized receptor activates the STAT (signal transducers and activators of transcription) class of transcription factors via activation of JAKs (Janus kinases). Ligand-induced dimerization of the receptor induces the reciprocal tyrosine phosphorylation of the associated JAKs, which, in turn, phosphorylates tyrosine residues on the cytoplasmic tail of the receptor.

bream (Santos *et al.*, 2001), Japanese flounder (*Paralichthys olivaceus*) (Higashimoto *et al.*, 2001), carp (San Martin *et al.*, 2004), and pufferfish (*Takifugu rubripes*) (Lee *et al.*, 2006). Multiple PRLR mRNAs were observed in goldfish (Tse *et al.*, 2000) and gilthead sea bream (Santos *et al.*, 2001). San Martin *et al.* (2007) first identified carp prolactin receptor gene expressing transcript isoforms encoding for short forms of the protein, but not associated with alternative promoters, unlike in humans and rodents (Bole-Feysot *et al.*, 1998). Recently Huang *et al.* (2007) identified two PRLR genes in the fugu and zebrafish (*Danio rerio*) genomes but not in the genomes of other vertebrates, and isolated two PRLR cDNAs from black sea bream (*Acanthopagrus schlegeli*) and Nile tilapia (*Oreochromis niloticus*). The classical one was named PRLR1, the newly identified one was named PRLR2. Both PRLRs resemble the long-form mammalian PRLRs, but have only about 30% similarity. These receptors differ mainly in box 2 of the intracellular domain and some intracellular Tyr residues are missing in PRLR2. Teleost PRLRs share important structural features for signal transduction in common with mammalian molecules, while the WS motif of fish PRLRs contains a substitution (Ser to Thr) in the fifth position of the goldfish (Tse *et al.*, 2000) and sea bream (Santos *et al.*, 2001) PRLRs.

The higher levels of PRLR expression were observed in osmoregulatory tissues, such as gill, kidney and intestine in tilapia (Sandra *et al.*, 1995, 2000; Prunet *et al.*, 2000; Pierce *et al.*, 2007), goldfish (Tse *et al.*, 2000), rainbow trout (Prunet *et al.*, 2000), gilthead sea bream (Santos *et al.*, 2001; Huang *et al.*, 2007), flounder (Higashimoto *et al.*, 2001), pufferfish (Lee *et al.*, 2006) and Atlantic salmon (*Salmo salar*) (Kiilerich *et al.*, 2007). Lower levels of expression were also detected in the brain, gonad, liver, muscle, skin, spleen, head kidney, lymphocytes and bone of some fish (Sandra *et al.*, 1995, 2000; Tse *et al.*, 2000; Higashimoto *et al.*, 2001; Santos *et al.*, 2001; Huang *et al.*, 2007). In black sea bream, the expression levels of PRLR1 were generally higher than PRLR2 in most tissues except for the gill. In the sea bream kidney, the PRLR1 expression was up-regulated by estradiol (E_2) and cortisol, but not testosterone, while the PRLR2 expression was down-regulated by E_2 and testosterone, but not cortisol (Huang *et al.*, 2007). The wide distribution of fish PRLRs in tissues and different gene expression patterns with steroid hormones correlates well with the multiple functions of PRL.

2.2. Functions of Prolactin

Water and electrolyte balance. Pickford and Phillips (1959) were the first to demonstrate that ovine PRL enabled hypophysectomized killifish (*Fundulus heteroclitus*) to survive in freshwater. Since then it has been shown that PRL plays a key role in the freshwater adaptation in many teleosts by stimulating ion retention and preventing the water influx to the osmoregulatory organs. The principal osmoregulatory organs are the gill, opercular membrane, skin, gastrointestinal tract, kidney and urinary bladder. Consistent with this action, PRL cell activity is higher in fish acclimated to freshwater than in those to seawater and PRL release from the pituitary *in vitro* increases as medium osmotic pressure is reduced (Nagahama *et al.*, 1975; Grau *et al.*, 1994). Because a large number of excellent reviews on this subject have been published, we will not repeat but will refer readers to these reviews: Hirano, 1986; Bern and Madsen, 1992; McCormick, 2001; Manzon, 2002; Sakamoto and McCormick, 2006. Recent studies showed higher PRLR expression levels in the primary osmoregulatory organs (Sandra *et al.*, 2000; Prunet *et al.*, 2000; Tse *et al.*, 2000; Higashimoto *et al.*, 2001; Santos *et al.*, 2001; Pierce *et al.*, 2007). PRLR gene expression in the gill was lower in seawater-acclimated tilapia than in freshwater fish (Shiraishi *et al.*, 1999; Prunet *et al.*, 2000; Pierce *et al.*, 2007). It is predicted that euryhaline teleosts may have the PRL and PRLR system to acclimatize themselves to sudden osmotic changes in the environment, ubiquitously expressing PRLRs in the osmoregulatory organs.

Growth and development. The somatotropic actions of PRL have been extensively reported in mammals, but much fewer studies are reported in fish. In tilapia (Shepherd *et al.*, 1997), PRL_{177}, but not PRL_{188}, has somatotropic activity as evidenced by an increase in $[H^3]$thymidine and $[S^{35}]$sulfate incorporation into branchial cartilage and also stimulation of hepatic IGF-I mRNA expression. In addition, PRL_{177} has the ability to displace tilapia GH from its receptor by radio receptor assay, but PRL_{188} does not. Sandra *et al.* (1995) demonstrated that a recombinant PRLR shows higher affinity for PRL_{188} than for PRL_{177}. Leena *et al.* (2001) showed that PRL inhibited several enzymes involved in biosynthesis of fatty acids in climbing perch (*Anabas testudineys*) liver, and PRL treatment in juvenile coho salmon (*Oncorhynchus kisutch*) resulted in pronounced lipid depletion (Sheridan, 1986).

The role of PRL in early development is well established in mammals. PRL may also play a role in the early development of teleosts. The PRL_{188} gene transcript was detected in newly hatched tilapia larvae, and the PRL_{177} gene transcript was detected in the pituitary of embryos one day before hatching (Ayson *et al.*, 1994). In rainbow trout, the PRL gene is detected during embryogenesis (prior to and after pituitary organogenesis), and after hatching (Yang *et al.*, 1999). In gilthead sea bream, PRL mRNA is detected in embryos and after hatching, and the PRLR gene is first detected in embryos at blastula, and higher levels were maintained until 2 days after hatching (Santos *et al.*, 2003). These studies have shown that PRL is in the developing pituitary gland in fish embryos and larvae, and PRLR mRNA and protein are also present in fish embryos and have a widespread tissue distribution in larvae (Power, 2005). These results support somatotropic action of PRL in teleosts.

Prolactin has an antimetamorphic action in amphibian tadpoles. In the Japanese flounder, ovine PRL antagonized the stimulatory effect of triiodothyronine on the resorption of the dorsal fin rays of prometamorphic larvae *in vitro* and also delayed the desorption of the dorsal fin rays without affecting the rates of eye migration and settling. However, ovine GH was without these effects. Both PRL and GH genes were increasingly expressed during successive metamorphic stages. These results suggest involvement of PRL with thyroid hormones in the control of development in the flounder (De Jesus *et al.*, 1994).

Immunoregulation. Endocrine–immune interactions have become an important research field in teleosts. It has been shown that cortisol, sex steroid and pituitary hormones including PRL influence immune function in several teleosts (Harris and Bird, 2000). PRL was shown to stimulate leukocyte mitogenesis (Sakai *et al.*, 1996a; Yada *et al.*, 2002a), respiratory burst activity (Sakai *et al.*, 1996b) and phagocytosis (Kajita *et al.*, 1992; Sakai *et al.*, 1995; Narnaware *et al.*, 1998) and to increase plasma IgM titers (Yada *et al.*, 2002a). In tilapia, PRLR mRNA was observed in the spleen and head kidney as well as the circulating lymphocytes (Sandra *et al.*, 2000). The expression of two tilapia PRLs and PRL receptors were detected in lymphoid tissues and cells, and the level of expression was higher in seawater-acclimated fish than that in freshwater fish. These results suggest that immunomodulatory actions of PRL appear to be independent of its osmoregulatory action.

Behavior. The behavioral effects of PRL in amphibians have been well studied, however much less is known in fish. During displaying parental fanning behavior, the rate of synthesis and release of the PRL cells in the pituitary of sexually mature male three-spined sticklebacks (*Gasterosteus aculeatus*), as estimated with quantitative electron microscopy, was enhanced considerably (Slijkhuis *et al.*, 1984) and PRL administration stimulated parental fanning behavior in sexually mature male three-spined sticklebacks (de Ruiter *et al.*, 1986). In tilapia, PRL induces transformation of skin gland to produce mucus for nourishment to the offspring (Ogawa, 1970). The injection of PRL into the fourth ventricle affects the drinking behavior of the seawater eel, inhibiting the water intake (Kozaka *et al.*, 2003).

Reproduction. It has been reported that PRL stimulates steroidogenesis in ovary and testis (Rubin and Specker, 1992). In salmonids, plasma PRL and pituitary mRNA levels increased during sexual maturation (Ogasawara *et al.*, 1996; Onuma *et al.*, 2003). An elevation of PRL levels during sexual maturation was also observed in tilapia (Tacon *et al.*, 2000). It has been well established that GnRH and E_2 stimulate PRL release in fish both *in vivo* and *in vitro* (Weber *et al.*, 1997; Kagabu *et al.*, 1998). Also, PRLR mRNA was detected in gonads in some teleosts, (Sandra *et al.*, 2000; Tse *et al.*, 2000; Higashimoto *et al.*, 2001; Santos *et al.*, 2001). These results suggest that PRL plays a role at various stages of the reproductive process in fish.

2.3. Prolactin Secretion by Hypothalamic Peptides

In teleosts, the hypothalamic neurohormones, plasma factors and plasma osmolality regulate the secretion of PRL from the pituitary gland. Plasma osmolality exerts direct and predominant regulatory actions on PRL secretion. Hyposmotic conditions stimulate PRL release (Kaneko and Hirano, 1993; Shepherd *et al.*, 1999; Weber *et al.*, 2004). Several plasma factors, which are secreted from other tissues, can also be classified into two categories as stimulating or inhibiting. The stimulating factors include E_2 (Barry and Grau 1986; Williams and Wigham 1994a; Weber *et al.*, 1997; Kagabu *et al.*, 1998), angiotensin II (Eckert *et al.*, 2003), natriuretic peptides

(Fox *et al.*, 2007) and IGFs (Fruchtman *et al.*, 2000). Inhibiting factors comprise cortisol (Borski *et al.*, 1991; Williams and Wigham, 1994a; Uchida *et al.*, 2004), ouabain (Kajimura *et al.*, 2005), urotensin II (Grau *et al.*, 1982; Leedom *et al.*, 2003) and vasoactive intestinal polypeptide (VIP) (Brinca *et al.*, 2003). It is well established that negative hypothalamic control of PRL release is exerted by dopamine (DA) and somatostatin (SS) in fish. On the other hand, several known hypothalamic neuropeptides such as GnRH, neuropeptides Y (NPY), copeptine and pituitary adenylate cyclase-activating polypeptide (PACAP), have been reported to enhance PRL-releasing activity. Recently, a novel neuropeptide, named PRL-releasing peptide (PrRP), was found in mammals and subsequently we identified teleost PrRP and confirmed the specific PRL-releasing action of this new hypothalamic neuropeptide (Moriyama *et al.*, 2002, 2007). Here we review the regulation of PRL secretion by hypothalamic neuropeptides including our new data on PrRP (Figure 5.2).

2.3.1. Inhibiting Factors

Dopamine (DA). DA belongs to a class of neurotransmitters in the central nervous system known as catecholamines. Since the recognition of its role as an inhibitor of the PRL release in 1970s, DA has been established as the primary regulator of PRL gene expression and PRL release in mammals and bird (Al Kahtane *et al.*, 2003). Two DA receptor subtypes, D_1 and D_2 receptors, are two distinct membrane proteins belonging to the GPCR superfamily (Oliveira *et al.*, 1994). D_1 receptor is linked to adenylate cyclase stimulation and mediates GH release from the goldfish pituitary (Wong *et al.*, 1993), while D_2 receptor inhibits adenylate cyclase activity and inhibits PRL release in mammals (Jose *et al.*, 1999).

In goldfish, the injection of DA agonists induced significant hypertrophy and nuclear enlargement in PRL cells. The numbers of cytoplasmic granules were similar in control and treated goldfish. In tilapia, the treatment with DA increased the quantities of rough endoplasmic reticulum in PRL cells and decreased the number of secretory granules (Hazineh *et al.*, 1997). In rainbow trout, incubation of the pituitary glands with DA reduced the release of PRL into the medium. The DA precursor, L-dopa, reduced pituitary PRL content *in vivo*, while the DA-receptor blocker increased total PRL content and amount released in the subsequent incubation (James and Wigham, 1984). These results suggest that DA inhibits PRL cell activity in teleosts.

Johnston and Wigham (1988) proposed involvement of the agonist-dependent G_i proteins, Ca^{2+}, calmodulin and cAMP in the mechanisms for regulation in the rainbow trout. GTP enhanced the inhibitory action of DA on PRL release *in vitro* and DA also reduced pituitary cAMP content. Forskolin increased both PRL release and cAMP content *in vitro*, but this

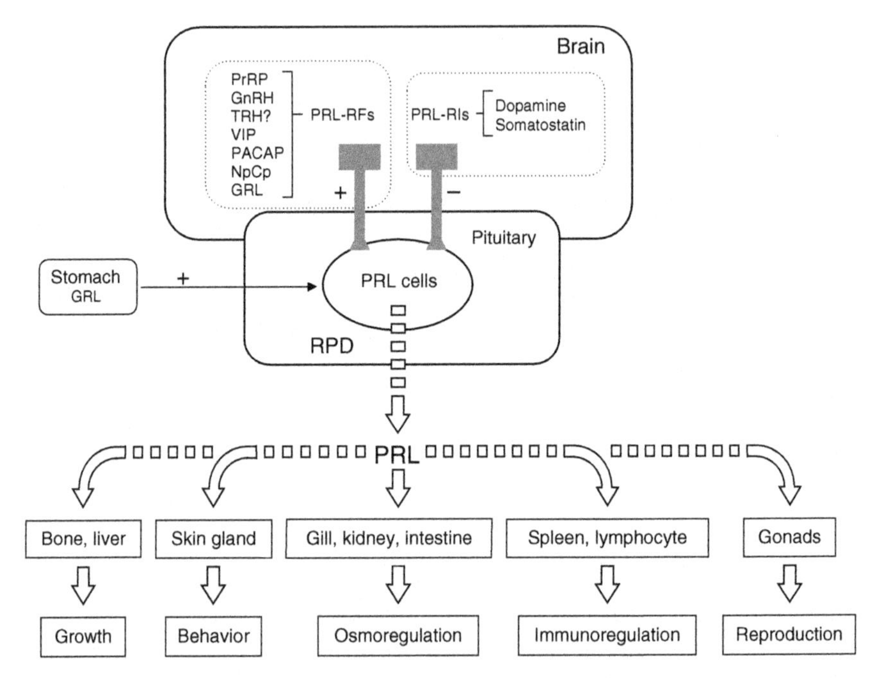

Fig. 5.2. Regulation of PRL release from the pituitary and functions of PRL in fish. PRL release from PRL-producing cells is under the multifunctional control of various neuropeptides and neurotransmitters. These stimulatory/inhibitory regulators from brain are delivered to PRL-producing cells in the RPD by direct innervation from the hypothalamus. PRL-RFs, prolactin-releasing factors; PRL-RIs, prolactin-release inhibitors; PrRP, prolactin-releasing peptide; GnRH, gonadotropin-releasing hormone; TRH, thyrotropin-releasing hormone; VIP, vasoactive intestinal peptide; PACAP, pituitary adenylate cyclase activating polypeptide; NpCp, copeptine; GRL, ghrelin; RPD, rostral pars distalis.

effect was prevented by DA and did not occur in Ca^{2+}-free medium. The cAMP analogue increased PRL synthesis in low Ca^{2+} medium, though release was not significantly affected. The calcium ionophore increased PRL release, but this effect was not seen with a voltage-dependent Ca^{2+} channel blocker. The calmodulin blocker increased PRL synthesis and pituitary PRL content *in vivo* and elevated pituitary cAMP levels (Johnston and Wigham, 1988). Thus, DA activates probably D_2 receptors in PRL cells, which causes a reduction in adenylate cyclase activity and thus inhibits cAMP production. The reduced amounts of cAMP induce inhibition of PRL secretion.

In goldfish, injection of DA [0.5–50 $\mu g\ g^{-1}$ body weight (BW)] and incubation of the pituitary cells with 1–100 μM exerted significant decrease of PRL gene expression in a dose-related manner. DA significantly inhibited the goldfish PRL promoter activity. A DA responsible region for the inhibition of promoter activity may exist within −188 bp, where two putative pit-1 binding sites are located in the goldfish PRL gene (Tse *et al.*, 2008). These results establish that DA acts as a potent negative regulator of PRL gene transcription and PRL secretion in teleosts.

Somatostatin (SS). The presence of multiple SS genes has been demonstrated in a number of fish species. SS-14 is conserved with identical primary structure in representative species across the vertebrates. SS receptors (SST) also belong to the GPCR superfamily. A total of four different SST subtypes are now known in fish (Zupanc *et al.*, 1999; Slagter *et al.*, 2004).

SS immunoreactivity was localized both in the brain and in the pituitary of teleosts (Dubois *et al.*, 1979; Olivereau *et al.*, 1984a,b; Grau *et al.*, 1985; Marchant *et al.*, 1989). In the pituitary, SS-ir fibers were found to extend to the GH cells in the PPD in the carps, goldfish, tilapia, killifish and mudskipper (*Periophthalmus modestus*) (Kah *et al.*, 1982; Olivereau *et al.*, 1984a,b; Grau *et al.*, 1985) and also to the ACTH cells in the RPD in the sea bass (*Dicentrarchus labrax*) (Power *et al.*, 1996). The mRNAs encoding SSTs in fish are broadly distributed in the pituitary (Sheridan *et al.*, 2000; Slagter *et al.*, 2004).

SS has been shown to inhibit PRL release from incubated pituitary tissue or RPD in the tilapia (Grau *et al.*, 1982, 1985; Helms *et al.*, 1991), the molly (*Poecilia latipinna*) (Wigham and Batten, 1984) and the rainbow trout (Williams and Wigham, 1994b). The exposure of organ-cultured pituitary gland or RPD of these teleosts to hyposmotic medium causes a rapid and steep rise in the release of newly synthesized PRL. When SS was added to the medium, SS quickly prevented and reduced the PRL release in a dose-related manner. SS antagonist blocked the inhibitory effect of SS in rainbow trout (William and Wigham, 1994b).

SS reduced the forskolin-stimulated increase in cAMP levels in a manner consistent with its rapid effects on PRL release. SS may act to render adenylate cyclase less responsive to direct stimulation by forskolin (Helms *et al.*, 1991). In tilapia, the inhibitory action of SS on PRL release was completely prevented by the presence of the calcium ionophore A23187. PRL release was also blocked when Ca^{2+} was excluded from the incubation medium, even in the presence of ionophore (Grau *et al.*, 1982).

2.3.2. Stimulating Peptides

Prolactin-releasing peptide (PrRP). Hinuma *et al.* (1998) identified a novel hypothalamic neuropeptide with a potent PRL-releasing activity from bovine hypothalamus extract as a ligand of an orphan GPCR. This neuropeptide, termed PrRP, possesses two molecular forms, a 31-amino acid peptide (PrRP31) and the C-terminal 20 residues (PrRP20), which belong to the so-called "arginyl-phenylalanyl-amide peptides" (RFa). To date, PrRP homologues have been identified from six classes of vertebrates (Figure 5.3).

In teleosts, a homologue of mammalian PrRP20 was first isolated from the brain of Japanese crucian carp (*Carassius auratus langsdorfi*) (Fujimoto *et al.*, 1998) and subsequently from chum salmon (*Oncorhynchus keta*) (Moriyama *et al.*, 2002) and Mozambique tilapia (Seale *et al.*, 2002). PreproPrRP cDNAs have been cloned from the brains of crucian carp (Satake *et al.*, 1999), chum salmon (Moriyama *et al.*, 2002), tilapia (Seale *et al.*, 2002) and Atlantic salmon (Montefusco-Siegmund *et al.*, 2006). They encode a single PrRP segment of 20 amino acid residues with identical sequence, so that it is designated as teleost PrRP (tPrRP) in this chapter (Figure 5.3).

PrRP-immunoreactive (tPrRP-ir) cell bodies were found in the posterior part of the hypothalamus, and PrRP-ir fibers were projected widely from the hypothalamus to the brain in goldfish (Wang *et al.*, 2000a), rainbow trout (Moriyama *et al.*, 2002) and guppy (*Poecilia reticulata*) (Amano *et al.*, 2007). Occurrence of PrRP mRNA in the hypothalamus has been detected by using RT-PCR in crucian carp (Satake *et al.*, 1999), mudskippers (Sakamoto *et al.*, 2005), Atlantic salmon (Montefusco-Siegmund *et al.*, 2006) and goldfish (Kelly and Peter, 2006). In rainbow trout (Moriyama *et al.*, 2002) and guppy (Amano *et al.*, 2007), a small number of PrRP-ir fibers projected to the pituitary and terminated close to the PRL cells in the RPDs and to the SL cells in the PI. In Atlantic salmon (Montefusco-Siegmund *et al.*, 2006),

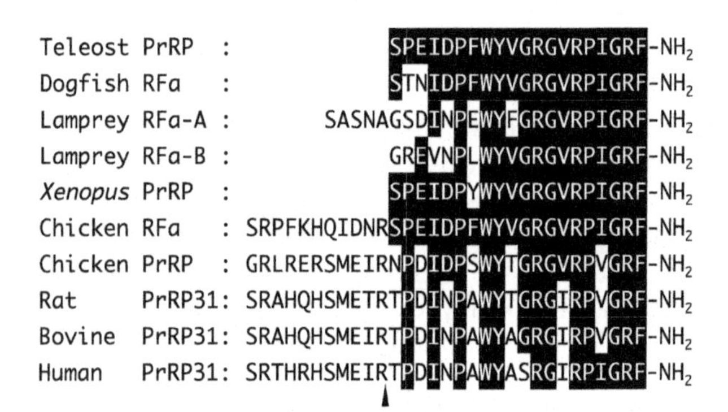

Fig. 5.3. Alignment of PrRP and its homologues. Highlighted amino acids are identical to those of teleost PrRP20. Teleost: crucian carp (AB020024), chum salmon (Moriyama *et al.*, 2002), Atlantic salmon (NM_001123641), tilapia (Seale *et al.*, 2002), zebrafish (EU117421). Dogfish RFa (AB433893), Lamprey (*Petromyzon marinus*) RFa-A and B (Moriyama *et al.*, 2007), *Xenopus laevis* PrRP (AB251344), Chicken PrRP (EF418015) and RFa (NM_001114503), Rat (NM_001101647), Bovine (NM_174790), Human (NM_015893).

although PrRP-ir fibers were not observed in the pituitary gland, PrRP-ir cell bodies and PrRP mRNA were also found in the RPD of the pituitary where PRL is synthesized. In addition the expression of PrRP mRNA was detected in the peripheral tissues such as in the liver, gut and ovary, and to a lesser extent, in the skin and kidney of mudskipper (Sakamoto *et al.*, 2005) and the retina of cyprinid fish (Wang *et al.,* 2000b). It is interesting to note that corresponding to the distribution of PrRP mRNA, PRL mRNA was also detectable in these organs of mudskipper (Sakamoto *et al.*, 2005). The widespread distribution of both tPrRP and PRL suggests involvement in a number of physiological functions in teleosts (Table 5.1).

As mentioned above, PrRP was originally identified as a ligand for the orphan GPCR (Hinuma *et al.*, 1998). The structural information of PrRP receptor gene homologues is now available from human, rat, mouse, guinea pig, chicken, pufferfish, zebrafish and sturgeon (Lagerström *et al.*, 2005). These are found as two separate subtypes and share a common ancestry with the NPY receptor. The distribution of PrRP receptor in teleost has not been reported as yet.

In tilapia, incubation of the tilapia pituitary with tPrRP (100 nM) significantly stimulated the release of two forms of tilapia PRL (PRL_{188} and PRL_{177}) but failed to stimulate GH release from the pituitary in organ culture. However, the effect of tPrRP on PRL release was less pronounced than the marked increase in PRL release in response to hyposmotic medium. In contrast, mammalian PrRPs had no effect on PRL release. tPrRP was

Table 5.1
Functions of teleost PrRP

Effect	Fish	Reference
PRL release	Rainbow trout	Moriyama *et al.* (2002); Sakamoto *et al.* (2003b)
	Tilapia	Seale *et al.* (2002)
	Mudskipper	Sakamoto *et al.* (2005)
	Goldfish	Kelly and Peter (2006)
SL release	Rainbow trout	Moriyama *et al.* (2002)
GH inhibition	Rainbow trout	Moriyama *et al.* (2002)
Osmoregulation	Mudskipper	Sakamoto *et al.* (2005)
	Goldfish	Fujimoto *et al.* (2006); Kelly and Peter (2006)
Terrestrial adaptation	Mudskipper	Sakamoto *et al.* (2005)
Visceral muscle contraction	Crucian carp	Fujimoto *et al.* (1998)
Cardiovascular regulation	Rainbow trout	Sakamoto *et al.* (2003)
Food intake	Goldfish	Kelly and Peter (2006)
Light response	Crucian carp	Wang *et al.* (2000b)

PRL, prolactin; SL, somatolactin; GH, growth hormone.

equipotent with chicken GnRH in stimulating PRL release in the pituitary preincubated with E_2. Circulating levels of PRL were significantly increased 1 h after intraperitoneal (ip) injection of 0.1 mg g^{-1} of tPrRP in female tilapia in freshwater but not in males (Seale *et al.*, 2002).

In rainbow trout, perfusion of the pituitaries with tPrRP at concentrations of 10 pM to 100 nM demonstrated maximum PRL release at 100 pM and maximum SL release at 10 and 100 nM. However, GH release was not affected. An ip injection of tPrRP into the rainbow trout increased the plasma PRL and SL levels at 3 and 9 h, respectively, at doses of 50 and 500 ng g^{-1} body weight. In contrast, plasma GH levels were decreased after 1 h at 500 ng g^{-1} body weight (Moriyama *et al.*, 2002). A single intra-arterial injection of tPrRP (40 nmol kg^{-1}) through a dorsal aorta catheter in rainbow trout increased plasma PRL levels rapidly at 2 min after injection and PRL mRNA levels in pituitaries were elevated at 8 h after the injection. In contrast, plasma SL levels were decreased and GH and SL mRNA levels were not significantly affected (Sakamoto *et al.*, 2003b). In goldfish, ip injection of tPrRP at doses of 25 ng g^{-1} significantly elevated pituitary PRL mRNA levels at 8 h after administration. Administration of 250 ng g^{-1} ip resulted in a significant elevation in pituitary PRL mRNA expression but not significantly different from that caused by the 25 ng g^{-1} ip injection (Kelly and Peter, 2006). Fujimoto *et al.* (2006) demonstrated that the pituitary mRNA levels of PRL were increased by tPrRP but decreased by anti-tPrRP. Furthermore, the rate of water inflow in the gills was decreased by tPrRP and increased by anti-tPrRP, and tPrRP expanded the mucosa cell layers on the scales.

These results suggest that tPrRP is physiologically involved in the regulation of PRL cell activity or PRL secretion and osmotic balance in teleosts, particularly freshwater fish. However, in mammals, the hypophysiotropic activity of PrRP to PRL cells remains unclear (Taylor and Samson, 2001). Instead, the widespread distribution of PrRP mRNA, peptide and receptor in the central nervous system suggests involvement in a number of brain functions (Sun *et al.*, 2005). Kelly and Peter (2006) found that tPrRP is involved in regulating appetite and hydromineral balance in goldfish. Goldfish exhibited a dose-dependent reduction in food intake in response to both ip and intracerebroventricular injections of PrRP. Further studies will be needed to clarify the putative brain functions of PrRP in fish.

Gonadotropin-releasing hormone (GnRH). GnRH is a decapeptide that controls reproduction via the hypothalamic–pituitary axis. GnRH has also been shown to be a potent stimulator of GH and PRL secretion in mammals (Blackwell *et al.*, 1986; Robberecht *et al.*, 1992). GnRH and its gene have been identified throughout the chordates (q.v. Chapter 3).

In many teleost fish, sea bream (sb) GnRH cell bodies are predominant in the preoptic area and sbGnRH immunoreactive fibers project mainly to the

PPD of the pituitary in sea bream (Gothilf *et al.,* 1996), tilapia (Parhar, 1997), sea bass and barfin flounder (Amano *et al.,* 2002). GnRH-ir fibers extend to the three areas of the pituitary, RPD, PPD and PI, and are found in close association in the pejerrey (Vissio *et al.,* 1999) and in the Nile perch (*Lates niloticus*) (Mousa and Mousa, 2003).

GnRH exerts its actions through binding to the GPCRs on the membrane (q.v. Chapter 3). They mediate their intracellular actions through the activation of one or more G-proteins (Sealfon *et al.*, 1997; Millar *et al.*, 2004). In the tilapia, in addition to LH and FSH cells, three GnRH-Rs were found in PRL, GH, thyrotropin (TSH), melanocyte-stimulating hormone, ACTH and SL cells (Parhar *et al.*, 2005). GnRH-R1 and -R2 transcripts were significantly higher in PRL cells of mature males (Parhar *et al.*, 2005). In the pejerrey, GnRH binding sites were also detected in PRL cells (Stefano *et al.*, 1999). These data on the localization of GnRH-ir fibers and GnRH-Rs suggest that GnRH is a possible hypothalamic neuropeptide that can modulate PRL.

In tilapia, three native forms of GnRH are shown to stimulate the release of PRL from the RPD of the pituitary *in vitro* with the following order of potency: cGnRHII>sGnRH>sbGnRH. In addition, a mammalian GnRH analogue stimulated the release of PRL from the RPD incubated in either iso-osmotic or hyperosmotic medium, the latter normally inhibiting PRL release. The response of the RPD to GnRH was augmented by co-incubation with testosterone or E_2 (Weber *et al.*, 1997). The mechanism by which GnRH regulates PRL cell function has also been examined in tilapia (Tipsmark *et al.*, 2005). GnRH stimulated PRL release through an increase in phospholipase C (PLC), inositol triphosphate (IP3) and intracellular calcium (Ca_i^{2+}) signaling in tilapia. cGnRH-II induced a rapid dose-dependent increase in Ca_i^{2+} in dispersed tilapia PRL cells. The Ca_i^{2+} signal was abolished by U-73122, an inhibitor of PLC-dependent phosphoinositide hydrolysis. Correspondingly, this inhibitor inhibited cGnRH-II-induced $tPRL_{188}$ secretion, suggesting that activation of PLC mediates cGnRH-II's stimulatory effect on PRL secretion. Pretreatment with Ca^{2+} antagonists, 8-(N, N-diethylamino) octyl-3,4,5-trimethoxybenzoate hydrochloride, an inhibitor of Ca^{2+} release from intracellular stores, impeded the effect of cGnRH-II on Ca_i^{2+}. Ca^{2+} antagonists suppressed secretion of $tPRL_{188}$ in response to cGnRH-II. These data suggest that GnRH may elicit its PRL-releasing effect by increasing Ca_i^{2+}. Furthermore, the rise in Ca_i^{2+} may be derived from PLC/IP3-induced mobilization of Ca^{2+} from intracellular stores along with influx through L-type voltage-gated Ca^{2+} channels.

In growing and maturing masu salmon (*Oncorhynchus masou*), GnRH increases the expression of genes encoding GH, PRL, and SL in the pituitary in particular seasons, probably in relation to physiological roles of the hormones (Bhandari *et al.*, 2003). Onuma *et al.* (2005) further suggested that sGnRH directly modulates synthesis of Pit-1, and expression of PRL and SL genes, but the regulation of GH/PRL/SL family hormone genes by GnRH is probably indirect, particularly in the late stage of gametogenesis.

Thyrotropin-releasing hormone (TRH). TRH exhibits, in addition to thyrotropin-releasing activity, a wide variety of hormonal and neurotransmitter/neuromodulator functions including GH- and PRL-releasing activity in many vertebrates (Nillni and Sevarino, 1999). In carp, TRH-ir fibers are present in the neuronal processes extending from the preoptic nucleus to the nucleus recessus lateralis of the hypothalamus and in the pituitary. The fibers in the pituitary were restricted mainly to the region of the neural lobe. Some fibers, localized in the neural lobe, were in close proximity to the anterior lobe (Hamano *et al.*, 1996).

TRH initiates all of its effects by interacting with receptors on the surfaces of cells, belonging to the GPCR superfamily (Sun *et al.*, 2003).

It has been well established that TRH stimulates PRL release *in vivo* and *in vitro* in mammals (Grosvenor and Mena, 1980). However, the physiological role of TRH on PRL cells in teleosts remains somewhat equivocal, i.e., species- or condition-specific. TRH was able to stimulate PRL release under certain conditions such as osmotic pressure and E_2 preincubation. In molly, TRH stimulated PRL release into a medium of hyperosmotic pressure, but not into that of hyposmotic pressure (Wigham and Batten, 1984). In tilapia, TRH could stimulate if the fish were primed with E_2 treatment (Barry and Grau, 1986). In rainbow trout, three different results have been reported; TRH exerts no effect (James and Wigham, 1984) or a stimulating effect (Prunet and Gonnet, 1986) or an inhibiting effect (Williams and Wigham, 1994a). In carp, TRH stimulated a dose-related increase in the release of newly synthesized PRL at 1–100 nM in the primary cultured pituitary cells of common carp (Kagabu *et al.*, 1998). In contrast, Tse *et al.* (2008) demonstrated that TRH decreased PRL mRNA levels in primary cultured pituitary cells of the goldfish through down-regulation of transcription for the PRL gene promoter. Due to these conflicting results TSH may not be a critical factor to control PRL release in teleosts.

Other hypothalamic peptides. *Vasoactive intestinal polypeptide (VIP)* is a neuropeptide synthesized in the cerebral cortex, hypothalamus and anterior pituitary, as well as in other tissues. It has been shown that hypothalamic VIP stimulated PRL release by activating the adenylate cyclase–cAMP pathway through GPCR (Robbercht *et al.*, 1979) *in vivo* and *in vitro* in mammals and birds (Mezey and Kiss, 1985). Secretion of the two tilapia PRLs, PRL_{177} and PRL_{188}, was significantly inhibited by VIP in both hyperosmotic and hyposmotic medium. In hyperosmotic medium, 300 nM VIP inhibited secretion of both PRLs by 47%, whereas in hyposmotic medium, 300 nM VIP inhibited their secretion by 27% (Kelly *et al.*, 1988). The inhibitory actions of VIP on

PRL secretion in tilapia are in contrast to the known stimulatory actions of VIP on PRL secretion in tetrapods.

Pituitary adenylate cyclase activating polypeptide (PACAP) was first isolated from ovine hypothalami as a novel hypophysiotropic peptide that could activate adenylate cyclase in cultured rat pituitary cells (Miyata *et al.*, 1989) and is involved in many physiological processes including hypophysiotropic actions on GH, gonadotropin (GTH) and PRL cells in mammals (q.v. Chapter 5). The primary structure of PACAP is highly conserved among vertebrates (Adams *et al.*, 2002) and PACAP receptors belonging to the GPCR superfamily were identified from the goldfish pituitary and brain (Chow *et al.*, 1997; Wong *et al.*, 1998). PACAP-ir cell bodies are distributed mainly in the diencephalon, and their fibers project into the PPD of goldfish (Wong *et al.*, 1998), stargazer *(Uranoscopus japonicus)* (Matsuda *et al.*, 1997) and European eel (*Anguilla anguilla*) (Montero *et al.*, 1998) and were found to stimulate GH and GTH release from cultured pituitary cells of the goldfish, European eel and salmon (Parker *et al.*, 1997; Wirachowsky *et al.*, 2000; Sawisky and Chang, 2005; Wong *et al.*, 2005). Moreover, Matsuda *et al.* (2008) identified PACAP-ir fibers in the neurohypophysis in close proximity to cells containing PRL and SL, and also where PACAP receptor immunoreactivity was localized. Treatment with PACAP increased the immunoblot area for PRL- and SL-ir cells in a dose-dependent manner (10^{-7}–10^{-9}M). The expression of SL mRNA, but not PRL mRNA, was increased significantly by the treatment with 10^{-7}M PACAP. Perfusion of 10^{-8}M PACAP to isolated cells from the goldfish pituitary increased intracellular calcium mobilization and made these cells immunoreactive for PRL and SL. These results indicate that PACAP functions as a hypophysiotropic factor, not only for GH and GTH, but also for PRL and SL in the goldfish pituitary.

Ghrelin (GRL) is a recently identified GH-releasing peptide isolated from stomach extracts as an endogenous ligand for the GH secretagogue receptor of the GPCR superfamily (Kojima *et al.*, 1999). GRL is primarily expressed in the intestinal tract and to a lesser extent in the brain and hypothalamus in fish species examined (Kojima *et al.*, 1999; Kaiya *et al.*, 2003a,b,c; Unniappan *et al.*, 2002). Together with stomach-derived GRL, hypothalamic GRL seems to participate in the regulation of GH secretion from the pituitary gland and also feeding behavior (Unniappan *et al.*, 2002). Stimulation of PRL release by homologous GRL has also been reported in bullfrog (Kaiya *et al.*, 2001) but not rat (Kojima *et al.*, 1999). Tilapia and eel GRL in a dose dependent manner stimulated the release of GH and PRL from *in vitro* cultured tilapia pituitary (Kaiya *et al.*, 2003b,c). However, rainbow trout GRL stimulated the release of GH but not the release of PRL *in vivo* and *in vitro* (Kaiya *et al.*, 2003a). These results suggest that the effect of GRL on the release of PRL may be species-specific.

Copeptine (NpCp). In rat, the 39-amino acid glycopeptide comprising the carboxyterminus of the neurohypophysial vasopressin-neurophysin precursor was reported to be a possible PRL-releasing factor (Nagy *et al.*, 1988). Recently, Flores *et al.* (2007) isolated the C-terminal peptide of isotocin precursor (cNpCp), which is uncleaved between the neurophysin (Np) and copeptin (Cp) domain from common carp. Carp NpCp-ir fibers were found to be abundant in hypothalamus and to directly contact PRL cells in the RPD. Incubation of the RPD of carp pituitary with carp NpCp (0.02~2 μg) for 30 min showed an increase of PRL in the culture media in a dose-related manner. However, a physiological significance of the peptide for PRL release needs further investigation.

Natriuretic peptides (NPs) consist of a family of hormones that is linked to the maintenance of cardiovascular and body fluid homeostasis in vertebrates (Takei, 2000; Toop and Donald, 2004; Potter *et al.*, 2006). Atrial NP (ANP), B-type NP (BNP), ventricular NP (VNP), and four C type NPs (CNP) have been identified in several teleost species (Inoue *et al.*, 2003, 2005). CNP gene was strongly expressed in the tilapia brain and slightly in the pituitary, whereas ANP and BNP genes were hardly expressed in either of these tissues. NP receptors (NPRs) have been identified in the hypothalamus and anterior pituitary of mammals (Potter *et al.*, 2006). The presence of NPRs in the pituitary suggests a role for circulating or locally produced NPs in the control of pituitary hormone secretion. However, none of the eel NPs had any effect on PRL release in eel (*Anguilla japonica*) (Eckert *et al.*, 2003). Tilapia ANP and BNP stimulated PRL and GH release from dispersed pituitary cells from freshwater-acclimated tilapia, while tilapia CNP was without effect on PRL release. Both ANP and BNP were effective in elevating intracellular cGMP accumulation, whereas no effect of CNP was observed. The brain CNP may not participate in PRL secretion in tilapia (Fox *et al.*, 2007).

3. THE NEUROENDOCRINE REGULATION OF SOMATOLACTIN SECRETION

3.1. Somatolactin and its Receptor

Somatolactin (SL) and gene. SL was originally found in the pituitary of Atlantic cod (*Gadus morhua*) (Rand-Weaver *et al.*, 1991a) and flounder (Ono *et al.*, 1990). SL cells are localized in the pituitary PI and periodic acid–Schiff (PAS)-positive cells in all the teleosts examined to date except salmonids (Rand-Weaver *et al.*, 1991b). The SL cDNA and genes have been isolated from many species of the Osteichthyes including surgeon and lungfish, but not from other classes of vertebrates. All SL genes consist of five exons as in all known PRL and GH genes excluding GH genes from the advanced teleost species as mentioned above. There are a TATA box and four Pit-1/GHF-1 binding sites. Immunohistochemical studies in the rainbow trout showed that Pit-1 protein is located in the nuclei of SL-producing cells in the well-developed intermediate lobe of the pituitary gland (Ono *et al.*, 1994).

Zhu *et al.* (2004) found two paralogous genes for zebrafish SL (named SLα and SLβ) that are expressed in different cells within the PI. On the basis of sequence similarity, SLβs include goldfish SL (CAU72940), catfish SL (*Ictalurus punctatus)* (AF062744), eel SL (AAU63884) and rainbow trout SLP (Yang and Chen, 2003). Other SLs, including two very similar SLs of sea bream (Y11144; L49205), belong to SLα. Two SL subtypes share 35–48% sequence identity. SLα is present in all fish species, but SLβ appears to be restricted to basal teleost species such as catfish, goldfish, salmonids and eel. The SLβ gene probably was lost in the diversification of teleosts (Lynn and Shepherd, 2007).

SLs have also been identified in sturgeon (AB017766) and lungfish (AB017200) (Amemiya *et al.*, 1999). Some cells in the PI of dogfish (*Triakis scyllium*) pituitary were stained with anti-salmon SL serum and its cDNA fragment has been cloned (unpublished data). However, there is no evidence that SL is present in tetrapod lineage and agnathans, suggesting that the SL gene emerged in the evolutionary process to gnathostomes and was lost during the evolution to tetrapods.

All SLs have at least 6 Cys residues, which are capable of forming three disulfide bonds in positions similar to those of tetrapod PRLs. SLs are glycoproteins with one to three *N*-glycosylation sites except SL of salmonids and goldfish, which possess no *N*-glycosylation sites (Takayama *et al.*, 1991; Yang *et al.*, 1997). The lack of glycosylation sites is probably the reason for the absence of PAS-positive cells in the PI of the pituitary gland in salmonids.

Somatolactin receptor (SLR). The identification of SLR in masu salmon (*Oncorhynchus masou*) (Fukada *et al.*, 2005), medaka and fugu (Fukamachi *et al.*, 2005), sea bream (AY573601) and eel (AB180476) demonstrated that SLR is a member of the cytokine receptor type I homodimeric group together with two GHRs (GHR1 and GHR2) and PRLR in teleost species as expected. SLRs show 38–58% and 28–33% sequence identity to vertebrate GHRs and PRLRs, respectively, and share common features of teleost GHR including FGEFS motif, six Cys residues in the extracellular domain, a single transmembrane region, and box 1 and box 2 regions in the intracellular domain. However, phylogenetic analysis revealed that salmon and medaka (*Oryzias latipes*) SLR belong to the GHR1 lineage. In addition, GHR2 binds GH, while GHR1 predominantly binds SL (Fukada *et al.*, 2004, 2005). These results suggest that sea bream and eel GHR1 have been improperly named and should be SLR (Fukamachi and Meyer, 2007).

In coho salmon, SLR mRNA was detected in the brain, pituitary, gill, heart, head kidney, posterior kidney, spleen, liver, muscle, fat and gonad, and the highest levels were observed in the liver and fat tissues (Fukada *et al.*, 2005). The higher levels of SLR mRNA were also observed in the liver of medaka (Fukamachi *et al.*, 2005). In tilapia, GHR1 was most highly expressed in fat, liver and muscle, suggesting a metabolic function. GHR1 expression was also high in skin, consistent with a function of SL in chromatophore regulation. These findings support the hypothesis that GHR1 is a receptor for SL.

3.2. Functions of Somatolactin

The physiological roles of SL have been proposed by morphological observation of SL cells by immunohistochemistry, by measuring the changes in plasma and pituitary SL levels by radioimmunoassay, and in expression levels of SL mRNA under different physiological and environmental conditions. It has been suggested that SL is involved in adaptation to environmental changes (Ono and Kawauchi, 1994), smoltification (Rand-Weaver and Swanson, 1993), adaptation to background and decreased illumination (Zhu and Thomas, 1995, 1998; Zhu *et al.*, 1999; Canepa *et al.*, 2006), stress responses (Kakizawa *et al.*, 1995a; Johnson *et al.*, 1997), control of some physiological aspects of reproduction (Planas *et al.,* 1992; Rand-Weaver *et al.,* 1992; Rand-Weaver and Swanson, 1993; Olivereau and Rand-Weaver, 1994a,b; Vissio *et al.,* 2002), acid-base balance (Kakizawa *et al.*, 1995a, 1996, 1997a,b) and the regulation of calcium (Kakizawa *et al.*, 1993, 1995a,b) and phosphate (Kakizawa *et al.*, 1995a) metabolism. It has also been implied that SL is involved in growth (Duan *et al.*, 1993; Company *et al.*, 2001) and energy metabolism (Rand-Weaver *et al.*, 1992; Kaneko *et al.*, 1993; Kakizawa *et al.*, 1995a; Company *et al.*, 2001; Mingarro *et al.*, 2002; Vega-Rubin de Celis *et al.*, 2004).

However, there are several conflicting results on these putative functions. Rainbow trout SLR (Fukada *et al.*, 2005) and tilapia GHR1 (Fukamachi *et al.*, 2006; Fukamachi and Meyer, 2007; Pierce *et al.*, 2007) expression was high in the skin, consistent with a SL functioning in chromophore regulation. The tilapia GHR1 from these studies was suggested to be a SLR based on analysis and binding studies. However, it has also been reported that background color does not affect SL levels in the plasma of rainbow trout (Kakizawa *et al.*, 1995a) and seasonal changes in SL secretion appear more closely linked to water temperature than to photoperiod (Rand-Weaver *et al.*, 1995). The stress-induced increase in circulating SL levels is also markedly different in two strains of rainbow trout (Rand-Weaver *et al.*, 1993) and no correlations were obtained between plasma SL concentrations and different salinities, external calcium concentrations or reproductive condition in red drum (*Sciaenops ocellatus*), Atlantic croaker (*Micropogonias undulatus*) (Zhu and Thomas, 1995), Atlantic halibut (*Hippoglossus hippoglossus*) and English sole (*Parophrys vetulus*) (Johnson *et al.*, 1997). There was also no evidence that SL was involved in an active role in response to fasting of rainbow trout (Pottinger *et al.*, 2003). However, these conflicting data should be reviewed in the light of the presence of two different SLs, SLα and SLβ, in the basal teleost species such as cyprinids, salmonids and eel. With the increasing number of different hormones and their receptors being identified, there are more challenges to fish endocrinologists to decipher the structure and function of these ligands/receptors.

It has been suggested that SL is involved in some aspects of reproduction as mentioned above. However, no further experimental evidence has been reported with the exception of the stimulatory effect of SL on gonadal steroidogenesis in coho salmon (Planas *et al.*, 1992).

Lu *et al.* (1995) examined the effects of salmon SL on renal transepithelial transport of P_i and Ca^{2+} by winter flounder (*Pseudopleuronectes americanus*) renal proximal tubule cells. SL stimulated P_i reabsorption in a dose-dependent manner at a physiological level of the hormone (12.5 ng mL^{-1}). Ca^{2+} fluxes were unchanged by the addition of 200 ng mL^{-1} SL. The SL-induced P_i reabsorption was abolished by a highly specific protein kinase A inhibitor. Moreover, the production and release of cAMP were significantly increased after 1 and 2 h exposure to SL. These data suggest that SL directly stimulates renal P_i reabsorption by a cAMP-dependent pathway.

In vitro effects of SL on melanosome aggregation were demonstrated with purified SL in red drum (Zhu and Thomas, 1997). The integument of the red drum became pale within 2 min following an intramuscular injection of SL (1 nmole g^{-1} BW) in fish held in a black-background aquarium, and gradually regained its black coloration during the subsequent 30 min. Melanosomes in the melanophores of scales were completely aggregated within 10 min of incubation with 1 μM SL *in vitro*. Furthermore, recombinant zebrafish SLβ was shown to induce melanosome aggregation in a concentration-related manner in skin of zebrafish at a concentration of 1 ng mL^{-1} (Nguyen *et al.*, 2006). Moreover, the identification of a gene responsible for medaka "color interfere" (*ci*) mutant has established that SL plays an essential role in body-color regulation in teleosts (Fukamachi *et al.*, 2004). These mutants do not show any obvious morphological and physiological defects other than defects in chromatophore proliferation and morphogenesis. The mutation has been identified as an 11-base deletion in *SL gene*, which causes truncation of 91-amino acids of the 230 amino acids of the SL hormone located upstream of the C-terminus. This medaka SL-deficient mutant color interfere exhibited a remarkable increase in the number of leukophores and a decreased number of visible xanthophores.

SL transcription changed dramatically during morphological body color adaptation to different backgrounds. These results demonstrated that SL plays a role in chromatophore development. The *ci* mutant fish possess constitutively increased numbers of leukophores and a concomitant decrease in visible xanthophores. Fukamachi *et al.* (2006) suggested a conserved function for SL in xanthophore regulation across species, rather than the evolution of a medaka-specific and leukophore-dependent role of SL in body-color regulation. This idea is supported by the remarkably paler body color in the "cobalt" variant of the rainbow trout, which has a defect in the pars intermedia of the pituitary in which SL-producing cells are distributed (Kaneko *et al.*, 1993), and by the expressional similarity of SLs in medaka, red drum and the Atlantic croaker, in which dark illumination or a black background dramatically increased SL expression (Zhu and Thomas, 1995; Fukamachi *et al.*, 2005).

Rainbow trout SLR and tilapia GHR1 were most highly expressed in fat, liver and muscle, suggesting a metabolic function. Vega-Rubin de Celis *et al.* (2003) found that a single ip injection of recombinant sea bass SL (0.1 $\mu g\ g^{-1}$ of body mass) to juvenile gilthead sea bream did not modify the circulating amount of IGF-I and nitrogen–ammonia excretion, but increased carbon dioxide output and oxygen uptake, which in turn decreased the respiratory quotient (CO_2 output per O_2 uptake). SL was also able to inhibit the hepatic activity of acetyl-coenzyme A carboxylase. These results support the involvement of SL in energy homeostasis, the enhancement of lipid metabolism. The involvement of SL in lipid metabolism has also been suggested by the study of cobalt rainbow trout, which lack SL-producing cells in the pituitary and accumulate a large amount of fat tissue (Kaneko *et al.*, 1993) and high contents of triglycerides and cholesterol in the liver and muscle (Yada *et al.*, 2002b).

3.3. Hypothalamic Control of Somatolactin Secretion

Together with the physiological functions of SL, the neuroendocrine regulatory mechanism for SL release from the pituitary remains largely unknown (Figure 5.4). Possible hypothalamic control of SL secretion has been supported by histological observations of several hypothalamic hormone neurons in the pituitary. The immunoreactive nerve fibers and endings of the following peptides were identified in close proximity to SL cells in the PI: GnRH in steelhead trout (Parhar and Iwata, 1994), Nile perch (Mousa and Mousa, 2003), pejerrey (*Odontesthes bonariensis*) (Vissio *et al.*, 1999; Stefano *et al.*, 1999) and sea bream (González-Martinez *et al.*, 2002); PrRP in goldfish and rainbow trout (Wang *et al.*, 2000b; Moriyama *et al.*, 2002); and PACAP in stargazer (*Uranoscopus japonicus*) (Matsuda *et al.*, 2005a,b) and goldfish (Matsuda *et al.*, 2008). In addition, GnRHR and PACAPR were detected in SL cells.

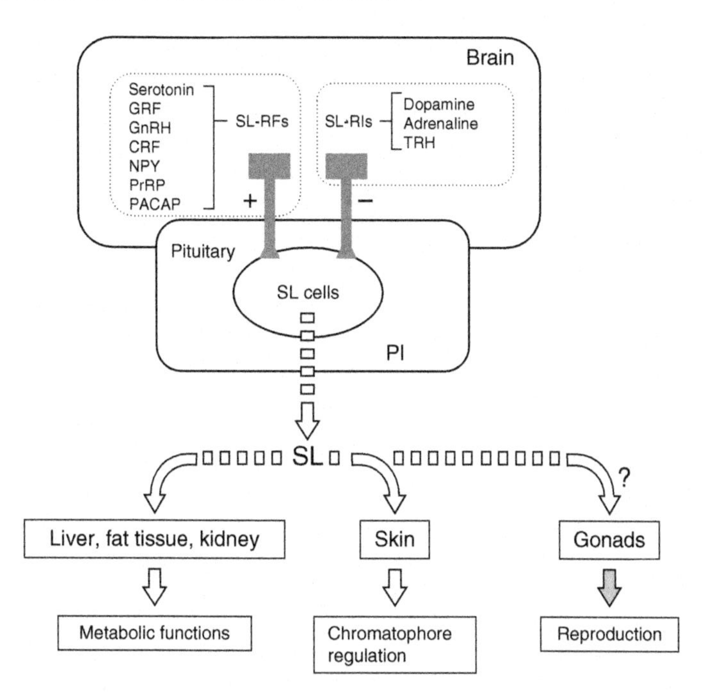

Fig. 5.4. Regulation of SL release from the pituitary and functions of SL in fish. SL release from SL-producing cells is under the multifunctional control of various neuropeptides and neurotransmitters. These stimulatory/inhibitory regulators from the brain are delivered to SL-producing cells in the PI by direct innervation from the hypothalamus. SL-RFs, somatolactin-releasing factors; SL-RIs, somatolactin-release inhibitors; GRF, growth hormone-releasing factor; GnRH, gonadotropin-releasing hormone; CRF, corticotropin-releasing factor; NPY, neuropeptide Y; PrRP, prolactin-releasing peptide; PACAP, pituitary adenylate cyclase activating polypeptide; TRH, thyrotropin-releasing hormone; PI, pars intermedia.

Inhibiting factors. Kakizawa *et al.* (1997a) first examined the effects of hypothalamic factors on SL secretion from the organ-cultured pituitary in rainbow trout. DA at concentrations of 30 and 300 μM, epinephrine at concentrations of 30 and 300 μM, and TRH at a concentration of 100 nM inhibited SL release.

Stimulating peptides. Kakizawa *et al.* (1997a) examined *in vitro* effects of hypothalamic factors on SL secretion from organ-cultured rainbow trout pituitaries in the absence or the presence of DA. GnRH, serotonin and corticotropin-releasing factor stimulate SL release in the DA-inhibited pituitary. Taniyama *et al.* (2000) demonstrated that GnRH analogue enhanced expression of SL gene in prespawning or maturing sockeye salmon. On the other hand, sbGnRH failed to affect basal SL release within 3 h of incubation *in vitro* (Peyon *et al.*, 2003).

In vertebrates including fish, NPY has been implicated in the regulation of various neuroendocrine axes. Porcine NPY alone has no effect on SL release from the European sea bass (*Dicentrachus labrax*) pituitary *in vitro*, whereas it dose-dependently enhanced SL release induced by leptin in the late pre-pubertal but not in the post-pubertal stage (Peyon *et al.*, 2001, 2003).

tPrRP is effective in stimulating SL release from the rainbow trout pituitary (Moriyama *et al.*, 2002). The elevation of plasma SL levels was observed after ip injection of tPrRP, but its levels did not show a dose dependency. The increase was observed later than the increase in PRL levels. It is possible that the effect of PrRP on SL release is mediated indirectly or by some other mechanism. Indeed, the stimulatory effect of tPrRP on SL release was less pronounced than that on PRL release *in vitro*. Sakamoto *et al.* (2003b) also reported that levels of SL in plasma and SL mRNA in the pituitary PrRP did not increase following intra-arterial injection of tPrRP.

PACAP is also effective in stimulating SL release from the goldfish pituitary (Matsuda *et al.*, 2008). In *in vitro* dispersed goldfish pituitary cells, treatment with PACAP in a dose-dependent manner increased the area of SL immunoreactivity. This peptide also stimulated the expression of SL mRNA, but not PRL mRNA. This indicates that PACAP can potentially act as a hypophysiotropic factor, not only for GH and GTH, but also for SL in goldfish.

Mammalian GnRHa significantly increased the level of pituitary SL mRNA in the males of prespawning sockeye salmon (*Oncorhynchus nerka*), but not in the females, whereas it did not induce significant increases in the levels of GH and PRL mRNAs in both the males and the females (Taniyama *et al.*, 2000).

4. CONCLUDING REMARKS

In this chapter, we present the state of the art on neuroendocrine control of synthesis and secretion of two homologous pituitary hormones, PRL and SL. In contrast to the rapid development of this area in rodents as model species of mammals, much less is known in fish, even though we now have identified much molecular information on the pituitary hormones and respective receptors. PRL and SL functions in teleosts have been investigated by morphological observation on hormone-producing cells, by changes of hormone levels in the pituitary and the plasma, and by changes of expression levels of the ligands and their respective receptors under given conditions, However, there have been few studies using homologous hormones and/or neurohormones. Therefore, conflicting results should be re-examined closely before conclusions can be made. The gene analyses of hormones, receptors and their mutants should bring new insights into the hormonal regulation of teleost physiology as described for the functions of SL, although such good fortune may not occur in every case. It should be noted that a general formula to obtain conclusive answers is the study on the protein levels, i.e., testing the effects of homologous neurohormones and pituitary hormones *in vivo* and *in vitro*.

It has been shown that many hypothalamic neurohormones participate in the regulation of PRL and SL. Moreover, these neurohormones and pituitary hormones exhibit versatile functions in teleosts as well (e.g., Table 5.1). An individual neurohormone appears to exert a different function or sometimes just opposite functions or no function, depending not only on species or ontogeny but many other factors including time of treatment, reproductive status, environmental conditions and route of administration. These all play important roles in determining the effects of neurohormones on PRL or SL synthesis and/or release. In fish, temperature should be considered a major environmental factor that can influence the expression and/or functions of hormones and respective receptors. These factors can thus activate or inhibit the cellular functions as a result of binding to the specific receptor. Many chemical messengers have kept the same or similar functions in phylogeny, but not always. In adapting to a new environment, some of the chemical messengers in an organism can be changed, modified or co-opted for other functions. Indeed, there are many examples to show that the ligand and its respective receptors exhibit considerable conservation despite strong mutation or selective pressures. On the other hand, their functions exhibit remarkable versatility in response to adaptive needs. This variety of actions has been made possible by differences in localization of tissue expressing specific receptors, by differentiation of alternative post-receptor intracellular phenomena and by regulation of hormone action by a wide variety of factors.

REFERENCES

Adams, B. A., Lescheid, D. W., Vickers, E. D., Crim, L. W., and Sherwood, N. M. (2002). Pituitary adenylate cyclase-activating polypeptide in sturgeon, whitefish, grayling, flounder and halibut: cDNA sequence, exon skipping and evolution. *Regul. Pept.* **109,** 27–37.

Al Kahtane, A., Chaiseha, Y., and El Halawani, M. (2003). Dopaminergic regulation of avian prolactin gene transcription. *J. Mol. Endocrinol.* **31,** 185–196.

Amano, M., Oka, Y., Yamanome, T., Okuzawa, K., and Yamamori, K. (2002). Three GnRH systems in the brain and pituitary of a pleuronectiform fish, the barfin flounder *Verasper moseri. Cell Tissue Res.* **309,** 323–329.

Amano, M., Oka, Y., Amiya, N., and Yamamori, K. (2007). Immunohistochemical localization and ontogenic development of prolactin-releasing peptide in the brain of the ovoviviparous fish species *Poecilia reticulata* (guppy). *Neurosci. Lett.* **413,** 206–209.

Amemiya, Y., Sogabe, Y., Nozaki, M., Takahashi, A., and Kawauchi, H. (1999). Somatolactin in the white sturgeon and African lungfish and its evolutionary significance. *Gen. Comp. Endocrinol.* **114,** 181–190.

Ayson, F. G., Kaneko, T., Hasegawa, S., and Hirano, T. (1994). Differential expression of two prolactin and growth hormone genes during early development of tilapia (*Oreochromis mossambicus*) in fresh water and seawater: Implications for possible involvement in osmoregulation during early life stages. *Gen. Comp. Endocrinol.* **95,** 143–152.

Barry, T. P., and Grau, E. G. (1986). Estradiol-17beta and thyrotropin-releasing hormone stimulate prolactin release from the pituitary gland of a teleost fish *in vitro*. *Gen. Comp. Endocrinol.* **62,** 306–314.

Bern, H. A., and Madsen, S. S. (1992). A selective survey of the endocrine system of the rainbow trout (*Oncorhynchus mykiss*) with emphasis on the hormonal regulation of ion balance. *Aquaculture* **100,** 237–262.

Bhandari, R. K., Taniyama, S., Kitahashi, T., Ando, H., Yamauchi, K., Zohar, Y., Ueda, H., and Uranob, A. (2003). Seasonal changes of responses to gonadotropin-releasing hormone analog in expression of growth hormone/prolactin/somatolactin genes in the pituitary of masu salmon. *Gen. Comp. Endocrinol.* **130,** 55–63.

Blackwell, R. E., Rodgers-Neame, N. T., Bradley, E. L., Jr., and Asch, R. H. (1986). Regulation of human prolactin secretion by gonadotropin-releasing hormone *in vitro*. *Fertil. Steril.* **46,** 26–31.

Bole-Feysot, C., Goffin, V., Edery, M., Binart, N., and Kelly, P. A. (1998). Prolactin (PRL) and its receptor: Actions, signal transduction pathways and phenotypes observed in PRL receptor knockout mice. *Endocr. Rev.* **19,** 225–268.

Borski, R. J., Helms, L. M., Richman, N. H., 3rd., and Grau, E. G. (1991). Cortisol rapidly reduces prolactin release and cAMP and $^{45}Ca^{2+}$ accumulation in the cichlid fish pituitary *in vitro*. *Proc. Natl. Acad. Sci. USA* **88,** 2758–2762.

Brinca, L., Fuentes, J., and Power, D. M. (2003). The regulatory action of estrogen and vasoactive intestinal peptide on prolactin secretion in sea bream (*Sparus aurata*, L.). *Gen. Comp. Endocrinol.* **131,** 117–125.

Canepa, M. M., Pandolfi, M., Maggese, M. C., and Vissio, P. G. (2006). Involvement of somatolactin in background adaptation of the cichlid fish *Cichlasoma dimerus*. *J. Exp. Zool.* **305,** 410–419.

Chow, B. K. C., Yuen, T. T. H., and Chan, K. W. (1997). Molecular evolution of vertebrate VIP receptors and functional characterization of a VIP receptor from goldfish *Carassius auratus*. *Gen. Comp. Endocrinol.* **105,** 176–185.

Company, R., Astola, A., Pendon, C., Valdivia, M. M., and Perez-Sanchez, J. (2001). Somatotropic regulation of fish growth and adiposity: Growth hormone (GH) and somatolactin (SL) relationship. *Comp. Biochem. Physiol.* **130C,** 435–445.

De Jesus, E. G., Hirano, T., and Inui, Y. (1994). The antimetamorphic effect of prolactin in the Japanese flounder. *Gen. Comp. Endocrinol.* **93,** 44–50.

de Ruiter, A. J., Wendelaar Bonga, S. E., Slijkhuis, H., and Baggerman, B. (1986). The effect of prolactin on fanning behavior in the male three-spined stickleback, *Gasterosteus aculeatus* L. *Gen. Comp. Endocrinol.* **64,** 273–283.

Duan, C., Dugua, S. J., and Plisetskaya, E. M. (1993). Insulin-like growth factor I (IGF-I) mRNA expression in coho salmon, *Oncorhynchus kisutch*: Tissue distribution and effects of growth hormone/ prolactin family proteins. *Fish Physiol. Biochem.* **11,** 371–379.

Dubois, M. P., Billard, R., Breton, B., and Peter, R. E. (1979). Comparative distribution of somatostatin, LH-RH, neurophysin, and alpha-endorphin in the rainbow trout: An immunocytological study. *Gen. Comp. Endocrinol.* **37,** 220–232.

Eckert, S. M., Hirano, T., Leedom, T. A., Takei, Y., and Grau, G. E. (2003). Effects of angiotensin II and natriuretic peptides of the eel on prolactin and growth hormone release in the tilapia. *Oreochromis mossambicus. Gen. Comp. Endocrinol.* **130,** 333–339.

Flores, C. M., Munoz, D., Soto, M., Kausel, G., Romero, A., and Figueroa, J. (2007). Copeptin, derived from isotocin precursor, is a probable prolactin releasing factor in carp. *Gen. Comp. Endocrinol.* **150,** 343–354.

Fox, B. K., Naka, T., Inoue, K., Takei, Y., Hirano, T., and Grau, E. G. (2007). *In vitro* effects of homologous natriuretic peptides on growth hormone and prolactin release in the tilapia, *Oreochromis mossambicus. Gen. Comp. Endocrinol.* **150,** 270–277.

Freeman, M. E., Kanyicska, B., Lerant, A., and Nagy, G. (2000). Prolactin: Structure, function, and regulation of secretion. *Physiol. Rev.* **80,** 1523–1631.

Fruchtman, S., Jackson, L., and Borski, R. (2000). Insulin-like growth factor I disparately regulates prolactin and growth hormone synthesis and secretion: Studies using the teleost pituitary model. *Endocrinology* **141,** 2886–2894.

Fujimoto, M., Takeshita, K., Wang, X., Takabatake, I., Fujisawa, Y., Teranishi, H., Ohtani, M., Muneoka, Y., and Ohta, S. (1998). Isolation and characterization of a novel bioactive peptide, Carassius RF-amide (C-RFa), from the brain of the Japanese crucian carp. *Biochem. Biophys. Res. Commun.* **242,** 436–440.

Fujimoto, M., Sakamoto, T., Kanetoh, T., Osaka, M., and Moriyama, S. (2006). Prolactin-releasing peptide is essential to maintain the prolactin levels and osmotic balance in freshwater teleost fish. *Peptides* **27,** 1104–1109.

Fukada, H., Ozaki, Y., Pierce, A. L., Adachi, S., Yamauchi, K., Hara, A., Swanson, P., and Dickhoff, W. W. (2004). Salmon growth hormone receptor: Molecular cloning, ligand specificity, and response to fasting. *Gen. Comp. Endocrinol.* **139,** 61–71.

Fukada, H., Ozaki, Y., Pierce, A. L., Adachi, S., Yamauchi, K., Hara, A., Swanson, P., and Dickhoff, W. W. (2005). Identification of the salmon somatolactin receptor, a new member of the cytokine receptor family. *Endocrinology* **146,** 2354–2361.

Fukamachi, S., and Meyer, A. (2007). Evolution of receptors for growth hormone and somatolactin in fish and land vertebrates: Lessons from the lungfish and sturgeon orthologues. *J. Mol. Evol.* **65,** 359–372.

Fukamachi, S., Sugimoto, M., Mitani, H., and Shima, A. (2004). Somatolactin selectively regulates proliferation and morphogenesis of neural-crest derived pigment cells in medaka. *Proc. Natl. Acad. Sci. USA* **101,** 10661–10666.

Fukamachi, S., Yada, T., and Mitani, H. (2005). Medaka receptors for somatolactin and growth hormone: Phylogenetic paradox among fish growth hormone receptors. *Genetics* **171,** 1875–1883.

Fukamachi, S., Wakamatsu, Y., and Mitani, H. (2006). Medaka double mutants for color interfere and leucophore free: Characterization of the xanthophore–somatolactin relationship using the leucophore free gene. *Dev. Genes Evol.* **216,** 152–157.

González-Martínez, D., Zmora, N., Mañanos, E., Saligaut, D., Zanuy, S., Zohar, Y., Elizur, A., Kah, O., and Muñoz-Cueto, J. A. (2002). Immunohistochemical localization of three different prepro-GnRHs in the brain and pituitary of the European sea bass (*Dicentrarchus labrax*) using antibodies to the corresponding GnRH-associated peptides. *J. Comp. Neurol.* **446,** 95–113.

Gothilf, Y., Muñoz-Cueto, J. A., Sagrillo, C. A., Selmanoff, M., Chen, T. T., Kah, O., Elizur, A., and Zohar, Y. (1996). Three forms of gonadotropin-releasing hormone in a perciform fish *(Sparus aurata)*: Complementary deoxyribonucleic acid characterization and brain localization. *Biol. Reprod.* **55,** 636–645.

Grau, E. G., Nishioka, R. S., and Bern, H. A. (1982). Effects of somatostatin and urotensin II on tilapia pituitary prolactin release and interactions between somatostatin, osmotic pressure,

Ca^{2+}, and adenosine 3,5-monophosphate in prolactin release *in vitro*. *Endocrinology* **110,** 910–914.

Grau, E. G., Nishioka, R. S., Young, G., and Bern, H. A. (1985). Somatostatin-like immunoreactivity in the pituitary and brain of three teleost fish species: Somatostatin as a potential regulator of prolactin cell function. *Gen. Comp. Endocrinol.* **59,** 350–357.

Grau, E. G., Richmann, N. H., III., and Borski, R. J. (1994). Osmoreception and a simple endocrine reflex of the prolactin cell of the tilapia *Oreochromis mossambicus*. *In* "Perspectives in Comparative Endocrinology" (Davey, K. G., Peter, R. E., and Tobe, S. S., Eds.), pp. 251–256. National Research Council of Canada, Ottawa.

Grosvenor, C. E., and Mena, F. (1980). Evidence that TRH and a hypothalamic prolactin-releasing factor may function in the release of prolactin in the lactating rat. *Endocrinology* **107,** 863–868.

Hamano, K., Yoshida, K., Suzuki, M., and Asahida, K. (1996). Changes in thyrotropin-releasing hormone concentration in the brain and levels of prolactin and thyroxin in the serum during spawning migration of the chum salmon, *Oncorhynchus keta*. *Gen. Comp. Endocrinol.* **101,** 275–281.

Harris, J., and Bird, D. J. (2000). Modulation of the fish immune system by hormones. *Vet. Immunol. Immunopathol.* **77,** 163–176.

Hazineh, A., Shin, S. H., Reifel, G., Pang, S. C., and Van der Kraak, G. J. (1997). Dopamine causes ultrastructural changes in prolactin cells of tilapia (*Oreochromis niloticus*). *Cell Mol. Life. Sci.* **53,** 452–458.

Helms, L. M., Grau, E. G., and Borski, R. J. (1991). Effects of osmotic pressure and somatostatin on the cAMP messenger system of the osmosensitive prolactin cell of a teleost fish, the tilapia (*Oreochromis mossambicus*). *Gen. Comp. Endocrinol.* **83,** 111–117.

Higashimoto, Y., Nakao, N., Ohkubo, T., Tanaka, M., and Nakashima, K. (2001). Structure and tissue distribution of prolactin receptor mRNA in Japanese flounder (*Paralichthys olivaceus*): Conserved and preferential expression in osmoregulatory organs. *Gen. Comp. Endocrinol.* **123,** 170–179.

Hinuma, S., Habata, Y., Fujii, R., Kawamata, Y., Hosoya, M., Fukusumi, S., Kitada, C., Masuo, Y., Asano, T., Matsumoto, H., Sekiguchi, M., Kurokawa, T., Nishimura, O., Onda, H., and Fujino, M. (1998). A prolactin-releasing peptide in the brain. *Nature* **393,** 272–276.

Hirano, T. (1986). The spectrum of prolactin action in teleosts. *In* "Comparative Endocrinology: Developments and Directions" (Ralph, C. L., Ed.), pp. 53–74. A. R. Liss, New York.

Huang, X., Jiao, B., Fung, C. K., Zhang, Y., Ho, W. K., Chan, C. B., Lin, H., Wang, D., and Cheng, C. H. (2007). The presence of two distinct prolactin receptors in sea bream with different tissue distribution patterns, signal transduction pathways and regulation of gene expression by steroid hormones. *J. Endocrinol.* **194,** 373–392.

Huising, M. O., Kruiswijk, C. P., and Flik, G. (2006). Phylogeny and evolution of class-I helical cytokines. *J. Endocrinol.* **189,** 1–25.

Imaoka, T., Matsuda, M., and Mori, T. (2000). Extrapituitary expression of the prolactin gene in the goldfish, African clawed frog and mouse. *Zool. Sci.* **17,** 791–796.

Inoue, K., Naruse, K., Yamagami, S., Mitani, H., Suzuki, N., and Takei, Y. (2003). Four functionally distinct C-type natriuretic peptides found in fish reveal evolutionary history of the natriuretic peptide system. *Proc. Natl. Acad. Sci. USA.* **100,** 10079–10084.

Inoue, K., Sakamoto, T., Yuge, S., Iwatani, H., Yamagami, S., Tsutsumi, M., Hori, H., Cerra, M. C., Tota, B., Suzuki, N., Okamoto, N., and Takei, Y. (2005). Structural and functional evolution of three cardiac natriuretic peptides. *Mol. Biol. Evol.* **2005,** 2428–2434.

James, V. A., and Wigham, Y. (1984). Evidence for dopaminergic and serotonergic regulation of prolactin cell activity in the trout *Salmo gairdneri*. *Gen. Comp. Endocrinol.* **56,** 231–239.

Johnson, L. L., Norberg, B., Willis, M. L., Zebroski, H., and Swanson, P. (1997). Isolation, characterization, and radioimmunoassay of Atlantic halibut somatolactin and plasma levels during stress and reproduction in flatfish. *Gen. Comp. Endocrinol.* **105,** 194–209.

Johnston, L. R., and Wigham, T. (1988). The intracellular regulation of prolactin cell function in the rainbow trout, *Salmo gairdneri*. *Gen. Comp. Endocrinol.* **71,** 284–289.

Jose, P. A., Yu, P. Y., Yamaguchi, I., Eisner, G. M., Mouradian, M. M., Felder, C. C., and Felder, R. A. (1999). Dopamine D1 receptor regulation of phospholipase C. *Hypertens. Res.* **18,** 39–42.

Kagabu, Y., Mishiba, T., Okino, T., and Yanagisawa, T. (1998). Effects of thyrotropin-releasing hormone and its metabolites, Cyclo(His-Pro) and TRH-OH, on growth hormone and prolactin synthesis in primary cultured pituitary cells of the common carp, *Cyprinus carpio*. *Gen. Comp. Endocrinol.* **111,** 395–403.

Kah, O., Chambolle, P., Dubourg, P., and Dubois, M. P. (1982). Immunocytochemical distribution of somatostatin in the forebrain of two teleosts, the goldfish (*Carassius auratus*) and Gambusia sp. *CR Seances Acad Sci III.* **294,** 519–524.

Kaiya, H., Kojima, M., Hosoda, H., Koda, A., Yamamoto, K., Kitajima, Y., Matsumoto, M., Minamitake, Y., Kikuyama, S., and Kangawa, K. (2001). Bullfrog ghrelin is modified by n-octanoic acid at its third threonine residue. *J. Biol. Chem.* **276,** 40441–40448.

Kaiya, H., Kojima, M., Hosoda, H., Moriyama, S., Takahashi, A., Kawauchi, H., and Kangawa, K. (2003a). Peptide purification, complementary deoxyribonucleic acid (DNA) and genomic DNA cloning, and functional characterization of ghrelin in rainbow trout. *Endocrinology* **144,** 5215–5226.

Kaiya, H., Kojima, M., Hosoda, H., Riley, L. G., Hirano, T., Grau, E. G., and Kangawa, K. (2003b). Identification of tilapia ghrelin and its effects on growth hormone and prolactin release in the tilapia, *Oreochromis mossambicus*. *Comp. Biochem. Physiol. B & Biochem. Mol. Biol.* **135,** 421–429.

Kaiya, H., Kojima, M., Hosoda, H., Riley, L. G., Hirano, T., Grau, E. G., and Kangawa, K. (2003c). Amidated fish ghrelin: Purification, cDNA cloning in the Japanese eel and its biological activity. *J. Endocrinol.* **176,** 415–423.

Kajimura, S., Seale, A. P., Hirano, T., Cooke, I. M., and Grau, E. G. (2005). Physiological concentrations of ouabain rapidly inhibit prolactin release from the tilapia pituitary. *Gen. Comp. Endocrinol.* **143,** 240–250.

Kajita, Y., Sakai, M., Kobayashi, M., and Kawauchi, H. (1992). Enhancement of non-specific cytotoxic activity of leucocytes in rainbow trout *Oncorhynchus mykiss* injected with growth hormone. *Fish Shellfish Immunol.* **2,** 155–157.

Kakizawa, S., Kaneko, T., Hasegawa, S., and Hirano, T. (1993). Activation of somalactin cells in the pituitary of the rainbow trout *Oncorhynchus mykiss* by low environmental calcium. *Gen. Comp. Endocrinol.* **91,** 298–306.

Kakizawa, S., Kaneko, T., Hasegawa, S., and Hirano, T. (1995a). Effects of feeding, fasting, background adaptation, acute stress, and exhaustive exercise on the plasma somatolactin concentrations in rainbow trout. *Gen. Comp. Endocrinol.* **98,** 137–146.

Kakizawa, S., Kaneko, T., Ogasawara, T., and Hirano, T. (1995b). Change in plasma somatolactin levels during spawning migration of chum salmon (*Oncorhynchus keta*). *Fish Physiol. Biochem.* **14,** 93–101.

Kakizawa, S., Kaneko, T., and Hirano, T. (1996). Elevation of plasma somatolactin concentrations during acidosis in rainbow trout (*Oncorhynchus mykiss*). *J. Exp. Biol.* **199,** 1043–1051.

Kakizawa, S., Kaneko, T., and Hirano, T. (1997a). Effects of hypothalamic factors on somatolactin secretion from the organ-cultured pituitary of rainbow trout. *Gen. Comp. Endocrinol.* **105,** 71–78.

Kakizawa, S., Ishimatsu, A., Takeda, T., Kaneko, T., and Hirano, T. (1997b). Possible involvement of somatolactin in the regulation of plasma bicarbonate for the compensation of acidosis in rainbow trout. *J. Exp. Biol.* **200,** 2675–2683.

Kaneko, T., and Hirano, T. (1993). Role of prolactin and somatolactin in calcium regulation in fish. *J. Exp. Biol.* **184,** 31–45.

Kaneko, T., Kakizawa, S., and Yada, T. (1993). Pituitary of "cobalt" variant of the rainbow trout separated from the hypothalamus lacks most pars intermedial and neurohypophysial tissue. *Gen. Comp. Endocrinol.* **92,** 31–40.

Kawauchi, H., Suzuki, K., Yamazaki, T., Moriyama, S., Nozaki, M., Yamaguchi, K., Takahashi, A., Youson, J., and Sower, S. A. (2002). Identification of growth hormone in the sea lamprey, an extant representative of a group of the most ancient vertebrates. *Endocrinology* **143,** 4916–4921.

Kelly, K. M., Nishioka, R. S., and Bern, H. A. (1988). Novel effect of vasoactive intestinal polypeptide and peptide histidine isoleucine: Inhibition of *in vitro* secretion of prolactin in the tilapia, *Oreochromis mossambicus. Gen. Comp. Endocrinol.* **72,** 98–106.

Kelly, S. P., and Peter, R. E. (2006). Prolactin-releasing peptide, food intake, and hydromineral balance in goldfish. *Am. J. Physiol. Regul. Integr. Comp. Physiol.* **291,** R1474–R1481.

Kiilerich, P., Kristiansen, K., and Madsen, S. S. (2007). Hormone receptors in gills of smolting Atlantic salmon, *Salmo salar:* expression of growth hormone, prolactin, mineralocorticoid and glucocorticoid receptors and 11beta-hydroxysteroid dehydrogenase type 2. *Gen. Comp. Endocrinol.* **152,** 295–303.

Kojima, M., Hosoda, H., Date, Y., Nakazato, M., Matsuo, H., and Kangawa, K. (1999). Ghrelin is a growth-hormone-releasing acylated peptide from stomach. *Nature* **402,** 656–660.

Kozaka, T., Fujii, Y., and Ando, M. (2003). Central effects of various ligands on drinking behavior in eels acclimated to seawater. *J. Exp. Biol.* **206,** 687–692.

Lagerström, M. C., Fredriksson, R., Bjarnadóttir, T. K., Fridmanis, D., Holmquist, T., Andersson, J., Yan, Y- L., Raudsepp, T., Zoorob, R., Kukkonen, J. P., Lundin, L-G., Klovins, J., *et al.* (2005). Origin of the prolactin-releasing hormone (PRLH) receptors: Evidence of coevolution between PRLH and a redundant neuropeptide Y receptor during vertebrate evolution. *Genomics* **85,** 688–703.

Lee, K. M., Kaneko, T., and Aida, K. (2006). Prolactin and prolactin receptor expressions in a marine teleost, pufferfish *Takifugu rubripes. Gen. Comp. Endocrinol.* **146,** 318–328.

Leedom, T. A., Hirano, T., and Grau, E. G. (2003). Effect of blood withdrawal and angiotensin II on prolactin release in the tilapia, *Oreochromis mossambicus. Comp. Biochem. Physiol. A & Mol. Integr. Physiol.* **135,** 155–163.

Leena, S., Shameena, B., and Oommen, O. V. (2001). *In vivo* and *in vitro* effects of prolactin and growth hormone on lipid metabolism in a teleost, *Anabas testudineus* (Bloch). *Comp. Biochem. Physiol. B* **128,** 761–766.

Lu, M., Swanson, P., and Renfro, J. L. (1995). Effect of somatolactin and related hormones on phosphate transport by flounder renal tubule primary cultures. *Am. J. Physiol.* **268,** R577–R582.

Lynn, S. G., and Shepherd, B. S. (2007). Molecular characterization and sex-specific tissue expression of prolactin, somatolactin and insulin-like growth factor-I in yellow perch *(Perca flavescens). Comp. Biochem. Physiol* B **147,** 412–427.

Manzon, L. A. (2002). The role of prolactin in fish osmoregulation. *Gen. Comp. Endocrinol.* **125,** 291–310.

Marchant, T. A., Dulka, J. G., and Peter, R. E. (1989). Relationship between serum growth hormone levels and the brain and pituitary content of immunoreactive somatostatin in the goldfish, *Carassius auratus* L. *Gen. Comp. Endocrinol.* **73,** 458–468.

Matsuda, K., Takei, Y., Katoh, J., Shioda, S., Arimura, A., and Uchiyama, M. (1997). Isolation and structural characterization of pituitary adenylate cyclase activating polypeptide (PACAP)-like peptide from the brain of a teleost, stargazer, *Uranoscopus japonicus*. *Peptides* **18,** 723–727.

Matsuda, K., Nagano, Y., Uchiyama, M., Onoue, S., Takahashi, A., Kawauchi, H., and Shioda, S. (2005a). Pituitary adenylate cyclase-activating polypeptide (PACAP)-like immunoreactivity in the brain of a teleost, *Uranoscopus japonicus*: Immunohistochemical relationship between PACAP and adenohypophysial hormones. *Regul. Pept.* **126,** 129–136.

Matsuda, K., Nagano, Y., Uchiyama, M., Takahashi, A., and Kawauchi, H. (2005b). Immunohistochemical observation of pituitary adenylate cyclase-activating polypeptide (PACAP) and adenohypophysial hormones in the pituitary of a teleost, *Uranoscopus japonicus*. *Zool. Sci.* **22,** 71–76.

Matsuda, K., Nejigaki, Y., Satoh, M., Shimaura, C., Tanaka, M., Kawamoto, K., Uchiyama, M., Kawauchi, H., Shioda, S., and Takahashi, A. (2008). Effect of pituitary adenylate cyclase-activating polypeptide (PACAP) on prolactin and somatolactin release from the goldfish pituitary *in vitro*. *Regul. Pept.* **145,** 72–79.

McCormick, S. D. (2001). Endocrine control of osmoregulation in teleost fish. *Am. Zool.* **41,** 781–794.

Mezey, E., and Kiss, J. Z. (1985). Vasoactive intestinal polypeptide-containing neurons in the paraventricular nucleus may participate in regulating prolactin secretion. *Proc. Natl. Acad. Sci. USA* **82,** 245–247.

Millar, R. P., Lu, Z.L, Pawson, A. J., Flanagan, C.A, Morgan, K., and Maudsley, S. R. (2004). Gonadotropin-releasing hormone receptors. *Endocrinol. Rev.* **25,** 235–275.

Mingarro, M., Vega-Rubín de Celis, S., Astola, A., Pendón, C., Valdivia, M. M., and Pérez-Sánchez, J. (2002). Endocrine mediators of seasonal growth in gilthead sea bream (*Sparus aurata*): The growth hormone and somatolactin paradigm. *Gen. Comp. Endocrinol.* **128,** 102–111.

Miyata, A., Arimura, A., Dahl, R. R., Minamino, N., Uehara, A., Jiang, L., Culler, M. D., and Coy, D. H. (1989). Isolation of a novel 38 residue-hypothalamic polypeptide which stimulates adenylate cyclase in pituitary cells. *Biochem. Biophys. Res. Commun.* **164,** 567–574.

Montefusco-Siegmund, R. A., Romero, A., Kausel, G., Muller, M., Fujimoto, M., and Figueroa, J. (2006). Cloning of the prepro C-RFa gene and brain localization of the active peptide in Salmo salar. *Cell Tissue Res.* **325,** 277–285.

Montero, M., Yon, L., Rousseau, K., Arimura, A., Fournier, A., Dufour, S., and Vaudry, H. (1998). Distribution, characterization, and growth hormone-releasing activity of pituitary adenylate cyclase-activating polypeptide in the European eel, *Anguilla anguilla*. *Endocrinology* **139,** 4300–4310.

Moriyama, S., Ito, T., Takahashi, A., Amano, M., Sower, S. A., Hirano, T., Yamamori, K., and Kawauch, H. (2002). A homolog of mammalian PRL-releasing peptide (fish arginyl-phenylalanyl-amide peptide) is a major hypothalamic peptide of PRL release in teleost fish. *Endocrinology* **143,** 2071–2079.

Moriyama, S., Kasahara, M., Amiya, N., Takahashi, A., Amano, M., Sower, S. A., Yamamori, K., and Kawauchi, H. (2007). RFamide peptides inhibit the expression of melanotropin and growth hormone genes in the pituitary of an Agnathan, the sea lamprey, *Petromyzon marinus*. *Endocrinology* **148,** 3740–3749.

Mousa, A., and Mousa, S. A. (2003). Immunohistochemical localization of gonadotropin releasing hormones in the brain and pituitary gland of the Nile perch, *Lates niloticus* (Teleostei, Centropomidae). *Gen. Comp. Endocrinol.* **130,** 245–255.

Nagahama, Y., Nishioka, R. S., Bern, H. A., and Gunther, R. L. (1975). Control of prolactin secretion in teleosts, with special reference to *Gillichthys mirabilis* and *Tilapia mossambica*. *Gen. Comp. Endocrinol.* **25,** 166–188.

Nagy, G., Mulchahey, J. J., Smyth, D. J., and Neull, J. D. (1988). The glycopeptide moiety of vasopressin-neurophysin precursor is neurohypophysial prolactin releasing factor. *Biochem. Biophys. Res. Commun.* **151,** 524–529.

Narnaware, Y. K., Kelly, S. P., and Woo, N. Y. (1998). Stimulation of macrophage phagocytosis and lymphocyte count by exogenous prolactin administration in silver sea bream (*Sparus sarba*) adapted to hyper- and hypo-osmotic salinities. *Vet. Immunol. Immunopathol.* **61,** 387–391.

Nguyen, N., Sugimoto, M., and Zhu, Y. (2006). Production and purification of recombinant somatolactin beta and its effects on melanosome aggregation in zebrafish. *Gen. Comp. Endocrinol.* **145,** 182–187.

Nicoll, C. S. (1993). Role of prolactin and placental lactogens in vertebrate growth and development. *In* "The Endocrinology of Growth, Development, and Metabolism in Vertebrates" (Schreibman, M. P., Scanes, C. G., and Pang, P. K. T., Eds.), pp. 183–219. Academic Press, New York.

Nillni, E. A., and Sevarino, K. A. (1999). The biology of pro-thyrotropin-releasing hormone-derived peptides. *Endo. Rev.* **20,** 599–648.

Ogasawara, T., Sakamoto, T., and Hirano, T. (1996). Prolactin kinetics during freshwater adaptation of maturing chum salmon. *Zool. Sci.* **13,** 443–447.

Ogawa, M. (1970). Effects of prolactin on epidermal mucous cells of goldfish, *Carassius auratus* L. *Can. J. Zool.* **48,** 501.

Oliveira, L., Paiva, A. C., Sander, C., and Vriend, G. (1994). A common step for signal transduction in G protein-coupled receptors. *Trends Pharmacol. Sci.* **15,** 170–172.

Olivereau, M., Ollevier, F., Vandesande, F., and Olivereau, J. (1984a). Somatostatin in the brain and the pituitary of some teleosts. Immunocytochemical identification and the effect of starvation. *Cell Tissue Res.* **238,** 289–296.

Olivereau, M., Ollevier, F., Vandesande, F., and Verdonck, W. (1984b). Immunocytochemical identification of CRF-like and SRIF-like peptides in the brain and the pituitary of cyprinid fish. *Cell Tissue Res.* **237,** 379–382.

Olivereau, M., and Rand-Weaver, M. (1994a). Immunoreactive somatolactin cells in the pituitary of young, migrating, spawning and spent chinook salmon, *Oncorhynchus tshawytscha*. *Fish Physiol. Biochem.* **13,** 141–151.

Olivereau, M., and Rand-Weaver, M. (1994b). Immunocytochemical study of the somatolactin cells in the pituitary of Pacific salmons, *Oncorhynchus nerka* and *O. keta,* at some stages of the reproductive cycle. *Gen. Comp. Endocrinol.* **93,** 28–35.

Ono, M., and Kawauchi, H. (1994). The somatolactin gene. *In* "Fish Physiology" (Farrell, A. P., Randall, D. J., Sherwood, N. M., and Hew, C. L., Eds.), Vol. 13, pp. 159–177. Academic Press, San Diego.

Ono, M., Takayama, Y., Rand-Weaver, M., Sakata, S., Yasunaga, T., Noso, T., and Kawauchi, H. (1990). cDNA cloning of somatolactin, a pituitary protein related to growth hormone and prolactin. *Proc. Natl. Acad. Sci. USA.* **87,** 4330–4334.

Ono, M., Harigai, T., Kaneko, T., Sato, Y., Ihara, S., and Kawauchi, H. (1994). Pit-1/GH factor-1 involvement in the gene expression of somatolactin. *Mol. Endocrinol.* **8,** 109–115.

Onuma, T., Kitahashi, T., Taniyama, S., Saito, D., Ando, H., and Urano, A. (2003). Changes in expression of genes encoding gonadotropin subunits and growth hormone/prolactin/somatolactin family hormones during final maturation and freshwater adaptation in prespawning chum salmon. *Endocrine* **20,** 23–34.

Onuma, T., Ando, H., Koide, N., Okada, H., and Urano, A. (2005). Effects of salmon GnRH and sex steroid hormones on expression of genes encoding growth hormone/prolactin/somatolactin family hormones and a pituitary-specific transcription factor in masu salmon pituitary cells *in vitro*. *Gen. Comp. Endocrinol.* **143,** 129–141.

Parhar, I. S. (1997). GnRH in tilapia: Three genes, three origins and their roles. *In* "GnRH Neurons: Gene to Behavior" (Parhar, I. S., and Sakuma, Y., Eds.), pp. 99–122. Brain Shuppan, Tokyo.

Parhar, I. S., and Iwata, M. (1994). Gonadotropin releasing hormone (GnRH) neurons project to growth hormone and somatolactin cells in the Steelhead trout. *Histochemistry* **102,** 195–203.

Parhar, I. S., Ogawa, S., and Sakuma, Y. (2005). Three GnRH receptor types in laser-captured single cells of the cichlid pituitary display cellular and functional heterogeneity. *Proc. Natl. Acad. Sci. USA* **102,** 2204–2209.

Parker, D. B., Power, M. E., Swanson, P., River, J., and Sherwood, N. M. (1997). Exon skipping in the gene encoding pituitary adenylate cyclase-activating polypeptide in salmon alters the expression of two hormones that stimulate growth hormone release. *Endocrinology* **138,** 414–423.

Peyon, P., Zanuy, S., and Carrillo, M. (2001). Action of leptin on *in vitro* luteinizing hormone release in the European sea bass (*Dicentrarchus labrax*). *Biol. Reprod.* **65,** 1573–1578.

Peyon, P., Vega-Rubin de Celis, S., Gomez-Requeni, P., Zanuy, S., Perez-Sanchez, J., and Carrillo, M. (2003). *In vitro* effect of leptin on somatolactin release in the European sea bass *(Dicentrarchus labrax)*: Dependence on the reproductive status and interaction with NPY and GnRH. *Gen. Comp. Endocrinol.* **132,** 284–292.

Pickford, G. E., and Phillips, J. G. (1959). Prolactin, a factor in promoting survival of hypophysectomized killifish in fresh water. *Science* **130,** 454–455.

Pierce, A. L., Fox, B. K., Davis, L. K., Visitacion, N., Kitahashi, T., Hirano, T., and Grau, E. G. (2007). Prolactin receptor, growth hormone receptor, and putative somatolactin receptor in Mozambique tilapia: Tissue specific expression and differential regulation by salinity and fasting. *Gen Comp. Endocrinol.* **154,** 31–40.

Planas, J. V., Swanson, P., Rand-Weaver, M., and Dickhoff, W. W. (1992). Somatolactin stimulates *in vitro* gonadal steroidogenesis in coho salmon, *Oncorhynchus kisutch*. *Gen. Comp. Endocrinol.* **87,** 1–5.

Potter, L. R., Abbey-Hosch, S., and Dickey, D. M. (2006). Natriuretic peptides, their receptors, and cyclic guanosine monophosphate-dependent signaling functions. *Endocrinol. Rev.* **27,** 47–72.

Pottinger, T. G., Rand-Weaver, M., and Sumpter, J. P. (2003). Overwinter fasting and re-feeding in rainbow trout: Plasma growth hormone and cortisol levels in relation to energy mobilisation. *Comp. Biochem. Physiol. B & Biochem. Mol. Biol.* **136,** 403–417.

Power, D. M. (2005). Developmental ontogeny of prolactin and its receptor in fish. *Gen. Comp. Endocrinol.* **142,** 25–33.

Power, D. M., Canario, A. V., and Ingleton, P. M. (1996). Somatotropin release-inhibiting factor and galanin innervation in the hypothalamus and pituitary of sea bream *(Sparus aurata)*. *Gen. Comp. Endocrinol.* **101,** 264–274.

Prunet, P., and Gonnet, F. (1986). Prolactin secretion in the rainbow trout: Effect of osmotic pressure and characterization of a hypothalamic PRL releasing activity *In* "Neuroendocrinology" 114, Abstract of the 1st International Congress of Neuroendocrinology. Karger, Basel.

Prunet, P., Sandra, O., Le Rouzic, P., Marchand, O., and Laudet, V. (2000). Molecular characterization of the prolactin receptor in two fish species, tilapia *Oreochromis niloticus* and rainbow trout, *Oncorhynchus mykiss*: A comparative approach. *Can. J. Physiol. Pharmacol.* **78,** 1086–1096.

Rand-Weaver, M., and Swanson, P. (1993). Plasma somatolactin levels in coho salmon (*Oncorhynchus kisutch*) during smoltification and sexual maturation. *Fish Physiol. Biochem.* **11,** 175–182.

Rand-Weaver, M., Noso, T., Muramoto, K., and Kawauchi, H. (1991a). Isolation and characterization of somatolactin, a new protein related to growth hormone and prolactin from Atlantic cod (*Gadus morhua*) pituitary glands. *Biochemistry* **30,** 1509–1515.

Rand-Weaver, M., Baker, B. I., and Kawauchi, H. (1991b). Cellular localization of somatolactin in the pars intermedia of some teleost fish. *Cell Tissue Res.* **263,** 207–215.

Rand-Weaver, M., Swanson, P., Kawauchi, H., and Dickhoff, W. W. (1992). Somatolactin, a novel pituitary protein: Purification and plasma levels during reproductive maturation of coho salmon. *J. Endocrinol.* **133,** 393–403.

Rand-Weaver, M., Kawauchi, H., and Ono, M. (1993). Evolution of the structure of the growth hormone and prolactin family. *In* "The Endocrinology of Growth, Development, and Metabolism in Vertebrates" (Schreibman, M. P., Scanes, C. G., and Pang, P. K. T., Eds.), pp. 13–42. Academic Press, New York.

Rand-Weaver, M., Pottinger, T. G., and Sumpter, J. P. (1995). Pronounced seasonal rhythms in plasma somatolactin levels in rainbow trout. *J. Endocrinol.* **146,** 113–119.

Robbercht, P., Deschodt-Lanckman, M., Camus, J. C., de-Neef, P., Lambert, M., and Christophe, M. (1979). VIP activation of rat anterior pituitary adenylate cyclase. *FEBS Lett.* **103,** 229–233.

Robberecht, W., Andries, M., and Denef, C. (1992). Stimulation of prolactin secretion from rat pituitary by luteinizing hormone-releasing hormone: Evidence against mediation by angiotensin II acting through a (Sar1-Ala8)-angiotensin II-sensitive receptor. *Neuroendocrinology* **56,** 185–194.

Rubin, D. A., and Specker, J. L. (1992). *In vitro* effects of homologous prolactins on testosterone production by testes of tilapia (*Oreochromis mossambicus*). *Gen. Comp. Endocrinol.* **87,** 189–196.

Sakai, M., Kobayashi, M., and Kawauchi, H. (1995). Enhancement of chemiluminescent responses of phagocytic cells from rainbow trout, *Oncorhynchus mykiss*, by injection of growth hormone. *Fish Shellfish Immunol.* **5,** 375–379.

Sakai, M., Kobayashi, M., and Kawauchi, H. (1996a). *In vitro* activation of fish phagocytic cells by GH, prolactin and somatolactin. *J. Endocrinol.* **151,** 113–118.

Sakai, M., Kobayashi, M., and Kawauchi, H. (1996b). Mitogenic effect of growth hormone and prolactin on chum salmon *Oncorhynchus keta* leukocytes *in vitro*. *Ver. Immunol. Immunopathol.* **53,** 185–189.

Sakamoto, T., and McCormick, S. D. (2006). Prolactin and growth hormone in fish osmoregulation. *Gen. Comp. Endocrinol.* **147,** 24–30.

Sakamoto, T., Fujimoto, M., and Ando, M. (2003a). Fishy tales of prolactin-releasing peptide. *Int. Rev. Cytol.* **225,** 91–130.

Sakamoto, T., Agustsson, T., Moriyama, S., Itoh, T., Takahashi, A., Kawauchi, H., Bjoarnsson, B. Th., and Ando, M. (2003b). Intra-arterial injection of prolactin-releasing peptide elevates prolactin gene expression and plasma prolactin levels in rainbow trout. *J. Comp. Physiol. B* **173,** 333–337.

Sakamoto, T., Amano, M., Hyodo, S., Moriyama, S., Takahashi, A., Kawauchi, H., and Ando, M. (2005). Expression of prolactin-releasing peptide and prolactin in the euryhaline mudskippers (*Periophthalmus modestus*): prolactin-releasing peptide as a primary regulator of prolactin. *J. Mol. Endocrinol.* **34,** 825–834.

Sandra, O., Sohm, F., De Luze, A., Prunet, P., Edery, M., and Kelly, P. A. (1995). Expression cloning of a cDNA encoding a fish prolactin receptor. *Proc. Natl. Acad. Sci. USA* **92,** 6037–6041.

Sandra, O., Le Rouzic, P., Cauty, C., Edery, M., and Prunet, P. (2000). Expression of the prolactin receptor (tiPRL-R) gene in tilapia *Oreochromis niloticus*: Tissue distribution and cellular localization in osmoregulatory organs. *J. Mol. Endocrinol.* **24,** 215–224.

San Martin, R., Caceres, P., Azocar, R., Alvarez, M., Molina, A., Vera, M. I., and Ktauskopf, M. (2004). Seasonal environmental changes modulate the prolactin receptor expression in an eurythermal fish. *J. Cell. Biochem.* **92,** 42–52.

San Martin, R., Hurtado, W., Quezada, C., Reyes, A. E., Vera, M. I., and Krauskopf, M. (2007). Gene structure and seasonal expression of carp fish prolactin short receptor isoforms. *J. Cell. Biochem.* **100,** 970–980.

Santos, C. R., Brinca, L., Ingleton, P. M., and Power, D. M. (1999). Cloning, expression, and tissue localisation of prolactin in adult sea bream (*Sparus aurata*). *Gen. Comp. Endocrinol.* **114,** 57–66.

Santos, C. R. A., Ingleton, P. M., Cavaco, J. E. B., Kelly, P. A., Edery, M., and Power, D. M. (2001). Cloning, characterization, and tissue distribution of prolactin receptor in the sea bream (*Sparus aurata*). *Gen. Comp. Endocrinol.* **121,** 32–47.

Santos, C. R., Cavaco, J. E., Ingleton, P. M., and Power, D. M. (2003). Developmental ontogeny of prolactin and prolactin receptor in the sea bream (*Sparus aurata*). *Gen. Comp. Endocrinol.* **132,** 304–314.

Satake, H., Minakata, H., Wang, X., and Fujimoto, M. (1999). Characterization of a cDNA encoding a precursor of Carassius RFamide, structurally related to a mammalian prolactin-releasing peptide. *FEBS Lett.* **446,** 247–250.

Sawisky, G. R., and Chang, J. P. (2005). Intracellular calcium involvement in pituitary adenylate cyclase-activating polypeptide stimulation of growth hormone and gonadotropin secretion in goldfish pituitary cells. *J. Neuroendocrinol.* **17,** 353–371.

Seale, A. P., Itoh, T., Moriyama, S., Takahashi, A., Kawauchi, H., Sakamoto, T., Fujimoto, M., Riley, L. G., Hirano, T., and Grau, E. G. (2002). Isolation and characterization of a homologue of mammalian prolactin-releasing peptide from the tilapia brain and its effect on prolactin release from the tilapia pituitary. *Gen. Comp. Endocrinol.* **125,** 328–339.

Sealfon, S. C., Weinstein, H., and Millar, R. P. (1997). Molecular mechanisms of ligand interaction with the gonadotropin-releasing hormone receptor. *Endocrinol. Rev.* **18,** 180–205.

Shepherd, B. S., Sakamoto, T., Nishioka, R. S., Richman, N. H., Mori, I., Madsen, S. S., Chen, T. T., Hirano, T., Bern, H. A., and Grau, E. G. (1997). Somatotropic actions of the homologous growth hormone and prolactins in the euryhaline teleost, tilapia, *Oreochromis mossambicus*. *Proc. Natl. Acad. Sci. USA* **94,** 2068–2072.

Shepherd, B. S., Sakamoto, T., Hyodo, S., Nishioka, R. S., Ball, C., Bern, H. A., and Grau, E. G. (1999). Is the primitive regulation of pituitary prolactin ($tPRL_{177}$ and $tPRL_{188}$) secretion and gene expression in the euryhaline tilapia (*Oreochromis mossambicus*) hypothalamic or environmental? *J. Endocrinol.* **161,** 121–129.

Sheridan, M. A. (1986). Effects of thyroxin, cortisol, growth hormone, and prolactin on lipid metabolism of coho salmon, *Oncorhynchus kisutch,* during smoltification. *Gen. Comp. Endocrinol.* **64,** 220–238.

Sheridan, M. A., Kittilson, J. D., and Slagter, B. J. (2000). Structure–function relationships of the signaling system for the somatostatin peptide hormone family. *Am. Zool.* **40,** 269–286.

Shiraishi, K., Matsuda, M., Mori, T., and Hirano, T. (1999). Changes in expression of prolactin and cortisol receptor genes during early life-stages of euryhaline tilapia (*Oreochromis mossambicus*) in fresh water and seawater. *Zool. Sci.* **16,** 139–146.

Slagter, B. J., Kittilson, J. D., and Sheridan, M. A. (2004). Somatostatin receptor subtype 1 and subtype 2 mRNA expression is regulated by nutritional state in rainbow trout (*Oncorhynchus mykiss*). *Gen. Comp. Endocrinol.* **139,** 236–244.

Slijkhuis, H., de Ruiter, A. J., Baggerman, B., and Wendelaar Bonga, S. E. (1984). Parental fanning behavior and prolactin cell activity in the male three-spined stickleback *Gasterosteus aculeatus* L. *Gen. Comp. Endocrinol.* **54,** 297–307.

Stefano, A. V., Vissio, P. G., Paz, D. A., Somoza, G. M., Maggese, M. C., and Barrantes, G. E. (1999). Colocalization of GnRH binding sites with gonadotropin-, somatotropin-, somatolactin-, and prolactin-expressing pituitary cells of the pejerrey, *Odontesthes bonariensis, in vitro*. *Gen. Comp. Endocrinol.* **116,** 133–139.

Sun, B., Fujiwara, K., Adachi, S., and Inoue, K. (2005). Physiological roles of prolactin-releasing peptide. *Regul. Pept.* **126,** 27–33.

Sun, Y., Lu, X., and Gershengorn, M. C. (2003). Thyrotropin-releasing hormone receptors – similarities and differences. *J. Mol. Endocrinol.* **30,** 87–97.

Tacon, P., Baroiller, J. F., Le Bail, P. Y., Prunet, P., and Jalabert, B. (2000). Effect of egg deprivation on sex steroids, gonadotropin, prolactin, and growth hormone profiles during the reproductive cycle of the mouthbrooding cichlid fish *Oreochromis niloticus*. *Gen. Comp. Endocrinol.* **117,** 54–65.

Takayama, Y., Ono, M., Rand-Weaver, M., and Kawauchi, H. (1991). Greater conservation of somatolactin, a presumed pituitary hormone of the growth hormone/prolactin family, than of growth hormone in teleost fish. *Gen. Comp. Endocrinol.* **83,** 366–374.

Takei, Y. (2000). Structural and functional evolution of the natriuretic peptide system in vertebrates. *Int. Rev. Cytol.* **194,** 1–66.

Taniyama, S., Kitahashi, T., Ando, H., Kaeriyama, M., Zohar, Y., Ueda, H., and Urano, A. (2000). Effects of gonadotropin-releasing hormone analog on expression of genes encoding the growth hormone/prolactin/somatolactin family and a pituitary-specific transcription factor in the pituitaries of prespawning sockeye salmon. *Gen. Comp. Endocrinol.* **118,** 418–424.

Taylor, M. M., and Samson, W. K. (2001). The prolactin releasing peptides: RFamide peptides. *Cell. Mol. Life Sci.* **58,** 1206–1215.

Tipsmark, C. K., Weber, G. M., Strom, C. N., Grau, E. G., Hirano, T., and Borski, R. J. (2005). Involvement of phospholipase C and intracellular calcium signaling in the gonadotropin-releasing hormone regulation of prolactin release from lactotrophs of tilapia *(Oreochromis mossambicus)*. *Gen. Comp. Endocrinol.* **142,** 227–233.

Toop, T., and Donald, J. A. (2004). Comparative aspects of natriuretic peptide physiology in non-mammalian vertebrates: A review. *J. Comp. Physiol. B* **174,** 189–204.

Tse, D. L. Y., Chow, B. K. C., Chan, C. B., Lee, L. T. O., and Cheng, C. H. K. (2000). Molecular cloning and expression studies of a prolactin receptor in goldfish (*Carassius auratus*). *Life Sci.* **66,** 593–605.

Tse, M. C. L., Wong, G. K. P., Xiao, P., Cheng, C. H. K., and Chan, K. M. (2008). Down-regulation of goldfish *(Carassius auratus)* prolactin gene expression by dopamine and thyrotropin releasing hormone. *Gen. Comp. Endocrinol.* **155,** 729–741.

Uchida, K., Yoshikawa-Ebesu, J. S., Kajimura, S., Yada, T., Hirano, T., and Grau, G. E. (2004). *In vitro* effects of cortisol on the release and gene expression of prolactin and growth hormone in the tilapia, *Oreochromis mossambicus*. *Gen. Comp. Endocrinol.* **135,** 116–125.

Unniappan, S., Lin, X., Cervini, L., Rivier, J., Kaiya, H., Kangawa, K., and Peter, R. E. (2002). Goldfish ghrelin: Molecular characterization of the complementary deoxyribonucleic acid, partial gene structure and evidence for its stimulatory role in food intake. *Endocrinology* **143,** 4143–4146.

Vega-Rubín de Celis, S., Gómez, P., Calduch-Giner, J. A., Médale, F., and Pérez-Sánchez, J. (2003). Expression and characterization of European sea bass (*Dicentrarchus labrax*) somatolactin: Assessment of *in vivo* metabolic effects. *Mar. Biotechnol.* **5,** 92–101.

Vega-Rubin de Celis, S., Rojas, P., Gomez-Requeni, P., Albalat, A., Gutierrez, J., Medale, F., Kaushik, S. J., Navarro, I., and Perez-Sanchez, J. (2004). Nutritional assessment of somatolactin function in gilthead sea bream *(Sparus aurata)*: Concurrent changes in somatotropic axis and pancreatic hormones. *Comp. Biochem. Biophys.* A **138,** 533–542.

Vissio, P. G., Stefano, A. V., Somoza, G. M., Maggese, M. C., and Paz, D. A. (1999). Close association among GnRH (Gonadotropin-releasing hormone) fibers and GtH, GH, SL and PRL expressing cells in pejerrey, *Odontesthes bonariensis* (Teleostei, Atheriniformes). *Fish Physiol. Biochem.* **21,** 121–127.

Vissio, P. G., Andreone, L., Paz, MD., Aaggese, M. C., Somoza, G. M., and Strussmann, C, A. (2002). Relation between the reproductive status and somatolactin cell activity in the pituitary of pejerrey, *Odontesthes bonariensis* (Atheriniformes). *J. Exp. Zool.* **293,** 492–499.

Wang, X., Morishita, F., Matsushima, O., and Fujimoto, M. (2000a). Immunohistochemical localization of C-RF-amide, a FMRF-related peptide, in the brain of the goldfish, *Carassius auratus*. *Zool. Sci.* **17,** 1067–1074.

Wang, X., Morishita, F., Matsushima, O., and Fujimoto, M. (2000b). Carassius RFamide, a novel FMRFa-related peptide, is produced within the retina and involved in retinal information processing in cyprinid fish. *Neurosci. Lett.* **289,** 115–118.

Weber, G. M., Powell, J. F. F., Park, M., Fischer, W. H., Craig, A. G., Rivier, J. E., Nanakorn, U., Parhar, I. S., Ngamvongchon, S., Grau, E. G., and Sherwood, N. M. (1997). Evidence that gonadotropinreleasing hormone (GnRH) functions as a prolactin-releasing factor in a teleost fish (*Oreochromis mossambicus*) and primary structures for three native GnRH molecules. *J. Endocrinol.* **155,** 121–132.

Weber, G. M., Seale, A. P., Richman, N. H., III, Stetson, M. H., Hirano, T., and Grau, E. G. (2004). Hormone release is tied to changes in cell size in the osmoreceptive prolactin cell of a euryhaline teleost fish, the tilapia, *Oreochromis mossambicus*. *Gen. Comp Endocrinol.* **138,** 8–13.

Wigham, T., and Batten, T. F. (1984). *In vitro* effects of thyrotropin-releasing hormone and somatostatin on prolactin and growth hormone release by the pituitary of *Poecilia latipinna*. I. An electrophoretic study. *Gen. Comp. Endocrinol.* **55,** 444–449.

Williams, A. J., and Wigham, T. (1994a). The regulation of prolactin cells in the rainbow trout (*Oncorhynchus mykiss*). I. Possible roles for thyrotropin-releasing hormone (TRH) and oestradiol. *Gen. Comp. Endocrinol.* **93,** 388–397.

Williams, A. J., and Wigham, T. (1994b). The regulation of prolactin cells in the rainbow trout *(Oncorhynchus mykiss)*. 2. Somatostatin. *Gen Comp. Endocrinol.* **93,** 398–405.

Wirachowsky, N. R., Kwong, P., Yunker, W. K., Johnson, J. D., and Chang, J. D. (2000). Mechanisms of action of pituitary adenylate cyclase-activating polypeptide (PACAP) on growth hormone release from dispersed goldfish pituitary cells. *Fish Physiol. Biochem.* **23,** 201–214.

Wong, A. O., Chang, J. P., and Peter, R. E. (1993). Characterization of D1 receptors mediating dopamine-stimulated growth hormone release from pituitary cells of the goldfish, *Carassius auratus*. *Endocrinology* **133,** 577–584.

Wong, A. O. L., Leung, M. Y., Shea, W. L. C., Tse, L. Y., Chang, J. P., and Chow, B. K. C. (1998). Hypophysiotropic action of pituitary adenylate cyclase-activating polypeptide (PACAP) in the goldfish: Immunohistochemical demonstration of PACAP in the pituitary, PACAP stimulation of growth hormone release from pituitary cells, and molecular cloning of pituitary type I PACAP receptor. *Endocrinology* **139,** 3465–3479.

Wong, A. O. L., Li, W., Leung, C. Y., Huo, L., and Zhou, H. (2005). Pituitary adenylate cyclase activating polypeptide (PACAP) and growth hormone (GH)-releasing factor in grass carp. I. functional coupling of cyclic adenosine 3′5′-monophosphate and Ca^{2+}/calmodulin-dependent signaling pathways in PACAP-induced GH secretion and GH gene expression in grass carp pituitary cells. *Endocrinology* **146,** 5407–5424.

Yada, T., Uchida, K., Kajimura, S., Azuma, T., Hirano, T., and Grau, E. G. (2002a). Immunomodulatory effects of prolactin and growth hormone in the tilapia, *Oreochromis mossambicus*. *J. Endocrinol.* **173,** 483–492.

Yada, T., Moriyama, S., Suzuki, Y., Azuma, T., Takahashi, A., Hirose, S., and Naito, N. (2002b). Relationships between obesity and metabolic hormones in the "cobalt" variant of rainbow trout. *Gen. Comp. Endocrinol.* **128,** 36–43.

Yang, B. Y., and Chen, T. T. (2003). Identification of a new growth hormone family protein, somatolactin-like protein, in the rainbow trout (*Oncorhyncus mykiss*) pituitary gland. *Endocrinology* **144,** 850–857.

Yang, B. Y., Arab, M., and Chen, T. T. (1997). Cloning and characterization of rainbow trout (*Oncorhynchus mykiss*) somatolactin cDNA and its expression in pituitary and nonpituitary tissues. *Gen. Comp. Endocrinol.* **106,** 271–280.

Yang, B. Y., Greene, M., and Chen, T. T. (1999). Early embryonic expression of the growth hormone family protein genes in the developing rainbow trout, *Oncorhynchus mykiss*. *Mol. Reprod. Dev.* **53,** 127–134.

Zhu, Y., and Thomas, P. (1995). Red drum somatolactin: Development of a homologous radioimmunoassay and plasma levels after exposure to stressors or various backgrounds. *Gen. Comp. Endocrinol.* **99,** 275–288.

Zhu, Y., and Thomas, P. (1997). Studies on the physiology of somatolactin secretion in red drum and Atlantic croaker. *Fish Physiol. Biochem.* **17,** 271–278.

Zhu, Y., and Thomas, P. (1998). Effects of light on plasma somatolactin levels in red drum (*Sciaenops ocellatus*). *Gen. Comp. Endocrinol.* **111,** 76–82.

Zhu, Y., Yoshiura, Y., Kikuchi, K., Aida, K., and Thomas, P. (1999). Cloning and phylogenetic relationship of red drum somatolactin cDNA and effects of light on pituitary somatolactin mRNA expression. *Gen. Comp. Endocrinol.* **113,** 69–79.

Zhu, Y., Stiller, J. W., Shaner, M. P., Baldini, A., Scemama, J. L., and Capehart, A. A. (2004). Cloning of somatolactin alpha and beta cDNAs in zebrafish and phylogenetic analysis of two distinct somatolactin subtypes in fish. *J. Endocrinol.* **182,** 509–518.

Zupanc, G. K., Siehler, S., Jones, E. M., Seuwen, K., Furuta, H., Hoyer, D., and Yano, H. (1999). Molecular cloning and pharmacological characterization of a somatostatin receptor subtype in the gymnotiform fish *Apteronotus albifrons*. *Gen. Comp. Endocrinol.* **115,** 333–345.

6

REGULATION AND CONTRIBUTION OF THE CORTICOTROPIC, MELANOTROPIC AND THYROTROPIC AXES TO THE STRESS RESPONSE IN FISHES

NICHOLAS J. BERNIER
GERT FLIK
PETER H.M. KLAREN

1. Introduction
2. Hypothalamic Regulation of Pituitary Cells
   2.1. Hypothalamic Regulation of Corticotropes
   2.2. Hypothalamic Regulation of Melanotropes
   2.3. Hypothalamic Regulation of Thyrotropes
3. Targets and Functions of the Corticotrope, Melanotrope and Thyrotrope Secretions
   3.1. Targets and Functions of Corticotrope Secretions: ACTH and Non-acetylated βEND
   3.2. Targets and Functions of Melanotrope Secretions: αMSH and *N*-acetylated βENDs
   3.3. Targets and Functions of Thyroid-stimulating Hormone
4. Targets, Functions and Feedback Effects of Cortisol and Thyroid Hormones
   4.1. Cortisol
   4.2. Thyroid Hormones
5. Contribution of the Corticotropic, Melanotropic, and Thyrotropic Axes to the Stress Response
   5.1. The Corticotropic Axis
   5.2. The Melanotropic Axis
   5.3. The Thyrotropic Axis
   5.4. Integrating at the Organismal Level: The Contribution of Hypophysiotropic Factors to the Stress Response *In Vivo*
6. Perspectives

Multiple hypothalamic factors are involved in the regulation of the secretions from the corticotrope, melanotrope and thyrotrope pituitary cells in fishes. Among these factors, corticotropin-releasing factor (CRF) and thyrotropin-releasing hormone (TRH) stimulate, and dopamine

Fish Neuroendocrinology: Volume 28
FISH PHYSIOLOGY

DOI: 10.1016/S1546-5098(09)28006-X

generally inhibits the secretions from all three hypothalamo-pituitary axes. The CRF system also plays a master role in the regulation of the endocrine response to stressors. In general, the contributions of the corticotropic, melanotropic and thyrotropic axes to the stress response are species-specific and depend on the challenge imposed on the system and its duration as well as the homeostatic resilience of the fish. Multiple interactions and feedback effects have been identified between these endocrine axes. We postulate that the extensive multidirectional communication as well as cross-talk between the corticotropic, melanotropic and thyrotropic axes forms a 'stress web' that exerts well-concerted actions on energy metabolism as its prime task. This chapter reviews the regulatory pathways, the targets, the functions and the interactions between the corticotropic, melanotropic and thyrotropic axes and the evidence that implicates each axis in the stress response.

1. INTRODUCTION

An "axis" is a conceptual approach to describe and model how an endocrine system is controlled. This chapter deals with hypothalamic–pituitary axes. The hypothalamic–pituitary–interrenal (HPI) axis and hypothalamic–pituitary–thyroid (HPT) axis comprise hypothalamic releasing factors that, via the pituitary pars distalis, modulate the output of peripheral cortisol and thyroid hormone producing glands. For lack of identification of a *bona fide* peripheral endocrine target, the melanotrope cell in the pituitary pars intermedia appears to be the ultimate target organ in the melanotropic axis.

As Ball *et al.* (1980) have suggested, the vertebrate hypothalamus–pituitary system may have evolved from an ancestral chordate structure in which the ventral floor of the brain was a site for systemic neuroendocrine secretions. Extant teleosts have a typical organization where a median eminence is absent and hypothalamic neurons, in particular those originating from the nucleus preopticus (NPO) and nucleus lateralis tuberis (NLT), directly project onto the pituitary pars distalis (Peter and Fryer, 1984). Multiple hypothalamic factors, which possibly once functioned as neurotransmitters, are involved in the regulation of the secretions from the corticotrope, melanotrope and thyrotrope pituitary cells. The contribution by some of these hypothalamic factors to the regulation of all three axes suggests the existence of a coordinated regulation.

In fact, multiple interactions and feedback effects have now been identified between the corticotropic, melanotropic and thyrotropic axes. These interactions mainly take place in the hypothalamus and peripheral target glands. Hence, it is difficult to adhere to a model consisting of three separate, parallel-running axes as control systems. While our appreciation of how this complex network contributes to maintaining homeostasis or to re-establishing steady state following a stressor is in its infancy, our ability to further understand the physiological significance of this communication and cross-talk will require a more comprehensive understanding of all three axes. Therefore, this chapter aims to review what is currently known in fish on the regulatory pathways and functions of the corticotropic, melanotropic and thyrotropic axes under basal conditions as well as in response to stressors. The chapter also highlights the physiological processes where the corticotropic, melanotropic and thyrotropic axes are already known to interact.

2. HYPOTHALAMIC REGULATION OF PITUITARY CELLS

2.1. Hypothalamic Regulation of Corticotropes

The corticotropes in the pituitary pars distalis produce the peptides adrenocorticotropic hormone (ACTH) and non-acetylated opioid β-endorphin (βEND) from the precursor proopiomelanocortin (POMC). While several studies have identified hypophysiotropic factors of ACTH secretion in fish, little is known about the regulatory mechanisms involved in the specific release of βEND from the pars distalis.

2.1.1. Stimulatory Factors

Among the multiple factors that affect the secretion of ACTH (Fig. 6.1), CRF is considered to be the major regulator of ACTH secretion in teleosts (Lederis *et al.*, 1994; Flik *et al.*, 2006). *In vitro* superfusion of pars distalis cells or whole pituitary glands with CRF dose-dependently stimulates ACTH secretion in several fish species (Table 6.1). CRF is expressed in the nucleus preopticus (NPO) of most teleosts examined to date as well as in several hypothalamic areas of the fish brain known to project to the pituitary gland (Okawara *et al.*, 1992; Pepels *et al.*, 2002; Alderman and Bernier, 2007). In the brown ghost knifefish (*Apteronotus leptorhynchus*), a combination of immunohistochemistry and neuronal tract tracing studies have provided direct evidence that a subpopulation of the CRF-ir cells of the NPO innervate the pituitary gland (Zupanc *et al.*, 1999). Moreover, immunohistochemical studies in over a dozen teleost species have shown that CRF-ir nerve fibers innervate the rostral pars distalis and are found in particular in the vicinity of ACTH cells (e.g., Pepels *et al.*, 2002; for review see Chapter 1, this volume).

Limited evidence also suggests that the non-acetylated βEND present in the corticotropes is under the stimulatory control of CRF. *In vitro*

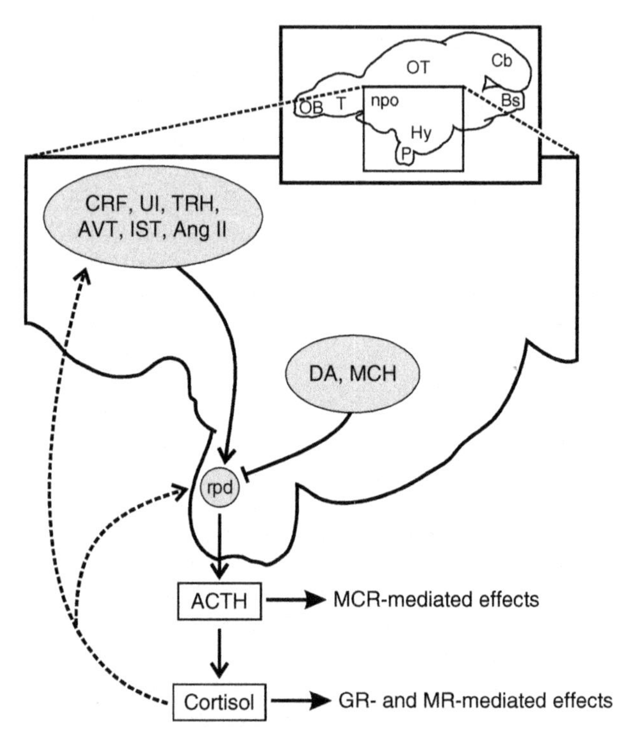

Fig. 6.1. Overview of the major factors that affect the activity of the corticotropic axis in teleosts. The mid-sagittal section in the inset shows the major divisions of the brain. Enlarged are the preoptic area and the hypothalamus, the primary regions involved in the hypophysiotropic regulation of the corticotropic axis. ACTH is the primary secretagogue of cortisol secretion, but several secondary factors may also play a role in the regulation of cortisol synthesis and secretion. The actions of both ACTH and cortisol are mediated by different receptor subtypes and cortisol can exert negative feedback effects on the corticotropes of the pituitary and on some of the brain's hypophysiotropic factors. Note that this overview is not a generalization, but a model constructed from experimental results obtained on different species (see text for detail). Solid arrow indicates stimulation, T-line indicates inhibition, and dashed arrow indicates negative feedback. Ang II, angiotensin II; AVT, arginine vasotocin; Bs, brainstem; Cb, cerebellum; CRF, corticotropin-releasing factor; DA, dopamine; GR, glucocorticoid receptor; Hy, hypothalamus; IST, isotocin; MCH, melanin-concentrating hormone; MCR, melanocortin receptors; MR, mineralocorticoid receptor; npo, nucleus preopticus; OB, olfactory bulb; OT, optic tectum; P, pituitary; rpd, rostral pars distalis;T, telencephalon; TRH, thyrotropin-releasing hormone; UI, urotensin I.

stimulation of common carp pituitary pars distalis with ovine CRF significantly increased βEND secretion (van den Burg *et al.*, 2001). Although non-acetylated βEND is an abundant form of circulating βEND and this opioid may play various physiological roles (see Section 3.1), to our

Table 6.1
Factors affecting ACTH secretion in teleost fish

Factor	Species	Reported effect	Effective or tested dose	Experimental observation/ comment	Reference
CRF	Goldfish (*Carassius auratus*)	Stimulates	2–100 nM	Ovine CRF; stimulation of ACTH secretion *in vitro* from superfused pars distalis cells	Fryer *et al.* (1983)
	Goldfish (*Carassius auratus*)	Stimulates	10–500 nM	White sucker and human/rat CRF; stimulation of ACTH secretion *in vitro* from superfused anterior pituitary cells	Lederis *et al.* (1994)
	Rainbow trout (*Oncorhynchus mykiss*)	Stimulates	0.1–1000 nM	Human/rat CRF; stimulation of ACTH secretion from pars distalis fragment incubated *in vitro* for 30 min	Baker *et al.* (1996)
	Rainbow trout (*Oncorhynchus mykiss*)	Stimulates	0.1 pM–0.1 μM ED_{50} 0.8 pM	Human/rat CRF; stimulation of ACTH secretion *in vitro* from superfused pituitary gland	Pierson *et al.* (1996)
	Gilthead sea bream (*Sparus auratus*)	Stimulates	0.01–1000 nM ED_{50} 1.5 nM	Human/rat CRF; stimulation of ACTH secretion *in vitro* from superfused pituitary gland	Rotllant *et al.* (2000b, 2001)
	Tilapia (*Oreochromis niloticus*)	Stimulates	100 nM	Tilapia and human/rat CRF; equipotent stimulation of ACTH secretion *in vitro* from superfused pituitary gland	van Enckevort *et al.* (2000)

(*continued*)

Table 6.1 *(continued)*

Factor	Species	Reported effect	Effective or tested dose	Experimental observation/ comment	Reference
	Common carp (*Cyprinus carpio*)	Stimulates	100 nM	Ovine CRF; stimulation of ACTH secretion *in vitro* from superfused pars distalis only in the presence of dopamine	Metz *et al.* (2004)
UI	Goldfish (*Carassius auratus*)	Stimulates	0.5–10 nM	White sucker UI; stimulation of ACTH secretion *in vitro* from superfused pars distalis cells. UI was 2–3 times more potent than ovine CRF or sauvagine in stimulating ACTH secretion	Fryer *et al.* (1983)
SVG	Goldfish (*Carassius auratus*)	Stimulates	2–100 nM	Stimulation of ACTH secretion *in vitro* from superfused pars distalis cells	Fryer *et al.* (1983)
AVT	Goldfish (*Carassius auratus*)	Stimulates	0.063–4 nM	Stimulation of ACTH secretion *in vitro* from superfused pars distalis cells	Fryer *et al.* (1985); Lederis *et al.* (1994)
	Rainbow trout (*Oncorhynchus mykiss*)	Stimulates	0.1 pM–1 μM ED_{50} 0.2 nM	Stimulation of ACTH secretion *in vitro* from superfused pituitary gland	Pierson *et al.* (1996)
	Rainbow trout (*Oncorhynchus mykiss*)	Stimulates	0.1–1000 nM	Stimulation of ACTH secretion from pars distalis fragment incubated *in vitro* for 30 min	Baker *et al.* (1996)

IST	Goldfish (*Carassius auratus*)	Stimulates	0.063–4 nM	Stimulation of ACTH secretion *in vitro* from superfused pars distalis cells	Fryer *et al.* (1985); Lederis *et al.* (1994)
	Rainbow trout (*Oncorhynchus mykiss*)	Stimulates	0.1 pM–1 μM ED_{50} 100 nM	Stimulation of ACTH secretion *in vitro* from superfused pituitary gland	Pierson *et al.* (1996)
AVP	Goldfish (*Carassius auratus*)	Stimulates	0.25–10 nM	Stimulation of ACTH secretion *in vitro* from superfused pars distalis cells	Fryer *et al.* (1985); Lederis *et al.* (1994)
Angiotensins	Goldfish (*Carassius auratus*)	Stimulates	0.5–200 nM	Salmon Ang I, human Ang I and Ang II; stimulation of ACTH secretion *in vitro* from superfused pars distalis cells. Ang I was about one-tenth as potent as Ang II	Weld and Fryer (1987, 1988)
TRH	Goldfish (*Carassius auratus*)	No effect	2.76 μM (1 μg mL^{-1})	TRH tested on ACTH secretion *in vitro* from superfused pars distalis cells	Fryer *et al.* (1983)
	Gilthead sea bream (*Sparus auratus*)	Stimulates	50 nM	Stimulation of ACTH secretion *in vitro* from superfused pituitary gland	Rotllant *et al.* (2000b)
MCH	Rainbow trout (*Oncorhynchus mykiss*)	Inhibits	0.1–100 nM	Salmon MCH; inhibition of basal and CRF-induced ACTH secretion from pars distalis incubated *in vitro* for 30 min	Baker *et al.* (1985)

(*continued*)

Table 6.1 (*continued*)

Factor	Species	Reported effect	Effective or tested dose	Experimental observation/ comment	Reference
DA	Goldfish (*Carassius auratus*)	Inhibits	Not applicable	Cytometric analysis: *in vivo* treatment with DA antagonists injected ip results in cellular and nuclear hypertrophy of corticotropic cells	Olivereau *et al.* (1988)
	Common carp (*Cyprinus carpio*)	Inhibits	10 μM	Suppression of ACTH secretion *in vitro* from superfused pars distalis fragment	Metz *et al.* (2004)

ACTH, adrenocorticotropic hormone; Ang, angiotensin; AVP, arginine vasopressin; AVT, arginine vasotocin; CRF, corticotropin-releasing factor; DA, dopamine; ip, intraperitoneal; IST, isotocin; MCH, melanin-concentrating hormone; TRH, thyrotropin-releasing hormone; SVG, sauvagine; UI, urotensin I.

knowledge no other study has examined the regulatory mechanisms involved in the secretion of this peptide.

Urotensin I (UI), another member of the CRF family of peptides, also stimulates ACTH secretion in fish. In fact, the potency of the heterologous peptides varies significantly. For example, white sucker (*Catostomus commersoni*) UI is significantly more potent than ovine CRF in stimulating ACTH secretion from superfused goldfish pars distalis cells (Fryer *et al.*, 1983). The corticotropic actions of UI, like those of CRF, are reversed by the non-selective CRF receptor antagonist, α-helical $CRF_{(9–41)}$ (Weld *et al.*, 1987). The limited studies describing the brain distribution of UI have shown expression in the NPO of zebrafish (*Danio rerio*; Alderman and Bernier, 2007), and in hypothalamic nuclei with known hypophysial projections in white sucker (Yulis *et al.*, 1986), goldfish (*Carassius auratus*; Fryer, 1989) and zebrafish (Alderman and Bernier, 2007). However, to date there is no clear evidence that UI-ir nerve fibers innervate the rostral pars distalis of fish. In mammals, urocortin (the orthologue to UI) has a potent ACTH-releasing activity *in vitro*, but it does not appear to be involved in the regulation of ACTH secretion *in vivo* (Oki and Sasano, 2004). Overall, more studies are needed to determine the relative importance of endogenous UI in fish as a corticotropic factor.

As recognized in mammals (Aguilera *et al.*, 2008), there is good evidence that neurohypophysial hormones also participate in the regulation of ACTH secretion in fish. *In vitro*, both arginine vasotocin (AVT) and isotocin (IST), the respective homologues to the mammalian hormones arginine vasopressin (AVP) and oxytocin, stimulate ACTH release from the pituitary of goldfish (Fryer *et al.*, 1985) and rainbow trout (*Oncorhynchus mykiss*; Baker *et al.*, 1996; Pierson *et al.*, 1996). In goldfish, the maximum ACTH-releasing activity of homologous AVT and IST is about half that of heterologous CRF and UI (Lederis *et al.*, 1994) and, unlike the situation in mammals (Aguilera *et al.*, 2008), neither AVT nor IST potentiates the corticotropic activity of CRF-related peptides (Fryer *et al.*, 1985). In rainbow trout, AVT and IST also have a lower ACTH-releasing efficacy than human/rat CRF, but here AVT clearly potentiates the ACTH-releasing properties of CRF (Baker *et al.*, 1996; Pierson *et al.*, 1996). Using a pharmacological approach, Pierson *et al.* (1996) also provided evidence that a V1-type vasopressin receptor is involved in mediating AVT-stimulated ACTH release in rainbow trout pituitary.

In the teleost brain, AVT and IST are almost exclusively expressed in the parvocellular and magnocellular neurons of the preoptic area (POA) and a good portion of these neurons project to the neurohypophysis (see Chapter 1, this volume). In some species, however, there is evidence that AVT-ir and IST-ir fibers make direct contact with the corticotropes in the adenohypophysis (Batten, 1986; Batten *et al.*, 1999). AVT-ir neurons also have been shown to colocalize with CRF neurons in the NPO of several teleost species (e.g., Ando *et al.*, 1999; Pepels *et al.*, 2002; Huising *et al.*, 2004). In mammals, while the secretion of magnocellular AVP into the peripheral circulation depends upon osmotic stimulation, the secretion of parvocellular AVP into the pituitary portal circulation is independent of the osmotic status and increases in response to specific types of acute stressors to potentiate the stimulatory effects of CRF on ACTH secretion (Aguilera *et al.*, 2008). Similarly, plasma osmolality is a potent stimulant of neurohypophysial AVT secretion in fish (see Chapter 8, this volume) and preoptic area AVT mRNA expression increases in response to diverse acute stressors which are not necessarily associated with an osmotic disturbance (see Section 5). While the pituitary content and the plasma levels of IST can also increase in response to specific stressors in some species (e.g., Mancera *et al.*, 2008), the evidence for a potential role of IST in the stress response in fish is still limited.

Angiotensins and thyrotropin-releasing hormone (TRH) are additional factors with ACTH-releasing activity in fish. Both angiotensin I (Ang I) and Ang II can stimulate ACTH secretion from superfused goldfish pars distalis cells (Weld and Fryer, 1987), but they do not potentiate the corticotropic actions of either CRF or UI (Weld and Fryer, 1988). Interestingly, many preoptic area AVT-ir neurosecretory neurons in the goldfish are also immunoreactive for Ang II, and some of these fibers reach the pars distalis of the adenohypophysis (Yamada *et al.*, 1990). While it remains to be seen whether angiotensins have a physiological role in the regulation of fish corticotropes, the stimulation of ACTH secretion by Ang II and the role of the renin–angiotensin system in the regulation of the stress response in mammals is well established (Saavedra and Benicky, 2007). Although TRH stimulates ACTH secretion from superfused gilthead sea bream (*Sparus auratus*) pituitary glands (Rotllant *et al.*, 2000b), it had no effect on the secretion of ACTH from superfused goldfish pars distalis cells (Fryer *et al.*, 1983). Moreover, to date, there is no clear evidence that TRH-ir fibers innervate the rostral pars distalis (RPD) in the teleost pituitary (for review see Cerdá-Reverter and Canosa, this volume). Finally, while both the monoamine serotonin (Winberg *et al.*, 1997; Höglund *et al.*, 2002b) and the cytokine interleukin 1β (Holland *et al.*, 2002; Metz *et al.*, 2006a; see Chapter 7, this volume) stimulate the HPI axis in fish, it remains to be determined whether these factors have a direct action on corticotropes. This question can be answered with *in situ* hybridization studies on receptor mRNAs.

2.1.2. Inhibitory Factors

Some hypothalamic factors can inhibit the release of ACTH from fish corticotropes. *In vitro*, melanin-concentrating hormone (MCH) is a potent inhibitor of basal and CRF-elicited ACTH secretion from the pars distalis of rainbow trout (Baker *et al.*, 1985, 1986). While most of the MCH-positive fibers that project to the pituitary are neurohypophysial and their product plays an important role in background color adaptation (Kawauchi, 2006), there is immunocytochemical evidence that some MCH fibers innervate the RPD (Naito *et al.*, 1986; Batten *et al.*, 1999). In addition, rainbow trout adapted to a light background have high circulating levels of MCH as well as lower plasma ACTH and cortisol levels than fish adapted to a dark background (Baker and Rance, 1981; Gilham and Baker, 1985).

Cytological and physiological evidence suggests that dopamine (DA) can inhibit basal ACTH secretion in fish. In goldfish, intraperitoneal (ip) injections of DA antagonists induce cytological changes in corticotropic cells consistent with an inhibitory effect of DA on ACTH synthesis and secretion (Olivereau *et al.*, 1988). In another cyprinid, the common carp (*Cyprinus carpio*), *in vitro* superfusion studies suggest that *in vivo* release of ACTH is under dopaminergic inhibitory control and that CRF can only stimulate ACTH secretion in the presence of a mild dopaminergic inhibitory tonus (Metz *et al.*, 2004). Dopaminergic innervation of the pituitary has been demonstrated in several teleost species (for review see Chapter 1, this volume). Whether the inhibitory effect of DA on common carp ACTH secretion is a general feature of teleost corticotrope secretion needs to be determined.

2.2. Hypothalamic Regulation of Melanotropes

The melanotropes in the pituitary pars intermedia produce α-melanocyte-stimulating hormone (αMSH) and βENDs from POMC. Both hormones are subject to post-translational acetylations that affect their biological activities profoundly (see Sections 3.1.2 and 3.2.1). Tables 6.2 and 6.3 show a consistent picture of stimulatory effects by TRH and members of the CRF family of neuropeptides (CRF, UI, sauvagine) on the release of αMSH and βENDs, and inhibitory effects exerted by DA and MCH (Fig. 6.2). It is insightful to first compare this picture with the multitude of factors that are involved in the regulation of the secretory activity of the amphibian melanotrope cell.

In the South African clawed toad *Xenopus laevis*, hypothalamic factors from the suprachiasmatic nucleus (NSC) and beyond [CRF, urocortin 1, TRH, DA, γ-aminobutyric acid (GABA), neuropeptide Y], adrenaline and serotonin (5-HT) from extrahypothalamic sites, and autocrine signals [purines, Ca^{2+}, brain-derived neurotrophic factor (BDNF), acetylcholine] all exert stimulatory or inhibitory effects on the release of αMSH (reviewed by Jenks *et al.*, 2007). It is perhaps difficult to reconcile the extensive regulation of the *Xenopus* melanotrope with the release of a single endocrine product that has, or appears to have, a simple peripheral role in melanosome dispersion in skin melanophores. Indeed, multiple regulatory pathways can

Table 6.2
Factors affecting αMSH secretion in teleost fish

Factor	Species	Reported effect	Effective or tested dose	Experimental observation/ comment	Reference
CRF	Goldfish (*Carassius auratus*)	Stimulates	1–64 nM	Ovine CRF; stimulation of αMSH secretion *in vitro* from superfused pars intermedia cells	Tran *et al.* (1990)
	Tilapia (*Oreochromis mossambicus*)	Stimulates	1–1000 nM EC_{50} 10 nM	Ovine CRF; stimulation of αMSH secretion *in vitro* from superfused pars intermedia fragment	Lamers *et al.* (1991)
	Gilthead sea bream (*Sparus auratus*)	Stimulates	0.01–1000 nM ED_{50} 12 nM	Human/rat CRF; stimulation of αMSH secretion *in vitro* from superfused pituitary glands	Rotllant *et al.* (2000b, 2001)
	Tilapia (*Oreochromis mossambicus*)	Stimulates	100 nM	Tilapia and rat CRF; equipotent stimulation of αMSH secretion *in vitro* from superfused pituitary glands	van Enckevort *et al.* (2000)
	Red porgy (*Pagrus pagrus*)	Stimulates	0.01–100 nM	Stimulation of αMSH secretion *in vitro* from superfused pituitary glands. Only the 100 nM dose was effective	van der Salm *et al.* (2004)
	Common carp (*Cyprinus carpio*)	Stimulates	0.01–1000 nM	Ovine CRF; stimulation of αMSH secretion *in vitro* from superfused pars intermedia fragment. Marked inter-individual variability in responsive state to CRF *in vitro*.	van den Burg *et al.* (2005)
UI	Goldfish (*Carassius auratus*)	Stimulates	1–64 nM	White sucker UI; stimulation of αMSH secretion *in vitro* from superfused pars intermedia cells. UI was 2–3 times more potent than oCRF or sauvagine in stimulating αMSH secretion	Tran *et al.* (1990)

SVG	Goldfish (*Carassius auratus*)	Stimulates	1–64 nM	Stimulation of αMSH secretion *in vitro* from superfused pars intermedia cells	Tran *et al.* (1990)
TRH	Goldfish (*Carassius auratus*)	Stimulates	0.5–250 nM	Stimulation of αMSH secretion *in vitro* from superfused pars intermedia cells	Tran *et al.* (1989)
	Goldfish (*Carassius auratus*)	Stimulates	1–10 000 nM EC_{50} 6.9 nM	Stimulation of αMSH secretion *in vitro* from superfused pars intermedia fragment	Omeljaniuk *et al.* (1989)
	Tilapia (*Oreochromis mossambicus*)	Stimulates	1–1000 nM EC_{50} 158 nM	Stimulation of αMSH secretion *in vitro* from superfused pars intermedia fragment	Lamers *et al.* (1991, 1994)
	Gilthead seabream (*Sparus auratus*)	Stimulates	50 nM	Stimulation of αMSH secretion *in vitro* from superfused pituitary glands	Rotllant *et al.* (2000b)
	Common carp (*Cyprinus carpio*)	Stimulates	0.01–10 nM	Stimulation of αMSH secretion *in vitro* from superfused pituitary parts	van den Burg *et al.* (2003)
	Red porgy (*Pagrus pagrus*)	Stimulates	0.01–100 nM EC_{50} 0.7 nM	Stimulation of αMSH secretion *in vitro* from superfused pituitary glands	van der Salm *et al.* (2004)
AVT	Goldfish (*Carassius auratus*)	No effect	19.0 nM (20 ng mL^{-1})	AVT tested on αMSH secretion *in vitro* from superfused pars intermedia cells	Tran *et al.* (1989, 1990)

(*continued*)

Table 6.2 (*continued*)

Factor	Species	Reported effect	Effective or tested dose	Experimental observation/ comment	Reference
IST	Goldfish (*Carassius auratus*)	No effect	20.7 nM (20 ng mL^{-1})	IST tested on αMSH secretion *in vitro* from superfused pars intermedia cells	Tran *et al.* (1989, 1990)
DA	European eel (*Anguilla anguilla*)	Inhibits	Not applicable	Cytometric analysis: *in vivo* treatment with the dopamine antagonist pimozide injected ip increased melanotrope activity	Olivereau (1978)
	Rainbow trout (*Oncorhynchus mykiss*)	Inhibits	10 μM	Suppression of αMSH secretion by pars intermedia incubated *in vitro* for 18 h	Barber *et al.* (1987)
	Goldfish (*Carassius auratus*)	Inhibits	Not applicable	Cytometric analysis: *in vivo* treatment with various dopamine antagonists injected ip increased melanotrope activity	Olivereau *et al.* (1987)
	Goldfish (*Carassius auratus*)	Inhibits	0.1 nM–100 μM	Suppression of basal and TRH-stimulated αMSH secretion *in vitro* from superfused pars intermedia fragment	Omeljaniuk *et al.* (1989)
	Tilapia (*Oreochromis mossambicus*)	Inhibits	3 nM–10 μM	Suppression of αMSH secretion *in vitro* from superfused pars intermedia fragment	Lamers *et al.* (1991)
	Tilapia (*Oreochromis mossambicus*)	Biphasic response	0.01 pM–10 μM	In low-pH acclimated fish, low DA levels stimulate and high levels suppress αMSH secretion *in vitro* from superfused pars intermedia fragment	Lamers *et al.* (1997)

	Red porgy (*Pagrus pagrus*)	Inhibits	0.01–100 nM	Suppression of αMSH secretion *in vitro* from superfused pituitary glands	van der Salm *et al.* (2004)
MCH	Rainbow trout (*Oncorhynchus mykiss*)	Inhibits	Not applicable	Cytometric and plasma analysis: *in vivo* treatment with salmon MCH injected ip depressed melanotrope activity and reduced plasma αMSH levels	Baker *et al.* (1986)
	European eel (*Anguilla anguilla*)	Inhibits	Not applicable	Immunoabsorption of endogenous MCH enhances αMSH secretion by pituitary halves incubated *in vitro* for 18 h	Barber *et al.* (1987)
	Rainbow trout (*Oncorhynchus mykiss*)	Inhibits	Not applicable	Immunoabsorption of endogenous MCH enhances αMSH secretion by pars intermedia incubated *in vitro* for 18 h	Barber *et al.* (1987)
	Tilapia (*Oreochromis mossambicus*)	Biphasic response	0.01–10 μM	Salmon MCH; 0.01–1 μM suppress αMSH and 10 μM stimulates the secretion *in vitro* from superfused pars intermedia fragment	Gröneveld *et al.* (1995b)
	Red porgy (*Pagrus pagrus*)	Inhibits	0.01–100 nM	Suppression of αMSH secretion *in vitro* from superfused pituitary glands	van der Salm *et al.* (2004)

αMSH, α-melanocyte-stimulating hormone; AVT, arginine vasotocin; CRF, corticotropin-releasing factor; DA, dopamine; ip, intraperitoneal; IST, isotocin; MCH, melanin-concentrating hormone; TRH, thyrotropin-releasing hormone; SVG, sauvagine; UI, urotensin I.

Table 6.3
Factors affecting β-endorphin secretion in teleost fish

Factor	Species	Reported effect	Effective or tested dose	Experimental observation/ comment	Reference
CRF	Common carp (*Cyprinus carpio*)	Stimulates	1 nM	Ovine CRF; stimulation of non-acetylated βEND secretion *in vitro* from superfused pars distalis	van den Burg *et al.* (2001)
	Common carp (*Cyprinus carpio*)	No effect	1 nM	Ovine CRF tested on βEND secretion *in vitro* from superfused pars intermedia	van den Burg *et al.* (2001)
	Common carp (*Cyprinus carpio*)	Stimulates	0.01–10 nM	Ovine CRF; stimulation of *N*-acetyl βEND secretion *in vitro* from superfused pars intermedia fragment	van den Burg *et al.* (2005)
	Gilthead sea bream (*Sparus auratus*)	Stimulates	0.01–1000 nM ED_{50} 16 nM	Human/rat CRF; stimulation of *N*-acetyl βEND secretion *in vitro* from superfused pituitary glands	Rotllant *et al.* (2001)
TRH	Common carp (*Cyprinus carpio*)	Stimulates	0.01–10 nM	Stimulation of *N*-acetyl βEND secretion *in vitro* from superfused pituitary parts	van den Burg *et al.* (2003)
Opioid βEND	Common carp (*Cyprinus carpio*)	Inhibits	2 nM	Ovine CRF inhibits *N*-acetyl βEND secretion *in vitro* from superfused pars intermedia fragment when pre-incubated with carp βEND[1–29] for 2 h	van den Burg *et al.* (2005)

CRF, corticotropin-releasing factor; βEND, β-endorphin; TRH, thyrotropin-releasing hormone.

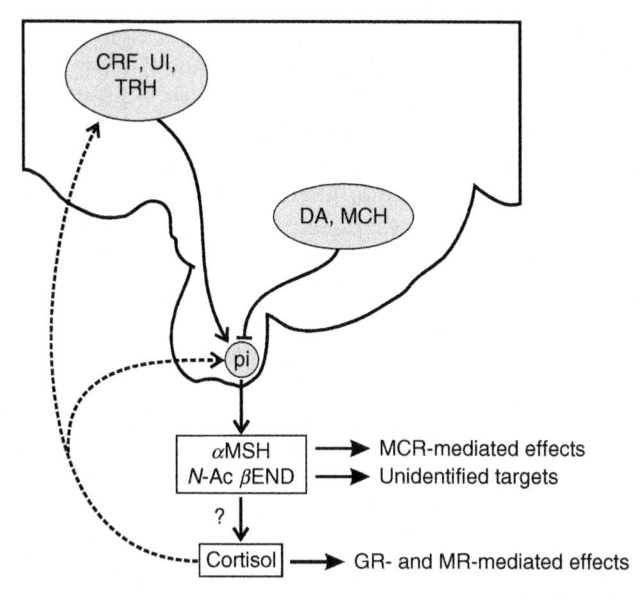

Fig. 6.2. Overview of the major factors that affect the activity of the melanotropic axis in teleosts. Shown is an enlargement of the preoptic area and hypothalamus, the primary regions involved in the hypophysiotropic regulation of the melanotropic axis. Adrenocorticotropic hormone (ACTH) is the primary secretagogue of cortisol secretion, but αMSH and *N*-Ac βEND may contribute to the regulation of cortisol release in some fish species (species-specificity indicated by "?"). While the actions of αMSH are mediated by melanocortin receptors (MCR) and those of cortisol are mediated by different corticosteroid receptor subtypes, the targets of *N*-Ac βEND have yet to be identified. Cortisol can exert negative feedback effects on the melanotropes of the pituitary and on some of the brain's hypophysiotropic factors. Note that this is not a generalization, but a model constructed from experimental results obtained on different species (see text for detail). Solid arrow indicates stimulation, T-line indicates inhibition, and dashed arrow indicates negative feedback. CRF, corticotropin-releasing factor; DA, dopamine; GR, glucocorticoid receptor; MCH, melanin-concentrating hormone; MR, mineralocorticoid receptor; pi, pars intermedia; TRH, thyrotropin-releasing hormone; UI, urotensin I.

indicate multiple functions of a hormone, and in teleost fish there is evidence, albeit equivocal, for pleiotropic actions of αMSH and βENDs (reviewed in Section 3.2). Alternatively, but not exclusively, we can conclude that the teleostean melanotrope cell is much less well investigated than its amphibian counterpart and the discovery of new regulatory factors can be awaited.

2.2.1. TRH and the CRF System

The hypothalamic factors TRH and CRF not only affect the corticotrope axis (Section 2.1), but also affect melanotrope secretions. The sensitivity of melanotropes to stimulation by TRH and CRF changes with exposure to a stressor (Lamers *et al.*, 1994; Rotllant *et al.*, 2000b; van den Burg *et al.*, 2003), indicating the involvement of these releasing hormones and their targets in a stress response. The biological activity of CRF is modulated by CRF-binding protein (CRF-BP), a phylogenetically highly conserved protein of hypothalamic origin (Huising and Flik, 2005; Westphal and Seasholtz, 2006). In common carp, nerve fibers containing either CRF-BP and CRF (present in separate axons originating from different neuronal populations) project from the NPO onto the pituitary pars intermedia, which indicates that melanotropes are affected by at least two components of the CRF system (Huising *et al.*, 2004). Interestingly, expression of hypothalamic CRF-BP mRNA had increased in common carp treated with thyroxine, which correlated predictably with decreased basal plasma cortisol levels (Geven *et al.*, 2006). Similar results were obtained *in vitro* (Geven *et al.*, 2009). These observations confirm the role of hypothalamic CRF-BP as a signaling molecule (or signal controlling molecule when it acts as chaperone) which allows the thyroid system to communicate with the melanotrope and corticotrope axes.

The study by van den Burg *et al.* (2005) provides the only data on the inter-individual variability for CRF effects on the *in vitro* release of αMSH and *N*-acetylated βEND (*N*-Ac βEND) (Table 6.2). CRF-stimulated pituitary release of αMSH and *N*-Ac βEND was assessed in carp acclimated to three different ambient temperatures (15, 22, and 29 °C). Not only did the magnitude of the maximal response and the CRF dose-response profile differ greatly with temperature, two-thirds of the animals tested did not respond to CRF at all. Since the release of αMSH and *N*-Ac βEND was consistently stimulated by TRH, and CRF consistently stimulated the release of opioid βEND from the pars distalis (van den Burg *et al.*, 2005), the possibility of an artifact was excluded. Interestingly, it appeared that TRH and CRF stimulated the release of *N*-Ac βEND from two different subcellular pools. Two different endocrine profiles could be distinguished in acutely stressed carp: fish that displayed elevated plasma levels of cortisol, αMSH and *N*-Ac βEND, as well as fish that only had elevated plasma cortisol levels (van den Burg *et al.*, 2005). Clearly, these observations suggest that the sensitivity of the melanotrope to CRF determines which stress response is launched. However, it has yet to be determined why the CRF-responsiveness should be different in the first place. In general, we should be alert for group effects that reflect responding as well as non-responding individuals; fish, like other vertebrates, have a continuum of shy to bold "personality traits" when responding to novelty and challenges (Wilson *et al.*, 1994).

2.2.2. MCH and DA

MCH can be viewed as an endogenous antagonist to αMSH, not only peripherally at the site of the skin melanophores, but also as an inhibitor of αMSH secretion by pituitary melanotropes. Apart from an inhibitory effect on the CRH-induced ACTH release in rainbow trout (Baker *et al.*, 1985), MCH appears to be mainly involved in the melanotropic axis. In fishes, MCH is localized in two hypothalamic nuclei, the lateral tuberal nucleus (NLT) and the lateral recess nucleus (NRL), of which only the former projects onto the pituitary (reviewed by Kawauchi and Baker, 2004). In Mozambique tilapia (*Oreochromis mossambicus*) acclimated to a white background, the levels of prepro-MCH (ppMCH) mRNA in the NLT, but not the NRL, had increased compared to animals on a black background. The expression of ppMCH in NLT neurons also increased upon exposure to very acid water (pH 3.5), which correlated positively with elevated plasma levels of ACTH and cortisol (Gröneveld *et al.*, 1995b). A milder pH stress (pH 4.0) and hyperosmotic conditions were without effect on ppMCH expression in the NLT and NRL, and did not produce significant changes in plasma ACTH and cortisol levels. In contrast, stress induced by repeated disturbance of the fishes resulted in a 45% increase in ppMCH mRNA expression in the NRL but not in the NLT (Gröneveld, *et al.*, 1995b). These observations reflect activities of stressor-specific hypothalamic MCH-neuron subpopulations in the regulation of skin pigmentation and the stress response.

In the last decade, an orexigenic role for hypothalamic MCH in the regulation of food intake and energy balance in mammals has become apparent (Qu *et al.*, 1996). While MCH also has an effect on feeding behavior in fish, central injection of either barfin flounder (*Verasper moseri*) or human MCH exerts an anorexigenic action in goldfish (Matsuda *et al.*, 2006) and immunoneutralization of brain MCH results in an increase in food intake (Matsuda *et al.*, 2009). Interestingly, studies into the pathways that mediate the anorexigenic action of MCH in goldfish suggest that MCH enhances the anorexigenic actions of αMSH via the MC4R signaling pathway and blocks the synthesis of NPY and ghrelin in the diencephalon (Shimakura *et al.*, 2008). MCH is not the sole piscine anorexigenic factor, however. In addition CRF, UI and TRH are established as regulatory peptides in the control of food intake and energy balance (Jensen, 2001; Joseph-Bravo, 2004; Bernier, 2006). Taken together, the hypothalamic control of the melanotrope axis by MCH, TRH and CRF-related peptides should be interpreted in a context of feeding and energy homeostasis, rather than physiological skin color changes.

DA typically plays an inhibitory role in all three axes discussed in this chapter. However, the plasticity of the dopaminergic system is reflected in the observation that, via the expression of a high-affinity D1 receptor on the tilapia melanotrope, DA can take on a stimulatory role in stressed fish (Lamers *et al.*, 1997). Interestingly, the low-affinity D2 receptor is not down-regulated in these animals, so that the net effect of the dopaminergic receptors is dependent on the local concentration of DA, which again is determined by the afferent input of higher brain centers.

Finally, the basal, unstimulated release of αMSH decreases over time during *in vitro* superfusion (Lamers *et al.*, 1991; van den Burg *et al.*, 2005). This is in contrast with the basal release of ACTH from the pituitary pars distalis and the activity of thyrotropes that become uninhibited and increase upon denervation in autotransplanted pituitary grafts (Metz *et al.*, 2004) (Section 2.3.1). This indicates that the melanotropes are under a net stimulatory control *in situ*, and the corticotropes and thyrotropes under a net inhibitory control. Perhaps this is related to differences between the nervous innervation of the pituitary pars intermedia and pars distalis, respectively. For now the physiological relevance, if any, of these observations is unknown.

2.3. Hypothalamic Regulation of Thyrotropes

Throughout the mammalian vertebrate class, the hypothalamic amidated tripeptide TRH (thyrotropin-releasing hormone) activates thyrotrope cells in the anterior pituitary gland to release thyrotropin (or thyroid-stimulating hormone, TSH). TSH stimulates multiple pathways involved in hormonogenesis in the thyroid gland (reviewed by Dunn and Dunn, 2001; Klaren *et al.*, 2007a) and increases the secretion of the prohormone thyroxine (T_4). The bioactive thyroid hormone 3,5,3′-triiodo-L-thyronine (T_3), derived from T_4, inhibits the expression and secretion of both pituitary TSH and hypothalamic TRH in a negative feedback loop. In the classical mammalian feedback system, negative thyroid hormone response elements in or near the promoter region allow for the T_3-suppressed expression of the genes for TSH α- and β-subunits, and TRH. In fishes the situation is definitely more complex, although much less well studied. Thyroid hormones and TSH negatively regulate TRH release (Gorbman *et al.*, 1983), but TRH cannot be assigned a general and unique thyrotropic role. Instead, multiple factors affect pituitary thyrotrope activity (Fig. 6.3), as discussed below.

2.3.1. Thyrotropin-Releasing Hormone

The role of TRH as a thyrotropic factor is well established for mammals (review by Nillni and Sevarino, 1999), but is still equivocal in fishes (see Table 6.4). "Classical" stimulatory thyrotropic effects of TRH have been observed in a number of teleost species, but inhibitory effects and negative results exist as well. This has turned attention to other potential hypothalamic thyrotropic factors, e.g., CRF, for which thyrotropic effects have been reported. It even appears that CRF, much more so than TRH, is considered to be the prime hypothalamic thyrotropic factor in early vertebrates (Lovejoy and Balment, 1999; Seasholtz *et al.*, 2002; De Groef *et al.*, 2006). However caution is needed when critically reviewing the evidence for or

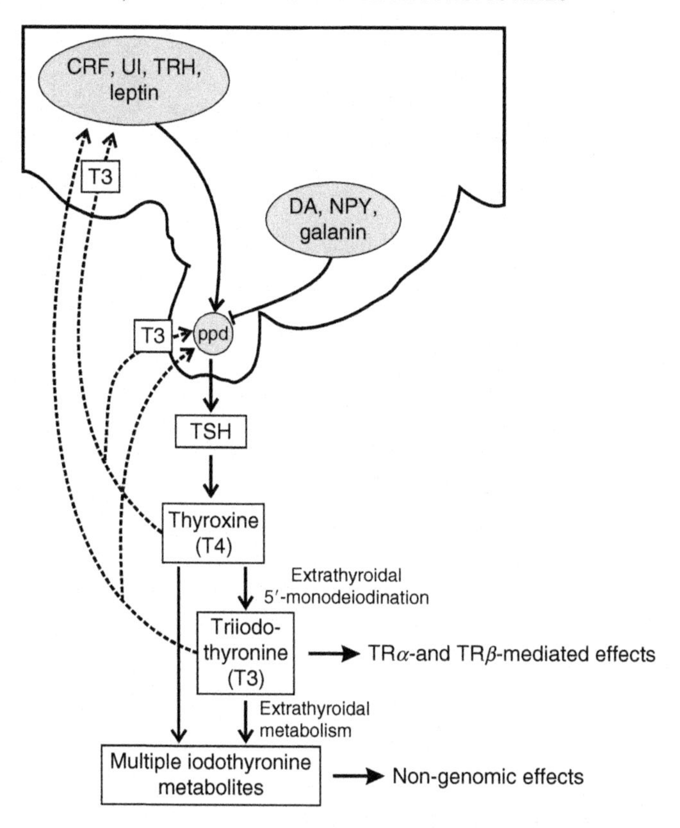

Fig. 6.3. Overview of the major factors that affect thyroxine secretion from the thyroid gland. Note that this is not a generalization, but a model constructed from experimental results obtained on different species. Thyroxine (T_4) is taken up by pituitary thyrotrope cells, but it is intracellularly deiodinated to T_3 that, ultimately, exerts a negative feedback. A similar mechanism is assumed for hypothalamic thyrotrope neurons. See Fig. 6.4 for a detailed overview of extrathyroidal metabolic pathways. Solid arrow indicates stimulation, T-line indicates inhibition, and dashed arrow indicates negative feedback. CRF, corticotropin-releasing factor; DA, dopamine; NPY, neuropeptide Y; ppd, proximal pars distalis; TR, thyroid hormone receptor; TRH, thyrotropin-releasing hormone; TSH, thyrotropin; UI, urotensin I.

against TRH as a piscine TSH-releasing hormone and a strong case can still be made for thyrotropic actions of TRH in fishes.

In clinical practice, the response of the pituitary to TRH is tested with a single intravenous dose (3–7 μg TRH per kg patient's body weight). Serum TSH is then measured 15–30 minutes later, and, ideally, at intervals up to 3 h post-injection. In healthy euthyroid patients, TRH administration typically elevates serum TSH levels fivefold within 15–40 min. Several hours later,

Table 6.4

Effects of TRH on thyrotropin (TSH) secretion in teleost fish. Superscripts indicate those cases where the effective dose was estimated using the median parameter values for the body length (L, cm) – body weight (W, g) relationship ($W = aL^b$) reported for the species or family (http://www.fishbase.org/search.php, version 04/2008)

Species	Reported effect	Effective or tested dose	Experimental observation/comment	Reference
Arctic charr (*Salvelinus alpinus*)	Stimulates	0.001–1 $\mu g\ g^{-1}$ BW	Single ip dose; increased plasma T_4 levels 1 to 5½ h post-injection in starved charr	Eales and Himick (1988)
Rainbow trout (*Oncorhynchus mykiss*)	Stimulates	0.7 $\mu g\ g^{-1}$ BW	Single ip dose; increased plasma T_4 levels 2 h post-injection	*Ibid.*
Goldfish (*Carassius auratus*)	Stimulates	2 × 2 $\mu g\ g^{-1}$ BW[a]	Two ip injections 1 h apart; degranulation of basophils in the pituitary pars distalis of fish 1 h post-injection (n=1, thiourea produced similar results)	Kaul and Vollrath (1974)
Chasmichthys dolichognathus (Gobiidae)	Stimulates	0.005–0.7 $\mu g\ g^{-1}$ BW[b]	Single ip dose, hypertrophy of basophils in the pituitary caudal pars distalis, "tentatively identified" as TSH cells, 4–10½ h post-injection	Tsuneki and Fernholm (1975)
Spotted snakehead (*Channa punctata*)	Stimulates	2–6 $\mu g\ g^{-1}$ BW	Single ip dose, increased thyroid peroxidase activity 6 h post-injection (higher doses inhibit)	Bhattacharya *et al.* (1979)
Japanese eel (*Anguilla japonica*)	Stimulates	10 nM	Increased TSH β-subunit mRNA expression in cultured pituitary cells incubated in vitro for 6 h	Han *et al.* (2004)
Bighead carp (*Aristichthys nobilis*)	Stimulates	0.01–10 nM	TRH dose-dependently increases TSH β-subunit mRNA expression in cultured pituitary cells incubated *in vitro* for 6–36 h	Chatterjee *et al.* (2001); Chowdhury *et al.* (2004)

Lungfish (*Protopterus ethiopicus*)	No effect	4×0.005–0.45 $\mu g\ g^{-1}$ BW	Four ip injections 24 h apart; no effect on subpharyngeal ("thyroidal") 24-h ^{131}I incorporation *in vivo* 24 h post-injection (stimulation upon βTSH treatment)	Gorbman and Hyder (1973)
Tilapia (*Oreochromis niloticus* × *O. aureus*)	No effect	5 ng g^{-1} BW	Single ip injection; no effect on plasma T_4 levels *in vivo* 4–24 h post-injection (stimulation upon βTSH treatment)	Melamed *et al.* (1995)
Goldfish (*Carassius auratus*)	No effect	6–250 $\mu g\ g^{-1}$ BW	Single ip dose, no induction of exophthalmos *in vivo* 4 h post-injection (TSH induces exophthalmos)	Wildmeister and Horster (1971)
	No effect	Not reported	Intraperitoneal administration of a "massive" TRH dose; no effect on plasma T_4 levels 1–24 h post-injection	Peter and McKeown (1975)
Common carp (*Cyprinus carpio*)	No effect	100 nM	No effect on the release of newly-synthesized TSH α- and β-subunits from cultured pituitary cells incubated *in vitro* for 24 h	Kagabu *et al.* (1998)
	No effect	100 nM	No effect on TSH β-subunit mRNA expression in intact pituitaries incubated *in vitro* for 6–36 h	Geven *et al.* (2009)
Coho salmon (*Oncorhynchus kisutch*)	No effect	0.01–100 nM	No effect on TSH secretion by cultured pituitary cells incubated for 6 h	Larsen *et al.* (1998)
Rainbow trout (*Oncorhynchus mykiss*)	Inhibits	3–7 $\mu g\ g^{-1}$ BW[c]	Mode of TRH administration not reported; "significant reduction" ($P < 0.001$) in serum T_4 levels *in vivo* (increased upon βTSH treatment)	Bromage *et al.* (1976)
Goldfish (*Carassius auratus*)	Inhibits	2×0.3–3 $\mu g\ g^{-1}$ BW	Two ip injections 12 h apart; 60% reduction in 24 h ^{131}I incorporation *in vivo* in the pharyngeal region and head kidney combined	Peter and McKeown (1975)

(*continued*)

Table 6.4 *(continued)*

Species	Reported effect	Effective or tested dose	Experimental observation/comment	Reference
	Inhibits	Not reported	Two "massive" ip TRH doses, delivered 12 h apart, caused "a significant suppression" of plasma T_4 levels in TSH-treated fish 12 h post-injection	*Ibid.*
Spotted snakehead (*Channa punctata*)	Inhibits	8–12 $\mu g\ g^{-1}$ BW	Single ip dose, decreased thyroid peroxidase activity 6 h post-injection (lower doses stimulate)	Bhattacharya *et al.* (1979)
Guppy (*Poecilia reticulata*)	Inhibits	0.2 $\mu g\ g^{-1}$ BW[d]	Three ip injections 24 h apart; 23% increase in dry weight per unit area of thyroid colloid (measured by interferometry), unvacuolated colloid and low epithelial cell height of thyroid follicles 24 h post-injection	Bromage (1975)
	Inhibits	Not defined	40% increase in dry weight per unit area of thyroid colloid (measured by inter–ferometry) in lower jaw incubated for 24 h in vitro in medium "containing the emanations" of Xiphophorus pituitary glands exposed to 28 μM TRH for 36 h	*Ibid.*

[a]Normalized dose estimated from the dose administered (2 × 300 μg/fish), reported body length (18–20 cm), and parameter values $a = 0.0295$, $b = 2.900$.

[b]Estimated from the dose administered (0.1–2 μg/fish), reported body length (6.6–13 cm), and parameter values $a = 0.0094$, $b = 3.0160$.

[c]Estimated from the dose administered (1–2 μg/fish), assuming a body length of adult males of 30 cm, and parameter values $a = 0.0118$, $b = 3.006$.

[d]Estimated from the dose administered (40 ng/fish), assuming a body length of 3 cm, and parameter values $a = 0.0081$, $b = 3.1492$.

BW, body weight; ip, intraperitoneal; TRH, thyrotropin-releasing hormone; TSH, thyrotropin.

serum T_3 and T_4 reach their peak concentrations, and TSH returns to preinjection levels (Faglia, 1998; van Tijn *et al.*, 2007). Because homologous serum-TSH assays are lacking for teleosts, indirect experimental readouts such as TSH β-subunit mRNA expression and thyroidal radioactive iodide incorporation, as well as plasma T_3 and T_4 concentrations are often used to assess the potency of TRH to stimulate TSH release in fish (see Table 6.4). However, results can be conflicting regarding the activity of the thyroid gland, and thus of the efficacy of administered TRH.

Important aspects to consider with respect to the use of such proxy parameters include the following. First, the elevated plasma thyroid hormone levels that result from an adequate TRH treatment *in vivo* will feed back on the pituitary, restoring the expression and secretion of TSH to normal levels. In particular when multiple doses, hours or days apart, are administered, not only negative feedback by thyroid hormones on pituitary TSH, but also depletion of the thyrotropes' TSH reserve can be a confounding factor. This can make it difficult to dissect a net TRH-induced effect from the combined stimulatory and inhibitory effects of, respectively, TRH and thyroid hormones on net TSH secretion (Rabello *et al.*, 1974; Staub *et al.*, 1978; Faglia, 1998). Second, not all experimental readouts, e.g., radioactive iodide uptake and exophthalmos[1], can be supposed to be equally indicative for TRH's putative thyrotropic action. Indeed, thyroidal radioiodine uptake in man has been reported to be quite insensitive to TRH treatment (Haigler *et al.*, 1972). Another consideration is that thyroidal radioiodine uptake is the net result of multiple, separate processes (reviewed by Dunn and Dunn, 2001). The thyroid gland very effectively clears iodide from the blood, and thus the incorporation of radioiodide measured at early time points following a bolus administration will mainly reflect a net transport capacity of the gland. Measurements at later time points, however, are the resultant of the enzymatic and cellular processes that are involved in thyroid hormone synthesis. Indeed, when Mozambique tilapia was injected intraperitoneally with a single dose of ^{125}I, the incorporation of radioiodide in the subpharyngeal thyroid gland was prominent within 24 h post-injection, but the appearance of radiolabeled thyroid hormones in plasma lagged behind and appeared at 48 h (Geven *et al.*, 2007). Likewise, in killifish (*Fundulus heteroclitus*) treated

[1]Exophthalmos is typically observed in hyperthyroid human patients with Graves' disease. In this autoimmune disease, the TSH receptor is activated by stimulating immunoglobulins. Symptomatically, over a period of weeks to months interstitial fluid accumulates in extraocular muscle which causes protrusion of the eyes. The appearance of exophthalmos in fishes only hours after experimental treatment (Table 6.4) indicates a very different etiology, and is even more surprising considering the anatomy of the orbital socket in the teleost skull that is much less confined by bony structures than in mammals. Exophthalmos in fishes was also discussed by Eales (1979).

with prolactin, plasma T_4 levels were significantly reduced without a concomitant change in thyroidal radioiodine uptake (Grau and Stetson, 1977a). Without additional information on iodide and thyroid hormone dynamics, a decreased thyroid gland radioiodine content can be interpreted as an inactive gland, or, equally likely, as an active gland with a high turnover of iodide.

Table 6.4 lists studies in which the thyrotropic effect of systemic TRH, administered intraperitoneally, was investigated. It is striking that only two teleost species, *viz.* goldfish and common carp, have been investigated separately by independent researchers. For seven out of 13 different investigated species, stimulatory thyrotropic effects of TRH are reported. These studies also fulfill at least two criteria of a clinical TRH test, in that a single dose (or two doses delivered with a short interval) was administered, and that the experimental readout was measured within a few hours post-treatment. Moreover, the effective doses applied here *in vivo* range from 0.001 to 6 $\mu g\ g^{-1}$ body weight, and are in the range of the plasma TRH concentrations (300 pM, or ca. 100 ng L^{-1}) that were measured in Mozambique tilapia (Lamers *et al.*, 1994).

TRH treatment was without effect in five species. The ip administration of a high dose of TRH in goldfish produced neither exophthalmos (Wildmeister and Horster, 1971) nor elevated plasma T_4 levels (Peter and McKeown, 1975). In contrast to these results, Kaul and Vollrath (1974) observed that lower TRH doses resulted in the degranulation of basophils, most likely thyrotropes, in the goldfish pituitary RPD. In their control experiments, TSH did induce exophthalmos and treatment with the thyrostatic thiourea mimicked the degranulation of the putative TSH cells. To reconcile these conflicting data, perhaps TRH-induced basophil degranulation should be considered, albeit that this was observed in a single goldfish specimen only (Kaul and Vollrath, 1974). A cytological response likely is a more direct indicator of TSH action than plasma T_4 concentrations and exophthalmos. It can then be argued that TRH is thyrotropic in goldfish, but that the resulting TSH release is inadequate to elicit changes downstream in the thyroid axis. We also have to be critical and acknowledge that the TRH dose used by Melamed *et al.* (1995) in tilapia hybrids is lower than those tested by most other authors and could have been inadequate to increase plasma T_4 levels *in vivo*. The unchanged thyroidal radioiodine uptake in lungfish reported by Gorbman and Hyder (1973) could suffer from the drawbacks of long-term TRH treatment mentioned above.

Only in common carp and coho salmon is the lack of thyrotropic effects induced by TRH well established. A gold standard was set by Larsen *et al.* (1998), who measured *in vitro* TSH secretion from coho salmon (*Oncorhynchus kisutch*) pituitary cells using a homologous immunoassay. Their observation that TRH did not stimulate TSH release was validated by positive results obtained with other endocrine factors. Moreover, no thyrotropic effect of TRH on the expression of TSH subunits was detected in two independent studies on common carp (Kagabu *et al.*, 1998; Geven *et al.*, 2009).

Not included in Table 6.4 are results from intracerebroventricular (icv) injections of TRH in goldfish brain, at doses of 0.2–1 μg per fish, that had no effect on plasma T_4 levels 5 min to 4 h post-injection (Crim *et al.*, 1978). It can be questioned whether the application of TRH in a brain ventricle will result in an adequate TRH activity in the pituitary. It also has to be stated that the experimental designs of those studies in which inhibitory thyrotropic effects of TRH were measured (Table 6.4) do not fit the criteria for a clinical TRH test and suffer from one or more of the drawbacks mentioned above. All in all, there is no convincing evidence for an inhibitory thyrotropic action of TRH in fishes.

2.3.2. Other Hypothalamic Factors

Table 6.5 displays the considerable number of additional (hypothalamic) thyrotropic factors besides TRH. The earliest investigations into the hypothalamic control of the teleostean thyroid axis were on fishes with autotransplanted, i.e., denervated, ectopic pituitary glands. These animals were shown to have an activated thyroid gland, as evidenced by histological criteria, increased radioiodide uptake and increased plasma T_4 levels. Moreover, thyrotropes from the pituitary grafts appeared more active than their intact non-transplanted counterparts (Ball *et al.*, 1963, 1965; Olivereau and Ball, 1966; Higgins and Ball, 1970; Peter, 1972; Grau and Stetson, 1977b). The notion that the secretion of TSH is under inhibitory control by a brain factor was established by electrolytic lesioning of either defined hypothalamic regions or the pituitary stalk. These procedures, which consistently activated the thyroid gland (Ball *et al.*, 1963; Peter, 1970, 1971; Ball *et al.*, 1972; Peter and McKeown, 1975; Pickford *et al.*, 1981), led to the conclusion that pituitary thyrotropes were tonically inhibited by a hypothalamic thyroid-inhibiting factor (Grau and Stetson, 1978; Olivereau *et al.*, 1988), most probably DA. This notion was supported by the expression of an inhibitory type-2 DA receptor in the pituitary pars distalis of rainbow trout (Vacher *et al.*, 2003). Somatostatin is also a well-established inhibitor of TRH-induced TSH release in vertebrates (Hedge *et al.*, 1981; De Groef *et al.*, 2005). However, experimental evidence for a similar role of somatostatin in teleosts is lacking (Byamungu *et al.*, 1991).

Several members of the CRF family, i.e., CRF, UI, sauvagine, had thyrotropic effects in salmonids and eels (Larsen *et al.*, 1998; Rousseau *et al.*, 1999). CRF's thyrotropic actions have been observed in other non-mammalian vertebrates (reviewed by De Groef *et al.*, 2006). Thyrotropes in

Table 6.5
Factors other than TRH affecting thyrotropin secretion in teleost fish

Factor	Species	Reported effect	Effective or tested dose	Experimental observation/comment	Reference
CRF	Coho salmon (*Oncorhynchus kisutch*)	Stimulates	0.01–100 nM	Ovine CRF; stimulation of TSH secretion by cultured pituitary cells incubated *in vitro* for 6 h	Larsen *et al.* (1998)
	European eel (*Anguilla Anguilla*)	Stimulates	Not reported	TSH β-subunit mRNA expression in cultured pituitary cells	Rousseau *et al.* (1999)
	Common carp (*Cyprinus carpio*)	No effect	100 nM	Ovine CRF; no effect on TSH β-subunit mRNA expression in intact pituitaries incubated *in vitro* for 6–36 h	Geven *et al.* (2009)
UI	Coho salmon (*Oncorhynchus kisutch*)	Stimulates	0.01–100 nM	Carp UI; stimulation of TSH secretion by cultured pituitary cells incubated *in vitro* for 6 h	Larsen *et al.* (1998)
SVG	Coho salmon (*Oncorhynchus kisutch*)	Stimulates	1–100 nM	Frog sauvagine; stimulation of TSH secretion by cultured pituitary cells incubated *in vitro* for 6 h	Larsen *et al.* (1998)
DA	Goldfish (*Carassius auratus*)	Inhibits	Not applicable	Cytometric analysis: *in vivo* treatment with DA antagonists injected ip results in cellular and nuclear hypertrophy of thyrotropic cells, and a slight increase in TSH immunoreactivity	Olivereau *et al.* (1988a)
	Killifish (*Fundulus heteroclitus*)	Inhibits	Not applicable	Intracranial injection of 6-OH DA, specifically destroying dopaminergic fibers, results in elevated serum T_4 levels	Grau and Stetson (1978)

βEND	Bighead carp (*Aristichthys nobilis*)	Stimulates		TSH β-subunit mRNA expression in cultured pituitary cells	Chowdhury *et al.* (2004)
PGE1, PGF2α	Singi fish (*Hetero–pneustes fossilis*)	Stimulates	100 μg per fish per day	Increased thyroidal ^{131}I incorporation and plasma TSH levels, decreased pituitary TSH content	Singh and Singh (1977)
Galanin	Bighead carp (*Aristichthys nobilis*)	Inhibits		TSH β-subunit mRNA expression in cultured pituitary cells	Chowdhury *et al.* (2004)
NPY	Bighead carp (*Aristichthys nobilis*)	Inhibits		TSH β-subunit mRNA expression in cultured pituitary cells	Chowdhury *et al.* (2004)
Leptin	Bighead carp (*Aristichthys nobilis*)	Stimulates		TSH β-subunit mRNA expression in cultured pituitary cells	Chowdhury *et al.* (2004)
GHRH	Coho salmon (*Oncorhynchus kisutch*)	No effect		Salmon GHRH, TSH secretion from cultured pituitary cells	Larsen *et al.* (1998)
GnRH	Killifish (*Fundulus heteroclitus*)	No effect		"No measurable effects on thyroid function"	Brown *et al.* (1982)
	Coho salmon (*Oncorhynchus kisutch*)	No effect		Salmon GnRH, TSH secretion from cultured pituitary cells	Larsen *et al.* (1998)
	Coho salmon (*Oncorhynchus kisutch*)	No effect		Salmon GnRH, plasma TSH levels *in vivo*	Moriyama *et al.* (1997)

CRF, corticotropin-releasing factor; DA, dopamine; βEND, β-endorphin; GHRH, growth hormone-releasing hormone; GnRH, gonadotropin-releasing hormone; ip, intraperitoneal; NPY, neuropeptide Y; PGE1, prostaglandin E1; PGF2α, prostaglandin F2α; SVG, sauvagine; TSH, thyrotropin; UI, urotensin I.

non-mammalian vertebrates express the type-2 CRF receptor (De Groef *et al.*, 2003, 2006), but experimental results recently obtained in mouse show that a small subset of thyrotropes express the type-1 CRF receptor (Westphal *et al.*, 2009). The thyrotropic effects of CRF are only partially confirmed by icv injections of CRF and its antagonist α-helical CRF_{9-41} in chinook salmon (*Oncorhynchus tshawytscha*) that produced no effect and a stimulatory effect, respectively, on plasma T_4 levels (Clements *et al.*, 2002). In common carp, the involvement of CRF in the regulation of TSH secretion could be indirectly inferred from the T_4-induced changes in the hypothalamic expression of CRF-binding protein (CRF-BP) *in vitro* and *in vivo*. However, the expression of TSHβ mRNA in cultured carp pituitary glands is refractory to both CRF and TRH (Kagabu *et al.*, 1998; Geven *et al.*, 2006, 2009), and the identification of a hypothalamic thyrotropic factor in this species is still awaited.

Putative thyrotropic factors other than CRF and TRH have been studied in single studies on a single species only. Chowdhury *et al.* (2004) observed stimulatory thyrotropic effects of NPY, galanin, leptin and βEND. The last factor is interesting as it is also a signaling molecule from the melanotrope axis (see Section 3.2.2). Still, too few studies have been done to unequivocally assign a thyrotropic function to these factors.

3. TARGETS AND FUNCTIONS OF THE CORTICOTROPE, MELANOTROPE AND THYROTROPE SECRETIONS

In all jawed vertebrates, the secretions from the corticotropes and the melanotropes are derived from a common precursor protein, POMC (Kawauchi and Sower, 2006). Proteolytic cleavage of the large POMC molecule is achieved by prohormone convertases (PCs) and is tissue-specific (Zhou *et al.*, 1993; Tanaka, 2003). In the corticotropes, the major products of POMC processing by PC1 are ACTH and β-lipotrophic hormone (β-LPH). β-LPH is further processed by PC2 to give rise to the opioid βEND. In the melanotropes, more extensive cleavage by both PC1 and PC2 results in the conversion of ACTH to αMSH and corticotropin-like intermediate lobe peptide (CLIP). β-LPH is also processed in the melanotropes by PC2 to form different βENDs which are mainly/to a large extent acetylated at the N-terminus. In contrast, the thyrotropes produce a single product, TSH. TSH belongs with follicle-stimulating hormone (FSH) and luteinizing hormone (LH) to the family of pituitary heterodimeric glycoproteins. These hormones consist of two chemically distinct subunits, of which the α-subunit is identical in all. The β-unit is unique and determines the protein signature and specifies its function (Kawauchi and Sower, 2006).

3.1. Targets and Functions of Corticotrope Secretions: ACTH and Non-acetylated βEND

3.1.1. ACTH

The first targets of ACTH and of the other POMC-derived melanocortins are the melanocortin receptors (MCR). MCRs belong to the superfamily of 7-transmembrane domain G-protein coupled receptors that stimulate adenylate cyclase and the cAMP signaling cascade (Mountjoy *et al.*, 1992). In several fish species, as in other vertebrates, five MCRs (MC1R–MC5R) have been identified (Logan *et al.*, 2003; Klovins *et al.*, 2004a; Flik *et al.*, 2006). While ACTH is the only natural ligand that binds and activates the MC2R in fish and mammals (Klovins *et al.*, 2004a), both ACTH and the MSHs bind to the other four MCRs (Schiöth *et al.*, 2005). However, unlike the mammalian situation where the MC1R, MC3R, MC4R and MC5R subtypes distinctly recognize MSHs, these four MCR subtypes in fish have a greater affinity for $ACTH_{1-24}$ than for the different MSHs (Haitina *et al.*, 2004; Klovins *et al.*, 2004a). Therefore, the binding properties of the different MCRs in fish, and in chicken as well (Barimo *et al.*, 2004), suggest that ACTH is a phylogenetically older ligand than the MSH peptides and that the selective recognition of the MSH peptides by the MCRs evolved in higher vertebrates (Cerdá-Reverter *et al.*, 2005; Schiöth *et al.*, 2005).

The primary function of ACTH in fish is the regulation of corticosteroid steroidogenesis in the interrenal cells of the head kidney (Donaldson, 1981; Wendelaar Bonga, 1997). While sympathetic nerve fibers (Arends *et al.*, 1999), αMSH (Lamers *et al.*, 1992), *N*-Ac βEND (Balm *et al.*, 1995), Ang II (Perrott and Balment, 1990), atrial natriuretic peptide (Arnold-Reed and Balment, 1991), UI (Kelsall and Balment, 1998) and several other factors stimulate cortisol secretion in some species (Schreck *et al.*, 1989; Mommsen *et al.*, 1999), ACTH is recognized as the principal stimulator of cortisol release during the acute phase of the stress response (Flik *et al.*, 2006). In return, the non-ACTH corticotropic signals likely enhance the steroidogenic action of ACTH in response to specific stressors and during the chronic phase of the stress response (see Section 3.2). At the cellular level, binding of ACTH to the MC2R in the head kidney activates the enzymatic pathways that convert cholesterol into corticosteroids (Huising *et al.*, 2005; Hagen *et al.*, 2006; Aluru and Vijayan, 2008). Specifically, ACTH enhances the steroidogenic capacity of the interrenal cells by stimulating steroidogenic acute regulatory protein (StAR; Hagen *et al.*, 2006; Aluru and Vijayan, 2008), cytochrome P450 side chain cleavage enzyme ($P450_{scc}$; Aluru and Vijayan, 2008) and 11β-hydroxylase ($P450_{c11}$; Hagen *et al.*, 2006). While StAR facilitates the transport of cholesterol across the mitochondrial membrane, $P450_{scc}$ converts cholesterol to pregnenolone (the first and rate-limiting enzymatic step of corticosteroid synthesis) and $P450_{c11}$ converts 11-deoxycortisol to cortisol (Mommsen *et al.*, 1999). Also, the stimulatory effect of ACTH on the expression of MC2R in the interrenal tissue likely contributes to the enhanced steroidogenic capacity of the interrenal tissue following an acute stressor (Aluru and Vijayan, 2008).

Beyond the MC2R-mediated effects of ACTH, the higher affinity of ACTH than that of the MSHs to all the MCR subtypes in fish (Schiöth *et al.*, 2005) suggests that the physiological functions of ACTH in fish may be more diverse than in mammals. For example, while αMSH is involved in melanosome dispersion in fish skin (see Section 3.2), MC1R, the main receptor for the melanotropic actions of αMSH on the skin melanophore, has a 10-times higher affinity for ACTH than for αMSH in pufferfish (*Takifugu rubripes*; Klovins *et al.*, 2004a). The melanosome-dispersing properties of ACTH in several fish species also suggest that ACTH, which contains the αMSH sequence, may play a role in the regulation of pigment dispersion in these animals (Fujii, 2000; van der Salm *et al.*, 2005). Peripherally, ACTH may also be involved in the control of catecholamine release from the chromaffin cells of the head kidney (Reid *et al.*, 1996), the stressor-mediated modulation of gonadal sex steroid production (Aluru and Vijayan, 2008) and the release of T_4 from renal thyroid follicles (Geven *et al.*, 2009). In the brain, the broad expression pattern of MC2R, MC4R and MC5R (Ringholm *et al.*, 2002; Cerdá-Reverter *et al.*, 2003a,b, 2005; Haitina *et al.*, 2004; Klovins *et al.*, 2004a) as well as the considerable ACTH immunoreactivity (Vallarino *et al.*, 1989; Olivereau and Olivereau, 1990b; Metz *et al.*, 2004) suggest that ACTH has various central effects. Whereas αMSH and βEND appear to be the predominant end products resulting from POMC processing in the brain of higher vertebrates, at least in rainbow trout brain the concentration of ACTH is in the same range as those of αMSH and βEND (Vallarino *et al.*, 1989). So while the specific functions of ACTH in the brain of fish remain unclear and species-specific differences likely exist, the central location of ACTH immunoreactivity and of MC4R- and MC5R-expressing neurons generally suggests that ACTH is involved both in hypophysiotropic regulation and, as in mammals, in a regulation of a variety of autonomic and behavioral functions (Bertolini *et al.*, 2009).

3.1.2. Non-acetylated βEND

Endorphins have been ascribed a remarkably diverse number of physiological roles. Although the functions of βENDs are generally poorly understood in fish, evidence exists to implicate these peptides in the control of pituitary hormone secretion, as well as in the regulation of the immune and stress responses. There are multiple chemical forms of βENDs with different biological activities. For instance, nascent β-endorphin is a powerful opiate, but loss of opioid activity follows acetylation of the peptide's *N*-terminus (*N*-Ac βEND) (Akil *et al.*, 1981). In common carp, the pituitary pars intermedia secretes full-length and a number of truncated *N*-acetylated βENDs (see Section 3.2.2), whereas opioid (non-acetylated) βEND is released from the pars distalis (van den Burg *et al.*, 2001). Pertinent data on the pituitary occurrence and release of βENDs in other teleost species are very scarce (Mosconi *et al.*, 1998).

Opioid peptides such as non-acetylated βEND can bind to four opioid receptor types [delta, kappa, mu, and the nociceptin receptor (NOP)] which arose early in vertebrate evolution as a result of gene duplications (Dreborg *et al.*, 2008). In mammals, βEND preferentially binds to delta and mu receptors (Przewlocki and Przewlocka, 2001). While βEND also binds to delta receptors in zebrafish, much work is still needed to accurately characterize the binding properties of opioid receptors in fish (Rodriguez *et al.*, 2000; Gonzalez-Nuñez *et al.*, 2006). To date, few studies have determined the expression pattern of opioid receptors in teleosts. In zebrafish, delta and mu receptors are widely distributed in the CNS and in peripheral organs throughout embryonic development (Sanchez-Simon and Rodriguez, 2008). In common carp, delta, mu and kappa receptors are constitutively expressed in the hypothalamus, the pars distalis and pars intermedia of the pituitary, the head kidney, kidney, thymus, spleen, and in leukocytes (Chadzinska *et al.*, 2009; see also Chapter 7, this volume).

A role for non-acetylated βEND in the control of pituitary hormone secretion is suggested by the distribution of this opiate peptide along the hypophysiotropic nerve fibers that run between the NPO and the pituitary gland. In common carp, the preoptic region shows prominent staining for non-acetylated βEND, and the nerve fibers of the pars nervosa that enter the pituitary gland are positive for non-acetylated βEND (Metz *et al.*, 2004; Flik *et al.*, 2006). Moreover, pre-incubation of carp melanotrope cells with homologous $\beta\text{END}_{(1-29)}$ reversed the stimulatory actions of CRF on *N*-Ac βEND secretion (van den Burg *et al.*, 2005).

In the immune system, non-acetylated βEND regulates diverse phagocyte functions in rainbow trout, common carp (Watanuki *et al.*, 2000) and spotted snakehead (*Channa punctatus*; Singh and Rai, 2008) that can be reversed by selective opioid receptor antagonists. There is also active regulation of opioid receptor gene expression on peritoneal leukocytes and head kidney phagocytes in response to both immune and stress challenges in common carp (Chadzinska *et al.*, 2009).

3.2. Targets and Functions of Melanotrope Secretions: αMSH and *N*-acetylated βENDs

3.2.1. αMSH

An important product of the precursor POMC in the melanotropes of the pituitary pars intermedia is αMSH. In vertebrates, post-translational acetylation of αMSH results in three αMSH forms: des-, mono- and di-acetylated αMSH. In fish, all three forms of αMSH are found in the plasma and in extracts of the neurointermediate lobe, and in Mozambique tilapia the release of the three forms may be differentially regulated (Lamers *et al.*, 1991). The functional significance of the different αMSH forms stems from their different potencies and half-lives (Rudman *et al.*, 1983).

In both fish and mammals, the melanocortin peptide αMSH can bind and activate four different melanocortin receptors: MC1R, MC3R, MC4R and MC5R (Metz *et al.*, 2006b). Among these, the MC1R has the highest affinity for αMSH and, while broadly expressed, it predominates in melanophore-rich organs (van der Salm *et al.*, 2005; Selz *et al.*, 2007). Expressed in many brain areas and peripheral organs in mammals, the distribution of MC3R in fish is not known except for its weak expression pattern in many brain areas of an elasmobranch, the spiny dogfish (*Squalus acanthias*; Klovins *et al.*, 2004b). MC4R is abundantly expressed in the fish brain and more sporadically reported in peripheral tissues (Metz *et al.*, 2006b). Detailed mapping of MC4R in goldfish revealed high expression levels in the neuroendocrine and food intake-controlling areas of the brain (Cerdá-Reverter *et al.*, 2003c). Finally, several studies have now demonstrated that MC5R is expressed in a variety of peripheral tissues in fish as well as in many different brain regions (Cerdá-Reverter *et al.*, 2003b; Haitina *et al.*, 2004; Metz *et al.*, 2005).

As the name implies, αMSH is involved in the regulation of skin coloration. In response to black background adaptation in fish, there is evidence that αMSH stimulates the dispersal of melanosomes in skin melanocytes that cover a large surface area and provide the fish with a darker appearance (van der Salm *et al.*, 2005; Amiya *et al.*, 2007). As in other vertebrates, the actions of αMSH in color adaptation in fish appear to be mediated by MC1R. Knockdown of MC1R expression in zebrafish embryos leads to aggregated melanosomes that do not disperse in dark conditions (Richardson *et al.*, 2008). Agouti-signaling peptide (ASP), an endogenous MC1R antagonist expressed almost exclusively in the ventral skin, also appears to be involved in determining the dorsal–ventral pigment pattern of fish (Cerdá-Reverter *et al.*, 2005).

While controversial, another physiological process in teleosts that may involve αMSH is the regulation of cortisol secretion from the interrenal cells. In Mozambique tilapia, a purified extract of pituitary PI known to contain αMSH was shown to have mild corticotropic actions *in vitro*

(Lamers *et al.*, 1992). Moreover, despite any corticotropic activity on its own, βEND significantly potentiated the corticotropic actions of an αMSH fraction from tilapia pituitary (Balm *et al.*, 1995). In contrast, while the pituitary PI of common carp contains at least one unidentified factor with corticotropic activity, synthetic αMSH and βEND, either alone or in combination, did not stimulate cortisol release in this species (Metz *et al.*, 2005). Similarly, despite the presence of MC4R in trout interrenals (Haitina *et al.*, 2004), neither αMSH nor NDP-MSH (an MC4R agonist) stimulated cortisol secretion from the interrenal tissue in rainbow trout, and SHU9119 (an MC4R/MC3R antagonist) did not modify ACTH-stimulated cortisol production *in vitro* (Aluru and Vijayan, 2008).

Recent evidence also implicates αMSH in the regulation of food intake and lipid metabolism in teleosts. In goldfish, while intracerebroventricular (icv) administration of the MC4R agonist NDP-MSH and of the non-specific agonist melanotan II (MT II) dose-dependently inhibited food intake, the specific MC4R antagonist HS024 stimulated appetite (Cerdá-Reverter *et al.*, 2003a,c). Similarly, in rainbow trout, while central administration of MTII decreased food intake, both HS024 and the MC3/4R antagonist SHU9119 had the opposite effect (Schjolden *et al.*, 2009). Interestingly, the orexigenic effects of MT II in goldfish are abolished by treatment with the CRF receptor antagonist, α-helical $CRF_{(9–41)}$, and there is immunohistological evidence that αMSH-containing nerve fibers or endings lay in close apposition to CRF-containing neurons in the hypothalamus (Matsuda *et al.*, 2008). Together, these results suggest that αMSH exerts a tonic inhibitory effect on food intake in fish through central MC4R signaling and that the anorexigenic actions of this melanocortin are mediated by the CRF-signaling pathway. Assuming αMSH crosses the blood–brain barrier in fish as it does in mammals (Strand, 1999), the increase in αMSH plasma levels that characterizes the response to some stressors (see Section 5) could contribute to the central regulation of food intake. The observation that αMSH stimulates hepatic triacylglycerol lipase activity and increases the circulating levels of fatty acids in rainbow trout (Yada *et al.*, 2000) suggests that circulating αMSH is involved in lipid mobilization. In agreement with the potential metabolic and anorexigenic roles attributed to αMSH in fish, the cobalt variant of rainbow trout, which lacks most of the PI and melanotropes, is characterized by a significant accumulation of fat, reduced hepatic triacylglycerol lipase activity and being hyperphagic (Yada *et al.*, 2002).

Finally, very recent evidence points to a thyrotropic action of αMSH in common carp. Here, the release of thyroxine (T_4) from the functional thyroid follicles in kidney and head kidney *in vitro* was stimulated by submicromolar concentrations of αMSH (Geven *et al.*, 2009). In T_4-treated carp, plasma αMSH levels had increased by 30% (Geven *et al.*, 2006). These observations suggest a positive feedback mechanism for the αMSH-induced release of T_4. It has been suggested that the increases in plasma αMSH observed in chronically stressed carp are involved in the regulation of food intake and energy balance (Metz *et al.*, 2005; Flik *et al.*, 2006), perhaps via thyroid hormones and the modulation of basal metabolic rate.

3.2.2. Acetylated β-endorphins

While the pituitary melanotropes of several fish species can produce and release several *N*-Ac-βENDs during stress responses (see Section 5.2), very little is known about the targets of these peptides and their functions. To date, specific stimulatory receptors for acetylated endorphins have not been identified (Flik *et al.*, 2006). In contrast, there is some evidence suggesting that the acetylation of opioid peptides can result in the production of opioid receptor antagonists (René *et al.*, 1998; Bennett *et al.*, 2005). Whether such a scenario actually applies to the *N*-Ac-βENDs of melanotropic origin is unknown but represents an interesting avenue for future research given the broad central and peripheral expression pattern of opioid receptors observed in fish (Sanchez-Simon and Rodriguez, 2008; Chadzinska *et al.*, 2009). Alternatively, one could argue that a cell that processes POMC say for "MSH-purposes" needs to inactivate the potent byproducts (opioids) that are inherent to POMC processing.

3.3. Targets and Functions of Thyroid-stimulating Hormone

3.3.1. Thyroid-stimulating Hormone: Structure and Functions

Thyroid-stimulating hormone (TSH, or thyrotropin) is the key regulating factor of thyroid gland function. TSH β-subunit has been characterized in species from different teleost taxa, and its mRNA is almost exclusively expressed in the pituitary gland (Ito *et al.*, 1993; Salmon *et al.*, 1993; Martin *et al.*, 1999; Yoshiura *et al.*, 1999; Chatterjee *et al.*, 2001; Han *et al.*, 2004; Lema *et al.*, 2008). An exception is the T_4-sensitive expression of TSHβ mRNA in Atlantic salmon (*Salmo salar*) liver and kidney (Mortensen and Arukwe, 2006). A large number of authors, using heterologous anti-human TSHβ antibodies, have described the distribution and topology of TSHβ in the teleostean pituitary pars distalis (reviewed by Kasper *et al.*, 2006).

Within the infraclass of Teleostei, sequence identities of the mature TSH β-subunit protein vary greatly from 97% to 53% (Marchelidon *et al.*, 1991; Lema *et al.*, 2008), and in some cases TSH β-subunits from two fish species share about as much identity as a fish and mammalian subunit. This diversity can explain, for instance, the reported 1000-fold difference between the potencies of bovine and salmon TSH to stimulate the release of T_4 from

Hawaiian parrotfish (*Scarus dubius*) thyroid tissue *in vitro* (Swanson *et al.*, 1988), and the thyroid-stimulating activities of heterologous gonadotropins in Nile tilapia (*Oreochromis niloticus*) and killifish (Brown *et al.*, 1985; Byamungu *et al.*, 1991).

Treatment with heterologous TSH consistently results in elevated plasma T_4 levels in fishes *in vivo* (Grau and Stetson, 1977a; Brown and Stetson, 1983; Specker and Richman, 1984; Brown *et al.*, 1985; Nishioka *et al.*, 1985; Leatherland, 1987; Inui *et al.*, 1989; Byamungu *et al.*, 1990; Bandyopadhyay and Bhattacharya, 1993) and an increased release of T_4 from cultured thyroid tissue *in vitro* (Bonnin, 1971; Jackson and Sage, 1973; Grau *et al.*, 1986; Swanson *et al.*, 1988; Okimoto *et al.*, 1991; Geven *et al.*, 2007). The cellular targets of TSH in the piscine thyroid follicle are less well investigated. It can be assumed that, for the large part, thyroid hormonogenesis in fishes follows the same biochemical and cellular routes as in mammals (Klaren *et al.*, 2007a). In coho salmon, TSH clearly has a trophic effect on the thyrocyte; treatment with heterologous TSH increased epithelial cell height of the thyroid follicle, a classic measure of enhanced thyroid activity. At the subcellular level, thyroid peroxidase (TPO) is an enzyme with a pivotal role in the synthesis of thyroid hormone, and its activity was stimulated by bovine TSH treatment *in vitro* in spotted snakehead (Chakraborti and Bhattacharya, 1978; Bhattacharya *et al.*, 1979).

Few data exist on the regulation of thyrotrope TSH expression and secretion, other than by hypothalamic factors and negative feedback by thyroid hormones. In salmonids the expression of pituitary TSHβ mRNA is highest in sexually immature animals compared to mature animals (Ito *et al.*, 1993; Martin *et al.*, 1999), indicating the involvement of thyroid hormones in sexual maturation. However, this correlation is much less conspicuous, if not absent, in eels (*Anguilla* species; Han *et al.*, 2004; Aroua *et al.*, 2005). Still, the notion of the involvement of thyroid hormones in sexual maturation is corroborated by the increased responsivity of pre-smolt coho salmon thyroid tissue compared to after smoltification (Specker and Kobuke, 1985) and the presence of estrogen and androgen receptors in tilapia pituitary thyrotropes (Arai *et al.*, 2001).

3.3.2. The Thyroid-stimulating Hormone Receptor

The TSH receptor (TSHR) is expressed in the basolateral membrane of the thyrocyte, but extrathyroidal locations exist. Indeed, following the elucidation of the primary sequence of its cDNA, the abundant TSHR mRNA expression in, in particular, male and female gonads in a number of teleost species was reported (Kumar *et al.*, 2000; Goto-Kazeto *et al.*, 2003; Vischer and Bogerd, 2003; Rocha *et al.*, 2007). Only in amago salmon (*Oncorhynchus rhodurus*) was the expression of TSHR found to be strictly confined to thyroidal tissue in the subpharyngeal region (Oba *et al.*, 2000). These observations were all obtained using reverse transcriptase PCR techniques, and it has been pointed out that this methodology is perhaps too sensitive in that it can pick up "illegitimate transcription" (*sic*) of the TSHR gene that does not result in physiologically relevant levels of protein expression (Rapoport *et al.*, 1998). However, the occurrence of TSHR in mammalian extrathyroidal tissues is well established because mRNA expression is corroborated by positive results from immunohistochemistry and ligand-binding assays (reviewed by Davies *et al.*, 2002).

The gonadal expression of TSHR changes with developmental stage and seasonally, and this can explain the involvement of thyroid hormones in the sexual maturation of some fish species (Ito *et al.*, 1993; Martin *et al.*, 1999; Rocha *et al.*, 2007). Still, care is needed in interpreting TSHR mRNA expression in organs and tissues other than the subpharyngeal region, which is the typical location of the thyroid gland in fishes. As stated earlier, functional heterotopic thyroid follicles occur in many organs, and the expression of TSHR mRNA could very well reflect the normal cellular location of TSHR in thyrocytes instead of other cell types.

4. TARGETS, FUNCTIONS AND FEEDBACK EFFECTS OF CORTISOL AND THYROID HORMONES

4.1. Cortisol

The principal corticosteroid secreted by the interrenal cells in teleost fishes is cortisol (reviewed by Mommsen *et al.*, 1999). While circulating levels of cortisone increase in response to different stressors, as does cortisol, cortisone is inactive and primarily the product of 11β-oxidation of cortisol in the target tissues (Patiño *et al.*, 1985, 1987). Moreover, in contrast to tetrapods, teleosts appear to lack the synthesizing capacity for producing the mineralocorticoid aldosterone (reviewed by Prunet *et al.*, 2006). As such, cortisol is considered the primary glucocorticoid *and* mineralocorticoid in teleosts. However, in rainbow trout 11-deoxycorticosterone (DOC) is a potent agonist of the mineralocorticoid receptor (Sturm *et al.*, 2005), and its plasma concentration increases 10–50 fold towards the end of the male reproductive cycle, reaching peak concentrations that are comparable to basal plasma cortisol levels (Milla *et al.*, 2008). Interestingly, DOC and its exogenous congener aldosterone were equally potent in activating the rainbow trout mineralocorticoid receptor, and it has been suggested that DOC is an ancestral endogenous ligand for the mineralocorticoid receptor (Bury and Sturm, 2007).

4.1.1. Targets: Glucocorticoid and Mineralocorticoid Receptors

The targets of corticosteroids in teleosts are ligand-activated transcription factors which include both glucocorticoid receptors (GR) and mineralocorticoid receptors (MR). GRs and MRs evolved from an ancestral corticosteroid receptor (CR) following a genome duplication event in the early gnathostomes (Thornton, 2001). A second whole genome duplication event dating back to some 335 million years led to two different GR proteins in most teleosts but only one of the duplicated MR appears to have been retained (Colombe *et al.*, 2000; Bury *et al.*, 2003; Greenwood *et al.*, 2003; Stolte *et al.*, 2006, 2008). A splice variant of one of the duplicate GR genes has also been identified in several teleost species (Bury and Sturm, 2007). Therefore, in most teleosts corticosteroid signaling is achieved through three GRs and one MR. In contrast, zebrafish have a single GR and MR gene and a GRβ splice variant that lacks transactivational activity but behaves as a dominant-negative inhibitor of GRα transactivation (Alsop and Vijayan, 2008; Schaaf *et al.*, 2008). While information on non-genomic signaling by corticosteroids in fish is lacking, there is evidence for membrane-associated glucocorticoid receptors in other vertebrates (Tasker *et al.*, 2006).

Transactivation assays demonstrate that the CRs of teleosts are selective for corticosteroid hormones and suggest that the multiple CRs in fish may mediate distinct physiological responses. In rainbow trout (Sturm *et al.*, 2005) and Burton's mouthbrooder (*Haplochromis burtoni*; Greenwood *et al.*, 2003), MRs are 10 to 100 times more sensitive to cortisol than are GRs. In contrast, in common carp the sensitivity of the MR is intermediate to that of the two GRs (Stolte *et al.*, 2008). Moreover, the GR2 isoform of rainbow trout (Bury *et al.*, 2003) and common carp (Stolte *et al.*, 2008), but not of Burton's mouthbrooder (Greenwood *et al.*, 2003), is significantly more sensitive to cortisol than the GR1 isoform. All in all, while the above differences in CR transactivation capacity may allow for differential regulation by basal and stress-induced cortisol levels, the selectivity of the CRs for steroids in teleosts is apparently species-specific.

The ubiquitous distribution of GRs and MRs in teleosts reveals the multiplicity of functions mediated and/or modulated by cortisol. The expression of GRs has been identified in the liver, gills, intestine, kidney, spleen, heart, skeletal muscles, gonads, leukocytes and erythrocytes (Mommsen *et al.*, 1999; Shrimpton and McCormick, 1999; Vijayan *et al.*, 2003; Takahashi *et al.*, 2006; Vazzana *et al.*, 2008). GR-ir and GR-mRNA expressing cells are also located in many regions of the CNS with predominant clusters in the dorsal telencephalon, and in nuclei with known neuroendocrine functions, including the NPO, the NLT, the inferior lobe of the hypothalamus and the caudal neurosecretory system (CNSS; Teitsma *et al.*,

1997, 1998; Pepels *et al.*, 2004; Marley *et al.*, 2008; Stolte *et al.*, 2008; for review see also Chapter 2, this volume). Both GRs and MRs are expressed in the pituitary pars distalis and pars intermedia (Pepels *et al.*, 2004; Stolte *et al.*, 2008). Specifically, GRs have been localized to the corticotropes of the RPD, and to the somatotropes and gonadotropes of the PPD (Teitsma *et al.*, 1998; Pepels *et al.*, 2004; Kitahashi *et al.*, 2007; Stolte *et al.*, 2008). While not as well characterized as the GRs, MRs are widely expressed in the brain and periphery of both rainbow trout and Burton's mouthbrooder (Greenwood *et al.*, 2003; Sturm *et al.*, 2005). GR and MR transcripts have also been observed throughout embryogenesis and both exhibit a differential expression pattern (Tagawa *et al.*, 1997; Alsop and Vijayan, 2008). Overall, although the specific physiological roles for each CR subtype are not known, the combination of multiple receptors that are differentially expressed and that show different transactivation capacity should facilitate a high degree of CR signaling flexibility in teleosts.

4.1.2. Functions

Cortisol in fish has been implicated in the regulation of a broad array of physiological functions that affect energy metabolism, growth, reproduction, hydromineral balance and the immune system (for detailed reviews see Wendelaar Bonga, 1997; Mommsen *et al.*, 1999; Norris and Hobbs, 2006). Specifically, cortisol affects energy metabolism in fish by increasing liver glucose production, and by stimulating proteolytic and lipolytic capacity (De Boeck *et al.*, 2001; Aluru and Vijayan, 2007). Key target genes mediating the metabolic effects of cortisol in fish include the hepatic gluconeogenic enzyme phosphoenolpyruvate carboxykinase (PEPCK), genes involved in protein metabolism (glutamine synthetase, arginase, aminotransferases), protein catabolism (cathepsin D, glutamate decarboxylase) and the lipolytic enzyme triacylglycerol lipase (Sheridan, 1986; Mommsen *et al.*, 1999; Aluru and Vijayan, 2007). Beyond its catabolic effects, cortisol suppresses somatic growth in fish through its actions on the growth hormone–insulin-like growth factor I axis (Kajimura *et al.*, 2003; Peterson and Small, 2005; Pierce *et al.*, 2005) and its effects on the neuroendocrine pathways that regulate food intake (Bernier 2006; see Chapter 9, this volume). Cortisol impairs sexual maturation and reproduction via multiple effects including the suppression of plasma gonadotropin levels, the inhibition of gonadal steroidogenesis and a reduction in liver vitellogenin synthesis (Contreras-Sanchez *et al.*, 1998; Lethimonier *et al.*, 2000; Pankhurst and Van Der Kraak, 2000; Consten *et al.*, 2001). Interestingly, as part of its key role in the restoration of internal fluid balance following a stressor, cortisol in teleosts can promote both ion uptake and ion secretion mechanisms (McCormick, 2001). In part, cortisol promotes ion uptake by stimulating the expression of the freshwater Na^+/K^+-ATPase isoform, NKA α1a, and promotes ion secretion by stimulating the expression of the seawater Na^+/K^+-ATPase isoform, NKA α1b, as well as the cystic fibrosis transmembrane conductance regulator (CFTR) anion channel (Kiilerich *et al.*, 2007; McCormick *et al.*, 2008). Clearly, cortisol induces subtypes of transporter enzymes, but one should realize that it is the epithelial electrochemical make-up (e.g., a negative transepithelial potential in a tight freshwater gill epithelium, a positive one in a leaky seawater gill) that determines net flow of ions. The effects of cortisol on the immune system are also complex. While cortisol generally suppresses the immune system, the response depends on the type of immune cell involved and the specific parameter considered (see Chapter 7, this volume).

The higher sensitivity of trout and carp MR for the mineralocorticoid DOC than for cortisol raises the possibility that DOC is a physiological ligand of piscine MRs (Sturm *et al.*, 2005; Stolte *et al.*, 2008). While the circulating levels of DOC under basal conditions are much lower than those of cortisol, several factors may be involved in reducing the bioavailability of cortisol (Prunet *et al.*, 2006). For example, the enzyme that converts cortisol into inactive cortisone, 11β-hydroxysteroid dehydrogenase, is widely expressed in fish tissues (Jiang *et al.*, 2003; Kusakabe *et al.*, 2003). In general, while the function of DOC in fish remains unclear, the plasma concentration of this steroid increases significantly at the end of the reproductive cycle in both male and female rainbow trout (Campbell *et al.*, 1980; Milla *et al.*, 2008) and there is evidence that DOC plays a role during spermiation (Milla *et al.*, 2008). In contrast, DOC does not appear to be involved either in oocyte hydration and maturation (Milla *et al.*, 2006), or in osmoregulation (McCormick *et al.*, 2008).

4.1.3. Negative Feedback

Through several feedback effects on multiple signaling factors and at different levels of the HPI axis, plasma cortisol also modulates the stress response. Intraperitoneal cortisol implants inhibit CRF expression in the NPO of goldfish (Bernier *et al.*, 1999, 2004). Similarly, removal of the negative feedback effects of cortisol by pharmacological adrenalectomy using metyrapone leads to an increase in NPO CRF gene expression in white sucker (Morley *et al.*, 1991) and goldfish (Bernier and Peter, 2001), and to an increase in CRF-ir in the brain and pituitary of the eel (*Anguilla anguilla*; Olivereau and Olivereau, 1990a). In contrast, despite clear evidence that NPO CRF-ir neurons co-express GRs in rainbow trout (Teitsma *et al.*, 1998; see also Chapter 2, this volume), the role of cortisol in the regulation of CRF gene expression in this species is equivocal. While cortisol implantation in rainbow trout reduced the stress-induced elevation in NPO CRF mRNA levels during isolation, it did not prevent the increase in CRF gene expression associated with confinement (Doyon *et al.*, 2006). Moreover, both isolated and confined trout treated with the GR antagonist, RU-486, were characterized by a reduction in NPO CRF mRNA levels (Doyon *et al.*, 2006). Lastly, chronically elevated plasma cortisol levels in rainbow trout can be associated with an increase in NPO CRF mRNA levels (Madison B. N., Tavakoli S. and Bernier N. J., unpublished data).

Moreover, while it is known that CRF expression in mammals (Shepard *et al.*, 2000) and *Xenopus* (Yao and Denver, 2007) is differentially regulated by corticosteroids in discrete regions of the brain, to date it is unclear whether similar cell-type specific effects of GCs on CRF expression exist in fish. Cortisol also inhibits the synthesis and pituitary release of AVT originating in the NPO in goldfish (Fryer and Lederis, 1988) and eels (Olivereau and Olivereau, 1990a), and the mRNA levels of UI in the hypothalamus of goldfish (Bernier and Peter, 2001). At the pituitary level, cortisol inhibits basal ACTH release and the ACTH-releasing activity of CRF and UI from superfused goldfish corticotropes (Fryer *et al.*, 1984). Exogenous corticosteroids can also suppress ACTH levels in gilthead sea bream (Rotllant *et al.*, 2000a) and brown trout (Pickering *et al.*, 1987), and dose-dependently decrease pituitary POMC expression and plasma ACTH levels in rainbow trout (Madison B.N., Tavakoli S. and Bernier N.J., unpublished data). Cortisol inhibits the *in vitro* CRF-stimulated release of αMSH from the melanotropes of Mozambique tilapia (Balm *et al.*, 1993), but has no effect on plasma αMSH levels in gilthead sea bream (Rotllant *et al.*, 2000a). In interrenal cells, cortisol may suppress its own secretion via a paracrine feedback loop in coho salmon (Bradford *et al.*, 1992), but not in gilthead sea bream (Rotllant *et al.*, 2000a). Finally, several studies in fish have shown that cortisol mediates GR protein degradation (e.g., Shrimpton and Randall, 1994; Aluru and Vijayan, 2007).

Overall, while there is clear evidence that cortisol can limit the magnitude and duration of the endocrine stress response in fish via several negative feedback pathways, the molecular mechanism by which cortisol and its receptors regulate the transcriptional machinery of their various targets within the HPI axis remains to be characterized.

4.2. Thyroid Hormones

4.2.1. Thyroxine and Extrathyroidal Deiodination

The thyroid gland or, more accurately, the gland's functional unit, the thyroid follicle, represents the only source of the prohormone thyroxine in vertebrates. Fishes generally do not have a compact thyroid gland, as other vertebrates do. Instead, they possess a loose, non-encapsulated organization of thyroid follicles. Thyroid follicles are typically scattered near the ventral aorta in the subpharyngeal region, but functional and endocrine active follicles are present in other anatomical areas as well in a large number of teleost species (reviewed by Baker-Cohen, 1959; Geven *et al.*, 2007). In particular renal tissues are preferential locations for heterotopic thyroid follicles, which in some species are the single site of the bioactive thyroid gland. In common carp (*Cyprinus carpio*), thyroid follicles in the subpharyngeal area, kidney and head kidney are all immunoreactive to T_4, but only the heterotopic thyroid follicles of the kidney and head kidney show their functionality by accumulating iodine and secreting T_4 *in vitro* upon stimulation by TSH and other endocrine factors (Geven *et al.*, 2007, 2009).

Thyroxine has no or few biological actions and is the major form in which thyroid hormones are secreted from the thyroid follicles under normal, non-fasting conditions (Eales and Brown, 1993). The chemical conversion to the potently bioactive 3,5,3′-triiodothyronine (T_3) that follows the secretion of T_4 is a pivotal step in thyroid hormone metabolism (reviewed by Köhrle, 1999, 2002) as the thyroid hormone receptor has an approximately 10-fold higher affinity for T_3 than for T_4. The conversion consists of the removal of a single iodine atom from the 5′-position of the prohormone molecule, and is catalyzed by two iodothyronine deiodinase isoforms, i.e., D1 and D2. Consequently, any cell or tissue that expresses a 5′-deiodinating activity is capable of producing bioactive thyroid hormone and, when the appropriate cellular extrusion pathway is present, can be considered a source of circulating T_3. This constitutes a key role for extrathyroidal tissues in the systemic supply of T_3.

The 5′-deiodinases D1 and D2 are widely expressed, and in particular fish liver and kidney display high T_4 5′-deiodination activities. These organs are therefore important determinants of thyroid status (Mol *et al.*, 1997; Van der Geyten *et al.*, 1998), and deiodinase activities are relevant experimental readouts. Deiodination can reliably be measured *in vitro* in whole organ homogenates or subcellular fractions using radiolabeled substrates, but care should be taken to validate assay conditions as the biochemical characteristics of teleost deiodinases can differ considerably from those of their mammalian orthologues (Klaren *et al.*, 2005; Arjona *et al.*, 2008). Since the end product of the hypothalamic–pituitary–thyroid axis in a strict sense is a prohormone, it follows that the conceptual model of the HPT axis is incomplete in that it neither includes the bioactive thyroid hormone nor appreciates the extrathyroidal conversion of T_4 to T_3.

In analogy to the post-translational modifications that control the function of some peptide hormones, thyroid hormones are subject to a wide variety of biochemical reactions (Fig. 6.4). The deiodinated, conjugated, decarboxylated and/or deaminated iodothyronine metabolites that are produced all have different biochemical and physiological properties (reviewed

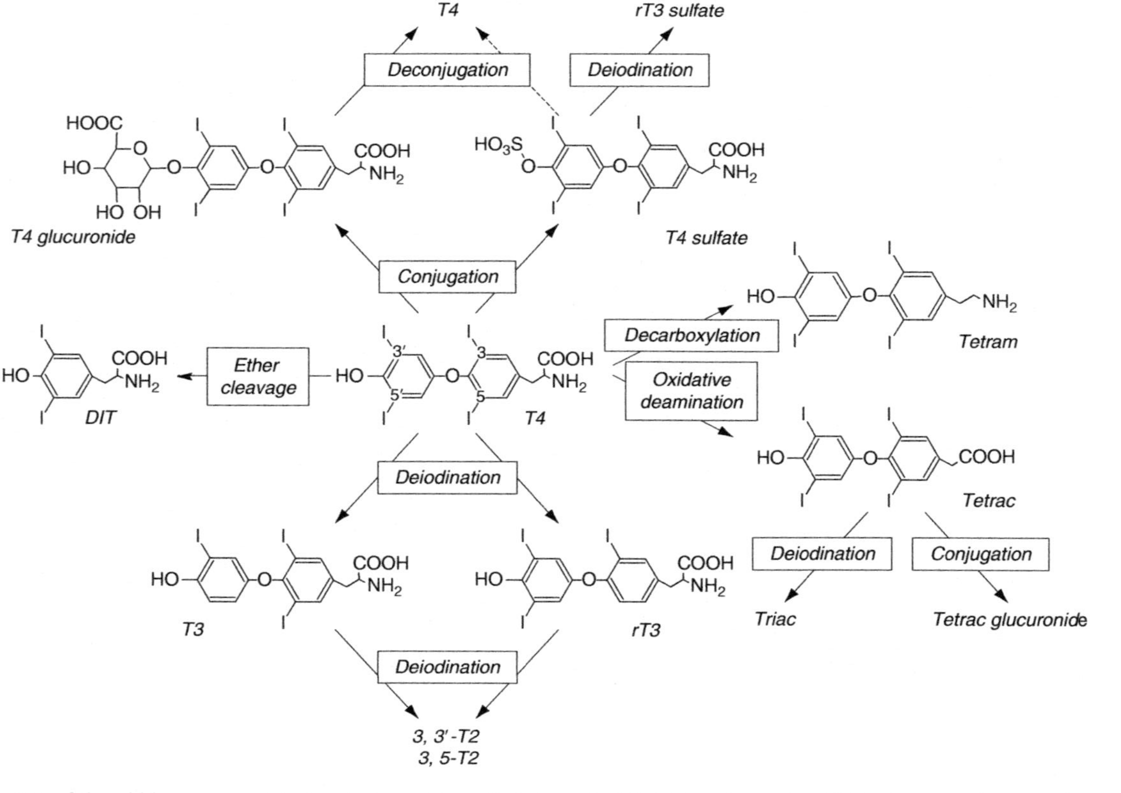

Fig. 6.4. Pathways of thyroid hormone metabolism (adapted from Köhrle *et al.*, 1987). Here, thyroxine (T_4) is chosen as the central metabolite, but most pathways are applicable to other iodothyronine species as well. Note: T_4 sulfate is not susceptible to deconjugation by sulfatase activity (as indicated by the dashed arrow) or 5′-deiodination by D1, but T_3 sulfate is. Abbreviations: DIT, diiodotyrosine; Tetrac, 3,3′5,5′-tetraiodothyroacetic acid; Tetram, 3,3′,5,5′-tetraiodothyronamine; Triac, 3,5,3′-triiodothyroacetic acid.

by Visser, 1994b; Wu *et al.*, 2005). Certain metabolic pathways, e.g., deiodination of T_3, and conjugation, are clearly involved in the termination of hormone signaling and systemic clearance, respectively. However, alternative actions, mostly non-genomic, can be attributed to multiple thyroid hormone metabolites. These are discussed in Section 4.2.4.

4.2.2. Thyroid Hormone Signaling is Dependent on Receptors and Cellular Uptake

Thyroid hormone receptors (TRs) are ligand-dependent transcription factors from the family of type-2 nuclear receptors that form dimers with the 9-*cis*-retinoic acid receptor, RXR (reviewed by Harvey and Williams, 2002; Tata, 2002).[2] In most vertebrates two separate genes code for the TRα and -β isoforms, but in fishes three genes, encoding two TRα and one TRβ receptors, respectively, have been identified (Yamano *et al.*, 1994; Yamano and Inui, 1995; Liu *et al.*, 2000; Tang *et al.*, 2008). Multiple functional receptors result from alternative mRNA splicing and alternative promoters (Lazar, 1993; Marchand *et al.*, 2001; Nelson and Habibi, 2009). Because of the particular genomic organization and transcriptional regulation, the expression of TRs can be accurately fine-tuned temporally and seasonally. Generally, TR mRNAs are expressed in virtually every tissue in a fish's body (Yamano and Miwa, 1998; Marchand *et al.*, 2004; Nelson and Habibi, 2006; Filby and Tyler, 2007), which accounts for the pleiotropic and pervasive actions of thyroid hormones.

Most molecular targets of T_3 are located within the genome. Genes that are responsive to T_3 carry a thyroid response element (TRE) that consists of a pair of DNA hexamer half-sites near the transcription initiation site in their promoter region. The TRE can bind TRs as monomer or homodimers, but the most active TR configuration is a heterodimer with RXR. Depending on the presence of a positive or negative TRE, T_3 can activate or suppress gene expression. Binding of T_3 results in the dissociation of a corepressor protein from the TR–RXR–TRE complex and the association of a coactivator. Many coactivators have an intrinsic histone acetyltransferase activity, and can modify the structure of chromatin to allow access for RNA polymerase and subsequent transcription (reviewed by Yen, 2001). The molecular mechanisms that underlie T_3-dependent transcriptional suppression are not yet fully elucidated.

Experimental data on the presence of (positive) TREs in piscine genes are scarce. They include growth hormone and the lipolytic enzyme lipoprotein lipase (Sternberg and Moav, 1999; Almuly *et al.*, 2000; Oku *et al.*, 2002).

[2]The Nuclear Receptor Signaling Atlas that is available via the portal: www.NURSA.org contains extensive and detailed information on thyroid hormone receptors (Margolis, 2008).

Besides one study that has attempted to identify target genes for T_3 in a metamorphosing flatfish species (Marchand *et al.*, 2004), very little is known about T_3-responsive genes in fishes. With the availability of the assembled fugu and zebrafish genomes, and the AGGT(C/A)A consensus sequence for the TRE half-site (Brent *et al.*, 1989), it should be feasible to locate TREs and the T_3-target genes they serve *in silico*.

Beyond the presence of thyroid hormone receptors, equally important is the repertoire of plasma membrane thyroid hormone transporters in target cells. Traditionally it was assumed that transmembrane transport occurred by diffusion due to the lipophilic nature of the iodothyronine molecule. However, even early studies indicated energy-dependent intracellular accumulation of thyroid hormones (reviewed by Hennemann, 2005). Moreover, nuclear magnetic resonance studies indicated preferential partition of iodothyronines within a lipid bilayer, and that passive transmembrane diffusion, if any, was greatly dependent on the phospholipid composition and cholesterol content of the membranes (Lai *et al.*, 1985; Chehín *et al.*, 1999). The permeability of a diffusional pathway is difficult to modulate, and most certainly is an inefficient way to regulate hormone entry. Over the past few years a number of thyroid hormone transporters have been characterized (reviewed by Jansen *et al.*, 2005) and their critical role in thyroid hormone signalling has been unequivocally demonstrated in patients with severe pathologies that could be attributed to dysfunctional mutant transporters (Friesema *et al.*, 2004; Jansen *et al.*, 2008). Undisturbed thyroid hormone signaling thus not only requires bioactive ligands and receptors, but also an intact transmembrane transport pathway (Hennemann *et al.*, 1998).

Members from the families of organic anion transporter polypeptides (OATP) and monocarboxylate transporters (MCT) possess specific thyroid hormone transport capacities (reviewed by Jansen *et al.*, 2005; Hagenbuch, 2007). We have virtually no pertinent information on the involvement of these transporters in thyroid hormone signaling in fish. Several MCT proteins, *viz.* MCT8 and MCT10 that specifically transport thyroid hormones in humans (Friesema *et al.*, 2008), have been identified in teleost genomes (Liu *et al.*, 2008), but we do not know what their role is in thyroid physiology *in vivo*. Interestingly, an organic anion transporter, fOat cloned from winter flounder (*Pseudopleuronectes americanus*) kidney, was shown to be homologous to rat and human Oat1 and Oat3 (Wolff *et al.*, 1997; Aslamkhan *et al.*, 2006), i.e., two members of the OATP family that mediate thyroid hormone transport in mammals (reviewed by Hagenbuch and Meier, 2004; Hagenbuch, 2007). It remains to be established whether the fOat protein is capable of thyroid hormone transport across a plasma membrane.

4.2.3. Targets and Functions

Mammalian genes that are positively regulated by T_3 are, for instance, deiodinase type 1, α-myosin heavy chain, several ATPases (Na^+/K^+-ATPase α-subunit, sarcoplasmic reticulum Ca^{2+}-ATPase), proteins that are involved in neuronal development (myelin basic protein, Purkinje cell protein *pcp2*), lipogenic enzymes (malic enzyme, glucose-6-phosphate dehydrogenase), the gluconeogenic enzyme phosphoenolpyruvate carboxykinase (PEPCK) and, in rat, growth hormone. Negatively regulated genes are those for the TSH α- and β-subunits, TRH, deiodinase type 2, prolactin and, surprisingly, human growth hormone (reviewed by Harvey and Williams, 2002; König and Neto, 2002). Only one study has attempted to identify target genes for T_3 in a teleost species (Marchand *et al.*, 2004), but very little else is known of T_3-responsive genes in fishes. Classically, the actions of thyroid hormone are considered to be the genomic actions of T_3 solely, but this view overlooks reported non-genomic effects of T_3 (reviewed by Davis and Davis, 1996; Bassett *et al.*, 2003; Goglia, 2005) and that of other iodothyronine metabolites (see Fig. 6.4), as discussed in the next section.

4.2.4. Biologically Active Thyroid Hormone Metabolites

In mammals and some fish species as well, biological effects of multiple thyroid hormone metabolites have been reported. Isolated nuclei from trout and salmon liver bind a wide array of thyroid hormone analogues, albeit with affinities that differ by three orders of magnitude (Darling *et al.*, 1982; Bres and Eales, 1986; Leeson *et al.*, 1988). Examples of effects which are not mediated via nuclear receptors and the genome include the induction of the mitogen-activated protein kinase (MAPK) pathway by T_4 and the T_3-isomer 3,3′,5′-triiodothyronine (rT_3) in cultured cell lines (Lin *et al.*, 1999), the stimulation of mammalian mitochondrial respiration and killifish deiodinase type 2 by diiodothyronine species 3,5-T_2 and 3,3′-T_2 (Lanni *et al.*, 1994; Goglia, 2005; García-G *et al.*, 2007), and the inotropic effects of the monoiodothyronamine T_1AM (Scanlan *et al.*, 2004). Sulfated and glucuronidated thyroid hormone conjugates have different reactivities in metabolic and transport pathways (Visser, 1994a; Wu *et al.*, 2005; van der Heide *et al.*, 2007). In gilthead sea bream the specific activities of enzymes that are involved in the conjugation and deconjugation of thyroid hormones were found to be responsive to an osmotic challenge (Klaren *et al.*, 2007b), hinting at a physiological role for conjugated iodothyronines.

There is no reason not to suspect novel physiological roles of iodothyronine metabolites other than native thyroid hormones in fish. Indeed, it can be envisaged that there is an evolutionary pressure favoring the assignment of biological functions to multiple metabolites of the iodothyronine molecule. In freshwater and seawater habitats, iodine and selenium (an essential ingredient of the deiodinase selenoprotein) are present in trace amounts, and this makes thyroid hormone action dependent on *two* trace elements. Monodeiodination of T_4's phenolic ring yields the potently bioactive T_3, and this already illustrates the impact of one iodine atom on the biological activity of a molecule. The iodine atoms in T_3 can be successively removed to yield di- and monoiodothyronines, as well as their derivates, all with different steric properties and (thus) different biological activities. The assignment of a biological activity to iodothyronine metabolites is a way to "economically" fully exploit trace elements.

4.2.5. Negative Feedback

A classical negative feedback exists between plasma thyroid hormones and TSH secretion by pituitary thyrotropes. Early studies already showed that experimentally induced hypothyroidism resulted in pituitary thyrotrope hypertrophy and degranulation (Barrington and Matty, 1955; Haider, 1975). More recently it has been shown in several teleost species that both T_4 and T_3 down-regulate the pituitary TSH β-subunit mRNA content *in vivo* and *in vitro* (Larsen *et al.*, 1997; Pradet-Balade *et al.*, 1997, 1999; Sohn *et al.*, 1999; Yoshiura *et al.*, 1999; Chatterjee *et al.*, 2001; Manchado *et al.*, 2008). Teleosts thus do not seem to conform to the mammalian feedback model in which only T_3, derived from the local pituitary deiodination of T_4, is the main suppressor of pituitary TSH secretion (Silva *et al.*, 1978; Christoffolete *et al.*, 2006). This likely reflects different properties of the thyroid hormone transport pathways in the plasma membrane of the pituitary thyrotrope.

5. CONTRIBUTION OF THE CORTICOTROPIC, MELANOTROPIC, AND THYROTROPIC AXES TO THE STRESS RESPONSE

The corticotropic, melanotropic and thyrotropic axes are all implicated in the endocrine response to stressors. While the large majority of the fish literature on this topic focuses on the direct contribution of the corticotropic axis to the regulation of cortisol secretion, evidence also exists that the melanotropic and thyrotropic axes interact centrally and peripherally with the corticotropic axis to coordinate the stress response. This section will summarize the evidence implicating each axis in the stress response, their interactions and their coordination.

5.1. The Corticotropic Axis

The basic function of the corticotropic axis in vertebrates is to stimulate the biosynthesis and release of corticosteroids from the steroidogenic tissue of the adrenal medulla or its equivalent. In return, as a result of their important roles in the stress response (see Section 4.1.2), elevated circulating levels of corticosteroids are a key indicator of stress. Accordingly, numerous studies on a variety of fish species have reported elevated plasma cortisol levels in response to diverse stressors (for reviews see Barton and Iwama, 1991; Mommsen *et al.*, 1999; Norris and Hobbs, 2006). Several studies have also implicated the corticotropic axis in the cortisol stress response by showing that the increase in circulating cortisol is preceded by a rise in plasma ACTH levels. For example, handling and confinement (Sumpter *et al.*, 1986; Rotllant *et al.*, 2001; Doyon *et al.*, 2006), crowding (Rotllant *et al.*, 2000b), thermal shock (Pickering *et al.*, 1986), osmotic shock (Craig *et al.*, 2005) and subordination (Höglund *et al.*, 2000) are characterized by increases in both plasma ACTH and cortisol levels. Similarly, the reduction in pituitary ACTH content in response handling, confinement and crowding (Rotllant *et al.*, 2000b, 2001), and the increase in pituitary pars distalis POMC gene expression associated with confinement (Gilchriest *et al.*, 2000) or restraint (Metz *et al.*, 2004) all suggest an increase in the release and synthesis of the POMC-derived peptide ACTH in response to stressors. In general, while a few acute stressors may be characterized by ACTH-independent stimulated cortisol release (Balm *et al.*, 1994; Arends *et al.*, 1999), the contribution of ACTH and of the corticotropic axis appears to be a common response to most stressors.

5.2. The Melanotropic Axis

Several studies have suggested that the melanotropic axis contributes to the endocrine stress response in fishes. For example, social subordination in Arctic charr (*Salvelinus alpinus*) results in elevated plasma levels of αMSH which are positively correlated with skin darkness (Höglund *et al.*, 2000, 2002a). Similarly, subordination in rainbow trout is associated with a sustained increase in pituitary POMC mRNA levels which is primarily of melanotropic origin (Winberg and Lepage, 1998). While increases in plasma αMSH and *N*-Ac βEND have also been observed in response to handling, confinement and chronic restraint (e.g., Sumpter *et al.*, 1985; Arends *et al.*, 1999; Rotllant *et al.*, 2000b; Metz *et al.*, 2005; van den Burg *et al.*, 2005), the same stressors have been shown to have either no effect or decrease the circulating levels of these peptides in other studies (e.g., Balm *et al.*, 1995; Balm and Pottinger, 1995; Ruane *et al.*, 1999; Rotllant *et al.*, 2001). In common carp, the variation in the contribution of αMSH and *N*-Ac βEND to the stress response depends on the responsive state of the melanotropes to CRF (van den Burg *et al.*, 2005) (discussed in Section 2.2.1). Overall, the contribution of the melanotropic axis to the stress response in fish depends on the nature and intensity of the stressor, the species involved and the responsive state of the melanotropes to hypophysiotropic factors.

The targets and actions of circulating αMSH and βEND within the corticotropic axis are still ill-defined in fish, but preliminary evidence suggests that multiple interactions are likely. While the corticotropic activity of αMSH observed in Mozambique tilapia may not generalize to other teleosts (see Section 3.2), the extensive gene expression of MC4R and MC5R in neurons of the preoptic area and the tuberal hypothalamus in goldfish (Cerdá-Reverter *et al.*, 2003b,c) suggests interactions between αMSH and the CRF system as evidenced by Matsuda *et al.* (2008). Peripheral αMSH can pass the blood–brain barrier (Banks and Kastin, 1995) and can probably reach these central targets. Still, more studies are needed to determine whether circulating αMSH exerts feedback effects on the hypophysiotropic regulators of the corticotropic axis (Metz *et al.*, 2005). Similarly, aside from a potentiating role in cortisol secretion in tilapia (Balm *et al.*, 1995), the contribution of *N*-Ac βEND to the stress response in fish remains elusive (Flik *et al.*, 2006). In contrast, several roles for non-acetylated βEND in the regulation of the corticotropic axis are suggested by the expression and stress-induced regulation of mu and delta opioid receptors in the hypothalamus, pituitary pars distalis and head kidney of common carp (Chadzinska *et al.* 2009).

5.3. The Thyrotropic Axis

Thyroid hormones have pervasive effects and virtually any cell in a fish body expresses thyroid hormone receptors. From this observation it can readily be envisaged that the thyrotropic axis is involved in the endocrine stress response in fishes. The involvement of the HPT axis is evidenced primarily through changes in plasma thyroid hormone levels. Still, it has to be stated that a general picture cannot be constructed. Some examples will illustrate this. Handling and blood sampling increased plasma T_3 and decreased plasma T_4 levels in rainbow trout (Todd and Eales, 2002). Feed restriction reduced plasma T_3 and T_4 levels in catfish (*Ictalurus punctatus*), rainbow trout and gilthead sea bream (Farbridge *et al.*, 1992; Power *et al.*, 2000; Gaylord *et al.*, 2001). Rainbow trout kept at high densities demonstrated decreased plasma T_3 levels (Leatherland, 1993). Brown trout (*Salmo trutta*) exposed to acid water displayed elevated T_4 but not T_3 levels, (Brown *et al.*, 1989). Osmotic challenges transiently increased plasma levels of total T_4 and total T_3 in rainbow trout (Orozco *et al.*, 2002), and of free T_4 in Senegal sole (*Solea senegalensis*; Arjona *et al.*, 2008). Typically plasma levels of T_4 and T_3 do not change in parallel, independent of whether total or free hormone concentrations are measured, which indicates the involvement of extrathyroidal deiodination pathways in the regulation of plasma thyroid hormone concentrations during stress (Waring and Brown, 1997; Orozco *et al.*, 2002; López-Bojorquez *et al.*, 2007; Arjona *et al.*, 2008). The vertebrate thyroid gland is the only endogenous source of the prohormone T_4, but peripheral tissues and organs are sources of circulating or locally bioactive T_3. This sets the HPT axis apart from the corticotropic and melanotropic axes as the regulation of the thyroid status of an animal extends well beyond a narrowly defined HPT axis.

The end products of at least two endocrine axes, *viz.* cortisol and thyroid hormones, have reciprocal effects, as is evident from experiments in which fishes were treated with either hormone (Young and Lin, 1988; Vijayan and Leatherland, 1992; Mustafa and MacKinnon, 1999; Walpita *et al.*, 2007). A bidirectional communication could take place via shared signaling molecules in the brain, as was shown in common carp (Geven *et al.*, 2006, 2009), but peripheral deiodination activities are a target as well (Vijayan and Leatherland, 1992; Walpita *et al.*, 2007).

Recently, efforts have been made to integrate central and peripheral regulation of the thyroid system in common carp (Geven *et al.*, 2006, 2007, 2009). A picture that emerges is that thyroid hormone reduces the *in vivo* activity of the corticotropic axis via CRF-BP and CRF, and activates the secretion of αMSH. Multiple peripheral factors, *viz.* TSH, ACTH, αMSH and cortisol, stimulate the release of T_4 from the renal thyroid gland in carp *in vitro*. Most probably a GR and MC2R are involved (Metz *et al.*, 2005a). An interesting notion is that the kidney and head kidney, which in common carp contain the functional thyroid gland, constitute a center where peripheral signals originating from the three endocrine axes are integrated.

5.4. Integrating at the Organismal Level: The Contribution of Hypophysiotropic Factors to the Stress Response *In Vivo*

While the results from *in vitro* studies suggest that multiple hypothalamic factors can stimulate and inhibit secretions from the corticotropic, melanotropic and thyrotropic axes in fish (see Section 2), *in vivo* studies provide evidence as to which of these potential hypophysiotropic signals actually contributes to regulating the secretion from the pituitary cells in response to diverse physiological conditions. The difference between *in vitro* and *in vivo* conditions is not trivial, and the translation of *in vitro* data to an organism can be hazardous. In previous sections we discussed changes in the basal activities of pituitary cells once the pituitary gland was denervated and devoid of (neuro)endocrine input, and presented data on the synergistic and permissive actions of combinations of endocrines. Moreover, since different types of stressors recruit separate stress-sensitive circuitries that converge on the hypothalamic-releasing and -inhibiting neurons of the diencephalon (Herman *et al.*, 2003), the relative contribution of the different hypophysiotropic signals is likely stressor-specific. This section will review the evidence from *in vivo* studies for the contribution of hypophysiotropic factors to the regulation of the corticotropic, melanotropic and thyrotropic axes in fish.

Beyond the studies which identified CRF and UI as important hypophysiotropic factors in the secretion of ACTH, TSH and αMSH, there is considerable evidence implicating the CRF system in the regulation of the corticotropic axis in fish. In rainbow trout, isolation and confinement (Ando *et al.*, 1999; Doyon *et al.*, 2005), social subordination (Doyon *et al.*, 2003), hyperammonemia (Ortega *et al.*, 2005), hypoxia (Bernier and Craig, 2005) and seawater transfer (Craig *et al.*, 2005) are characterized by stressor-specific and time-dependent increases in plasma cortisol and in preoptic area CRF mRNA levels (Bernier *et al.*, 2008). All of the above stressors in rainbow trout, with the exception of social subordination, which paradoxically elicits the most pronounced cortisol response, are also associated with an increase in forebrain UI mRNA levels (Craig *et al.*, 2005; Bernier *et al.*, 2008). In common carp, the stress of a 24-h restraint increased plasma cortisol and hypothalamic CRF gene expression, leaving UI mRNA levels unchanged (Huising *et al.*, 2004). In contrast, a 30-min restraint stress in flounder (*Platichthys flesus*) increased hypothalamic UI mRNA levels 3 h later, but left CRF gene expression unchanged (Lu *et al.*, 2004). In masu salmon (*Oncorhynchus masou*), the seasonal changes in plasma cortisol and in forebrain CRF and UI mRNA levels suggest that UI may be of greater importance than CRF in regulating the corticotropic axis during the spawning period (Westring *et al.*, 2008). Unfortunately, none of the above studies quantified whether the stressors had an impact on the circulating levels of αMSH, *N*-Ac βEND, TSH or thyroid hormones, thus leaving us to question whether CRF or UI contribute to the regulation of the melanotropic or thyrotropic axes in response to these specific stressors. In contrast, the increase in *in vitro* melanotrope sensitivity to CRF following a crowding stress in gilthead sea bream (Rotllant *et al.*, 2000b) suggests a potential involvement of the CRF system in the regulation of αMSH release.

In addition to the stimulatory role of CRF-related peptides in the regulation of the corticotropic axis, other components of the CRF system may be involved in dampening the endocrine response to stressors and regulating its duration. For example, the 24-h restraint stress in common carp is associated with a down-regulation of pituitary CRF type 1 receptor (CRFR1) mRNA levels and an increase in hypothalamic CRF-BP gene expression (Huising *et al.*, 2004). In rainbow trout, the increase in preoptic area CRF mRNA following repeated chasing events increased pituitary CRF-BP gene expression (Doyon *et al.*, 2005), and both hypoxia and subordination stressors produced region-specific increases in brain CRF-BP mRNA levels (Alderman *et al.*, 2008). Similarly, one month of social stress in Burton's mouthbrooder decreased mRNA levels of brain CRF and pituitary CRFR1, and increased pituitary CRF-BP (Chen and Fernald, 2008). The stimulatory effects of thyroxine on hypothalamic CRF-BP mRNA levels in common carp also suggest *in vivo* interactions between axes (Geven *et al.*, 2006, 2009) (see Section 2.2.1). Interestingly, in zebrafish, CRF-BP shows a pronounced regional co-expression with CRF and comparatively little UI co-expression (Alderman and Bernier, 2007). Overall, while specific experiments are needed to provide direct evidence for an involvement of CRF-BP in CRF-related peptide signaling activities in fish and to identify the specific CRF receptor subtypes mediating the hypophysiotropic effects of CRF and UI, results from the above studies identify CRF-BP and CRF receptors as potentially important players in the overall coordination of the three hypothalamo-pituitary axes being considered.

Studies aimed at characterizing the impact of stressors on the mRNA expression of preoptic area AVT have also provided evidence for the multiplicity of hypophysiotropic factors involved in the regulation of the corticotropic axis in fish. In rainbow trout, the increase in plasma cortisol caused by acute confinement stress, but not chronic stress, is associated with an increase in AVT mRNA levels in the parvocellular neurons of the NPO (Gilchriest *et al.*, 2000). Confinement stress is also characterized by increases in plasma cortisol and NPO AVT mRNA expression in the flounder, but in this species the stressor was associated with magnocellular rather than parvocellular changes in AVT transcripts (Bond *et al.*, 2007). A role for AVT in the activation of the corticotropic axis is also suggested in zebrafish where social subordination markedly increased parvocellular preoptic area AVT immunoreactivity (Larson *et al.*, 2006).

To date, beyond the *in vitro* evidence implicating TRH as a stimulator of corticotrope, melanotrope and thyrotrope secretions in fishes, direct *in vivo* evidence that this hypophysiotropic factor contributes to the regulation of these three axes has been scarce. Nevertheless, the increase in *in vitro* melanotrope sensitivity to TRH following a chronic acidification stress in tilapia (Lamers *et al.*, 1994) and a crowding stress in gilthead sea bream (Rotllant *et al.*, 2000b) suggests that TRH is involved in the stimulation of αMSH release in response to stressors. Moreover, in common carp, hypothalamic expression of TRH is increased by more than twofold following 24 h of restraint stress (Metz J. and Flik G., unpublished observation), a stressor also known to be associated with increases in plasma cortisol and hypothalamic CRF mRNA levels (Huising *et al.*, 2004).

The potential contribution of MCH in the regulation of ACTH and αMSH secretion during the stress response in fish is suggested by the observation that various stressors are associated with changes in hypothalamic MCH perikarya content, whole pituitary MCH content, plasma MCH levels (Green and Baker, 1991; Green *et al.*, 1991) and hypothalamic MCH mRNA levels (Gröneveld *et al.*, 1995a). However, whether and to what extent the inhibitory effects attributed to MCH *in vitro* (see Sections 2.1 and 2.2) modulate the corticotropic and melanotropic axes under conditions of stress remains to be demonstrated. Similarly, while diverse stressors in fish are associated with an increase in hypothalamic (Øverli *et al.*, 1999, 2001; Gesto *et al.*, 2008) and pituitary (Gesto *et al.*, 2008) dopaminergic activity, and DA can inhibit the secretion of ACTH, αMSH and TSH *in vitro* (see Section 2), future studies are needed to determine the extent to which the inhibitory effects of DA are a factor in determining the stressor-specific contribution of the pituitary to the endocrine stress response.

In addition to their role in pituitary secretion, the key hypophysiotropic factors of the corticotropic, melanotropic and thyrotropic axes are involved in multiple regulatory functions. In the zebrafish brain, for example, the broad expression patterns of CRF, UI, CRF-BP (Alderman and Bernier, 2007) and TRH (Díaz *et al.*, 2002) suggest multiple roles in sensory, autonomic, behavioral and neuroendocrine functions. The actions attributed to CRF and UI in the regulation of the cardiovascular system (Le Mével *et al.*, 2006), ventilation (Le Mével *et al.*, 2009), locomotion (Clements *et al.*, 2002; Carpenter *et al.*, 2007) and appetite (Bernier, 2006) in fish also support a much broader role for the CRF system in the coordination of the stress response. In addition to the forebrain, fish are unique in having a second major source of CRF-related peptides, the CNSS (McCrohan *et al.*, 2007), that can be released to the circulation and that is responsive to stressors (Bernier *et al.*, 2008). In common carp, both CRF and CRF-BP are abundantly expressed within the chromaffin compartment of the head kidney, suggesting that like mammals a local CRF system may play a role in the regulation of cortisol release (Huising *et al.*, 2007). Integrating the hypophysiotropic functions of CRF and TRH with the broader physiological roles of the peptides will be an important avenue of future investigations.

6. PERSPECTIVES

If one aspect emerges from our collation and discussion of data, it is that the same hypothalamic signaling molecules, *viz.* TRH and peptides from the CRF-family, play prominent roles in the central regulation of the corticotrope, melanotrope and thyrotrope axes, other hypothalamic factors notwithstanding. The end products of the three axes also cross-communicate peripherally as well as centrally. It is difficult to continue to hold to a conceptual model of the regulation of stress responses as parallel-running endocrine axes. We provide ample documentation that CRF and TRH both can stimulate the release of the major pituitary hormones ACTH, αMSH and TSH, that again all can stimulate the release of the peripherally released hormones cortisol and T_4/T_3. The roles of αMSH as an anorexigenic and lipolytic factor, cortisol as a glucocorticoid, and the effects of thyroid hormone on basal metabolic rate all appear to converge on the common denominator of energy homeostasis. To reconcile the apparent paradox of only two hypothalamic factors regulating so basic a process via multiple intermediate signaling molecules, we could be more aware of the specific hypothalamic locations of these factors, and in particular the afferent input from higher/other brain centers that they connect with/are connected with. The notion that the hypothalamus is an integrating center for central signals on the physiological state of an animal is real, but needs to be explicated by stating what those signals then are. Our future investigations should take us upstream from the hypothalamus to identify new players – not to extend endocrine axes, but to better understand the stress web here constructed at a basic level. The pivotal importance of proper regulation of energy homeostasis/allostasis through this web for the evolutionary success of vertebrates is substantiated by the extreme conservation of its key signaling molecules as well as its multilevel and highly integrative character providing the organism with a great flexibility to cope with an ever changing environment, a key requirement for survival.

ACKNOWLEDGMENT

We thank Ian Smith and Sarah Alderman for help with figure formatting and presentation. Supported by a NSERC DG to N.J.B.

REFERENCES

Aguilera, G., Subburaju, S., Young, S., and Chen, J. (2008). The parvocellular vasopressinergic system and responsiveness of the hypothalamic pituitary adrenal axis during chronic stress. *Prog. Brain Res.* **170,** 29–39.

Akil, H., Young, E., Watson, S. J., and Coy, D. H. (1981). Opiate binding properties of naturally occurring N- and C-terminal modified β-endorphins. *Peptides* **2,** 289–292.

Alderman, S. L., and Bernier, N. J. (2007). Localization of corticotropin-releasing factor, urotensin I, and CRF-binding protein gene expression in the brain of the zebrafish, *Danio rerio*. *J. Comp. Neurobiol.* **502,** 783–793.

Alderman, S. L., Raine, J. C., and Bernier, N. J. (2008). Distribution and regional stressor-induced regulation of corticotrophin-releasing factor binding protein in rainbow trout (*Oncorhynchus mykiss*). *J. Neuroendocrinol.* **20,** 347–358.

Almuly, R., Cavari, B., Ferstman, H., Kolodny, O., and Funkenstein, B. (2000). Genomic structure and sequence of the gilthead seabream (*Sparus aurata*) growth hormone-encoding gene: Identification of minisatellite polymorphism in intron I. *Genome* **43,** 836–845.

Alsop, D., and Vijayan, M. M. (2008). Development of the corticosteroid stress axis and receptor expression in zebrafish. *Am. J. Physiol.* **294,** R711–719.

Aluru, N., and Vijayan, M. M. (2007). Hepatic transcriptome response to glucocorticoid receptor activation in rainbow trout. *Physiol. Genomics* **31,** 483–491.

Aluru, N., and Vijayan, M. M. (2008). Molecular characterization, tissue-specific expression, and regulation of melanocortin 2 receptor in rainbow trout. *Endocrinology* **149,** 4577–4588.

Amiya, N., Amano, M., Takahashi, A., Yamanome, T., and Yamamori, K. (2007). Profiles of α-melanocyte-stimulating hormone in the Japanese flounder as revealed by a newly developed time-resolved fluoroimmunoassay and immunohistochemistry. *Gen. Comp. Endocrinol.* **151,** 135–141.

Ando, H., Hasegawa, M., Ando, J., and Urano, A. (1999). Expression of salmon corticotropin-releasing hormone precursor gene in the preoptic nucleus in stressed rainbow trout. *Gen. Comp. Endocrinol.* **113,** 87–95.

Arai, M., Assil, I. Q., and Abou-Samra, A. B. (2001). Characterization of three corticotropin-releasing factor receptors in catfish: A novel third receptor is predominantly expressed in pituitary and urophysis. *Endocrinology* **142,** 446–454.

Arends, R. J., Mancera, J. M., Muñoz, J. L., Wendelaar Bonga, S. E., and Flik, G. (1999). The stress response of the gilthead sea bream (*Sparus aurata* L.) to air exposure and confinement. *J. Endocrinol.* **163,** 149–157.

Arjona, F. J., Vargas-Chacoff, L., Martín del Río, M. P., Flik, G., Mancera, J. M., and Klaren, P. H. M. (2008). The involvement of thyroid hormones and cortisol in the osmotic acclimation of *Solea senegalensis*. *Gen. Comp. Endocrinol.* **155,** 796–803.

Arnold-Reed, D. E., and Balment, R. J. (1991). Atrial natriuretic factor stimulates *in-vivo* and *in-vitro* secretion of cortisol in teleosts. *J. Endocrinol.* **128,** R17–R20.

Aroua, S., Schmitz, M., Baloche, S., Vidal, B., Rousseau, K., and Dufour, S. (2005). Endocrine evidence that silvering, a secondary metamorphosis in the eel, is a pubertal rather than a metamorphic event. *Neuroendocrinology* **82,** 221–232.

Aslamkhan, A. G., Thompson, D. M., Perry, J. L., Bleasby, K., Wolff, N. A., Barros, S., Miller, D. S., and Pritchard, J. B. (2006). The flounder organic anion transporter fOat has sequence, function, and substrate specificity similarity to both mammalian Oat1 and Oat3. *Am. J. Physiol. Regul. Integr. Comp. Physiol.* **291,** R1773–1780.

Baker-Cohen, K. F. (1959). Renal and other heterotopic thyroid tissue in fishes. *In* "Proceedings of the Columbia University Symposium on Comparative Endocrinology" (Gorbman, A., Ed.), pp. 283–301. Wiley, Cold Spring Harbor.

Baker, B. I., and Rance, T. A. (1981). Differences in concentrations of plasma-cortisol in the trout and the eel following adaptation to black or white backgrounds. *J. Endocrinol.* **89,** 135–140.

Baker, B. I., Bird, D. J., and Buckingham, J. C. (1985). Salmonid melanin-concentrating hormone inhibits corticotropin release. *J. Endocrinol.* **106,** R5–R8.

Baker, B. I., Bird, D. J., and Buckingham, J. C. (1986). Effects of chronic administration of melanin-concentrating hormone on corticotrophin, melanotrophin, and pigmentation in the trout. *Gen. Comp. Endocrinol.* **63,** 62–69.

Baker, B. I., Bird, D. J., and Buckingham, J. C. (1996). In the trout, CRH and AVT synergize to stimulate ACTH release. *Regul. Peptides* **67,** 207–210.

Ball, J. N., Olivereau, M., and Kallman, K. D. (1963). Secretion of thyrotrophic hormone by pituitary transplants in a teleost fish. *Nature* **199,** 618–620.

Ball, J. N., Olivereau, M., Slicher, A. M., and Kallman, K. D. (1965). Functional capacity of ectopic pituitary transplants in the teleost *Poecilia formosa*, with a comparative discussion on the transplanted pituitary. *Phil. Trans. R. Soc. London B* **249,** 69–99.

Ball, J. N., Baker, B. I., Olivereau, M., and Peter, R. E. (1972). Investigations on hypothalamic control of adenohypophysial functions in teleost fishes. *Gen. Comp. Endocrinol. Suppl.* **3,** 11–21.

Ball, J. N., Batten, T. F. C., and Young, G. (1980). Evolution of hypothalamo-adenohypophysial systems in lower vertebrates. *In* "Hormones, Adaptation and Evolution" (Ishii, S., Hirano, T., and Wada, M., Eds.). Japan Scientific Societies Press/Springer-Verlag, Tokyo/Berlin.

Balm, P. H. M., and Pottinger, T. G. (1995). Corticotrope and melanotrope POMC-derived peptides in relation to interrenal function during stress in rainbow trout (*Oncorhynchus mykiss*). *Gen. Comp. Endocrinol.* **98,** 279–288.

Balm, P. H. M., Gröneveld, D., Lamers, A. E., and Wendelaar Bonga, S. E. (1993). Multiple actions of melanotropic peptides in the teleost *Oreochromis mossambicus* (Tilapia). *Annals N. Y. Acad. Sci.* **680,** 448–450.

Balm, P. H. M., Pepels, P., Helfrich, S., Hovens, M. L. M., and Wendelaar Bonga, S. E. (1994). Adrenocorticotropic hormone in relation to interrenal function during stress in tilapia (*Oreochromis mossambicus*). *Gen. Comp. Endocrinol.* **96,** 347–360.

Balm, P. H. M., Hovens, M. L., and Wendelaar Bonga, S. E. (1995). Endorphin and MSH in concert form the corticotropic principle released by tilapia (*Oreochromis mossambicus*; Teleostei) melanotropes. *Peptides* **16,** 463–469.

Bandyopadhyay, S., and Bhattacharya, S. (1993). Purification and properties of an Indian major carp (*Cirrhinus mrigala,* Ham.) pituitary thyrotropin. *Gen. Comp. Endocrinol.* **90,** 192–204.

Banks, W. A., and Kastin, A. J. (1995). Permeability of the blood–brain-barrier to melanocortins. *Peptides* **16,** 1157–1161.

Barber, L. D., Baker, B. I., Penny, J. C., and Eberle, A. N. (1987). Melanin concentrating hormone inhibits the release of αMSH from teleost pituitary glands. *Gen. Comp. Endocrinol.* **65,** 79–86.

Barimo, J. F., Steele, S. L., Wright, P. A., and Walsh, P. J. (2004). Dogmas and controversies in the handling of nitrigenous wastes: Ureotely and ammonia tolerance in early life stages of the gulf toadfish, *Opsanus beta*. *J. Exp. Biol.* **207,** 2011–2020.

Barrington, E. J. W., and Matty, A. J. (1955). The identification of thyrotrophin-secreting cells in the pituitary gland of the minnow (*Phoxinus phoxinus*). *Q. J. Microsc. Sci.* **96,** 193–201.

Barton, B. A., and Iwama, G. K. (1991). Physiological changes in fish from stress in aquaculture with emphasis on the response and effects of corticosteroids. *Ann. Rev. Fish Dis.* **1,** 3–26.

Bassett, J. H. D., Harvey, C. B., and Williams, G. R. (2003). Mechanisms of thyroid hormone receptor-specific nuclear and extra nuclear actions. *Mol. Cell. Endocrinol.* **213,** 1–11.

Batten, T. F. (1986). Ultrastructural characterization of neurosecretory fibres immunoreactive for vasotocin, isotocin, somatostatin, LHRH, and CRF in the pituitary of a teleost fish, *Poecilia latipinna*. *Cell Tissue Res.* **244,** 661–672.

Batten, T. F. C., Moons, L., and Vandesande, F. (1999). Innervation and control of the adenohypophysis by hypothalamic peptidergic neurons in teleost fishes: EM immunohistochemical evidence. *Microsc. Res. Techn.* **44,** 19–35.

Bennett, M. A., Murray, T. F., and Aldrich, J. V. (2005). Structure-activity relationships of arodyn, a novel acetylated kappa opioid receptor antagonist. *J. Pept. Res.* **65,** 322–332.

Bernier, N. J. (2006). The corticotropin-releasing factor system as a mediator of the appetite-suppressing effects of stress in fish. *Gen. Comp. Endocrinol.* **146,** 45–55.

Bernier, N. J., and Craig, P. M. (2005). CRF-related peptides contribute to stress response and regulation of appetite in hypoxic rainbow trout. *Am. J. Physiol.* **289,** R982–R990.

Bernier, N. J., and Peter, R. E. (2001). Appetite-suppressing effects of urotensin I and corticotropin-releasing hormone in goldfish (*Carassius auratus*). *Neuroendocrinology* **73,** 248–260.
Bernier, N. J., Lin, X., and Peter, R. E. (1999). Differential expression of corticotropin-releasing factor (CRF) and urotensin I precursor genes, and evidence of CRF gene expression regulated by cortisol in goldfish brain. *Gen. Comp. Endocrinol.* **116,** 461–477.
Bernier, N. J., Bedard, N., and Peter, R. E. (2004). Effects of cortisol on food intake, growth, and forebrain neuropeptide Y and corticotropin-releasing factor gene expression in goldfish. *Gen. Comp. Endocrinol.* **135,** 230–240.
Bernier, N. J., Alderman, S. L., and Bristow, E. N. (2008). Heads or tails? Stressor-specific expression of corticotropin-releasing factor and urotensin I in the preoptic area and caudal neurosecretory system of rainbow trout. *J. Endocrinol.* **196,** 637–648.
Bertolini, A., Tacchi, R., and Vergoni, A. V. (2009). Brain effects of melanocortins. *Pharmacol. Res.* **59,** 13–47.
Bhattacharya, S., Mukherjee, D., and Sen, S. (1979). Role of synthetic mammalian thyrotropin releasing hormone on fish thyroid peroxidase activity. *Indian J. Exp. Biol.* **17,** 1041–1043.
Bond, H., Warne, J. M., and Balment, R. J. (2007). Effect of acute restraint on hypothalamic pro-vasotocin mRNA expression in flounder, *Platichthys flesus*. *Gen. Comp. Endocrinol.* **153,** 221–227.
Bonnin, J. P. (1971). Cultures organotypiques de thyroïdes d'un poisson Téléostéen marin: *Gobius niger* L. Effets de la TSH et de la prolactine. *Compt. Rend. Seanc. Soc. Biol. Fil.* **165,** 1284–1291.
Bradford, C. S., Fitzpatrick, M. S., and Schreck, C. B. (1992). Evidence for ultra-short-loop feedback in ACTH-induced interrenal steroidogenesis in coho salmon: Acute self-suppression of cortisol secretion *in vitro*. *Gen. Comp. Endocrinol.* **87,** 292–299.
Brent, G. A., Harney, J. W., Chen, Y., Warne, R. L., Moore, D. D., and Larsen, P. R. (1989). Mutations of the rat growth hormone promoter which increase and decrease response to thyroid hormone define a consensus thyroid hormone response element. *Mol. Endocrinol.* **3,** 1996–2004.
Bres, O., and Eales, J. G. (1986). Thyroid hormone binding to isolated trout (*Salmo gairdneri*) liver nuclei *in vitro*: Binding affinity, capacity, and chemical specificity. *Gen. Comp. Endocrinol.* **61,** 29–39.
Bromage, N. R. (1975). The effects of mammalian thyrotropin-releasing hormone on the pituitary–thyroid axis of teleost fish. *Gen. Comp. Endocrinol.* **25,** 292–297.
Bromage, N. R., Whitehead, C., and Brown, T. J. (1976). Thyroxine secretion in teleosts and effects of TSH, TRH and other peptides. *Gen. Comp. Endocrinol.* **29,** 246 (conference abstract).
Brown, C. L., and Stetson, M. H. (1983). Prolactin – thyroid interaction in *Fundulus heteroclitus*. *Gen. Comp. Endocrinol.* **50,** 167–171.
Brown, C. L., Grau, E. G., and Stetson, M. H. (1982). Endogenous gonadotropin does not have heterothyrotropic activity in *Fundulus heteroclitus*. *Am. Zool.* **22,** 854 (conference abstract).
Brown, C. L., Grau, E. G., and Stetson, M. H. (1985). Functional specificity of gonadotropin and thyrotropin in *Fundulus heteroclitus*. *Gen. Comp. Endocrinol.* **58,** 252–258.
Brown, J. A., Edwards, D., and Whitehead, C. (1989). Cortisol and thyroid hormone responses to acid stress in the brown trout, *Salmo trutta* L. *J. Fish Biol.* **35,** 73–84.
Bury, N. R., and Sturm, A. (2007). Evolution of the corticosteroid receptor signalling pathway in fish. *Gen. Comp. Endocrinol.* **153,** 47–56.
Bury, N. R., Sturm, A., LeRouzic, P., Lethimonier, C., Ducouret, B., Guiguen, Y., Robinson-Rechavi, M., Laudet, V., Rafestin-Oblin, M. E., and Prunet, P. (2003). Evidence for two distinct functional glucocorticoid receptors in teleost fish. *J. Mol. Endocrinol.* **31,** 141–156.

Byamungu, N., Corneillie, S., Mol, K., Darras, V., and Kühn, E. R. (1990). Stimulation of thyroid function by several pituitary hormones results in an increase in plasma thyroxine and reverse triiodothyronine in tilapia (*Tilapia nilotica*). *Gen. Comp. Endocrinol.* **80,** 33–40.

Byamungu, N., Mol, K., and Kühn, E. R. (1991). Somatostatin increases plasma T_3 concentrations in *Tilapia nilotica* in the presence of increased plasma T_4 levels. *Gen. Comp. Endocrinol.* **82,** 401–406.

Campbell, C. M., Fostier, A., Jalabert, B., and Truscott, B. (1980). Identification and quantification of steroids in the serum of rainbow trout during spermiation and oocyte maturation. *J. Endocrinol.* **85,** 371–378.

Carpenter, R. E., Watt, M. J., Forster, G. L., Øverli, Ø., Bockholt, C., Renner, K. J., and Summers, C. H. (2007). Corticotropin releasing factor induces anxiogenic locomotion in trout and alters serotonergic and dopaminergic activity. *Horm. Behav.* **52,** 600–611.

Cerdá-Reverter, J. M., Schiöth, H. B., and Peter, R. E. (2003a). The central melanocortin system regulates food intake in goldfish. *Regul. Pept.* **115,** 101–113.

Cerdá-Reverter, J. M., Ling, M. K., Schiöth, H. B., and Peter, R. E. (2003b). Molecular cloning, characterization and brain mapping of the melanocortin 5 receptor in the goldfish. *J. Neurochem.* **87,** 1354–1367.

Cerdá-Reverter, J. M., Ringholm, A., Schiöth, H. B., and Peter, R. E. (2003c). Molecular cloning, pharmacological characterization and brain mapping of the melanocortin 4 receptor in the goldfish: Involvement in the control of food intake. *Endocrinology* **144,** 2336–2349.

Cerdá-Reverter, J. M., Haitina, T., Schiöth, H. B., and Peter, R. E. (2005). Gene structure of the goldfish agouti-signaling protein: A putative role in the dorsal-ventral pigment pattern of fish. *Endocrinology* **146,** 1597–1610.

Chadzinska, M., Hermsen, T., Savelkoul, H. F. J., and Verburg-van Kemenade, B. M. L. (2009). Cloning of opioid receptors in common carp (*Cyprinus carpio* L.) and their involvement in regulation of stress and immune response. *Brain Behav. Immun.* **23,** 257–266.

Chakraborti, P., and Bhattacharya, S. (1978). Bovine TSH-stimulation of fish thyroid peroxidase activity and role of thyroxine thereon. *Experientia* **34,** 136–137.

Chatterjee, A., Hsieh, Y.-L., and Yu, J. Y. L. (2001). Molecular cloning of cDNA encoding thyroid stimulating hormone β subunit of bighead carp *Aristichthys nobilis* and regulation of its gene expression. *Mol. Cell. Endocrinol.* **174,** 1–9.

Chehín, R. N., Issé, B. G., Rintoul, M. R., and Farías, R. N. (1999). Differential transmembrane diffusion of triiodothyronine and thyroxine in liposomes: Regulation by lipid composition. *J. Membrane Biol.* **167,** 251–256.

Chen, C. C., and Fernald, R. D. (2008). Sequences, expression patterns and regulation of the corticotropin-releasing factor system in a teleost. *Gen. Comp. Endocrinol.* **157,** 148–155.

Chowdhury, I., Chien, J. T., Chatterjee, A., and Yu, J. Y. L. (2004). *In vitro* effects of mammalian leptin, neuropeptide-Y, β-endorphin and galanin on transcript levels of thyrotropin β and common α subunit mRNAs in the pituitary of bighead carp (*Aristichthys nobilis*). *Comp. Biochem. Physiol. B* **139,** 87–98.

Christoffolete, M. A., Ribeiro, R., Singru, P., Fekete, C., da Silva, W. S., Gordon, D. F., Huang, S. A., Crescenzi, A., Harney, J. W., Ridgway, E. C., Larsen, P. R., Lechan, R. M., *et al.* (2006). Atypical expression of type 2 iodothyronine deiodinase in thyrotrophs explains the thyroxine-mediated pituitary TSH feedback mechanism. *Endocrinology* **147,** 1735–1743.

Clements, S., Schreck, C. B., Larsen, D. A., and Dickhoff, W. W. (2002). Central administration of corticotropin-releasing hormone stimulates locomotor activity in juvenile chinook salmon (*Oncorhynchus tshawytscha*). *Gen. Comp. Endocrinol.* **125,** 319–327.

Colombe, L., Fostier, A., Bury, N., Pakdel, F., and Guiguen, Y. (2000). A mineralocorticoid-like receptor in the rainbow trout, *Oncorhynchus mykiss*: Cloning and characterization of its steroid binding domain. *Steroids* **65,** 319–328.

Consten, D., Bogerd, J., Komen, J., Lambert, J. G. D., and Goos, H. J. T. (2001). Long-term cortisol treatment inhibits pubertal development in male common carp, *Cyprinus carpio* L. *Biol. Reprod.* **64,** 1063–1071.

Contreras-Sanchez, W. M., Schreck, C. B., Fitzpatrick, M. S., and Pereira, C. B. (1998). Effects of stress on the reproductive performance of rainbow trout (*Oncorhynchus mykiss*). *Biol. Reprod.* **58,** 439–447.

Craig, P. M., Al-Timimi, H., and Bernier, N. J. (2005). Differential increase in forebrain and caudal neurosecretory system corticotropin-releasing factor and urotensin I gene expression associated with seawater transfer in rainbow trout. *Endocrinology* **146,** 3851–3860.

Crim, J. W., Dickhoff, W. W., and Gorbman, A. (1978). Comparative endocrinology of piscine hypothalamic hypophysiotropic peptides: Distribution and activity. *Am. Zool.* **18,** 411–424.

Darling, D. S., Dickhoff, W. W., and Gorbman, A. (1982). Comparison of thyroid hormone binding to hepatic nuclei of the rat and a teleost (*Oncorhynchus kisutch*). *Endocrinology* **111,** 1936–1943.

Davies, T., Marians, R., and Latif, R. (2002). The TSH receptor reveals itself. *J. Clin. Invest.* **110,** 161–164.

Davis, P. J., and Davis, F. B. (1996). Nongenomic actions of thyroid hormone. *Thyroid* **6,** 497–504.

De Boeck, G., Alsop, D., and Wood, C. (2001). Cortisol effects on aerobic and anaerobic metabolism, nitrogen excretion, and whole-body composition in juvenile rainbow trout. *Physiol. Biochem. Zool.* **74,** 858–868.

De Groef, B., Goris, N., Arckens, L., Kühn, E. R., and Darras, V. M. (2003). Corticotropin-releasing hormone (CRH)-induced thyrotropin release is directly mediated through CRH receptor type 2 on thyrotropes. *Endocrinology* **144,** 5537–5544.

De Groef, B., Vandenborne, K., Van As, P., Darras, V. M., Kühn, E. R., Decuypere, E., and Geris, K. L. (2005). Hypothalamic control of the thyroidal axis in the chicken: Over the boundaries of the classical hormonal axes. *Domest. Anim. Endocrinol.* **29,** 104–110.

De Groef, B., Van der Geyten, S., Darras, V. M., and Kühn, E. R. (2006). Role of corticotropin-releasing hormone as a thyrotropin-releasing factor in non-mammalian vertebrates. *Gen. Comp. Endocrinol.* **146,** 62–68.

Díaz, M. L., Becerra, M., Manso, M. J., and Anadón, R. (2002). Distribution of thyrotropin-releasing hormone (TRH) immunoreactivity in the brain of the zebrafish (*Danio rerio*). *J. Comp. Neurol.* **450,** 45–60.

Donaldson, E. M. (1981). The pituitary-interrenal axis as an indicator of stress in fish. *In* "Stress and Fish" (Pickering, A. D., Ed.), pp. 11–47. Academic Press, London.

Doyon, C., Gilmour, K. M., Trudeau, V. L., and Moon, T. W. (2003). Corticotropin-releasing factor and neuropeptide Y mRNA levels are elevated in the preoptic area of socially subordinate rainbow trout. *Gen. Comp. Endocrinol.* **133,** 260–271.

Doyon, C., Trudeau, V. L., and Moon, T. W. (2005). Stress elevates corticotropin-releasing factor (CRF) and CRF-binding protein mRNA levels in rainbow trout (*Oncorhynchus mykiss*). *J. Endocrinol.* **186,** 123–130.

Doyon, C., Leclair, J., Trudeau, V. L., and Moon, T. W. (2006). Corticotropin-releasing factor and neuropeptide Y mRNA levels are modified by glucocorticoids in rainbow trout, *Oncorhynchus mykiss*. *Gen. Comp. Endocrinol.* **146,** 126–135.

Dreborg, S., Sundström, G., Larsson, T. A., and Larhammar, D. (2008). Evolution of vertebrate opioid receptors. *Proc. Natl. Acad. Sci. USA* **105,** 15487–15492.

Dunn, J. T., and Dunn, A. D. (2001). Update on intrathyroidal iodine metabolism. *Thyroid* **11,** 407–414.

Eales, J. G. (1979). Thyroid functions in cyclostomes and fishes. *In* "Hormones and Evolution" (Barrington, E. J. W., Ed.), Vol. 1, pp. 341–436. Academic Press, New York.

Eales, J. G., and Brown, S. B. (1993). Measurement and regulation of thyroidal status in teleost fish. *Rev. Fish Biol. Fish.* **3,** 299–347.

Eales, J. G., and Himick, B. A. (1988). The effects of TRH on plasma thyroid hormone levels of rainbow trout (*Salmo gairdneri*) and arctic charr (*Salvelinus alpinus*). *Gen. Comp. Endocrinol.* **72,** 333–339.

Faglia, G. (1998). The clinical impact of the thyrotropin-releasing hormone test. *Thyroid* **8,** 903–908.

Farbridge, K. J., Flett, P. A., and Leatherland, J. F. (1992). Temporal effects of restricted diet and compensatory increased dietary intake on thyroid function, plasma growth hormone levels and tissue lipid reserves of rainbow trout *Oncorhynchus mykiss. Aquaculture* **104,** 157–174.

Filby, A. L., and Tyler, C. R. (2007). Cloning and characterization of cDNAs for hormones and/ or receptors of growth hormone, insulin-like growth factor-I, thyroid hormone, and corticosteroid and the gender-, tissue-, and developmental-specific expression of their mRNA transcripts in fathead minnow (*Pimephales promelas*). *Gen. Comp. Endocrinol.* **150,** 151–163.

Flik, G., Klaren, P. H. M., van den Burg, E. H., Metz, J. R., and Huising, M. O. (2006). CRF and stress in fish. *Gen. Comp. Endocrinol.* **146,** 36–44.

Friesema, E. C. H., Grueters, A., Biebermann, H., Krude, H., von Moers, A., Reeser, M., Barrett, T. G., Mancilla, E. E., Svensson, J., Kester, M. H. A., Kuiper, G. G. J. M., Balkassmi, S., *et al.* (2004). Association between mutations in a thyroid hormone transporter and severe X-linked psychomotor retardation. *Lancet* **364,** 1435–1437.

Friesema, E. C. H., Jansen, J., Jachtenberg, J. W., Visser, W. E., Kester, M. H. A., and Visser, T. J. (2008). Effective cellular uptake and efflux of thyroid hormone by human monocarboxylate transporter 10. *Mol. Endocrinol.* **22,** 1357–1369.

Fryer, J., Lederis, K., and Rivier, J. (1983). Urotensin I, a CRF-like neuropeptide, stimulates ACTH release from the teleost pituitary. *Endocrinology* **113,** 2308–2310.

Fryer, J., Lederis, K., and Rivier, J. (1984). Cortisol inhibits the ACTH-releasing activity of urotensin I, CRF and sauvagine observed with superfused goldfish pituitary cells. *Peptides* **5,** 925–930.

Fryer, J., Lederis, K., and Rivier, J. (1985). ACTH-releasing activity of urotensin I and ovine CRF: Interactions with arginine vasotocin, isotocin and arginine vasopressin. *Regul. Peptides* **11,** 11–15.

Fryer, J. N. (1989). Neuropeptides regulating the activity of goldfish corticotropes and melanotropes. *Fish Physio. Biochem.* **7,** 21–27.

Fryer, J. N., and Lederis, K. (1988). Comparison of actions of posterior pituitary hormones in corticotropin secretion in mammals and fishes. *In* "Recent Progress in Posterior Pituitary Hormones" (Yoshida, S., and Share, L., Eds.), pp. 337–344. Elsevier, Amsterdam.

Fujii, R. (2000). The regulation of motile activity in fish chromatophores. *Pigment Cell Res.* **13,** 300–319.

García-G, C., López-Bojorquez, L., Nunez, J., Valverde-R, C., and Orozco, A. (2007). 3,5-Diiodothyronine *in vivo* maintains euthyroidal expression of type 2 iodothyronine deiodinase, growth hormone, and thyroid hormone receptor $\beta 1$ in the killifish. *Am. J. Physiol. Regul. Integr. Comp. Physiol.* **293,** R877–R883.

Gaylord, T. G., MacKenzie, D. S., and Gatlin, D. M., III (2001). Growth performance, body composition and plasma thyroid hormone status of channel catfish (*Ictalurus punctatus*) in response to short-term feed deprivation and refeeding. *Fish Physiol. Biochem.* **24,** 73–79.

Gesto, M., Soengas, J. L., and Miguez, J. M. (2008). Acute and prolonged stress responses of brain monoaminergic activity and plasma cortisol levels in rainbow trout are modified by PAHs (naphthalene, beta-naphthoflavone and benzo(a)pyrene) treatment. *Aquat. Toxicol.* **86,** 341–351.

Geven, E. J. W., Verkaar, F., Flik, G., and Klaren, P. H. M. (2006). Experimental hyperthyroidism and central mediators of stress axis and thyroid axis activity in common carp (*Cyprinus carpio* L.). *J. Mol. Endocrinol.* **37,** 443–452.

Geven, E. J. W., Nguyen, N.-K., van den Boogaart, M., Spanings, F. A. T., Flik, G., and Klaren, P. H. M. (2007). Comparative thyroidology: Thyroid gland location and iodothyronine dynamics in Mozambique tilapia (*Oreochromis mossambicus* Peters) and common carp (*Cyprinus carpio* L.). *J. Exp. Biol.* **210,** 4005–4015.

Geven, E. J. W., Flik, G., and Klaren, P. H. M. (2009). Central and peripheral integration of interrenal and thyroid axes signals in common carp (*Cyprinus carpio* L.). *J. Endocrinol.* **200,** 117–123.

Gilchriest, B. J., Tipping, D. R., Hake, L., Levy, A., and Baker, B. I. (2000). The effects of acute and chronic stresses on vasotocin gene transcripts in the brain of the rainbow trout (*Oncorhynchus mykiss*). *J. Neuroendocrinol.* **12,** 795–801.

Gilham, I. D., and Baker, B. I. (1985). A black background facilitates the response to stress in teleosts. *J. Endocrinol.* **105,** 99–105.

Goglia, F. (2005). Biological effects of 3,5-diiodothyronine (T_2). *Biochemistry (Mosc.)* **70,** 164–172.

Gonzalez-Nuñez, V., Barrallo, A., Traynor, J. R., and Rodriguez, R. E. (2006). Characterization of opioid-binding sites in zebrafish brain. *J. Pharmacol. Exp. Ther.* **316,** 900–904.

Gorbman, A., and Hyder, M. (1973). Failure of mammalian TRH to stimulate thyroid function in the lungfish. *Gen. Comp. Endocrinol.* **20,** 588–589.

Gorbman, A., Dickhoff, W. W., Vigna, S. R., Clark, N. B., and Ralph, C. L. (1983). *In* "Comparative Endocrinology" John Wiley & Sons, Inc., New York.

Goto-Kazeto, R., Kazeto, Y., and Trant, J. M. (2003). Cloning and seasonal changes in ovarian expression of a TSH receptor in the channel catfish, *Ictalurus punctatus*. *Fish Physiol. Biochem.* **28,** 339–340.

Grau, E. G., and Stetson, M. H. (1977a). The effects of prolactin and TSH on thyroid function in *Fundulus heteroclitus*. *Gen. Comp. Endocrinol.* **33,** 329–335.

Grau, E. G., and Stetson, M. H. (1977b). Pituitary autotransplants in *Fundulus heteroclitus*: Effect on thyroid function. *Gen. Comp. Endocrinol.* **32,** 427–431.

Grau, E. G., II, and Stetson, M. H. (1978). Dopaminergic neurons and TSH release in *Fundulus heteroclitus*. *Am. Zool.* **18,** 651 (conference abstract).

Grau, E. G., Helms, L. M. H., Shimoda, S. K., Ford, C.-A., LeGrand, J., and Yamauchi, K. (1986). The thyroid gland of the Hawaiian parrotfish and its use as an *in vitro* model system. *Gen. Comp. Endocrinol.* **61,** 100–108.

Green, J. A., and Baker, B. I. (1991). The influence of repeated stress on the release of melanin-concentrating hormone in the rainbow trout. *J. Endocrinol.* **428,** 261–266.

Green, J. A., Baker, B. I., and Kawauchi, H. (1991). The effect of rearing rainbow trout on black or white backgrounds on their secretion of melanin-concentrating hormone and their sensitivity to stress. *J. Endocrinol.* **128,** 267–274.

Greenwood, A. K., Butler, P. C., White, R. B., DeMarco, U., Pearce, D., and Fernald, R. D. (2003). Multiple corticosteroid receptors in a teleost fish: Distinct sequences, expression patterns, and transcriptional activities. *Endocrinology* **144,** 4226–4236.

Gröneveld, D., Balm, P. H. M., Martens, G. J. M., and Wendelaar Bonga, S. E. (1995a). Differential melanin-concentrating hormone gene expression in two hypothalamic nuclei of the teleost tilapia in response to environmental changes. *J. Neuroendocrinol.* **7,** 527–533.

Gröneveld, D., Balm, P. H. M., and Wendelaar Bonga, S. E. (1995b). Biphasic effect of MCH on α-MSH release from the tilapia (*Oreochromis mossambicus*) pituitary. *Peptides* **16,** 945–949.
Hagen, I. J., Kusakabe, M., and Young, G. (2006). Effects of ACTH and cAMP on steroidogenic acute regulatory protein and P450 11β-hydroxylase messenger RNAs in rainbow trout interrenal cells: Relationship with *in vitro* cortisol production. *Gen. Comp. Endocrinol.* **145,** 254–262.
Hagenbuch, B. (2007). Cellular entry of thyroid hormones by organic anion transporting polypeptides. *Best Pract. Res. Clin. Endocrinol. Metab.* **21,** 209–221.
Hagenbuch, B., and Meier, P. J. (2004). Organic anion transporting polypeptides of the OATP/*SLC21* family: Phylogenetic classification as OATP/*SLCO* superfamily, new nomenclature and molecular/functional properties. *Pflügers Arch.* **447,** 653–665.
Haider, S. (1975). Pituitary cytology of radiothyroidectomised teleost *Heteropneustes fossilis* (Bloch.). *Endokrinologie* **65,** 300–307.
Haigler, E. D., Jr., Hershman, J. M., and Pittman, J. A., Jr. (1972). Response to orally administered synthetic thyrotropin-releasing hormone in man. *J. Clin. Endocrinol. Metab.* **35,** 631–635.
Haitina, T., Klovins, J., Andersson, J., Fredriksson, R., Lagerström, M. C., Larhammar, D., Larson, E. T., and Schiöth, H. B. (2004). Cloning, tissue distribution, pharmacology and three-dimensional modelling of melanocortin receptors 4 and 5 in rainbow trout suggest close evolutionary relationship of these subtypes. *Biochem. J.* **380,** 475–486.
Han, Y.-S., Liao, I.-C., Tzeng, W.-N., and Yu, J. Y.-L. (2004). Cloning of the cDNA for thyroid stimulating hormone β subunit and changes in activity of pituitary–thyroid axis during silvering of the Japanese eel, *Anguilla japonica. J. Mol. Endocrinol.* **32,** 179–194.
Harvey, C. B., and Williams, G. R. (2002). Mechanism of thyroid hormone action. *Thyroid* **12,** 441–446.
Hedge, G. A., Wright, K. C., and Judd, A. (1981). Factors modulating the secretion of thyrotropin and other hormones of the thyroid axis. *Environ. Health Perspect.* **38,** 57–63.
Hennemann, G. (2005). Notes on the history of cellular uptake and deiodination of thyroid hormone. *Thyroid* **15,** 753–756.
Hennemann, G., Everts, M. E., de Jong, M., Lim, C.-F., Krenning, E. P., and Docter, R. (1998). The significance of plasma membrane transport in the bioavailability of thyroid hormone. *Clin. Endocrinol.* **48,** 1–8.
Herman, J. P., Figueiredo, H., Mueller, N. K., Ulrich-Lai, Y., Ostrander, M. M., Choi, D. C., and Cullinan, W. E. (2003). Central mechanisms of stress integration: Hierarchical circuitry controlling hypothalamo-pituitary-adrenocortical responsiveness. *Front. Neuroendocrinol.* **24,** 151–180.
Higgins, K. M., and Ball, J. N. (1970). Investigations on the hypothalamic control of thyroid-stimulating hormone secretion in the teleost *Poecilia latipinna. J. Endocrinol.* **48,** xxix.
Höglund, E., Balm, P. H. M., and Winberg, S. (2000). Skin darkening, a potential social signal in subordinate arctic charr (*Salvelinus alpinus*): The regulatory role of brain monoamines and pro-opiomelanocortin-derived peptides. *J. Exp. Biol.* **203,** 1711–1721.
Höglund, E., Balm, P. H. M., and Winberg, S. (2002a). Behavioural and neuroendocrine effects of environmental background colour and social interaction in Arctic charr (*Salvelinus alpinus*). *J. Exp. Biol.* **205,** 2535–2543.
Höglund, E., Balm, P. H. M., and Winberg, S. (2002b). Stimulatory and inhibitory effects of 5-HT_{1A} receptors on adrenocorticotropic hormone and cortisol secretion in a teleost fish, the Arctic charr (*Salvelinus alpinus*). *Neurosci. Lett.* **324,** 193–196.
Holland, J. W., Pottinger, T. G., and Secombes, C. J. (2002). Recombinant interleukin-1β activates the hypothalamic-pituitary-interrenal axis in rainbow trout, *Oncorhynchus mykiss. J. Endocrinol.* **175,** 261–267.

Huising, M. O., and Flik, G. (2005). The remarkable conservation of corticotropin-releasing hormone-binding protein (CRH-BP) in the honeybee (*Apis mellifera*) dates the CRH system to a common ancestor of insects and vertebrates. *Endocrinology* **146,** 2165–2170.

Huising, M. O., Metz, J. R., van Schooten, C., Taverne-Thiele, A. J., Hermsen, T., Verburg-van Kemenade, B. M. L., and Flik, G. (2004). Structural characterisation of a cyprinid (*Cyprinus carpio* L.) CRH, CRH-BP and CRH-R1, and the role of these proteins in the acute stress response. *J. Mol. Endocrinol.* **32,** 627–648.

Huising, M. O., Metz, J. R., de Mazon, A. F., Verburg-van Kemenade, B. M. L., and Flik, G. (2005). Regulation of the stress response in early vertebrates. *Ann. N.Y. Acad. Sci.* **1041,** 345–347.

Huising, M. O., van der Aa, L. M., Metz, J. R., de Mazon, A. F., Verburg-van Kemenade, B. M. L., and Flik, G. (2007). Corticotropin-releasing factor (CRF) and CRF-binding protein expression in and release from the head kidney of common carp: Evolutionary conservation of the adrenal CRF system. *J. Endocrinol.* **193,** 349–357.

Inui, Y., Tagawa, M., Miwa, S., and Hirano, T. (1989). Effects of bovine TSH on the tissue thyroxine level and metamorphosis in prometamorphic flounder larvae. *Gen. Comp. Endocrinol.* **74,** 406–410.

Ito, M., Koide, Y., Takamatsu, N., Kawauchi, H., and Shiba, T. (1993). cDNA cloning of the β subunit of teleost thyrotropin. *Proc. Natl. Acad. Sci. USA* **90,** 6052–6055.

Jackson, R. G., and Sage, M. (1973). A comparison of the effects of mammalian TSH on the thyroid glands of the teleost *Galeichthys felis* and the elasmobranch *Dasyatis sabina*. *Comp. Biochem. Physiol. A* **44,** 867–870.

Jansen, J., Friesema, E. C. H., Milici, C., and Visser, T. J. (2005). Thyroid hormone transporters in health and disease. *Thyroid* **15,** 757–768.

Jansen, J., Friesema, E. C. H., Kester, M. H. A., Schwartz, C. E., and Visser, T. J. (2008). Genotype–phenotype relationship in patients with mutations in thyroid hormone transporter MCT8. *Endocrinology* 2184–2190.

Jenks, B. G., Kidane, A. H., Scheenen, W. J. J. M., and Roubos, E. W. (2007). Plasticity in the melanotrope neuroendocrine interface of *Xenopus laevis*. *Neuroendocrinology* **85,** 177–185.

Jensen, J. (2001). Regulatory peptides and control of food intake in non-mammalian vertebrates. *Comp. Biochem. Physiol. A* **128,** 471–479.

Jiang, J. Q., Wang, D. S., Senthilkumaran, B., Kobayashi, T., Kobayashi, H. K., Yamaguchi, A., Ge, W., Young, G., and Nagahama, Y. (2003). Isolation, characterization and expression of 11 β-hydroxysteroid dehydrogenase type 2 cDNAs from the testes of Japanese eel (*Anguilla japonica*) and Nile tilapia (*Oreochromis niloticus*). *J. Mol. Endocrinol.* **31,** 305–315.

Joseph-Bravo, P. (2004). Hypophysiotropic thyrotropin-releasing hormone neurons as transducers of energy homeostasis. *Endocrinology* **145,** 4813–4815.

Kagabu, Y., Mishiba, T., Okino, T., and Yanagisawa, T. (1998). Effects of thyrotropin-releasing hormone and its metabolites, cyclo(His-Pro) and TRH-OH, on growth hormone and prolactin synthesis in primary cultured pituitary cells of the common carp, *Cyprinus carpio*. *Gen. Comp. Endocrinol.* **111,** 395–403.

Kajimura, S., Hirano, T., Visitacion, N., Moriyama, S., Aida, K., and Grau, E. G. (2003). Dual mode of cortisol action on GH/IGF-I/IFG binding proteins in the tilapia, *Oreochromis mossambicus*. *J. Endocrinol.* **178,** 91–99.

Kasper, R. S., Shved, N., Takahashi, A., Reinecke, M., and Eppler, E. (2006). A systematic immunohistochemical survey of the distribution patterns of GH, prolactin, somatolactin, β-TSH, β-FSH, β-LH, ACTH, and α-MSH in the adenohypophysis of *Oreochromis niloticus*, the Nile tilapia. *Cell Tissue Res.* **325,** 303–313.

Kaul, S., and Vollrath, L. (1974). The goldfish pituitary. I. Cytology. *Cell Tissue Res.* **154,** 211–230.

Kawauchi, H. (2006). Functions of melanin-concentrating hormone in fish. *J. Exp. Zool. A* **305,** 751–760.

Kawauchi, H., and Baker, B. I. (2004). Melanin-concentrating hormone signaling systems in fish. *Peptides* **25,** 1577–1584.

Kawauchi, H., and Sower, S. A. (2006). The dawn and evolution of hormones in the adenohypophysis. *Gen. Comp. Endocrinol.* **148,** 3–14.

Kelsall, C. J., and Balment, R. J. (1998). Native urotensins influence cortisol secretion and plasma cortisol concentration in the euryhaline flounder, *Platichthys flesus*. *Gen. Comp. Endocrinol.* **112,** 210–219.

Kiilerich, P., Kristiansen, K., and Madsen, S. S. (2007). Cortisol regulation of ion transporter mRNA in Atlantic salmon gill and the effect of salinity on the signaling pathway. *J. Endocrinol.* **194,** 417–427.

Kitahashi, T., Ogawa, S., Soga, T., Sakuma, Y., and Parhar, I. (2007). Sexual maturation modulates expression of nuclear receptor types in laser-captured single cells of the cichlid (*Oreochromis niloticus*) pituitary. *Endocrinology* **148,** 5822–5830.

Klaren, P. H. M., Haasdijk, R., Metz, J. R., Nitsch, L. M. C., Darras, V. M., Van der Geyten, S., and Flik, G. (2005). Characterization of an iodothyronine 5′-deiodinase in gilthead seabream (*Sparus auratus*) that is inhibited by dithiothreitol. *Endocrinology* **146,** 5621–5630.

Klaren, P. H. M., Geven, E. J. W., and Flik, G. (2007a). The involvement of the thyroid gland in teleost osmoregulation. *In* "Fish Osmoregulation" (Baldisserotto, B., Mancera, J. M., and Kapoor, B. G., Eds.), pp. 35–65. Science Publishers, Inc., Enfield, USA.

Klaren, P. H. M., Guzmán, J. M., Reutelingsperger, S. J., Mancera, J. M., and Flik, G. (2007b). Low salinity acclimation and thyroid hormone metabolizing enzymes in gilthead seabream (*Sparus auratus*). *Gen. Comp. Endocrinol.* **152,** 215–222.

Klovins, J., Haitina, T., Fridmanis, D., Kilianova, Z., Kapa, I., Fredriksson, R., Gallo-Payet, N., and Schiöth, H. B. (2004a). The melanocortin system in fugu: Determination of POMC/AGRP/MCR gene repertoire and synteny, as well as pharmacology and anatomical distribution of the MCRs. *Mol. Biol. Evol.* **21,** 563–579.

Klovins, J., Haitina, T., Ringholm, A., Löwgren, M., Fridmanis, D., Slaidina, M., Stier, S., and Schiöth, H. B. (2004b). Cloning of two melanocortin (MC) receptors in spiny dogfish: MC3 receptor in cartilaginous fish shows high affinity to ACTH-derived peptides while it has lower preference to α-MSH. *Eur. J. Biochem.* **271,** 4320–4331.

Köhrle, J. (1999). Local activation and inactivation of thyroid hormones: The deiodinase family. *Mol. Cell. Endocrinol.* **151,** 103–119.

Köhrle, J. (2002). Iodothyronine deiodinases. *Methods Enzymol.* **347,** 125–167.

Köhrle, J., Brabant, G., and Hesch, R. D. (1987). Metabolism of the thyroid hormones. *Horm. Res.* **26,** 58–78.

König, S., and Neto, V. M. (2002). Thyroid hormone actions on neural cells. *Cell. Mol. Neurobiol.* **22,** 517–544.

Kumar, R. S., Ijiri, S., Kight, K., Swanson, P., Dittman, A., Alok, D., Zohar, Y., and Trant, J. M. (2000). Cloning and functional expression of a thyrotropin receptor from the gonads of a vertebrate (bony fish): Potential thyroid-independent role for thyrotropin in reproduction. *Mol. Cell. Endocrinol.* **167,** 1–9.

Kusakabe, M., Nakamura, I., and Young, G. (2003). 11β-Hydroxysteroid dehydrogenase complementary deoxyribonucleic acid in rainbow trout: Cloning, sites of expression, and seasonal changes in gonads. *Endocrinology* **144,** 2534–2545.

Lai, C.-S., Korytowski, W., Niu, C.-H., and Cheng, S.-Y. (1985). Transverse motion of spin-labeled 3,3′,5-triiodo-L-thyronine in phospholipid bilayers. *Biochem. Biophys. Res. Commun.* **131,** 408–412.

Lamers, A. E., Balm, P. H. M., Haenen, H. E. M. G., Jenks, B. G., and Wendelaar Bonga, S. E. (1991). Regulation of differential release of α-melanocyte-stimulating hormone forms from the pituitary of a teleost fish, *Oreochromis mossambicus. J. Endocrinol.* **129,** 179–187.

Lamers, A. E., Flik, G., Atsma, W., and Wendelaar Bonga, S. E. (1992). A role for di-acetyl α-melanocyte-stimulating hormone in the control of cortisol release in the teleost Oreochromis mossambicus. *J. Endocrinol.* **135,** 285–292.

Lamers, A. E., Flik, G., and Wendelaar Bonga, S. E. (1994). A specific role for TRH in release of diacetyl α-MSH in tilapia stressed by acid water. *Am. J. Physiol.* **267,** R1302–R1308.

Lamers, A. E., ter Brugge, P. J., Flik, G., and Wendelaar Bonga, S. E. (1997). Acid stress induces a D1-like dopamine receptor in pituitary MSH cells of *Oreochromis mossambicus. Am. J. Physiol.* **273,** R387–R392.

Lanni, A., Moreno, M., Lombardi, A., and Goglia, F. (1994). Rapid stimulation *in vitro* of rat liver cytochrome oxidase activity by 3,5-diiodo-L-thyronine and by 3,3′-diiodo-L-thyronine. *Mol. Cell. Endocrinol.* **99,** 89–94.

Larsen, D. A., Dickey, J. T., and Dickhoff, W. W. (1997). Quantification of salmon α- and thyrotropin (TSH) β-subunit messenger RNA by an RNase protection assay: Regulation by thyroid hormones. *Gen. Comp. Endocrinol.* **107,** 98–108.

Larsen, D. A., Swanson, P., Dickey, J. T., Rivier, J., and Dickhoff, W. W. (1998). *In vitro* thyrotropin-releasing activity of corticotropin-releasing hormone-family peptides in coho salmon, *Oncorhynchus kisutch. Gen. Comp. Endocrinol.* **109,** 276–285.

Larson, E. T., O'Malley, D. M., and Melloni, R. H., Jr. (2006). Aggression and vasotocin are associated with dominant-subordinate relationships in zebrafish. *Behav. Brain Res.* **167,** 94–102.

Lazar, M. A. (1993). Thyroid hormone receptors: multiple forms, multiple possibilities. *Endocrine Rev.* **14,** 184–193.

Le Mével, J.-C., Mimassi, N., Lancien, F., Mabin, D., and Conlon, J. M. (2006). Cardiovascular actions of the stress-related neurohormonal peptides, corticotropin-releasing factor and urotensin-I in the trout *Oncorhynchus mykiss. Gen. Comp. Endocrinol.* **146,** 56–61.

Le Mével, J. C., Lancien, F., Mimassi, N., and Conlon, J. M. (2009). Central hyperventilatory action of the stress-related neurohormonal peptides, corticotropin-releasing factor and urotensin-I in the trout *Oncorhynchus mykiss. Gen Comp Endocrinol* doi:10.1016/j.ygen.2009.03.019.

Leatherland, J. F. (1987). Thyroid response to ovine thyrotropin challenge in cortisol- and dexamethasone-treated rainbow trout, *Salmo gairdneri. Comp. Biochem. Physiol. A* **86,** 383–387.

Leatherland, J. F. (1993). Stocking density and cohort sampling effects on endocrine interactions in rainbow trout. *Aquacult. Int.* **1,** 137–156.

Lederis, K., Fryer, J. N., Okawara, Y., Schönrock, C., and Richter, D. (1994). Corticotropin-releasing factors acting on the fish pituitary: Experimental and molecular analysis. *In* "Fish Physiology" (Sherwood, N. M., and Hew, C. L., Eds.), Vol. XIII, pp. 67–100. Academic Press, San Diego.

Leeson, P. D., Ellis, D., Emmett, J. C., Shah, V. P., Showell, G. A., and Underwood, A. H. (1988). Thyroid hormone analogs. Synthesis of 3′-substituted 3,5-diiodo-L-thyronines and quantitative structure-activity studies of *in vitro* and *in vivo* thyromimetic activities in rat liver and heart. *J. Med. Chem.* **31,** 37–54.

Lema, S. C., Dickey, J. T., and Swanson, P. (2008). Molecular cloning and sequence analysis of multiple cDNA variants for thyroid-stimulating hormone β subunit (TSHβ) in the fathead minnow (*Pimephales promelas*). *Gen. Comp. Endocrinol.* **155,** 472–480.

Lethimonier, C., Flouriot, G., Valotaire, Y., Kah, O., and Ducouret, B. (2000). Transcriptional interference between glucocorticoid receptor and estradiol receptor mediates the inhibitory effect of cortisol on fish vitellogenesis. *Biol. Reprod.* **62,** 1763–1771.

Lin, H.-Y., Davis, F. B., Gordinier, J. K., Martino, L. J., and Davis, P. J. (1999). Thyroid hormone induces activation of mitogen-activated protein kinase in cultured cells. *Am. J. Physiol.* **276,** C1014–C1024.

Liu, Q. P., Dou, S. J., Wang, G. E., Li, Z. M., and Feng, Y. (2008). Evolution and functional divergence of monocarboxylate transporter genes in vertebrates. *Gene* **423,** 14–22.

Liu, Y.-W., Lo, L.-J., and Chan, W.-K. (2000). Temporal expression and T3 induction of thyroid hormone receptors $\alpha 1$ and $\beta 1$ during early embryonic and larval development in zebrafish, *Danio rerio*. *Mol. Cell. Endocrinol.* **159,** 187–195.

Logan, D. W., Bryson-Richardson, R. J., Pagan, K. E., Taylor, M. S., Currie, P. D., and Jackson, I. J. (2003). The structure and evolution of the melanocortin and MCH receptors in fish and mammals. *Genomics* **81,** 184–191.

López-Bojorquez, L., Villalobos, P., García-G, C., Orozco, A., and Valverde-R, C. (2007). Functional identification of an osmotic response element (ORE) in the promoter region of the killifish deiodinase 2 gene (*FhDio2*). *J. Exp. Biol.* **210,** 3126–3132.

Lovejoy, D. A., and Balment, R. J. (1999). Evolution and physiology of the corticotropin-releasing factor (CRF) family of neuropeptides in vertebrates. *Gen. Comp. Endocrinol.* **115,** 1–22.

Lu, W., Dow, L., Gumusgoz, S., Brierly, M. J., Warne, J. M., McCrohan, C. R., Balment, R. J., and Riccardi, D. (2004). Coexpression of corticotropin-releasing hormone and urotensin I precursor genes in the caudal neurosecretory system of the Euryhaline flounder (*Platichthys flesus*): A possible shared role in peripheral regulation. *Endocrinology* **145,** 5786–5797.

Mancera, J. M., Vargas-Chacoff, L., Garcia-Lopez, A., Kleszczynska, A., Kalamarz, H., Martinez-Rodriguez, G., and Kulczykowska, E. (2008). High density and food deprivation affect arginine vasotocin, isotocin and melatonin in gilthead sea bream (*Sparus auratus*). *Comp. Biochem. Physiol. A* **149,** 92–97.

Manchado, M., Infante, C., Asensio, E., Planas, J. V., and Cañavate, J. P. (2008). Thyroid hormones down-regulate thyrotropin β subunit and thyroglobulin during metamorphosis in the flatfish Senegalese sole (*Solea senegalensis* Kaup). *Gen. Comp. Endocrinol.* **155,** 447–455.

Marchand, O., Safi, R., Escriva, H., Van Rompaey, E., Prunet, P., and Laudet, V. (2001). Molecular cloning and characterization of thyroid hormone receptors in teleost fish. *J. Mol. Endocrinol.* **26,** 51–65.

Marchand, O., Duffraisse, M., Triqueneaux, G., Safi, R., and Laudet, V. (2004). Molecular cloning and developmental expression patterns of thyroid hormone receptors and T3 target genes in the turbot (*Scophtalmus maximus*) during post-embryonic development. *Gen. Comp. Endocrinol.* **135,** 345–357.

Marchelidon, J., Huet, J. C., Salmon, C., Pernollet, J. C., and Fontaine, Y. A. (1991). Purification and characterization of the putative thyrotropic hormone subunits of a teleost fish, the eel (*Anguilla anguilla*). *C. R. Acad. Sci. Ser. III* **313,** 253–258.

Margolis, R. N. (2008). The nuclear receptor signaling atlas: Catalyzing understanding of thyroid hormone signaling and metabolic control. *Thyroid* **18,** 113–122.

Marley, R., Lu, W., Balment, R. J., and McCrohan, C. R. (2008). Cortisol and prolactin modulation of caudal neurosecretory system activity in the euryhaline flounder *Platichthys flesus*. *Comp. Biochem. Physiol. A Mol. Integr. Physiol.* **151,** 71–77.

Martin, S. A. M., Wallner, W., Youngson, A. F., and Smith, T. (1999). Differential expression of Atlantic salmon thyrotropin β subunit mRNA and its cDNA sequence. *J. Fish Biol.* **54,** 757–766.

Matsuda, K., Shimakura, S., Maruyama, K., Miura, T., Uchiyama, M., Kawauchi, H., Shioda, S., and Takahashi, A. (2006). Central administration of melanin-concentrating hormone (MCH) suppresses food intake, but not locomotor activity, in the goldfish, *Carassius auratus*. *Neurosci. Lett.* **399,** 259–263.

Matsuda, K., Kojima, K., Shimakura, S., Wada, K., Maruyama, K., Uchiyama, M., Kikuyama, S., and Shioda, S. (2008). Corticotropin-releasing hormone mediates α-melanocyte-stimulating hormone-induced anorexigenic action in goldfish. *Peptides* **29,** 1930–1936.

Matsuda, K., Kojima, K., Shimakura, S., and Takahashi, A. (2009). Regulation of food intake by melanin-concentrating hormone in goldfish. *Peptides* doi: 10.1016/j.peptides.2009.02.015.

McCormick, S. D. (2001). Endocrine control of osmoregulation in teleost fish. *Am. Zool.* **41,** 781–794.

McCormick, S. D., Regish, A., O'Dea, M. F., and Shrimpton, J. M. (2008). Are we missing a mineralocorticoid in teleost fish? Effect of cortisol, deoxycorticosterone and aldosterone on osmoregulation, gill Na^+,K^+-ATPase activity and isoform mRNA levels in Atlantic salmon. *Gen. Comp. Endocrinol.* **157,** 35–40.

McCrohan, C. R., Lu, W., Brierley, M. J., Dow, L., and Balment, R. J. (2007). Fish caudal neurosecretory system: A model for the study of neuroendocrine secretion. *Gen. Comp. Endocrinol.* **153,** 243–250.

Melamed, P., Eliahu, N., Levavi-Sivan, B., Ofir, M., Farchi-Pisanty, O., Rentier-Delrue, F., Smal, J., Yaron, Z., and Naor, Z. (1995). Hypothalamic and thyroidal regulation of growth hormone in tilapia. *Gen. Comp. Endocrinol.* **97,** 13–30.

Metz, J. R., Huising, M. O., Meek, J., Taverne-Thiele, A. J., Wendelaar Bonga, S. E., and Flik, G. (2004). Localisation, expression and control of adrenocorticotropic hormone in the nucleus preopticus and pituitary gland of common carp (*Cyprinus carpio* L.). *J. Endocrinol.* **182,** 23–31.

Metz, J. R., Geven, E. J. W., van den Burg, E. H., and Flik, G. (2005). ACTH, α-MSH and control of cortisol release: Cloning, sequencing and functional expression of the melanocortin-2 and melanocortin-5 receptor in *Cyprinus carpio*. *Am. J. Physiol.* **289,** R814–R826.

Metz, J. R., Huising, M. O., Leon, K., Verburg-van Kemenade, B. M. L., and Flik, G. (2006a). Central and peripheral interleukin-1β and interleukin-1 receptor I expression and their role in the acute stress response of common carp, *Cyprinus carpio* L. *J. Endocrinol.* **191,** 25–35.

Metz, J. R., Peters, J. J. M., and Flik, G. (2006b). Molecular biology and physiology of the melanocortin system in fish: A review. *Gen. Comp. Endocrinol.* **148,** 150–162.

Milla, S., Jalabert, B., Rime, H., Prunet, P., and Bobe, J. (2006). Hydration of rainbow trout oocyte during meiotic maturation and *in vitro* regulation by 17,20β-hydroxy-4-pregnen-3-one and cortisol. *J. Exp. Biol.* **209,** 1147–1156.

Milla, S., Terrien, X., Sturm, A., Ibrahim, F., Giton, F., Fiet, J., Prunet, P., and Le Gac, F. (2008). Plasma 11-deoxycorticosterone (DOC) and mineralocorticoid receptor testicular expression during rainbow trout *Oncorhynchus mykiss* spermiation: Implication with 17α,20β-dihydroxyprogesterone on the milt fluidity? *Reprod. Biol. Endocrinol.* **6,** 19.

Mol, K. A., Van der Geyten, S., Darras, V. M., Visser, T. J., and Kühn, E. R. (1997). Characterization of iodothyronine outer ring and inner ring deiodinase activities in the blue tilapia, *Oreochromis aureus*. *Endocrinology* **138,** 1787–1793.

Mommsen, T. P., Vijayan, M. M., and Moon, T. W. (1999). Cortisol in teleosts: Dynamics, mechanisms of action, and metabolic regulation. *Rev. Fish Biol. Fish.* **9,** 211–268.

Moriyama, S., Swanson, P., Larsen, D. A., Miwa, S., Kawauchi, H., and Dickhoff, W. W. (1997). Salmon thyroid-stimulating hormone: Isolation, characterization, and development of a radioimmunoassay. *Gen. Comp. Endocrinol.* **108,** 457–471.

Morley, S. D., Schonrock, C., Richter, D., Okawara, Y., and Lederis, K. (1991). Corticotropin-releasing factor (CRF) gene family in the brain of the teleost fish *Catostomus commersoni* (white sucker): Molecular analysis predicts distinct precursors for two CRFs and one urotensin I peptide. *Mol. Mar. Biol. Biotechnol.* **1,** 48–57.

Mortensen, A. S., and Arukwe, A. (2006). The persistent DDT metabolite, 1,1-dichloro-2,2-bis (*p*-chlorophenyl)ethylene, alters thyroid hormone-dependent genes, hepatic cytochrome P4503A, and pregnane X receptor gene expressions in Atlantic salmon (*Salmo salar*) parr. *Environ. Toxicol. Chem.* **25,** 1607–1615.

Mosconi, G., Gallinelli, A., Polzonetti-Magni, A. M., and Facchinetti, F. (1998). Acetyl salmon endorphin-like and interrenal stress response in male gilthead sea bream, *Sparus aurata. Neuroendocrinology* **68,** 129–134.

Mountjoy, K. G., Robbins, L. S., Mortrud, M. T., and Cone, R. D. (1992). The cloning of a family of genes that encode the melanocortin receptors. *Science* **257,** 1248–1251.

Mustafa, A., and MacKinnon, B. M. (1999). Atlantic salmon, *Salmo salar* L., and Arctic char, *Salvelinus alpinus* (L.): Comparative correlation between iodine-iodide supplementation, thyroid hormone levels, plasma cortisol levels, and infection intensity with the sea louse *Caligus elongatus. Can. J. Zool.* **77,** 1092–1101.

Naito, N., Kawazoe, I., Nakai, Y., Kawauchi, H., and Hirano, T. (1986). Coexistence of immunoreactivity for melanin-concentrating hormone and α-melanocyte-stimulating hormone in the hypothalamus of the rat. *Neurosci. Lett.* **70,** 81–85.

Nelson, E. R., and Habibi, H. R. (2006). Molecular characterization and sex-related seasonal expression of thyroid receptor subtypes in goldfish. *Mol. Cell. Endocrinol.* **253,** 83–95.

Nelson, E. R., and Habibi, H. R. (2009). Thyroid receptor subtypes: Structure and function in fish. *Gen. Comp. Endocrinol.* **161,** 90–96.

Nillni, E. A., and Sevarino, K. A. (1999). The biology of pro-thyrotropin-releasing hormone-derived peptides. *Endocrine Rev.* **20,** 599–648.

Nishioka, R. S., Grau, E. G., Lai, K. V., and Bern, H. A. (1985). Normal and induced development of the thyroid gland of coho salmon. *Aquaculture* **45,** 384–385.

Norris, D. O., and Hobbs, S. L. (2006). The HPA axis and functions of corticosteroids in fishes. *In* "Fish Endocrinology" (Reinecke, M., Zaccone, G., and Kapoor, B. G., Eds.), Vol. 2, pp. 721–765. Science Publishers, Enfield, USA.

Oba, Y., Hirai, T., Yoshiura, Y., Kobayashi, T., and Nagahama, Y. (2000). Cloning, functional characterization, and expression of thyrotropin receptors in the thyroid of amago salmon (*Oncorhynchus rhodurus*). *Biochem. Biophys. Res. Commun.* **276,** 258–263.

Okawara, Y., Ko, D., Morley, S. D., Richter, D., and Lederis, K. P. (1992). *In situ* hybridization of corticotropin-releasing factor-encoding messenger RNA in the hypothalamus of the white sucker, *Catostomus commersoni. Cell Tissue Res.* **267,** 545–549.

Oki, Y., and Sasano, H. (2004). Localization and physiological roles of urocortin. *Peptides* **25,** 1745–1749.

Okimoto, D. K., Tagawa, M., Koide, Y., Grau, E. G., and Hirano, T. (1991). Effects of various adenohypophyseal hormones of chum salmon on thyroxine release *in vitro* in the medaka, *Oryzias latipes. Zool. Sci.* **8,** 567–573.

Oku, H., Ogata, H. Y., and Liang, X. F. (2002). Organization of the lipoprotein lipase gene of red sea bream *Pagrus major. Comp. Biochem. Physiol. B* **131,** 775–785.

Olivereau, M. (1978). Effect of pimozide on the cytology of the eel pituitary. *Cell Tiss. Res.* **189,** 231–239.

Olivereau, M., and Ball, J. N. (1966). Histological study of functional ectopic pituitary transplants in a teleost fish (*Poecilia formosa*). *Proc. R. Soc. Lond. B* **164,** 106–129.

Olivereau, M., and Olivereau, J. (1990a). Effect of pharmacological adrenalectomy on corticotropin-releasing factor-like and arginine vasotocin immunoreactivities in the brain and pituitary of the eel: immunocytochemical study. *Gen. Comp. Endocrinol.* **80,** 199–215.

Olivereau, M., and Olivereau, J. M. (1990b). Corticotropin-like immunoreactivity in the brain and pituitary of three teleost species (goldfish, trout and eel). *Cell Tissue Res.* **262,** 115–123.

Olivereau, M., Olivereau, J., and Lambert, J. (1987). *In vivo* effect of dopamine antagonists on melanocyte-stimulating hormone cells of the goldfish (*Carassius auratus* L.) pituitary. *Gen. Comp. Endocrinol.* **68,** 12–18.

Olivereau, M., Olivereau, J.-M., and Lambert, J.-F. (1988). Cytological responses of the pituitary (rostral pars distalis) and immunoreactive corticotropin-releasing factor (CRF) in the goldfish treated with dopamine antagonists. *Gen. Comp. Endocrinol.* **71,** 506–515.

Omeljaniuk, R. J., Tonon, M., and Peter, R. E. (1989). Dopamine inhibition of gonadotropin and α-melanocyte stimulating hormone release *in vitro* from the pituitary of the goldfish (*Carassius auratus*). *Gen. Comp. Endocrinol.* **74,** 451–467.

Orozco, A., Villalobos, P., and Valverde-R, C. (2002). Environmental salinity selectively modifies the outer-ring deiodinating activity of liver, kidney and gill in the rainbow trout. *Comp. Biochem. Physiol. A* **131,** 387–395.

Ortega, V. A., Renner, K. J., and Bernier, N. J. (2005). Appetite-suppressing effects of ammonia exposure in rainbow trout associated with regional and temporal activation of brain monoaminergic and CRF systems. *J. Exp. Biol.* **208,** 1855–1866.

Øverli, Ø., Harris, C. A., and Winberg, S. (1999). Short-term effects of fights for social dominance and the establishment of dominant-subordinate relationships on brain monoamines and cortisol in rainbow trout. *Brain Behav. Evol.* **54,** 263–275.

Øverli, Ø., Pottinger, T. G., Carrick, T. R., Øverli, E., and Winberg, S. (2001). Brain monoaminergic activity in rainbow trout selected for high and low stress responsiveness. *Brain Behav. Evol.* **57,** 214–224.

Pankhurst, N. W., and Van Der Kraak, G. (2000). Evidence that acute stress inhibits ovarian steroidogenesis in rainbow trout *in vivo*, through the action of cortisol. *Gen. Comp. Endocrinol.* **117,** 225–237.

Patiño, R., Schreck, C. B., and Redding, J. M. (1985). Clearance of plasma corticosteroids during smoltification of coho salmon, *Oncorhynchus kisutch*. *Comp. Biochem. Physiol. A* **82,** 531–535.

Patiño, R., Redding, J. M., and Schreck, C. B. (1987). Interrenal secretion of corticosteroids and plasma cortisol and cortisone concentrations after acute stress and during seawater acclimation in juvenile coho salmon (*Oncorhynchus kisutch*). *Gen. Comp. Endocrinol.* **68,** 431–439.

Pepels, P. P. L. M., Meek, J., Wendelaar Bonga, S. E., and Balm, P. H. M. (2002). Distribution and quantification of corticotropin-releasing hormone (CRH) in the brain of the teleost fish *Oreochromis mossambicus* (Tilapia). *J. Comp. Neurol.* **453,** 247–268.

Pepels, P. P. L. M., van Helvoort, H., Wendelaar Bonga, S. E., and Balm, P. H. M. (2004). Corticotropin-releasing hormone in the teleost stress response: Rapid appearance of the peptide in plasma of tilapia (*Oreochromis mossambicus*). *J. Endocrinol.* **180,** 425–438.

Perrott, M. N., and Balment, R. J. (1990). The renin-angiotensin system and the regulation of plasma cortisol in the flounder, *Platichthys flesus*. *Gen. Comp. Endocrinol.* **78,** 414–420.

Peter, R. E. (1970). Hypothalamic control of thyroid gland activity and gonadal activity in the goldfish, *Carassius auratus*. *Gen. Comp. Endocrinol.* **14,** 334–356.

Peter, R. E. (1971). Feedback effects of thyroxine on the hypothalamus and pituitary of goldfish, *Carassius auratus*. *J. Endocrinol.* **51,** 31–39.

Peter, R. E. (1972). Feedback effects of thyroxine in goldfish *Carassius auratus* with an autotransplanted pituitary. *Neuroendocrinology* **10,** 273–281.

Peter, R. E., and Fryer, J. N. (1984). Endocrine function of the hypothalamus of actinopterygians. *In* "Fish neurobiology" (Davis, R. E., and Northcutt, R. G., Eds.), pp. 165–201. University of Michigan Press, Ann Arbor.

Peter, R. E., and McKeown, B. A. (1975). Hypothalamic control of prolactin and thyrotropin secretion in teleosts, with special reference to recent studies on the goldfish. *Gen. Comp. Endocrinol.* **25,** 153–165.

Peterson, B. C., and Small, B. C. (2005). Effects of exogenous cortisol on the GH/IGF/IGFBP network in channel catfish. *Dom. Anim. Endocrinol.* **28,** 391–404.

Pickering, A. D., Pottinger, T. G., and Sumpter, J. P. (1986). Independence of the pituitary–interrenal axis and melanotroph activity in the brown trout, *Salmo trutta* L., under conditions of environmental stress. *Gen. Comp. Endocrinol.* **64,** 206–211.

Pickering, A. D., Pottinger, T. G., and Sumpter, J. P. (1987). On the use of dexamethasone to block the pituitary–interrenal axis in the brown trout, *Salmo trutta* L. *Gen. Comp. Endocrinol.* **65,** 346–353.

Pickford, G. E., Knight, W. R., Knight, J. N., Gallardo, R., and Baker, B. I. (1981). Long-term effects of hypothalamic lesions on the pituitary and its target organs in the killifish *Fundulus heteroclitus*. 1. Effects on the gonads, thyroid, and growth. *J. Exp. Zool.* **217,** 341–351.

Pierce, A. L., Fukada, H., and Dickhoff, W. W. (2005). Metabolic hormones modulate the effect of growth hormone (GH) on insulin-like growth factor-I (IGF-I) mRNA level in primary culture of salmon hepatocytes. *J. Endocrinol.* **184,** 341–349.

Pierson, P. M., Guibbolini, M. E., and Lahlou, B. (1996). A V1-type receptor for mediating the neurohypophysial hormone-induced ACTH release in trout pituitary. *J. Endocrinol.* **149,** 109–115.

Power, D. M., Melo, J., and Santos, C. R. A. (2000). The effect of food deprivation and refeeding on the liver, thyroid hormones and transthyretin in sea bream. *J. Fish Biol.* **56,** 374–387.

Pradet-Balade, B., Schmitz, M., Salmon, C., Dufour, S., and Querat, B. (1997). Down-regulation of TSH subunit mRNA levels by thyroid hormones in the European eel. *Gen. Comp. Endocrinol.* **108,** 191–198.

Pradet-Balade, B., Burel, C., Dufour, S., Boujard, T., Kaushik, S. J., Querat, B., and Boeuf, G. (1999). Thyroid hormones down-regulate thyrotropin beta mRNA level *in vivo* in the turbot (*Psetta maxima*). *Fish Physiol. Biochem.* **20,** 193–199.

Prunet, P., Sturm, A., and Milla, S. (2006). Multiple corticosteroid receptors in fish: From old ideas to new concepts. *Gen. Comp. Endocrinol.* **147,** 17–23.

Przewlocki, R., and Przewlocka, B. (2001). Opioids in chronic pain. *Eur. J. Pharmacol.* **429,** 79–91.

Qu, D., Ludwig, D. S., Gammeltoft, S., Piper, M., Pelleymounter, M. A., Cullen, M. J., Mathes, W. F., Przypek, J., Kanarek, R., and Maratos-Flier, E. (1996). A role for melanin-concentrating hormone in the central regulation of feeding behaviour. *Nature* **380,** 243–247.

Rabello, M. M., Snyder, P. J., and Utiger, R. D. (1974). Effects on pituitary–thyroid axis and prolactin secretion of single and repetitive oral doses of thyrotropin-releasing hormone (TRH). *J. Clin. Endocrinol. Metab.* **39,** 571–578.

Rapoport, B., Chazenbalk, G. D., Jaume, J. C., and McLachlan, S. M. (1998). The thyrotropin (TSH) receptor: interaction with TSH and autoantibodies. *Endocrine Rev.* **19,** 673–716.

Reid, S. G., Vijayan, M. M., and Perry, S. F. (1996). Modulation of catecholamine storage and release by the pituitary–interrenal axis in the rainbow trout (*Oncorhynchus mykiss*). *J. Comp. Physiol. B* **165,** 665–676.

René, F., Muller, A., Jover, E., Kieffer, B., Koch, B., and Loeffler, J.-P. (1998). Melanocortin receptors and δ-opioid receptor mediate opposite signalling actions of POMC-derived peptides in CATH.a cells. *Eur. J. Neurosci.* **10,** 1885–1894.

Richardson, J., Lundegaard, P. R., Reynolds, N. L., Dorin, J. R., Porteous, D. J., Jackson, I. J., and Patton, E. E. (2008). mc1r Pathway regulation of zebrafish melanosome dispersion. *Zebrafish* **5,** 289–295.

Ringholm, A., Fredriksson, R., Poliakova, N., Yan, Y., Postlethwait, J. H., Larhammar, D., and Schiöth, H. B. (2002). One melanocortin 4 and two melanocortin 5 receptors from zebrafish show remarkable conservation in structure and pharmacology. *J. Neurochem.* **82,** 6–18.

Rocha, A., Gómez, A., Galay-Burgos, M., Zanuy, S., Sweeney, G. E., and Carrillo, M. (2007). Molecular characterization and seasonal changes in gonadal expression of a thyrotropin receptor in the European sea bass. *Gen. Comp. Endocrinol.* **152,** 89–101.

Rodriguez, R. E., Barrallo, A., Garcia-Malvar, F., McFadyen, I. J., Gonzalez-Sarmiento, R., and Traynor, J. R. (2000). Characterization of ZFOR1, a putative delta-opioid receptor from the teleost zebrafish (*Danio rerio*). *Neurosci. Lett.* **288,** 207–210.

Rotllant, J., Arends, R. J., Mancera, J. M., Flik, G., Wendelaar Bonga, S. E., and Tort, L. (2000a). Inhibition of HPI axis response to stress in gilthead sea bream (*Sparus aurata*) with physiological plasma levels of cortisol. *Fish Physiol. Biochem.* **23,** 13–22.

Rotllant, J., Balm, P. H. M., Ruane, N. M., Perez-Sanchez, J., Wendelaar Bonga, S. E., and Tort, L. (2000b). Pituitary proopiomelanocortin-derived peptides and hypothalamus–pituitary–interrenal axis activity in gilthead sea bream (*Sparus aurata*) during prolonged crowding stress: Differential regulation of adrenocorticotropin hormone and α-melanocyte-stimulating hormone release by corticotropin-releasing hormone and thyrotropin-releasing hormone. *Gen. Comp. Endocrinol.* **119,** 152–163.

Rotllant, J., Balm, P. H. M., Perez-Sanchez, J., Wendelaar Bonga, S. E., and Tort, L. (2001). Pituitary and interrenal function in gilthead sea bream (*Sparus aurata* L., Teleostei) after handling and confinement stress. *Gen. Comp. Endocrinol.* **121,** 333–342.

Rousseau, K., Le Belle, N., Marchelidon, J., and Dufour, S. (1999). Evidence that corticotropin-releasing hormone acts as a growth hormone-releasing factor in a primitive teleost, the European eel (*Anguilla anguilla*). *J. Neuroendocrinol.* **11,** 385–392.

Ruane, N. M., Wendelaar Bonga, S. E., and Balm, P. H. M. (1999). Differences between rainbow trout and brown trout in the regulation of the pituitary–interrenal axis and physiological performance during confinement. *Gen. Comp. Endocrinol.* **115,** 210–219.

Rudman, D., Hollins, B. M., Kutner, M. H., Moffitt, S. D., and Lynn, M. J. (1983). Three types of α-melanocyte-stimulating hormone: Bioactivities and half-lives. *Am. J. Physiol.* **245,** E47–54.

Saavedra, J. M., and Benicky, J. (2007). Brain and peripheral angiotensin II play a major role in stress. *Stress* **10,** 185–193.

Salmon, C., Marchelidon, J., Fontaine, Y. A., Huet, J. C., and Querat, B. (1993). Cloning and sequence of thyrotropin beta subunit of a teleost fish: The eel (*Anguilla anguilla* L.). *C. R. Acad. Sci. Ser. III* **316,** 749–753.

Sanchez-Simon, F. M., and Rodriguez, R. E. (2008). Developmental expression and distribution of opioid receptors in zebrafish. *Neuroscience* **151,** 129–137.

Scanlan, T. S., Suchland, K. L., Hart, M. E., Chiellini, G., Huang, Y., Kruzich, P. J., Frascarelli, S., Crossley, D. A., Bunzow, J. R., Ronca-Testoni, S., Lin, E. T., Hatton, D., *et al.* (2004). 3-Iodothyronamine is an endogenous and rapid-acting derivative of thyroid hormone. *Nature Med.* **10,** 638–642.

Schaaf, M. J. M., Champagne, D., van Laanen, I. H. C., van Wijk, D. C. W. A., Meijer, A. H., Meijer, O. C., Spaink, H. P., and Richardson, M. K. (2008). Discovery of a functional glucocorticoid receptor β-isoform in zebrafish. *Endocrinology* **149,** 1591–1599.

Schiöth, H. B., Haitina, T., Ling, M. K., Ringholm, A., Fredriksson, R., Cerdá-Reverter, J. M., and Klovins, J. (2005). Evolutionary conservation of the structural, pharmacological, and genomic characteristics of the melanocortin receptor subtypes. *Peptides* **26,** 1886–1900.

Schjolden, J., Schiöth, H. B., Larhammar, D., Winberg, S., and Larson, E. T. (2009). Melanocortin peptides affect the motivation to feed in rainbow trout (*Oncorhynchus mykiss*). *Gen. Comp. Endocrinol.* **160,** 134–138.

Schreck, C. B., Bradford, C. S., Fitzpatrick, M. S., and Patiño, R. (1989). Regulation of the interrenal of fishes: Non-classical control mechanisms. *Fish Physiol. Biochem.* **7,** 259–265.

Seasholtz, A. F., Valverde, R. A., and Denver, R. J. (2002). Corticotropin-releasing hormone-binding protein: Biochemistry and function from fishes to mammals. *J. Endocrinol.* **175,** 89–97.

Selz, Y., Braasch, I., Hoffmann, C., Schmidt, C., Schultheis, C., Schartl, M., and Volff, J. N. (2007). Evolution of melanocortin receptors in teleost fish: The melanocortin type 1 receptor. *Gene* **401,** 114–122.

Shepard, J. D., Barron, K. W., and Myers, D. A. (2000). Corticosterone delivery to the amygdala increases corticotropin-releasing factor mRNA in the central amygdaloid nucleus and anxiety-like behavior. *Brain Res.* **861,** 288–295.

Sheridan, M. A. (1986). Effects of thyroxin, cortisol, growth hormone, and prolactin on lipid metabolism of coho salmon, *Oncorhynchus kisutch*, during smoltification. *Gen. Comp. Endocrinol.* **64,** 220–238.

Shimakura, S., Miura, T., Maruyama, K., Nakamachi, T., Uchiyama, M., Kageyama, H., Shioda, S., Takahashi, A., and Matsuda, K. (2008). Alpha-melanocyte-stimulating hormone mediates melanin-concentrating hormone-induced anorexigenic action in goldfish. *Horm. Behav.* **53,** 323–328.

Shrimpton, J. M., and McCormick, S. D. (1999). Responsiveness of gill Na^+/K^+-ATPase to cortisol is related to gill corticosteroid receptor concentration in juvenile rainbow trout. *J. Exp. Biol.* **202,** 987–995.

Shrimpton, J. M., and Randall, D. J. (1994). Downregulation of corticosteroid receptors in gills of coho salmon due to stress and cortisol treatment. *Am. J. Physiol.* **267,** R432–R438.

Silva, J. E., Dick, T. E., and Larsen, P. R. (1978). The contribution of local tissue thyroxine monodeiodination to the nuclear 3,5,3′-triiodothyronine in pituitary, liver, and kidney of euthyroid rats. *Endocrinology* **103,** 1196–1207.

Singh, A. K., and Singh, T. P. (1977). Thyroid activity and TSH levels in pituitary gland and blood serum in response to clomid, sexovid and prostaglandins treatment in *Heteropneustes fossilis* (Bloch). *Endokrinologie* **70,** 69–76.

Singh, R., and Rai, U. (2008). β-Endorphin regulates diverse functions of splenic phagocytes through different opioid receptors in freshwater fish *Channa punctatus* (Bloch): An *in vitro* study. *Dev. Comp. Immunol.* **32,** 330–338.

Sohn, Y. C., Yoshiura, Y., Suetake, H., Kobayashi, M., and Aida, K. (1999). Isolation and characterization of the goldfish thyrotropin β subunit gene including the 5′-flanking region. *Gen. Comp. Endocrinol.* **115,** 463–473.

Specker, J. L., and Kobuke, L. (1985). Thyroid physiology of juvenile coho salmon. *Aquaculture* **45,** 389–390.

Specker, J. L., and Richman, N. H., 3rd (1984). Environmental salinity and the thyroidal response to thyrotropin in juvenile coho salmon (*Oncorhynchus kisutch*). *J. Exp. Zool.* **230,** 329–333.

Staub, J. J., Girard, J., Mueller-Brand, J., Noelpp, B., Werner-Zodrow, I., Baur, U., Heitz, P., and Gemsenjaeger, E. (1978). Blunting of TSH response after repeated oral administration of TRH in normal and hypothyroid subjects. *J. Clin. Endocrinol. Metab.* **46,** 260–266.

Sternberg, H., and Moav, B. (1999). Regulation of the growth hormone gene by fish thyroid retinoid receptors. *Fish Physiol. Biochem.* **20,** 331–339.

Stolte, E. H., Verburg-van Kemenade, B. M. L., Savelkoul, H. F. J., and Flik, G. (2006). Evolution of glucocorticoid receptors with different glucocorticoid sensitivity. *J. Endocrinol.* **190,** 17–28.

Stolte, E. H., de Mazon, A. F., Leon-Koosterziel, K. M., Jesiak, M., Bury, N. R., Sturm, A., Savelkoul, H. F. J., van Kemenade, B. M. L., and Flik, G. (2008). Corticosteroid receptors involved in stress regulation in common carp, *Cyprinus carpio*. *J. Endocrinol.* **198,** 403–417.

Strand, F. L. (1999). New vistas for melanocortins. Finally, an explanation for their pleiotropic functions. *Ann. N. Y. Acad. Sci.* **897,** 1–16.
Sturm, A., Bury, N., Dengreville, L., Fagart, J., Flouriot, G., Rafestin-Oblin, M. E., and Prunet, P. (2005). 11-Deoxycorticosterone is a potent agonist of the rainbow trout (*Oncorhynchus mykiss*) mineralocorticoid receptor. *Endocrinology* **146,** 47–55.
Sumpter, J. P., Pickering, A. D., and Pottinger, T. G. (1985). Stress-induced elevation of plasma α-MSH and endorphin in brown trout, *Salmo trutta* L. *Gen. Comp. Endocrinol.* **59,** 257–265.
Sumpter, J. P., Dye, H. M., and Benfey, T. J. (1986). The effects of stress on plasma ACTH, α-MSH, and cortisol levels in salmonid fishes. *Gen. Comp. Endocrinol.* **62,** 377–385.
Swanson, P., Grau, E. G., Helms, L. M., and Dickhoff, W. W. (1988). Thyrotropic activity of salmon pituitary glycoprotein hormones in the Hawaiian parrotfish thyroid *in vitro*. *J. Exp. Zool.* **245,** 194–199.
Tagawa, M., Hagiwara, H., Takemura, A., Hirose, S., and Hirano, T. (1997). Partial cloning of the hormone-binding domain of the cortisol receptor in Tilapia, *Oreochromis mossambicus*, and changes in the mRNA levels during embryonic development. *Gen. Comp. Endocrinol.* **108,** 132–140.
Takahashi, H., Sakamoto, T., Hyodo, S., Shepherd, B. S., Kaneko, T., and Grau, E. G. (2006). Expression of glucocorticoid receptor in the intestine of a euryhaline teleost, the Mozambique tilapia (*Oreochromis mossambicus*): Effect of seawater exposure and cortisol treatment. *Life Sci.* **78,** 2329–2335.
Tanaka, S. (2003). Comparative aspects of intracellular proteolytic processing of peptide hormone precursors: Studies of proopiomelanocortin processing. *Zool. Sci.* **20,** 1183–1198.
Tang, X., Liu, X., Zhang, Y., Zhu, P., and Lin, H. (2008). Molecular cloning, tissue distribution and expression profiles of thyroid hormone receptors during embryogenesis in orange-spotted grouper (*Epinephelus coioides*). *Gen. Comp. Endocrinol.* **159,** 117–124.
Tasker, J. G., Di, S., and Malcher-Lopes, R. (2006). Minireview: Rapid glucocorticoid signaling via membrane-associated receptors. *Endocrinology* **147,** 5549–5556.
Tata, J. R. (2002). Signalling through nuclear receptors. *Nature Rev. Mol. Cell Biol.* **3,** 702–710.
Teitsma, C. A., Bailhache, T., Tujague, M., Balment, R. J., Ducouret, B., and Kah, O. (1997). Distribution and expression of glucocorticoid receptor mRNA in the forebrain of the rainbow trout. *Neuroendocrinology* **66,** 294–304.
Teitsma, C. A., Anglade, I., Toutirais, G., Muñoz-Cueto, J.-A., Saligaut, D., Ducouret, B., and Kah, O. (1998). Immunohistochemical localization of glucocorticoid receptors in the forebrain of the rainbow trout (*Oncorhynchus mykiss*). *J. Comp. Neurol.* **401,** 395–410.
Thornton, J. W. (2001). Evolution of vertebrate steroid receptors from an ancestral estrogen receptor by ligand exploitation and serial genome expansions. *Proc. Natl. Acad. Sci. USA* **98,** 5671–5676.
Todd, K. J., and Eales, J. G. (2002). The effect of handling and blood removal on plasma levels and hepatic deiodination of thyroid hormones in adult male and female rainbow trout, *Oncorhynchus mykiss*. *Can. J. Zool.* **80,** 372–375.
Tran, T. N., Fryer, J. N., Bennett, H. P., Tonon, M. C., and Vaudry, H. (1989). TRH stimulates the release of POMC-derived peptides from goldfish melanotropes. *Peptides* **10,** 835–841.
Tran, T. N., Fryer, J. N., Lederis, K., and Vaudry, H. (1990). CRF, urotensin I, and sauvagine stimulate the release of POMC-derived peptides from goldfish neurointermediate lobe cells. *Gen. Comp. Endocrinol.* **78,** 351–360.
Tsuneki, K., and Fernholm, B. (1975). Effect of thyrotropin-releasing hormone on the thyroid of a teleost, *Chasmichthys dolichognathus*, and a hagfish, *Eptatretus burgeri*. *Acta Zool.* **56,** 61–65.
Vacher, C., Pellegrini, E., Anglade, I., Ferriére, F., Saligaut, C., and Kah, O. (2003). Distribution of dopamine D_2 receptor mRNAs in the brain and the pituitary of female rainbow trout: An in situ hybridization study. *J. Comp. Neurol.* **458,** 32–45.

Vallarino, M., Delbende, C., Bunel, D. T., Ottonello, I., and Vaudry, H. (1989). Proopiomelanocortin (POMC)-related peptides in the brain of the rainbow trout, *Salmo gairdneri*. *Peptides* **10,** 1223–1230.

van den Burg, E. H., Metz, J. R., Arends, R. J., Devreese, B., Vandenberghe, I., Van Beeumen, J., Wendelaar Bonga, S. E., and Flik, G. (2001). Identification of β-endorphins in the pituitary gland and blood plasma of the common carp (*Cyprinus carpio*). *J. Endocrinol.* **169,** 271–280.

van den Burg, E. H., Metz, J. R., Ross, H. A., Darras, V. M., Wendelaar Bonga, S. E., and Flik, G. (2003). Temperature-induced changes in thyrotropin-releasing hormone sensitivity in carp melanotropes. *Neuroendocrinology* **77,** 15–23.

van den Burg, E. H., Metz, J. R., Spanings, F. A. T., Wendelaar Bonga, S. E., and Flik, G. (2005). Plasma α-MSH and acetylated β-endorphin levels following stress vary according to CRH sensitivity of the pituitary melanotropes in common carp, *Cyprinus carpio*. *Gen. Comp. Endocrinol.* **140,** 210–221.

Van der Geyten, S., Mol, K. A., Pluymers, W., Kühn, E. R., and Darras, V. M. (1998). Changes in plasma T_3 during fasting/refeeding in tilapia (*Oreochromis niloticus*) are mainly regulated through changes in hepatic type II iodothyronine deiodinase. *Fish Physiol. Biochem.* **19,** 135–143.

van der Heide, S. M., Joosten, B. J. L. J., Dragt, B. S., Everts, M. E., and Klaren, P. H. M. (2007). A physiological role for glucuronidated thyroid hormones: Preferential uptake by H9c2(2-1) myotubes. *Mol. Cell. Endocrinol.* **264,** 109–117.

van der Salm, A. L., Pavlidis, M., Flik, G., and Wendelaar Bonga, S. E. (2004). Differential release of α-melanophore stimulating hormone isoforms by the pituitary gland of red porgy, *Pagrus pagrus*. *Gen. Comp. Endocrinol.* **135,** 126–133.

van der Salm, A. L., Metz, J. R., Wendelaar Bonga, S. E., and Flik, G. (2005). Alpha-MSH, the melanocortin-1 receptor and background adaptation in the Mozambique tilapia, *Oreochromis mossambicus*. *Gen. Comp. Endocrinol.* **144,** 140–149.

van Enckevort, F. H. C., Pepels, P. P. L. M., Leunissen, J. A. M., Martens, G. J. M., Wendelaar Bonga, S. E., and Balm, P. H. M. (2000). *Oreochromis mossambicus* (tilapia) corticotropin-releasing hormone: cDNA sequence and bioactivity. *J. Neuroendocrinol.* **12,** 177–186.

van Tijn, D. A., de Vijlder, J. J. M., and Vulsma, T. (2007). Role of the thyrotropin-releasing hormone stimulation test in diagnosis of congenital central hypothyroidism in infants. *J. Clin. Endocrinol. Metab.* **93,** 410–419.

Vazzana, M., Vizzini, A., Salerno, G., Di Bella, M. L., Celi, M., and Parrinello, N. (2008). Expression of a glucocorticoid receptor (DlGR1) in several tissues of the teleost fish *Dicentrarchus labrax*. *Tissue Cell* **40,** 89–94.

Vijayan, M. M., and Leatherland, J. F. (1992). *In vivo* effects of the steroid analogue RU486 on some aspects of intermediary and thyroid metabolism of brook charr, *Salvelinus fontinalis*. *J. Exp. Zool.* **263,** 265–271.

Vijayan, M. M., Raptis, S., and Sathiyaa, R. (2003). Cortisol treatment affects glucocorticoid receptor and glucocorticoid-responsive genes in the liver of rainbow trout. *Gen. Comp. Endocrinol.* **132,** 256–263.

Vischer, H. F., and Bogerd, J. (2003). Cloning and functional characterization of a testicular TSH receptor cDNA from the African catfish (*Clarias gariepinus*). *J. Mol. Endocrinol.* **30,** 227–238.

Visser, T. J. (1994a). Role of sulfation in thyroid hormone metabolism. *Chem.-Biol. Interact.* **92,** 293–303.

Visser, T. J. (1994b). Sulfation and glucuronidation pathways of thyroid hormone metabolism. *In* "Thyroid Hormone Metabolism: Molecular Biology and Alternate Pathways" (Wu, S. Y., and Visser, T. J., Eds.), pp. 85–117. CRC Press, Boca Raton.

Walpita, C. N., Grommen, S. V. H., Darras, V. M., and Van der Geyten, S. (2007). The influence of stress on thyroid hormone production and peripheral deiodination in the Nile tilapia (*Oreochromis niloticus*). *Gen. Comp. Endocrinol.* **150,** 18–25.

Waring, C. P., and Brown, J. A. (1997). Plasma and tissue thyroxine and triiodothyronine contents in sublethally stressed, aluminum-exposed brown trout (*Salmo trutta*). *Gen. Comp. Endocrinol.* **106,** 120–126.

Watanuki, H., Gushiken, Y., Takahashi, A., Yasuda, A., and Sakai, M. (2000). *In vitro* modulation of fish phagocytic cells by β-endorphin. *Fish Shellfish Immunol.* **10,** 203–212.

Weld, M. M., and Fryer, J. N. (1987). Stimulation by angiotensins I and II of ACTH release from goldfish pituitary cell columns. *Gen. Comp. Endocrinol.* **68,** 19–27.

Weld, M. M., and Fryer, J. N. (1988). Angiotensin II stimulation of teleost adrenocorticotropic hormone release: Interactions with urotensin I and corticotropin-releasing factor. *Gen. Comp. Endocrinol.* **69,** 335–340.

Weld, M. M., Fryer, J. N., Rivier, J., and Lederis, K. (1987). Inhibition of CRF- and urotensin I-stimulated ACTH release from goldfish pituitary cell columns by the CRF analogue α-helical CRF(9-41). *Regul. Peptides* **19,** 273–280.

Wendelaar Bonga, S. E. (1997). The stress response in fish. *Physiol. Rev.* **77,** 591–625.

Westphal, N. J., and Seasholtz, A. F. (2006). CRH-BP: The regulation and function of a phylogenetically conserved binding protein. *Front. Biosci.* **11,** 1878–1891.

Westphal, N. J., Evans, R. T., and Seasholtz, A. F. (2009). Novel expression of type 1 corticotropin-releasing hormone receptor in multiple endocrine cell types in the murine anterior pituitary. *Endocrinology* **150,** 260–267.

Westring, C. G., Ando, H., Kitahashi, T., Bhandari, R. K., Ueda, H., Urano, A., Dores, R. M., Sher, A. A., and Danielson, P. B. (2008). Seasonal changes in CRF-I and urotensin I transcript levels in masu salmon: Correlation with cortisol secretion during spawning. *Gen. Comp. Endocrinol.* **155,** 126–140.

Wildmeister, W., and Horster, F. A. (1971). Die Wirkung von synthetischem Thyrotropin Releasing Hormone auf die Entwicklung eines experimentellen Exophthalmos beim Goldfish. *Acta Endocrinol.* **68,** 363–366.

Wilson, D. S., Clark, A. B., Coleman, K., and Dearstyne, T. (1994). Shyness and boldness in humans and other animals. *Trends Ecol. Evol.* **9,** 442–446.

Winberg, S., and Lepage, O. (1998). Elevation of brain 5-HT activity, POMC expression, and plasma cortisol in socially subordinate rainbow trout. *Am. J. Physiol. Regul.* **274,** R645–R654.

Winberg, S., Nilsson, A., Hylland, P., Soderstöm, V., and Nilsson, G. E. (1997). Serotonin as a regulator of hypothalamic–pituitary–interrenal activity in teleost fish. *Neurosci. Lett.* **230,** 113–116.

Wolff, N. A., Werner, A., Burkhardt, S., and Burkhardt, G. (1997). Expression cloning and characterization of a renal organic anion transporter from winter flounder. *FEBS Lett.* **417,** 287–291.

Wu, S.-y., Green, W. L., Huang, W.-s., Hays, M. T., and Chopra, I. J. (2005). Alternate pathways of thyroid hormone metabolism. *Thyroid* **15,** 943–958.

Yada, T., Azuma, T., Takahashi, A., Suzuki, Y., and Hirose, S. (2000). Effects of desacetyl-α-MSH on lipid mobilization in the rainbow trout, *Oncorhynchus mykiss*. *Zool. Sci.* **17,** 1123–1127.

Yada, T., Moriyama, S., Suzuki, Y., Azuma, T., Takahashi, A., Hirose, S., and Naito, N. (2002). Relationships between obesity and metabolic hormones in the "cobalt" variant of rainbow trout. *Gen. Comp. Endocrinol.* **128,** 36–43.

Yamada, C., Noji, S., Shioda, S., Nakai, Y., and Kobayashi, H. (1990). Intragranular colocalization of arginine vasopressin- and angiotensin II-like immunoreactivity in the hypothalamo-neurohypophysial system of the goldfish, *Carassius auratus*. *Zool. Sci.* **7,** 257–263.

Yamano, K., and Inui, Y. (1995). cDNA cloning of thyroid hormone receptor β for the Japanese flounder. *Gen. Comp. Endocrinol.* **99,** 197–203.

Yamano, K., and Miwa, S. (1998). Differential gene expression of thyroid hormone receptor α and β in fish development. *Gen. Comp. Endocrinol.* **109,** 75–85.

Yamano, K., Araki, K., Sekikawa, K., and Inui, Y. (1994). Cloning of thyroid hormone receptor genes expressed in metamorphosing flounder. *Dev. Genet.* **15,** 378–382.

Yao, M., and Denver, R. J. (2007). Regulation of vertebrate corticotropin-releasing factor genes. *Gen. Comp. Endocrinol.* **153,** 200–216.

Yen, P. M. (2001). Physiological and molecular basis of thyroid hormone action. *Physiol. Rev.* **81,** 1097–1142.

Yoshiura, Y., Sohn, Y. C., Munakata, A., Kobayashi, M., and Aida, K. (1999). Molecular cloning of the cDNA encoding the β subunit of thyrotropin and regulation of its gene expression by thyroid hormones in the goldfish, *Carassius auratus*. *Fish Physiol. Biochem.* **21,** 201–210.

Young, G., and Lin, R. J. (1988). Response of the interrenal to adrenocorticotropic hormone after short-term thyroxine treatment of coho salmon (*Oncorhynchus kisutch*). *J. Exp. Zool.* **245,** 53–58.

Yulis, C. R., Lederis, K., Wong, K.-L., and Fisher, A. W. F. (1986). Localization of urotensin I- and corticotropin-releasing factor-like immunoreactivity in the central nervous system of *Catostomus commersoni*. *Peptides* **7,** 79–86.

Zhou, A., Bloomquist, B. T., and Mains, R. E. (1993). The prohormone convertases PC1 and PC2 mediate distinct endoproteolytic cleavages in a strict temporal order during proopiomelanocortin biosynthetic processing. *J. Biol. Chem.* **268,** 1763–1769.

Zupanc, G. K. H., Horschke, I., and Lovejoy, D. A. (1999). Corticotropin-releasing factor in the brain of the gymnotiform fish, *Apteronotus leptorhynchus*: Immunohistochemical studies combined with neuronal tract tracing. *Gen. Comp. Endocrinol.* **114,** 349–364.

7

NEUROENDOCRINE–IMMUNE INTERACTIONS IN TELEOST FISH

B. M. LIDY VERBURG-VAN KEMENADE
ELLEN H. STOLTE
JURIAAN R. METZ
MAGDALENA CHADZINSKA

1. Neuroendocrine–Immune Interaction
   1.1. Introduction to Neuroendocrine–Immune Interaction
   1.2. Immunity in Fishes
   1.3. Cytokine Families and Phylogenetic Relationship between Cytokines and Hormones
2. Immune Modulation by Neuroendocrine Factors
   2.1. Sympathetic Innervation
   2.2. Hypothalamus–Pituitary–Interrenal (HPI) Axis
   2.3. Opioids
   2.4. Members of the Type I Cytokine Family: Growth Hormone, Prolactin, Leptin
   2.5. The Hypothalamic–Pituitary–Thyroid (HPT) Axis
   2.6. The Brain–Pituitary–Gonadal (BPG) Axis
3. Neuroendocrine Modulation by Cytokines
   3.1. A Stress Response Follows the Immune Response
   3.2. From Periphery to Brain
   3.3. How Are Peripheral Immune Signals Conveyed to the Brain?
   3.4. Cytokine Production in the Brain
   3.5. Cytokine-induced Activation of Hypothalamo–Pituitary–Interrenal Axis
4. Conclusions and Future Perspectives

1. NEUROENDOCRINE–IMMUNE INTERACTION

Physical, chemical and biological disturbances evoke an incredible repertoire of physiological, endocrinological and immunological responses. It is now recognized that the neuroendocrine and immune systems interact in a bi-directional fashion. In this way the status of pathogen recognition is

Fish Neuroendocrinology: Volume 28
FISH PHYSIOLOGY

DOI: 10.1016/S1546-5098(09)28007-1

communicated to the brain and the immune response is influenced by physiological changes. This explicit communication consequently needs a common language of signaling molecules and receptors. The network includes corticosteroids, classical pituitary hormones, cytokines and neuropeptides, as well as neural pathways. Understanding the basic biological significance of this dialogue may provide therapeutic benefit for treatment of pathologies. This review focuses on the pathways, receptors and mechanisms involved in teleost fish which form an excellent model to reveal phylogenetically old and original mechanisms of stress physiology and immunology.

1.1. Introduction to Neuroendocrine–Immune Interaction

Maintenance of a balanced internal milieu is based on a dynamic equilibrium of bi-directional physiological processes. This equilibrium is constantly threatened by actions of intrinsic and extrinsic stimuli or stressors of physical, chemical or biological nature, such as pathogens.

Some twenty years ago, Blalock proposed that the immune system served a sensory role, a "sixth sense" to detect pathogens and tumors that the body otherwise could not hear, feel, smell or taste (Blalock, 1984). To launch an appropriate neuroendocrine response, this pathogen detection needs to be signaled to the nervous system. Reciprocally it was found that stress (Fast *et al.*, 2008) and subordinate hierarchy (Faisal *et al.*, 1989) cause decreased immune functions in fish, implying that signals from the endocrine system elicit effects on the immune system (Ader *et al.*, 1995). In fact, some signals appear to be variants of an evolutionary conserved communication system.

Fish are in intimate contact with their aqueous environment that can harbor high numbers of pathogens. The extant teleost fishes are among the evolutionary oldest vertebrates and are equipped with a well-developed and effective immune system. An elaborate communication system of cell–cell contacts and humoral factors (cytokines) exists to communicate pathogen recognition and to coordinate appropriate response measures of the different types of leukocytes. Cytokines are polypeptides or glycoproteins that generally have a very low constitutive expression, their production is transient and actions are usually local (auto- and paracrine rather than endocrine). And whereas hormones are usually produced by a limited number of cell types and directed to restricted target cells to elicit specific responses, cytokines are produced by different cell types, elicit multiple effects in different target cells (pleiotropy) and may show functional redundancy (Vilcek, 2003). Interestingly, many signaling molecules referred to as either cytokine or hormone belong to the same molecular family of structurally related proteins, the members of which have evolved from a common ancestor. Moreover peptide hormones appear to be more pleiotropic than originally hypothesized. The common use of intracellular pathways, like the Janus kinase (JAK), and signal inducer and activator of transcription (STAT) pathway, by hormones and cytokines may explain the potential redundancy in their actions.

The need for a balanced communication in optimal stress coping and wound healing is illustrated when keeping in mind that although a powerful inflammatory cytokine response is essential to overcome a bacterial infection, this may cut as a double-edged sword as a too strong inflammatory response may lead to destruction of host tissues. Moreover, stress (pathogen challenge is a crucial extrinsic stressor) induced energy redistribution as a result of enhanced cortisol activity may result in suppression of (immune) functions that are of second order importance when survival is at stake (Wendelaar Bonga, 1997). Also the fact that immune responses may be related to smolting (Maule *et al.*, 1987), temperature or season (Nakanishi, 1986), gender or circadian rhythm (Nevid and Meier, 1995) underlines the physiological importance of interaction.

The worldwide growing demand for protein has resulted in intensive aquaculture practices that have rapidly expanded over the last few decades. Unfavorable conditions that come with intensive fish farming include high rearing densities and unnatural housing and social situations, frequent handling (e.g., during vaccination procedures), transport, overfeeding and suboptimal water quality (e.g., variable temperatures, waste products). Such conditions will interfere with proper immune function (Engelsma *et al.*, 2003; Terova *et al.*, 2005) and facilitate rapid pathogen expansion. Prevention and/or effective vaccination and treatment of infectious diseases require an understanding of the elaborate bi-directional communication between the neuroendocrine and immune system.

1.2. Immunity in Fishes

Teleost fish have developed an effective protective immune system. First, they produce a biochemical barrier to prevent pathogen invasion, viz. the integumental mucus layer, which contains many anti-bacterial peptides: lysozyme, lectins and proteases (Palaksha *et al.*, 2008). Second, when pathogens pass this barrier, an array of soluble and cellular defense mechanisms is activated. Fish are the earliest vertebrates that have developed both arms of the immune system reflected by the innate and the adaptive immune response. The innate immune system functions in "generic" recognition of pathogens and prevention of pathogen dispersal, whereas the adaptive immune system specifically recognizes pathogens on the basis of specific surface antigens and clears an infection via production of specific antibodies and cytotoxic lymphocytes. Importantly, the adaptive immune system will generate a memory. Third, the combined immune responses are regulated to be rapidly terminated after pathogen clearance to prevent collateral damage to the host.

1.2.1. Immune Organs and Communication

Fish lack bone marrow as a primary immune organ, but their hematopoietic cells reside in the head kidney and give rise to both myeloid and lymphoid immune cells (van Muiswinkel, 1995). Fish lack lymph nodes as secondary immune organs. The spleen and the head kidney are sites for interaction of the immune system with antigens and harbor the antibody-producing lymphocytes. The thymus, as in mammals, acts as a center of T-lymphocyte maturation. The mucosal immune system associated with epithelial surfaces of the gut, gills and skin forms the first line of defense and the connective tissues associated are densely populated with immune cells to attack any penetrating infectious agent.

As mentioned above, immune cells communicate via cytokines and associated receptors and this is of critical importance for a fast and effective attack on pathogens. Minute amounts of cytokines can generate strong inflammatory responses, and tight control over this production is required to prevent cell damage. Dependent on the stage of infection and function of particular cell types, immune cells produce different types of cytokines. Among these cytokines we distinguish interleukins (ILs), substances that communicate between leukocytes, and chemokines, an acronym for chemoattractant cytokines which supervise cell migration (chemotaxis). Cytokines can elicit a broad range of actions on cell growth and differentiation. The kinetics of cytokine expression after detection of the pathogen ensures effective clearing of the pathogen whilst minimizing the damage to the host.

1.2.2. Innate Immune Response; Cell Types and Pathogen Recognition

The principal cell types of the teleost fish innate immune system are the myeloid cells, neutrophilic granulocytes and macrophages (from a monocyte precursor), cells that all have phagocytosis as a major function. These cells migrate to a site of infection/inflammation and can kill bacteria and infected cells (Gomez and Balcazar, 2008). Killing of infected cells can also be performed by non-specific cytotoxic cells (NCC), the fish equivalent of the mammalian natural killer (NK) cells (Fischer *et al.*, 2006), or by production of bactericidal or lytic proteins/enzymes (Magnadottir, 2006).

Pathogens have particular characteristics important for virulence, which are usually not expressed in the host. These characteristics are collectively called pathogen-associated molecular patterns (PAMPs) such as fungal β1,3-glucan, viral double stranded RNA or bacterial cell wall products such as lipopolysaccharide (LPS). They are recognized by pathogen recognition receptors (PRR). PRRs exist as soluble, humoral variants, like the complement protein C3, or expressed as membrane receptors on cells of the immune system, such as many of the Toll-like receptors (TLRs) (Magnadottir, 2006). In mammals particular TLRs recognize a separate group of infectious agents (e.g., TLR1 and 2 principally recognize lipoprotein and peptidoglycans from gram-positive bacteria, TLR3 recognizes double stranded RNA from viruses, TLR4 recognizes LPS of gram-negative bacteria, etc.; Purcell *et al.*, 2006). Analysis of the zebrafish genomic database reveals orthologues for most TLRs expressed in mammals.

When activated, these recognition molecules induce a number of responses directed towards killing of the pathogen and infected cells, prevention of pathogen dispersal, and communication of type and severity of infection via release of cytokines and pro-inflammatory molecules. Killing of pathogens is increased by opsonization (i.e., a process by which a pathogen is marked for ingestion and destruction by phagocytic cells) and subsequent phagocytosis of the pathogen. Moreover, microbicidal reactive oxygen species (ROS) are produced by phagocytic cells. Additionally, natural cytotoxic cells are stimulated to assist in killing of infected cells and the complement system is activated to attack pathogen membranes (Magnadottir, 2006). Moreover, the acute phase response is activated, which comprises a number of plasma proteins [e.g., serum amyloid A (SAA), transferrins, lysozyme, and alpha-2-macroglobulin (α2M)] and limits dispersal of pathogens but also initiates tissue repair (Magnadottir, 2006).

TLR activation will induce pro-inflammatory cytokine expression via intracellular pathways that are presumed to be analogous to those described in mammalian systems. Intriguingly, due to TLR specificity the resulting cytokine profile will communicate the type of infection (Purcell *et al.*, 2006). For instance, a viral attack will induce the production of type I interferon-α and -β that will "warn" other cells to increase their antiviral defenses and stimulate cytotoxic cell activity to lyse infected cells and prevent further multiplication of the virus (Robertsen, 2006).

1.2.3. Innate Immune Response; Cytokine Signaling

As with the TLR genes, for many of the cytokines produced in mammals, an orthologous fish gene has been demonstrated (Figure 7.1). Despite the very low sequence homologies found, essential motifs and three-dimensional structures are conserved. In the vertebrate immune system, pro-inflammatory cytokines including tumor necrosis factor alpha (TNF-α), interleukin 1 beta (IL-1β) and interleukin 6 (IL-6) induce the acute phase response and chemokine release (Roitt *et al.*, 2006). Next, interleukin 12 (IL-12) is released, which in turn stimulates the release of type II interferon: interferon gamma (IFN-γ). In comparison to mammals, the fish pro-inflammatory response is characterized by a first wave of TNF-α -and IL-1β expression followed by chemokine expression and the peak of IL-12 expression (Chadzinska *et al.*, 2008). *TNF-α* is constitutively produced in head kidney and gills and can be induced by LPS stimulation of head kidney macrophages. It mediates resistance to infections by controlling intracellular pathogen replication and induces cell

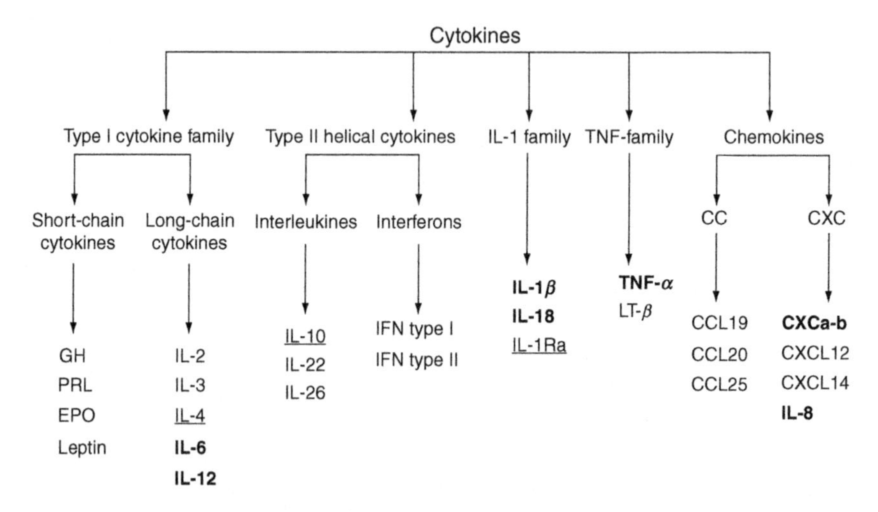

Fig. 7.1. Cytokine family members identified in several teleost fish species (see references in text). Pro-inflammatory cytokines are in bold, anti-inflammatory underlined. GH, growth hormone; PRL, prolactin; EPO, erythropoietin; IL, interleukin; IFN, interferon; IL-1Ra, interleukin 1 receptor antagonist; TNF, tumor necrosis factor; LT, lymphotoxin.

proliferation after an immune stimulus. TNF-α is also involved in the production of microbicidal nitric oxide (NO), presumably by regulating inducible nitric oxide synthase (iNOS) expression (Saeij *et al.*, 2003a). Macrophage-secreted *IL-1β* stimulates thymocyte (T-lymphocyte) proliferation, initiates the acute phase response and activates macrophages and T-lymphocytes (Buonocore *et al.*, 2004). *IL-12* expression can be induced in head kidney macrophages after bacterial and viral infections and is suggested to increase the cytolytic properties of NCC and T-lymphocytes to ensure clearing of virus-infected cells (Huising *et al.*, 2006c; Nascimento *et al.*, 2007; Forlenza *et al.*, 2008). *IFN-γ* is expressed by natural cytotoxic cells and T-lymphocytes and activates macrophages to increase microbicidal activity and stimulates antigen presentation (Zou *et al.*, 2005). Chemokines are expressed by all classical immune organs such as head kidney, spleen, thymus and kidney, and are involved in directing leukocytes to the site of inflammation (Huising *et al.*, 2003c).

1.2.4. Adaptive Immune Response: Cell Types, Response and Memory

The adaptive immune response provides the vertebrate immune system with the ability to recognize and remember specific pathogens and to mount a stronger and faster response in a subsequent encounter with the same pathogen. To mount such an immune response, a specific pathogen or antigen needs to be recognized. The major histocompatibility complex (MHC) proteins act as "signposts" and display fragmented pieces of an antigen on the cell surface of antigen-presenting cells (Stet *et al.*, 2003). The T-cell receptor (TCR) is a molecule found on the surface of T-lymphocytes (T-cells) that is, in general, responsible for recognition of antigens bound to MHC molecules. CD4 is a co-receptor that assists TCR to activate its T-cell following an interaction with an antigen-presenting cell (Roitt *et al.*, 2006). Antigen presentation is mediated by macrophages (Martin *et al.*, 2007) and B-lymphocytes produce antibodies and are characterized by expression of immunoglobulin at their cell surface (Magnadottir *et al.*, 2005). Mammals display a division of T-lymphocyte functions, executed by two different types of T-lymphocytes, helper T-lymphocytes (Th) that support B-lymphocyte function and cytotoxic T-lymphocytes (Tc) that will attack and kill virus-infected cells. Th-cells are characterized by co-receptor CD4 that binds MHC class II, whereas Tc-cells are characterized by CD8 that binds MHC I. Such a division of functions is not fully established in fish, although fish leukocytes do display cytotoxic activity (reviewed by Fischer *et al.*, 2006). Moreover, teleostean T-lymphocytes express CD4 (Suetake *et al.*, 2004), CD8 (Hansen and Strassburger, 2000) and MHC class I (Stet *et al.*, 2003).

Recognition of a particular antigen (e.g., small pathogenic fragments) presented on the MHC of antigen-presenting cells will lead to clonal expansion of specific B- and T-lymphocytes. B-lymphocytes will secrete specific antibodies (humoral immunity). These antibodies will attach to the pathogen which results in increased opsonization and phagocytosis of the pathogens (Sanmartin *et al.*, 2008). A first antigen encounter will induce long-lived memory cells that retain the capacity to be stimulated by the antigen for a faster and more vigorous secondary response (Kaattari *et al.*, 2002). It should be kept in mind that teleost fish have different immunoglobulin (Ig) isotypes compared to mammals. The main Ig found after an immune challenge is a tetrameric IgM molecule. Additionally an IgD molecule and Ig heavy chain variations such as IgZ (*Danio rerio* immunoglobulin), IgT (tau; for teleostean), and the chimera IgM–IgZ have also been described (Hansen *et al.*, 2005; Savan *et al.*, 2005). The function of these various Ig molecules in fish is under current investigation.

Antigen-specific T-lymphocytes will either kill infected host cells or assist antibody production of the antigen-specific B-lymphocytes (by release of cytokines such as IFN-γ) (Cain *et al.*, 2002). This cellular immunity is also more efficient upon a second stimulation as was shown by faster rejection of second-set grafts compared to first-set grafts (Grether *et al.*, 2004).

Although this adaptive immune response is rapid and strong upon a second encounter with the same pathogen, in ectotherms such as fishes, depending on the ambient temperature, it takes days to weeks to be fully effective after a first encounter (Rijkers *et al.*, 1980; van Muiswinkel, 1995). Therefore the innate immune system is of crucial importance. Without a proper initial killing and prevention of dispersion of the pathogen, the host may succumb to the infection long before an adaptive immune response has been mounted (van Muiswinkel, 1995; Raida and Buchmann, 2007).

1.2.5. Termination of the Immune Response

Production of microbicidal proteins, reactive oxygen species, and high circulating numbers of cytotoxic cells are potentially harmful to the host and may even become fatal. Therefore, the anti-inflammatory cytokines IL-10 and TGF beta (TGF-β) are produced to inhibit excessive activation of the immune response and initiate processes of wound healing, tissue remodeling and recovery (Figure 7.1). The peak of IL-10 expression occurs during the late phase of an inflammatory response (Pinto *et al.*, 2007) and correlates with inhibited expression of the pro-inflammatory cytokine IL-1β and chemokines (Chadzinska *et al.*, 2008; Seppola *et al.*, 2008). Moreover TGF-β, which also increases at the late stage of infection (Tafalla *et al.*, 2005), down-regulates the NO response of TNF-α-activated macrophages (Haddad *et al.*, 2008). Apoptosis is required to maintain a balance in dynamic cell populations and can initially be inhibited to increase the life-span and effectiveness of phagocytic cells, but needs to swiftly increase to terminate the initial task (Elenkov and Chrousos, 2006).

1.2.6. Prerequisites for Interaction with the Neuroendocrine System

The widespread presence of hormones and receptors in/on immune cells is hypothesized to create the delicate balance between immune stimulation and suppression and contribute to neuroendocrine functioning. For the neuroendocrine system to reciprocally receive signals from the immune system endocrine cells need specific cytokine receptors and cytokines need to be produced close to endocrine cells, or should be released into the circulation.

Direct (para) sympathetic innervation of the teleost immune system is possible as catecholamine receptors of the sympatho-adrenomedullary system are present on immune cells (Roy and Rai, 2008). Hormone receptors were found on/in immune cells of higher and lower vertebrates, or their presence was corroborated by a functional response after hormone stimulation (see Section 2). Cytokines that affect neuroendocrine functions are mostly produced by immune-related cells in the brain, e.g., astrocytes or glia cells (Kelley *et al.*, 2003; Churchill *et al.*, 2006; Garden and Moller, 2006). When produced by peripheral immune organs, cytokines have to be able to pass the circumventricular organ (CVO) (see Section 3).

In addition to endocrine effects, paracrine and autocrine effects within the immune system are possible if immune cells are capable of producing hormones. Traditional hypothalamic and pituitary hormones are modestly but widely expressed in mammalian immune cells (Elenkov and Chrousos, 2006 and refs therein). This phenomenon is evolutionarily old. In teleost fish, corticotropin-releasing factor (CRF) immunoreactivity was found in macrophage-like cells of the gill and skin (Mazon *et al.*, 2006). The pro-opiomelanocortin (POMC)-derived peptides adrenocorticotropin (ACTH), β-endorphin and alpha-melanocyte-stimulating hormone (α-MSH) were demonstrated in the thymus of goldfish (Ottaviani *et al.*, 1995). In leukocytes, growth hormone and prolactin expression has been confirmed (Yada, 2007). Other classical hormones such as leptin I and II were found in the thymus and spleen (Huising *et al.*, 2006a).

The teleost head kidney is an immune and endocrine organ: the hematopoietic tissue is situated adjacent to the endocrine tissue, creating the exemplary location for neuroendocrine–immune interaction. The chromaffin cells of the head kidney produce catecholamines, whereas the interrenal cells produce cortisol, making the head kidney the functional analogue of the mammalian adrenal gland. At the same time, both lymphoid (B- and T-lymphocytes) and myeloid (phagocytic) cells are produced in the head kidney hematopoietic tissue, making it the functional analogue of the mammalian bone marrow (van Muiswinkel, 1995).

1.3. Cytokine Families and Phylogenetic Relationship between Cytokines and Hormones

Although convenient to define a particular field of research, the separation of hormones and cytokines seems arbitrary as cytokines are also released into the circulation and signal in an autocrine, paracrine and even endocrine fashion. With the recent availability of whole genome databases of several teleost and various other vertebrate species it is now possible to analyze phylogenetic relationships of gene families. These analyses show that many signaling molecules referred to as cytokines, hormones or growth factors belong to the same family of structurally related proteins. All members of such a family probably have evolved from one common ancestor through a series of successive duplication events, although the level of sequence conservation varies greatly (Huising *et al.*, 2006b). The type I helical cytokine family and also the chemokines form interesting examples of this co-evolution and will therefore be discussed in more detail in this chapter.

1.3.1. Classification of Cytokines and their Receptors

Many cytokines and cytokine receptors have been grouped into families (Figure 7.1). Cytokine families may have a high degree of sequence homology, but the tertiary structure and presence of distinct family motifs are also important to achieve this classification. The most important cytokine families are a) the type I cytokine family, b) the type II helical cytokines, and c) the IL-1 family.

Among the families of cytokines, the type I cytokine family of ligands and receptors is the most extensive. The type I cytokine receptor family, with a conserved WSXWS motif in the extracellular domain, comprises receptors for many crucial hematopoietic cytokines (e.g., IL-2, IL-3, IL-4, IL-5, IL-6, IL-7, IL-12 and IL-13), hematopoietic growth factors [e.g., colony-stimulating factors (CSFs), and erythropoietin], growth factors [e.g., growth hormone (GH) and prolactin (PRL)], and the satiety factor leptin. This family of ligands (with four tightly packed α-helices) and receptors comprises an interesting palette of "endocrine" and "immune" signals that have probably co-evolved from the same ancestral precursor molecule (Boulay *et al.*, 2003; Huising *et al.*, 2006b). Their role in neuroendocrine–immune interaction will be elaborated in Section 2.4.

Type II helical cytokines (HCII) such as IL-10, IL-26, IL-22, and interferons (both type I and type II) are homodimers composed of two cytokine molecules that contain six α-helices each. The evolution of HCII and their receptors suggests that diversification of the ancestral mechanism necessary for host defense against infections probably followed different pathways in tetrapods and teleost fish (Lutfalla *et al.*, 2003).

The IL-1 family, which comprises pro-inflammatory cytokines such as IL-1β and IL-18, but also receptor antagonists, is characterized by a three-dimensional structure with a fold rich in beta strands. Few IL-1 family members and their receptors have unambiguous orthologues in fish. However, phylogenetic analyses support the concept of shared signaling mechanisms in the neuroendocrine and immune systems (Huising *et al.*, 2004b).

1.3.2. Chemokines and Chemokine Receptors: An Example of Co-evolution of Immune and Neuroendocrine Signals

Chemokines are important mediators to regulate leukocyte trafficking by chemotaxis. Different leukocytes display chemotaxis to different types of chemotactic agents for various reasons. Chemokines form a superfamily of small, structurally and functionally related cytokines. Based on their pattern of conserved cysteine residues near the NH2 terminus, four classes of chemokines are distinguished: CXC, CC, CX3C and C chemokines (Laing and Secombes, 2004). Interestingly the CC chemokines of teleost fish are characterized by an enormous redundancy. While 30 CC chemokines were characterized in mammals, over 100 different CC chemokines with different expression patterns could be identified in teleost fish (DeVries *et al.*, 2006; Peatman and Liu, 2007). As such, it is not possible to discuss all chemokines and their known actions in teleost immunity. Since chemotaxis by chemokines is a very old phenomenon and CXC chemokines form a beautiful example of the co-evolution of neuroendocrine and immune signaling molecules, it is these which will be discussed below.

CXC chemokines are characterized by the first two cysteine residues being separated by one intervening amino acid. CXC chemokines are further classified according to the presence of the tri-peptide motif ELR. This motif is found directly preceding the CXC motif. Compared to bony fish, mammals have an elaborate CXC chemokine family. There are 16 CXC chemokines identified in mammals to date, systematically named CXCL1–CXCL16. The ELR motif, an evolutionarily recent phenomenon exclusive to mammals, is present in CXCL1–CXCL8. Collectively, these chemokines are implicated in the chemoattraction of polymorphonuclear neutrophils. The ELR^{-}CXC chemokines are implicated in numerous different processes, mostly concerning lymphocyte chemoattraction.

Genomic data mining and molecular analyses revealed five teleost CXC cytokines. Three of these (CXCa–c) do not resemble any of the mammalian CXC chemokine clusters. Spacing of cysteine residues as well as gene structure of CXCa and CXCb are typical for CXC chemokines and they are functionally analogous to some of the mammalian CXC chemokines (Huising *et al.*, 2003c). In contrast, the other two fish CXC chemokines are evolutionarily old and orthologous to mammalian CXCL12 and CXCL14. Interestingly they are prominently expressed in the brain (Huising *et al.*, 2004b,c). The good conservation and the expression patterns of both CXCL12 and its receptor CXCR4 (and also CXCL14) suggest that they are the modern descendants of the primordial CXC chemokine/receptor pair and also suggest an important function for this ligand/receptor pair. This is confirmed by the fact that knock-out mice for either CXCL12 or CXCR4 display an embryonically lethal phenotype (Ma *et al.*, 1998). The disruption of CXCL12/CXCR4 signaling however causes a chemotactic defect in a subset of cerebellar neurons not in leukocytes. Reconstruction of the genesis of the CXC chemokine family in a cladogram, as inferred from structural and functional features, led us to hypothesize that CXC chemokines originated in the brain and not in the immune system and that the CXC chemokine family was recruited to perform chemotactic immune functions only after the vertebrate immune system began to take shape as a specialized organized system (Huising *et al.*, 2003b).

CXC chemokine receptors belong to the large family of 7-helix G-protein coupled receptors. In mammals, there are six CXC chemokine receptors identified (CXCR1–CXCR6). A key feature of CXC chemokines and their receptors is their promiscuous interaction, meaning that several chemokines share one receptor and that most receptors are able to recognize multiple chemokines. Moreover, as will be discussed later, G-protein coupled chemokine receptors may heterodomerize with similar receptors for neuropeptides, e.g., opioid receptors (Suzuki *et al.*, 2002). Bony fish contain homologues to CXCR4 and CXCR1/2, but not to the other mammalian CXCRs (Huising *et al.*, 2003b). Judged by the evolutionary conservation and the relatively short branch lengths in the phylogenetic tree throughout the cluster, CXCR4 is the oldest CXC chemokine receptor.

2. IMMUNE MODULATION BY NEUROENDOCRINE FACTORS

For decades the effect of stress on the immune system has been widely investigated in human studies and animal models (Elenkov and Chrousos, 2006). Although acute stress can have beneficial effects, chronic stress was found to inhibit an optimal immune response in both mammals and teleost fish (Weyts *et al.*, 1998b; Viveros-Paredes *et al.*, 2006; Edwards *et al.*, 2007; Fast *et al.*, 2008).

Teleost fish are especially intriguing animals in which to study the cooperation between the immune and neuroendocrine systems because the head kidney combines immune and endocrine functions. This organization suggests that the end-products of the brain–sympathetic–chromaffin cell axis, adrenaline (AD) and noradrenaline (NA), and of the hypothalamic–pituitary–interrenal axis, cortisol, have direct paracrine access to the cells of the immune system and vice versa. Therefore, chemical signals from the immune system could also exert direct paracrine action on the chromaffin and interrenal cells.

2.1. Sympathetic Innervation

Mammalian lymphoid organs appear to be innervated by both parasympathetic and sympathetic nerve fibers, which have been speculated to respectively stimulate and inhibit an immune response. Moreover, mammalian leukocytes express most of the adrenergic and cholinergic components found in the nervous system, including muscarinic and nicotinic acetylcholine (ACh) receptors (Kawashima and Fujii, 2003). As there is still very little information available on whether piscine immune organs and cells possess ACh receptors or are directly innervated by cholinergic fibers, we will focus on the communication between the immune and adrenergic systems.

2.1.1. Catecholaminergic Innervation and Catecholamine Receptors in the Immune System

The initial descriptions of adrenergic innervation of the thymus and spleen in mice were subsequently extended to include other species, as well as organs like lymph nodes, bone marrow and gut (Felten *et al.*, 1985).

In coho salmon (*Oncorhynchus kisutch*), for example, the spleen is richly innervated by adrenergic neurons. While this innervation enters the spleen and remains largely associated with the splenic vasculature, fibers can also be observed in spleen parenchyma (Flory, 1989). There are two types of receptors that bind catecholamines: alpha (α-AR) and beta (β-AR) adrenergic receptors. Based on binding affinities they may be further divided into several α and β subtypes. In mammals innate immune cells express both α- and β-ARs, while the β2-AR is the predominant AR expressed on T- and B-lymphocytes, except for murine Th2 cells that lack expression of any subtype (Nance and Sanders, 2007). Experiments with specific radioligands confirm the expression of β-ARs on goldfish (*Carassius auratus*) head kidney, spleen and peritoneal leukocytes, but the rank order of ligand potency does not allow classification into any of the known mammalian subtypes (Jozefowski and Plytycz, 1998). Similarly, the presence of β-AR was demonstrated in membranes isolated from head kidney and spleen leukocytes of channel catfish (*Ictalurus punctatus*) (Finkenbine *et al.*, 2002). Recent sequencing and characterization of teleost β-ARs in different tissues (Dugan *et al.*, 2003; Fabbri *et al.*, 2008) will facilitate further identification of these on immune cells or tissues and will enable a study towards their regulation and interaction with other G-protein coupled receptors (Heijnen, 2007).

2.1.2. Effects of Catecholamines on Innate and Adaptive Immunity

In vitro studies with specific β-AR agonists and antagonists showed significant inhibitory effects on innate and adaptive immune parameters in several species of teleost fish, while α-AR agonists showed predominantly stimulation of respiratory burst and antibody production (Table 7.1). The differential effects measured with AD and NA therefore seem to be generated via different receptor profiles.

These *in vitro* effects were corroborated *in vivo.* In tilapia (*Oreochromis aureus*), Chen *et al.* (2002) showed that cold stress-induced changes in catecholamines and cortisol depressed phagocytic activity of leukocytes and plasma immunoglobulin M (IgM) level, while *in vitro,* a combination of cortisol and isoproterenol had an additive effect in reducing phagocytosis. Finally, propranolol (a β-AR antagonist) prolonged scale allograft survival in gulf killifish *(Fundulus grandis)* (Nevid and Meier, 1995).

2.2. Hypothalamus–Pituitary–Interrenal (HPI) Axis

Release of the stress hormone cortisol is under control of the HPI or stress axis which is the functional analogue of the mammalian hypothalamus–pituitary–adrenal (HPA) axis (see Chapter 6 for details). When stress signals are perceived, the hypothalamic region of the nucleus

Table 7.1
In vitro effects of adrenergic receptor agonists on immune parameters in fish

Agonist	Effect	Species	Reference
NA	Reduction of macrophage phagocytosis	Spotted murrel	Roy and Rai, 2008
	Stimulation of respiratory burst in macrophages		
AD	Reduction of macrophage phagocytosis	Spotted murrel	Roy and Rai, 2008
	Stimulation of respiratory burst in macrophages		
	Reduction of respiratory burst in phagocytes	Rainbow trout	Flory and Bayne, 1991
Isoproterenol	Reduction of respiratory burst in phagocytes	Rainbow trout	Flory and Bayne, 1991
	Reduction of leukocyte proliferation after stimulation with LPS, ConA, PHA		
	Reduction of leukocyte phagocytosis		Nanaware *et al.*, 1994
	Reduction of antibody response		Flory, 1990
Phenylephrine	Reduction of leukocyte phagocytosis	Rainbow trout	Nanaware *et al.*, 1994
	Stimulation of respiratory burst in phagocytes		Flory and Bayne, 1991
	Stimulation of antibody response		Flory, 1990

NA, noradrenaline; AD adrenaline; LPS, lipopolysaccharide; ConA, concavalin A; PHA, phytohaemagglutinin.

pre-opticus in the brain responds by releasing CRF into the pituitary. This CRF signal is received by the CRF receptor subtype 1 (CRF_1) of the corticotropes in the pars distalis of the pituitary. CRF binding protein (CRF-BP) regulates jointly with CRF the release of ACTH into the circulation (Huising *et al.*, 2004a; Metz *et al.*, 2004). ACTH subsequently induces cortisol release by the interrenal cells of the head kidney into the circulation. The importance of corticosteroids for the mammalian immune balance is well described (Elenkov and Chrousos, 2006). Examples include the mostly anti-inflammatory actions of corticosteroids and the fact that adrenalectomy or treatment with the glucocorticoid antagonist RU468 prevents the thymic atrophy that occurs during a *Trypanosoma* infection (Perez *et al.*, 2007).

2.2.1. Cortisol Receptor Types and their Presence in Immune Cells

After passing through the cell membrane, cortisol is bound by an intracellular receptor, the glucocorticoid receptor (GR). For this binding to occur, two heat shock proteins (Hsp), Hsp70 and Hsp90, are essential chaperones (Pratt and Toft, 1997). This hormone–receptor complex then translocates into the nucleus where it binds to a glucocorticoid responsive element (GRE) to activate or repress transcription of effector genes (Stolte *et al.*, 2006). Interestingly, GR is not the only receptor capable of binding cortisol, nor is it present as a single type of receptor. As in mammals, both the GR and the mineralocorticoid receptor (MR) are capable of binding cortisol (Bridgham *et al.*, 2006). Additionally, in contrast to mammals, fish have duplicate GR genes (GR1 and GR2) that both transcribe into functional proteins (Stolte *et al.*, 2006). This is likely a result of an early genome duplication, before the radiation of the teleosts, but after the divergence of tetrapods from the fish lineage. GR1 also exists in two alternative splice variants (GR1a and GR1b), leading to the translation of an extra exon, and therefore of an extra nine amino acids in the DNA binding region of GR1a (Ducouret *et al.*, 1995; Stolte *et al.*, 2008c). Thus there are four receptors capable of binding cortisol in fish: GR1a, GR1b, GR2, and MR. However, their ability to induce activation of downstream genes is dependent on the cortisol concentration. In both rainbow trout and carp the GR2 can be activated at low cortisol concentrations (present in non-stressed animals) whereas the GR1 receptor is only sensitive to high levels of cortisol as observed in stressed animals (Stolte *et al.*, 2008b).

In mammals, receptors for cortisol have been described in the cells of the immune system and these play a role in determining the final immune response (Heijnen, 2007). Similarly, in fish, cortisol receptors have been described in immune cells, that affect the immune response (Maule and Schreck, 1991). In common carp (*Cyprinus carpio*), all four cortisol receptors are widely expressed throughout the body, but in the cells and tissues of the immune system the MR is hardly expressed. Carp lymphocytes show a higher relative expression of the "sensitive" GR2 (Stolte *et al.*, 2008c). As in other organs, expression and functional properties of these receptors can be modulated by cortisol stimulation. In rainbow trout, binding affinity of GR was decreased after acute stress in splenic leukocytes, although numbers of GR were increased (Maule and Schreck, 1991). In contrast, seawater acclimation of rainbow trout, which increased cortisol levels, decreased both peripheral blood and head kidney leukocyte GR1, GR2 and MR expression (Yada *et al.*, 2008). Interestingly, GR expression is also affected by immune stimulation. Injection of the endotoxin LPS (cell wall component of gram-negative bacteria) increased expression of GR in the spleen of gilthead sea bream (Acerete *et al.*, 2007). Similarly in common carp, GR1 expression in peritoneal leukocytes was increased after 1 and 2 days of zymosan (yeast cell wall component)-induced peritonitis, whereas GR2 expression was unaffected (Stolte *et al.*, 2008a). *In vitro* stimulation of head kidney phagocytes with LPS quickly and transiently induced expression of GR1, whereas GR2 expression was reduced, which implies an immune-dependent GR response

(Stolte *et al.*, 2008c). This might reflect a temporal surge of the "stress" GR1 receptor expression to increase sensitivity for feedback control in order to prevent a too fierce pro-inflammatory response.

2.2.2. Effects of Cortisol on Immune Responses

Glucocorticoids regulate multiple aspects of the immune defense in mammals and influence the balanced successive secretion of pro- and anti-inflammatory cytokines and balanced apoptosis (Elenkov and Chrousos, 2006). They therefore are widely used as an anti-inflammatory drug. Similarly, cortisol has profound and differential effects on the teleostean immune system, mediated by one or several of the GRs present in immune cells.

In the innate arm of the immune system, *in vitro* cortisol stimulation of common carp, tilapia and silver sea bream (*Sparus sarba*) head kidney leukocytes significantly depressed phagocytosis (Law *et al.*, 2001). Similarly, cortisol dose-dependently inhibited chemotaxis and phagocytosis in a goldfish macrophage cell line and strongly inhibited respiratory burst activity (Wang and Belosevic, 1995). Decreased respiratory burst activity was also shown for sea bream head kidney leukocytes (Esteban *et al.*, 2004) and this response was shown to be mediated by GR (Vizzini *et al.*, 2007). Additionally, cortisol affects the cytokine production of immune cells. Indeed, in several *in vitro* studies it was demonstrated that cortisol inhibits LPS-induced expression of acute phase protein serum amyloid S (SAA) and pro-inflammatory cytokines IL-1β, TNF-α, IL-12p35, IL-11 and iNOS (Saeij *et al.*, 2003b; Huising *et al.*, 2005; Fast *et al.*, 2008; Stolte *et al.*, 2008c).

Besides modulation of cytokine responses, cortisol was shown to affect apoptosis and proliferation of immune cells for effective activation and deactivation of the teleostean immune response. Cortisol inhibited proliferation of a rainbow trout monocyte/macrophage cell line and these responses were GR dependent (Pagniello *et al.*, 2002). Similarly, cortisol-induced apoptosis was found in silver sea bream macrophages, and Atlantic salmon (*Salmo salar*) macrophages isolated from stressed fish showed decreased survival when exposed to the bacterium *Aeromonas salmonicida* (Fast *et al.*, 2008). Interestingly, neutrophilic granulocytes are protected by stress. Cortisol stimulation inhibited (GR dependently) apoptosis in neutrophilic granulocytes of common carp (Weyts *et al.*, 1998a). *In vivo* results corroborated these *in vitro* experiments. Stress reduced the number of circulating B-lymphocytes, whereas the relative percentage of circulating granulocytes nearly doubled (Engelsma *et al.*, 2003). As neutrophilic granulocytes are of great importance to the first line of defense, it may be beneficial in situations of acute stress (and possible injury) to prolong their life-span and maintain higher numbers in circulation.

Although less extensively studied, stress-induced immune modulation has also been found in the adaptive arm of the immune system. In common carp, *in vitro* stimulation with cortisol reduced the IgM secretion by lymphocytes from spleen, head kidney and blood (Saha *et al.*, 2004). An *in vivo* temperature stress resulted in decreased antibody responses after immunization (Verburg-van Kemenade *et al.*, 1999). Reduced IgM secretion and antibody production could result from cortisol-induced apoptosis. Inhibition of proliferation and an induction of apoptosis in lymphocytes from blood, spleen, head kidney and thymus was reported (Saha *et al.*, 2003). This is a GR-dependent process, as addition of the specific GR blocker RU486 rescued B-lymphocytes (Weyts *et al.*, 1997).

2.2.3. Pro-opiomelanocortin-related Peptides

The POMC-derived peptides ACTH, MSH, endorphin and the N-terminal peptide of POMC (NPP, or pro-gamma-MSH) are primarily produced and released from the cells of the pars distalis and the pars intermedia, respectively, and may reach immune cells in the blood stream or immune tissues. Local production of POMC-derived peptides within immune tissues has also been reported. Although POMC-derived peptides are expressed in low amounts by immune cells, the localized contribution of POMC-derived peptides may be significant as the number of cells involved is large.

Research on stress-induced immune modulation has focused primarily on cortisol, and information about other HPI axis factors is scarce. Interestingly, ACTH and ACTH receptor-like molecules were found during early development in immune tissues such as the interrenal tissue, the thymus and spleen (Mola *et al.*, 2005). LPS treatment induced production of ACTH-like molecules in the interrenal area and liver, suggesting a role for this peptide before the lymphoid cells have reached complete maturation. In tilapia (*Oreochromis mossambicus*), LPS treatment inhibited both *in vitro* pituitary ACTH and α-MSH release and head kidney responsiveness to ACTH (Balm *et al.*, 1995). Reciprocally, ACTH has been implicated in increased oxidative burst activity in phagocytes (Bayne and Levy, 1991).

Besides stimulating ACTH release from the cells in the pars distalis, CRF also induces α-MSH release from cells in the pars intermedia of the pituitary and stress can thus lead to concomitant release of α-MSH and endorphins. Both ACTH and α-MSH have been shown to have immunomodulatory effects. In thymic areas POMC-derived peptides and cytokines were found to be co-localized with apoptotic cells in the thymus, which could suggest that neuroendocrine cells might play a role in the selection and apoptosis of thymic lymphocytes (Ottaviani *et al.*, 1995). α-MSH was found to induce production of superoxide anions in macrophages and to increase phagocytosis. Also, carp lymphocytes treated with α-MSH increased the mitogenic responses to phytohaemagglutinin (PHA) (Watanuki *et al.*, 2003). Furthermore it was found that α-MSH was ineffective on isolated phagocytes but supernatants derived from mixed leukocytes treated with α-MSH increased phagocytosis. This suggests that more than one type of immune cell is required for this type of immune modulation to take effect (Harris and Bird, 2000). NPP increased phagocytosis *in vitro* in rainbow trout and carp macrophages. These results were corroborated *in vivo* where administration of NPP significantly increased superoxide anion production as well as phagocytosis in macrophages, thus showing that NPP in lower vertebrates activates the function of the phagocytic cells (Sakai *et al.*, 2001).

2.3. Opioids

Investigation of the opioid system in phylogenetically more primitive animals resulted in the intriguing hypothesis that this system arose in invertebrates as a part of the immunomodulatory system and that its analgesic properties were developed later, when pain became an alerting process (Stefano *et al.*, 1998). This theory is supported by the fact that both in invertebrates and vertebrates antimicrobial peptides (e.g., ekelytin) originate from opioid prohormones (Salzet *et al.*, 2006). Moreover, leukocytes can synthesize opioid peptides and posses opioid receptors (Sharp, 2006). Therefore, opioids can affect immune processes directly through activation of opioid receptors on immunocytes. However, indirect modulation of immune responses by opioids is also possible, when opioid activation of the stress axis results in increased production of corticosteroids or in activation of the sympathetic nervous system with subsequent release of catecholamines (Mellon and Bayer, 1998).

2.3.1. Opioid Peptides and their Synthesis in Immune Cells

The main groups of opioid peptides are produced as a result of post-translatory processing of opioid prohormones: pro-enkephalin (PENK), pro-dynorphin (PDYN) and POMC (Figure 7.2). POMC is the precursor of α- and β-endorphin (END) and several non-opioid peptides (Laurent *et al.*, 2004). PENK is the source of leu-enkephalin (LE), met-enkephalin (ME), met-enkephalin-Arg^6-Gly^7-Leu^8, met-enkephalin-Arg^6-Phe^7, peptides E, B and F, and also for antibacterial enkelytin (Laurent *et al.*, 2004; Brogden *et al.*, 2005). PDYN gives rise to dynorphin A, B (rimorphin, DYN), α- and β-neoendorphin (NEO) (Laurent *et al.*, 2004). Recently, novel groups of opioid peptides have been discovered: endomorphins (1 and 2, precursor unknown) and nociceptin/orphanin FQ (cleaved from pro-nociceptin) (Przewlocki and Przewlocka, 2005). All opioid prohormone genes probably derived from a common ancestor and during the course of chordate evolution the PENK gene underwent a series of duplication events to give rise to POMC, pro-nociceptin and PDYN (Danielson and Dores, 1999).

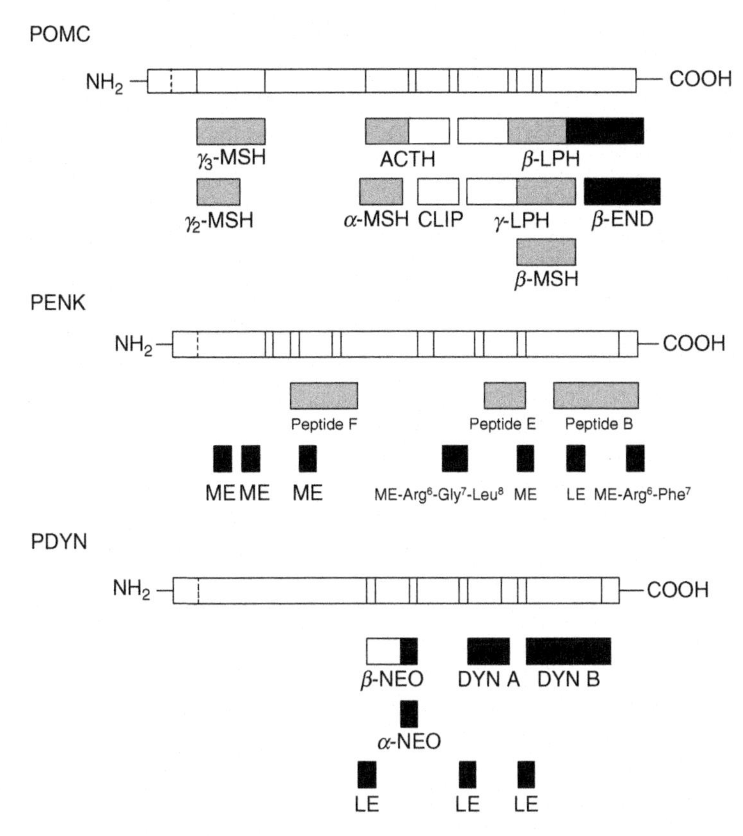

Fig. 7.2. Prohormones as a source of opioid peptides and hormones. ACTH, adrenocorticotropic hormone; CLIP, corticotropin-like intermediate lobe peptide; DYN, dynorphin; END, endorphin; LE, leu-enkephalin; LPH, lipotropin; ME, met-enkephalin; MSH, melanocytes-stimulating hormone; NEO, neoendorphin; POMC, proopiomelanocortin; PENK, proenkephalin; PDYN, prodynorphin.

Opioid prohormones have been cloned and characterized in several fish species (e.g., Gonzalez-Nunez *et al.*, 2003a,b; Gonzalez-Nunez *et al.*, 2003).

2.3.2. Opioid Receptors and their Function in the Immune System

Three major types of classical opioid receptors have been cloned and pharmacologically characterized in mammals: the morphine-like receptor, mu (MOR, μ, oprm), a delta receptor (DOR, δ, oprd) and a kappa receptor (KOR, κ, oprk). Moreover there is one non-classical opioid-like receptor, ORL-1 (NOR, oprn). Based on radioligand binding, several subtypes of the opioid receptors are defined (Kieffer and Gaveriaux-Ruff, 2002;

Corbett *et al.*, 2006). The endogenous ligands of the different opioid receptors are given in Table 7.2.

All known opioid receptors are G-protein coupled receptors, and agonist binding results in inhibition of adenylyl cyclase activity and intracellular cAMP via activation of pertussis toxin-sensitive G_i protein (Kieffer and Gaveriaux-Ruff, 2002). Opioid receptors may also form homo- as well as heterodimeric receptor complexes with any other opioid receptor, but also with non-opioid receptors such as the β2-adrenergic receptor and even chemokine receptors. Homodimerization or heterodimerization of different types of opioid receptors has been shown to change their pharmacology as well as signal transduction (Jordan *et al.*, 2001; Gomes *et al.*, 2003; Corbett *et al.*, 2006).

The predicted protein structures for opioid receptors are highly homologous, and the almost identical genomic organization suggests a common ancestor gene (Stevens, 2004). Putative opioid receptors with high homology to their mammalian counterparts were cloned in a few fish species (Darlison *et al.*, 1997; Barrallo *et al.*, 1998a,b, 2000; Pinal-Seoane *et al.*, 2006), and recently in carp (Chadzinska *et al.*, 2009a).

In mammals, mu, delta and kappa, as well as non-classical opioid binding sites were found on leukocytes (Sharp, 2006). Characterization of opioid receptors on goldfish head kidney leukocytes indicated the presence of at least two different opiate-binding sites: a [^{3}H]naloxone binding site and a [^{3}H] naltrindole binding site (Chadzinska *et al.*, 1997; Jozefowski and Plytycz, 1997). The naloxone-binding site is similar to the morphine-sensitive and opioid peptide-insensitive mu3 receptor. Low affinity binding of selective ligands excluded the presence of neuronal-type mu and delta opioid receptors on goldfish leukocytes (Chadzinska *et al.*, 1997; Jozefowski and Plytycz, 1997). In carp, low constitutive expression of mu, delta and kappa genes was observed in immune organs. Both during *in vivo* zymosan-induced peritonitis and after *in vitro* LPS-induced stimulation, expression of the opioid receptor genes in leukocytes was up-regulated (Chadzinska *et al.*, 2009a).

Table 7.2
Opioid receptor types and ligands in mammals

Receptor	Subtypes	Endogenous ligands
Mu	1–3	Endomorphins, β-END
Delta	1–2	Enkephalins, β-END
Kappa	1–3	Dynorphins
ORL-1		Nociceptin

In mammals, opioids modulate both innate and acquired immune responses, altering resistance to a variety of infectious agents (McCarthy *et al.*, 2001; Eisenstein *et al.*, 2006; Singh and Rai, 2008). In fact, all of the major properties of cytokines are shared by opioids, i.e., production by immune cells with paracrine, autocrine and endocrine sites of action, functional redundancy, pleiotropy and effects that are both dose and time dependent (Peterson *et al.*, 1998). Moreover, opioid receptor agonists affect *in vitro* phagocytosis and respiratory burst activity (for review see Eisenstein *et al.*, 2006). Similarly, prominent *in vitro* effects on important innate immune parameters have been observed in teleosts (Table 7.3). Strikingly, opposite effects of β-END on phagocytes at either low or high concentrations are antagonized by mu and delta receptor agonists, respectively. The effects of opioids on leukocyte chemotaxis are especially interesting. Mu and delta opioid receptor agonists act as potent *in vitro* chemoattractants for different leukocyte populations (Grimm *et al.*, 1998a,b; Chadzinska *et al.*, 1999; Miyagi *et al.*, 2000; Chadzinska and Plytycz, 2004), but the same opioids can also inhibit cell migration towards chemoattractants via a process of heterologous desensitization (Grimm *et al.*, 1998a,b; Choi *et al.*, 1999; Miyagi *et al.*, 2000). In general, activation of one type of G-protein coupled receptor (GPCR) can result in phosphorylation of C-terminal cytosolic tails of other GPCR by second messenger-mediated kinases such as protein kinase A (PKA) or PKC. A phosphorylated receptor loses its capacity to couple to the downstream heterotrimeric G protein and therefore becomes insensitive to any stimulation (Rogers *et al.*, 2000).

In goldfish, leukocyte pre-incubation with mu and delta (but not kappa) agonists enhanced random migration, while it inhibited migration towards zymosan-stimulated serum. The latter was reversed by specific antagonists of mu and delta opioid receptors (Chadzinska and Plytycz, 2004). Moreover, both *in vitro* and *in vivo* morphine down-regulated expression of chemokines (CXCa and CXCb) as well as chemokine receptors (CXCR1 and CXCR2) on carp leukocytes (Chadzinska *et al.*, 2009b). *In vitro* incubation of mammalian macrophages and monocytes with morphine and other mu (but not delta) opioid receptor agonists induced apoptosis (Malik *et al.*, 2002; Bhat *et al.*, 2004). Our preliminary results show that pre-incubation of carp phagocytes either with morphine or with deltorphine II does not induce cell apoptosis (Chadzinska *et al.*, 2009b; Verburg-van Kemenade *et al.*, in press).

Since the impact of opioids on leukocytes suggests a crucial change of the inflammatory response, *in vivo* experiments with a pro-inflammatory agent (zymosan or thioglycollate) were designed to corroborate this hypothesis. Morphine injected together with zymosan or thioglycollate reduced the number of inflammatory leukocytes in fish (goldfish, salmon, carp) (Chadzinska *et al.*, 1999, 2000). This gave strong evidence that the decreased

Table 7.3
Effects of opioid receptor agonists on immune parameters in fish

Agonist	Treatment	Effect	Species	Reference
Morphine	*In vitro*	Decreased expression of genes for IL-1β, TNF-α, CXCa, CXCR1 and iNOS and inhibition of NO production in LPS-stimulated phagocytes	Common carp	Chadzinska *et al.*, 2009b
		Reduction of leukocyte chemotaxis towards chemoattractants	Common carp Wild goldfish	Chadzinska and Plytycz, 2004; Chadzinska *et al.*, 2009b
	In vivo	Reduction of cell influx and level of chemoattractants during peritonitis	Atlantic salmon Wild goldfish Common carp	Chadzinska *et al.*, 1999, 2000, 2009b
		Stimulation of respiratory burst activity in exudatory leukocytes	Atlantic salmon	Chadzinska *et al.*, 1999
		Decreased expression of genes for TNF-α, CXCa, CXCb, CXCR1, CXCR2, iNOS and Arginase2 in exudatory leukocytes	Common carp	Chadzinska *et al.*, 2009b
β-END	*In vitro*	Stimulation of superoxide anion production and phagocytosis	Rainbow trout Common carp	Watanuki *et al.*, 2000
		Stimulation of phagocytosis and NO production Stimulation (at low doses) and inhibition (at high doses) of respiratory burst activity	Spotted murrel	Singh and Rai, 2008
		Stimulation of bactericidal activity of leukocytes	Rainbow trout	Watanuki *et al.*, 2000
		Reduction of leukocyte cytotoxicity and proliferation	Tilapia	Faisal *et al.*, 1989
	In vivo	Stimulation of respiratory burst activity and phagocytosis in phagocytes. Stimulation of leukocyte chemotaxis towards opsonized bacteria	Rainbow trout	Watanuki *et al.*, 1999

Endomorphin 2	*In vitro*	Reduction of phagocyte chemotaxis towards chemokine. Decreased expression of genes for TNF-α, CXCb and CXCR1 in LPS-stimulated phagocytes	Common carp	Chadzinska *et al.*, 2009a; Verburg-van Kemenade *et al.*, in press
Deltorphine II	*In vitro*	Reduction of leukocyte chemotaxis towards chemoattractants	Common carp Wild goldfish	Chadzinska *et al.*, 2009a; Chadzinska and Plytycz, 2004; Verburg-van Kemenade *et al.*, in press
		Decreased expression of genes for CXCa, CXCb, CXCR1 and CXCR2 in LPS-stimulated phagocytes	Common carp	Chadzinska *et al.*, 2009a; Verburg-van Kemenade *et al.*, in press
U50,488H	*In vitro*	Reduction of phagocyte chemotaxis towards chemokine. Decreased expression of genes for IL-10 and CXCR1 in LPS-stimulated phagocytes	Common carp	Chadzinska *et al.*, 2009a; Verburg-van Kemenade *et al.*, in press

IL, interleukin; TNF-α, tumor necrosis factor alpha; CXC, chemokine; CXCR, chemokine receptor; iNOS, inducible nitric oxide synthase; NO, nitric oxide; LPS, lipopolysaccharide.

number of peritoneal leukocytes in morphine-treated animals can be due to reduced levels of chemoattractants and chemokine receptors as well as heterologous desensitization of chemokine receptors (Chadzinska *et al.*, 1999, 2000, 2009a,b). Interestingly, the results of Metz and co-workers (Metz *et al.*, 2006) indicate that the communication between the opioid system and inflammatory response is bi-directional (see Section 3).

Many studies of the effects of opioids on immune function have focused on perturbation of natural killer and lymphocyte functions (Shavit *et al.*, 1984; Yeager *et al.*, 1995). In tilapia, social confrontation-induced immunosuppression was reversed by the opioid receptor antagonist naltrexone (Faisal *et al.*, 1989). Indirectly, these results demonstrate that the immunosuppression associated with this social stressor is at least partly mediated via the endogenous opioid system. However, it is important to note that the naltrexone effect was limited to the cytotoxic and T-lymphocytes. Moreover, exogenous β-END induced significant suppression of cytotoxicity as well as proliferation, which could be reversed by naltrexone (Faisal *et al.*, 1989, 1992). The longer allograft survival in gulf killifish injected with naloxone (opioid antagonist) points in the same direction (Nevid and Meier, 1995).

2.4. Members of the Type I Cytokine Family: Growth Hormone, Prolactin, Leptin

2.4.1. Type I Cytokines and their Linkage to the Immune System

The type I cytokines comprise a family of "endocrine" and "immune" signals that probably have co-evolved from the same ancestral precursor molecule (Huising *et al.*, 2006b). They are characterized by a bundle of four tightly packed α-helices, although their primary amino acid sequences share little sequence similarity. Members of this cytokine family include GH and PRL, two pleiotropic hormones that among other functions contribute to the restoration of cell depletion induced by corticosteroids or stress. For example, a state of hyperprolactinemia is associated with active stages of autoimmune diseases such as rheumatoid arthritis and systemic lupus erythematosus and stress (Orbach and Shoenfeld, 2007). In healthy conditions glucocorticoids act to inhibit excessive synthesis and release of pro-inflammatory cytokines. Inflammation disrupts the cytokine balance in either the brain or the periphery and the body responds with a rise in hormone release to restore this balance (Kelley *et al.*, 2007). In teleost fish two forms of PRL, PRL_{177} and PRL_{188}, have been identified (Swennen *et al.*, 1991; Flik *et al.*, 1994). PRL concentrations are related to different conditions of stress and infection (Wendelaar Bonga *et al.*, 1984). Both PRL and GH are, together with cortisol, involved in osmoregulation and their expression and/or plasma protein levels change after adaptation to either fresh or salt water. The reductions in immune parameters during this transformation are therefore assumed to be associated with the changes in hormone balance.

The classical role of the type I cytokine leptin as a satiety hormone is well known and has been extensively studied in order to understand weight regulation and its related disorders (Schwartz *et al.*, 2000). In addition, there is now great interest in the potential roles of leptin as a type I cytokine in immunity (Lago *et al.*, 2008). Recently leptin sequences were described for pufferfish and carp (Huising *et al.*, 2006a), opening this area of research for poikilothermic animals. We will therefore now focus on the mechanisms linking type I cytokines/hormones to immune defense in teleosts.

2.4.2. Production of GH, PRL, SL and Leptin in Immune Tissues

Peripheral expression of GH, insulin-like growth factor I (IGF-I), PRL and leptin has been found in lymphoid tissues and leukocytes of mammals (Kelley *et al.*, 2007). The extrapituitary sites of PRL production include neurons, prostate, epithelium and endothelium (Ben-Jonathan *et al.*, 1996). Multiple immunoreactive PRL molecules (24, 21 and 11 kDa in thymocytes and a 27 kDa form in mononuclear cells) are detected, including a pituitary-like PRL with biological activity (Montgomery, 2001). Information on the regulation of expression of these immune cell related factors is still scarce. PRL expression in mammalian T-lymphocytes is regulated by IL-2, IL-4 and IL-1β (Gerlo *et al.*, 2005). Moreover, PRL gene expression in lymphoid tissue is regulated independently of the pituitary transcription factor Pit-1 (Berwaer *et al.*, 1994; Gellersen *et al.*, 1994) and it is now hypothesized that PRL expression in human lymphocytes and pituitary is under the control of a different promoter (Gerlo *et al.*, 2006). Similarly, expression of GH in mammalian pituitary or lymphocytes may require different transcription factors (Kooijman *et al.*, 2000; Weigent *et al.*, 2000).

In various teleost species, expression of type I cytokines was found in leukocytes and lymphoid tissue (Table 7.4). Interestingly, GH synthesis in rainbow trout leukocytes is stimulated by cortisol (Yada *et al.*, 2005), which corroborates an evolutionary conserved balance of pro- and anti-inflammatory signals maintained by cortisol versus GH/PRL.

2.4.3. Receptors for GH, PRL and Leptin on Leukocytes

GH, PRL, SL and leptin receptors belong to a large and heterogeneous family of receptors, the class I cytokine receptor family and they show strong homology, which indicates an evolutionary association with the immune system (Heinrich *et al.*, 2003). After cytokine binding, tyrosine phosphorylation is mediated by JAK family tyrosine kinases. Subsequently, JAK phosphorylates intracellular tyrosine residues in the cytoplasmic domain of the

Table 7.4
Expression of type I cytokines in lymphoid tissues of fish

Cytokine	Site of expression	Species	Reference
PRL	Leukocytes	Rainbow trout Tilapia	Yada and Azuma, 2002
	Head kidney, spleen, intestine, PBL and head kidney leukocytes	Rainbow trout Tilapia	Yada *et al.*, 2002
SL	Spleen	Rainbow trout	Yang *et al.*, 1997
GH	PBL	Rainbow trout Tilapia *Not in catfish*	Yada *et al.*, 2001b, 2002, 2005
	Head kidney	Sea bream	Calduch-Giner and Perez-Sanchez, 1999
	Spleen, kidney, intestine	Salmon	Mori and Devlin, 1999
Leptin	Thymus	Common carp	Huising *et al.*, 2006a

PRL, prolactin; PBL, peripheral blood leukocytes; SL, somatolactin; GH, growth hormone.

receptor chain. This complex activates STAT by phosphorylation. A STAT dimer is then formed that translocates to the nucleus and initiates transcription (Heinrich *et al.*, 2003). Which gene is transcribed depends on the different members of the STAT family that form the dimer. The mammalian STAT family consists of seven members, and in teleost fish several members have been described (Lewis and Ward, 2004). Many cytokines show pleiotropy and redundancy in their biological action which is a likely result of the shared use of the GP130 signaling chain and promiscuity of components of the JAK/STAT pathway (Huising *et al.*, 2006b). Induction of suppressor of cytokine signaling (SOCS) molecules is initiated upon JAK/STAT activation to ensure a mechanism of negative feedback. This indicates a lively interaction with immune functions as SOCS proteins are also up-regulated through activation of various cytokines (e.g., type I cytokines and CXCL12 stimulation of the CXCR4 receptor; Redelman *et al.*, 2008).

In mammals, several isoforms of the PRL receptor (PRLR) are described, one of which may act as a decoy receptor. Binding to these receptors may be promiscuous as high concentrations of GH may bind and signal through the PRLR to influence human neutrophils (Fu *et al.*, 1992; Soares, 2004). Signaling of the JAK–STAT pathway after ligand binding to the PRLR or GHR occurs after homodimerization. The PRLR in fish has structural similarity to its mammalian counterpart although the overall sequence is poorly conserved (26–37%). The intracellular domain (ICD) contains a region rich in proline (box 1) in the membrane proximal region, which in mammals is responsible for signal transmission through the JAK2–STAT signaling pathway. The activation of the fish PRLR appears to employ a

Table 7.5
Receptors for type I cytokines in lymphoid tissues of fish

Receptor	Expression or specific binding	Species	Reference
PRLR	Head kidney	Sea bream	Santos *et al.*, 2001
	Lymphocytes	Tilapia	Sandra *et al.*, 2000; Yada and Azuma, 2002
		Rainbow trout	Prunet *et al.*, 2000
GHR	Lymphocytes	Sea bass	Calduch-Giner *et al.*, 1995
	Spleen	Rainbow trout	Very *et al.*, 2005
LeptinR	leukocytes	Common carp	Unpublished observation

PRLR, prolactin receptor; GHR, growth hormone receptor; LeptinR, leptin receptor.

similar mechanism because the Nile tilapia PRLR is capable of activating transcription of a STAT-5-responsive reporter gene (Sohm *et al.*, 1998; Tse *et al.*, 2000). The expression of type I cytokine receptors on teleost immune cells is summarized in Table 7.5.

2.4.4. Effects of PRL, GH and Leptin on the Immune Response

The immune deficiencies of hypophysectomized dwarf mammals originally focused attention on the potential immune functions of hormones. Similarly, hypophysectomy in teleosts leads to decreased innate and adaptive immune functions (Hull and Harvey, 1997; Yada *et al.*, 1999, 2001a; Yada and Azuma, 2002). A role for prolactin was subsequently confirmed for mammals *in vitro* and *in vivo* (reviewed by Kelley *et al.*, 2007). GH or PRL therapy could also reverse the effects of hypophysectomy in fish. Yet PRL does not seem to be a prerequisite for immune function as PRL and PRLR deficient mice have a proper immune defense (Horseman *et al.*, 1997; Bouchard *et al.*, 1999). The specific effects of PRL and GH on teleost innate and adaptive immune parameters are, as in mammals, generally stimulatory (reviewed by Redelman *et al.*, 2008) and are summarized in Table 7.6.

Higher occurrence of autoimmune diseases in obese individuals is hypothesized to be leptin-related (Harle and Straub, 2006). This is based on prominent stimulatory effects of leptin on innate responses like chemotaxis and secretion of oxygen radicals (La Cava and Matarese, 2004), phagocytosis and secretion of pro-inflammatory mediators of the acute phase (Sanchez-Margalet *et al.*, 2003) and NCC cytotoxic ability (Zhao *et al.*, 2003). To date, partly because leptin sequences in poikolotherms have only recently been identified, it is not known whether leptin affects teleost innate immunity. In the adaptive immunity of mammals, leptin affects the generation,

Table 7.6

Effects of PRL and GH on innate and adaptive immune parameters in fish

Cytokine	Effect	Species	Reference
PRL, but not SL			
In vitro	Stimulation of phagocytosis in leukocytes	Atlantic salmon	Sakai *et al.*, 1996
	Stimulation of leukocyte proliferation		Yada *et al.*, 2004
In vivo	Stimulation of Ig production	Rainbow trout	Yada et al., 1999, 2002
GH			
In vivo and *in vitro*	Stimulation of superoxide anion production and phagocytosis	Rainbow trout	Sakai *et al.*, 1995, 1996; Yada *et al.*, 2001b
		Tilapia	Yada *et al.*, 2002
		Sea bass	Munoz *et al.*, 1998
	NCC activity of leukocytes from spleen, head kidney and peripheral blood leukocytes	Rainbow trout	Sakai *et al.*, 1995
	Hemolytic activity of serum	Rainbow trout	Sakai *et al.*, 1995
	Stimulation of Ig production	Rainbow trout	Yada *et al.*, 1999, 2002

PRL, prolactin; SL, somatolactin; GH, growth hormone.

maturation and survival of thymic T-cells (Howard *et al.*, 1999), and increases the proliferation and IL-2 secretion in naïve T-lymphocytes (Martin-Romero and Sanchez-Margalet, 2001). In memory T-cells, leptin promotes the switch towards helper 1 (Th1)-cell immune responses by increasing IFN-γ and TNF-α secretion and the production of IgG2a by B-cells. Preliminary data indicate increased expression of leptin in PBLs and thymocytes upon immune stimuli and increased expression of IFN-γ upon administration of human leptin (Verburg-van Kemenade, unpublished results).

GH, PRL and leptin are not the only type I cytokines that communicate with the immune system. In mammals there is now increasing evidence for a role of other family members that certainly deserve further investigation. IL-6 is involved in antibody production, the acute phase response and stimulation of T-cells (Bird *et al.*, 2005). IL-6 has been reported in two fish species, flounder (*Paralichthys olivaceus*) and pufferfish (*Takifugu rubripes*). IL-6 exhibits a conserved gene structure and shows an expression pattern similar to mammalian orthologues, which suggests conservation of gene function (Bird *et al.*, 2005; Nam *et al.*, 2007). Leukemia inhibitory factor (LIF) and oncostatin M (OSM) are involved in the regulation of tumor cell proliferation (Abe *et al.*, 2007). Recently LIF orthologues have been found in zebrafish (*Danio rerio*) and green spotted pufferfish (*Tetraodon nigroviridis*).

2.5. The Hypothalamic–Pituitary–Thyroid (HPT) Axis

In mammals, the interaction between the HPT axis and the immune system is quite well established both *in vitro* and with hyper- and hypothyroid models *in vivo* (Bagriacik and Klein, 2000; Dorshkind and Horseman, 2001). These effects are generally considered stimulatory for the adaptive response where especially B-cell development is influenced (Dorshkind and Horseman, 2000). Halabe Bucay (2007) recently reviewed the effects of thyroid hormones (THs) on immune parameters, including activation of proliferation and cytokine production (e.g., IFN-γ) in leukocytes. In contrast to results for adaptive responses, Dorshkind and Horseman (2000) suggest impairment of the innate response. Furthermore, thyroid-stimulating hormone (TSH) may also be produced within the immune system by leukocytes (reviewed by Weigent and Blalock, 1995).

In teleost fishes the thyroid function is also regulated by a comparable HPT axis, although differences in structure and regulation have been noted. For instance it was found that CRF rather than TRH forms the main thyrotropin-releasing hormone (Flik *et al.*, 2006; Geven *et al.*, 2006). Interestingly, in some species thyroid follicles have been found within the hematopoietic tissue of the head kidney (Geven *et al.*, 2007). Moreover THs are suggested to have an important regulatory role in stress axis function with experimentally induced hyperthyroidism in carp leading to a strong decrease in plasma cortisol levels (Geven *et al.*, 2006). Early studies observed a significant decrease in the number of circulating leukocytes in hypothyroid fish (Slicher, 1961), which could be restored by addition of thyroxine (T_4) or mammalian TSH (Ball and Hawkins, 1976). Lam *et al.* (2005) were the first to find evidence that the thyroid relationship with thymus development is conserved from fish to higher vertebrates as thymus development and thymopoiesis (thymus size, recombination activating gene RAG-1 positivity, expression of T-cell antigen receptors) were found to be affected by T_4. It is of special interest to study the thyroid–immune interaction in teleosts as a period of smolting, preparing the juvenile salmon to entry into saltwater, is accompanied by significant increases in plasma T_4 and cortisol. Upon immunization with *Vibrio anguillarum* antigen decreased numbers of peripheral blood lymphocytes and splenic plaque forming cells were noted together with relative increases in small lymphocytes. The contribution of cortisol versus T_4 is, however, not yet elucidated (Maule *et al.*, 1987).

2.6. The Brain–Pituitary–Gonadal (BPG) Axis

Along with the recognition of the potential impact of stress on immunity, stressful situations are often accompanied by changes in sex steroid levels and gonadal development or function. Moreover, changes in the activity of immune cells and apoptosis in mammals can be induced by estrogens and immune cells do possess estrogen receptors (Ahmed, 2000). Endocrine disrupting compounds (EDCs) in the environment have also been shown to interfere with reproduction and disease resistance of several species. Important EDCs are natural and synthetic estrogens and androgens (Segner *et al.*, 2006) and aquatic ecosystems may be under the influence of natural estrogens such as 17β-estradiol (E2) and synthetic estrogens like 17α-ethinylestradiol (EE2) (Cargouet *et al.*, 2004; Kolodziej *et al.*, 2004). Overall, our knowledge on the influence of steroids on teleost immunity is fragmentary. Gender and species differences, and differences in developmental stage or spawning season must be taken into account. Moreover, potential interference of estrogens with the GH/IGF-I system may complicate results (Riley *et al.*, 2004; Filby *et al.*, 2006).

2.6.1. Estrogen/androgen Receptors and Effects of Estrogens/androgens on Innate and Adaptive Immunity

The fact that the androgen receptor is detected in salmonid lymphocytes and estrogen receptor-α (ER-α) is reported in catfish leukocytes (Patino *et al.*, 2000) indicates involvement of the sex steroids in teleost immunity. Moreover, increased phagocytosis and increased resistance to *Vibrio* infection in sea bream were related to testosterone (not E2) levels (Deane *et al.*, 2001). *In vivo* application of E2 in goldfish led to increased susceptibility to *Trypanosoma danilewskyi* and lower proliferation of leukocytes (Wang and Belosevic, 1994). Injection of carp with E2, 11 ketotestosterone (11KT) and progesterone reduced NO and phagocytosis in a dose-dependent manner (Watanuki *et al.*, 2002). *In vitro* assays with carp head kidney leukocytes showed impaired phagocytosis by nanomolar concentrations of E2, progesterone and 11KT, suppression of NO by 11KT and progesterone, but no reduction in respiratory burst by any of the steroids (Yamaguchi *et al.*, 2001). Wang and Belosevic (1995) also tested E2 on cells of a goldfish macrophage derived cell line and measured suppression of chemotaxis and phagocytosis, but unlike cortisol no effect could be found on NO production. Saha *et al.* (2003, 2004) measured no apoptosis of carp thymocytes and no reduction in IgM levels with the gonadal steroids, although high IgM levels are usually found during the spawning season.

3. NEUROENDOCRINE MODULATION BY CYTOKINES

As a result of technical limitations, the majority of cytokine research in fish has concentrated on the role of cytokines in immune responses and as yet there are only a handful of studies that have focused on neuroendocrine modulation. Nonetheless, there is increasing evidence that supports the existence of bi-directional immune–endocrine interactions. Clearly, for a physiological connection between the immune and neuroendocrine system, there must be a mechanism by which signals released peripherally can induce central responses. Indeed, it took scientists more than two decades to gain acceptance for the concept of communication between the immune system and the brain (Quan and Banks, 2007). In this section it is our goal to review and discuss the current knowledge of complex cytokine signaling in neuro–immune–endocrine interactions, where possible with special reference to fish.

3.1. A Stress Response Follows the Immune Response

In mammals and fish alike, the response to a systemic infection or inflammatory reaction is generally characterized by three phases. First, the acute phase response (see Section 1.2) is characterized by a trigger of the innate immune response. Second, there is an activation of the HPA axis and an involvement of the brain (see below). To date, however, only a handful of studies in fish have characterized the impact of an immunological challenge on plasma cortisol concentrations. In both mammals and fish, an immune challenge results in increased activity of the neuroendocrine pathway leading to elevation of plasma cortisol (Camp *et al.*, 2000; Haukenes and Barton, 2004). Third, there is a strong feedback action by glucocorticoids, balanced by GH and PRL, to suppress potentially excessive cytokine production and pro-inflammatory response (see Section 2.2).

3.2. From Periphery to Brain

As described earlier, the innate immune response is based on the recognition of PAMPs by TLRs which induces an increase in the expression of pro-inflammatory cytokine genes. Typically, IL-1β, IL-6 and TNF-α are the first pro-inflammatory cytokines expressed. In addition to their "classical" role in mediating and coordinating the local and systemic inflammatory response to pathogens, these cytokines are responsible for the induction of the sickness syndrome; a term that collectively describes all endocrine, autonomic and behavioral changes that occur in response to an infection (Dantzer, 2006). For example, the above-mentioned cytokines induce among others fever, loss of appetite, withdrawal from social activities and fatigue. This set of physiological adjustments is regarded as an adaptive response of the host to infectious microorganisms which together positively affects recovery and survival (Dantzer, 2006). Indeed, disturbances of the sickness syndrome can lead to the development of neural disorders like depression and Alzheimer's disease. Central to our understanding of immune system–CNS communication is to identify how the CNS gets "informed" about an immune response that takes places peripherally.

3.3. How Are Peripheral Immune Signals Conveyed to the Brain?

Currently, there are four proposed mechanisms that link peripheral inflammation signals to CNS responses (Correa *et al.*, 2007). The first pathway comprises activation of afferent fibers in the vagal nerve during infection (Bluthé *et al.*, 1994; Goehler *et al.*, 1999). These fibers project to several brain nuclei, including the paraventricular nucleus (PVN) of the hypothalamus. Second, circulating PAMPs trigger macrophages in the CVOs (blood–brain barrier-deficient areas that surround the brain ventricles) to produce and release pro-inflammatory cytokines (Quan *et al.*, 1998). These cytokines are then believed to reach receptors via volume diffusion (Vitkovic *et al.*, 2000). Neurons of CVOs were found to express pro-inflammatory cytokine receptors, including IL-1RI and TNFR (Bette *et al.*, 2003; Nadjar *et al.*, 2003, 2005). These neurons have direct projections to the PVN of the hypothalamus (Rivest *et al.*, 2000). The third proposed mechanism is via active transport by specific carriers to allow passage through the blood–brain barrier (Banks *et al.*, 1989). Interestingly, permeability of the blood–brain barrier is under the influence of IL-1β (Blamire *et al.*, 2000). The fourth mode of action is based on local synthesis of prostaglandin E2, induced by pro-inflammatory cytokines, in macrophages and endothelial cells at the interface of blood and brain (Konsman *et al.*, 2004). Prostaglandin E2 can cross the blood–brain barrier and once in the brain induces fever (Engblom *et al.*, 2002).

Although it is not yet known whether peripheral immune signals can reach the brain in fish, available evidence suggests that mechanisms similar to those identified in mammals may be present in fish. First, TLRs activated by recognized PAMPs have recently been identified in flounder, zebrafish and catfish (see Section 1.2; Hirono *et al.*, 2004; Meijer *et al.*, 2004). These fish TLRs were found to be well conserved and were shown to be involved in the responses to infection (Baoprasertkul *et al.*, 2007). Second, fish do have CVOs such as the saccus vasculosus (Sueiro *et al.*, 2007). Anatomically and functionally, the CVOs of fish are similar to mammalian CVOs: their vessels bear the same functional difference with other brain areas, including a concomitant lack of a blood–brain barrier (Jeong *et al.*, 2008). Third, despite the lack of a specific study on the passage of cytokines through the blood–brain barrier in fish, there is recent evidence suggesting that the blood–brain barrier of zebrafish is molecularly and functionally highly similar to that of higher vertebrates (Jeong *et al.*, 2008). Lastly, intraperitoneal injection of IL-1β induced an increase in cyclooxygenase-2 (COX-2), the enzyme responsible for prostaglandin synthesis (Hong *et al.*, 2003). Together these observations strongly indicate that in fish, like mammals, the brain is accessible for, or sensitive to, peripherally released cytokines. Obviously, whether and how accessibility and sensitivity are regulated needs further study.

Among the key functions of pro-inflammatory cytokines is induction of fever. At first glance this function may appear not applicable to ectothermic fish. However, regulated elevation of body temperature has been reported in all vertebrate groups, including fish (Reynolds *et al.*, 1976). Although it was originally reported that injection of LPS failed to induce an increase in body temperature in pumpkinseed (*Lepomis gibbosus*; Marx *et al.*, 1984), this was contradicted in a later study using goldfish where both LPS and IL-2 injections did induce fever (Cabanac and Laberge, 1998).

3.4. Cytokine Production in the Brain

It is important to realize that in the CNS itself, including the hypothalamus, neurons constitutively express cytokines, where they contribute to normal brain function (Shintani *et al.*, 1995). In addition, astrocytes and glia cells are immunocompetent cells and respond like immune cells within the CNS. Reportedly, cytokines and chemokines including IL-1 (Giulian *et al.*, 1986), IL-6 (Frei *et al.*, 1989) and TNF-α (Chung and Benveniste, 1990) are produced and secreted in amounts significant enough to have autocrine actions (Aloisi, 2001). Inflammatory cytokines are also expressed in the brain of fish. For example, IL-1β expression was demonstrated in the preoptic nucleus of carp (Metz *et al.*, 2006). Central IL-1 and IL-6 expression have also been reported in rainbow trout (Iliev *et al.*, 2007) and sea bream (Castellana *et al.*, 2008).

3.5. Cytokine-induced Activation of Hypothalamo–Pituitary–Interrenal Axis

Among the first indications in mammals that cytokines influence neuroendocrine functions was the finding that corticosterone levels in mice markedly increased during the primary immune response (Shek and Sabiston, 1983). Nowadays the cytokinergic activation of the HPA axis is the most firmly established type of immune–neuroendocrine modulation (Berkenbosch *et al.*, 1987; Sapolsky *et al.*, 1987; Goshen & Yirmiya, 2009). In mammals, IL-1, IL-2, IL-6, IL-10, IFN-β, IFN-γ, LIF, G-CSF and TNF-α have all been shown to influence HPA axis activity (Hermus and Sweep, 1990; Harbuz *et al.*, 1992; Mastorakos *et al.*, 1994; Shintani *et al.*, 1995; Kim and Melmed, 1999; Smith *et al.*, 1999; Zylinska *et al.*, 1999; Dunn, 2000). Most cytokines activate the HPA axis in mammals via stimulation of hypothalamic CRF and AVP secretion. Indeed, the IL-1RI is expressed in the mammalian PVN (Yabuuchi *et al.*, 1994). For IL-1, IL-2, IL-6 and IL-10, it has been shown that these cytokines can stimulate ACTH release from pituitary cells (Woloski *et al.*, 1985; Prickett *et al.*, 2000). Interestingly, IL-2 is even more potent than CRF in stimulating ACTH release (Karanth and McCann, 1991). There is ample evidence as well that IL-1, IL-2 and IL-6, G-CSF and TNF-α have, both *in vitro* and *in vivo,* stimulatory effects on glucocorticoid release directly from the adrenal gland (Turnbull and Rivier, 1999).

Besides their actions on HPA axis activity, cytokines have important effects on several other endocrine processes. For example, many cytokines, including IL-1, have been shown to affect pituitary secretion of gonadotropins and steroidogenesis by testes and ovaries (Cannon, 1998). Also, IL-6 decreases secretion of GH and PRL from rat pituitary cells (Tomida *et al.*, 2001).

To date only two studies have investigated the potential roles of cytokines in the regulation of the HPI axis in fish and both studies focused on IL-1β. In parallel with what we know for mammals, both *in vivo* and *in vitro,* IL-1β activates the fish HPI axis. In rainbow trout, plasma cortisol levels were elevated following injection with IL-1β (Holland *et al.*, 2002). In contrast, in *in vitro* superfusion experiments on common carp, while IL-1β did not affect cortisol production, it stimulated pituitary release of α-MSH and β-endorphin (Metz *et al.*, 2006). So far a possible action via stimulation of CRF secretion has not been investigated, but IL-1RI mRNA was found in the preoptic area of the common carp brain (Metz *et al.*, 2006). In addition, as observed in mammals, IL-1β expression appears to be under the influence of HPI axis activity. In carp, restraint stress was associated with a 2.5-fold up-regulation of preoptic area IL-1β expression (Metz *et al.*, 2006). It was thus concluded that hypothalamic IL-1β fulfils a similar function in fish as it does in mammals and that this mechanism is phylogenetically conserved.

4. CONCLUSIONS AND FUTURE PERSPECTIVES

Communication between the neuroendocrine and immune system in teleosts is now firmly established. Although many of the key players have now been identified, its fine details of regulation, especially its bi-directional nature, and the implication for physiological homeostasis and health are far from elucidated.

1. Research into neuroendocrine–immune interaction is relatively young, but the existence of this interaction now receives general acceptance.
2. Its evolutionary conservation from fish to humans underlines the physiological importance of this communication.
3. As in mammals, situations of stress not only inhibit, but also enhance specific immune function to ensure adaptive responses.
4. The evolutionary conserved critical balance between the stress axis hormones and the release of GH/PRL is pivotal for an adaptive response. The potential impact of other pathways may however not be underestimated as the HPT and BPG axes until now have received much less attention throughout species from all evolutionary stages.

The great progress in genomic sequencing projects and the sensitivity of detection not only ensured an immense step forward in finding the array of potentially active regulatory factors with their respective receptors, but also revealed the vast spread of these factors in different organs, tissues and cell types. Thereby it became clear that the pleiotropic and redundant characteristics of these factors are not limited to the immune system. The neuroendocrine system may also be characterized by regulatory factors with pleiotropy and redundancy. This pleiotropic character certainly adds a few dimensions of complexity to the solution of this issue as ligands from both neuroendocrine and immune sources will be regulated by multiple factors from multiple origins. Moreover, the signal will be detected in multiple places by multiple receptors that will take care of detection and physiological action in endocrine and paracrine fashion. Therefore simple stress or infection model does not exist and even the simple administration of one factor will have multiple consequences. Furthermore, we have to take into account that 300 million years ago teleost fish underwent extra genome duplications and many of these duplicated genes survived evolution during which many copies of a single pair have obtained a different receptor binding capacity or even a totally different function. Moreover, in both fish and mammals we find many splice variants and binding proteins that considerably alter the bioactive ligand or ligand concentration or that may strongly affect the number of available receptors.

To this complexity, the complexity of the immune system as such should be added, in which the cells of different types are constantly traveling throughout the body and upon immune stimuli vigorously respond with fast proliferation or apoptosis. Stress and infection in mammals and in fish have extensive influence and cause immense redistribution which in the case of infection is of course needed to induce a strong and local response (Dhabhar, 2003; Engelsma *et al.*, 2003; Huising *et al.*, 2003a).

For the future it is very important to not only direct our focus on the ligands; the importance of the receptor and selective receptor regulation should not be neglected as this may form an important factor to determine the outcome of the bi-directional communication. Multiple subtypes of receptors exist that may have a completely different profile of ligand specificity and sensitivity. It is becoming increasingly evident that receptors may be influenced by stress or infection. Regulation of glucocorticoid and opioid receptor expression or heterodimerization as described for opioid receptors and chemokine receptors are good examples. Recently it was elucidated that cytokines, neurotransmitters and oxygen radicals are indeed capable of influencing the responsiveness of G-protein coupled kinases. Changes in one kinase may alter the sensitivity of multiple receptors and thereby result in a different "signalosome." Heijnen (2007) launched an interesting hypothesis for which clear indications have been found in mammals. This hypothesis implicates not only the activity of the neuroendocrine or immune ligands but also the expression levels of the different kinases involved in signaling. G-protein coupled receptor kinase 2 (GRK2) is a key kinase in leukocytes which is coupled to receptors for important neuroendocrine ligands like NA, dopamine, serotonin, opioids and adenosine. In patients suffering from infection or autoimmune disease the levels of GRK2 expression may be significantly altered. As these neuroendocrine ligands and cytokines are of course present in large amounts within the brain, this cannot be without significant effect at this level. Lombardi *et al.* (2002) demonstrated that in brain, stress will induce decreases in GRK2 expression levels.

Although the complexity of the problem is increasing, our possibilities for research in teleost species are also growing. The whole genome of zebrafish and pufferfish was rapidly sequenced and more species are now being added. Microarray analyses with full annotation are now available and will help us to distinguish the different pathways involved. In mammals, models for autoimmune disease or cancer have already delivered interesting results. For fishes we will have to choose models of infection and stress and a combination of these two. This combination is very important as often the influence of neuroendocrine factors on the immune system was investigated without immune stimuli involved. Under "normal" physiological conditions in the absence of serious threat of infection, the immune system was investigated displays low activity which is essential to prevent tissue damage due to, e.g., multiple microbicidal factors. As discussed, neuroendocrine factors will often manipulate or stabilize an immune response but will not themselves trigger the cells to proliferation or increased activity. Research in fish has the advantage over human research that one may examine cell populations or tissues from different organs at different time points of experimental stimulation in stress or infection models with which we can search not only for ligands and receptors but also for the promoters involved. These patterns may become clearer with the use of the available isogenic lines.

As will be clear from this review, extensive genomics research has already delivered and will continue to deliver intriguing results on the co-evolution of the immune and endocrine systems. Moreover, strong conservation over millions of years of evolution provides indications for its physiological impact.

Finally, a solid proteomics program will without doubt be indispensible. For research in teleosts it is crucial that homologous proteins become available. Sequence homology between orthologue cytokines of men and fish is often between 20 and 30%. This implies important differences in binding or affinity to antibodies, receptors or binding proteins. Moreover, when non-homologous proteins are used, the danger of immune activity against these proteins cannot be excluded.

A bi-directional communication between the neuroendocrine system and the immune system of fish, long neglected, is now generally accepted and the evolutionary conserved nature of this interaction clarifies its physiological importance. The greatest challenge, however, will be the analysis of the processes that take place at a central level and the determination of the direct or indirect influence of peripheral factors. Until now, little evidence was available to indicate that the neuroendocrine system responds equally to stressors as to an immune challenge. The practical implications of these studies are numerous and of great importance for successful aquaculture practice.

REFERENCES

Abe, T., Mikekado, T., Haga, S., Kisara, Y., Watanabe, K., Kurokawa, T., and Suzuki, T. (2007). Identification, cDNA cloning, and mRNA localization of a zebrafish ortholog of leukemia inhibitory factor. *Comp. Biochem. Physiol. B* **147,** 38–44.

Acerete, L., Balasch, J. C., Castellana, B., Redruello, B., Roher, N., Canario, A. V., Planas, J. V., MacKenzie, S., and Tort, L. (2007). Cloning of the glucocorticoid receptor (GR) in gilthead seabream (*Sparus aurata*). Differential expression of GR and immune genes in gilthead seabream after an immune challenge. *Comp. Biochem. Physiol. B* **148,** 32–43.

Ader, R., Cohen, N., and Felten, D. (1995). Psychoneuroimmunology – Interactions between the nervous system and the immune system. *Lancet* **345,** 99–103.

Ahmed, S. R. (2000). The immune system as a potential target for environmental estrogens (endocrine disrupters): A new emerging field. *Toxicology* **150,** 191–206.

Aloisi, F. (2001). Immune function of microglia. *Glia* **36,** 165–179.

Bagriacik, E. U., and Klein, J. R. (2000). The thyrotropin (thyroid-stimulating hormone) receptor is expressed on murine dendritic cells and on a subset of CD45RB(high) lymph node T cells: Functional role for thyroid-stimulating hormone during immune activation. *J. Immunol.* **164,** 6158–6165.

Ball, J. N., and Hawkins, E. F. (1976). Adrenocortical (interrenal) responses to hypophysectomy and adenohypophyseal hormones in teleost *Poecilia latipinna. Gen. Comp. Endocrinol.* **28,** 59–70.

Balm, P. H., van Lieshout, E., Lokate, J., and Wendelaar Bonga, S. E. (1995). Bacterial lipopolysaccharide (LPS) and interleukin-1 (IL-1) exert multiple physiological effects in the tilapia *Oreochromis mossambicus* (Teleostei). *J. Comp. Physiol. B* **165,** 85–92.

Banks, W. A., Kastin, A. J., and Durham, D. A. (1989). Bidirectional transport of interleukin-1-alpha across the blood-brain barrier. *Brain Research Bulletin* **23,** 433–437.

Baoprasertkul, P., Peatman, E., Abernathy, J., and Liu, Z. J. (2007). Structural characterisation and expression analysis of toll-like receptor 2 gene from catfish. *Fish Shellfish Immunol.* **22,** 418–426.

Barrallo, A., Gonzalez-Sarmiento, R., Porteros, A., Garcia-Isidoro, M., and Rodriguez, R. E. (1998a). Cloning, molecular characterization, and distribution of a gene homologous to delta opioid receptor from zebrafish (*Danio rerio*). *Biochem. Biophys. Res. Commun.* **245,** 544–548.

Barrallo, A., Malvar, F. G., Gonzalez, R., Rodriguez, R. E., and Traynor, J. R. (1998b). Cloning and characterization of a delta opioid receptor from zebrafish. *Biochem. Soc. Trans.* **26,** S360.

Barrallo, A., Gonzalez-Sarmiento, R., Alvar, F., and Rodriguez, R. E. (2000). ZFOR2, a new opioid receptor-like gene from the teleost zebrafish (*Danio rerio*). *Brain Res. Mol. Brain Res.* **84,** 1–6.

Bayne, C. J., and Levy, S. (1991). Modulation of the oxidative burst in trout myeloid cells by adrenocorticotropic hormone and catecholamines: Mechanisms of action. *J. Leukoc. Biol.* **50,** 554–560.

Ben-Jonathan, N., Mershon, J. L., Allen, D. L., and Steinmetz, R. W. (1996). Extrapituitary prolactin: Distribution, regulation, functions, and clinical aspects. *Endocr. Rev.* **17,** 639–669.

Berkenbosch, F., Vanoers, J., Delrey, A., Tilders, F., and Besedovsky, H. (1987). Corticotropin-releasing factor producing neurons in the rat activated by interleukin-1. *Science* **238,** 524–526.

Berwaer, M., Martial, J. A., and Davis, J. R. (1994). Characterization of an up-stream promoter directing extrapituitary expression of the human prolactin gene. *Mol. Endocrinol.* **8,** 635–642.

Bette, M., Kaut, O., Schafer, M. K. H., and Weihe, E. (2003). Constitutive expression of p55TNFR mRNA and mitogen-specific up-regulation of TNF alpha and p75TNFR mRNA in mouse brain. *J. Comp. Neurol.* **465,** 417–430.

Bhat, R. S., Bhaskaran, M., Mongia, A., Hitosugi, N., and Singhal, P. C. (2004). Morphine-induced macrophage apoptosis: Oxidative stress and strategies for modulation. *J. Leukoc. Biol.* **75,** 1131–1138.

Bird, S., Zou, J., Savan, R., Kono, T., Sakai, M., Woo, J., and Secombes, C. (2005). Characterisation and expression analysis of an interleukin-6 homologue in the Japanese pufferfish, *Fugu rubripes*. *Dev. Comp. Immunol.* **29,** 775–789.

Blalock, J. E. (1984). The immune system as a sensory organ. *J. Immunol.* **132,** 1067–1070.

Blamire, A. M., Anthony, D. C., Rajagopalan, B., Sibson, N. R., Perry, V. H., and Styles, P. (2000). Interleukin-1 beta-induced changes in blood-brain barrier permeability, apparent diffusion coefficient, and cerebral blood volume in the rat brain: A magnetic resonance study. *J. Neurosci.* **20,** 8153–8159.

Bluthé, R. M., Walter, V., Parnet, P., Laye, S., Lestage, J., Verrier, D., Poole, S., Stenning, B. E., Kelley, K. W., and Dantzer, R. (1994). Lipopolysaccharide induces sickness behavior in rats by a vagal mediated mechanism. *C. R. Acad. Sci. III* **317,** 499–503.

Bouchard, B., Ormandy, C. J., Di Santo, J. P., and Kelly, P. A. (1999). Immune system development and function in prolactin receptor-deficient mice. *J. Immunol.* **163,** 576–582.

Boulay, J. L., O'Shea, J. J., and Paul, W. E. (2003). Molecular phylogeny within type I cytokines and their cognate receptors. *Immunity* **19,** 159–163.

Bridgham, J. T., Carroll, S. M., and Thornton, J. W. (2006). Evolution of hormone-receptor complexity by molecular exploitation. *Science* **312,** 97–101.

Brogden, K. A., Guthmiller, J. M., Salzet, M., and Zasloff, M. (2005). The nervous system and innate immunity: The neuropeptide connection. *Nat. Immunol.* **6,** 558–564.

Buonocore, F., Mazzini, M., Forlenza, M., Randelli, E., Secombes, C. J., Zou, J., and Scapigliati, G. (2004). Expression in *Escherichia coli* and purification of sea bass (*Dicentrarchus labrax*) interleukin-1beta, a possible immunoadjuvant in aquaculture. *Mar. Biotechnol.* **6,** 53–59.

Cabanac, M., and Laberge, F. (1998). Fever in goldfish is induced by pyrogens but not by handling. *Physiol. Behav.* **63,** 377–379.

Cain, K. D., Jones, D. R., and Raison, R. L. (2002). Antibody-antigen kinetics following immunization of rainbow trout (*Oncorhynchus mykiss*) with a T-cell dependent antigen. *Dev. Comp. Immunol.* **26,** 181–190.

Calduch-Giner, J. A., and Perez-Sanchez, J. (1999). Expression of growth hormone gene in the head kidney of gilthead sea bream (*Sparus aurata*). *J. Exp. Zool.* **283,** 326–330.

Calduch-Giner, J. A., Sitja-Bobadilla, A., Alvarez-Pellitero, P., and Perez-Sanchez, J. (1995). Evidence for a direct action of GH on haemopoietic cells of a marine fish, the gilthead sea bream (*Sparus aurata*). *J. Endocrinol.* **146,** 459–467.

Camp, K. L., Wolters, W. R., and Rice, C. D. (2000). Survivability and immune responses after challenge with *Edwardsiella ictaluri* in susceptible and resistant families of channel catfish, *Ictalurus punctatus*. *Fish Shellfish Immunol.* **10,** 475–487.

Cannon, J. G. (1998). Adaptive interactions between cytokines and the hypothalamic-pituitary-gonadal axis. *Ann. N. Y. Acad. Sci.* **856,** 234–242.

Cargouet, M., Perdiz, D., Mouatassim-Souali, A., Tamisier-Karolak, S., and Levi, Y. (2004). Assessment of river contamination by estrogenic compounds in Paris area (France). *Sci. Total Environ.* **324,** 55–66.

Castellana, B., Iliev, D. B., Sepulcre, M. P., MacKenzie, S., Goetz, F. W., Mulero, V., and Planas, J. V. (2008). Molecular characterization of interleukin-6 in the gilthead seabream (*Sparus aurata*). *Mol. Immunol.* **45,** 3363–3370.

Chadzinska, M., and Plytycz, B. (2004). Differential migratory properties of mouse, fish, and frog leukocytes treated with agonists of opioid receptors. *Dev. Comp. Immunol.* **28,** 949–958.

Chadzinska, M., Jozefowski, S., Bigaj, J., and Plytycz, B. (1997). Morphine modulation of thioglycollate-elicited peritoneal inflammation in the goldfish, *Carassius auratus*. *Arch. Immunol. Ther. Exp. (Warsz.)* **45,** 321–327.

Chadzinska, M., Kolaczkowska, E., Seljelid, R., and Plytycz, B. (1999). Morphine modulation of peritoneal inflammation in Atlantic salmon and CB6 mice. *J. Leukoc. Biol.* **65,** 590–596.

Chadzinska, M., Scislowska-Czarnecka, A., and Plytycz, B. (2000). Inhibitory effects of morphine on some inflammation-related parameters in the goldfish *Carassius auratus* L. *Fish Shellfish Immunol.* **10,** 531–542.

Chadzinska, M., Leon-Kloosterziel, K. M., Plytycz, B., and Verburg-van Kemenade, B. M. L. (2008). *In vivo* kinetics of cytokine expression during peritonitis in carp: Evidence for innate and alternative macrophage polarization. *Dev. Comp. Immunol.* **32,** 509–518.

Chadzinska, M., Hermsen, T., Savelkoul, H. F., and Verburg-van Kemenade, B. M. (2009a). Cloning of opioid receptors in common carp (*Cyprinus carpio* L.) and their involvement in regulation of stress and immune response. *Brain Behav. Immun.* **23,** 257–266.

Chadzinska, M., Savelkoul, H. F. J., and Verburg-van Kemenade, B. M. L. (2009b). Morphine affects the inflammatory response in carp by impairment of leukocyte migration. *Dev. Comp. Immunol.* **33,** 88–96.

Chen, W. H., Sun, L. T., Tsai, C. L., Song, Y. L., and Chang, C. F. (2002). Cold-stress induced the modulation of catecholamines, cortisol, immunoglobulin M, and leukocyte phagocytosis in tilapia. *Gen. Comp. Endocrinol.* **126,** 90–100.

Choi, Y., Chuang, L. F., Lam, K. M., Kung, H. F., Wang, J. M., Osburn, B. I., and Chuang, R. Y. (1999). Inhibition of chemokine-induced chemotaxis of monkey leukocytes by mu-opioid receptor agonists. *In Vivo* **13,** 389–396.

Chung, I. Y., and Benveniste, E. N. (1990). Tumor necrosis factor-alpha production by astrocytes–induction by lipopolysaccharide, IFN-gamma, and IL-1-beta. *J. Immunol.* **144,** 2999–3007.

Churchill, L., Taishi, P., Wang, M. F., Brandt, J., Cearley, C., Rehman, A., and Krueger, J. M. (2006). Brain distribution of cytokine mRNA induced by systemic administration of interleukin-1 beta or tumor necrosis factor alpha. *Brain Res.* **1120,** 64–73.

Corbett, A. D., Henderson, G., McKnight, A. T., and Paterson, S. J. (2006). 75 years of opioid research: the exciting but vain quest for the Holy Grail. *Br. J. Pharmacol.* **147**(Suppl. 1), S153–S162.

Correa, S. G., Maccioni, M., Rivero, V. E., Iribarren, P., Sotomayor, C. E., and Riera, C. M. (2007). Cytokines and the immune-neuroendocrine network: What did we learn from infection and autoimmunity? *Cytokine Growth Factor Rev.* **18,** 125–134.

Danielson, P. B., and Dores, R. M. (1999). Molecular evolution of the opioid/orphanin gene family. *Gen. Comp. Endocrinol.* **113,** 169–186.

Dantzer, R. (2006). Cytokine, sickness behavior, and depression. *Neurol. Clin.* **24,** 441–460.

Darlison, M. G., Greten, F. R., Harvey, R. J., Kreienkamp, H. J., Stuhmer, T., Zwiers, H., Lederis, K., and Richter, D. (1997). Opioid receptors from a lower vertebrate (*Catostomus commersoni*): Sequence, pharmacology, coupling to a G-protein-gated inward-rectifying potassium channel (GIRK1), and evolution. *Proc. Natl. Acad. Sci. USA* **94,** 8214–8219.

Deane, E. E., Li, J., and Woo, N. Y. S. (2001). Hormonal status and phagocytic activity in sea bream infected with vibriosis. *Comp. Biochem. Physiol. B* **129,** 687–693.

DeVries, M. E., Kelvin, A. A., Xu, L. L., Ran, L. S., Robinson, J., and Kelvin, D. J. (2006). Defining the origins and evolution of the chemokine/chemokine receptor system. *J. Immunol.* **176,** 401–415.

Dhabhar, F. S. (2003). Stress, leukocyte trafficking, and the augmentation of skin immune function. *Ann. N. Y. Acad. Sci.* **992,** 205–217.

Dorshkind, K., and Horseman, N. D. (2000). The roles of prolactin, growth hormone, insulin-like growth factor-I, and thyroid hormones in lymphocyte development and function: Insights from genetic models of hormone and hormone receptor deficiency. *Endocr. Rev.* **21,** 292–312.

Dorshkind, K., and Horseman, N. D. (2001). Anterior pituitary hormones, stress, and immune system homeostasis. *Bioessays* **23,** 288–294.

Ducouret, B., Tujague, M., Ashraf, J., Mouchel, N., Servel, N., Valotaire, Y., and Thompson, E. B. (1995). Cloning of a teleost fish glucocorticoid receptor shows that it contains a deoxyribonucleic acid-binding domain different from that of mammals. *Endocrinology* **136,** 3774–3783.

Dugan, S. G., Lortie, M. B., Nickerson, J. G., and Moon, T. W. (2003). Regulation of the rainbow trout (*Oncorhynchus mykiss*) hepatic beta(2)-adrenoceptor by adrenergic agonists. *Comp. Biochem. Physiol. B* **136,** 331–342.

Dunn, A. J. (2000). Cytokine activation of the HPA axis. *Ann. N. Y. Acad. Sci.* **917,** 608–617.

Edwards, K. M., Burns, V. E., Carroll, D., Drayson, M., and Ring, C. (2007). The acute stress-induced immunoenhancement hypothesis. *Exerc. Sport Sci. Rev.* **35,** 150–155.

Eisenstein, T. K., Rahim, R. T., Feng, P., Thingalaya, N. K., and Meissler, J. J. (2006). Effects of opioid tolerance and withdrawal on the immune system. *J. Neuroimmune Pharmacol.* **1,** 237–249.

Elenkov, I. J., and Chrousos, G. P. (2006). Stress system – organization, physiology and immunoregulation. *Neuroimmunomodulation* **13,** 257–267.

Engblom, D., Ek, M., Saha, S., Ericsson-Dahlstrand, A., Jakobsson, P. J., and Blomqvist, A. (2002). Prostaglandins as inflammatory messengers across the blood-brain barrier. *J. Mol. Med.* **80,** 5–15.

Engelsma, M. Y., Hougee, S., Nap, D., Hofenk, M., Rombout, J. H., van Muiswinkel, W. B., and Verburg-van Kemenade, B. M. L. (2003). Multiple acute temperature stress affects leucocyte populations and antibody responses in common carp, *Cyprinus carpio* L. *Fish Shellfish Immunol.* **15,** 397–410.

Esteban, M. A., Rodriguez, A., Ayala, A. G., and Meseguer, J. (2004). Effects of high doses of cortisol on innate cellular immune response of seabream (*Sparus aurata* L.). *Gen. Comp. Endocrinol.* **137,** 89–98.

Fabbri, E., Chen, X., Capuzzo, A., and Moon, T. W. (2008). Binding kinetics and sequencing of hepatic alpha(1)-adrenergic receptors in two marine teleosts, mackerel (*Scomber scombrus*) and anchovy (*Engraulis encrasicolus*). *J. Exp. Zool. A* **309A,** 157–165.

Faisal, M., Chiappelli, F., Ahmed, I., Cooper, E. L., and Weiner, H. (1989). Social confrontation "stress" in aggressive fish is associated with an endogenous opioid-mediated suppression of proliferative response to mitogens and nonspecific cytotoxicity. *Brain Behav. Immun.* **3,** 223–233.

Faisal, M., Chiappelli, F., Ahmed, II, Cooper, E. L., and Weiner, H. (1992). [The role of endogenous opioids in modulation of immunosuppression in fish]. *Schriftenr. Ver. Wasser Boden. Lufthyg.* **89,** 785–799.

Fast, M. D., Hosoya, S., Johnson, S. C., and Afonso, L. O. B. (2008). Cortisol response and immune-related effects of Atlantic salmon (*Salmo salar* Linnaeus) subjected to short- and long-term stress. *Fish Shellfish Immunol.* **94,** 194–204.

Felten, D. L., Felten, S. Y., Carlson, S. L., Olschowka, J. A., and Livnat, S. (1985). Noradrenergic and peptidergic innervation of lymphoid tissue. *J. Immunol.* **135,** 755s–765s.

Filby, A. L., Thorpe, K. L., and Tyler, C. R. (2006). Multiple molecular effect pathways of an environmental oestrogen in fish. *J. Mol. Endocrinol.* **37,** 121–134.

Finkenbine, S. S., Gettys, T. W., and Burnett, K. G. (2002). Beta-adrenergic receptors on leukocytes of the channel catfish, *Ictalurus punctatus*. *Comp. Biochem. Physiol. C* **131,** 27–37.

Fischer, U., Utke, K., Somamoto, T., Kollner, B., Ototake, M., and Nakanishi, T. (2006). Cytotoxic activities of fish leucocytes. *Fish Shellfish Immunol.* **20,** 209–226.

Flik, G., Rentier-Delrue, F., and Wendelaar Bonga, S. E. (1994). Calcitropic effects of recombinant prolactins in *Oreochromis mossambicus*. *Am. J. Physiol.* **266,** R1302–R1308.

Flik, G., Klaren, P. H. M., Van den Burg, E. H., Metz, J. R., and Huising, M. O. (2006). CRF and stress in fish. *Gen. Comp. Endocrinol.* **146,** 36–44.

Flory, C. M. (1989). Autonomic innervation of the spleen of the coho salmon, *Oncorhynchus kisutch*. A histochemical demonstration and preliminary assessment of its immunoregulatory role. *Brain Behav. Immun.* **3,** 331–344.

Flory, C. M. (1990). Phylogeny of neuroimmunoregulation: effects of adrenergic and cholinergic agents on the in vitro antibody response of the rainbow trout, *Onchorynchus mykiss*. *Dev. Comp. Immunol.* **14,** 283–294.

Flory, C. M., and Bayne, C. J. (1991). The influence of adrenergic and cholinergic agents on the chemiluminescent and mitogenic responses of leukocytes from the rainbow trout, *Oncorhynchus mykiss*. *Dev. Comp. Immunol.* **15,** 135–142.

Forlenza, M., de Carvalho Dias, J. D., Vesely, T., Pokorova, D., Savelkoul, H. F., and Wiegertjes, G. F. (2008). Transcription of signal-3 cytokines, IL-12 and IFN alpha beta, coincides with the timing of CD8 alpha beta up-regulation during viral infection of common carp (*Cyprinus carpio* L.). *Mol. Immunol.* **45,** 1531–1547.

Frei, K., Malipiero, U. V., Leist, T. P., Zinkernagel, R. M., Schwab, M. E., and Fontana, A. (1989). On the cellular source and function of interleukin-6 produced in the central nervous system in viral diseases. *Eur. J. Immunol.* **19,** 689–694.

Fu, Y. K., Arkins, S., Fuh, G., Cunningham, B. C., Wells, J. A., Fong, S., Cronin, M. J., Dantzer, R., and Kelley, K. W. (1992). Growth hormone augments superoxide anion secretion of human neutrophils by binding to the prolactin receptor. *J. Clin. Invest.* **89,** 451–457.

Garden, G. A., and Moller, T. (2006). Microglia biology in health and disease. *J. Neuroimmune Pharmacol.* **1,** 127–137.

Gellersen, B., Kempf, R., Telgmann, R., and DiMattia, G. E. (1994). Nonpituitary human prolactin gene transcription is independent of Pit-1 and differentially controlled in lymphocytes and in endometrial stroma. *Mol. Endocrinol.* **8,** 356–373.

Gerlo, S., Verdood, P., Hooghe-Peters, E. L., and Kooijman, R. (2005). Modulation of prolactin expression in human T lymphocytes by cytokines. *J. Neuroimmunol.* **162,** 190–193.

Gerlo, S., Davis, J. R., Mager, D. L., and Kooijman, R. (2006). Prolactin in man: A tale of two promoters. *Bioessays* **28,** 1051–1055.

Geven, E. J. W., Verkaar, F., Flik, G., and Klaren, P. H. M. (2006). Experimental hyperthyroidism and central mediators of stress axis and thyroid axis activity in common carp (*Cyprinus carpio* L.). *J. Mol. Endocrinol.* **37,** 443–452.

Geven, E. J. W., Nguyen, N. K., van den Boogaart, M., Spanings, F. A. T., Flik, G., and Klaren, P. H. M. (2007). Comparative thyroidology: thyroid gland location and iodothyronine dynamics in Mozambique tilapia (*Oreochromis mossambicus* Peters) and common carp (*Cyprinus carpio* L.). *J. Exp. Biol.* **210,** 4005–4015.

Giulian, D., Baker, T. J., Shih, L. C. N., and Lachman, L. B. (1986). Interleukin-1 of the central nervous system is produced by ameboid microglia. *J. Exp. Med.* **164,** 594–604.

Goehler, L. E., Gaykema, R. P. A., Nguyen, K. T., Lee, J. E., Tilders, F. J. H., Maier, S. F., and Watkins, L. R. (1999). Interleukin-1 beta in immune cells of the abdominal vagus nerve: A link between the immune and nervous systems? *J. Neurosci.* **19,** 2799–2806.

Gomes, I., Filipovska, J., and Devi, L. A. (2003). Opioid receptor oligomerization. Detection and functional characterization of interacting receptors. *Methods Mol. Med.* **84,** 157–183.

Gomez, G. D., and Balcazar, J. L. (2008). A review on the interactions between gut microbiota and innate immunity of fish. *FEMS Immunol. Med. Microbiol.* **52,** 145–154.

Gonzalez-Nunez, V., Gonzalez-Sarmiento, R., and Rodriguez, R. E. (2003). Characterization of zebrafish proenkephalin reveals novel opioid sequences. *Brain Res. Mol. Brain Res.* **114,** 31–39.

Gonzalez-Nunez, V., Gonzalez-Sarmiento, R., and Rodriguez, R. E. (2003a). Cloning and characterization of a full-length pronociceptin in zebrafish: evidence of the existence of two different nociceptin sequences in the same precursor. *Biochim. Biophys. Acta* **1629,** 114–118.

Gonzalez-Nunez, V., Gonzalez-Sarmiento, R., and Rodriguez, R. E. (2003b). Identification of two proopiomelanocortin genes in zebrafish (*Danio rerio*). *Brain Res. Mol. Brain Res.* **120,** 1–8.

Goshen, I., and Yirmiya, R. (2009) Interleukin-1 (IL-1): A central regulator of stress responses. Frontiers in Neuroendocrinol. **30,** 30–45.

Grether, G. F., Kasahara, S., Kolluru, G. R., and Cooper, E. L. (2004). Sex-specific effects of carotenoid intake on the immunological response to allografts in guppies (*Poecilia reticulata*). *Proc. Biol. Sci.* **271,** 45–49.

Grimm, M. C., Ben-Baruch, A., Taub, D. D., Howard, O. M., Resau, J. H., Wang, J. M., Ali, H., Richardson, R., Snyderman, R., and Oppenheim, J. J. (1998a). Opiates transdeactivate chemokine receptors: delta and mu opiate receptor-mediated heterologous desensitization. *J. Exp. Med.* **188,** 317–325.

Grimm, M. C., Ben-Baruch, A., Taub, D. D., Howard, O. M., Wang, J. M., and Oppenheim, J. J. (1998b). Opiate inhibition of chemokine-induced chemotaxis. *Ann. N. Y. Acad. Sci.* **840,** 9–20.

Haddad, G., Hanington, P. C., Wilson, E. C., Grayfer, L., and Belosevic, M. (2008). Molecular and functional characterization of goldfish (*Carassius auratus* L.) transforming growth factor-beta. *Dev. Comp. Immunol.* **32,** 654–663.

Halabe Bucay, A. (2007). Clinical hypothesis: Application of AIDS vaccines together with thyroid hormones to increase their immunogenic effect. *Vaccine* **25,** 6292–6293.

Hansen, J. D., and Strassburger, P. (2000). Description of an ectothermic TCR coreceptor, CD8 alpha, in rainbow trout. *J. Immunol.* **164,** 3132–3139.

Hansen, J. D., Landis, E. D., and Phillips, R. B. (2005). Discovery of a unique Ig heavy-chain isotype (IgT) in rainbow trout: Implications for a distinctive B cell developmental pathway in teleost fish. *Proc. Natl. Acad. Sci. USA* **102,** 6919–6924.

Harbuz, M. S., Stephanou, A., Sarlis, N., and Lightman, S. L. (1992). The effects of recombinant human interleukin (IL)-1-alpha, IL-1-beta or IL-6 on hypothalamo-pituitary-adrenal axis activation. *J. Endocrinol.* **133,** 349–355.

Harle, P., and Straub, R. H. (2006). Leptin is a link between adipose tissue and inflammation. *Ann. N. Y. Acad. Sci.* **1069,** 454–462.

Harris, J., and Bird, D. J. (2000). Supernatants from leucocytes treated with melanin-concentrating hormone (MCH) and alpha-melanocyte stimulating hormone (alpha-MSH) have a stimulatory effect on rainbow trout (*Oncorhynchus mykiss*) phagocytes *in vitro*. *Vet. Immunol. Immunopathol.* **76,** 117–124.

Haukenes, A. H., and Barton, B. A. (2004). Characterization of the cortisol response following an acute challenge with lipopolysaccharide in yellow perch and the influence of rearing density. *J. Fish Biol.* **64,** 851–862.

Heijnen, C. J. (2007). Receptor regulation in neuroendocrine-immune communication: Current knowledge and future perspectives. *Brain Behav. Immun.* **21,** 1–8.

Heinrich, P. C., Behrmann, I., Haan, S., Hermanns, H. M., Muller-Newen, G., and Schaper, F. (2003). Principles of interleukin (IL)-6-type cytokine signalling and its regulation. *Biochem. J.* **374,** 1–20.

Hermus, A. R. M. M., and Sweep, C. G. J. (1990). Cytokines and the hypothalamic pituitary adrenal axis. *J. Ster. Biochem. Mol. Biol.* **37,** 867–871.

Hirono, I., Takami, M., Miyata, M., Miyazaki, T., Han, H. J., Takano, T., Endo, M., and Aoki, T. (2004). Characterization of gene structure and expression of two toll-like receptors from Japanese flounder, *Paralichthys olivaceus. Immunogenetics* **56,** 38–46.

Holland, J. W., Pottinger, T. G., and Secombes, C. J. (2002). Recombinant interleukin-1 beta activates the hypothalamic-pituitary-interrenal axis in rainbow trout, *Oncorhynchus mykiss. J. Endocrinol.* **175,** 261–267.

Hong, S. H., Peddie, S., Campos-Perez, J. J., Zou, J., and Secombes, C. J. (2003). The effect of intraperitoneally administered recombinant IL-1 beta on immune parameters and resistance to *Aeromonas salmonicida* in the rainbow trout (*Oncorhynchus mykiss*). *Dev. Comp. Immunol.* **27,** 801–812.

Horseman, N. D., Zhao, W., Montecino-Rodriguez, E., Tanaka, M., Nakashima, K., Engle, S. J., Smith, F., Markoff, E., and Dorshkind, K. (1997). Defective mammopoiesis, but normal hematopoiesis, in mice with a targeted disruption of the prolactin gene. *EMBO J.* **16,** 6926–6935.

Howard, J. K., Lord, G. M., Matarese, G., Vendetti, S., Ghatei, M. A., Ritter, M. A., Lechler, R. I., and Bloom, S. R. (1999). Leptin protects mice from starvation-induced lymphoid atrophy and increases thymic cellularity in ob/ob mice. *J. Clin. Invest.* **104,** 1051–1059.

Huising, M. O., Guichelaar, T., Hoek, C., Verburg-van Kemenade, B. M. L., Flik, G., Savelkoul, H. F. J., and Rombout, J. H. W. M. (2003a). Increased efficacy of immersion vaccination in fish with hyperosmotic pretreatment. *Vaccine* **21,** 4178–4193.

Huising, M. O., Stet, R. J. M., Kruiswijk, C. P., Savelkoul, H. F. J., and Verburg-van Kemenade, B. M. L. (2003b). Response to Shields: Molecular evolution of CXC chemokines and receptors. *Trends Immunol.* **24,** 356–357.

Huising, M. O., Stolte, E. H., Flik, G., Savelkoul, H. F. J., and Verburg-van Kemenade, B. M. L. (2003c). CXC chemokines and leukocyte chemotaxis in common carp (*Cyprinus carpio* L.). *Dev. Comp. Immunol.* **27,** 875–888.

Huising, M. O., Metz, J. R., van Schooten, C., Taverne-Thiele, A. J., Hermsen, T., Verburg-van Kemenade, B. M. L., and Flik, G. (2004a). Structural characterisation of a cyprinid (*Cyprinus carpio* L.) CRH, CRH-BP and CRH-R1, and the role of these proteins in the acute stress response. *J. Mol. Endocrinol.* **32,** 627–648.

Huising, M. O., Stet, R. J. M., Savelkoul, H. F. J., and Verburg-van Kemenade, B. M. L. (2004b). The molecular evolution of the interleukin-1 family of cytokines; IL-18 in teleost fish. *Dev. Comp. Immunol.* **28,** 395–413.

Huising, M. O., van der Meulen, T., Flik, G., and Verburg-van Kemenade, B. M. L. (2004c). Three novel carp CXC chemokines are expressed early in ontogeny and at nonimmune sites. *Eur. J. Biochem.* **271,** 4094–4106.

Huising, M. O., Kruiswijk, C. P., van Schijndel, J. E., Savelkoul, H. F. J., Flik, G., and Verburg-van Kemenade, B. M. L. (2005). Multiple and highly divergent IL-11 genes in teleost fish. *Immunogenetics* **57,** 432–443.

Huising, M. O., Geven, E. J. W., Kruiswijk, C. P., Nabuurs, S. B., Stolte, E. H., Spanings, F. A. T., Verburg-van Kemenade, B. M. L., and Flik, G. (2006a). Increased leptin expression in common carp (*Cyprinus carpio*) after food intake but not after fasting or feeding to satiation. *Endocrinology* **147,** 5786–5797.

Huising, M. O., Kruiswijk, C. P., and Flik, G. (2006b). Phylogeny and evolution of class-I helical cytokines. *J. Endocrinol.* **189,** 1–25.

Huising, M. O., van Schijndel, J. E., Kruiswijk, C. P., Nabuurs, S. B., Savelkoul, H. F. J., Flik, G., and Verburg-van Kemenade, B. M. L. (2006c). The presence of multiple and differentially regulated interleukin-12p40 genes in bony fishes signifies an expansion of the vertebrate heterodimeric cytokine family. *Mol. Immunol.* **43,** 1519–1533.

Hull, K. L., and Harvey, S. (1997). Growth hormone: An immune regulator in vertebrates. *In* "Advances in Comparative Endocrinology" (Kawashima, S., and Kikuyama, S., Eds.), pp. 565–572. Monduzzi Editore, Bologna.

Iliev, D. B., Castellana, B., MacKenzie, S., Planas, J. V., and Goetz, F. W. (2007). Cloning and expression analysis of an IL-6 homolog in rainbow trout (*Oncorhynchus mykiss*). *Mol. Immunol.* **44,** 1803–1807.

Jeong, J. Y., Kwon, H. B., Ahn, J. C., Kang, D., Kwon, S. H., Park, J. A., and Kim, K. W. (2008). Functional and developmental analysis of the blood-brain barrier in zebrafish. *Brain Res. Bull.* **75,** 619–628.

Jordan, B. A., Trapaidze, N., Gomes, I., Nivarthi, R., and Devi, L. A. (2001). Oligomerization of opioid receptors with beta 2-adrenergic receptors: A role in trafficking and mitogen-activated protein kinase activation. *Proc. Natl. Acad. Sci. USA* **98,** 343–348.

Jozefowski, S., and Plytycz, B. (1997). Characterization of opiate binding sites on the goldfish (*Carassius auratus* L.) pronephric leukocytes. *Pol. J. Pharmacol.* **49,** 229–237.

Jozefowski, S. J., and Plytycz, B. (1998). Characterization of beta-adrenergic receptors in fish and amphibian lymphoid organs. *Dev. Comp. Immunol.* **22,** 587–603.

Kaattari, S. L., Zhang, H. L., Khor, I. W., Kaattari, I. M., and Shapiro, D. A. (2002). Affinity maturation in trout: Clonal dominance of high affinity antibodies late in the immune response. *Dev. Comp. Immunol.* **26,** 191–200.

Karanth, S., and McCann, S. M. (1991). Anterior pituitary hormone control by interleukin-2. *Proc. Natl. Acad. Sci. USA* **88,** 2961–2965.

Kawashima, K., and Fujii, T. (2003). The lymphocytic cholinergic system and its contribution to the regulation of immune activity. *Life Sci.* **74,** 675–696.

Kelley, K. W., Bluthe, R. M., Dantzer, R., Zhou, J. H., Shen, W. H., Johnson, R. W., and Broussard, S. R. (2003). Cytokine-induced sickness behavior. *Brain Behav. Immun.* **17,** S112–S118.

Kelley, K. W., Weigent, D. A., and Kooijman, R. (2007). Protein hormones and immunity. *Brain Behav. Immun.* **21,** 384–392.

Kieffer, B. L., and Gaveriaux-Ruff, C. (2002). Exploring the opioid system by gene knockout. *Prog. Neurobiol.* **66,** 285–306.

Kim, D. S., and Melmed, S. (1999). Stimulatory effect of leukemia inhibitory factor on ACTH secretion of dispersed rat pituitary cells. *Endocr. Res.* **25,** 11–19.

Kolodziej, E. P., Harter, T., and Sedlak, D. L. (2004). Dairy wastewater, aquaculture, and spawning fish as sources of steroid hormones in the aquatic environment. *Environ. Sci. Technol.* **38,** 6377–6384.

Konsman, J. P., Vigues, S., Mackerlova, L., Bristow, A., and Blomqvist, A. (2004). Rat brain vascular distribution of interleukin-1 type-1 receptor immunoreactivity: Relationship to patterns of inducible cyclooxygenase expression by peripheral inflammatory stimuli. *J. Comp. Neurol.* **472,** 113–129.

Kooijman, R., Gerlo, S., Coppens, A., and Hooghe-Peters, E. L. (2000). Growth hormone and prolactin expression in the immune system. *Ann. N. Y. Acad. Sci.* **917,** 534–540.

La Cava, A., and Matarese, G. (2004). The weight of leptin in immunity. *Nat. Rev. Immunol.* **4,** 371–379.

Lago, R., Gómez, R., Lago, F., Gómez-Reino, J., and Gualillo, O. (2008). Leptin beyond body weight regulation – current concepts concerning its role in immune function and inflammation. *Cell. Immunol.* **252,** 139–145.

Laing, K. J., and Secombes, C. J. (2004). Chemokines. *Dev. Comp. Immunol.* **28,** 443–460.

Lam, S. H., Sin, Y. M., Gong, Z., and Lam, T. J. (2005). Effects of thyroid hormone on the development of immune system in zebrafish. *Gen. Comp. Endocrinol.* **142,** 325–335.

Laurent, V., Jaubert-Miazza, L., Desjardins, R., Day, R., and Lindberg, I. (2004). Biosynthesis of proopiomelanocortin-derived peptides in prohormone convertase 2 and 7B2 null mice. *Endocrinology* **145,** 519–528.

Law, W. Y., Chen, W. H., Song, Y. L., Dufour, S., and Chang, C. F. (2001). Differential *in vitro* suppressive effects of steroids on leukocyte phagocytosis in two teleosts, tilapia and common carp. *Gen. Comp. Endocrinol.* **121,** 163–172.

Lewis, R. S., and Ward, A. C. (2004). Conservation, duplication and divergence of the zebrafish stat5 genes. *Gene* **338,** 65–74.

Lombardi, M. S., Kavelaars, A., Penela, P., Scholtens, E. J., Roccio, M., Schmidt, R. E., Schedlowski, M., Mayor, F., and Heijnen, C. J. (2002). Oxidative stress decreases G protein-coupled receptor kinase 2 in lymphocytes via a calpain-dependent mechanism. *Mol. Pharmacol.* **62,** 379–388.

Lutfalla, G., Crollius, H. R., Stange-Thomann, N., Jaillon, O., Mogensen, K., and Monneron, D. (2003). Comparative genomic analysis reveals independent expansion of a lineage-specific gene family in vertebrates: The class II cytokine receptors and their ligands in mammals and fish. *BMC Genomics* **4,** 29.

Ma, Q., Jones, D., Borghesani, P. R., Segal, R. A., Nagasawa, T., Kishimoto, T., Bronson, R. T., and Springer, T. A. (1998). Impaired B-lymphopoiesis, myelopoiesis, and derailed cerebellar neuron migration in CXCR4- and SDF-1-deficient mice. *Proc. Natl. Acad. Sci. USA* **95,** 9448–9453.

Magnadottir, B. (2006). Innate immunity of fish (overview). *Fish Shellfish Immunol.* **20,** 137–151.

Magnadottir, B., Lange, S., Gudmundsdottir, S., Bogwald, J., and Dalmo, R. A. (2005). Ontogeny of humoral immune parameters in fish. *Fish Shellfish Immunol.* **19,** 429–439.

Malik, A. A., Radhakrishnan, N., Reddy, K., Smith, A. D., and Singhal, P. C. (2002). Morphine-induced macrophage apoptosis modulates migration of macrophages: Use of *in vitro* model of urinary tract infection. *J. Endourol.* **16,** 605–610.

Martin-Romero, C., and Sanchez-Margalet, V. (2001). Human leptin activates PI3K and MAPK pathways in human peripheral blood mononuclear cells: Possible role of Sam68. *Cell. Immunol.* **212,** 83–91.

Martin, S. A. M., Zou, J., Houlihan, D. F., and Secombes, C. J. (2007). Directional responses following recombinant cytokine stimulation of rainbow trout (*Oncorhynchus mykiss*) RTS-II macrophage cells as revealed by transcriptome profiling. *BMC Genomics* **8,** 150.

Marx, J., Hilbig, R., and Rahmann, H. (1984). Endotoxin and prostaglandin E1 fail to induce fever in a teleost fish. *Comp. Biochem. Physiol. A* **77,** 483–487.

Mastorakos, G., Weber, J. S., Magiakou, M. A., Gunn, H., and Chrousos, G. P. (1994). Hypothalamic pituitary adrenal axis activation and stimulation of systemic vasopressin secretion by recombinant interleukin-6 in humans – potential implications for the syndrome of inappropriate vasopressin secretion. *J. Clin. Endocrinol. Met.* **79,** 934–939.

Maule, A. G., and Schreck, C. B. (1991). Stress and cortisol treatment changed affinity and number of glucocorticoid receptors in leukocytes and gill of coho salmon. *Gen. Comp. Endocrinol.* **84,** 83–93.

Maule, A. G., Schreck, C. B., and Kaattari, S. L. (1987). Changes in the immune system of coho salmon (*Oncorhynchus kisutch*) during the parr to smolt transformation and after implantation of cortisol. *Can. J. Fish. Aquat. Sci.* **44,** 161–166.

Mazon, A. F., Verburg-van Kemenade, B. M. L., Flik, G., and Huising, M. O. (2006). Corticotropin-releasing hormone-receptor 1 (CRH-R1) and CRH-binding protein (CRH-BP) are expressed in the gills and skin of common carp *Cyprinus carpio* L. and respond to acute stress and infection. *J. Exp. Biol.* **209,** 510–517.

McCarthy, L., Wetzel, M., Sliker, J. K., Eisenstein, T. K., and Rogers, T. J. (2001). Opioids, opioid receptors, and the immune response. *Drug Alcohol Depend.* **62,** 111–123.

Meijer, A. H., Gabby Krens, S. F., Medina Rodriguez, I. A., He, S., Bitter, W., Ewa Snaar-Jagalska, B., and Spaink, H. P. (2004). Expression analysis of the Toll-like receptor and TIR domain adaptor families of zebrafish. *Mol. Immunol.* **40,** 773–783.

Mellon, R. D., and Bayer, B. M. (1998). Evidence for central opioid receptors in the immunomodulatory effects of morphine: Review of potential mechanism(s) of action. *J. Neuroimmunol.* **83,** 19–28.

Metz, J. R., Huising, M. O., Meek, J., Taverne-Thiele, A. J., Wendelaar Bonga, S. E., and Flik, G. (2004). Localization, expression and control of adrenocorticotropic hormone in the nucleus preopticus and pituitary gland of common carp (*Cyprinus carpio* L.). *J. Endocrinol.* **182,** 23–31.

Metz, J. R., Huising, M. O., Leon, K., Verburg-van Kemenade, B. M. L., and Flik, G. (2006). Central and peripheral interleukin-1beta and interleukin-1 receptor I expression and their role in the acute stress response of common carp, *Cyprinus carpio* L. *J. Endocrinol.* **191,** 25–35.

Miyagi, T., Chuang, L. F., Lam, K. M., Kung, H., Wang, J. M., Osburn, B. I., and Chuang, R. Y. (2000). Opioids suppress chemokine-mediated migration of monkey neutrophils and monocytes – an instant response. *Immunopharmacology* **47,** 53–62.

Mola, L., Gambarelli, A., Pederzoli, A., and Ottaviani, E. (2005). ACTH response to LPS in the first stages of development of the fish *Dicentrarchus labrax* L. *Gen. Comp. Endocrinol.* **143,** 99–103.

Montgomery, D. W. (2001). Prolactin production by immune cells. *Lupus* **10,** 665–675.

Mori, T., and Devlin, R. H. (1999). Transgene and host growth hormone gene expression in pituitary and nonpituitary tissues of normal and growth hormone transgenic salmon. *Mol. Cell. Endocrinol.* **149,** 129–139.

Munoz, P., Calduch-Giner, J. A., Sitja-Bobadilla, A., Alvarez-Pellitero, P., and Perez-Sanchez, J. (1998). Modulation of the respiratory burst activity of Mediterranean sea bass (*Dicentrarchus labrax* L.) phagocytes by growth hormone and parasitic status. *Fish Shellfish Immunol.* **8,** 25–36.

Nadjar, A., Combe, C., Laye, S., Tridon, V., Dantzer, R., Amedee, T., and Parnet, P. (2003). Nuclear factor kappa B nuclear translocation as a crucial marker of brain response to interleukin-1. A study in rat and interleukin-1 type I deficient mouse. *J. Neurochem.* **87,** 1024–1036.

Nadjar, A., Combe, C., Busquet, P., Dantzer, R., and Parnet, P. (2005). Signaling pathways of interleukin-1 actions in the brain: Anatomical distribution of phospho-ERK1/2 in the brain of rat treated systemically with interleukin-1 beta. *Neuroscience* **134,** 921–932.

Nakanishi, T. (1986). Seasonal changes in the humoral immune response and the lymphoid tissues of the marine teleost, *Sebastiscus marmoratus*. *Vet. Immunol. Immunopathol.* **12,** 213–221.

Nam, B. H., Byon, J. Y., Kim, Y. O., Park, E. M., Cho, Y. C., and Cheong, J. (2007). Molecular cloning and characterisation of the flounder (*Paralichthys olivaceus*) interleukin-6 gene. *Fish Shellfish Immunol.* **23,** 231–236.

Nanaware, Y. K., Baker, B. I., and Tomlinson, M. G. (1994). The effect of various stresses, corticosteroids and adrenergic agents on phagocytosis in the rainbow trout, *Oncorhynchus mykiss*. *Fish Physiol. Biochem.* **13,** 31–40.

Nance, D. M., and Sanders, V. M. (2007). Autonomic innervation and regulation of the immune system (1987–2007). *Brain Behav. Immun.* **21,** 736–745.

Nascimento, D. S., do Vale, A., Tomas, A. M., Zou, J., Secombes, C. J., and dos Santos, N. M. (2007). Cloning, promoter analysis and expression in response to bacterial exposure of sea bass (*Dicentrarchus labrax* L.) interleukin-12 p40 and p35 subunits. *Mol. Immunol.* **44,** 2277–2291.

Nevid, N. J., and Meier, A. H. (1995). Timed daily administrations of hormones and antagonists of neuroendocrine receptors alter day-night rhythms of allograft rejection in the gulf killifish, *Fundulus grandis*. *Gen. Comp. Endocrinol.* **97,** 327–339.

Orbach, H., and Shoenfeld, Y. (2007). Hyperprolactinemia and autoimmune diseases. *Autoimmun. Rev.* **6,** 537–542.

Ottaviani, E., Franchini, A., and Franceschi, C. (1995). Evidence for the presence of immunoreactive POMC-derived peptides and cytokines in the thymus of the goldfish (*Carassius auratus*). *Histochem. J.* **27,** 597–601.

Pagniello, K. B., Bols, N. C., and Lee, L. E. (2002). Effect of corticosteroids on viability and proliferation of the rainbow trout monocyte/macrophage cell line, RTS11. *Fish Shellfish Immunol.* **13,** 199–214.

Palaksha, K. J., Shin, G. W., Kim, Y. R., and Jung, T. S. (2008). Evaluation of non-specific immune components from the skin mucus of olive flounder (*Paralichthys olivaceus*). *Fish Shellfish Immunol.* **24,** 479–488.

Patino, R., Xia, Z. F., Gale, W. L., Wu, C. F., Maule, A. G., and Chang, X. T. (2000). Novel transcripts of the estrogen receptor alpha gene in channel catfish. *Gen. Comp. Endocrinol.* **120,** 314–325.

Peatman, E., and Liu, Z. J. (2007). Evolution of CC chemokines in teleost fish: A case study in gene duplication and implications for immune diversity. *Immunogenetics* **59,** 613–623.

Perez, A. R., Roggero, E., Nicora, A., Palazzi, J., Besedovsky, H. O., del Rey, A., and Bottasso, O. A. (2007). Thymus atrophy during *Trypanosoma cruzi* infection is caused by an immuno-endocrine imbalance. *Brain Behav. Immun.* **21,** 890–900.

Peterson, P. K., Molitor, T. W., and Chao, C. C. (1998). The opioid–cytokine connection. *J. Neuroimmunol.* **83,** 63–69.

Pinal-Seoane, N., Martin, I. R., Gonzalez-Nunez, V., de Velasco, E. M., Alvarez, F. A., Sarmiento, R. G., and Rodriguez, R. E. (2006). Characterization of a new duplicate delta-opioid receptor from zebrafish. *J. Mol. Endocrinol.* **37,** 391–403.

Pinto, R. D., Nascimento, D. S., Reis, M. I. R., do Vale, A., and dos Santos, N. M. S. (2007). Molecular characterization, 3D modelling and expression analysis of sea bass (*Dicentrarchus labrax* L.) interleukin-10. *Mol. Immunol.* **44,** 2056–2065.

Pratt, W. B., and Toft, D. O. (1997). Steroid receptor interactions with heat shock protein and immunophilin chaperones. *Endocr. Rev.* **18,** 306–360.

Prickett, T. C. R., Inder, W. J., Evans, M. J., and Donald, R. A. (2000). Interleukin-1 potentiates basal and AVP-stimulated ACTH secretion *in vitro* – The role of CRH pre-incubation. *Horm. Metab. Res.* **32,** 350–354.

Prunet, P., Sandra, O., Le Rouzic, P., Marchand, O., and Laudet, V. (2000). Molecular characterization of the prolactin receptor in two fish species, tilapia *Oreochromis niloticus* and rainbow trout, *Oncorhynchus mykiss*: A comparative approach. *Can. J. Physiol. Pharmacol.* **78,** 1086–1096.

Przewlocki, R., and Przewlocka, B. (2005). Opioids in neuropathic pain. *Curr. Pharm. Des.* **11,** 3013–3025.

Purcell, M. K., Smith, K. D., Aderem, A., Hood, L., Winton, J. R., and Roach, J. C. (2006). Conservation of Toll-like receptor signaling pathways in teleost fish. *Comp. Biochem. Physiol. D* **1,** 77–88.

Quan, N., and Banks, W. A. (2007). Brain–immune communication pathways. *Brain Behav. Immun.* **21,** 727–735.

Quan, N., Whiteside, M., and Herkenham, M. (1998). Time course and localization patterns of interleukin-1 beta messenger RNA expression in brain and pituitary after peripheral administration of lipopolysaccharide. *Neuroscience* **83,** 281–293.

Raida, M. K., and Buchmann, K. (2007). Temperature-dependent expression of immune-relevant genes in rainbow trout following *Yersinia ruckeri* vaccination. *Dis. Aquat. Organ.* **77,** 41–52.

Redelman, D., Welniak, L. A., Taub, D., and Murphy, W. J. (2008). Neuroendocrine hormones such as growth hormone and prolactin are integral members of the immunological cytokine network. *Cell. Immunol.* **252,** 111–121.

Reynolds, W. W., Casterlin, M. E., and Covert, J. B. (1976). Behavioural fever in teleost fishes. *Nature* **259,** 41–42.

Rijkers, G. T., Frederix-Wolters, E. M. H., and Van Muiswinkel, W. B. (1980). The immune system of cyprinid fish – kinetics and temperature dependence of antibody-producing cells in carp (*Cyprinus carpio*). *Immunology* **41,** 91–97.

Riley, L. G., Hirano, T., and Grau, E. G. (2004). Estradiol-17 beta and dihydrotestosterone differentially regulate vitellogenin and insulin-like growth factor-I production in primary hepatocytes of the tilapia *Oreochromis mossambicus. Comp. Biochem. Physiol. C* **138,** 177–186.

Rivest, S., Lacroix, S., Vallieres, L., Nadeau, S., Zhang, J., and Laflamme, N. (2000). How the blood talks to the brain parenchyma and the paraventricular nucleus of the hypothalamus during systemic inflammatory and infectious stimuli. *Proc. Soc. Exp. Biol. Med.* **223,** 22–38.

Robertsen, B. (2006). The interferon system of teleost fish. *Fish Shellfish Immunol.* **20,** 172–191.

Rogers, T. J., Steele, A. D., Howard, O. M., and Oppenheim, J. J. (2000). Bidirectional heterologous desensitization of opioid and chemokine receptors. *Ann. N. Y. Acad. Sci.* **917,** 19–28.

Roitt, I., Delves, P., Martin, S., and Burton, D. (2006). "Roitt's Essential Immunology." Blackwell Publishing, London.

Roy, B., and Rai, U. (2008). Role of adrenoceptor-coupled second messenger system in sympatho-adrenomedullary modulation of splenic macrophage functions in live fish *Channa punctatus. Gen. Comp. Endocrinol.* **155,** 298–306.

Saeij, J. P., Stet, R. J., de Vries, B. J., van Muiswinkel, W. B., and Wiegertjes, G. F. (2003a). Molecular and functional characterization of carp TNF: A link between TNF polymorphism and trypanotolerance? *Dev. Comp. Immunol.* **27,** 29–41.

Saeij, J. P., Verburg-van Kemenade, B. M. L., van Muiswinkel, W. B., and Wiegertjes, G. F. (2003b). Daily handling stress reduces resistance of carp to *Trypanoplasma borreli*: *In vitro* modulatory effects of cortisol on leukocyte function and apoptosis. *Dev. Comp. Immunol.* **27,** 233–245.

Saha, N. R., Usami, T., and Suzuki, Y. (2003). A double staining flow cytometric assay for the detection of steroid induced apoptotic leucocytes in common carp (*Cyprinus carpio*). *Dev. Comp. Immunol.* **27,** 351–363.

Saha, N. R., Usami, T., and Suzuki, Y. (2004). *In vitro* effects of steroid hormones on IgM-secreting cells and IgM secretion in common carp (*Cyprinus carpio*). *Fish Shellfish Immunol.* **17,** 149–158.

Sakai, M., Kobayashi, M., and Kawauchi, H. (1995). Enhancement of chemiluminescent responses of phagocytic cells from rainbow trout, *Oncorhynchus mykiss*, by injection of growth hormone. *Fish Shellfish Immunol.* **5,** 375–379.

Sakai, M., Kobayashi, M., and Kawauchi, H. (1996). *In vitro* activation of fish phagocytic cells by GH, prolactin and somatolactin. *J. Endocrinol.* **151,** 113–118.

Sakai, M., Yamaguchi, T., Watanuki, H., Yasuda, A., and Takahashi, A. (2001). Modulation of fish phagocytic cells by N-terminal peptides of proopiomelanocortin (NPP). *J. Exp. Zool.* **290,** 341–346.

Salzet, M., Tasiemski, A., and Cooper, E. (2006). Innate immunity in lophotrochozoans: The annelids. *Curr. Pharm. Des.* **12,** 3043–3050.

Sanchez-Margalet, V., Martin-Romero, C., Santos-Alvarez, J., Goberna, R., Najib, S., and Gonzalez-Yanes, C. (2003). Role of leptin as an immunomodulator of blood mononuclear cells: mechanisms of action. *Clin. Exp. Immunol.* **133,** 11–19.

Sandra, O., Le Rouzic, P., Cauty, C., Edery, M., and Prunet, P. (2000). Expression of the prolactin receptor (tiPRL-R) gene in tilapia *Oreochromis niloticus*: Tissue distribution and cellular localization in osmoregulatory organs. *J. Mol. Endocrinol.* **24,** 215–224.

Sanmartin, M. L., Parama, A., Castro, R., Cabaleiro, S., Leiro, J., Lamas, J., and Barja, J. L. (2008). Vaccination of turbot, *Psetta maxima* (L.), against the protozoan parasite *Philasterides dicentrarchi*: Effects on antibody production and protection. *J. Fish Dis.* **31,** 135–140.

Santos, C. R., Ingleton, P. M., Cavaco, J. E., Kelly, P. A., Edery, M., and Power, D. M. (2001). Cloning, characterization, and tissue distribution of prolactin receptor in the sea bream (*Sparus aurata*). *Gen. Comp. Endocrinol.* **121,** 32–47.

Sapolsky, R., Rivier, C., Yamamoto, G., Plotsky, P., and Vale, W. (1987). Interleukin-1 stimulates the secretion of hypothalamic corticotropin-releasing factor. *Science* **238,** 522–524.

Savan, R., Aman, A., Nakao, M., Watanuki, H., and Sakai, M. (2005). Discovery of a novel immunoglobulin heavy chain gene chimera from common carp (*Cyprinus carpio* L.). *Immunogenetics* **57,** 458–463.

Schwartz, M. W., Woods, S. C., Porte, D. J., Seeley, R. J., and Baskin, D. G. (2000). Central nervous system control of food intake. *Nature* **6,** 661–671.

Segner, H., Eppler, E., and Reinecke, M. (2006). The impact of environmental hormonally active substances on the endocrine and immune systems of fish. *In* "Fish Endocrinology" (Reinecke, M., Zaccone, G., and Kapoor, B. G., Eds.), pp. 809–865. Science Publishers, Enfield.

Seppola, M., Larsen, A. N., Steiro, K., Robertsen, B., and Jensen, I. (2008). Characterisation and expression analysis of the interleukin genes, IL-1beta, IL-8 and IL-10, in Atlantic cod (*Gadus morhua* L.). *Mol. Immunol.* **45,** 887–897.

Sharp, B. M. (2006). Multiple opioid receptors on immune cells modulate intracellular signaling. *Brain Behav. Immun.* **20,** 9–14.

Shavit, Y., Lewis, J. W., Terman, G. W., Gale, R. P., and Liebeskind, J. C. (1984). Opioid peptides mediate the suppressive effect of stress on natural killer cell cytotoxicity. *Science* **223,** 188–190.

Shek, P. N., and Sabiston, B. H. (1983). Neuroendocrine regulation of immune processes – change in circulating corticosterone levels induced by the primary antibody response in mice. *Int. J. Immunopharmacol.* **5,** 23–33.

Shintani, F., Nakaki, T., Kanba, S., Kato, R., and Asai, M. (1995). Role of interleukin-1 in stress responses – a putative neurotransmitter. *Mol. Neurobiol.* **10,** 47–71.

Singh, R., and Rai, U. (2008). beta-Endorphin regulates diverse functions of splenic phagocytes through different opioid receptors in freshwater fish *Channa punctatus* (Bloch): An *in vitro* study. *Dev. Comp. Immunol.* **32,** 330–338.

Slicher, A. M. (1961). Endocrinological and hematological studies in *Fundulus heteroclitus* (Linn.). *Bull. Bingham Oceanogr. Coll.* **17,** 1–55.

Smith, E. M., Cadet, P., Stefano, G. B., Opp, M. R., and Hughes, T. K. (1999). IL-10 as a mediator in the HPA axis and brain. *J. Neuroimmunol.* **100,** 140–148.

Soares, M. J. (2004). The prolactin and growth hormone families: Pregnancy-specific hormones/cytokines at the maternal-fetal interface. *Reprod. Biol. Endocrinol.* **2,** 51.

Sohm, F., Pezet, A., Sandra, O., Prunet, P., de Luze, A., and Edery, M. (1998). Activation of gene transcription by tilapia prolactin variants tiPRL188 and tiPRL177. *FEBS Lett.* **438,** 119–123.

Stefano, G. B., Salzet, B., and Fricchione, G. L. (1998). Enkelytin and opioid peptide association in invertebrates and vertebrates: Immune activation and pain. *Immunol. Today* **19,** 265–268.

Stet, R. J. M., Kruiswijk, C. P., and Dixon, B. (2003). Major histocompatibility lineages and immune gene function in teleost fishes: The road not taken. *Crit. Rev. Immunol.* **23,** 441–471.

Stevens, C. W. (2004). Opioid research in amphibians: An alternative pain model yielding insights on the evolution of opioid receptors. *Brain Res. Brain Res. Rev.* **46,** 204–215.

Stolte, E. H., Verburg-van Kemenade, B. M. L., Savelkoul, H. F. J., and Flik, G. (2006). Evolution of glucocorticoid receptors with different glucocorticoid sensitivity. *J. Endocrinol.* **190,** 17–28.

Stolte, E. H., Chadzinska, M., Przybylska, D., Flik, G., Savelkoul, H. F., and Verburg-van Kemenade, B. M. L. (2008a). The immune response differentially regulates Hsp70 and glucocorticoid receptor expression *in vitro* and *in vivo* in common carp (*Cyprinus carpio* L.). *Fish Shellfish Immunol.* [Epub ahead of print].

Stolte, E. H., de Mazon, A. F., Leon-Kloosterziel, K. M., Jesiak, M., Bury, N. R., Sturm, A., Savelkoul, H. F., Verburg-van Kemenade, B. M. L., and Flik, G. (2008b). Corticosteroid receptors involved in stress regulation in common carp, *Cyprinus carpio*. *J. Endocrinol.* **198,** 403–417.

Stolte, E. H., Nabuurs, S. B., Bury, N. R., Sturm, A., Flik, G., Savelkoul, H. F., and Verburg-van Kemenade, B. M. L. (2008c). Corticosteroid receptors and pro-inflammatory cytokines. *Mol. Immunol.* **46,** 70–79.

Sueiro, C., Carrera, I., Ferreiro, S., Molist, P., Adrio, F., Anadon, R., and Rodriguez-Moldes, I. (2007). New insights on saccus vasculosus evolution: A developmental and immunohistochemical study in elasmobranchs. *Brain Behav. Evol.* **70,** 187–204.

Suetake, H., Araki, K., and Suzuki, Y. (2004). Cloning, expression, and characterization of fugu CD4, the first ectothermic animal CD4. *Immunogenetics* **56,** 368–374.

Suzuki, S., Chuang, L. F., Yau, P., Doi, R. H., and Chuang, R. Y. (2002). Interactions of opioid and chemokine receptors: Oligomerization of mu, kappa, and delta with CCR5 on immune cells. *Exp. Cell Res.* **280,** 192–200.

Swennen, D., Rentier-Delrue, F., Auperin, B., Prunet, P., Flik, G., Wendelaar Bonga, S. E., Lion, M., and Martial, J. A. (1991). Production and purification of biologically active recombinant tilapia (*Oreochromis niloticus*) prolactins. *J. Endocrinol.* **131,** 219–227.

Tafalla, C., Coll, J., and Secombes, C. J. (2005). Expression of genes related to the early immune response in rainbow trout (*Oncorhynchus mykiss*) after viral haemorrhagic septicemia virus (VHSV) infection. *Dev. Comp. Immunol.* **29,** 615–626.

Terova, G., Gornati, R., Rimoldi, S., Bernardini, G., and Saroglia, M. (2005). Quantification of a glucocorticoid receptor in sea bass (*Dicentrarchus labrax*, L.) reared at high stocking density. *Gene* **357,** 144–151.

Tomida, M., Yoshida, U., Mogi, C., Maruyama, M., Goda, H., Hatta, Y., and Inoue, K. (2001). Leukaemia inhibitory factor and interleukin-6 inhibit secretion of prolactin and growth hormone by rat pituitary MtT/SM cells. *Cytokine* **14,** 202–207.

Tse, D. L., Chow, B. K., Chan, C. B., Lee, L. T., and Cheng, C. H. (2000). Molecular cloning and expression studies of a prolactin receptor in goldfish (*Carassius auratus*). *Life Sci.* **66,** 593–605.

Turnbull, A. V., and Rivier, C. L. (1999). Regulation of the hypothalamic-pituitary-adrenal axis by cytokines: actions and mechanisms of action. *Physiol. Rev.* **79,** 1–71.

van Muiswinkel, W. B. (1995). "The Piscine Immune System: Innate and Acquired Immunity." CAB International, Wallingford, UK.

Verburg-van Kemenade, B. M. L., Nowak, B., Engelsma, M. Y., and Weyts, F. A. A. (1999). Differential effects of cortisol on apoptosis and proliferation of carp B-lymphocytes from head kidney, spleen and blood. *Fish Shellfish Immunol.* **9,** 405–415.

Verburg-van Kemenade, B. M. L., Savelkoul, H. F., and Chadzinska, M. (2009). Function for the opioid system during inflammation in carp. *Ann. NY Acad Sci.* in press.

Very, N. M., Kittilson, J. D., Norbeck, L. A., and Sheridan, M. A. (2005). Isolation, characterization, and distribution of two cDNAs encoding for growth hormone receptor in rainbow trout (*Oncorhynchus mykiss*). *Comp. Biochem. Physiol. B* **140,** 615–628.

Vilcek, J. (2003). "The Cytokines; An Overview." Academic Press, London.

Vitkovic, L., Konsman, J. P., Bockaert, J., Dantzer, R., Homburger, V., and Jacque, C. (2000). Cytokine signals propagate through the brain. *Mol. Psychiatry* **5,** 604–615.

Viveros-Paredes, J. M., Puebla-Perez, A. M., Gutierrez-Coronado, O., Sandoval-Ramirez, L., and Villasenor-Garcia, M. M. (2006). Dysregulation of the Th1/Th2 cytokine profile is associated with immunosuppression induced by hypothalamic-pituitary-adrenal axis activation in mice. *Int. Immunopharmacol.* **6,** 774–781.

Vizzini, A., Vazzana, M., Cammarata, M., and Parrinello, N. (2007). Peritoneal cavity phagocytes from the teleost sea bass express a glucocorticoid receptor (cloned and sequenced) involved in genomic modulation of the *in vitro* chemiluminescence response to zymosan. *Gen. Comp. Endocrinol.* **150,** 114–123.

Wang, R., and Belosevic, M. (1994). Estradiol increases susceptibility of goldfish to *Trypanosoma danilewskyi*. *Dev. Comp. Immunol.* **18,** 377–387.

Wang, R., and Belosevic, M. (1995). The *in vitro* effects of estradiol and cortisol on the function of a long term goldfish macrophage cell line. *Dev. Comp. Immunol.* **19,** 327–336.

Watanuki, N., Takahashi, A., Yasuda, A., and Sakai, M. (1999). Kidney leucocytes of rainbow trout, *Oncorhynchus mykiss*, are activated by intraperitoneal injection of beta-endorphin. *Vet. Immunol. Immunopathol.* **71,** 89–97.

Watanuki, H., Gushiken, Y., Takahashi, A., Yasuda, A., and Sakai, M. (2000). *In vitro* modulation of fish phagocytic cells by beta-endorphin. *Fish Shellfish Immunol.* **10,** 203–212.

Watanuki, H., Yamaguchi, T., and Sakai, M. (2002). Suppression in function of phagocytic cells in common carp *Cyprinus carpio* L. injected with estradiol, progesterone or 11-ketotestosterone. *Comp. Biochem. Physiol. C. Toxicol. Pharmacol.* **132,** 407–413.

Watanuki, H., Sakai, M., and Takahashi, A. (2003). Immunomodulatory effects of alpha melanocyte stimulating hormone on common carp (*Cyprinus carpio* L.). *Vet. Immunol. Immunopathol.* **91,** 135–140.

Weigent, D. A., and Blalock, J. E. (1995). Associations between the neuroendocrine and immune systems. *J. Leukoc. Biol.* **58,** 137–150.

Weigent, D. A., Vines, C. R., Long, J. C., Blalock, J. E., and Elton, T. S. (2000). Characterization of the promoter-directing expression of growth hormone in a monocyte cell line. *Neuroimmunomodulation* **7,** 126–134.

Wendelaar Bonga, S. E. (1997). The stress response in fish. *Physiol. Rev.* **77,** 591–625.

Wendelaar Bonga, S. E., van der Meij, J. C., and Flik, G. (1984). Prolactin and acid stress in the teleost *Oreochromis* (formerly *Sarotherodon*) *mossambicus*. *Gen. Comp. Endocrinol.* **55,** 323–332.

Weyts, F. A., Verburg-van Kemenade, B. M., Flik, G., Lambert, J. G., and Wendelaar Bonga, S. E. (1997). Conservation of apoptosis as an immune regulatory mechanism: Effects of cortisol and cortisone on carp lymphocytes. *Brain Behav. Immun.* **11,** 95–105.

Weyts, F. A., Flik, G., and Verburg-van Kemenade, B. M. (1998a). Cortisol inhibits apoptosis in carp neutrophilic granulocytes. *Dev. Comp. Immunol.* **22,** 563–572.

Weyts, F. A. A., Flik, G., Rombout, J. H., and Verburg-van Kemenade, B. M. L. (1998b). Cortisol induces apoptosis in activated B cells, not in other lymphoid cells of the common carp, *Cyprinus carpio* L. *Dev. Comp. Immunol.* **22,** 551–562.

Woloski, B. M. M. J., Smith, E. M., Meyer, W. J., Fuller, G. M., and Blalock, J. E. (1985). Corticotropin-releasing activity of monokines. *Science* **230,** 1035–1037.

Yabuuchi, K., Minami, M., Katsumata, S., and Satoh, M. (1994). Localization of type-I interleukin-1 receptor messenger RNA in the rat brain. *Mol. Brain Res.* **27,** 27–36.

Yada, T. (2007). Growth hormone and fish immune system. *Gen. Comp. Endocrinol.* **152,** 353–358.

Yada, T., and Azuma, T. (2002). Hypophysectomy depresses immune functions in rainbow trout. *Comp. Biochem. Physiol. C* **131,** 93–100.

Yada, T., Nagae, M., Moriyama, S., and Azuma, T. (1999). Effects of prolactin and growth hormone on plasma immunoglobulin M levels of hypophysectomized rainbow trout, *Oncorhynchus mykiss*. *Gen. Comp. Endocrinol.* **115,** 46–52.

Yada, T., Azuma, T., Hirano, T., and Grau, E. G. (2001a). Effects of hypophysectomy on immune functions in channel catfish. *In* "Perspective in Comparative Endocrinology: Unity and Diversity" (Goos, H. J. T., Rastogi, R. K., Vaudry, H., and Pierantoni, R., Eds.), pp. 369–376. Monduzzi Editore, Bologna.

Yada, T., Azuma, T., and Takagi, Y. (2001b). Stimulation of non-specific immune functions in seawater-acclimated rainbow trout, *Oncorhynchus mykiss*, with reference to the role of growth hormone. *Comp. Biochem. Physiol. B* **129,** 695–701.

Yada, T., Uchida, K., Kajimura, S., Azuma, T., Hirano, T., and Grau, E. G. (2002). Immunomodulatory effects of prolactin and growth hormone in the tilapia, *Oreochromis mossambicus*. *J. Endocrinol.* **173,** 483–492.

Yada, T., Misumi, I., Muto, K., Azuma, T., and Schreck, C. B. (2004). Effects of prolactin and growth hormone on proliferation and survival of cultured trout leucocytes. *Gen. Comp. Endocrinol.* **136,** 298–306.

Yada, T., Muto, K., Azuma, T., Hyodo, S., and Schreck, C. B. (2005). Cortisol stimulates growth hormone gene expression in rainbow trout leucocytes *in vitro*. *Gen. Comp. Endocrinol.* **142,** 248–255.

Yada, T., Hyodo, S., and Schreck, C. B. (2008). Effects of seawater acclimation on mRNA levels of corticosteroid receptor genes in osmoregulatory and immune systems in trout. *Gen. Comp. Endocrinol.* **156,** 622–627.

Yamaguchi, T., Watanuki, H., and Sakai, M. (2001). Effects of estradiol, progesterone and testosterone on the function of carp, *Cyprinus carpio*, phagocytes *in vitro*. *Comp. Biochem. Physiol. C* **129,** 49–55.

Yang, B. Y., Arab, M., and Chen, T. T. (1997). Cloning and characterization of rainbow trout (*Oncorhynchus mykiss*) somatolactin cDNA and its expression in pituitary and nonpituitary tissues. *Gen. Comp. Endocrinol.* **106,** 271–280.

Yeager, M. P., Colacchio, T. A., Yu, C. T., Hildebrandt, L., Howell, A. L., Weiss, J., and Guyre, P. M. (1995). Morphine inhibits spontaneous and cytokine-enhanced natural killer cell cytotoxicity in volunteers. *Anesthesiology* **83,** 500–508.

Zhao, Y. R., Sun, R., You, L., Gao, C. Y., and Tian, Z. G. (2003). Expression of leptin receptors and response to leptin stimulation of human natural killer cell lines. *Biochem. Biophys. Res. Commun.* **300,** 247–252.

Zou, J., Carrington, A., Collet, B., Dijkstra, J. M., Yoshiura, Y., Bols, N., and Secombes, C. J. (2005). Identification and bioactivities of IFN-gamma in rainbow trout *Oncorhynchus mykiss*: The first Th1-type cytokine characterized functionally in fish. *J. Immunol.* **175,** 2484–2494.

Zylinska, K., Mucha, S., Komorowski, J., Korycka, A., Pisarek, H., Robak, T., and Stepien, H. (1999). Influence of granulocyte-macrophage colony stimulating factor on pituitary–adrenal axis (PAA) in rats *in vivo*. *Pituitary* **2,** 211–216.

8

THE NEUROENDOCRINE REGULATION OF FLUID INTAKE AND FLUID BALANCE

YOSHIO TAKEI
RICHARD J. BALMENT

1. Introduction
   1.1. Fluid Exchange and Balancing Mechanisms
   1.2. Volemic versus Osmotic Mechanisms for Body Fluid Regulation
2. Regulation of Fluid Intake
   2.1. Unique Regulatory Mechanism in Fish
   2.2. Hormonal Regulation of Drinking in Fish
   2.3. Neural Mechanisms to Elicit Drinking (Swallowing) in Fish
   2.4. Regulation of Sodium Appetite (Preference) in Fish
3. Regulation of Fluid Balance: Renal and Extrarenal Mechanisms
   3.1. Arginine Vasotocin
   3.2. Renin–angiotensin System (RAS)
   3.3. Prolactin, Growth Hormone and Cortisol
   3.4. Urotensins
4. Perspectives

Fishes are unique among vertebrates in that their body fluids directly contact the environmental water across thin respiratory epithelia. To maintain body fluid homeostasis (both osmolality and volume) in various salinities, loss and accumulation of ions and water are balanced through dietary uptake and osmoregulatory organ activities, processes regulated and integrated by hormones, including many from the brain. Drinking plays an essential role in body fluid balance in marine teleost fish, as the sole means to compensate for osmotic water loss. While several neuropeptides may act centrally in a paracrine/autocrine fashion to regulate drinking, peripheral hormones also act through structures devoid of the blood–brain barrier. Reflecting their aquatic life, mechanisms controlling drinking in fishes appear to differ considerably

Fish Neuroendocrinology: Volume 28
FISH PHYSIOLOGY

DOI: 10.1016/S1546-5098(09)28008-3

from those of terrestrial animals. Fluid balance also relies upon integration and control of transporting epithelia of the gill, gut, kidney and rectal gland of chondrichthyes by endocrine and neuroendocrine factors, some of which are secreted and act locally in these peripheral osmoregulatory organs.

1. INTRODUCTION

1.1. Fluid Exchange and Balancing Mechanisms

Fish body fluids are separated by only a thin respiratory epithelium from their surrounding media. Representatives of all the major groups occupy both freshwater (FW) and seawater (SW) environments, which have an osmolality range from a few to 1000 mOsm kg^{-1}. Accordingly, an ability to regulate body fluids independently of the external environment is essential for their survival. Most species are restricted to one or other medium (stenohaline) but a few exhibit extreme plasticity in their osmoregulatory capacity and migrate between these media (euryhaline). Although a minority group, and possibly not truly representative of fish osmoregulatory physiology, these latter species have been the major focus for researchers as euryhalinity is evident in the cyclostomes (lampreys), elasmobranchs and teleosts.

For fish in FW the major challenges are the need to maintain extracellular fluid composition and volume in the face of continuous osmotic gain of water from the very dilute medium, coupled with a steady diffusional loss of major body fluid ions such as Na^+ and Cl^- (Fig 8.1). The glomerular kidney provides the essential means for excreting the excess water through filtration of blood and tubular reabsorption of ions and other solutes to produce large volumes of dilute urine. This process is likely further supported by the ion reabsorptive capacity of the fish urinary bladder. The ion losses may be balanced through dietary intake and gut absorption, though this may also be supplemented by active uptake of ions from the media by the mitochondria-rich cells (MRCs, often referred to as chloride cells or ionocytes) in the gill leaflets (see review by Evans *et al.*, 2005). The active uptake is thought to be facilitated by H^+-ATPase with synergistic action of Na/K-ATPase (Lin and Randall, 1995).

Marine cyclostome species such as the hagfish (*Myxine glutinosa*) have a plasma osmolality close to that of SW, and body fluid ion concentrations are also similar. There thus appears to be limited exchange of ions and water with the medium and little need to osmoregulate. Indeed, these fish appear to be like osmoconforming marine invertebrates as their body fluid osmolality falls in line with dilution of the surrounding medium (Foster and Forster, 2007). The glomerular kidney is, however, present in these ancient species and may support body volume and some specific ion regulation such as Mg^{2+} and $SO_4{}^{2-}$.

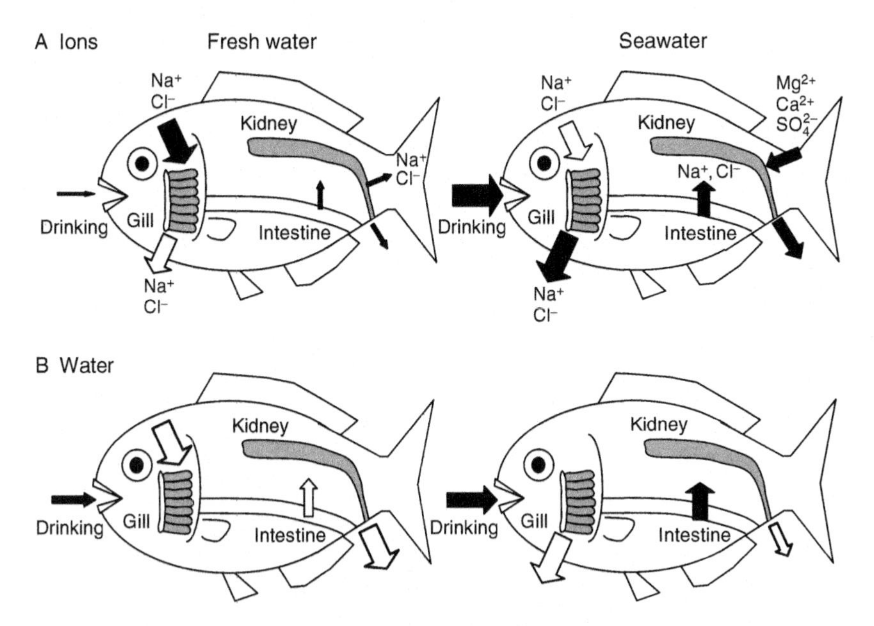

Fig. 8.1. Budget of ions (A) and water (B) between body and environment in starved teleost fish in fresh water or seawater. Open arrows show passive movement and filled arrows active movement.

In contrast, marine teleosts maintain a body fluid composition comparable with but slightly higher than that of FW species, at around 300–400 mOsm kg^{-1}. Accordingly, they are considerably hypotonic to the surrounding SW (ca. 1000 mOsm kg^{-1}) and thus continuously experience osmotic water loss to the medium (Fig 8.1). This water loss is restored by regulated drinking of SW. After some initial desalination in the esophageal tract, the remaining salt is actively absorbed in the intestine accompanied by the water. Clearly, although this provides for restoration of overall fluid balance and body fluid volume, this mechanism results in a very large salt load for the body, which must be eliminated to sustain the body fluid composition. Many studies have confirmed that it is the MRCs in the skin and gill epithelia which are responsible for the active secretion of excess Na^+ and Cl^- ions in SW. This involves the cooperative actions of three ion transporters: basolaterally located Na/K-ATPase and Na/K/2Cl cotransporter (NKCC) and an apical cystic fibrosis transmembrane regulator (CFTR)-type Cl channel (Hiroi and McCormick, 2007). The salt load to be excreted by this mechanism is further exacerbated by the diffusional gain of salt as a result of the standing ion gradients between fish and the medium (e.g., Na^+ and Cl^- are at 2–3 times higher concentration in SW compared with teleost blood). Accordingly, in marine teleosts body fluid balancing mechanisms are dominated by SW drinking and gut absorption coupled with active gill ion extrusion. The kidney in marine teleosts produces little urine to further conserve water and this is rich in divalent ions such as Mg^{2+} and $SO_4{}^{2+}$, inevitably accumulated as a result of SW imbibition (Beyenbach, 1995). In contrast to FW species, urine formation relies rather less on blood filtration and more upon tubular secretion. Indeed, in a number of SW teleosts the kidney glomeruli are greatly reduced in size and they are completely absent in some species like the toadfish (*Opsanus tau*) (Beyenbach, 2004).

Elasmobranchs are predominantly marine and characteristically accumulate and retain urea and trimethylamine oxide in their plasma. This physiological uremia coupled with blood Na^+ and Cl^- concentrations higher than those of marine teleosts maintains blood osmolality slightly higher than that of SW, accordingly reducing the osmoregulatory demands in this group. Elasmobranchs show regulated drinking of SW if they become dehydrated though basal levels are low by comparison with marine teleosts (Anderson *et al.*, 2002c). The much reduced standing ion gradients between fish and surrounding SW also result in lower rates of diffusional gain of Na^+ and Cl^- and consequent need for active extrusion. There is evidence to support gill NaCl secretion but elasmobranchs also exhibit a further salt secreting tissue, the rectal gland, which affords rapid regulated excretion of concentrated NaCl and may be of particular importance in eliminating salt after meal ingestion (Anderson *et al.*, 2002a). Through experimental manipulation of environmental salinity it has been shown that blood Na^+, Cl^- and urea levels are regulated downwards as salinity decreases (Hazon and Henderson, 1984). The salinity threshold at which osmotic homeostasis is maintained varies between species, which appears to reflect differing capacities of urea, Na^+ and Cl^- retention (Hazon *et al.*, 2003). The few species that live whole or part of their life in FW show sustained reductions in blood Na^+, Cl^- and urea concentrations, reducing the inevitable osmotic water accumulation. Nonetheless, the euryhaline bullshark (*Carcharhinus leucas*), still maintains a plasma osmolality that is twice that of FW teleost fish (650–700 mOsm kg^{-1}) when it is in FW (Pillans and Franklin, 2004). The regulation of body fluid volume, salt and urea concentrations in elasmobranchs involves the integration of adjustments in renal excretion (glomerular filtration, tubular secretion and reabsorption), branchial fluxes, drinking and gut absorption, rectal gland secretion and liver urea biosynthesis (see Anderson *et al.*, 2006).

Acclimation to altered external salinity in all fish groups involves permeability changes in the gill and other osmoregulatory epithelia, in association with close regulation of specific transporter activities. There is increasing evidence to support roles for claudins, occludin and aquaporins (AQPs) in these regulated changes in epithelial permeability. In sea bass cDNA sequences encoding homologues of mammalian AQP1 and AQP3 have been isolated and characterized (Giffard-Mena *et al.*, 2007). Changing salinity influenced aquaporin expression levels in gill, kidney and digestive tract. These authors suggest a major role for AQP1 in water transport in kidney and gut of SW-acclimated fish, while AQP3 was implicated in gill water transport in FW-acclimated animals. In teleost gill AQP3 is expressed in MRCs and in some species in other epithelial cells. Most studies of ionoregulatory epithelia in fish have focused on the transcellular mechanisms of ion movement, driven by active transport. Less attention has been given to those mechanisms which enhance ion retention by limiting paracellular losses. This paracellular pathway, controlled by the tight junction complex, accordingly remains rather poorly understood. Tight junctions comprise transmembrane and cytosolic protein complexes around the apical domain of epithelial cells, which associate between neighboring cells to form a semipermeable paracellular seal that restricts solute movement. More than 40 tight junction and associated proteins have been identified. Claudins are a large family of tight junction proteins involved in determining ion selectivity and general permeability of the paracellular pathway in epithelia. Tipsmark and colleagues (2008a) identified five isoforms (10e, 27a, 28a, 28b and 30) of the claudin family in Atlantic salmon gill EST (Expressed Sequence Tag) libraries. The acclimation of salmon to SW was associated with reduced expression of claudin 27a and 30 with no clear effect on claudin 28a or 28b, while there was a fourfold increase in the expression of claudin 10e. These observations were in accordance with an observed peak in branchial claudin 10e expression during smoltification, coincident with optimal SW tolerance. These authors suggest that claudin 10e could be important in cation selective channels, while the reduction in claudin 27a and 30 might alter gill permeability in favor of the ion secretory mode of the SW gill. In tilapia claudin 3- and claudin 4-like immunoreactive proteins have been identified in the gill (Tipsmark *et al.*, 2008a). Gill claudin 3- and 4-like proteins were reduced when fish were exposed to SW, while FW transfer was associated with increased protein content. Tilapia claudin 28a and 30 mRNA expression was also increased during FW acclimation. These findings imply that claudins contribute to the gill permeability changes associated with salinity acclimation possibly through the formation of deeper tight junctions in the FW gill. This would be expected to lower ion permeability which is essential for FW survival. Gill claudin 3- and 4-like immunoreactive proteins were also shown to be elevated in FW compared with SW acclimated southern flounder (Tipsmark *et al.*, 2008b). The dynamics of occludin, one of the first transmembrane tight junction proteins identified, have recently been examined in goldfish exposed to ion-poor water (Chasiotis *et al.*, 2009). After abrupt exposure to ion-poor water, gill tissue exhibited increased Na/K-

ATPase activity and occludin expression. Occludin expression was also elevated after 14 and 28 days in ion-poor water, implying a role in the acclimation of fish to these ion-poor conditions. This is an important emerging field of study for understanding the complexity of integrated responses to challenges to body fluid and ionic balance. Thus far there has been little advance in revealing the endocrine regulation of claudin, occludin or AQP expression in fish transporting epithelia. This will be a very fruitful area of future research and likely involves many of the factors already established in the regulation of specific transporter activities discussed below.

As outlined briefly above, the osmoregulatory demands (hemodilution in FW and dehydration and salt loading in SW) and thus the evident physiological responses are greatly affected by environmental salinity. While the responses for survival in FW appear more consistent, different strategies (hypo- vs. hyper-osmoregulation) have emerged between the major marine fish groups to secure survival in SW. The uremia and development of the secretory rectal gland clearly sets the elasmobranchs apart from teleost fish, though both groups exploit renal, branchial and drinking mechanisms. These diverse mechanisms are employed ultimately to achieve the homeostatic balance of water and solute influx and efflux between the fish and its surrounding medium. This dynamic balance is in turn achieved by the integrated regulation of these mechanisms by a cocktail of neural and hormonal inputs. In this chapter we focus on the endocrine contributions to the regulation. These can be considered in two broad categories: those which afford rapid adjustments in existing transport/flux mechanisms (acute regulatory response), and those which enable substantive change in transport/flux including the generation of new proteins/cells (acclimation response). The latter may take hours or days and are often part of the responses to an anticipated (migratory) or enforced change in the external medium. See the recent review by McCormick and Bradshaw (2006) for further consideration of these principles.

1.2. Volemic versus Osmotic Mechanisms for Body Fluid Regulation

Although blood volume and plasma osmolality are closely linked with each other for body fluid regulation, the relative importance of volume and salt regulation appears to differ depending on the animal's habitat. In addition to the diverse body fluid regulation within fish species (cyclostomes, elasmobranchs and teleosts), fish appear to be unique in the major mechanisms for body fluid regulation compared with terrestrial animals (mammals, birds and reptiles). As mentioned above, teleost fish in FW lose ions, particularly Na^+ and Cl^-, via the gills and in the large amount of urine as teleost renal tubules have limited abilities to reabsorb these ions compared with mammals and birds (Brown *et al.*, 1993). Thus the ability to gain ions through active uptake by the gills is essential for teleost fish to survive in FW if they have no access to Na^+ and Cl^- from food. By contrast, teleost fish in SW suffer from excess Na^+ and Cl^- that enter the body via the gills and are absorbed by the intestine from the ingested SW, as SW fish drink copiously to compensate for the water lost osmotically. Therefore, they can survive in SW only if they can excrete excess ions that enter the body. For these reasons, it appears that ion regulation, rather than water regulation, is of primary importance for teleost fish to adapt in either FW or SW (McCormick and Bradshaw, 2006; Takei, 2008). This is in contrast to the terrestrial animals in which water regulation (retention) dominates over ion regulation as exemplified by the essential role of antidiuretic hormone in body fluid regulation, while volume control is achieved by the retention of ions (Takei *et al.*, 2007). This difference may be explained by the fact that fish have access to water irrespective of the environmental salinities because of their aquatic life.

Although fish can survive in diverse environmental salinities only if they are capable of ion regulation, they seem to maintain blood volume more strictly over plasma osmolality when they are exposed to body fluid disturbances (Takei, 2000). When euryhaline eels are transferred from FW to SW, plasma Na^+ concentration and osmolality increase immediately and the increase continues for a week (Takei *et al.*, 1998). The blood volume decreases after SW transfer, but the change is much smaller and transient compared with the change in osmolality as measured by the body weight changes (Oide and Utida, 1968; Kirsch and Mayer-Gostan, 1973), hematocrit changes (Okawara *et al.*, 1987; Takei *et al.*, 1998), or the dye dilution method (Takei, 1988). Since cellular fluids exude into the extracellular space because of the hyperosmoremia, the decrease in blood volume may be further ameliorated. The fact that plasma angiotensin II (ANG II) concentration increases only mildly and transiently after SW transfer supports this small decrease in blood volume (Okawara *et al.*, 1987). Hypovolemia is a potent stimulus for renin secretion in teleost fish (Nishimura *et al.*, 1979).

Ken Olson and his colleagues, through their extensive studies on cardiovascular regulation in the trout (Olson, 1992), have proposed that the primary target of regulation in fish is blood volume and not salt balance. In their recent experiments (Olson and Hoagland, 2008), they prepared trout in three different body fluid balances, FW fish (volume loading, salt depleting), SW fish (volume depleting, salt loading), and FW fish with high salt diet (volume loading, salt loading), and measured blood volume and arterial and venous pressure and showed that blood pressure is intimately related to blood volume but not to salt balance, although they did not measure ion concentrations in body fluids in each fish group. These results show that blood volume is the primary determinant of blood pressure in trout. The trout were acclimated in each condition for more than 2 weeks, so that the results show that they cannot maintain blood volume in different osmotic conditions. In the eel, both blood volume (Nishimura *et al.*, 1976) and arterial pressure (Nobata *et al.*, 2008) are maintained constant in both FW and SW, but plasma osmolality is higher in SW fish than FW fish (Tsuchida and Takei, 1998). Therefore, blood volume is more strictly regulated in the eel than in the trout when exposed to body fluid challenges. The trout is basically a FW fish while the eel is a SW fish (Tsukamoto *et al.*, 1998) though both are migratory species. It is likely that the small but significant differences in the body fluid regulation between the two species may originate from such differences in their evolutionary history.

The terrestrial animals appear to maintain plasma Na^+ concentration and osmolality primarily over blood volume after water deprivation, which mimics SW challenges in teleost fish. When drinking water is withdrawn from the dog, plasma osmolality is maintained while extracellular fluid volume decreases in the initial phase of water deprivation (Elkinton and Taffel, 1942). Accordingly, a sequence of physiological responses to water deprivation is an initial loss of Na^+ in the urine to protect plasma tonicity, which is followed by a vigorous defense of blood volume in the later phase first at the expense of interstitial fluid and then of cellular fluid, which protects the circulatory system from embarrassment. The initial protection of plasma tonicity can be accounted for by the fact that increased plasma Na^+ concentration hyperpolarizes the cells and diminishes excitability of neurons and muscles. Similar types of regulation, i.e., an initial protection of plasma tonicity followed by the volume protection, is also observed in birds after water deprivation (Takei *et al.*, 1988a). Although a significant difference is evident in the regulation of volume or osmolality between fish and tetrapods (mammals and birds), these two parameters are inseparable and it is suggested that mammals primarily maintain total body sodium for the maintenance of body fluid volume (Reinhardt and Seeliger, 2000).

2. REGULATION OF FLUID INTAKE

Animals gain body fluids by oral intake of free water, preformed water from the food, and from oxidation water produced by cell metabolism of carbohydrates, lipids and proteins (Schmidt-Nielsen, 1997). As fish live in water, their requirements for oral water intake differ greatly from terrestrial animals that abandoned aquatic life. The requirements are also different among fishes depending on the environmental salinities, particularly in teleost fish that have plasma osmolality ca. one third of SW. Therefore, FW teleost fish usually drink little water to avoid overhydration as they gain sufficient water via the gills by osmosis, although FW fish have been shown to drink at the larval stage (Flik *et al.*, 2002). In particular, a substantial amount of water is ingested to moisten the feed (Ruohonen *et al.*, 1997). On the other hand, marine teleost fish must ingest water orally to compensate for the osmotic loss, a situation similar to terrestrial animals. Since euryhaline fish live in a habitat where salinity changes from time to time, they have developed mechanisms to regulate drinking depending on the environmental salinities. Accordingly, they serve as excellent models that have been heavily exploited to investigate the mechanism regulating drinking in fish.

2.1. Unique Regulatory Mechanism in Fish

Although oral drinking is essential for marine teleost fish to survive in hyperosmotic SW as in terrestrial animals, the regulatory mechanism appears to differ substantially between these animals because of the difference in their habitats. The unique mechanisms in fish appear mostly to stem from their aquatic life. Because water is always available in the mouth of fish, they can drink water by only swallowing without searching for water that is obligatory for terrestrial animals. Furthermore, as fish are exposed to possible overdrinking of surrounding water, mechanisms to suppress drinking may be more important than those that induce drinking. This is in contrast to the condition of terrestrial animals where thirst, which motivates the water searching behavior, is indispensable. In addition, the difference in the relative importance of volemic and osmotic mechanisms in body fluid regulation between fishes and terrestrial animals may impose some differences in the mechanisms controlling fluid intake.

2.1.1. Reflex Drinking versus Thirst-motivated Drinking

Drinking behavior consists of a series of processes that result in the intake of fluid from the environment into the digestive tracts, which ultimately becomes body fluids after absorption by the intestine. The drinking behavior of terrestrial animals consists of a series of events, i.e., searching for water, taking water into the mouth, and finally swallowing. The former two behaviors are motivated by the sensation of thirst, while swallowing is a reflex that is induced automatically when water reaches the pharynx (Doty, 1968). There has been a long history of research on the mechanism arousing thirst in mammals, which is summarized in a number of extensive monographs (Fitzsimons, 1979; Rolls and Rolls, 1982; de Caro *et al.*, 1986; Grossman, 1990; Ramsay and Booth, 1991). It is apparent that the sensation of thirst is required for terrestrial animals to drink water, but fishes can swallow water by a reflex without thirst as water is usually present in the mouth. Therefore, it is not known whether fish feel thirsty when they drink. Since marine fish drink copiously and constantly almost at the same rate when drinking rate is measured through a catheter in the esophagus, water seems to be imbibed automatically without time-to-time regulation. Furthermore, dipsogenic hormones are effective in teleost fish even after removal of the forebrain (Hirano *et al.*, 1972; Takei *et al.*, 1979), where arousal of thirst is integrated in mammals (Fitzsimons, 1998). The fact that fish can drink without the thirst component of behavior is the first feature that characterizes drinking in fish. However, it should be determined more clearly whether the sensation of thirst is in fact absent in fish or not, using for example terrestrial fish that move into water when they are thirsty. If dipsogenic stimuli urge them to enter water and drink, "thirst-motivated behavior" is also likely generated by these stimuli.

2.1.2. Acceleration versus Inhibition of Drinking

Because of the importance of water drinking, terrestrial animals have developed mechanisms for acceleration of drinking to ensure survival in the water-deficient environment. It is now known that multiple mechanisms work in concert to arouse thirst, and these mechanisms compensate for each other to guarantee drinking even if one is impaired (Fitzsimons, 1979). By contrast, fish can ingest water at any time just by swallowing. Therefore, fish are always exposed to the danger of excess drinking, which seems to have led to fish developing more elaborate mechanisms to inhibit drinking. As discussed in detail in the subsequent section, this notion coincides well with the fact that antidipsogenic hormones are more dominant in teleost fish for their action and in number than in terrestrial animals, while dipsogenic hormones are more evident in mammals and birds than in fishes (Fitzsimons, 1998; Takei, 2002; Kozaka *et al.*, 2003). For instance, bradykinin is dipsogenic in mammals but potently antidipsogenic in fishes (Takei *et al.*, 2001). The dominance of the inhibitory mechanism also stems from the finding that the effect of ANG II on acceleration of drinking was even enhanced after removal of the forebrain in the eel (Takei *et al.*, 1979). This result suggests that some inhibitory signals originating in the forebrain are continuously transmitted to the hindbrain to suppress swallowing, and the forebrain ablation removed the inhibition, resulting in an increased sensitivity to the dipsogenic hormone. Furthermore, an osmotic stimulus or cellular dehydration, the most potent stimulus for arousal of thirst in mammals and birds (Fitzsimons, 1979; Kaufman and Peters, 1980; Takei *et al.*, 1988b), is an inhibitory stimulus for drinking in teleosts (Takei *et al.*, 1988c) and elasmobranchs (Anderson *et al.*, 2002b). These data also illustrate more dominant inhibitory mechanisms for regulation of drinking in fishes. The dominance of the antidipsogenic mechanism is the second interesting feature that characterizes the drinking regulatory mechanisms in fish.

2.1.3. Volemic versus Osmotic Factors as Driving Stimuli

Consistent with the precise regulation of blood volume in teleost fish, a volemic stimulus (hypovolemia) caused by blood withdrawal is highly potent and consistent for eliciting drinking in the eel (Hirano, 1974). The hypovolemia is accompanied by increased plasma ANG II, a potent dipsogenic hormone (Fitzsimons, 1998), which may contribute to the increased drinking induced by the volemic stimulus. By contrast, osmotic stimuli (hyperosmoremia) caused by injections of hypertonic NaCl, mannitol or sucrose solution inhibit drinking in the eel (Takei *et al.*, 1988c). The inhibition of drinking after an osmotic stimulus is not due to hypervolemia caused by increased influx of water across the gills, because the inhibition is seen also in SW fish where osmotic water influx may not occur even after such an osmotic stimulus. Interestingly, the osmotic stimuli increase plasma ANG II concentration in the eel. This is contrary to the decrease in mammals and birds, where increased Na^+ and Cl^- load to the macula densa of the renal distal tubule inhibits renin release and thus ANG II production (Takei *et al.*, 1988b; Komlosi *et al.*, 2004). As the teleost nephron lacks the macula densa, this result indicates that an osmotic stimulus itself is stimulatory to the juxtaglomerular cells for renin secretion. Collectively, osmotic stimuli inhibit drinking in the eel even though plasma ANG II increases.

In elasmobranchs, hypovolemia caused by hemorrhage is a potent stimulus for acceleration of drinking but increases in plasma osmolality after injection of various osmolytes inhibit drinking in two species of dogfish, *Scyliorhinus canicula* and *Triakis scyllia* (Anderson *et al.*, 2002b). The hypovolemia was accompanied by an increase in plasma ANG II, but hypernatremia caused by a bolus injection of hypertonic NaCl solution failed to increase it. Since hypertonic sucrose solution increased plasma ANG II concentration, the failure of NaCl solution may be due to the presence of the macula densa in the elasmobranch kidney (Lacy and Reale, 1990). Drinking rate was closely correlated with plasma osmolality when the river lamprey, *Lampetra fluviatilis,* was kept in media of different salinities (Rankin, 2002). However, it is not certain whether osmotic stimulus itself or hypovolemia caused by increased medium osmolality is responsible for the enhanced drinking. It is of interest to examine the effects of hypertonic solutions on drinking in this species. Therefore, the third feature that characterizes the drinking regulatory mechanisms in fish is predominance of the volemic over osmotic mechanisms, which of course is in line with the more precise regulation of blood volume than plasma osmolality in fish.

Indeed, in contrast to fish, osmotic stimuli are much more potent than volemic stimuli to promote drinking in mammals and birds (Fitzsimons, 1979), even though plasma ANG II is depressed by these treatments.

Hypovolemia caused by blood withdrawal or intraperitoneal injection of hyperoncotic polyethylene glycol induces only weak drinking in mammals and birds even with a concomitant increase in plasma ANG II (Fitzsimons, 1979; Takei *et al.*, 1989). These data are consistent with the more precise regulation of plasma osmolality than blood volume in terrestrial animals.

2.2. Hormonal Regulation of Drinking in Fish

As drinking behavior is integrated in the brain, neural mechanisms appear to play a pivotal role in its regulation. Since the late 1960s, however, involvement of the endocrine system has been implicated in the induction of drinking; ANG II induces vigorous drinking in the rat in normal water balance (Epstein *et al.*, 1970). This report opened a new field in endocrinology. The new field, so-called "behavioral endocrinology", attracted the attention of many endocrinologists because it was innovative compared with traditional hormone effects in that hormones were shown to act on the brain to modulate behavior. The hormonal regulation of drinking has been investigated in some piscine species including teleosts (Takei, 2002), elasmobranchs (Anderson *et al.*, 2002c) and cyclostomes (Rankin, 2002).

2.2.1. Hormones that Induce Drinking

Angiotensin II (ANG II). The best known dipsogenic hormone is ANG II, which is an active component of the renin–angiotensin system (RAS) (Figure 8.2). ANG II induces copious drinking in almost all vertebrate classes thus far examined, although the potency differs greatly among species in different habitats or demands for drinking (Kobayashi *et al.*, 1979, 1983). In fishes, for example, ANG II increased drinking in euryhaline species that live in estuarine brackish waters and start drinking on moving into hypertonic waters. However, ANG II did not induce drinking in several stenohaline species that are confined in either FW or SW. It is known that marine teleost fish drink copiously but at a constant rate, which indicates that water passively enters the esophagus through the loosened upper esophageal sphincter. By contrast, ANG II induces a burst of drinking probably through coordinated movement of muscles concerned in swallowing. Therefore, ANG II may not be involved in the constant drinking of SW fish. In fact, removal of free ANG II from plasma by infusion of antiserum raised against ANG II did not suppress drinking rate in SW eels (Takei and Tsuchida, 2000). However, captopril, a converting enzyme inhibitor that blocks the conversion of inactive ANG I to active ANG II, profoundly suppressed drinking in the eel (Tierney *et al.*, 1995). This effect of captopril may be non-specific because the inhibition lasted after termination of captopril infusion, when plasma ANG II concentration rebounded above the

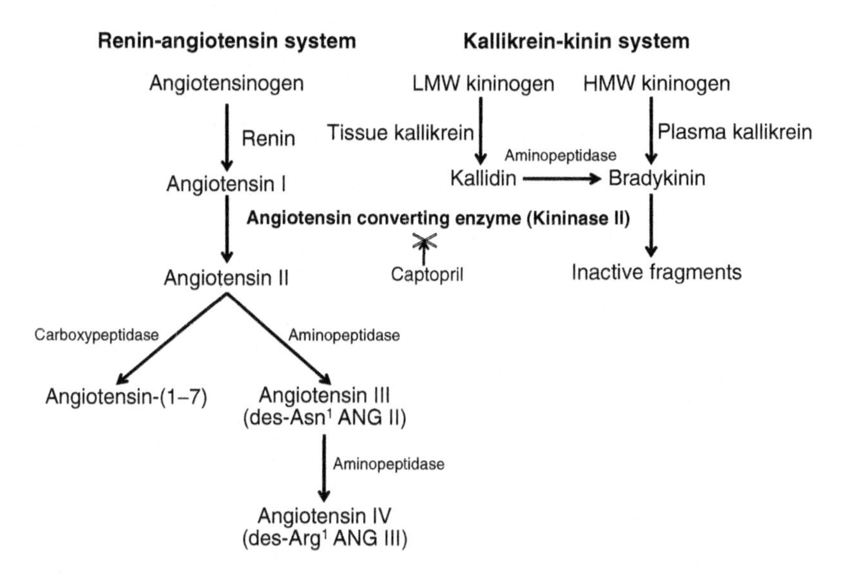

Fig. 8.2. Components of the renin–angiotensin and kallikrein–kinin cascade and their relationship via angiotensin-converting enzyme. Captopril inhibits angiotensin-converting enzyme and thus ANG I to ANG II conversion but captopril inhibition of this enzyme also reduces the truncation of bradykinin into its inactive fragments. HMW, high molecular weight; LMW, low molecular weight.

pre-infusion level (Takei and Tsuchida, 2000). The inhibition of drinking may be due to the increased bradykinin in plasma, which also results from captopril treatment because the converting enzyme is also a major degrading enzyme of bradykinin and thus named kininase II (Figure 8.2), and bradykinin is potently antidipsogenic in the eel (Takei *et al.*, 2001).

Exogenous ANG II induces drinking when administered into the periphery of the eel (Takei *et al.*, 1979), killifish (Malvin *et al.*, 1980) and flounder (Carrick and Balment, 1983). Treatments that activate the endogenous RAS, such as the smooth muscle relaxant papaverine, also increase drinking rates in the flounder (Balment and Carrick, 1985). ANG II is dipsogenic in two species of dogfish, *S. canicula* and *T. scyllium*, and activation of the RAS also enhances drinking rate (Anderson *et al.*, 2001b). However, exogenous ANG II failed to induce drinking in the migratory river lamprey, *L. fluviatilis* (Rankin, 2002). Increased ANG II in the circulation likely acts on target sites in the brain that lack the blood–brain barrier (BBB) to induce drinking, such as the subfornical organ (SFO) and/or organon vasculosum of the lamina terminalis (OVLT) as shown in mammals (Simpson and Routtenberg, 1973; Nicolaidis and Fitzsimons, 1975) and birds (Takei, 1977) (Figure 8.3). In mammals and birds, such vascular-rich circumventricular organs directly

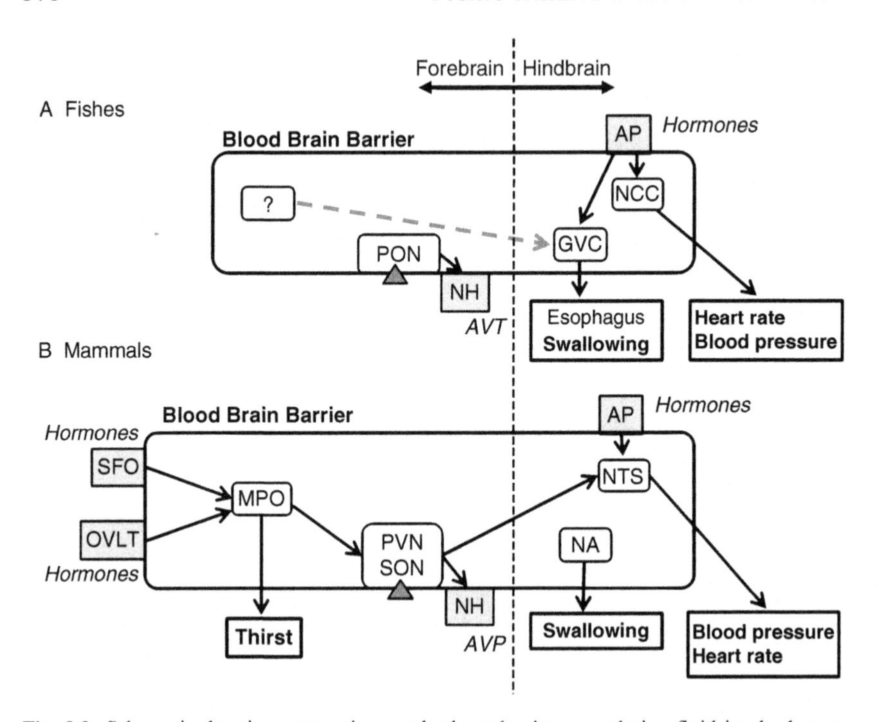

Fig. 8.3. Schematic drawing comparing cerebral mechanisms regulating fluid intake between fishes and mammals. Circulating hormones act on the circumventricular organs that are located outside the blood–brain barrier such as area postrema (AP), organon vasculosum of the lamina terminalis (OVLT), subfornical organ (SFO) and neurohypophysis (NH). Neurohypophysial hormone-producing preoptic nucleus (PON), paraventricular nucleus (PVN) and supraoptic nucleus (SON) extend dendrites outside the blood–brain barrier and directly receive information from blood. In mammals, thirst-regulating hormones, such as angiotensin II, are synthesized locally in the SFO, OVLT, median preoptic nucleus (MPO) and other brain sites and regulate fluid intake. AVP, arginine vasopressin; AVT, arginine vasotocin; GVC, glossopharyngeal–vagal motor complex; NA, nucleus ambiguus; NCC, commissural nucleus of Cajal; NTS, nucleus tractus solitarius.

respond to ANG II from blood (McKinley *et al.*, 2003b), as ablation of these structures abolishes drinking caused by peripheral injection of ANG II. Therefore, they are named sensory circumventricular organs (Johnson and Thunhorst, 1997). These forebrain structures may transmit the dipsogenic information generated by ANG II to the inferred thirst center to motivate a series of drinking behaviors in mammals and birds. However, these circumventricular organs in the forebrain have not thus far been identified in fish (Mukuda *et al.*, 2005).

Unlike mammals and birds, the forebrain does not seem to be necessary for ANG II to initiate drinking in teleost fish (Takei, 2002). As mentioned above, injection of ANG II into the circulation induces copious drinking in the eel (Takei *et al.*, 1979). However, ANG II was still effective, or rather more effective, after "decerebration" that removes the whole forebrain and most of the midbrain. The "decerebrated" eel may not feel thirsty as the sensation of thirst is integrated in the forebrain including amygdala (Fitzsimons, 1998). Therefore, ANG II in blood may act on the hindbrain at the level of medulla oblongata to induce reflex swallowing, resulting in drinking of environmental water (Figure 8.3). In fact, infusion of ANG II into the fourth ventricle above the medulla oblongata induced drinking in the eel (Kozaka *et al.*, 2003). The most likely site of action of ANG II is the area postrema (AP), which is another circumventricular organ in the hindbrain facing the fourth ventricle (Mukuda *et al.*, 2005). However, the exact site of action of ANG II has not yet been determined.

In addition to the dipsogenic effect, centrally administered ANG II increases arterial blood pressure and heart rate in mammals principally through sympathetic activation (Reid, 1992) (Figure 8.3). In teleost fish, Le Mevel and his colleagues have shown that central ANG II increases arterial pressure and heart rate, and decreases heart rate variability and baroreflex sensitivity when injected into the third cerebral ventricle of the conscious trout *Oncorhynchus mykiss* (Le Mevel *et al.*, 1994, 2002). In the trout, however, the effect on the heart rate is more potent than on the arterial pressure, suggesting that the vasopressor effect is mediated by tachycardia (Figure 8.3). The topical injection of ANG II to the dorsal vagal motor nucleus is particularly sensitive (Pamantung *et al.*, 1997), and the cardiovascular effects of central ANG II disappeared in atropinized fish (Lancien and Le Mevel, 2007). Therefore, ANG II effects may be mediated by the parasympathetic inhibition rather than by the sympathetic activation in the trout. Central cardiovascular actions have also been reported for arginine vasotocin, urotensin II and other hormones in the trout (Le Mevel *et al.*, 2008a,b), and these are considered later.

Adrenomedullins (AMs). AM was first isolated from pheochromocytoma cells of adrenal medulla origin (Kitamura *et al.*, 1993). AM was previously thought to be a member of the calcitonin gene-related peptide (CGRP) family that consists of CGRP, AM and amylin. However five AMs form an independent subfamily in teleost fish and are named AM1 through AM5 (Ogoshi *et al.*, 2003). It has now been shown that mammalian AM is an orthologue of teleost AM1, and that AM1/4 and AM2/3 were generated as a result of a whole genome duplication that occurred only in the teleost lineage (Ogoshi *et al.*, 2006). Based on these findings, AM2 and AM5 have been newly identified in mammals (Takei *et al.*, 2004a, 2008). AM(1) binds to calcitonin receptor-like receptor (CLR) associated with receptor activity-modifying protein (RAMP) in teleost fish and mammals (Hay *et al.*, 2004; Nag *et al.*, 2006) as does CGRP. However novel receptor(s) specific for AM2 and/or AM5 may well exist because AM2 and AM5 are more potent than AM(1) for the central actions despite lower affinity to the known AM receptors, and because their actions are only partially blocked by the antagonist specific for the AM receptor in the rat (Hashimoto *et al.*, 2007b). The AM gene is expressed in various tissues and thus is basically a local paracrine factor in mammals (Lopez and Martinez, 2001). AM has extremely low immunogenicity and thus it is difficult to raise good antibodies; however, plasma AM concentration has been measured in healthy human volunteers (3.3 ± 0.4 fmol mL^{-1}, $n = 8$), and the level increases in hypertensive patients according to the severity of hypertension (Kitamura *et al.*, 1994). The origin of plasma AM is not known. The AM genes are expressed in various tissues in the pufferfish (Ogoshi et al., 2003) and the eel (Nobata *et al.*, 2008), except for the AM2 and AM3 genes in the pufferfish which are expressed almost exclusively in the brain.

Intra-arterial bolus injections of AM2 and AM5 induced profound hypotension (Nobata *et al.*, 2008) and copious drinking (Ogoshi *et al.*, 2008) in the eel. The decrease in arterial pressure was more than 50% at 1 nmol kg^{-1}, which is the most efficacious vasodepressor hormone in fishes known thus far. Although vasodepressor substances are generally dipsogenic in fish (Hirano and Hasegawa, 1984), the AM-induced drinking was not due to hypotension because slow infusion of AM2 and AM5 at low, non-depressor doses was also dipsogenic. The dipsogenic effect of AM2 was dose-dependent and even more profound than that of ANG II. However, intracerebroventricular (icv) injection of AM2 and AM5 did not induce drinking in the eel, although ANG II was strongly dipsogenic when injected into the same site through a permanently implanted cannula. On the other hand, icv injection of AM2 at the same site increased arterial pressure as did ANG II, suggesting different sites of action of AM2 in the brain for the dipsogenic and vasopressor effects. Peripherally administered AMs should act on target sites in the brain that lack the blood–brain barrier such as the AP located in the hindbrain. It is likely that AMs injected into the cerebral ventricle also act on the circumventricular structure. Since AM molecules are fivefold larger in molecular mass than ANG II, it is possible that AM injected into the cerebral ventricle cannot reach its target neurons across the cerebrospinal fluid–brain barrier (McKinley *et al.*, 2003b). It is also possible that peripherally and centrally administered AM and ANG II act on different target sites.

2.2.2. Hormones that Inhibit Drinking

Natriuretic peptides (NPs). The NP family consists of seven members in teleost fish: atrial, B-type, ventricular NP (ANP, BNP and VNP) that are synthesized mostly in the heart, and four C-type NPs (CNP1, 2, 3 and 4) that have the common structural characteristic of lacking a C-terminal "tail" sequence extending from the intramolecular ring and are most abundantly synthesized in the brain (Inoue *et al.*, 2003). All seven members have been identified in the chondrostean and some primitive teleostean fishes such as eel and salmonids, but ANP and/or VNP seem to have been silenced in some species during teleost evolution, suggesting that BNP is a basic cardiac natriuretic peptide in vertebrates (Inoue *et al.*, 2005).

ANP and VNP are potent antidipsogenic hormones in the eel (Tsukada and Takei, 2006). BNP was identified in the eel only recently and thus its effect on drinking has not yet been examined. When infused into the circulation of the SW eel, ANP dose-dependently decreased drinking rate at 0.3–3 pmol kg min^{-1} and the copious drinking of SW eels was almost abolished at 3 pmol kg min^{-1}. This dose was still non-hypotensive and probably represents physiologically relevant effects (Tsuchida and Takei, 1998). In fact, osmotic stimuli increase plasma endogenous ANP concentration to the level achieved during infusion at 3 pmol kg min^{-1} (Kaiya and Takei, 1996b). ANP is known as a potent natriuretic hormone in mammals, but the natriuretic effect was appreciable at 50 pmol kg min^{-1} in the dog where significant hypotension occurred at the same time (Seymour *et al.*, 1986). ANP is also antidipsogenic in mammals (Antunes-Rodrigues *et al.*, 1985), but the minimum effective dose of ANP was much greater than the dose required for the dipsogenic effect of ANG II when injected icv. By contrast, the antidipsogenic effect of ANP was two orders of magnitude more potent than dipsogenic effect of ANG II in the eel when administered into the circulation (Tsuchida and Takei, 1998). This is also true for elasmobranchs that have only CNP as a cardiac hormone, in which CNP was shown to be 50 times more potent in its inhibition of drinking than the stimulatory effect of ANG II (Anderson *et al.*, 2001c). The difference in the potency of dipsogenic and antidipsogenic hormones in fish and mammals may be due to the relative importance of the inhibitory mechanism for drinking in aquatic fishes as mentioned above. ANP infusion also dose-dependently decreased plasma ANG II concentration in SW eels, which may enhance its antidipsogenic effect (Tsuchida and Takei, 1998).

The antidipsogenic effect of ANP may be advantageous for SW adaptation (Takei and Hirose, 2002). As mentioned above, to drink SW is indispensable for teleost fish to survive in the hypertonic SW environment. As fish are in water, however, they are prone to overdrinking and to resultant hypernatremia when they are in SW. After transfer of eels from FW to SW, robust drinking occurs within 1 min in response to the high Cl^- ion concentration in SW (Hirano, 1974; Takei *et al.*, 1998). The acute and excessive intake ceased in 15 min and suppressed drinking continued for a few hours at a level lower than the constant SW drinking rate. Such transient inhibition may be due to an inhibitory signal, such as stomach distension after the initial robust drinking (Hirano, 1974), but the time course of this inhibition coincides well with the transient increase in plasma ANP concentration after SW transfer (Kaiya and Takei, 1996a). Therefore, it is likely that ANP suppresses excessive drinking when encountering SW, to ameliorate sudden increases in plasma Na^+ concentration and to promote SW adaptation (Takei and Hirose, 2002). Plasma ANP concentration does not differ between FW and SW-adapted eels (Kaiya and Takei, 1996a). However, ANP seems to chronically inhibit excessive drinking in SW-adapted eels to maintain plasma Na^+ concentration at a level slightly higher than in FW, because infusion of ANP antiserum to remove circulating ANP results in increased drinking and increased plasma Na^+ concentration in SW eels (Tsukada and Takei, 2006).

ANP administered peripherally most probably acts on the BBB-deficient structures in the brain and diminishes thirst or inhibits reflex swallowing. The most likely site of action is the AP based on the following evidence: 1) ANP injected into the fourth ventricle near the AP inhibited drinking in SW eel (Kozaka *et al.*, 2003); 2) Evans blue injected into the circulation stained the AP and some other circumventricular structures without staining other brain parenchyma (Mukuda *et al.*, 2005); 3) ANP receptor (A-type natriuretic peptide receptor) has been localized at the AP by immunohistochemistry (Tsukada *et al.*, 2007); and 4) heat coagulation or local neuronal lesioning of the AP by kinic acid significantly attenuated the inhibitory effect of ANP on drinking in SW eels (Tsukada *et al.*, 2007). However, direct injection of ANP into the ventricle is scarcely antidipsogenic compared with systemic administration in our experiment, although icv ANG II is more effective than peripheral injection of ANG II (Nobata, S. and Takei, Y., unpublished data). It is possible that ANP is less penetrative across the ependymal barrier compared with ANG II because of its larger molecular mass (McKinley *et al.*, 2003b).

As discussed in Section 1.2, there are interesting differences in volume versus osmotic regulation between trout and eels and also for ANP actions. In the trout, ANP secretion is primarily regulated by the increased blood volume and ANP acts to decrease blood volume as observed in mammals (Johnson and Olson, 2008). The volume decrease leads to a decrease in blood pressure and to cardioprotection. In the eel, however, osmolality is the major stimulant for ANP secretion, and ANP acts to decrease body sodium and chloride and thus promote SW adaptation (Takei and Hirose, 2002). Indeed, ANP secretion occurs *in vivo* soon after SW transfer even with the decreased blood volume described above and occurs *in vitro* from isolated eel atria in response to increased medium osmolality (Kaiya, H. and Takei, Y., unpublished data). Although heterologous ANP and radioimmunoassay have been used in the trout studies, it is likely that ANP actions are relatively stronger on volume regulation in trout and on salt regulation in eels. This again illustrates the diversity of major regulatory systems within the teleost species.

Ghrelin. Ghrelin was first isolated from the stomach as a ligand for the growth hormone secretagogue receptor (Kojima *et al.*, 1999). Ghrelin is a linear peptide consisting of 19–28 amino acid residues with addition of various middle-chain fatty acids such as octanoic, decanoic, or decenoic acid at the third Ser or Thr residue (Kaiya *et al.*, 2008). Ghrelin has been identified in two species of elasmobranchs (Kawakoshi *et al.*, 2007) and eleven species of teleosts, and two different genes have been identified in the rainbow trout and channel catfish (Kaiya *et al.*, 2008). When administered into the periphery, ghrelin actually stimulated growth hormone secretion, but the hormone attracted the attention of many researchers more widely when it was shown to be a potent orexigenic hormone acting on the brain of rats (Nakazato *et al.*, 2001).

With regard to effects on body fluid regulation, ghrelin was found to have potent antidipsogenic actions when injected into the brain and periphery of the eel, the first such study in vertebrates (Kozaka *et al.*, 2003). Based on these findings, it was later found that ghrelin is a potent antidipsogenic hormone in the rat (Hashimoto *et al.*, 2007a) and in the chick (Tachibana *et al.*, 2006). The antidipsogenic effect was as potent as the orexigenic effect in the rat, so ghrelin inhibits drinking even though food intake is enhanced simultaneously. It is generally recognized that drinking usually occurs when animals eat, which is called prandial drinking (Fitzsimons, 1979). The antidipsogenic effect of ghrelin has been observed in all species thus far examined including teleosts, mammals and birds. However, while ghrelin is orexigenic in mammals and teleost fishes (Unniappan *et al.*, 2004; Riley *et al.*, 2005; Matsuda *et al.*, 2006), it is anorectic in the chicken (Saito *et al.*, 2005).

Ghrelin is even more potent than ANP when injected into the fourth ventricle of SW-adapted eels (Kozaka *et al.*, 2003). When injected or infused into the circulation, however, antidipsogenic effects of the two peptides are comparable (Kaiya, H., Nobata, S. and Takei, Y., unpublished data). ANP seems to act on the AP to inhibit swallowing (see above and Tsukada *et al.*, 2007), but nothing is known about the site of action of ghrelin in the brain. The difference in the relative potency of the two peptides after central and peripheral injections may be due to the difference in the ease of access to parenchymal neurons from the ventricle. Eel ghrelin (21 amino acids) and eel ANP (27 amino acids) have similar molecular mass, but ghrelin is acylated with fatty acid, which may change the ability to cross the ependymal layer of the ventricular surface. The site of action of ghrelin secreted from the stomach should be the BBB-insufficient circumventricular organs such as the AP. Thus it will be intriguing to examine whether the AP serves as both a window for various hormones produced in the blood and the target of locally synthesized hormones in the brain to integrate the regulation of drinking in fishes.

Ghrelin has been investigated primarily for its orexigenic effects and stimulatory actions on growth hormone secretion in many vertebrate species including fishes (Kaiya *et al.*, 2008). Consistently, changes in plasma ghrelin concentration are measured in relation to fasting and energy metabolism. Two forms of ghrelins, the acylated active form and the non-acylated inactive form, exist in plasma. A radioimmunoassay using antiserum directed to the N-terminus measures the active form, and one using that directed to the C-terminus measures both active and inactive forms (Hosoda *et al.*, 2000). In the rat, acylated active ghrelin concentrations in plasma are 5–80 fmol mL^{-1}, which are approximately 10-fold lower than the total ghrelin concentrations. In eels, the active form of ghrelin circulates at 50 fmol mL^{-1}, and the concentration increases 6 h after transfer of eels from FW to SW (Kaiya *et al.*, 2006). Furthermore, homologous ghrelin increases growth hormone secretion *in vitro* and stimulates growth hormone/insulin-like growth factor-I (IGF-I) axis in the tilapia (Fox *et al.*, 2007). The growth hormone/IGF-I axis is implicated in SW adaptation (Sakamoto and McCormick, 2006). Together with the potent antidipsogenic effect, it is apparent that ghrelin may be involved more widely in body fluid regulation in teleost fishes.

Bradykinin. Bradykinin, a linear nonapeptide hormone, is the final and active product of the kallikrein–kinin system (KKS) that is known to have potent cardiovascular and inflammatory actions (Regoli and Barabe, 1980). Two KKSs, plasma KKS and tissue KKS, are known to exist; plasma kallikrein acts on the high molecular weight kininogen to produce bradykinin, while tissue kallikrein acts on the low molecular weight kininogen to produce kallidin, [Lys^0]-bradykinin (Figure 8.3). Bradykinin has been identified from bony fish to mammals, but not in cartilaginous fish and cyclostomes. In fishes, incubation of porcine tissue kallikrein with fish plasma usually produces kallidin, and it has been sequenced in all groups of ray-finned fish (chondrostei, holostei and teleostei) and lobe-finned fish (lungfish) (Conlon, 1999). It is not yet known whether bradykinin is circulating in fish blood because of its extremely short half life. Kallidin ([Lys^0]- or [Arg^0]-bradykinin) exhibited potent cardiovascular actions in the cod (Platzack and Conlon, 1997), sturgeon (Li *et al.*, 1998), rainbow trout (Olson *et al.*, 1997), eel (Takei *et al.*, 2001), and lungfish (Balment *et al.*, 2002), and the effect of kallidin was more potent than homologous bradykinin in the eel. Bradykinin receptor (B2-like receptor) has been cloned in the zebrafish (Duner *et al.*, 2002).

The KKS is closely related to the RAS because both systems share angiotensin-converting enzyme (ACE) for inactivation and activation of the system, respectively (Skidgel and Erdos, 2004). The ACE is also named kininase II and degrades bradykinin (Figure 8.3). Therefore, treatment of captopril to inhibit ACE also inhibits degradation of bradykinin.

Homologous [Arg^0]-bradykinin is a potent antidipsogenic hormone in the eel when injected as a bolus or infused at a rate that does not change arterial pressure (Takei *et al.*, 2001). [Arg^0]-bradykinin injected into the circulation was antidipsogenic even with a concomitant increase in plasma ANG II concentration. The increase in plasma ANG II may be due to an increase in the ACE activity caused by bradykinin injection. [Arg^0]-bradykinin was more potent than bradykinin or [Arg^0]-des-Arg^9-bradykinin for this antidipsogenic effect. It is not known whether the antidipsogenic effect was a direct action or mediated by the increase of other potent antidipsogenic hormones such as ANP and ghrelin. Therefore, icv injection of homologous [Arg^0]-bradykinin into the brain is needed to solve this problem and to identify the target site. As bradykinin induces drinking in the rat (Fregly and Rowland, 1991), the antidipsogenic effect in the eel also supports the notion that inhibitory mechanisms for drinking predominate in aquatic fish.

2.2.3. Other Hormones that Regulate Drinking

Hormones that modulate drinking after administration in fish are listed in Table 8.1. It is generally considered that hypotensive substances are dipsogenic, and that hypertensive substances are antidipsogenic (Hirano and Hasegawa, 1984). This idea originated from the fact that vasopressor α-adrenergic agonist, adrenaline or noradrenaline, arginine vasotocin, oxytocin, and urophysial extract (probably urotensin II), are all antidipsogenic, while vasodepressor β-adrenergic agonist, isoproterenol, histamine and acetylcholine are dipsogenic in the eel. Among the above-mentioned hormones, this theory is applicable to AM and bradykinin, but ANG II and ANP are exceptions; ANG II is vasopressor but dipsogenic, and ANP is vasodepressor but antidipsogenic in the eel. In addition, mammalian substance P and serotonin increased water intake, while vasoactive intestinal peptide, eel intestinal pentapeptide and cholecystokinin depressed drinking in SW-adapted eels (Ando *et al.*, 2000). There are other hormones that changed drinking rate slightly (Table 8.1) but the cardiovascular actions of these hormones have not yet been examined in the eel.

When injected centrally, acetylcholine, isoproterenol and substance P also enhanced drinking in addition to ANG II, while serotonin, γ-amino butyric acid (GABA), prolactin, arginine vasotocin, vasoactive intestinal peptide, and noradrenaline inhibited drinking in addition to ANP and ghrelin in SW eels (Kozaka *et al.*, 2003). It is not known how icv injections of these substances alter arterial pressure, but ANG II elevates arterial pressure when injected icv (Le Mevel *et al.*, 2008a). Cortisol and growth hormone are important SW-adapting hormones in teleost fish, and these

Table 8.1
Hormones that regulate drinking in fishes and mammals thus far examined

Fishes		Mammals	
Stimulatory	Inhibitory	Stimulatory	Inhibitory
Angiotensin II	Noradrenaline	Angiotensin II	Atrial natriuretic peptide
Adrenomedullin 2	Atrial natriuretic peptide	Noradrenaline	Adrenomedullin
Substance P	Ghrelin	Bradykinin	Ghrelin
Prolactin (iv)	Bradykinin	Tachykinins	Tachykinins (rat)
	Arginine vasotocin	Arginine vasopressin	Glucagon-like peptide
	Vasoactive intestinal peptide	PACAP	Bombesin-like peptide (rat)
	Prolactin (icv)	Prolactin (icv)	Endothelin
	Eel intestinal peptide	Bombesin-like peptide	Opioids (rat)
	Isotocin	Relaxin	
	Cholecystokinin	Neurotensin	
		Neuropeptide Y	
		Opioids	
		Insulin	

For original reference, see Fitzsimons (1998), Ando *et al.* (2000), and Kozaka *et al.* (2003). Some hormones have opposite effects when injected peripherally and centrally or among different species.

icv, intracerebroventricular; iv, intravenous; PACAP, pituitary adenylate cyclase-activating peptide.

hormones appear to have a permissive role in enhancing drinking rate in salmonids when fish migrate to hyperosmotic media (Fuentes *et al.*, 1996; Nielsen *et al.*, 1999).

Collecting all the information about drink-regulating hormones obtained thus far, it is apparent that antidipsogenic hormones exceed in number dipsogenic hormones in teleost fish, probably to avoid over-drinking of surrounding water as mentioned above. In the future, it is necessary to examine how these peripheral and central regulators for drinking interact in the brain to optimize the amount of water intake in teleost fish. In addition, there are a number of other candidate hormones to be investigated that may participate in body fluid regulation in fishes such as relaxin as suggested for mammals (Fitzsimons, 1998; McKinley *et al.*, 2003b).

2.2.4. Central versus Peripheral Hormones

It is most likely that peripherally generated hormones act on the brain to regulate drinking in fishes through the BBB-deficient target sites, such as the SFO, AP and OVLT, that possess their receptors (Ferguson and Bains, 1996; Johnson and Thunhorst, 1997; McKinley *et al.*, 2003b). In addition, most peripheral hormones are also synthesized locally in the brain and act on central target sites in a paracrine or autocrine fashion. For example, blood-borne ANG II acts on the circumventricular organs on the lamina terminalis to induce drinking in mammals (Johnson and Thunhorst, 1997) and birds (Takei, 1977), while ANG II is also synthesized locally in the brain by the brain RAS and acts on the neurons within the BBB independently of the peripheral RAS (Bader and Ganten, 2002; McKinley *et al.*, 2003a). In fact, the responses of ANG II concentration in plasma and brain are not parallel after alteration of water and electrolyte balance. Little is yet known about the interaction of central and peripheral hormones to ensure appropriate thirst and drinking responses in terrestrial or aquatic animals.

In fishes, immunohistochemical localization of drink-regulating hormones has not yet been examined even for ANG II and ANP. However, it is most probable that these hormones are synthesized in the fish brain and act on the regulatory sites in a paracrine/autocrine fashion. Physiological data obtained thus far indicate that the AP is the site of action of circulating hormones that regulate swallowing in fish. However, a possibility remains that other circumventricular organs also exist in the forebrain that respond to the circulating hormone and send the regulatory signal to the AP. In the rat, the SFO and OVLT are connected with nerve fibers and communicate with each other (Fitzsimons, 1998; McKinley *et al.*, 2003b). Further analysis of the cerebral mechanisms may elucidate how peripheral and central hormones interact and how the inhibitory mechanism dominates the stimulatory mechanism in the fish brain. Although less is known about the structure and function of the fish brain, it has a possibility to serve as a good model for analysis of cerebral mechanisms because of its simpler architecture.

2.3. Neural Mechanisms to Elicit Drinking (Swallowing) in Fish

Although drinking is essential for SW adaptation, studies on the regulatory mechanisms of drinking in fish are rather scanty compared with terrestrial animals. Reflecting their easy access to water, however, the mechanisms regulating drinking behavior may be less complicated in fish than in terrestrial animals. Since removal of the forebrain and most of the midbrain does not influence the copious drinking of SW eels (Hirano *et al.*, 1972), drinking of fish appears to be basically regulated by the hindbrain. The ANG II-induced drinking still survives in such "decerebrated" eels, so that the site of action of ANG II should exist in the hindbrain also (Takei *et al.*, 1979). On the other hand, the tenth cranial nerve, the vagus nerve, is important for the act of drinking, as bilateral transection of the vagus nerve abolished both SW-induced drinking (Hirano *et al.*, 1972) and ANG II-induced drinking (Takei *et al.*, 1979).

Mukuda and Ando (2003) showed that different skeletal muscles, the sternohyoid, third branchial, fourth branchial, opercular, pharyngeal, upper esophageal sphincter, and esophageal body muscles, which are concerned in a series of swallowing movements are innervated by the vagus and/or glossopharyngeal nerves. It was also found that each muscle originates from different parts of the glossopharyngeal–vagal motor complex (GVC) that extends anterio-posteriorly along both sides of the medulla oblongata. Among the muscles involved in swallowing, the upper esophageal sphincter muscle (UES), which consists of skeletal muscle, acts as a gate for ingestion and is usually constricted under cholinergic control (Kozaka and Ando, 2003). They also suggested that the activity of neurons in the GVC that are concerned in swallowing is inhibited by catecholamines, indicating adrenergic innervation probably from the commissural nucleus of Cajal and/or the AP, the former of which is equivalent to the nucleus tractus solitarius of mammals (Ito *et al.*, 2006). Based on these data, it is hypothesized that tonic stimulatory signals through the vagus constantly constrict the UES and block water entering into the esophagus, but when adrenergic innervation to the GVC is activated, it inhibits the GVC neurons that send stimulatory signals to the UES, resulting in relaxation of UES and ingestion of water. However, this model does not fit with the result that bilateral vagotomy abolishes SW- and ANG II-induced drinking in the eel, as the transection may release the UES from the tonic constricting signal. However, it is likely that a series of muscle contractions for swallowing is governed by the vagus nerve, so that inability of coordinated movement does not elicit drinking even though the UES is opened.

2.4. Regulation of Sodium Appetite (Preference) in Fish

Sodium appetite is an important regulator for body fluid balance and has a long history of research in mammals alongside thirst (Fitzsimons, 1979; Denton, 1982; de Caro *et al.*, 1986; Grossman, 1990; Ramsay and Booth, 1991). The behavioral urge for salt is innate and specific to NaCl, and the hunger for salt was demonstrable 3 days postnatal when renin was administered into the cerebral ventricle of newborn rats (Leshem, 1999). Therefore, it is possible that a sensation similar to the sodium appetite in mammals also exists in fishes. However, nothing is known about the hormones that potentially induce sodium appetite in fishes.

Fish migration between FW and SW is regulated by many factors including gonadal maturation, etc., but sodium preference could also be involved in the motivation to final movement to waters of different salinities. There are several reports on the mechanisms for the onset of downstream migration in the fry of salmonid fish (Iwata, 1995). It is known that smoltification occurs in salmonid fish (parr–smolt transformation) and in eels (yellow to silver transformation) before the fish start seaward migration (Folmar and Dickhoff, 1980). In particular, salinity tolerance (Parry, 1960) and salinity preference (McInerney, 1964) increase according to the progress of smoltification in salmonids. Several hormones, thyroxine, cortisol, insulin, growth hormone, etc., increase during the course of smoltification, of which thyroid hormones, thyroxine (T_4) and T_3, exhibit the most dramatic surge during smoltification (Hoar, 1988).

Using a preference tank that has FW and SW compartments, Iwata *et al.* (1986) examined the time course of changes in seawater preference in the chum salmon (*O. keta*) fry that start downstream migration soon after hatching. They found that the SW preference was gradually lost when the fry were kept in FW. A close relationship between plasma thyroxine level and seaward migration has been demonstrated (Ojima and Iwata, 2007), but it is not known whether the seaward migration is motivated by salinity preference or not. Furthermore, it remains to be determined whether hormones that promote smoltification and SW adaptation in fish and those that induce sodium appetite in mammals, enhance salt preference in teleost fish or not. There is a single report showing that stress prevents SW entry in the juvenile chinook salmon, *O. tshawytscha* (Price and Schreck, 2003), suggesting that cortisol inhibits salt preference in this salmonid. There are migratory species in the cartilaginous fish, e.g., the bullshark, *Carcharhinus leucas,* and in cyclostomes, the sea lamprey, *Petromyzon marinus*, and river lamprey, *L. fluviatilis,* but no studies seem to have been performed on the sodium preference in these vertebrate groups.

3. REGULATION OF FLUID BALANCE: RENAL AND EXTRARENAL MECHANISMS

As outlined in Section 1, fish maintain body fluids by balancing the loss and accumulation of ions and water through dietary uptake and exchanges with the environment. In this process there is close coupling of ion balance with body water content to ensure stability of body fluid osmolality and total volume of intra- and extra-cellular fluid compartments. In addition to the fluid intake mechanisms discussed above, fluid balance also relies upon contributions from the transporting epithelia of the gill, gut, kidney and rectal gland. The activities of these organs are in turn regulated and integrated by a plethora of endocrine and neuroendocrine factors which are now considered below.

3.1. Arginine Vasotocin

Two neurohypophysial hormones (homologues of mammalian vasopressin and oxytocin) are present in all jawed vertebrates. Arginine vasotocin (AVT) is the common basic peptide for all non-mammalian species, replaced by arginine vasopressin (AVP) in mammals (Acher, 1996). Throughout the vertebrate series AVT/AVP is intimately involved in the physiological regulation of body fluid volume and composition. Extracellular dehydration (volume reduction) and cellular dehydration (raised osmolality) are both stimulants of AVP secretion in mammals, though increased plasma osmolality is more potent (Balment, 2002). Studies in the euryhaline flounder, *Platichthys flesus*, indicate that raised osmolality is also a much more potent stimulant for AVT secretion than a reduction in blood volume (Warne and Balment, 1995). Plasma AVT concentrations are strongly related to coincident plasma osmolality in the flounder irrespective of whether fish are exposed to SW or FW (Balment *et al.*, 2006). AVT would appear to contribute to both the acute response to osmotic challenge as well as longer term acclimation responses. Hypothalamic AVT mRNA expression increases, pituitary AVT content falls and plasma AVT levels are raised in the initial hours and days after transfer of fish between SW and FW (Bond *et al.*, 2002; Warne *et al.*, 2005). Plasma AVT concentration also falls in the initial hours after entry into FW and rises acutely in association with the increased plasma osmolality on transfer of fish to SW. The salt loading and potential cellular dehydration accompanying the early phase of SW exposure can be mimicked by intraperitoneal injection of hypertonic saline. The acute rise in plasma osmolality 60 min after hypertonic saline injection in flounder was associated with raised plasma AVT levels (Warne and Balment, 1995). More recent studies by Hyodo and colleagues revealed a similar picture of AVT secretion stimulated in response to cellular dehydration in the marine elasmobranch *T. scyllium* (Hyodo *et al.*, 2004). Plasma AVT and hypothalamic AVT mRNA expression levels were significantly elevated after fish were held in concentrated SW (130%) for 2 days.

The primary target for AVP in mammals is the kidney, more specifically the distal tubule (collecting duct) where it regulates the insertion of aquaporins to adjust the water permeability and thus volume and osmotic concentration of final urine. It should be noted, however, that even in mammals physiologically relevant levels of AVP can influence not only water excretion but also renal NaCl loss (see Balment *et al.*, 2006). Three major types of AVP receptors have been described in mammals – V1a associated with vascular smooth muscle, V1b associated with pituitary corticotroph cells and V2 linked to regulation of renal tubule function. Thus far only a V1 type vasotocin receptor has been cloned in fish (Mahlmann *et al.*, 1994; Warne, 2001) and using cell expression systems this has been shown to be coupled to the phospholipase C-inositol triphosphate signaling pathway (Warne, 2001). Injections of AVT produce a systemic pressor response in teleost fish (Le Mevel *et al.*, 1993) alongside changes in cardiac output and stroke volume (Oudit and Butler, 1995). Typically there is a transient fall in dorsal aortic blood pressure, which reflects the initial profound constriction of the branchial vasculature (Bennett and Rankin, 1986). The slower development of increase in post-branchial systemic blood pressure results from a more sustained peripheral vasoconstriction. Exploiting agonists of mammalian defined V1 and V2 receptors, it appears that different receptor types may mediate the actions of AVT at the gill and peripheral blood vessels (Warne and Balment, 1997). The AVT receptors mediating the gill vasoconstriction responded to a V2 agonist, which had no apparent effect on the peripheral vascular tone. Although these observations imply a role for AVT in the regulation of some cardiovascular parameters, many of the reported effects are only achieved at concentrations of AVT that exceed the physiological range (Warne and Balment, 1997). In most teleost species so far studied, circulating AVT concentration falls within the range 1–20 fmol mL^{-1}, while recently reported levels in the elasmobranch, *T. scyllium*, appear a little higher at 50–120 fmol mL^{-1} (Hyodo *et al.*, 2004). Therefore, as is the case for AVP in mammals, AVT probably does not directly contribute to the maintenance of systemic blood pressure or cardiovascular function in fish, though it is likely to contribute to adjustments in regional blood flow distribution and blood flow rates in key tissues. The contributions of AVT to fluid and electrolyte balance are also clearly important indirectly to the security of cardiovascular function and blood pressure regulation.

The AVT actions contributing to body fluid balance in fish involve both renal and extrarenal target tissues. The renal effects appear to rely less upon tubular effects, as occurs in mammals, and more upon adjustments in the rates of glomerular filtration. Early studies in FW eels reported a diuresis alongside a pressor response to high doses of AVT (Henderson and Wales, 1974; Babiker and Rankin, 1978), while lower, non-pressor doses of AVT produced a clear fall in urine flow or antidiuresis. It is these lower doses that we now understand will have produced physiologically relevant plasma concentrations of the hormone (Warne and Balment, 1997). Using the *in situ* trout-trunk preparation, which allows systemic pressure to be maintained at a consistent level, Amer and Brown (1995) were able to show that physiologically relevant concentrations of AVT are indeed antidiuretic. This appears to be largely driven by a reduction in the glomerular filtration rate (GFR), with little evidence for an additional tubular action. This AVT induced fall in GFR appears to be the result of a reduction in the number of filtering nephrons (derecruitment), when glucose transport maxima were used as a marker for the number of filtering nephrons (Henderson and Wales, 1974; Amer and Brown, 1995). Wells *et al.* (2002, 2005) have shown more recently, using a similar *in situ* perfused kidney preparation of the dogfish, *S. canicula*, that AVT also induces a glomerular antidiuresis in elasmobranchs, again involving glomerular intermittency. Immunostaining for the V1 receptor, using an homologous specific antibody for flounder, indicated an entirely renal vascular distribution (Warne *et al.*, 2005). Immunostaining for the receptor was evident on the afferent and efferent arterioles of the glomerulus and extended to the small blood vessels in the smooth muscle layer surrounding the collecting duct. Clearly AVT-induced contraction of the afferent or efferent glomerular arterioles would provide a mechanism to either decrease or increase GFR respectively. However, the AVT induced glomerular intermittency involves at least two populations of non-filtering nephrons, one in which the glomeruli are perfused with blood and one which is not. This implies an additional more subtle action of AVT, which could be possibly afforded by the receptors in the smooth muscle surrounding the collecting duct. It has been suggested that contraction of this smooth muscle could increase intratubular pressure and thus reduce filtration rate (Tsuneki *et al.*, 1984). Recent assessment of global renal V1 receptor mRNA expression by quantitative PCR indicated lower expression levels in flounder adapted to FW compared with SW acclimated fish. There was also a significant reduction in relative AVT V1 receptor mRNA expression at 24 h following transfer of fish from SW to FW (see Balment *et al.*, 2006). Accordingly, it is likely that part of the altered actions of AVT in FW and SW fish involve changes in tissue receptor expression and thus sensitivity to the neurohypohysial secreted peptide.

Although the V2 receptor is generally considered to have emerged in vertebrate evolution at the level of tetrapods, with their emergence on land (Pang, 1983), there is some indirect evidence which implies the presence of an AVT receptor coupled to the cAMP signaling pathway in fish kidney (Perrott *et al.*, 1993; Warne *et al.*, 2002). *In vitro* trout renal tubule preparations showed concentration-dependent stimulation of accumulated cAMP production. Significant cAMP generation was observed when tissue was exposed to 10^{-12} M AVT, which accords well with reported circulating levels. The presence of a V2 receptor, typically linked to tubular water reabsorption through insertion of AQP2 in the apical membrane of distal tubular cells in tetrapods, would clearly require a major reassessment of how AVT might modulate urine production in fish. This would imply a combination of both glomerular and tubular components, comparable with those seen in non-mammalian tetrapods, fundamentally changing our current view of the evolutionary development of AVT regulation of renal function in vertebrates. It would clearly be very rewarding to positively identify and characterize this putative AVT V2 receptor.

In fish the gills are an integral part of the extrarenal mechanisms supporting ion and water balance and, as indicated above, AVT receptors are present in the gill. Iodinated vasotocin binding studies showed the presence of a single population of receptors in the eel gill, with the number of binding sites being higher in cells from SW compared with FW fish (Guibbolini *et al.*, 1988). Vasotocin V1 receptor mRNA is expressed in the gill (Mahlmann *et al.*, 1994; Warne, 2001) and while there was no difference between relative receptor mRNA expression levels in SW and FW acclimated flounder, there was a significant fall at 24 h after transfer of fish from SW to FW (Balment *et al.*, 2006). As indicated earlier, AVT can profoundly affect blood flow through the gills, decreasing lamellar perfusion, which can indirectly influence both passive and active ion fluxes across the epithelia (Bennett and Rankin, 1986; Olson, 2002). In trout the vascular receptor would appear to be pharmacologically comparable with the mammalian V1 type and oxytocin receptor (Conklin *et al.*, 1999).

There is also a suggestion that AVT can directly affect gill ion and water transport (Maetz *et al.*, 1964; Marshall, 2003). Guibbolini and Avella (2003) using sea bass gill respiratory cells in culture showed a direct effect of AVT on monolayer short-circuit current. This was interpreted as AVT stimulation of Cl^- secretion mediated by a V1-type receptor, a response that would be potentially important for survival of fish in SW. Interestingly, these responses were also shared by the other teleost neurohypophysial peptide, isotocin. While this work highlights actions for AVT on gill respiratory cells, Guibbolini and Avella (2003) did not rule out additional AVT actions on the gill MRCs. Indeed, V1 receptor immunolocalization in SW adapted flounder gill does support the presence of receptors in presumptive MRCs (see Balment *et al.*, 2006). Taken together the scattered observations of AVT and the gill currently support likely roles in ion extrusion, which are likely, if the changes in receptor mRNA expression are associated with altered protein, to be of greater importance in SW than FW environments.

In recent years there has been increasing evidence of AVT involvement in various behaviors, including those related to reproduction (see Balment *et al.*, 2006), so it is not surprising that spawning migrations of fish like chum salmon (Hiraoka *et al.*, 1997) are associated with altered hypothalamic AVT mRNA levels. Kulczykowska (1995) has proposed that melatonin and AVT interact to control adaptations to seasonal and daily environmental changes (light, temperature and salinity). In fish, as in other vertebrates, plasma cortisol concentrations show a diurnal rhythm. In the rainbow trout hypothalamic parvocellular AVT mRNA expression shows a daily rhythm which is the inverse of that for cortisol (Gilchriest *et al.*, 1998). In flounder there is a very clear diurnal variation in plasma melatonin levels which peak at the onset of the dark phase, a pattern which remains entrained even when animals are maintained in continuous darkness (Kulczykowska *et al.*, 2001). Plasma AVT concentrations show an inverse relationship with melatonin with highest AVT levels in the daylight hours, implying a functional relationship between melatonin and AVT, perhaps comparable with that for melatonin and AVP in mammals, in which AVT/AVP is consistently inhibitory to pineal melatonin secretion (Olcese *et al.*, 1993).

The innervation of pituitary corticotroph cells by AVT neurons (Batten *et al.*, 1990) affords AVT influence over adrenocorticotropin (ACTH) and thus interrenal cortisol secretion. Using *in vitro* rainbow trout anterior pituitary cell incubations, Baker *et al.* (1996) showed AVT dose-dependent stimulation of ACTH secretion and synergistic stimulatory actions with corticotropin-releasing factor (CRF), as reported for AVP and CRF in mammals (Rivier and Vale, 1983). In rainbow trout *in situ* hybridization analysis of the preoptic neurons indicated increased AVT mRNA expression in parvocellular but not magnocellular neurons in response to confinement stress (Gilchriest *et al.*, 2000). This contrasts with observations in the flounder where altered magnocellular rather than parvocellular AVT mRNA expression was evident 3, 24 and 48 h following confinement stress (Bond *et al.*, 2007). These neurons also expressed the glucocorticoid receptor which would potentially support negative feedback of enhanced plasma cortisol levels. It appears there is likely variation between species in which subsets of the hypothalamic AVT neurons may modulate the hypothalamic–pituitary axis, nonetheless this does allow for AVT influence over cortisol secretion not only in relation to specific stressors but also to osmoregulatory challenges (see below for more detail).

3.2. Renin–angiotensin System (RAS)

The RAS is one of the major endocrine systems modulating osmoregulatory mechanisms throughout the vertebrates (Kobayashi and Takei, 1996). The kidney as the primary source of renin and the gill as the major source of converting enzyme not only play central roles in the control of this system's activity but are themselves target tissues for the active ANG II. Salinity transfer or hemorrhage and other challenges that are likely to produce hypotension or hypovolemia usually activate the RAS (Butler and Brown, 2007). A number of studies have shown that the RAS is activated in fish held in hyperosmotic media, where its actions afford maintenance of blood volume and pressure in a dehydrating environment (see Wong *et al.*, 2006). In addition to the ANG II pressor and dipsogenic actions described above, further direct and indirect effects of ANG II are important to sustain body fluid balance. Cortisol from the interrenal is important for many of the longer term acclimatory adjustments as teleost fish accommodate to the movement between media of different salinities, including MRC functions and gut transport capacity. In flounder ANG II administration stimulates raised plasma cortisol levels (Perrott and Balment, 1990), while plasma cortisol and ANG II changed in parallel as sea bream were adapated to different salinities (Wong *et al.*, 2006). The cortisol rise following transfer of eels from FW to SW can be blocked by administration of the converting enzyme inhibitor, captopril (Kenyon *et al.*, 1985). Accordingly, a number of effects associated with altered RAS activity are likely to include indirect actions as a result of modulating cortisol secretion either at the level of the pituitary ACTH secretion (Weld and Fryer, 1987) or directly at the interrenal. More recent studies have also indicated a direct effect of the RAS to regulate Na/K-ATPase activity in osmoregulatory tissues of the eel including gill (Marsigliante *et al.*, 1997), kidney (Marsigliante *et al.*, 2000) and intestine (Marsigliante *et al.*, 2001). In sea bream intraperitoneal injection of ANG II produced dose-dependent increases in branchial Na/K-ATPase activity (Wong *et al.*, 2006). Clearly such effects of ANG II would enhance potential ion transport activity in these tissues.

In accordance with the apparent activation of the RAS by exposure of fish to dehydrating conditions the direct renal effect of ANG II is to induce vasoconstriction of renal microvasculature leading to a fall in GFR and an antidiuresis (Gray and Brown, 1985; Olson *et al.*, 1986). Obviating the potential confounding effects of systemic blood pressure changes induced by ANG II *in vivo*, this antidiuretic action of ANG II has been confirmed *in vitro* using the perfused trout trunk and *in situ* kidney preparations (Dunne and Rankin, 1992; Brown *et al.*, 1993). While the renal antidiuretic response is dominated by change in GFR, there is also some evidence for possible tubular antidiuretic actions of ANG II (see Brown *et al.*, 2000). A number of studies have shown the presence of ANG II receptors along the teleost renal tubule (Cobb and Brown, 1992; Marsigliante *et al.*, 1997). The renal actions of ANG II *in vivo* are a likely result of a combination of ANG II delivered via the renal arterial supply and also ANG II generated intrarenally. There is growing evidence that, as for mammals, a local intrarenal RAS system exists and that may contribute subtle control of renal function and may contribute to the glomerular intermittency that appears to be central to the control of urine production in fish (Brown *et al.*, 2000).

The unique structure of their ANG II and its lack of activity in mammalian bioassays contributed to an initial view that the elasmobranchs did not possess an RAS, though it is now clear that all components exist as in other vertebrates (Takei *et al.*, 2004b; see Anderson *et al.*, 2001c). Pressor activity for the peptide is evident, part of which may rely upon catecholamine stimulation (Tierney *et al.*, 1997a), and the RAS is considered to contribute to blood pressure regulation during acute blood volume loss and/or fall in blood pressure. Indirect osmoregulatory actions involve stimulation of interrenal secretion of 1α-hydroxycorticosterone, the unique elasmobranch mineralocortocoid (Armour *et al.*, 1993a,b). Receptors for this steroid have been identified in the rectal gland, kidney and gills of the ray, *Raja ocellata* (Mood and Idler, 1974; Idler and Kane, 1980). The presence of juxtaglomerular cells and macula densa in elasmobranchs (Lacy and Reale, 1990) suggests that the RAS may be involved in the regulation of glomerular filtration rate, as in teleosts. Indeed, ANG II induced an antidiuretic effect in the isolated trunk preparation of *S. canicula* (Anderson *et al.*, 2001c). More recent observations confirm an antidiuretic response to ANG II based upon a fall in glomerular filtration rate and the proportion of glomeruli filtering. This contrasted with the diuretic actions of CNP (defined above) in the same preparation (Wells *et al.*, 2006). Elasmobranch gills possess MRCs but Na/K-ATPase activity is 10–15 times less than that of the marine teleost gill (Jampol and Epstein, 1970). Specific ANG II binding is present in both gill cell membranes and branchial blood vessels in elasmobranchs (Tierney *et al.*, 1997b; Anderson *et al.*, 2001a). The physiological role of ANG II in elasmobranch gill awaits clear definition but is likely to contribute to body fluid ion and water balance. The rectal gland is a highly vascularized secretory tissue that is active intermittently to permit excretion of excess NaCl (Shuttleworth, 1988). Volume expansion will directly stimulate rectal gland secretion (Solomon *et al.*, 1985) and it is considered that this affords removal of excess salts during periods of acute volume/salt loading, such as feeding events. In accordance with other ANG II effects which support salt and fluid retention, ANG II appears to inhibit rectal gland secretion (Anderson *et al.*, 2001c). Administration of ANG II in the isolated, perfused rectal gland of *S. canicula* produced increased vascular resistance contrasting with the relaxant actions of CNP, affording a potential dynamic balance of control of gland secretion.

In the last 5 years ANG I has been isolated and characterized from two cyclostomes: the sea lamprey *P. marinus* (Takei *et al.*, 2004b) and the river lamprey *L. fluviatilis* (Rankin *et al.*, 2004). Rankin *et al.* (2001) using a heterologous assay showed that plasma ANG II and III concentrations were higher in SW than FW acclimated river lampreys. So for such an anadromous species migrating between rivers and SW the RAS is likely to be involved in body fluid regulation similar to that seen in euryhaline teleosts. Recently Brown *et al.* (2005) have shown that reduced blood volume is an effective stimulus for RAS activation and that rapid transfer of fish between SW and FW and FW and SW leads to a fall and rise in plasma ANG respectively. These authors suggest that volume/pressure receptors and osmoreceptors interact in the regulation of lamprey RAS activity. This volume/pressure sensitivity is in accord with evidence of a vasoconstrictor action of homologous lamprey ANG II (Rankin *et al.*, 2004) and supports a fundamental role for the RAS in maintaining volume and pressure conserved over 500 million years for vertebrate evolution. It will be fascinating in future studies to learn of the other direct and indirect actions of ANG in this intriguing group of fish.

3.3. Prolactin, Growth Hormone and Cortisol

Prolactin and growth hormone (GH) belong to a family of pituitary polypeptide hormones that are thought to have arisen along with their receptors from a gene duplication event and subsequent divergence early in vertebrate evolution (Forsyth and Wallis, 2002; Kawauchi and Sower, 2006). Both hormones are considered to play a significant role in the acclimation process as fish enhance their ion and water transport capacities to meet the demands of new environments. It was Grace Pickford who provided the first convincing evidence that prolactin had a role in the ion uptake mechanisms in FW teleost fish (Pickford and Phillips, 1959). Since then a very large mass of evidence has accumulated in a number of fish to confirm prolactin's vital role in supporting survival in FW. Following exposure to FW gene expression, synthesis, secretion and plasma prolactin levels increase (see review by Manzon, 2002). Plasma osmolality and cortisol appear to directly influence prolactin secretion (Seale *et al.*, 2006) but an additional prolactin-releasing peptide (PrRP) has also been recently identified in teleosts. Sakamoto *et al.* (2003) have shown that PrRP specifically promotes prolactin transcription and secretion and that PrRP nerve terminals are localized close to the pituitary prolactin cells. It is worth noting that there is variation among species in the ability to survive in FW without pituitary hormones (possible for salmonids but not killifish or tilapia).

Prolactin primarily reduces the ion and water permeability of osmoregulatory epithelia (Hirano, 1986). In euryhaline species prolactin reduces gut water and ion reabsorption, though there are some species variations (Manzon, 2002). This contrasts with the actions of cortisol which tends to increase gut epithelial ion and water permeability and active uptake of ions to increase osmotic water uptake (Loretz, 1995). Although prolactin is associated with Na and Cl retention in a number of FW and euryhaline teleosts, the precise target tissue actions are often rather poorly understood. In the gill prolactin has been shown to affect MRC development, inhibiting the emergence of SW MRC (Herndon *et al.*, 1991) and promoting the development of the FW ion uptake MRC morphology (Pisam *et al.*, 1993). Cortisol may also promote the development of ion uptake MRCs in some species (Perry and Goss, 1994), though cortisol is most obviously recognized for its role in promoting the SW acclimation and the emergence of SW MRCs (see Sakamoto and McCormick, 2006). There is a suggestion of an interaction between cortisol and prolactin for the FW acclimation process but there is little direct evidence to support this (McCormick, 2001).

GH was first recognized as having a capacity to improve brown trout (*Salmo trutta*) tolerance of exposure to SW (Smith, 1956). This has subsequently been shown to be due to GH-induced increase in the number and size of gill MRCs, Na/K-ATPase and NKCC enhancing the gill salt secretory capacity (McCormick, 2001; Pelis and McCormick, 2001). There is an important additive/synergistic interaction between GH and cortisol in these effects (Madsen, 1990). Some of the actions of GH are mediated through insulin-like growth factor (IGF-I). IGF-I treatment increases salinity tolerance in Atlantic salmon, killifish and rainbow trout (Mancera and McCormick, 1998). There is evidence for both endocrine and autocrine/paracrine actions of IGF in the teleost gill (see Sakamoto and McCormick, 2006). This GH/IGF-I axis is clearly of great importance in the physiological adaptations that characterize the parr–smolt transformation of anadromous salmonids.

There are species variations in the actions of GH and prolactin; for example GH appears to be without effect on osmoregulatory parameters in the gilthead sea bream (*Sparus auratus*) (Mancera *et al.*, 2002) or the eel (Sakamoto *et al.*, 1993). The actions of prolactin would also not be appropriate for many stenohaline marine teleosts, though we have no studies yet to assess this. Another consideration is that there are high levels of extrapituitary prolactin and GH reported in teleosts (Imaoka *et al.*, 2000; Sakamoto *et al.*, 2005a,b). It is thus tempting to consider, as suggested by Sakamoto and McCormick (2006), whether prolactin secretion only became centralized into the pituitary with the emergence of the tetrapods.

As indicated above, cortisol appears to be integral to the complex osmoregulatory processes for teleost survival in FW and SW. Cortisol secretion from the interrenal is under hypothalamic–pituitary control via the CRF and ACTH axis, as in other vertebrates. However, further neuroendocrine input to cortisol secretion may emanate from hypothalamic AVT neurons (see above) and from the caudal neurosecretory system (CNSS), which provides potential stimulatory input via its secretion of CRF and urotensins I (UI) and II (UII) (see Lu *et al.*, 2004). Indeed, recent studies in masu salmon suggest that UI, which is also generated in the hypothalamus, may be of greater importance than CRF in regulating cortisol secretion in the spawning period (Westring *et al.*, 2008). Furthermore it is apparent that ANP (Arnold-Reed and Balment, 1991) and ANG II (Perrott and Balment, 1990) can influence the interrenal to ensure that cortisol secretion is integrated with many other aspects of the endocrine response to osmotic and volume challenge. This is of particular importance as cortisol in fish subserves not only osmoregulation but several other important aspects of physiology, including reproduction, immune responses, growth and metabolism (Mommsen *et al.*, 1999). Unlike tetrapods teleost fish do not secrete a separate mineralocorticoid like aldosterone, rather these types of actions as described above are mediated by cortisol. The actions of corticosteroid hormones rely upon intracellular receptors that act as ligand-dependent transcription factors. A single class of high affinity, low capacity binding sites for cortisol has been identified in many tissues including gill and intestine. The number of binding sites and affinity can be altered by changes in external salinity (see Prunet *et al.*, 2006). A glucocorticoid receptor (GR) was first cloned for the rainbow trout (Ducouret *et al.*, 1995) and later in several other teleost fish. However, recent studies indicate the existence of a second GR isoform in trout (Bury *et al.*, 2003) and also the presence of a mineralocorticoid receptor (MR) (Sturm *et al.*, 2005). In the absence of a separate mineralocorticoid secreted from the teleost interrenal this was surprising, though evidence that deoxycorticosterone (DOC) was a potent stimulator of the MR and its presence in fish has raised the possibility that DOC could be the physiological ligand for teleost MR (Prunet *et al.*, 2006). Information on DOC actions in fish is scarce and further work is required to elucidate the validity of this proposal. However, a recent study in Atlantic salmon comparing the effects of exogenous cortisol, DOC and aldosterone on salinity tolerance provides strong evidence that it is cortisol, probably acting through the GR, that is responsible for the increased salinity tolerance and increased gill Na/K-ATPase activity. There was no clear effect of DOC or aldosterone, and the putative MR blocker, spironolactone, did not alter the increased salinity tolerance induced by cortisol (McCormick *et al.*, 2008). A study by Shaw and colleagues (2007) in killifish (*Fundulus heteroclitus*) similarly supports GR rather than MR mediation of the corticosteroid promotion of acclimation to SW. Inhibition of the GR by RU-486 prevented killifish from acclimating to increased salinity, while spironolactone (MR blocker) had no effect on SW acclimation.

3.4. Urotensins

Urotensin I (UI) is a 41-amino acid peptide with close structural similarities to CRF. Although UI was originally identified in fish it is now apparent that it is present as a neurotransmitter/neuromodulator throughout vertebrates and is a significant contributor to stress responses (see Chapter 6). Urotensin II (UII) is a cyclic peptide originally isolated from the CNSS of the teleost, *Gillichthys mirabilis* (Pearson *et al.*, 1980), on the basis of its smooth muscle-stimulating activity (Figure 8.4). Recently it has become evident that UII is the natural ligand of the GPR14 orphan receptor (Ames *et al.*, 1999), now renamed the UII receptor or UT receptor. In mammals UII has been shown to have extremely potent cardiovascular actions and is apparently

Human	ETPDCFWKYCV	AA535545
Flounder	QFAGTTECFWKYCV	P21857
Takifugu	TGNNECFWKYCV	Scaf. 60
Tetraodon	HGNDECFWKYCV	Chr. 9
Goby	AGTADCFWKYCV	P01147
Carp-a	GGGADCFWKYCV	P04560
Carp-b	GGNTECFWKYCV	P04561
Zebrafish-a	GGGADCFWKYCV	Chr. 7
Zebrafish-b	GSNTECFWKYCV	EH608865
White sucker-a	GSGADCFWKYCV	P04558
White sucker-b	GSNTECFWKYCV	P04559
Medaka	SGNTECFWKYCV	Chr. 7
Stickleback	AGNSECFWKYCV	Contig 2571
Paddlefish	GSTSECFWKYCV	P81022
Dogfish	NNFSDCFWKYCV	P35490
River lamprey	NNFSDCFWKYCV	Waugh et al. (1995)
Sea lamprey	NNFSDCFWKYCV	Waugh et al. (1995)

Fig. 8.4. Urotensin II sequences identified thus far in fishes. The human sequence is shown for reference. Conserved amino acids across species are shadowed. Accession number of protein and EST database, and chromosome number on which the gene localizes, are noted on the right.

associated with a number of related diseases such as essential hypertension (Balment *et al.*, 2005).

Both UI and UII are present in specific brain regions in both fish and mammals but in fish a major source of circulating peptides is proposed to be the CNSS. This unique fish neuroendocrine structure is present in the spinal cord of the terminal vertebral segments and comprises large peptide synthesizing neurons, Dahlgren cells, which in teleosts project axons to a neuroheamal organ, the urophysis (Lu *et al.*, 2004, 2006; McCrohan *et al.*, 2007). Dahlgren cells are also present in elasmobranchs but the CNSS is less well organized and lacks a clear urophysis. Recent studies in a number of fish have provided detailed information regarding the structural organization of the CNSS (Parmentier *et al.*, 2006). These underline the potential value of this discrete structure in teleosts for investigation of the fundamental mechanisms of neuroendocrine secretion (see review by McCrohan *et al.*, 2007). Unlike the hypothalamic– neurohypohysial system, with which it shares many parallels, the CNSS is easily accessed for both *in vivo* and *in vitro* electrophysiological studies (Brierley *et al.*, 2001; Ashworth *et al.*, 2005). This has enabled recent advances in our understanding of the neurophysiology of this system, together with factors that modulate its output (Lu *et al.*, 2007; McCrohan *et al.*, 2007; Marley *et al.*, 2007).

Urotensins have been implicated in various aspects of fish physiology including osmoregulation, reproduction and stress responses. It has been shown that both UI and UII modulate cortisol secretion from the interrenal suggesting that the CNSS affords stress-specific stimulation of cortisol secretion (Kelsall and Balment, 1998) independent of hypothalamic–pituitary input (Winter *et al.*, 2000). Clearly such effects of urotensins on cortisol secretion are likely to have an impact on body fluid balance as outlined above. Indeed, secretion of urophysial peptides is sensitive to external salinity. Reduction in medium tonicity increased Dahlgren cell UI immunoreactivity in two species of teleosts (Minniti and Minniti, 1995). In *G. mirabilis* urophysial UI content was increased 24 h after transfer of fish to FW by comparison with SW maintained animals (Larson and Madani, 1991). Recent studies in the rainbow trout have shown complex changes in forebrain and CNSS CRF and UI mRNA expression in response to transfer of fish from FW to SW (Craig *et al.*, 2005). There was a transient rise in plasma ACTH and cortisol alongside an increase in hypothalamic and preoptic CRF mRNA expression at 24 h, while there was a delayed increase in hypothalamic UI mRNA expression. In contrast, both UI and CRF mRNA expression in the CNSS increased in parallel 24 h after SW transfer. The authors suggest that specific forebrain and CNSS populations of neurons have different roles in the coordinated acute and chronic responses to fluid balance disturbances.

Direct actions of UI on transporting epithelia have been described. Significant reductions in water and NaCl absorption were induced by UI in isolated intestinal segments of FW- but not SW-adapted tilapia (Mainoya and Bern, 1982). Active chloride secretion by the MRCs of the opercular skin of *G. mirabilis* was stimulated by UI (Marshall and Bern, 1981), representing an important aspect of ion balance in marine fish. In estuarine fish such as the killifish, UI may contribute along with AVT to the complex and rapid onset of gill salt secretion as fish move between dilute and hypertonic microenvironments (see Marshall, 2003). The icv injection of UI in unanesthetized trout induced an increase in dorsal aortic blood pressure, while intra-arterial injection of UI produced a transient fall in blood pressure reversed to hypertension by the release of catecholamines (Le Mevel *et al.*, 2006). In the elasmobranch, *S. canicula*, bolus injection of UI again produced a transient fall and then a sustained rise in blood pressure (Platzack *et al.*, 1998). Clearly such effects are likely to impinge upon renal excretion and in elasmobranchs possibly rectal gland secretion but this awaits further study.

The icv injection of UII in trout also produced a long-lasting hypertensive response though at higher doses compared with UI (Le Mevel *et al.*, 2008b). Intra-arterial UII administration produced a clear dose-dependent increase in dorsal aortic pressure and a decrease in heart rate of long duration. The effects of UII on blood flow and blood pressure in renal and gill tissues are also likely to alter ion and water fluxes across these epithelia and mediate in part the observed effects of UII administration or urophysis removal on plasma composition (see Lederis, 1977). There is, nonetheless, evidence of direct UII effects on ion and water transport. Eliminating any effects of UII on blood flow, Marshall and Bern (1979) used the isolated opercular skin of *G. mirabilis* to show that UII inhibited the short circuit current (SCC), a measure of active chloride secretion across the epithelium. Similarly, using an isolated *G. mirabilis* bladder preparation, Loretz and Bern (1981) showed that UII stimulated active Na uptake. More recently, high doses of UII were shown to stimulate SCC in isolated hind gut preparations of FW adapted *A. anguilla*, potentially reflecting increased absorption of Na across the gut wall (Baldisserotto and Mimura, 1997). These effects on isolated epithelia would provide fish in hypotonic media with an increased ability to conserve NaCl, which represents one of their major osmoregulatory challenges. In fish the major source of circulating UII appears to be the CNSS; surgical removal of the urophysis greatly reduces plasma UII levels (Winter *et al.*, 1999; Lu *et al.*, 2006). Urophysial peptide content varies in accordance with external salinity in euryhaline fish like the flounder, *P. flesus* (Arnold-Reed *et al.*, 1991) and plasma UII levels are also rapidly responsive to transfer of fish between SW and FW (Bond *et al.*, 2002). Plasma UII levels fell and urophysial UII storage rose in fish transferred to FW relative to SW maintained animals.

The first fish UII receptor was cloned recently for the flounder (Lu *et al.*, 2006) and according to its widespread tissue expression is likely to mediate diverse actions of UII. However, it is now clear that UII receptor mRNA is present in all the key osmoregulatory tissues, gill, kidney, bladder and gut, which also express UII mRNA, affording the potential for both paracrine and endocrine UII control of ion and water transport. Using a heterologous UII receptor antibody, UII receptor immunoreactivity appeared to be localized to vascular elements in both kidney and gill (Lu *et al.*, 2006). Interestingly, UII receptor mRNA expression was lower in the kidney and gill of FW compared with SW adapted fish, which may provide a basis for altered target tissue sensitivity (Lu *et al.*, 2006). Direct transfer of flounder from SW to FW resulted in increased CNSS UII mRNA expression at 8 h, while the reverse transfer FW to SW was associated with a fall in CNSS UII mRNA after 8 h relative to FW maintained animals, underlining the capacity for rapid responses of the CNSS neuroendocrine system to altered external salinity (Lu *et al.*, 2006). Kidney UII receptor mRNA expression was reduced 8 h and 24 h after transfer of fish from SW to FW, while the reverse transfer was associated with markedly increased UII receptor mRNA levels at 24 h. A similar pattern of change was also evident for gill UII receptor mRNA levels, implying considerable plasticity in target tissue responsiveness to UII under these osmoregulatory challenges.

The recent extensive mammalian UII research has revealed a second gene encoding a precursor of a UII paralogue, called UII-related peptide (URP). This octapeptide, identified in humans, mice and rats (Sugo *et al.*, 2003), shares the cyclic hexapeptide sequence with UII. Both UII and URP have limited structural identity to somatostatin 1 (SS1) and UII and SS1 share some functional characteristics (see Conlon *et al.*, 1997). However, cDNA analysis of prepro-UII and prepro-SS1 indicates little sequence identity (Coulouarn *et al.*, 1999) and probably that these were not derived from a common ancestral gene. However, using a comparative genomics approach, Tostivint *et al.* (2006) have provided convincing evidence that the UII and URP genes and the SSI and its related peptide genes arose through segmental duplication of two ancestral genes that were linked to each other. Accordingly, it appears that UII- and somatostatin-encoding genes do ultimately belong to one superfamily. Tostivint *et al.* (2006) were unable to detect a URP-like gene in the zebrafish database and suggested that the gene has been lost in this species. A URP-like sequence was, however, evident in the genome of *Tetraodon* and *Takifugu*, though the predicted peptide appears atypical compared with tetrapod URP and the phylogenetic and functional significance of URP in fish groups awaits further study.

4. PERSPECTIVES

Fishes are in a unique position within vertebrate species in terms of body fluid regulation because their blood is in direct contact with environmental water of varying salinities via only a monolayer respiratory epithelium. Accordingly, regulation of water and ions is profoundly influenced by the environmental salinity and reversed when fish migrate between FW and SW. Marine teleost fish lose water by osmosis at a rate even greater than terrestrial animals. Thus, oral fluid intake is essential for their survival as in terrestrial animals. Furthermore, mechanisms regulating drinking are expected to be not as complicated in fishes as in terrestrial animals because of easier access to water. Thus fish may be an excellent model to analyze these mechanisms. However, research on this subject is still in its infancy compared with studies in mammals. It is important to determine whether fishes feel thirsty or not when they drink, as they can ingest surrounding water simply by reflex swallowing. Frogs are thought to feel thirsty when they absorb water by the ventral skin, although they do not drink orally (Hillyard, 1999). The presence of the sensation of thirst is determined by the extended time in water with an assumption that thirst has motivated frogs to stay in water for absorption by the skin. A similar criterion can be used for the amphibious fish that mostly stay on the land.

Another important field in fluid intake in fishes is the cerebral mechanisms that lead to drinking behavior. As fishes drink surrounding water by only swallowing, coordinated muscular movements may be regulated by the GVC through the vagus nerve. Therefore, the regulation of GVC activities from upper neuronal systems will be the next target. There is some evidence that neurons in the AP send their axons to the GVC directly or through neurons in other brain nuclei (Mukuda *et al.*, 2005). There is circumstantial evidence suggesting that inhibitory signals are sent from the forebrain or midbrain to prevent excess drinking. The presence and localization of drink-regulating hormones in fish brain should also be investigated to analyze the interaction of intrinsic brain hormones with peripheral ones to optimize the intake. In any case, less complicated fish brains appear to serve as excellent tools for the analysis.

Neuroendocrine secretions from the brain and CNSS contribute directly and indirectly to the regulation of ion and water fluxes at the gills, gut and kidney. Observations of the direct actions of AVT are still very limited and future work should include assessment of whether V2 receptor mediated effects are present in fish. It seems unlikely that this receptor class and linked responses are limited to the terrestrial tetrapods as currently accepted. As a result of the recent rapid growth in mammalian UII research it is likely that there will also be impetus to reassess UII actions in fish. For these and other intriguing apparent gaps or contradictions between knowledge of regulatory mechanisms in fish and tetrapods, amphibious fish species, such as the mudskipper, may afford valuable models for future investigation. Recent rapid advances in ligand and receptor characterization in cyclostomes and elasmobranchs provide the basis for growth in functional studies in these groups which is critical for the development of a better understanding of the evolution of body fluid regulatory mechanisms in aquatic species. Finally it is becoming increasingly apparent that regulatory pathways have representation in the brain, the periphery as circulating hormones as well as local osmoregulatory tissue expression (ligand and receptor). Future work will need to consider how these central, endocrine and autocrine/paracrine components interact to achieve appropriate adjustments in fluid and ion balances.

As this chapter has emphasized, body fluid homeostasis is underpinned by complex and highly interactive regulatory pathways. We have only just begun to understand some of the cross-talk among these pathways but with technical advances which allow multicomponent analysis and real-time coincident measures, we can look forward to significant advances in this area. This is critical if we are to establish how this complexity is integrated to secure both rapid and longer term adjustments in physiological systems as fish experience challenges to the security of fluid balance.

REFERENCES

Acher, R. (1996). Molecular evolution of fish neurohypophysial hormones: Neutral and selective evolutionary mechanisms. *Gen. Comp. Endocrinol.* **102,** 157–172.

Amer, S., and Brown, J. A. (1995). Glomerular actions of arginine vasotocin in the in situ perfused trout kidney. *Am. J. Physiol.* **269,** R775–R780.

Ames, R. S., Sarau, H. M., Chambers, J. K., Willette, R. N., Aiyar, N. V., Romanic, A. M., Louden, C. S., Foley, J. J., Sauermelch, C. F., Coatney, R. W., Ao, Z., Disa, J., *et al.* (1999). Human urotensin-II is a potent vasoconstrictor and agonist for the orphan receptor GPR14. *Nature* **401,** 282–286.

Anderson, W. G., Cerra, M. C., Wells, A., Tierney, M. L., Tota, B., Takei, Y., and Hazon, N. (2001a). Angiotensin and angiotensin receptors in cartilaginous fishes. *Comp. Biochem. Physiol.* **128A,** 31–40.

Anderson, W. G., Takei, Y., and Hazon, N. (2001b). The dipsogenic effect of the renin–angiotensin system in elasmobranch fish. *Gen. Comp. Endocrinol.* **125,** 300–307.

Anderson, W.G, Takei, Y., and Hazon, N. (2001c). Possible interaction between the renin–angiotensin system and natriuretic peptides on drinking in elasmobranch fish. *In* "Perspectives in Comparative Endocrinology" (Goos, H. J. T., Rastoi, R. K., Vaudry, H., and Perantoni, R., Eds.), pp. 753–758. Monduzzi Editore, Bologna.

Anderson, W. G., Good, J. P., and Hazon, N. (2002a). Changes in secretion rate and vascular perfusion in the rectal gland of the European lesser-spotted dogfish (*Scyliorhnus canicula* L.) in response to environmental and hormonal stimuli. *J. Fish Biol.* **60,** 1580–1590.

Anderson, W. G., Takei, Y., and Hazon, N. (2002b). Osmotic and volemic effects on drinking rate in elasmobranch fish. *J. Exp. Biol.* **205,** 1115–1122.

Anderson, W. G., Wells, A., Takei, Y., and Hazon, N. (2002c). The control of drinking in elasmobranch fish with special reference to the renin–angiotensin system. *In* "Osmoregulation and Drinking in Vertebrates" (Hazon, N., and Flik, G., Eds.), pp. 19–30. BIOS Scientific Publishers Ltd, Oxford.

Anderson, W. G., Pillans, R. D., Hyodo, S, Tsukada, T., Good, J. P., Takei, Y., Franklin, C. E., and Hazon, N. (2006). The effects of freshwater to seawater transfer on circulating levels of angiotensin II, C-type natriuretic peptide and arginine vasotocin in the euryhaline elasmobranch, *Carchahinus leucas. Gen. Comp. Endocrinol.* **147,** 39–46.

Ando, M., Fujii, Y., Kadota, T., Kozaka, T., Mukuda, T., Takase, I., and Kawahara, A. (2000). Some factors affecting drinking behavior and their interactions in seawater-acclimated eels, *Anguilla japonica. Zool. Sci.* **17,** 171–178.

Antunes-Rodrigues, J., McCann, S. M., Rogers, L. C., and Samson, W. K. (1985). Atrial natriuretic factor inhibits dehydration- and angiotensin-induced water intake in the conscious unrestrained rat. *Proc. Natl. Acad. Sci. USA* **82,** 8720–8723.

Armour, K. J., O'Toole, L. B., and Hazon, N. (1993a). The effect of dietary protein restriction on the secretory dynamics of 1α-hydroxycorticosterone and urea in the dogfish *Scyliorhinus canicula*: a possible role for 1α-hydroxycorticosterone in sodium retention. *J. Endocrinol.* **138,** 275–282.

Armour, K. J., O'Toole, L. B., and Hazon, N. (1993b). Mechanisms of ACTH- and angiotensin II-stimulated 1α-hydroxycorticosterone secretion in the dogfish, *Scyliorhinus canicula. J. Mol. Endocrinol.* **19,** 235–242.

Arnold-Reed, D. E., and Balment, R. J. (1991). Atrial naturetic factor stimulates in-vivo and in-vitro secretion of cortisol in teleosts. *J. Endocrinol.* **128,** R17–R20.

Arnold-Reed, D. E., Balment, R. J., McCrohan, C. R., and Hackney, C. M. (1991). The caudal neurosecretory system of *Platichthys flesus*: General morphology and responses to altered salinity. *Comp. Biochem. Physiol.* **99A,** 137–143.

Ashworth, A. J., Banks, J., Brierley, M. J., Balment, R. J., and McCrohan, C. R. (2005). Electrical activity of caudal neurosecretory neurons in seawater and freshwater-adapted *Platichthys flesus, in vivo*. *J. Exp. Biol.* **208,** 267–275.

Babiker, M. M., and Rankin, J. C. (1978). Neurohypophysial hormone control of kidney function in the European eel (*Anguilla anguilla*) adapted to sea water or fresh water. *J. Endocrinol.* **78,** 347–358.

Bader, M., and Ganten, D. (2002). It's renin in the brain: Transgenic animals elucidate the brain renin–angiotensin system. *Circ. Res.* **90,** 8–10.

Baker, B. I., Bird, D. J., and Buckingham, J. C. (1996). In the trout, CRH and AVT synergize to stimulate ACTH release. *Regul. Pept.* **67,** 207–210.

Baldisserotto, B., and Mimura, O. M. (1997). Changes in the electrophysiological parameters of the posterior intestine of *Anguilla anguilla* (Pisces) induced by oxytocin, urotensin II and aldosterone. *Braz. J. Med. Biol. Res.* **30,** 35–39.

Balment, R. J. (2002). Control of water balance in mammals. *In* "Osmoregulation and Drinking in Vertebrates" (Hazon, N., and Flik, G., Eds.), pp. 153–168. Bios Scientific Publishers Ltd, Oxford.

Balment, R. J., and Carrick, S. (1985). Endogenous renin–angiotensin system and drinking behavior in flounder. *Am. J. Physiol.* **248,** R157–R160.

Balment, R. J., Masini, M. A., Vallarino, M., and Conlon, J. M. (2002). Cardiovascular actions of lungfish bradykinin in the unanesthetised African lungfish, *Protopterus annectens*. *Comp. Biochem. Physiol.* **131A,** 467–474.

Balment, R. J., Song, W., and Ashton, N. (2005). Urotensin II – ancient hormone with new functions in vertebrate body fluid regulation. *Ann. N.Y. Acad. Sci.* **1040,** 66–73.

Balment, R. J., Lu, W., Weybourne, E., and Warne, J. M. (2006). Arginine vasotocin a key hormone in fish physiology and behaviour: A review with insights from mammalian models. *Gen. Comp. Endocrinol.* **147,** 9–16.

Batten, T. F. C., Cambre, M. L., Moons, L., and Vandesande, F. (1990). Comparative distribution of neuropeptides – immunoreactive systems in the brain of the green molly (*Poecilia latipinna*). *J. Comp. Neurol.* **302,** 893–919.

Bennett, M. B., and Rankin, J. C. (1986). The effect of neurohypophysial hormones on the vascular resistance of the isolated perfused gill of the European eel *Anguilla anguilla*. *Gen. Comp. Endocrinol.* **64,** 60–66.

Beyenbach, K. W. (1995). Secretory electrolyte transport in renal proximay tubules of fish. *In* "Cellular and Molecular Approaches to Fish Ionic Regulation" (Wood, C. M., and Shuttleworth, T. J., Eds.), pp. 85–105. Academic Press, San Diego.

Beyenbach, K. W. (2004). Kidneys sans glomeruli. *Am. J. Physiol.* **286,** F811–F827.

Bond, H., Winter, M., Warne, J. M., and Balment, R. J. (2002). Plasma concentrations of arginine vasotocin and urotensin II are reduced following transfer of flounder (*Platichthys flesus*) from sea water to fresh water. *Gen. Comp. Endocrinol.* **125,** 113–120.

Bond, H., Warne, J. M., and Balment, R. J. (2007). Effect of acute restraint on hypothalamic provasotocin mRNA expression in flounder, *Platichthys flesus*. *Gen Comp. Endocrinol.* **153,** 221–227.

Brierley, M. J., Ashworth, A. J., Banks, R. J., Balment, R. J., and McCrohan, C. R. (2001). Bursting properties of caudal neurosecretory cells in the flounder, *Platichthys flesus, in vitro*. *J. Exp. Biol.* **204,** 2733–2739.

Brown, J. A., Rankin, J. C., and Yokota, S. D. (1993). Glomerular heamodynamics of filtration in single nephrons of non-mammalian vertebrates. *In* "New Insights in Vertebrate Kidney

Function" (Brown, J. A., Balment, R.J, and Rankin, J. C., Eds.), pp. 1–44. Cambridge University Press, Cambridge.

Brown, J. A., Paley, R. K., Amer, S., and Aves, S. J. (2000). Evidence for an intrarenal renin–angiotensin system in the rainbow trout, *Oncorhynchus mykiss*. *Am. J. Physiol.* **278,** R1685–R1691.

Brown, J. A., Cobb, C. S., Frankling, S. C., and Rankin, J. C. (2005). Activation of the newly discovered cyclostome renin–angiotensin system in the river lamprey *Lampetra fluviatilis*. *J. Exp. Biol.* **208,** 223–232.

Bury, N. R., Sturm, A., Rouzic, P. Le, Lethimonier, C., Ducouret, B., Guiguen, Y., Robinson-Rechavi, M., Laudet, V., Rafestin-Oblin, M. E., and Prunet, P. (2003). Evidence for two distinct functional glucocorticoid receptors in teleost fish, *J. Mol. Endocrinol.* **31,** 141–156.

Butler, D. G., and Brown, J. A. (2007). Stanniectomy attenuates the renin–angiotensin response to hypovolemic hypotension in freshwater eels (*Anguilla rostrata*) but not blood pressure. *J. Comp. Physiol.* **177B,** 143–151.

Carrick, S., and Balment, R. J. (1983). The renin–angiotensin system and drinking in the euryhaline flounder, *Platichthys flesus*. *Gen. Comp. Endocrinol.* **51,** 423–433.

Chasiotis, H., Effendi, J. C., and Kelly, S. P. (2009). Occludin expression in goldfish held in ion-poor water. *J. Comp. Physiol. B.* **179,** 145–154.

Cobb, C. S., and Brown, J. A. (1992). Localisation of angiotensin II binding to tissues of the rainbow trout, *Oncorhynchus mykiss*, adapted to freshwater and seawater: An autoradiographic study. *J. Comp. Physiol. B* **162,** 197–202.

Conklin, D. J., Smith, M. P., and Olson, K. R. (1999). Pharmacological characterisation of arginine vasotocin vascular smooth muscle receptors in the trout (*Onorhynchus mykiss*) *in vitro*. *Gen. Comp. Endocrinol.* **114,** 36–46.

Conlon, J. M. (1999). Bradykinin and its receptors in non-mammalian vertebrates. *Regul. Pept.* **79,** 1–81.

Conlon, J. M., Tostivint, H., and Vaudry, H. (1997). Somatostatin- and urotensin II-related peptides: Molecular diversity and evolutionary perspectives. *Regul. Pept.* **69,** 95–103.

Coulouarn, Y., Jégou, S., Tostivint, H., Vaudry, H., and Lihrmann, I. (1999). Cloning, sequence analysis and tissue distribution of the mouse and rat urotensin II precursors. *FEBS Lett.* **457,** 28–32.

Craig, P. M., Al-Timimi, H., and Bernier, N. J. (2005). Differential increase in forebrain and caudal neurosecretory system corticotrophin-releasing factor and urotensin I gene expression associated with seawater transfer in rainbow trout. *Endocrinology* **146,** 3851–3860.

de Caro, G., Epstein, A. N., and Massi, M. (1986). "The Physiology of Thirst and Sodium Appetite." Plenum Press, New York.

Denton, D. (1982). "The Hunger for Salt." Springer-Verlag, Berlin.

Doty, R. W. (1968). Neural organization of deglutition. *In* "Handbook of Physiology: Alimentary Canal" (Code, C. F., Ed.), pp. 1861–1902. American Physiological Society, Washington DC.

Ducouret, B., Tujague, M., Ashraf, J., Mouchel, N., Servel, N., Valotaire, Y., and Thompson, E. B. (1995). Cloning of a teleost fish glucocorticoid receptor shows that it contains a deoxyribonucleic acid-binding domain different from that of mammals. *Endocrinology* **136,** 3774–3783.

Duner, T., Conlon, J. M., Kukkonen, J. P., Akerman, K. E. O., Yan, Y., Postlethwait, J. H., and Larhammer, D. (2002). Cloning, structural characterization and functional expression of a zebrafish bradykinin B2-related receptor. *Biochem. J.* **364,** 817–824.

Dunne, J. B., and Rankin, J. C. (1992). Effects of atrial natriuretic peptide and angiotensin II on salt and water excretion by the perfused rainbow trout kidney. *J. Physiol. (Lond.)* **446,** 92P.

Elkinton, J. R., and Taffel, M. (1942). Prolonged water deprivation in the dog. *J. Clin. Invest.* **21,** 787–794.

Epstein, A. N., Fitzsimons, J. T., and Rolls, B. J. (1970). Drinking induced by injection of angiotensin into the brain of the rat. *J. Physiol. (London)* **210,** 457–474.

Evans, D. H., Piermarini, P. M., and Choe, K. P. (2005). The multifunctional fish gill: Dominant site of gas exchange, osmoregulation, acid-base regulation and excretion of nitrogenous waste. *Physiol. Rev.* **85,** 97–177.

Ferguson, A. V., and Bains, J. S. (1996). Electrophysiology of the circumventricular organs. *Frontiers Neuroendocrinol.* **17,** 440–475.

Fitzsimons, J. T. (1979). "The Physiology of Thirst and Sodium Appetite." Cambridge University Press, Cambridge.

Fitzsimons, J. T. (1998). Angiotensin, thirst, and sodium appetite. *Physiol. Rev.* **78,** 583–686.

Flik, G., Varsamos, S., Guerreiro, P. M. G., Fuentes, X., Huising, M. O., and Fenwick, J. C. (2002). Drinking in (very young) fish. *In* "Osmoregulation and Drinking in Vertebrates" (Hazon, N., and Flik, G., Eds.), pp. 31–47. BIOS Scientific Publishers Ltd, Oxford.

Folmar, L. C., and Dickhoff, W. W. (1980). The Parr-smalt transformation (smoltification) and seawater adaptation in salmonids. *Aquaculture* **21,** 1–37.

Forsyth, I. A., and Wallis, M. (2002). Growth hormone and prolactin – molecular and functional evolution. *J. Mammary Gland Biol. Neoplasia.* **7,** 291–312.

Foster, J. M., and Forster, M. E. (2007). Effects of salinity manipulations on blood pressures in an osmoconforming chordate, the hagfish, *Eptatretus cirrhatus. J. Comp. Physiol. B* **177,** 31–39.

Fox, B. K., Riley, L. G., Dorough, C., Kaiya, H., Hirano, T., and Grau, E. G. (2007). Effects of homologous ghrelins on the growth hormone/ insulin-like growth factor-I axis in the tilapia, *Oreochromis mossambicus. Zool. Sci.* **24,** 391–400.

Fregly, M. J., and Rowland, N. E. (1991). Bradykinin-induced dipsogenesis in captopril-treated rats. *Brain Res. Bull.* **26,** 169–172.

Fuentes, J., Bury, N. R., Carroll, S., and Eddy, F. B. (1996). Drinking in Atlantic salmon presmolts (*Salmo salar* L.) and juvenile rainbow trout (*Oncorhynchus mykiss* Walbaum) in response to cortisol and seawater challenge. *Aquaculture* **141,** 129–137.

Giffard-Mena, I., Boulo, V., Aujoulat, F., Fowden, H., Castille, R., Charmantier, G., and Cramb, G. (2007). Aquaporin molecular characterization in the sea-bass (*Dicentrarchus labrax*): The effect of salinity on AQP1 and AQP3 expression. *Comp. Biochem. Physiol. A.* **148,** 430–444.

Gilchriest, B. J., Tipping, D. R., Levy, A., and Baker, B. I. (1998). Diurnal changes in the expression of genes encoding for arginine vasotocin and pituitary proopiomelanocortin in the rainbow trout (*Oncorhynchus mykiss*): Correlation with changes in plasma hormones. *J. Neuroendocrinol.* **10,** 937.

Gilchriest, B. J., Tipping, D. R., Hake, L., Levy, A., and Baker, B. I. (2000). The effects of acute and chronic stresses on vasotocin gene transcripts in the brain of the rainbow trout (*Oncorhynchus mykiss*). *J. Neuroendocrinol.* **12,** 795–801.

Gray, J. C., and Brown, J. A. (1985). Renal and cardiovascular effects of angiotensin II in the rainbow trout, *Salmo gairdneri. Gen. Comp. Endocrinol.* **59,** 375–381.

Grossman, S. P. (1990). "Thirst and Sodium Appetite. Physiological Basis." Academic Press, San Diego.

Guibbolini, M. E., and Avella, M. (2003). Neurohypophysial hormone regulation of Cl- secretion: Physiological evidence for V1-type receptors in sea bass gill respiratory cells in culture. *J. Endocrinol.* **176,** 111–119.

Guibbolini, M. E., Henderson, I. W., Mosley, W., and Lahlou, B. (1988). Arginine vasotocin binding to isolated branchial cells of the eel: Effect of salinity. *J. Mol. Endocrinol.* **1,** 125–130.

Hashimoto, H., Fujihara, H., Kawasaki, M., Saito, T., Shibata, M., Takei, Y., and Ueta, Y. (2007a). Centrally and peripherally administered ghrelin potently inhibits water intake in rats. *Endocrinology* **148,** 1638–1647.

Hashimoto, H., Hyodo, S., Kawasaki, M., Shibata, M., Saito, T., Suzuki, H., Ohtsubo, H., Yokoyama, T., Fujihara, H., Higuchi, T., Takei, Y., and Ueta, Y. (2007b). Adrenomedullin 2 is a more potent activator of hypothalamic oxytocin-secreting neurons than adrenomedullin in rats, and its effects are only partially blocked by antagonists for adrenomedullin and calcitonin gene-related peptide receptors. *Peptides* **28,** 1104–1112.

Hay, D. L., Conner, A. C., Howitt, S. G., Smith, D. M., and Poyner, D. R. (2004). The pharmacology of adrenomedullin receptors and their relationship to CGRP receptors. *J. Mol. Neurosci.* **22,** 105–114.

Hazon, N., and Henderson, I. W. (1984). Secretory dynamics of 1α-hydroxycorticosterone in the elasmobranch fish, *Scyliorhinus canicula. J. Endocrinol.* **103,** 205–211.

Hazon, N., Wells, A., Pillans, R. D., Good, J. P., Anderson, W. G., and Franklin, C. E. (2003). Urea based osmoregulation and endocrine control in elasmobranch fish with special reference to euryhalinity. *Comp. Biochem. Physiol.* **136B,** 685–700.

Henderson, I. W., and Wales, N. A. M. (1974). Renal diuresis and antidiuresis after injections of arginine vasotocin in the fresh water eel *Anguilla anguilla. J. Endocrinol.* **61,** 487–500.

Herndon, T. M., McCormick, S. D., and Bern, H. A. (1991). Effects of prolactin on chloride cells in opercular membrane of seawater-adapted tilapia. *Gen. Comp. Endocrinol.* **83,** 283–289.

Hillyard, S. D. (1999). Behavioral, molecular and integrative mechanisms of amphibian osmoregulation. *J. Comp. Zool.* **283,** 662–674.

Hirano, T. (1974). Some factors regulating drinking by the eel, *Anguilla japonica. J. Exp. Biol.* **61,** 737–747.

Hirano, T. (1986). The spectrum of prolactin action in teleosts. *Prog. Clin. Biol. Res.* **205,** 53–74.

Hirano, T., and Hasegawa, S. (1984). Effects of angiotensins and other vasoactive substances on drinking in the eel *Anguilla japonica. Zool. Sci.* **1,** 106–113.

Hirano, T., Satou, M., and Utida, S. (1972). Central nervous system control of osmoregulation in the eel (*Anguilla japonica*). *Comp. Biochem. Physiol.* **43A,** 537–544.

Hiraoka, S., Ando, H., Ban, M., Ueda, H, and Urano, A. (1997). Changes in expression of neurohypophysial hormone genes during spawning migration in chum salmon, *Ocorhynchus keta. J. Mol. Endocrinol.* **18,** 49–55.

Hiroi, J., and McCormick, S. D. (2007). Variation in salinity tolerance, gill $Na^+/K^+/2Cl^-$ cotransporter and mitochondrial-rich cell distribution in three salmonids *Salvelinus namaycush, Salvelinus fontinalis* and *Salmo salar. J. Exp. Biol.* **210,** 1015–1024.

Hoar, W. S. (1988). The physiology of smolting salmonids. *In* "Fish Physiology" (Hoar, W. S., and Randall, D. J., Eds.), Vol. 11B, pp. 275–344. Academic Press, San Diego.

Hosoda, H., Kojima, M., Matsuo, H., and Kangawa, K. (2000). Ghrelin and des-acyl ghrelin: Two major forms of rat ghrelin peptide in gastrointestinal tissues. *Biochem. Biophys. Res. Commun.* **279,** 909–913.

Hyodo, S., Tsukada, T., and Takei, Y. (2004). Neurohypophysial hormones of dogfish, *Triakis scyllium*: Structures and salinity-dependent secretion. *Gen. Comp. Endocrinol.* **138,** 97–104.

Idler, D. R., and Kane, K. M. (1980). Cytosol receptor glycoprotein for 1α-hydroxycorticosterone in tissues of elasmobranch fish, *Raja ocellata. Gen. Comp. Endocrinol.* **42,** 259–266.

Imaoka, T., Matsuda, M., and Mori, T. (2000). Extrapituitary expression of the prolactin gene in the goldfish, African clawed frog and mouse. *Zool. Sci.* **17,** 791–796.

Inoue, K., Naruse, K., Yamagami, S., Mitani, H., Suzuki, N., and Takei, Y. (2003). Four functionally distinct C-type natriuretic peptides found in fish reveal new evolutionary history of the natriuretic system. *Proc. Natl. Acad. Sci. USA* **100,** 10079–10084.

Inoue, K., Sakamoto, T., Yuge, S., Iwatani, H., Yamagami, S., Tsutsumi, M., Hori, H., Cerra, M. C., Tota, B., Suzuki, N., Okamoto, N., and Takei, Y. (2005). Structural and functional evolution of three cardiac natriuretic peptides. *Mol. Biol. Evol.* **22,** 2428–2434.

Ito, S., Mukuda, T., and Ando, M. (2006). Catecholamines inhibit neuronal activity in the glossopharyngeal-vagal motor complex of the Japanese eel: Significance for controlling swallowing water. *J. Exp. Zool.* **305A,** 499–506.

Iwata, M. (1995). Downstream migratory behavior of salmonids and its relationship with cortisol and thyroid hormones: A review. *Aquaculture* **135,** 131–139.

Iwata, M., Ogura, H., Komatsu, S., and Suzuki, K. (1986). Loss of seawater preference in chum salmon (*Oncorhynchus keta*) fry retained in fresh water after migration season. *J. Exp. Zool.* **240,** 369–376.

Jampol, L. M., and Epstein, F. M. (1970). Sodium-potassium-activated adenosine triphosphatase and osmotic regulation by fishes. *Am. J. Physiol.* **218,** 607–611.

Johnson, A. K., and Thunhorst, R. L. (1997). The neuroendocrinology of thirst and salt appetite: Visceral sensory signals and mechanisms of central integration. *Frontiers Neuroendocrinol.* **18,** 292–353.

Johnson, K. R., and Olson, K. R. (2008). Comparative physiology of the piscine natriuretic peptide system. *Gen. Comp. Endocrinol.* **157,** 21–26.

Kaiya, H., and Takei, Y. (1996a). Changes in plasma atrial and ventricular natriuretic peptide concentrations after transfer of eels from fresh water and seawater or *vice versa*. *Gen. Comp. Endocrinol.* **104,** 337–345.

Kaiya, H., and Takei, Y. (1996b). Osmotic and volaemic regulation of atrial and ventricular natriuretic peptide secretion in conscious eels. *J. Endocrinol.* **149,** 441–447.

Kaiya, H., Tsukada, T., Yuge, S., Mondo, H., Kangawa, K., and Takei, Y. (2006). Identification of eel ghrelin in plasma and stomach by raidoimmunoassay and histochemistry. *Gen. Comp. Endocrinol.* **148,** 375–382.

Kaiya, H., Miyazato, M., Kangawa, K., Peter, R. E., and Unniappan, S. (2008). Ghrelin: A multifunctional hormone in non-mammalian vertebrates. *Comp. Biochem. Physiol.* **149A,** 109–128.

Kaufman, S., and Peters, G. (1980). Regulatory drinking in the pigeon, *Columba livia*. *Am. J. Physiol.* **239,** R219–R225.

Kawakoshi, A., Kaiya, H., Riley, L. G., Hirano, T., Grau, E. G., Miyazato, M., Hosoda, H., and Kangawa, K. (2007). Identification of ghrelin-like peptide in two species of shark, *Sphyrna lewini* and *Carcharhinus melanopterus*. *Gen. Comp. Endocrinol.* **151,** 259–268.

Kawauchi, H., and Sower, S. A. (2006). The dawn and evolution of hormones in the adenohypophysis. *Gen. Comp. Endocrinol.* **148,** 3–14.

Kelsall, C. J., and Balment, R. J. (1998). Native urotensins influence cortisol secretion and plasma cortisol concentration in the euryhaline flounder, *Platichthys flesus*. *Gen. Comp. Endocrinol.* **112,** 210–219.

Kenyon, C. J., McKeever, A., Oliver, J. A., and Henderson, I. W. (1985). Control of renal and adrenocortical function by the renin–angiotensin system in two euryhaline teleost fishes. *Gen. Comp. Endocrinol.* **58,** 93–100.

Kirsch, R., and Mayer-Gostan, N. (1973). Kinetics of water and chloride exchanges during adaptation of the European eel to sea water. *J. Exp. Biol.* **58,** 105–121.

Kitamura, K., Kangawa, K., Kawamoto, M., Ichiki, Y., Nakamura, S., Matsuo, H., and Eto, T. (1993). Adrenomedullin: A novel hypotensive peptide isolated from human pheochromocytoma. *Biochem. Biophys. Res. Commun.* **192,** 553–560.

Kitamura, K., Ichiki, Y., Tanaka, M., Kawamoto, M., Emura, J., Sakakibara, S., Kangawa, K., Matsuo, H., and Eto, T. (1994). Immunoreactive adrenomedullin in human plasma. *FEBS Lett.* **341,** 288–290.

Kobayashi, H., and Takei, Y. (1996). "The Renin–Angiotensin System: Comparative Aspects." Springer International, Berlin.

Kobayashi, H., Uemura, H., Wada, M., and Takei, Y. (1979). Ecological adaptation of angiotensin II-induced thirst mechanism in tetrapods. *Gen. Comp. Endocrinol.* **38,** 93–104.

Kobayashi, H., Uemura, H., Takei, Y., Itazu, N., Ozawa, M., and Ichinohe, K. (1983). Drinking induced by angiotensin II in fishes. *Gen. Comp. Endocrinol.* **49,** 295–306.

Kojima, M., Hosoda, H., Date, Y., Nakazato, M., Matsuo, H., and Kangawa, K. (1999). Ghrelin is a growth-hormone-releasing acylated peptide from stomach. *Nature* **402,** 656–660.

Komlosi, P., Fintha, A., and Bell, P. D. (2004). Current mechanisms of macula densa signaling. *Acta Physiol. Scand.* **181,** 463–469.

Kozaka, T., and Ando, M. (2003). Cholinergic innervations to the upper esophageal sphincter muscle in the eel, with special reference to drinking behavior. *J. Comp. Physiol. B* **173,** 135–140.

Kozaka, T., Fujii, Y., and Ando, M. (2003). Central effects of various ligands on drinking behavior in eels acclimated to seawater. *J. Exp. Biol.* **206,** 687–692.

Kulczykowska, E. (1995). Arginine vasotocin–melatonin interactions in fish: A hypothesis. *Fish Biol. Fish* **5,** 96–102.

Kulczykowska, E., Warne, J. M., and Balment, R. J. (2001). Day–night variations in plasma melatonin and arginine vasotocin concentrations in chronically cannulated flounder (*Platichthys flesus*). *Comp. Biochem. Physiol.* **130A,** 827–834.

Lacy, E. R., and Reale, E. (1990). The presence of juxtaglomerular apparatus in elasmobranch fish. *Anat. Embryol.* **182,** 249–262.

Lancien, F., and Le Mevel, J.-C. (2007). Central actions of angiotensin II on spontaneous baroreflex sensitivity in the trout *Oncorhynchus mykiss. Regul. Pept.* **138,** 94–102.

Larson, B. A., and Madani, Z. (1991). Increased urotensin I and II immunoreactivity in the urophysis of *Gillichthys mirabilis* transferred to low salinity. *Gen. Comp. Endocrinol.* **83,** 379–387.

Lederis, K. (1977). Chemical properties and the physiological and pharmacological actions of urophysial peptides. *Am. Zool.* **17,** 823–832.

Le Mével, J.-C., Pamantung, T. F., Mabin, D., and Vaudrey, H. (1993). Effects of central and peripheral administration of arginine vasotocin and related neuropeptides on blood pressure and heart rate in the conscious trout. *Brain Res.* **610,** 82–89.

Le Mével, J.-C., Pamantung, T. F., Mabin, D., and Vaudry, H. (1994). Intracerebroventricular administration of angiotensin II increases heart rate in the conscious trout. *Brain Res.* **654,** 216–222.

Le Mével, J.-C., Mimassi, N., Lancien, F., Mabin, D., Boucher, J. M., and Blanc, J. J. (2002). Heart rate variability, a target for the effects of angiotensin II in the brain of the trout *Oncorhynchus mykiss. Brain Res.* **947,** 34–40.

Le Mével, J. C., Mimassi, N., Lancien, F., Mabin, D., and Conlon, J. M. (2006). Cardiovascular actions of the stress-related neurohormonal peptides corticotropin- releasing factor and urotensin-I in the trout *Oncorhynchus mykiss. Gen. Comp. Endocrinol.* **146,** 56–61.

Le Mével, J.-C., Lancien, F., and Mimassi, N. (2008a). Central cardiovascular actions of angiotensin in trout. *Gen. Comp. Endocrinol.* **157,** 27–34.

Le Mével, J.-C., Lancien, F., Mimassi, N., Leprince, J., Conlon, J. M., and Vaudry, H. (2008b). Central and peripheral cardiovascular, ventilatory, and motor effects of trout urotensin-II in the trout. *Peptides* **29,** 830–837.

Leshem, M. (1999). The ontogeny of salt hunger in the rat. *Neurosci. Biobehav. Rev.* **23,** 649–659.

Li, Z., Smith, M. P., Duff, D. W., Barton, B. A., Olson, K. R., and Conlon, J. M. (1998). Isolation and cardiovascular activity of [Met^1, Met^1] bradykinin from the plasma of a sturgeon (Acipenseriformes). *Peptides* **19,** 635–641.

Lin, H., and Randall, D. (1995). Proton pump in fish gills. *In* "Cellular and Molecular Approaches to Fish Ionic Regulation" (Wood, C. M., and Shuttleworth, T. J., Eds.), pp. 229–256. Academic Press, San Diego.

López, J., and Martínez, A. (2001). Cell and molecular biology of the multifunctional peptide, adrenomedullin. *Int. Rev. Cytol.* **221,** 1–92.

Loretz, C. A. (1995). Electrophysiology of ion transport in teleost intestinal cells. *In* "Cellular and Molecular Approaches to Fish Ionic Regulation" (Wood, C. M., and Shuttleworth, T. J., Eds.), pp. 25–56. Academic Press, San Diego.

Loretz, C. A., and Bern, H. A. (1981). Stimulation of sodium transport across the teleost urinary bladder by Urotensin II. *Gen. Comp. Endocrinol.* **43,** 325–330.

Lu, W., Gumusgoz, S., Dow, L., Brierley, M. J., Warne, J. M., McCrohan, C. R., Balment, R. J., and Riccardi, D. (2004). Co-expression of corticotrophin-releasing hormone (CRH) and urotensin I (UI) precursor genes in the caudal neurosecretory system of the euryhaline flounder (*Platichthys flesus*): A possible shared role in phenotypic plasticity. *Endocrinology* **145,** 5786–5797.

Lu, W., Greenwood, M., Dow, L., Yuill, J., Worthington, J., Brierley, M. J., McCrohan, C. R., Riccardi, D., and Balment, R. J. (2006). Molecular characterization and expression of urotensin II and its receptor in the flounder (*Platichthys flesus*): A hormone system supporting body fluid homeostasis in euryhaline fish. *Endocrinology* **147,** 3692–3708.

Lu, W., Worthington, J., Riccardi, D., Balment, R. J., and McCrohan, C. R. (2007). Seasonal changes in peptide, receptor and ion channel mRNA expression in the caudal neurosecretory system of the European flounder (*Platichthys flesus*). *Gen Comp. Endocrinol.* **153,** 262–267.

Madsen, S. S. (1990). The role of cortisol and growth hormone in seawater adaptation and development of hypoosmoregulatory mechanisms in sea trout parr (*Salmo trutta trutta*). *Gen. Comp. Endocrinol.* **79,** 1–11.

Maetz, J., Bourguet, J., Lahlou, B., and Hourdry, J. (1964). Peptides neurohypophysiares et osmoregulation chez *Carassius auratus*. *Gen. Comp. Endocrinol.* **4,** 508–522.

Mahlmann, S., Meyerhof, W., Hausmann, H., Heierhorst, J., Schönrock, C., Zwiers, H., Lederis, K., and Richter, D. (1994). Structure, function, and phylogeny of [Arg^8] vasotocin receptors from teleost fish and toad. *Proc. Natl. Acad. Sci. USA.* **91,** 1342–1345.

Mainoya, J. R., and Bern, H. A. (1982). Effects of teleost urotensins on intestinal absorption of water and NaCl in Tilapia, *Saratherodon moassambicus*, adapted to fresh water or sea water. *Gen. Comp. Endocrinol.* **47,** 54–58.

Malvin, R. L., Schiff, D., and Eiger, S. (1980). Angiotensin and drinking rates in the euryhaline killifish. *Am. J. Physiol.* **239,** R31–R34.

Mancera, J. M., and McCormick, S. D. (1998). Osmoregulatory actions of the GH/IGF axis in non-salmonid teleosts. *Comp. Biochem. Physiol.* **121B,** 43–48.

Mancera, J. M., Carrion, R. L., and del Rio, M. D. M. (2002). Osmoregulatory action of PRL, GH, and cortisol in the gilthead seabream (*Sparus aurata* L.). *Gen. Comp. Endocrinol.* **129,** 95–103.

Manzon, L. A. (2002). The role of prolactin in fish osmoregulation: A review. *Gen. Comp. Endocrinol.* **125,** 291–310.

Marley, R., Lu, W, Balment, R. J., and McCrohan, C. R. (2007). Evidence for nitric oxide role in the caudal neurosecretory system of the European flounder, *Platichthys flesus*. *Gen. Comp. Endocrinol.* **153,** 251–261.

Marshall, W. S. (2003). Rapid regulation of NaCl secretion by estuarine teleost fish: Coping strategies for short-duration fresh water exposures. *Biochim. Biophys. Acta* **1618,** 95–105.

Marshall, W. S., and Bern, H. A. (1979). Teleostean urophysis: Urotensin II and ion transport across the isolated skin of a marine teleost. *Science* **204,** 519–520.

Marshall, W. S., and Bern, H. A. (1981). Active chloride transport by the skin of a marine teleost is stimulated by urotensin I and inhibited by urotensin II. *Gen. Comp. Endocrinol.* **43,** 484–491.

Marsigliante, S., Muscella, A., Vinson, G. P., and Storelli, C. (1997). Angiotensin II receptors in the gill of sea water- and freshwater-adapted eel. *J. Mol. Endocrinol.* **18,** 67–76.

Marsigliante, S., Muscella, A., Barker, S., and Storelli, C. (2000). Angiotensin II modulates the activity of the Na^+/K^+ATPase in eel kidney. *J. Endocrinol.* **165,** 147–156.

Marsigliante, S., Muscella, A., Greco, S., Elia, M. G., Vilella, S., and Storelli, C. (2001). Na^+/K^+ATPase activity inhibition and isoform-specific translocation of protein kinase C following angiotensin II administration in isolated eel enterocytes. *J. Endocrinol.* **168,** 339–346.

Matsuda, K., Miura, T., Kaiya, H., Maruyama, K., Shimakura, S., Uchiyama, M., Kangawa, K., and Shioda, S. (2006). Regulation of food intake by acyl and des-acyl ghrelins in the goldfish. *Peptides* **27,** 2321–2325.

McCormick, S. D. (2001). Endocrine control of osmoregulation in teleost fish. *Am. Zool.* **41,** 781–794.

McCormick, S. D., and Bradshaw, D. (2006). Hormonal control of salt and water balance in vertebrates. *Gen. Comp. Endocrinol.* **147,** 3–8.

McCormick, S. D., Regish, A., O'Dea, M. F., and Shrimpton, J. M. (2008). Are we missing a mineralocorticoid in teleost fish? Effects of cortisol, deoxycorticosterone and aldosterone on osmoregulation, gill Na^+, K^+-ATPase activity and isoform mRNA levels in Atlantic salmon. *Gen. Comp. Endocrinol.* **157,** 35–41.

McCrohan, C. R., Lu, W., Brierley, M. J., Dow, L., and Balment, R. J. (2007). Fish caudal neurosecretory system: A model for the study of neuroendocrine secretion. *Gen. Comp. Endocrinol.* **153,** 243–250.

McInerney, J. E. (1964). Salinity preference: An orientation mechanism in salmon migration. *J. Fish. Res. Bd. Canada* **21,** 995–1018.

McKinley, M. J., Albiston, A. L., Allen, A. M., Mathai, M. L., May, C. N., McAllen, R. M., Oldfield, B. J., Mendelsohn, F. A. O., and Chai, S. Y. (2003a). The brain renin–angiotensin system: Location and physiological roles. *Int. J. Biochem. Cell Biol.* **35,** 901–918.

McKinley, M. J., McAllen, R. M., Davern, P., Giles, M. E., Penschow, J., Sunn, N., Uschakov, A., and Oldfield, B. J. (2003b). "The Sensory Circumventricular Organs of the Mammalian Brain. Advances in Anatomy, Embryology and Cell Biology 172." Springer, Berlin.

Minniti, F., and Minniti, G. (1995). Immunocytochemical and ultrastructural changes in the caudal neurosecretory system of a sea water fish *Boops boops* L (teleostei: Sparidae) in relation to the osmotic stress. *Eur. J. Morphol.* **33,** 473–483.

Mommsen, T. P., Vijayan, M. M., and Moon, T. W. (1999). Cortisol in teleosts: Dynamics, mechanisms of action, and metabolic regulation. *Rev. Fish Biol. Fish.* **9,** 211–268.

Mood, T. W., and Idler, D. R. (1974). The binding of 1α-hydroxycorticosterone to tissue soluble proteins in the skate, *Raja ocellata. Comp. Biochem. Physiol.* **48,** 499–500.

Mukuda, T., and Ando, M. (2003). Medullary motor neurons associated with drinking behavior of Japanese eels. *J. Fish Biol.* **62,** 1–12.

Mukuda, T., Matsunaga, Y., Kawamoto, K., Yamaguchi, K., and Ando, M. (2005). "Blood-contacting neurons" in the brain of the Japanese eel, *Anguilla japonica. J. Exp. Zool.* **303A,** 366–376.

Nag, K., Kato, A., Nakada, T., Hoshijima, K., Mistry, A. C., Takei, Y., and Hirose, S. (2006). Molecular and functional characterization of adrenomedullin receptors in pufferfish. *Am. J. Physiol.* **290,** R467–R478.

Nakazato, M., Murakami, N., Date, Y., Kojima, M., Matsuo, H., Kangawa, K., and Matsukura, S. (2001). A role for ghrelin in the central regulation of feeding. *Nature* **409,** 194–198.

Nicolaidis, S., and Fitzsimons, J. T. (1975). La dependence de la prise d'eau induite par l'angiotensine II envers la function vasomotrice cerebrale chez le rat. *C. R. Acad. Sci. Paris* **281,** 1417–1420.

Nielsen, C., Madsen, S. S., and Bjornsson, T. B. (1999). Changes in branchial and intestinal osmoregulatory mechanisms and growth hormone levels during smolting in hatchery-reared and wild brown trout. *J. Fish. Biol.* **54,** 799–818.

Nishimura, H., Sawyer, W. H., and Nigrelli, R. F. (1976). Renin, cortisol and plasma volume in marine teleost fishes adapted to dilute media. *J. Endocrinol.* **70,** 47–59.

Nishimura, H., Lunde, L. G., and Zucker, A. (1979). Renin response to hemorrhage and hypotension in the aglomerular toadfish *Opsanus tau. Am. J. Physiol.* **237,** H105–H111.

Nobata, S., Ogoshi, M., and Takei, Y. (2008). Potent cardiovascular actions of homologous adrenomedullins in eel. *Am. J. Physiol.* **294,** R1544–R1553.

Ogoshi, M., Inoue, K., and Takei, Y. (2003). Identification of a novel adrenomedullin gene family in teleost fish. *Biochem. Biophys. Res. Commun.* **311,** 1072–1077.

Ogoshi, M., Inoue, K., Naruse, K., and Takei, Y. (2006). Evolutionary history of the calcitonin gene-related peptide family in vertebrates revealed by comparative genomic analyses. *Peptides* **27,** 3154–3164.

Ogoshi, M., Nobata, S., and Takei, Y. (2008). Potent osmoregulatory actions of peripherally and centrally administered homologous adrenomedullins in eels. *Am. J. Physiol.* **295,** R2075–R2083.

Oide, H., and Utida, S. (1968). Changes in intestinal absorption and renal excretion of water during adaptation to sea-water in the Japanese eel. *Marine Biol.* **1,** 172–177.

Ojima, D., and Iwata, M. (2007). The relationship between thyroxine surge and onset of downstream migration in chum salmon *Oncorhynchus keta* fry. *Aquaculture* **273,** 185–193.

Okawara, Y., Karakida, T., Aihara, M., Yamaguchi, K., and Kobayashi, H. (1987). Involvement of angiotensin II in water intake in the Japanese eel, *Anguilla japonica. Zool. Sci.* **4,** 523–528.

Olcese, J., Sinemus, C., and Ivell, R. (1993). Vasopressinergic innervation of the bovine pineal gland: Is there a local source for arginine vasopressin? *Mol. Cell. Neurosci.* **4,** 47–54.

Olson, K. R. (1992). Blood and extracellular fluid volume regulation: Role of the renin–angiotensin kallikrein–kinin systems and atrial natriuretic peptides. *In* "Fish Physiology" (Hoar, W. S., Randall, D. J., and Farrell, A. P., Eds.), Vol. XIIB, pp. 135–254. Academic Press, San Diego.

Olson, K. R. (2002). Gill circulation: Regulation of perfusion distribution and metabolism of regulatory molecules. *J. Exp. Zool.* **293,** 320–335.

Olson, K. R., and Hoagland, T. M. (2008). Effects of freshwater/saltwater adaptation and dietary salt on fluid compartments, blood pressure and venous capacitance in trout. *Am. J. Physiol.* **294,** R1061–R1067.

Olson, K. R., Kullman, D., Nakartes, A. J., and Oparil, S. (1986). Angiotensin extraction by trout tissues *in vivo* and metabolism by the perfused gill. *Am. J. Physiol.* **250,** R532–R538.

Olson, K. R., Conklin, D. J., Weaver, L., Jr., Duff, D. W., Herman, C. A., Wang, X., and Conlon, J. M. (1997). Cardiovascular effects of homologous bradykinin in rainbow trout. *Am. J. Physiol.* **272,** R1112–R1120.

Oudit, G. Y., and Butler, D. G. (1995). Cardiovascular effects of arginine vasotocin, atrial natriuretic peptide, and epinephrine in freshwater eels. *Am. J. Physiol.* **268,** R1273–R1280.

Pamantung, T. F., Leroy, J. P., Mabin, D., and Le Mével, J.-C. (1997). Role of dorsal vagal motor nucleus in angiotensin II-mediated tachycardia in the conscious trout *Oncorhynchus mykiss. Brain Res.* **777,** 167–175.

Pang, P. K. T. (1983). Evolution of control of epithelial transport in vertebrates. *J. Exp. Biol.* **106,** 549–556.

Parmentier, C., Taxi, J., Balment, R. J., Nicolas, G., and Calas, A. (2006). Caudal neurosecretory system of the zebrafish: Ultrastructural organization and immunocytochemical detection of urotensins. *Cell Tiss. Res.* **325,** 111–124.

Parry, G. (1960). The development of salinity tolerance in the salmon, *Salmo salar* (L.) and some related species. *J. Exp. Biol.* **37,** 425–434.

Pearson, D., Shively, J. E., Clark, B. R., Geschwind, I. I., Barkley, M., Nishioka, R. S., and Bern, H. A. (1980). Urotensin II: A somatostatin-like peptide in the caudal neurosecretory system of fishes. *Proc. Natl. Acad. Sci. USA* **77,** 5021–5024.

Pelis, R. M., and McCormick, S. D. (2001). Effects of growth hormone and cortisol on Na^+-K^+-$2Cl^-$ cotransporter localization and abundance in the gills of Atlantic salmon. *Gen. Comp. Endocrinol.* **124,** 134–143.

Perrott, M. N., and Balment, R. J. (1990). The renin–angiotensin system and the regulation of plasma cortisol in the flounder, *Platichthys flesus*. *Gen. Comp. Endcorinol.* **78,** 414–420.

Perrott, M. N., Sainsbury, R. J., and Balment, R. J. (1993). Peptide hormone stimulated second messenger production in the teleoston nephron. *Gen. Comp. Endocrinol.* **89,** 387–395.

Perry, R. M., and Goss, G. G. (1994). The effects of experimentally altered gill chloride cell surface area on acid-base regulation in rainbow trout during metabolic alkalosis. *J. Comp. Physiol. B.* **164,** 327–336.

Pickford, G. E., and Phillips, J. G. (1959). Prolactin, a factor promoting survival of hypophysectomized killifish in freshwater. *Science* **130,** 454–455.

Pillans, R. D., and Franklin, C. E. (2004). Plasma osmolyte concentrations and rectal gland mass of bull sharks, *Carcharhinus leucas*, captured along a salinity gradient. *Comp. Biochem. Physiol.* **138A,** 363–371.

Pisam, M., Auperin, B., Prunet, P., Rentierdelrue, F., Martial, J., and Rambourg, A. (1993). Effects of prolactin on alpha and beta chloride cells in the gill epithelium of the saltwater adapted tilapia *Orecochromis niloticus*. *Anat. Rec.* **235,** 275–284.

Platzack, B., and Conlon, J. M. (1997). Purification, structural characterization and cardiovascular activity of cod bradykinins. *Am. J. Physiol.* **272,** R710–R717.

Platzack, B., Schaffert, C., Hazon, N., and Conlon, J. M. (1998). Cardiovascular actions of dogfish urotensin-I in the dogfish, *Scyliorhinus canicula*. *Gen. Comp. Endocrinol.* **109,** 269–275.

Price, C. S., and Schreck, C. B. (2003). Stress and saltwater-entry behavior of juvenile chinook salmon (*Oncorhynchus tshawytscha*): Conflicts in physiological motivation. *Can. J. Fish. Aquat. Sci.* **60,** 910–918.

Prunet, P., Sturm, A., and Milla, S. (2006). Multiple corticosteroid receptors in fish: From old ideas to new concepts. *Gen. Comp. Endocrinol.* **147,** 17–23.

Ramsay, D. J., and Booth, D. A. (1991). "Thirst. Physiological and Psychological Aspects." Springer-Verlag, Berlin.

Rankin, J. C. (2002). Drinking in hagfishes and lampreys. *In* "Osmoregulation and Drinking in Vertebrates" (Hazon, N., and Flik, G., Eds.), pp. 1–18. BIOS Scientific Publishers Ltd., Oxford.

Rankin, J. C., Cobb, C. S., Frankling, S. C., and Brown, J. A. (2001). Circulating angiotensins in the river lamprey, *Lampetra fluviatilis*, acclimated to freshwater and seawater: Possible involvement in the regulation of drinking. *Comp. Biochem. Physiol.* **129B,** 311–318.

Rankin, J. C., Watanabe, T. X., Nakajima, K., Broadhead, C., and Takei, Y. (2004). Identification of angiotensin I in a cyclostome, *Lampetra fluviatilis*. *Zool. Sci.* **21,** 173–179.

Regoli, D., and Barabe, J. (1980). Pharmacology of bradykinin and related kinins. *Pharmacol. Rev.* **32,** 1–46.

Reid, I. A. (1992). Interactions between ANG II, sympathetic nervous system, and baroreceptor reflexes in regulation of blood pressure. *Am. J. Physiol.* **262,** E763–E778.

Reinhardt, H. W., and Seeliger, E. (2000). Toward an integrative concept of control of total body sodium. *News Physiol. Sci.* **15,** 319–325.

Riley, L. G., Fox, B. K., Kaiya, H., Hirano, T., and Grau, E. G. (2005). Long-term treatment of ghrelin stimulates feeding, fat deposition and alters the GH/IGF-I axis in the tilapia, *Oreochromis mossambicus*. *Gen. Comp. Endocrinol.* **142,** 234–240.

Rivier, C., and Vale, W. (1983). Modulation of stress-induced ACTH release by corticotrophin-releasing factor, catecholamines and vasopressin. *Nature* **305,** 325–327.

Rolls, B. J., and Rolls, E. T. (1982). "Thirst," Cambridge University Press, Cambridge.

Ruohonen, K., Grove, D. J., and McIloy, J. T. (1997). The amount of food ingested in a single meal by rainbow trout offered chopped herring, dry and wet diets. *J. Fish. Biol.* **51,** 93–105.

Saito, E., Kaiya, H., Tachibana, T., Tomonaga, S., Denbow, D. M., Kangawa, K., and Furuse, M. (2005). Inhibitory effect of ghrelin on food intake is mediated by the corticotropin-releasing factor system in neonatal chicks. *Regul. Pept.* **251,** 201–208.

Sakamoto, T., and McCormick, S. D. (2006). Prolactin and growth hormone in fish osmoregulation. *Gen. Comp. Endocrinol.* **147,** 24–30.

Sakamoto, T., McCormick, S. D., and Hirano, T. (1993). Osmoregulatory actions of growth hormone and its mode of action in salmonids: A review. *Fish Physiol. Biochem.* **11,** 155–164.

Sakamoto, T., Fujimoto, M., and Ando, M. (2003). Fishy tales of prolactin-releasing peptide. *Int. Rev. Cytol.* **225,** 91–130.

Sakamoto, T., Amano, M., Hyodo, S., Moriyama, S., Takahashi, A., Kawauchi, H., and Ando, M. (2005a). Expression of prolactin-releasing peptide and prolactin in the euryhaline mudskippers (*Periophthalmus modestus*): prolactin-releasing peptide as a primary regulator of prolactin. *J. Mol. Endocrinol.* **34,** 825–834.

Sakamoto, T., Oda, A., Narita, K., Takahashi, H., Oda, T., Fujiwara, J., and Godo, W. (2005b). Prolactin: Fishy tales of its primary regulator and function. *Ann. NY Acad. Sci.* **1040,** 184–188.

Seale, A. P., Fiess, J. C., Hirano, T., Cooke, I. M., and Grau, E. G. (2006). Disparate release of prolactin and growth hormone from the tilapia pituitary in response to osmotic stimulation. *Gen. Comp. Endocrinol.* **145,** 222–231.

Schmidt-Nielsen, K. (1997). "Animal Physiology. Adaptation and Environment. Fifth Ed.," pp. 301–394. Cambridge University Press, Cambridge.

Seymour, A. A., Smith, S. G., Mazack, E. K., and Blaine, E. H. (1986). A comparison of synthetic rat and human atrial natriuretic factor in conscious dogs. *Hypertension* **8,** 211–216.

Shaw, J. R., Gabor, K., Hand, H., Lankowski, A., Durant, L., Thibodeau, R., Stanton, C. R., Barnaby, R., Coutermarsh, B., Karlson, K. H., Sato, J. D., Hamilton, J. W., *et al.* (2007). Role of glucocorticoid receptor in acclimation of killifish (*Fundulus heteroclitus*) to seawater and effects of arsenic. *Am. J. Physiol.* **292,** R1052–R1060.

Shuttleworth, T. J. (1988). Salt and water balance. *In* "Physiology of Elasmobranch Fishes" (Shuttleworth, T. J., Ed.), pp. 171–199. Springer Verlag, Berlin.

Simpson, J. B., and Routtenberg, A. (1973). Subfornical organ: Site of drinking elicitation by angiotensin II. *Science* **181,** 1172–1175.

Skidgel, R. A., and Erdos, E. G. (2004). Angiotensin converting enzyme (ACE) and neprilysin hydrolyze neuropeptides: A brief history, the beginning and follow-ups to early studies. *Peptides* **25,** 521–525.

Smith, D. C. W. (1956). The role of the endocrine organs in the salinity tolerance of trout. *Mem. Soc. Endocrinol.* **5,** 83–101.

Solomon, R., Taylor, M., Sheth, S., Silva, P., and Epstein, F. H. (1985). Primary role of volume expansion in stimulation of rectal gland function. *Am. J. Physiol.* **248,** R638–R640.

Sturm, A., Bury, N., Dengreville, L., Fagart, J., Flouriot, G., Rafestin-Oblin, M. E., and Prunet, P. (2005). 11-deoxycorticosterone is a potent agonist of the rainbow trout (*Oncorhynchus mykiss*) mineralocorticoid receptor. *Endocrinology* **146,** 47–55.

Sugo, T., Murakami, Y., Shimomura, Y., Harada, M., Abe, M., Ishibashi, Y., Kitada, C., Miyajima, N., Suzuki, N., Mori, M., and Fujino, M. (2003). Identification of urotensin II-related peptide as the urotensin II-immunoreactive molecule in the rat brain. *Biochem. Biophys. Res. Commun.* **310,** 860–868.

Tachibana, T., Kaiya, H., Denbow, D. M., Kangawa, K., and Furuse, M. (2006). Central ghrelin acts as an anti-dipsogenic peptide in chicks. *Neurosci. Lett.* **405,** 241–245.

Takei, Y. (1977). The role of subfornical organ in drinking induced by angiotensin in the Japanese quail, *Coturnix coturnix japonica*. *Cell Tiss. Res.* **185,** 175–185.

Takei, Y. (1988). Changes in blood volume after alteration of hydromineral balance in conscious eels, *Anguilla japonica*. *Comp. Biochem. Physiol.* **91A,** 293–297.

Takei, Y. (2000). Comparative physiology of body fluid regulation in vertebrates with special reference to thirst regulation. *Jpn. J. Physiol.* **50,** 171–186.

Takei, Y. (2002). Hormonal control of drinking in the eel: An evolutionary approach. *In* "Osmoregulation and Drinking in Vertebrates" (Hazon, N., and Flik, G., Eds.), pp. 61–82. BIOS Scientific Publishers Ltd., Oxford.

Takei, Y. (2008). Exploring novel hormones essential for seawater adaptation in teleost fish. *Gen. Comp. Endocrinol.* **157,** 3–13.

Takei, Y., and Hirose, S. (2002). The natriuretic peptide system in eel: A key endocrine system for euryhalinity? *Am. J. Physiol.* **282,** R940–R951.

Takei, Y., and Tsuchida, T. (2000). Role of the renin–angiotensin system in drinking of seawater-adapted eels *Anguilla japonica*. *Am. J. Physiol.* **279,** R1105–R1111.

Takei, Y., Hirano, T., and Kobayashi, H. (1979). Angiotensin and water intake in the Japanese eel, *Anguilla japonica*. *Gen. Comp. Endocrinol.* **38,** 446–475.

Takei, Y., Okawara, Y., and Kobayashi, H. (1988a). Water intake caused by water deprivation in the quail, *Coturnix coturnix japonica*. *J. Comp. Physiol. B* **158,** 519–525.

Takei, Y., Okawara, Y., and Kobayashi, H. (1988b). Drinking induced by cellular dehydration in the quail, *Coturnix coturnix japonica*. *Comp. Biochem. Physiol.* **90A,** 291–296.

Takei, Y., Okubo, J., and Yamaguchi, K. (1988c). Effect of cellular dehydration on drinking and plasma angiotensin II level in the eel, *Anguilla japonica*. *Zool. Sci.* **5,** 43–51.

Takei, Y., Okawara, Y., and Kobayashi, H. (1989). Control of drinking in birds. *In* "Progress in Avian Osmoregulation" (Hughes, M. R., and Chadwick, A. C., Eds.), pp. 1–12. Leeds Philosophical and Literary Society Ltd, Leeds.

Takei, Y., Tsuchida, T., and Tanakadate, A. (1998). Evaluation of water intake in seawater adaptation in eels using a synchronized drop counter and pulse injector system. *Zool. Sci.* **15,** 677–682.

Takei, Y., Tsuchida, T., Li, Z., and Conlon, J. M. (2001). Antidipsogenic effect of eel bradykinin in the eel, *Anguilla japonica*. *Am. J. Physiol.* **281,** R1090–R1096.

Takei, Y., Inoue, K., Ogoshi, M., Kawahara, T., Bannai, H., and Miyano, S. (2004a). Mammalian homolog of fish adrenomedullin 2: Identification of a novel cardiovascular and renal regulator. *FEBS Lett.* **556,** 53–58.

Takei, Y., Joss, J. M. P., Kloas, W., and Rankin, J. C. (2004b). Identification of angiotensin I in several vertebrate species: Its structural and functional evolution. *Gen. Comp. Endocrinol.* **135,** 286–292.

Takei, Y., Ogoshi, M., and Inoue, K. (2007). A 'reverse' phylogenetic approach for identification of novel osmoregulatory and cardiovascular hormones in vertebrates. *Frontiers Neuroendocrinol.* **28,** 143–160.

Takei, Y., Hashimoto, H., Inoue, K., Osaki, T., Yoshizawa-Kumagaye, K., Watanabe, T. X., Minamino, N., and Ueta, Y. (2008). Central and peripheral cardiovascular actions of adrenomedullin 5, a novel member of the calcitonin gene-related peptide family, in mammals. *J. Endocrinol.* **197,** 391–400.

Tierney, M. L., Luke, G., Cramb, G., and Hazon, N. (1995). The role of the renin–angiotensin system in the control of blood pressure and drinking in the European eel, *Anguilla anguilla. Gen. Comp. Endocrinol.* **100,** 39–48.

Tierney, M. L., Hamano, K., Anderson, G., Takei, Y., Ashida, K., and Hazon, N. (1997a). Interactions between the renin–angiotensin system and catecholamines on the cardiovascular system of elasmobranchs. *Fish. Physiol. Biochem.* **17,** 333–337.

Tierney, M. L., Takei, Y., and Hazon, N. (1997b). The presence of angiotensin II receptors in elasmobranchs. *Gen. Comp. Endocrinol.* **105,** 9–17.

Tipsmark, C. K., Kiilerich, P., Nilsen, T. O., Ebbesson, L. O., Stefansson, S. O., and Madsen, S. S. (2008a). Branchial expression patterns of claudin isoforms in Atlantic salmon during seawater acclimation and smoltification. *Am. J. Physiol.* **294,** R1563–R1574.

Tipsmark, C. K., Luckenbach, J. A., Madsen, S. S., Kiilerich, P., and Borski, R. J. (2008b). Osmoregulation and expression of ion transport proteins and putative claudins in the gill of southern flounder (*Paralichthys lethostigma*). *Comp. Biochem. Physiol. A* **150,** 265–273.

Tostivint, H., Joly, L., Lihrmann, I., Parmentier, C., Lebon, A., Morisson, M., Calas, A., Ekker, M., and Vaudry, H. (2006). Comparative genomics provides evidence for close evolutionary relationships between the urotensin II and somatostatin gene families. *Proc. Natl. Acad. Sci. USA* **103,** 2237–2242.

Tsuchida, T., and Takei, Y. (1998). Effects of homologous atrial natriuretic peptide on drinking and plasma angiotensin II level in eels. *Am. J. Physiol.* **275,** R1605–R1610.

Tsukada, T., and Takei, Y. (2006). Integrative approach to osmoregulatory action of atrial natriuretic peptide in seawater eels. *Gen. Comp. Endocrinol.* **147,** 31–38.

Tsukada, T., Nobata, S., Hyodo, S., and Takei, Y. (2007). Area postrema, a brain circumventricular organ, is the site of antidipsogenic action of circulating atrial natriuretic peptide in eels. *J. Exp. Biol.* **210,** 3970–3978.

Tsukamoto, K., Nakai, I., and Tesch, F.-W. (1998). Do all freshwater eels migrate? *Nature* **396,** 635–636.

Tsuneki, K., Kobayashi, H., and Pang, P. K. T. (1984). Electron microscope study of innervation of smooth muscle cells surrounding collecting tubules of the fish kidney. *Cell Tiss. Res.* **238,** 307–312.

Unniappan, S., Canosa, L. F., and Peter, R. E. (2004). Orexigenic actions of ghrelin in goldfish: feeding-induced changes in rain and gut mRNA expression and serum levels, and responses to central and peripheral injections. *Neuroendocrinology* **79,** 100–108.

Warne, J. M. (2001). Cloning and characterisation of an arginine vasotocin receptor from the euryhaline flounder *Platichthys flesus. Gen. Comp. Endocrinol.* **122,** 312–319.

Warne, J. M., and Balment, R. J. (1995). Effect of acute manipulation of blood volume and osmolality on plasma [AVT] in seawater flounder. *Am. J. Physiol.* **269,** R1107–R1112.

Warne, J. M., and Balment, R. J. (1997). Changes in plasma arginine vasotocin (AVT) concentration and dorsal aortic blood pressure following AVT injection in the teleost *Platichthys flesus. Gen. Comp. Endocrinol.* **105,** 358–364.

Warne, J. M., Harding, K. E., and Balment, R. J. (2002). Neurohypohysial hormones and renal function in fish and mammals. *Comp. Biochem. Physiol.* **132B,** 231–237.

Warne, J. M., Bond, H., Weybourne, E., Sahajpal, V., Lu, W., and Balment, R. J. (2005). Altered plasma and pituitary arginine vasotocin and hypothalamic provasotocin expression in flounder (*Platichthys flesus*) following hypertonic challenge and distribution of vasotocin receptors within the kidney. *Gen. Comp. Endocrinol.* **144,** 240–247.

Waugh, D., Youson, J., Mims, S. D., Sower, S., and Conlon, J. M. (1995). Urotensin II from the river lamprey (Lampetra fluviatilis), the sea lamorey (Petromyzon marinus) and the paddlefish (Polyodon spathula). *Gen. Comp. Endocrinol.* **99,** 323–332.

Weld, M. M., and Fryer, J. N. (1987). Stimulation by angiotensin I and II of ACTH release from goldfish pituitary cell columns. *Gen. Comp. Endocrinol.* **68,** 19–27.

Wells, A., Anderson, W. G., and Hazon, N. (2002). Development of an in situ perfused kidney preparation for elasmobranch fish: Action of arginine vasotocin. *Am. J. Physiol.* **282,** R1636–R1642.

Wells, A., Anderson, W. G., and Hazon, N. (2005). Glomerular effects of AVT on the *in situ* perfused trunk preparation of the dogfish. *Ann. N. Y. Acad. Sci.* **1040,** 515–517.

Wells, A., Anderson, W. G., Cains, J. E., Cooper, M. W., and Hazon, N. (2006). Effects of angiotensin II and C-type natriuretic peptide on the in situ perfused trunk preparation of the dogfish, *Scyliorhinus canicula*. *Gen. Comp. Endocrinol.* **145,** 109–115.

Westring, C. G., Ando, H., Kitahashi, T., Bhandari, R. K., Ueda, H., Urano, A., Dores, R. M., Sher, A. A., and Danielson, P. B. (2008). Seasonal changes in CRF-1 and urotensin I transcript in masu salmon: Correlation with cortisol secretion during spawning. *Gen. Comp. Endocrinol.* **155,** 126–140.

Winter, M. J., Hubbard, P. C., McCrohan, C. R., and Balment, R. J. (1999). A homologous radioimmunoassay for the measurement of urotensin II in the euryhaline flounder, *Platichthys flesus*. *Gen.Comp. Endocrinol.* **114,** 249–256.

Winter, M. J., Ashworth, A., Bond, H., Brierley, M. J., McCrohan, C. R., and Balment, R. J. (2000). The caudal neurosecretory system; control and function of a novel neuroendocrine system in fish. *Biochem. Cell Biol.* **78,** 1–11.

Wong, M. K. S., Takei, Y., and Woo, N. Y. S. (2006). Differential status of the renin–angiotensin system of the silver bream (*Sparus sarba*) in different salinities. *Gen. Comp. Endocrinol.* **149,** 81–89.

9

THE ENDOCRINE REGULATION OF FOOD INTAKE

HÉLÈNE VOLKOFF
SURAJ UNNIAPPAN
SCOTT P. KELLY

1. Introduction
   1.1. Overview of Feeding in Vertebrates
   1.2. Fishes as Models for the Study of Feeding Regulation
2. Endocrine Regulation
   2.1. Hunger Signals
   2.2. Satiation Signals
   2.3. The Growth Hormone System
   2.4. The Melanocortin System and Melanin-concentrating Hormone
3. Influence of Intrinsic Factors
   3.1. Metabolic Signals/Energy Reserves
   3.2. Ontogeny
   3.3. Gender and Reproductive Status
   3.4. Genetic Influence
4. Influence of Extrinsic Factors
   4.1. Temperature
   4.2. Photoperiod
   4.3. Salinity
   4.4. Hypoxia
   4.5. Pollutants and Health Status
5. Proposed Model of Endocrine Circuitries Involved in Feeding in Fishes and Concluding Remarks

In vertebrates, the regulation of food intake is a complex phenomenon involving numerous central and peripheral endocrine factors. The actions of these factors are modulated by both intrinsic and extrinsic variables, including energy reserves, metabolic fuel usage, ontogeny, reproductive status and environmental conditions. To date, molecular and behavioral evidence indicates that the regulation of food intake has been relatively well conserved along the vertebrate lineage, at least between piscine and mammalian groups.

Fish Neuroendocrinology: Volume 28
FISH PHYSIOLOGY

DOI: 10.1016/S1546-5098(09)28009-5

However, fundamental differences in physiology between homeothermic mammals and poikilothermic fishes suggest that the endocrine control of feeding behavior might involve specific mechanisms in each of the two groups. This review highlights our current knowledge on the endocrine regulation of feeding in fishes, with an emphasis on how our knowledge of the piscine system integrates with a broader vertebrate picture.

1. INTRODUCTION

1.1. Overview of Feeding in Vertebrates

In all vertebrates, energy balance is regulated through multiple pathways involving key appetite- stimulating (orexigenic) or appetite-inhibiting (anorexigenic) factors. The regulation of feeding involves not only the central nervous system (CNS), but also peripheral organs such as the gastrointestinal (GI) tract and adipose tissue. The brain, particularly the hypothalamus, plays a critical role in the regulation of energy homeostasis (Valassi *et al.*, 2008). Originally thought to be under the influence of specific hypothalamic nuclei, feeding is now believed to be regulated by several neuronal circuits that integrate peripheral metabolic, endocrine and neuronal signals that reflect an animal's energy status. Peripheral hormones convey information to central feeding centers either via the vagus nerve or by crossing the blood–brain barrier and acting directly in the brain (Brightman and Broadwell, 1976). These peripheral hormones are in part released in response to the presence of food within the digestive tract (Holmgren and Olsson, Chapter 10 of this volume). In fishes as in mammals, several brain regions have been shown to be involved in the control of food intake. These include not only the hypothalamus but also extrahypothalamic regions such as the ventral telencephalon and olfactory tract (Peter, 1979). In recent years, a growing number of homologues of mammalian appetite-regulating peptides have been characterized in fishes (Volkoff *et al.*, 2005; Gorissen *et al.*, 2006). These peptides, which include both central and peripheral factors, appear to have similar appetite-regulating effects (Table 9.1) in fishes as in mammals, suggesting that, although group-specific differences might exist, the regulation of food intake has been relatively well conserved along the vertebrate lineage. Table 9.1 lists the fish homologues known to date and summarizes their effects on feeding and their response to nutritional status. This review highlights our current understanding on the endocrine regulation of feeding in fishes, with an emphasis on how our knowledge of the piscine system integrates with a broader vertebrate picture.

Table 9.1

Appetite regulatory endocrine factors, their tissue distribution, effects on feeding and mRNA expression in response to variations in nutritional status in fishes. Included are synopses of: 1) major tissues expressing mRNAs encoding appetite regulatory peptides; 2) known effects of appetite regulatory factors on feeding following central or peripheral administration; 3) fasting-induced changes in mRNA expression; and 4) postprandial changes in mRNA expression. For detailed information and further references, please refer to the subsections on specific factors within the text

Endocrine factors	Tissue(s) of mRNA expression	Appetite regulatory effects and lowest reported effective dose		Fasting-induced changes in mRNA expression	Postprandial changes in mRNA expression
		Intracerebroventricular administration	Intraperitoneal administration		
Neuropeptide Y (NPY)	Brain Gut Pituitary	Stimulation Goldfish 0.5 ng g^{-1} gfNPY[1] 100 ng g^{-1} pNPY[2] Trout 400 ng g^{-1} pNPY[3] Catfish 50 ng g^{-1} pNPY [4]	No effect Goldfish 0.1 or 0.33 μg g^{-1} pNPY[2]	Yes +ve: gut, brain[5–7]	Yes +ve: brain[5,8]
Peptide Y	Gut Brain	ND	ND	Yes +ve: gut[9]	ND
Orexins (OX)	Brain Gut	Stimulation Goldfish 1 ng g^{-1} hOX-A[10] 10 ng g^{-1} hOX-B[10]	ND	Yes +ve: brain[11,12]	Yes +ve: brain[13]

(continued)

Table 9.1 *(continued)*

Endocrine factors	Tissue(s) of mRNA expression	Appetite regulatory effects and lowest reported effective dose		Fasting-induced changes in mRNA expression	Postprandial changes in mRNA expression
		Intracerebroventricular administration	Intraperitoneal administration		
Galanin (GAL)	Brain Gut	Stimulation Goldfish/tench 10 ng g^{-1} rGAL[14,15] 100 ng g^{-1} pGAL[14]	None Goldfish/tench 100 ng g^{-1} pGAL[14]	Yes +ve: brain[16]	No brain[16]
Ghrelin	Gut Brain	Stimulation Goldfish 10 ng g^{-1} gfGhrelin ([1–12][17] 1 ng g^{-1} hGhrelin[17] ~0.5 ng g^{-1} gfAcyl-Ghrelin[18]	Stimulation Goldfish ~5 ng g^{-1} gfAcyl-Ghrelin[17] Trout ~3 ng g^{-1} rGrhelin[19] Tilapia ~1 ng g h^{-1} tilGhrelin-C10[20]	Yes +ve: gut, brain[17,21,22]	Yes −ve: brain, gut[17,23]
Cocaine- and amphetamine-regulated transcript (CART)	Brain Gut Gonad	Inhibition Goldfish 1 ng g^{-1} hCART 55–102[24] 10 ng g^{-1} CART 62–76[24]	ND	Yes −ve: brain[8,25,26]	Yes +ve: brain[8] −ve: brain[25]
Corticotropin-releasing factor (CRF)	Brain	Inhibition Goldfish 2 ng g^{-1} rCRF[27] Tench ~200 ng/g oCRF[28]	ND	Yes +ve: brain[29]	ND
Urotensin I (UI)	Brain	Inhibition Goldfish 2 ng g^{-1} gfUI[27]	ND	Yes +ve: brain[29]	ND

Cortisol	Interrenal	ND	Inhibition Trout 250 μg g^{-1} [30]	ND	ND
Prolactin-releasing peptide (PrRP)	Brain	Inhibition Goldfish 10 ng g^{-1} gfPrPR[31]	Inhibition Goldfish 25 ng g^{-1} gfPrPR[31]	Yes +ve: brain[31]	Yes +ve: brain[31]
Tachykinins	Brain Gut neurons	ND	ND	ND	Yes +ve: brain[32]
Endocannabinoids	Brain	ND	Stimulation/ Inhibition Goldfish +ve, 1 pg g^{-1} anandamide[33] −ve, 10 pg g^{-1} anandamide[33]	ND	ND
Cholecystokinin (CCK)	Gut Brain	Inhibition Goldfish 5 ng g^{-1} CCK-8[34]	Inhibition Goldfish 50 ng g^{-1} CCK-8[34] Catfish 12.5 ng g^{-1} CCK-8[4]	Yes +ve: gut[7,35]	Yes +ve: gut, brain[35,36]
Gastrin-releasing peptide/ bombesin (BBS)	Gut Brain	Inhibition Goldfish 5 ng g^{-1} BBS[37]	Inhibition Goldfish 50 ng g^{-1} BBS[37]	Yes +ve: gut[38]	ND
Amylin	Brain Gonad	Inhibition Goldfish 10 ng g^{-1} rAmylin[39]	Inhibition Goldfish 100 ng g^{-1} rAmylin[39]	ND	ND
Calcitonin gene-related peptide (CGRP)	Brain Pituitary Gut Gonad	Inhibition Goldfish 10 ng g^{-1} hCGRP[40]	ND	ND	ND

(*continued*)

Table 9.1 *(continued)*

Endocrine factors	Tissue(s) of mRNA expression	Appetite regulatory effects and lowest reported effective dose		Fasting-induced changes in mRNA expression	Postprandial changes in mRNA expression
		Intracerebroventricular administration	Intraperitoneal administration		
Intermedin (IM)	Brain Pituitary Gut Gonad	Inhibition Goldfish 10 ng g^{-1} pfIM[40]	ND	ND	ND
Neuromedin U (NMU)	Brain Gut Gonad	Inhibition Goldfish ~5 ng g^{-1} NMU-21[41]	ND	Yes −ve: brain[41]	ND
Leptin	Liver	Inhibition Goldfish 100 ng g^{-1} hLeptin[42]	Inhibition Goldfish 300 ng g^{-1} hLeptin[42] Trout 720 ng g^{-1} trLeptin[43]	ND	Yes +ve: liver[44]
Glucagon-like peptide -1 (GLP-1)	Pancreas Gut	Inhibition Catfish 0.25 ng g^{-1} cfGLP-1[45] 0.25 ng g^{-1} hGLP-1[45]	Inhibition Catfish 150 ng g^{-1} cfGLP-1, iv[45]	ND	ND
Insulin	Pancreas	Inhibition/none Trout −ve: 9 ng g^{-1} [46] Catfish None: 25 and 50 ng g^{-1} [4]	Inhibition Trout 4 μg g^{-1} [46]	ND	ND

Insulin-like growth factors (IGFs)		ND	ND		
	Liver Muscle			Yes −ve: muscle, liver[47–49]	ND
Growth hormone (GH)		ND	Stimulation		
	Pituitary		Trout 5 μg g^{-1} oGH[50] 25 μg g^{-1} oGH implant[51]	Yes +ve: pituitary[47,49]	ND
Pituitary adenylate cyclase activating polypeptide (PACAP)		Inhibition	Inhibition		
	Brain	Goldfish 30 ng g^{-1} fPACAP[52]	Goldfish 140 ng g^{-1} fPACAP[52]	None brain[52,53]	ND
Somatostatin		ND	No effect		
	Brain Gut Pancreas		Trout ~1 ng g h^{-1} for 20 days, implant[54]	Yes +ve: brain, pancreas[53,55]	ND
Melanocyte-stimulating hormone (MSH)		Inhibition	ND		
	Brain	Goldfish ~20 ng g^{-1} NDP-α-MSH[56] ~15–20 ng g^{-1} MT II[57]		ND	ND
Agouti-related protein	Brain	ND	ND	Yes +ve: brain[58,59]	ND
Melanin-concentrating hormone (MCH)		Inhibition	ND		
	Brain	Goldfish ~0.5 ng g^{-1} flMCH[60] ~10 ng g^{-1} hMCH[60]		Yes +ve: brain[61]	ND

(*continued*)

Table 9.1 *(continued)*

Endocrine factors	Tissue(s) of mRNA expression	Appetite regulatory effects and lowest reported effective dose: Intracerebroventricular administration	Intraperitoneal administration	Fasting-induced changes in mRNA expression	Postprandial changes in mRNA expression
Gonadotropin-releasing hormone (GnRH)	Brain	Inhibition Goldfish 0.1–0.5 ng g^{-1} cGnRH[62,63]	ND	ND	ND
Testosterone	Gonad	ND	Inhibition Sea bass 60 μg g^{-1} implants[64]	ND	ND
Estradiol	Gonad	ND	Inhibition Sea bass 60 μg g^{-1} implants[64]	ND	ND

+ve, positive changes in mRNA expression (up-regulation); −ve, negative changes in mRNA expression (down-regulation); ND, not determined; iv, intravenous.

Prefixes: h, human; p, porcine; o, ovine; r, rat; c, chicken; f, frog; gf, goldfish; til, tilapia; tr, trout; cf, catfish; fl, flounder.

References: [1] Narnaware *et al.* (2000); [2] Lopez-Patino *et al.* (1999); [3] Aldegunde and Mancebo (2006); [4] Silverstein and Plisetskaya (2000); [5] Narnaware and Peter (2001); [6] Silverstein *et al.* (1999a); [7] MacDonald and Volkoff (2009); [8] Kehoe and Volkoff (2007); [9] Murashita *et al.* (2007); [10] Volkoff *et al.* (1999); [11] Nakamachi *et al.* (2006); [12] Novak *et al.* (2005); [13] Xu and Volkoff (2007); [14] de Pedro *et al.* (1995a); [15] Volkoff and Peter (2001b); [16] Unniappan *et al.* (2004b); [17] Unniappan *et al.* (2004a); [18] Matsuda *et al.* (2006a); [19] Shepherd *et al.* (2007); [20] Riley *et al.* (2005); [21] Terova *et al.* (2008); [22] Amole and Unniappan (2009); [23] Peddu *et al.* (2009); [24] Volkoff and Peter (2000); [25] Volkoff and Peter (2001a); [26] Kobayashi *et al.* (2008a); [27] Bernier and Peter (2001); [28] De Pedro *et al.* (1995); [29] Bernier and Craig (2005); [30] Gregory and Wood (1999); [31] Kelly and Peter (2006); [32] Peyon *et al.* (2000); [33] Valenti *et al.* (2005); [34] Himick and Peter (1994b); [35] Murashita *et al.* (2006); [36] Peyon *et al.* (1999); [37] Himick and Peter (1994a); [38] Xu and Volkoff (2009); [39] Thavanathan and Volkoff (2006); [40] Martinez-Alvarez *et al.* (2008); [41] Maruyama *et al.* (2008); [42] Volkoff *et al.* (2003); [43] Murashita *et al.* (2008); [44] Huising *et al.* (2006a); [45] Silverstein *et al.* (2001); [46] Soengas and Aldegunde (2004); [47] Ayson *et al.* (2007); [48] Terova *et al.* (2007); [49] Pedroso *et al.* (2006); [50] Johnsson and Bjornsson (1994); [51] Johansson *et al.* (2005); [52] Matsuda *et al.* (2005); [53] Xu and Volkoff (2008); [54] Very *et al.* (2001); [55] Ehrman *et al.* (2002); [56] Cerdá-Reverter *et al.* (2003); [57] Matsuda *et al.* (2008b); [58] Cerdá-Reverter *et al.* (2003); [59] Song *et al.* (2003); [60] Matsuda *et al.* (2006); [61] Takahashi *et al.* (2004); [62] Hoskins *et al.* (2008); [63] Matsuda *et al.* (2008); [64] Leal *et al.* (2009).

1.2. Fishes as Models for the Study of Feeding Regulation

1.2.1. Comparison to Mammalian Models

Increasing evidence indicates that basic mechanisms controlling feeding behavior are generally conserved among vertebrates. As experimental models, fishes are valuable as they can usually be submitted to invasive protocols (e.g., injections) or repeated sampling with few or no effects on feeding behavior. Also, fishes show a remarkable level of diversity with regards to morphology, ecology, behavior and genomes (Volff, 2004), which makes them attractive for the study of the evolution of appetite-regulating systems in vertebrates. This diversity and the existence of major anatomical and physiological differences between homeothermic mammals and ectothermic fishes suggest that the endocrine control of feeding involves molecules and mechanisms that might be specific to certain species or groups. Obvious anatomical differences between mammals and fishes include differences in brain and gut morphologies as well as the existence in fishes of specific organs not present in mammals (e.g., caudal neurosecretory organ). Anatomical and functional differences are also seen within fish species. For example, fishes feeding habits range from omnivore to carnivore, which translate into differences in gastrointestinal tract morphology and gut hormone profiles (Holmgren and Olsson, Chapter 10 of this volume). The brain distribution of neuropeptides also varies within fish species (Cerdá-Reverter and Canosa, Chapter 1 of this volume). Although energy metabolism in fishes is similar to that in mammals and birds, fishes do not expend energy to maintain a body temperature different from that of their environment and require less energy for the excretion of waste nitrogenous products. Also, as opposed to most mammals, which have a finite growth, the growth of many fish species is indeterminate, in which there is no fixed size and some growth may continue throughout the life. Fishes not only need to feed to maintain their basal metabolism but also experience a trade-off in resource allocation between reproduction and growth throughout their lives (Heino and Kaitala, 1999). Fishes are also submitted to a number of environmental challenges, which has led to a number of feeding adaptations, such as coping with long-term fasting.

1.2.2. Methods for Studying Feeding in Fishes

Although simple in concept, measuring food intake in fishes can be challenging and has been performed using a number of approaches. In wild animals, estimates of food intake can be obtained by examining stomach or gut contents or by using models that take into account variables such as prey taxa proportions (Godby *et al.*, 2007). In laboratory conditions, the more accurate assessment of food intake is that obtained by quantifying the number of food pellets eaten by individual fish (Himick and Peter, 1994a; Volkoff *et al.*, 1999). Food intake can also be estimated by assessing differences in weight between full and empty guts (Delicio and Vicentini-Paulino, 1993) or between food left in tanks and food offered to fish (Bernier *et al.*, 2004), by quantifying radio-labeled food (Ronnestad *et al.*, 2007) or by the X-ray detection of pellets containing opaque beads (Craig *et al.*, 2005) in fishes. When dealing with large numbers of animals, feeding of individual fish can be inferred by dividing the amount of food eaten by a group of fish by the number of fish and the average weight of fish (Kehoe and Volkoff, 2008), by using underwater sensors to detect uneaten food (Noble *et al.*, 2007) or by counting feed demands in fish trained to use self-feeding devices (Millot *et al.*, 2008). The effects of peptides on feeding and metabolism in fishes can be assessed directly by using various hormone treatments, where the peptides are delivered via peripheral (intra-peritoneal, ip) or brain injections (intracerebroventricular, icv) (Silverstein *et al.*, 2001; Volkoff *et al.*, 2003), by oral administration (Bernier, 2006), via ip-implanted pellets (Johansson *et al.*, 2005) or via osmotic mini-pumps (Riley *et al.*, 2005). The role of hormones can also be assessed indirectly by examining blood levels or protein/mRNA expression levels of these peptides in fish submitted to various conditions (e.g., starvation, environmental stresses) (see subsequent sections). It is noteworthy that the advent of fish genome projects and new technologies (e.g., transgenesis and morpholino "knock-downs") (Volff, 2004) will likely allow for the development of valuable genetic piscine models for the study of feeding regulation.

2. ENDOCRINE REGULATION

Appetite-regulating factors can be categorized in hunger signals versus satiation signals and central versus peripheral signals. This section provides a description of all the known major appetite regulators in fish (also summarized in Table 9.1 and Figure 9.1).

2.1. Hunger Signals

2.1.1. Brain Signals

Neuropeptide Y family of peptides. The neuropeptide Y (NPY) family of peptides consists of NPY and peptide YY (PYY) (found in all vertebrate classes), pancreatic polypeptide (PP, found only in the pancreas of tetrapods), and peptide Y (PY, found only in some teleost fishes) (Hoyle, 1999), which have five known receptors, Y1, Y2, Y4, Y5 and Y6 (Kamiji and Inui, 2007). In mammals, NPY is abundant in the CNS, particularly in

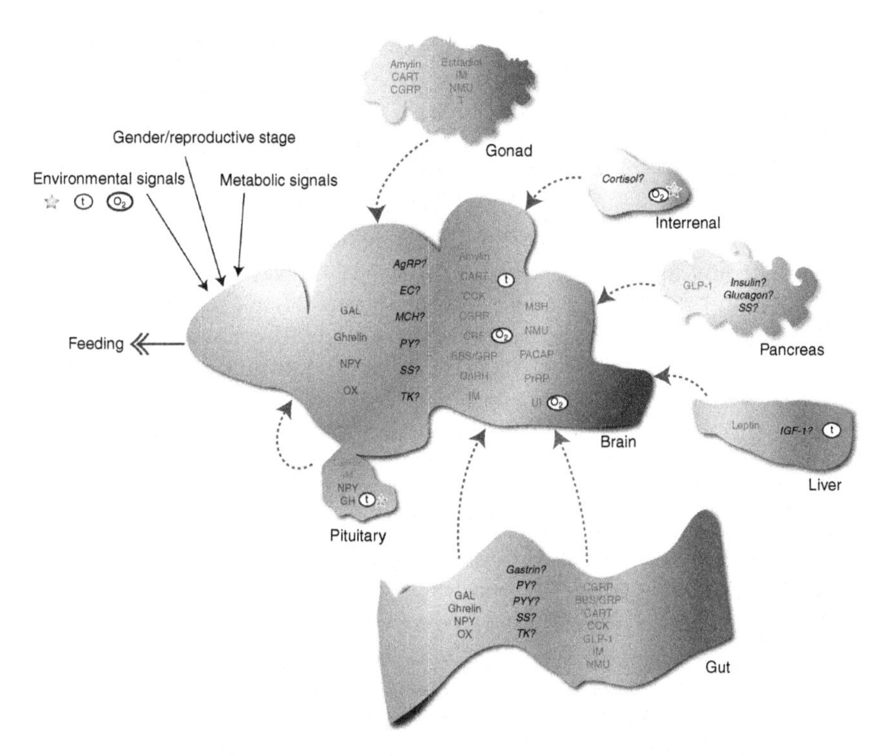

Fig. 9.1. Summary of our current knowledge of the regulation of feeding in fishes by hormonal and other factors. Note that this figure represents neither the species specificity associated with many hormones, with regards to their presence, locations or actions nor the interactions between hormones. Peptides in blue characters are orexigenic. Peptides in red characters are anorexigenic. Peptides in black italics and followed by a "?" are peptides for which the role in feeding is unknown or unclear. Dashed curved red arrow indicates a blood connection whereas the gray dashed arrow between gut and brain indicates a nervous (vagal) connection. Hormones adjacent to a circled t, a star or an "O_2" represent hormones affected by temperature, photoperiod or oxygen levels, respectively. AgRP, agouti-related peptide; BBS/GRP, bombesin/gastrin-releasing peptide; CART, cocaine- and amphetamine-regulated transcript; CCK, cholecystokinin; CGRP, calcitonin gene-related peptide; CRF, corticotropin-releasing factor; EC, endocannabinoids; GAL, galanin; GH, growth hormone; GLP-1, glucagon-like peptide 1; GnRH, gonadotropin-releasing hormone; IGF-1, insulin-like growth factor 1; IM, intermedin; MCH, melanin-concentrating hormone; MSH, melanocyte-stimulating hormone; NMU, neuromedin U; NPY, neuropeptide Y; OX, orexins; PACAP, pituitary adenylate cyclase activating polypeptide; PrRP, prolactin-releasing peptide; PY, peptide Y; PYY, peptide YY; SS, somatostatin; T, testosterone; TK, tachykinins; UI, urotensin I. (See Color Insert.)

hypothalamic nuclei involved in the regulation of feeding, and is one of the most potent orexigenic peptides known, this effect being mediated by Y1 and Y5 receptor subtypes (Gao and Horvath, 2007). All fishes produce NPY and PYY whereas only some teleosts produce PY (Cerdá-Reverter *et al.*, 2000a).

NPY sequences have been determined for several fish species and NPY neurons are widely distributed in fish CNS (Sundstrom *et al.*, 2005; Kehoe and Volkoff, 2007; Sueiro *et al.*, 2007; MacDonald and Volkoff, 2009) as well as pituitary and gut (Cerdá-Reverter *et al.*, 2000a; Rodriguez-Gomez *et al.*, 2001). To date, seven NPY receptor subtypes that bind both NPY and PYY have been identified in fishes (Y1, Y2, Y4–Y8) (Salaneck *et al.*, 2008). Icv, but not ip (Lopez-Patino *et al.*, 1999), injections of mammalian or fish NPY increase feeding in fish (Lopez-Patino *et al.*, 1999; Narnaware *et al.*, 2000; Silverstein and Plisetskaya, 2000; Aldegunde and Mancebo, 2006) but it is unclear which NPY receptor subtype may mediate this response (Salaneck *et al.*, 2008). Brain NPY mRNA levels increase following fasting in goldfish, *Carassius auratus* (Narnaware and Peter, 2001), winter skate, *Raja ocellata* (MacDonald and Volkoff, 2009), and chinook and coho salmon, *Oncorhynchus tshawytscha* and *Oncorhynchus kisutch* (Silverstein *et al.*, 1999a), undergo periprandial variations in goldfish (Narnaware and Peter, 2001), Atlantic cod, *Gadus morhua* (Kehoe and Volkoff, 2007), and tilapia, *Oreochromis mossambicus* (Peddu *et al.*, 2009), and are influenced by macronutrient intake in goldfish (Narnaware and Peter, 2002). In fishes as in mammals, the actions of NPY on feeding occur in part by the modulation of other appetite regulators, e.g., corticotrophin-releasing factor (CRF) and cortisol (Bernier *et al.*, 2004), cocaine- and amphetamine-regulated transcript (CART) (Volkoff and Peter, 2000), leptin (Volkoff *et al.*, 2003), melanin-concentrating hormone (MCH) (Matsuda *et al.*, 2008b), orexins (OXs) and galanin (GAL) (Volkoff and Peter, 2001b), growth hormone (GH) (Mazumdar *et al.*, 2006) and ghrelin (Miura *et al.*, 2006). Although PY and/or PYY have been identified in several species (Cerdá-Reverter *et al.*, 2000a), their physiological role in fishes remains unclear. PYY gene expression is present in sea bass, *Dicentrarchus labrax* (Cerdá-Reverter *et al.*, 2000b), and pufferfish, *Takifugu rubripes* (Sundstrom *et al.*, 2005), brains. PY gene expression is present in brain of sea bass (Cerdá-Reverter *et al.*, 2000b) and in intestine of yellowtail, *Seriola quinqueradiata* (Murashita *et al.*, 2007), where its mRNA expression increases with fasting, suggesting PY is a feeding regulator.

Orexins. OXs (hypocretins) consist of two peptides, orexin A (OX-A) and orexin B (OX-B) produced by cleavage of a single precursor, preproorexin. In mammals, OXs are produced mainly in the lateral hypothalamus and act through two G-protein coupled receptors (OX_1 and OX_2) to stimulate feeding, control gastric secretion (Korczynski *et al.*, 2006) and regulate sleep and wakefulness (Ohno and Sakurai, 2007). In fishes, mRNAs encoding for prepro-OX have been reported for six species (Faraco *et al.*, 2006; Xu and Volkoff, 2007), where it is present in several brain regions including the hypothalamus (Amiya *et al.*, 2007; Xu and Volkoff, 2007), in pituitary and in peripheral tissues, including gut (Xu and Volkoff, 2007). In sea perch, *Lateolabrax japonicus,* OX-A-immunoreactive cells in the pituitary co-localize with GH-secreting cells (Suzuki *et al.*, 2007), suggesting a role of OXs in the control of pituitary secretion in fishes. OXs seem to regulate feeding in fishes, as icv injections of OXs stimulate appetite in goldfish (Volkoff *et al.*, 1999; Nakamachi *et al.*, 2006), OX-like immunoreactive (ir) cells in the hypothalamus of goldfish brain increase in fasted fish and decrease in glucose-injected fish (Nakamachi *et al.*, 2006), food deprivation increases brain prepro-OX mRNA levels in goldfish (Nakamachi *et al.*, 2006) and zebrafish, *Danio rerio* (Novak *et al.*, 2005). In Atlantic cod, prepro-OX brain mRNA expression levels display periprandial changes and are high in fish fed low rations (Xu and Volkoff, 2007). Icv injections of OXs increase locomotor activity in goldfish (Nakamachi *et al.*, 2006), and zebrafish lacking the OX receptor (Yokogawa *et al.*, 2007) or OX-overexpressing larvae (Prober *et al.*, 2006) both display abnormal sleeping patterns, suggesting that OXs are involved in the regulation of states of wakefulness in fish, which might indirectly affect feeding and metabolism. In fishes, as in mammals (Saper, 2006; Nishino, 2007), OXs appear to interact with other appetite regulators. In goldfish, blocking of OX receptors results in a decrease in NPY-, GAL- (Volkoff and Peter, 2001b) and ghrelin-induced (Miura *et al.*, 2007) feeding, and blocking NPY, GAL or ghrelin receptors inhibits OX-induced feeding. In addition, central administration of OX results in increases in both NPY and ghrelin mRNA expression (Volkoff and Peter, 2001b; Miura *et al.*, 2007) and ghrelin treatment increases brain OX expression (Miura *et al.*, 2007). In goldfish, CART (Volkoff and Peter, 2000) and leptin (Volkoff *et al.*, 2003) both inhibit OX-induced feeding. In medaka, MCH-immunoreactive fibers are in close contact with the OX-producing cells (Amiya *et al.*, 2007). This evidence points to a functional interdependence between OX and other peptidergic systems in the control of energy balance in goldfish.

Galanin. In mammals, GAL is widely expressed in the CNS and intestine and stimulates food intake and weight gain through GAL1R, GAL2R, and GAL3R receptors (Walton *et al.*, 2006; Lang *et al.*, 2007). GAL has been isolated from several fishes (Volkoff *et al.*, 2005) and preproGAL mRNAs encoding five putative GAL peptides have been identified in goldfish (Unniappan *et al.*, 2003). PreproGAL mRNA (Unniappan *et al.*, 2004b) and GAL-like immunoreactivity (Jadhao and Meyer, 2000; Jadhao and Pinelli, 2001; Riley *et al.*, 2005) have a widespread distribution in fish brain. Icv, but not ip, administration of mammalian GAL stimulates food intake in goldfish (De Pedro *et al.*, 1995a; Volkoff and Peter, 2001b) and tench, *Tinca tinca* (Guijarro *et al.*, 1999), and fasting increases brain preproGAL mRNA expressions in goldfish (Unniappan *et al.*, 2004b), suggesting an orexigenic role of GAL in fishes. Galanin-like peptide (GALP) is a recently discovered second member of the GAL peptide family that displays similarities in amino acid composition with GAL and binds to GAL receptors (Lang *et al.*, 2007). The appetite regulating effects of GALP are controversial, with both stimulatory and inhibitory effects reported in rodents (Lang *et al.*, 2007). The structure and functions of GALP in fish are currently unknown.

2.1.2. Peripheral Signals

Ghrelin. Ghrelin is a 28-amino acid acylated peptide predominantly secreted by the stomach but also by the brain. In mammals, ghrelin stimulates both GH secretion and appetite, and it is the only known GI hormone with confirmed orexigenic properties (Olszewski *et al.*, 2008). Ghrelin has been identified from several teleost (Kaiya *et al.*, 2008; Manning *et al.*, 2008; Miura *et al.*, 2008; Olsson *et al.*, 2008; Terova *et al.*, 2008; Xu and Volkoff, 2009) and elasmobranch (Kawakoshi *et al.*, 2007) fishes. In fishes as in mammals, ghrelin mRNA is mostly expressed in stomach with low levels found in the brain (Kaiya *et al.*, 2008). Injections of either goldfish or human ghrelin stimulate food intake in goldfish (Unniappan and Peter, 2005; Matsuda *et al.*, 2006a; Miura *et al.*, 2007, 2008) and continuous infusion of tilapia ghrelin causes an increase in food intake in tilapia (Riley *et al.*, 2005). Ip ghrelin injections in trout cause an increase in food intake of rainbow trout, *Oncorhynchus mykiss* (Shepherd *et al.*, 2007), while no effect for ghrelin on food intake in trout was found by another group (Jonsson *et al.*, 2007). This variation in the appetite regulatory effects of ghrelin in the same species could be due to differences in the doses of peptide tested and methods used for peptide administration and food intake quantification. Both hypothalamic and gut preproghrelin mRNA expression as well as serum ghrelin levels display periprandial variations in goldfish (Unniappan *et al.*, 2004a) and tilapia (Peddu *et al.*, 2009). Fasting induces an up-regulation of ghrelin mRNA expression in sea bass stomach (Terova *et al.*, 2008), goldfish hypothalamus and gut (Unniappan *et al.*, 2004a) and zebrafish brain and gut (Amole and Unniappan, 2009) but does not affect gut ghrelin mRNA levels in Nile tilapia, *Oreochromis niloticus* (Parhar *et al.*, 2003), and Atlantic cod (Xu and Volkoff, 2009). Fasting results in a similar decrease in plasma ghrelin levels in trout (Jonsson *et al.*, 2007) and burbot, *Lota lota* (Nieminen *et al.*, 2003). In goldfish, ghrelin has been shown to interact with other appetite-related peptides, including NPY (Miura *et al.*, 2006), OX (Miura *et al.*, 2007) and GRP/BBS (Canosa *et al.*, 2005). Icv injections of ghrelin result in an increase in prepro-OX and NPY mRNA expression in the brain. Pre-injection of orexin receptor 1A antagonist, SB334867, abolishes the appetite stimulatory effects of ghrelin, while the NPY Y1 receptor antagonist attenuated appetite stimulatory effects of ghrelin. These results indicate that the orexigenic effects of ghrelin are mediated via NPY and OX-dependent pathways (Miura *et al.*, 2006, 2007). The ghrelin receptor (GRLR) has been characterized in several fishes (Kaiya *et al.*, 2008), where it is highly expressed in the brain (Chan *et al.*, 2004), suggesting ghrelin has direct central actions. The presence of ghrelin receptor in the gut and brain provides additional support for the appetite regulatory role of the ghrelin–GRLR system in fishes (Kaiya *et al.*, 2008). In tilapia, significant elevations in preprandial brain mRNA levels of the GRLR are observed (Peddu *et al.*, 2009). Ghrelin causes contraction of zebrafish intestinal muscles *in vitro* suggesting a role in modulating gut motility (Olsson *et al.*, 2008), which could contribute to the orexigenic actions of ghrelin in fish.

2.2. Satiation Signals

2.2.1. Brain Signals

Cocaine- and amphetamine-regulated transcript (CART) peptide was originally isolated as an mRNA up-regulated in rat brain following acute administration of psychomotor stimulants and later shown to be expressed in brain, gut and pancreas (Ekblad, 2006; Vicentic and Jones, 2007). CART treatment inhibits feeding in mammals (Vicentic and Jones, 2007) and birds (Tachibana *et al.*, 2003), although orexigenic actions have also been reported in rats following injections into specific hypothalamic areas (Abbott *et al.*, 2001). In fishes, CART mRNA sequences have been published for goldfish (Volkoff and Peter, 2001b), channel catfish, *Ictalurus punctatus* (Kobayashi *et al.*, 2008a), winter skate (MacDonald and Volkoff, 2009) and cod (Kehoe and Volkoff, 2007), with two forms of CART peptide precursors, CART I and CART II, identified in goldfish (Volkoff and Peter, 2001b). In fish, CART mRNA is present in brain and peripheral tissues including gonads and kidney. CART immunoreactivity has also been detected throughout brain and pituitary of catfish, *Clarias batrachus* (Singru *et al.*, 2007), and, interestingly, in the venom gland of niquim, *Thalassophryne nattereri* (Magalhaes *et al.*, 2006). Icv injections of human CART inhibit feeding in goldfish (Volkoff and Peter, 2000) and CART mRNA levels decline following fasting in goldfish (Volkoff and Peter, 2001b), channel catfish (Kobayashi *et al.*, 2008a) and Atlantic cod (Kehoe and Volkoff, 2007), suggesting that CART act as an anorexigenic factor in fishes. Postprandial hypothalamic CART expression increases in goldfish (Volkoff and Peter, 2001a) but decreases in cod (Kehoe and Volkoff, 2007). Administration of human CART inhibits both NPY- and OX A-stimulated feeding in goldfish, suggesting an inhibitory action of

CART on both NPY and OX-A systems (Volkoff and Peter, 2000). In addition, in catfish brain, CART- and NPY-ir axons were closely associated (Singru *et al.*, 2008). A synergistic interaction between leptin and CART has also been suggested in goldfish (Volkoff *et al.*, 2003).

Corticotropin-releasing factor (CRF)-related neuropeptides include CRF, several urocortins (UCN), fish urotensin I (UI), and amphibian sauvagine, which differ in their affinities for binding to two CRF receptor subtypes (CRF1 and CRF2) (Bernier, 2006; Bernier *et al.*, Chapter 6 of this volume). The transcripts encoding CRF, UI and UCNs as well as CRF1 and CRF2 have been reported for several fish species (Alderman and Bernier, 2007). In fishes as in all vertebrates, CRF has an important role in the stress response, but CRF-expressing neurons are widely distributed in the brain, suggesting that CRF has diverse physiological and behavioral functions (Bernier *et al.*, Chapter 6 of this volume), including the regulation of feeding. Ip injections of CRF in goldfish do not affect food intake (De Pedro *et al.*, 1993). Icv treatments with CRF in tench (De Pedro *et al.*, 1995b) and with either CRF (De Pedro *et al.*, 1993; Bernier and Peter, 2001) or UI (Bernier and Peter, 2001) in goldfish inhibit feeding. Increases in both UI and CRF brain mRNA levels are seen in anorexic goldfish treated with ip implants of a glucocorticoid receptor antagonist, or a cortisol synthesis inhibitor in goldfish (Bernier and Peter, 2001) and in anorexic trout submitted to hypoxia (Bernier and Craig, 2005). The role of UCNs in the regulation of food intake in fishes is not known and that of cortisol is unclear. Ip injections of cortisol do not affect food intake in goldfish (De Pedro *et al.*, 1997) but have an anorexigenic effect in rainbow trout (Gregory and Wood, 1999). In channel catfish (Peterson and Small, 2005) and in goldfish, moderate increases in plasma cortisol stimulate food intake and decrease CRF mRNA expression, whereas high doses of cortisol decrease CRF mRNA levels but have no effects on food intake (Bernier *et al.*, 2004). It has been suggested that CRF-related peptides interact with other appetite-regulating systems in fishes, such as serotonin (De Pedro *et al.*, 1998) and NPY (Bernier *et al.*, 2004).

RFamides belong to a diverse group of peptides that are characterized by a common N-terminal sequence. First discovered in venus clam, RFamides were subsequently isolated from chicken and mammalian brains and include prolactin-releasing peptide (PrRP), RFamide-related peptides (RFRPs), metastin and kisspeptin(s). RFamide receptors are yet to be fully characterized (Dockray, 2004). Of the vertebrate RFamides identified to date, all except for members of the kisspeptin family have been demonstrated to modulate food intake in mammals (Bechtold and Luckman, 2007). Little is known about the structure and the functions of RFamides in fishes.

RFRP-expressing neurons have been localized in the midbrain of goldfish, suggesting a role of these peptides in neuroendocrine function (Sawada *et al.*, 2002) and orthologues to mammalian PrRP have been identified in a few fishes (Moriyama *et al.*, 2002; Seale *et al.*, 2002; Moriyama *et al.*, 2007). In goldfish, hypothalamic PrRP mRNA expression significantly increases after feeding, and following 7 days of food deprivation (Kelly and Peter, 2006). In goldfish, PrRP appears to have anorexigenic actions but is also intimately involved in the regulation of salt and water balance (Kelly and Peter, 2006), which might be the major role of this peptide in fishes (Takei and Balment, Chapter 8 of this volume).

Tachykinins (TKs) constitute a family of small neuropeptides defined structurally by a common C-terminal amino acid sequence and distributed in the nervous system of chordates. The TK family comprises several members including substance P (in mammals, birds and fishes), neurokinins, and carassin (in goldfish) (Conlon and Larhammar, 2005). TK immunoreactivity has been demonstrated in the CNS and in enteric neurons of several fish species (Bermudez *et al.*, 2007; Pinuela and Northcutt, 2007). TKs have been shown to induce contractions of gut smooth muscle of several fishes (Liu *et al.*, 2002; Holmberg *et al.*, 2004; Kim *et al.*, 2005), and hypothalamic mRNA expression of preproTK displays postprandial changes (Peyon *et al.*, 2000), suggesting TKs are involved in feeding and digestion processes in fishes (Holmgren and Olsson, Chapter 10 of this volume).

Endocannabinoid system. Endocannabinoids (ECs) are endogenous phospholipid derivatives, which activate two cannabinoid receptor subtypes (CB1 and CB2 receptors). In mammals, ECs have central orexigenic effects through hypothalamic CB1 receptors (Despres, 2007). Orthologues of the mammalian cannabinoid CB1 and CB2 receptors have recently been identified in fishes (Elphick and Egertova, 2001; McPartland *et al.*, 2007) but the role of the EC system in the regulation of food intake in fishes remains unclear. In goldfish, ECs (anandamide and arachidonoylglycerol) and CB1-like immunoreactivity are distributed throughout the brain (Valenti *et al.*, 2005). Ip administration of anandamide either increases or reduces food intake, at low (1 pg g^{-1} body weight) or intermediate (10 pg g^{-1}) doses, respectively, and food deprivation induces an increase in anandamide (but not arachidonoylglycerol) mRNA levels in the telencephalon (Valenti *et al.*, 2005).

Other peptides. Neuromedin U (NMU), first isolated from the porcine spinal cord, is an anorexigenic neuropeptide in mammals. cDNAs encoding three NMU orthologues (NMU-21, NMU-25 and NMU-38) have been isolated in goldfish, where NMU mRNAs are expressed in brain, gut and gonads (Maruyama *et al.*, 2008). In goldfish, fasting induces a decrease in brain proNMU mRNA levels in the brain and icv administration of NMU-21 suppresses food intake (Maruyama *et al.*, 2008), suggesting that NMU regulates appetite in fish.

Calcitonin gene-related peptide (CGRP), adrenomedullin (AM) and intermedin (IM), along with amylin (see below) are structurally related peptides, members of the calcitonin/CGRP peptide family. All three peptides have been shown to be involved in the control of food intake in mammals. In goldfish, CGRP, IM and AM mRNAs are expressed throughout the brain, in pituitary and in several peripheral tissues, including gut and gonad (Martinez-Alvarez *et al.*, 2008). Icv injections of either CGRP or IM, but not AM, significantly decrease food intake in goldfish (Martinez-Alvarez *et al.*, 2009).

2.2.2. Peripheral Signals

Cholecystokinin/gastrin. The GI peptides cholecystokinin (CCK) and gastrin are characterized by a common C-terminal tetrapeptide sequence. In mammals, CCK is found in both brain and GI tract, whereas gastrin is produced only by gastric endocrine cells. Both peptides have multiple biologically active forms that bind to two receptor subtypes (CCK1R and CCK2R). CCK acts peripherally via vagal pathways to decrease gastric emptying, stimulate gastric secretions and reduce food intake, but binding of CCK to brain receptors also results in satiety (Raybould, 2007).

CCK/gastrin-like immunoreactivity has been shown in the nervous system and gut of several fish species (Aldman *et al.*, 1989; Himick and Peter, 1994b; Bermudez *et al.*, 2007). mRNAs encoding for CCK have been determined for several species (Peyon *et al.*, 1998; Jensen *et al.*, 2001; Kurokawa *et al.*, 2003) whereas gastrin mRNAs have only been identified in spotted river puffer, *Tetraodon nigroviridis,* Japanese flounder, *Paralichthys olivaceus* (Kurokawa *et al.*, 2003), and two shark species, the spiny dogfish, *Squalus acanthias,* and the porbeagle, *Lamna cornubica* (Johnsen *et al.*, 1997). CCK mRNA expression (Peyon *et al.*, 1998; Jensen *et al.*, 2001; Kurokawa *et al.*, 2003; MacDonald and Volkoff, 2009) and binding sites (Himick *et al.*, 1996; Oliver and Vigna, 1996) are present in both brain and intestine whereas gastrin mRNA is only seen in intestine (Kurokawa *et al.*, 2003). CCK-related peptides influence digestion and feeding processes in fishes. They are released when food is present in the intestine, inhibit gastric emptying and increase gut motility (Forgan and Forster, 2007; Holmgren and Olsson, Chapter 10 of this volume). CCK also acts as a satiety factor in fishes. CCK suppresses food intake following both icv and ip INJECTIONS in goldfish (Himick and Peter, 1994b; Thavanathan and Volkoff, 2006) and following icv INJECTIONS in channel catfish (Silverstein and Plisetskaya, 2000) and treatment with CCK antagonists induces an increase in food intake in rainbow trout (Gelineau and Boujard, 2001). In addition, CCK mRNA levels increase following a meal in both goldfish brain (Peyon *et al.*, 1999) and in yellowtail pyloric ceca (Murashita *et al.*, 2007). Fasting increases mRNA levels of CCK in gut of yellowtail (Murashita *et al.*, 2006) and winter skate (MacDonald and Volkoff, 2009). Rainbow trout fed high fat diets have higher plasma CCK levels compared with fish fed high protein diets (Jonsson *et al.*, 2006), suggesting that the release of CCK is influenced by diet in fishes. CCK might mediate in part the effects of both leptin (Volkoff *et al.*, 2003) and amylin (Thavanathan and Volkoff, 2006) on food intake in goldfish. Interactions between CCK and somatostatin (SS) (Eilertson *et al.*, 1996; Canosa and Peter, 2004) and GH (Himick *et al.*, 1993) have also been suggested. Although the role of gastrin in the regulation of feeding is not known, gastrin treatment elicits contraction of chinook salmon gut rings *in vitro*, suggesting it might have a role in gut motility in fishes (Forgan and Forster, 2007).

Bombesin/GRP. Bombesin (BBS), first isolated from the skin of amphibians, and gastrin-releasing peptide (GRP) are structurally related peptides that share a similar C-terminal sequence and have similar biological actions. In vertebrates, BBS/GRP peptides are widely distributed in the GI tract and CNS (McCoy and Avery, 1990). BBS/GRP-like peptides have been detected in GI tract and brain of fishes, including elasmobranchs (Bjenning *et al.*, 1991) and teleosts (Volkoff *et al.*, 2000; Bosi *et al.*, 2004; Xu and Volkoff, 2009). In fishes, BBS-like peptides regulate gastric acid secretion and motility (Thorndyke *et al.*, 1990) and might act as satiety factors, as BBS injections suppress food intake in goldfish (Himick and Peter, 1994a). In Atlantic cod, GRP gut mRNA expression is higher in fish fed high rations as compared to fish fed low rations (Xu and Volkoff, 2009). In rainbow trout, however, plasma GRP levels are not influenced by feeding or diet composition (Jonsson *et al.*, 2006). In goldfish, BBS treatment stimulates GH release and reduces pro-SS expression in forebrain (Canosa and Peter, 2004; Canosa *et al.*, 2005) and these effects are partially blocked by ghrelin, suggesting an interaction between BBS, SS and ghrelin (Canosa *et al.*, 2005).

Amylin. In mammals, amylin is co-secreted with insulin from pancreatic β-cells in response to meals. It binds to specific receptors in the CNS to suppress nutrient-stimulated glucagon secretion, slow gastric emptying and reduce food intake (Lutz, 2006). In fishes, amylin has been identified and its deduced amino-acid sequence partially characterized in zebrafish, Atlantic salmon (*Salmo salar*), sculpin (*Myoxocephalus scorpius*) (Westermark *et al.*, 2002), goldfish (Martinez-Alvarez *et al.*, 2008) and pufferfish (Chang *et al.*, 2004). Amylin immunoreactivity has been observed in insulin-producing cells of sculpin Brockmann bodies and in Atlantic salmon endocrine pancreas, suggesting that in fishes as in mammals, amylin is produced in the periphery

(Westermark *et al.*, 2002). In goldfish, amylin mRNA is mostly expressed in the brain but is also present in pituitary, gonad, kidney and muscle (Martinez-Alvarez *et al.*, 2008). Although little is known about the physiological role of amylin in fishes, both icv and ip injections of human amylin decrease food intake in goldfish (Thavanathan and Volkoff, 2006), suggesting that it has an anorexigenic role in fishes.

Leptin. In mammals, leptin is produced by adipocytes and regulates food intake, energy balance and reproduction. Although leptin-like ir material had previously been detected in the GI tract (Muruzabal *et al.*, 2002; Bosi *et al.*, 2004), blood (Mustonen *et al.*, 2002; Nieminen *et al.*, 2003; Nagasaka *et al.*, 2006), brain, liver (Johnson *et al.*, 2000) and fat (Yaghoubian *et al.*, 2001; Vegusdal *et al.*, 2003) of several fish species, fish leptins have only recently been characterized (Kurokawa *et al.*, 2005; Huising *et al.*, 2006a). Low amino acid identities are seen between fish leptins and between fish and mammalian leptins (Kurokawa *et al.*, 2005). In contrast to mammals, where leptin is predominantly expressed in adipose tissue, the liver appears to be the major site for leptin expression in fishes (Kurokawa *et al.*, 2005; Huising *et al.*, 2006a,b). cDNA encoding a leptin receptor has recently been reported in medaka, *Oryzias melastigma*, where it is abundantly expressed in brain and peripheral tissues (Wong *et al.*, 2007), and in pufferfish (Kurokawa *et al.*, 2008), where it is abundantly expressed in the pituitary and ovary. The role of leptin in regulation of feeding in fishes is still controversial. Although mammalian leptin treatments have no marked effects in coho salmon (Baker *et al.*, 2000) or channel catfish (Silverstein and Plisetskaya, 2000), they decrease food intake in goldfish (Volkoff *et al.*, 2003; De Pedro *et al.*, 2006). In goldfish, leptin seems to potentiate the actions of CART and CCK, and to inhibit the actions of NPY and OX-A (Volkoff and Peter, 2000; Volkoff *et al.*, 2003). In trout, ip injections of recombinant trout leptin suppress food intake, reduce NPY mRNA expression and increase POMC mRNA levels (Murashita *et al.*, 2008). High blood ir-leptin levels are associated with a decrease in appetite in spawning ayu, *Plecoglossus altivelis* (Nagasaka *et al.*, 2006), fasting induces a decrease in plasma leptin-ir peptides levels in burbot (Nieminen *et al.*, 2003), liver leptin mRNA expression peaks in response to a meal in carp, *Cyprinus carpio* (Huising *et al.*, 2006b), and treatment with mammalian leptin increases fat metabolism in green sunfish, *Lepomis cyanellus* (Londraville and Duvall, 2002). However, fasting does not affect liver leptin expression in carp (Huising *et al.*, 2006b).

Glucagon/glucagon-like peptide-1 and insulin/insulin-like growth factors. Glucagon and glucagon-like peptides (GLP) are peptides of the secretin family, which also includes gastric inhibitory peptide (GIP), secretin, vasoactive intestinal peptide (VIP) and oxyntomodulin (Nelson and Sheridan, 2006). The vertebrate proglucagon gene encodes three glucagon-like sequences [glucagon, glucagon-like peptide-1 (GLP-1), and glucagon-like peptide 2 (GLP-2)] that have distinct functions in regulating metabolism (Irwin and Wong, 2005). In fishes, a pancreatic glucagon gene produces glucagon and GLP-1, whereas an intestinal gene encodes for all three glucagon peptides as well as oxyntomodulin (Navarro *et al.*, 1999). Although there is no evidence for a role of glucagon in regulating feeding in fishes, it appears to affect metabolism. Glucagon treatment elevates plasma glucose (Irwin and Wong, 2005; Nelson and Sheridan, 2006) and plasma fatty acid levels (Albalat *et al.*, 2005), and high glucose diets induce a decrease in glucagon circulating levels (del Sol Novoa *et al.*, 2004). GLP-1 has been identified in several fish species, where it is produced by both pancreas and intestine (Nelson and Sheridan, 2006). In channel catfish, both icv and ip injections of GLP-1 decrease food intake (Silverstein *et al.*, 2001) and affect metabolism by promoting glycogenolysis and gluconeogenesis (Silverstein *et al.*, 2001). As opposed to mammals where GLP-1 affects gastric motility, human GLP-1 has no effect on motility of chinook salmon gut rings (Forgan and Forster, 2007). Of the other members of the secretin family, only VIP has been shown to affect feeding as both ip and icv injections of VIP decrease feeding in goldfish (Matsuda *et al.*, 2005a).

Insulin has been identified (Irwin, 2004) and insulin-ir cells have been observed in pancreatic islets of several fishes (Nelson and Sheridan, 2006). In fishes, as in other vertebrates, insulin acts as an anabolic factor and stimulates glycogen synthesis, lipogenesis and protein synthesis (Nelson and Sheridan, 2006). However, fishes have a low capacity to utilize carbohydrates (Albalat *et al.*, 2007) and insulin secretion is only weakly stimulated by glucose, suggesting that other compounds, possibly amino acids, might be the major insulinotropic factors (Andoh, 2007). The role of insulin in the regulation of feeding in fishes is unclear: fasted fish usually have lower plasma insulin levels than fed fish (Navarro *et al.*, 2006; Montserrat *et al.*, 2007) and administration of insulin inhibits food intake in rainbow trout (Soengas and Aldegunde, 2004), but not in channel catfish (Silverstein and Plisetskaya, 2000). Although fasting has been shown to decrease the expression of insulin-like growth factors (IGFs) mRNA in the liver and muscle of several fishes (Pedroso *et al.*, 2006; Ayson *et al.*, 2007; Terova *et al.*, 2007), there is no evidence for a role of IGFs in the control of feeding in fishes.

2.3. The Growth Hormone System

Growth hormone (GH) plays a key role in growth and metabolism of all vertebrates, including fishes (Canosa *et al.*, 2007; Rousseau and Dufour, 2007; Chang and Wong, Chapter 4 of this volume) and GH-treated (Mclean *et al.*, 1997) and transgenic GH fish (Fu *et al.*, 2007; Hallennan *et al.*, 2007) both display increased growth. GH also appears to be involved in the regulation of feeding in fishes, as feeding behavior is stimulated in both GH-treated teleosts [via ip injections (Johnsson and Bjornsson, 1994) or ip pellet implantation (Johansson *et al.*, 2005)] and GH transgenic coho salmon (Stevens and Devlin, 2005) and starvation induces changes in both pituitary GH mRNA and plasma GH levels (Small, 2005; Pedroso *et al.*, 2006; Ayson *et al.*, 2007; Chang and Wong, Chapter 4 of this volume). Higher growth in transgenic animals appears to be due not only to increased food consumption but also to an increase in feed conversion efficiency (Raven *et al.*, 2006).

In vertebrates, growth hormone (GH) secretion is regulated by hypothalamic hormones, in particular growth hormone-releasing hormone (GHRH)/ pituitary adenylate cyclase activating polypeptide (PACAP) and somatostatin (SS). GHRH-like peptides and PACAP have been identified in a number of fish species (Chang and Wong, Chapter 4 of this volume). In fishes as in mammals, PACAP stimulates GH release from fish pituitary cells *in vitro* (Wong *et al.*, 2005; Sze *et al.*, 2007) and also appears to be involved in the regulation of feeding. When injected either centrally or peripherally, PACAP suppresses food intake in goldfish (Matsuda *et al.*, 2005b), and these actions might be mediated by pro-opiomelanocortin (POMC) and corticotropin-releasing factor (CRF) (Matsuda and Maruyama, 2007). In goldfish (Matsuda *et al.*, 2005a) and cod (Xu and Volkoff, 2008), PACAP mRNA expression is not affected by food deprivation but increases with excessive feeding or re-feeding. PACAP potentiates both meal-induced pancreatic insulin release and the lipid-storing actions of insulin on adipocytes (Nakata and Yada, 2007), suggesting that fish PACAP regulates energy metabolism in several tissues.

In mammals, somatostatin (SS) exists in two biologically active forms, SS-14 and SS-28, that are produced by alternative cleavage from the same precursor (preprosomatostatin I, PPS I), and inhibits pituitary GH secretion (Patel, 1999). SS proteins or mRNAs encoding PPSs have been isolated from over 20 fish species. Three PPS genes have been identified: PPS I, which encodes for SS-14, and PPSII and PPSIII, which produce SS14 variants. SS-14 has been shown to inhibit GH secretion in several teleosts, both *in vitro* and *in vivo* (Klein and Sheridan, 2008; Chang and Wong, Chapter 4 of this volume). Little is known about the effects of SS on food intake in fishes (Klein and Sheridan, 2008). Implantation of rainbow trout with SS-14-I inhibits growth but has no effect on food intake (Very *et al.*, 2001). However, SS has been shown to affect metabolism and energy homeostasis. Not only SS promotes lipid mobilization and hyperglycemia (Eilertson and Sheridan, 1993), but also SS biosynthesis and secretion are both regulated by nutrients, such as lipid and glucose (Sheridan and Kittilson, 2004), and by appetite regulators, such as CCK, NPY and GAL (Eilertson *et al.*, 1996). In addition, fasting induces increases in brain PPS I expression levels in cod (Xu and Volkoff, 2008) and in pancreatic PPS I mRNA in rainbow trout (Ehrman *et al.*, 2002).

2.4. The Melanocortin System and Melanin-concentrating Hormone

The melanocortin (MC) system plays a key role in regulating body weight and pigmentation (Millington, 2007). It is composed of melanocyte-stimulating hormone (MSH) adrenocorticotropic hormone (ACTH), melanocortin receptors (MCRs) and two endogenous antagonists, agouti-related protein (AgRP) and the agouti-signaling peptide (ASP). ACTH and MSH are produced from a common precursor, proopiomelanocortin (POMC), for which the predominant sites of expression are corticotrope and melanotrope cells of the pituitary gland. Corticotrope cells mainly produce ACTH, whereas MSH is produced by melanotrope cells (Metz *et al.*, 2006). In fishes, MCRs have been identified in teleosts, agnathans and elasmobranchs (Haitina *et al.*, 2007; Selz *et al.*, 2007) and consist of at least six subtypes (Schioth *et al.*, 2005). In both mammals and fishes, skin coloration is mediated via MC1R while MC4R receptors are thought to play a key role in the central regulation of food intake and energy balance (Metz *et al.*, 2006). Whereas in mammals the MC4R is exclusively expressed in the CNS, it is also peripherally expressed in fishes (Metz *et al.*, 2006). Central administration of MC4R agonists inhibits food intake in goldfish, whereas treatments with MC4R antagonists stimulate feeding (Cerdá-Reverter *et al.*, 2003) and food deprivation results in an increase in the number of MC4R transcripts in the liver, but not in brain, of flounder (Kobayashi *et al.*, 2008b), suggesting that both central and peripheral MC4Rs might be in part involved in the regulation of feeding in fishes. Evidence also supports a role for AgRP and MSH in the regulation of feeding and metabolism of fishes. In zebrafish, α-MSH- and AgRP-ir cells in the ventral periventricular hypothalamus are more pronounced at 5 days post fertilization when larvae begin to feed actively, which suggests a role of these peptides in regulating feeding behavior (Forlano and Cone, 2007). Fasting up-regulates hypothalamic AgRP mRNA levels in both goldfish and zebrafish (Cerdá-Reverter *et al.*, 2003; Song *et al.*, 2003), and transgenic zebrafish overexpressing AgRP exhibit obesity, increased growth and adipocyte hypertrophy (Song and Cone, 2007), suggesting an orexigenic role for AgRP. In goldfish, central injections of either NDP-α-MSH (an MC4R agonist) (Cerdá-Reverter *et al.*, 2003) or MT II (an MSH agonist) (Matsuda *et al.*, 2008a) inhibit feeding. The anorexigenic action of MSH might be mediated by the CRF-signaling pathway (Matsuda *et al.*, 2008a). In salmonids, α-MSH stimulates hepatic lipase activity and increases circulating levels of fatty acids, and fish with defective α-MSH synthesis are hyperphagic, have enlarged livers and accumulation of abdominal fat (Yada *et al.*, 2002). These data suggest that α-MSH inhibits feeding and affects energy balance in fishes. In goldfish, fasting does not significantly modify hypothalamic mRNA levels of POMC (Cerdá-Reverter *et al.*, 2003), suggesting that POMC is not directly involved in the regulation of feeding. ASP has been identified in fishes, and appears to act as a MC1R and MC4R melanocortin antagonist and a melanization inhibition factor (Cerdá-Reverter *et al.*, 2005; Klovins and Schioth, 2005), but its role in regulating feeding in fishes is not known.

Melanin-concentrating hormone (MCH) plays a major role in the control of feeding and energy homeostasis in mammals. In fishes, MCH regulates color by antagonizing the actions of α-MSH (Pissios *et al.*, 2006). Two fish MCH genes, MCH1 and MCH2, and fish MCH receptors have been characterized and their role in the regulation of food intake in fishes remains controversial. In goldfish, icv treatment with either mammalian or flounder MCH decreases feeding (Matsuda *et al.*, 2006b; Shimakura *et al.*, 2008) and fasting induces a decrease in the number of hypothalamic MCH-like ir neurons (Matsuda *et al.*, 2007). In goldfish, the anorexigenic actions of MCH appear to be mediated by NPY and MSH but not by CRF, PACAP or CCK (Matsuda *et al.*, 2008b; Shimakura *et al.*, 2008). However, fasting induces increases in hypothalamic MCH expression in barfin flounder, *Verasper moseri* (Takahashi *et al.*, 2004), and transgenic medaka overexpressing the MCH gene display changes in body color, but have normal growth or feeding behavior (Kinoshita *et al.*, 2001). To our knowledge, the role of ACTH in fishes has not been examined to date.

3. INFLUENCE OF INTRINSIC FACTORS

3.1. Metabolic Signals/Energy Reserves

Circulating metabolite levels alter food intake in fishes. Glucose administration induces hyperglycemia and a decrease in food intake in trout (Banos *et al.*, 1998). Conversely, 2-deoxyglucose administration (a non-metabolizable analogue of glucose) stimulates food intake (Delicio and Vicentini-Paulino, 1993; Soengas and Aldegunde, 2004). Hyperglycemia also increases feeding latency time in fishes (Kuz'mina and Garina, 2001; Kuz'mina *et al.*, 2002; Kuz'mina, 2005). In addition to carbohydrates, protein and lipid metabolites have also been demonstrated to alter feeding in fishes. Ip administration of amino acids, either singly or as a mixture, decreases food intake in carp (Kuz'mina, 2005). In chinook salmon and channel catfish, fat fish eat less than lean fish, suggesting a lipostatic control of food intake (Shearer *et al.*, 1997;

Silverstein and Plisetskaya, 2000). However, in sea bream, a diet rich in linoleic acid induces a decrease in food intake and a lower total body fat content (Diez *et al.*, 2007).

3.2. Ontogeny

Developmental profiles have been examined for several appetite-regulating factors in fishes and shown to exhibit factor-specific as well as species-specific variations in both distribution patterns and timing of appearance. In general, appetite-regulating factors appear early in development, suggesting that they play a role in embryogenesis (pre-hatch) as well as nutrient absorption/acquisition (pre- and post-hatch) in fish larvae. A later appearance of appetite-regulating factors, especially when associated with metamorphic events or mixed feeding phases (e.g., onset of exogenous feeding), may reflect ontogenic shifts in feeding. GI hormones have been extensively studied due to their important role in the digestive physiology of larval fishes (Ronnestad *et al.*, 2007). From studies carried out on the endocrine factors and fishes examined to date, it can be hypothesized that CCK and ghrelin make an early appearance in larval development, at least in certain fish species (Kamisaka *et al.*, 2003; Parhar *et al.*, 2003; Kamisaka *et al.*, 2005; Manning *et al.*, 2008). GI CCK immunoreactivity has been detected upon first hatch in herring, *Clupea harengus* (Kamisaka *et al.*, 2005), and ghrelin has been detected 48 h post fertilization in zebrafish (Pauls *et al.*, 2007). In contrast, gastrin and GLP-1 are detected later in development (Kurokawa *et al.*, 2003; Navarro *et al.*, 2006). A number of other signals involved in the regulation of feeding in fishes appear to be present prior to hatching, such as PACAP (Krueckl *et al.*, 2003; Xu and Volkoff, 2008), somatostatin (Xing *et al.*, 2005; Xu and Volkoff, 2008), NPY (Xu and Volkoff, unpubl), OX (Xu and Volkoff, 2007), α-MSH and AgRP (Forlano and Cone, 2007).

3.3. Gender and Reproductive Status

Gender-specific differences in both distribution and expression levels have been reported for several appetite regulators, including TKs (Peyon *et al.*, 2000), GAL (Rao *et al.*, 1996; Jadhao and Meyer, 2000) and ghrelin (Parhar *et al.*, 2003). These observations would suggest that gender, or more specifically sex steroids, might influence feeding. It seems likely that under natural conditions, these influences would be more pronounced when fish are either breeding or preparing to breed (e.g., migrating). In this regard, there is a well-documented cessation or decline in feeding during spawning migrations or other reproductive behaviors (e.g., courtship, spawning, territoriality, guarding) (van Ginneken and Maes, 2005). Evidence that appetite-regulating factors are involved in this spawning-induced anorexia awaits further study. However, strong interactions between appetite regulators and reproduction have been suggested in goldfish. In this species, icv treatment with gonadotrophin-releasing hormone (GnRH) induces a decrease in food intake (Hoskins *et al.*, 2008; Matsuda *et al.*, 2008c), which is in part due to down-regulation of brain OX mRNA expression (Hoskins *et al.*, 2008). Conversely, icv OX-A injections induce a decrease in spawning behavior and a decrease in GnRH mRN expression levels in the brain (Hoskins *et al.*, 2008). In addition, seasonal changes correlated to gonadal cycles have been shown for NPY in ayu (Chiba *et al.*, 1996) and catfish (Mazumdar *et al.*, 2007) as well as for UI mRNA and circulating cortisol levels in masu salmon, *Oncorhynchus masou* (Westring *et al.*, 2008). To date, the effects of sex steroids on food intake are unclear. Testosterone (T) treatment decreases food intake in male perch (Mandiki *et al.*, 2005) and elevates MCH mRNA expression in the hypothalamus of both male and female goldfish (Cerdá-Reverter *et al.*, 2006), and castration reduces the density of NPY-ir fibers in the forebrain of tilapia (Sakharkar *et al.*, 2005), suggesting an anorexic role for T. However, T treatments have been shown to increase feeding in sea bream (Woo *et al.*, 1993). Estradiol treatment stimulates feeding in female, but not male, perch (Mandiki *et al.*, 2005) but has no effects on food intake of sea bream (Woo *et al.*, 1993), and elevates hypothalamic MCH (a putative feeding inhibitor) mRNA expression in goldfish (Cerdá-Reverter *et al.*, 2006). In sea bass, ip implantation of either 17-β-estradiol or T decreases food intake, the effects of T probably being mediated by central aromatization to estradiol (Leal *et al.*, 2009).

3.4. Genetic Influence

Food intake has been shown to vary among fish strains of different genetic backgrounds. In captive trout, different strains display variations in feeding activities and growth (Mambrini *et al.*, 2004), feeding patterns (Boujard *et al.*, 2007) and nutrient utilization (Quillet *et al.*, 2007). Individuals from different phenotypes may also display different feeding and growth patterns when exposed to specific environments or treatments. For example, in channel catfish, anorexigenic compounds inhibit feeding in some strains more than others and low temperature exposure affects the strains differently with respect to feed efficiency (Silverstein *et al.*, 1999b). Variations in microsatellites (tandem repeats) have been detected in NPY genes of both channel catfish and rainbow trout (Silverstein, 2002), suggesting that genetic variations in appetite-regulating peptides may account in part for these divergent phenotypes.

4. INFLUENCE OF EXTRINSIC FACTORS

4.1. Temperature

A relationship between temperature and food intake has been demonstrated in a number of fishes. Fish tend to decrease their food intake when placed in extreme temperatures but usually increase their food consumption and growth rates with rising temperatures within a "tolerable" range (Russell *et al.*, 1996; Guijarro *et al.*, 1999; Bendiksen *et al.*, 2002; Sunuma *et al.*, 2007; Kehoe and Volkoff, 2008). Little is known about the endocrine mechanisms regulating these temperature-induced changes in feeding. Increased temperatures usually lead to increased plasma GH and IGF-I levels (Gabillard *et al.*, 2005). In Atlantic cod, brain NPY mRNA expression does not seem to be affected by temperature, but brain CART mRNA expression levels are higher in fish held at 2°C than in fish held at either 11°C or 15°C. This suggests that CART, but not NPY, may contribute to temperature-induced changes in appetite in this species (Kehoe and Volkoff, 2008).

4.2. Photoperiod

Feeding activity has been shown to be affected by photoperiod and light regimens in a number of fish species (Noble *et al.*, 2005; Tucker *et al.*, 2006; Sunuma *et al.*, 2007), although exceptions have been reported (Canavate *et al.*, 2006). It has been suggested that photoperiod may indirectly modify growth by increasing food intake and/or muscle mass by exercise (Boeuf and Le Bail, 1999). The specific endocrine mechanisms behind these changes are unclear and will require further investigation. Plasma GH, thyroid hormone and cortisol concentrations are all affected by light regimens in rainbow trout (Reddy and Leatherland, 2003), and Atlantic salmon respond to increased photoperiod by elevating plasma GH (Nordgarden *et al.*, 2007). In contrast, photoperiod does not affect circulating IGF-I levels in sunshine bass (Davis and McEntire, 2006). These data suggest a role for appetite-regulating factors that stimulate the GH related pathways, at least in salmonids. The hormone melatonin, synthesized in the pineal gland, acts as a neuroendocrine signal that coordinates the environment and rhythmic physiological processes. Melatonin decreases feeding and body weight when administered ip in goldfish (Pinillos *et al.*, 2001; De Pedro *et al.*, 2008) and tench (Lopez-Olmeda *et al.*, 2006) and when given orally to sea bass (Rubio *et al.*, 2004). In goldfish, plasma leptin and ghrelin as well as hypothalamic NPY levels are not affected by melatonin ip treatment, suggesting that these feeding regulators may not be involved in the effects of melatonin on energy homeostasis in fish (De Pedro *et al.*, 2008).

4.3. Salinity

Salinity has been shown to affect feeding in several fish species (De Boeck *et al.*, 2000) but little is known about the appetite-regulating factors that may be involved. The somatotropic axis is a major regulator of growth as well as salt water tolerance in salmonids, which suggests that this system may play a role in salinity-related alterations in feeding (Boeuf and Payan, 2001). Studies have also linked food intake and hydromineral balance in the goldfish, where PrRP appears to play a role in both the regulation of food intake and ionoregulatory homeostasis, but participates in either process according to systemic needs (Kelly and Peter, 2006). Lastly, under conditions of acute osmoregulatory disturbance, fishes often transiently reduce food intake, which has been shown to be concomitant with increases in brain CRF and UI mRNA expression in trout (Craig *et al.*, 2005), suggesting a role for these compounds.

4.4. Hypoxia

In the species examined to date, hypoxia has distinct appetite suppressive effects (Buentello *et al.*, 2000; Ripley and Foran, 2007). In rainbow trout, exposure to low oxygen levels increases forebrain CRF and UI mRNA levels as well as plasma cortisol, suggesting that CRF-related peptides play a physiological role in mediating at least a portion of the reduction in food intake under hypoxic conditions (Bernier and Craig, 2005).

4.5. Pollutants and Health Status

Contaminants and disease agents can also influence feeding but the responses vary depending on the species considered as well as on the agent and the method of exposure. For example, rainbow trout exposed to elevated waterborne concentrations of metals feed less (Todd *et al.*, 2007), whereas coho salmon fed a high zinc diet display increased feeding rates (Bowen *et al.*, 2006). An immune challenge produced by LPS treatment decreases feeding and induces changes in CART, NPY and CRF gene expression in the goldfish brain (Volkoff and Peter, 2004) and in CRF ir in tilapia brain (Pepels *et al.*, 2004). In lake trout, *Salvelinus namaycush*, exposed to low doses of the insecticide tebufenozide, although food intake and brain NPY and CRF mRNAs are unaffected, brain CART mRNA expression levels are significantly higher than in control fish (Volkoff *et al.*, 2007). These changes in CART and perhaps other appetite regulators might be linked to tebufenozide-induced immunostimulation in this species (Hamoutene *et al.*, 2008).

5. PROPOSED MODEL OF ENDOCRINE CIRCUITRIES INVOLVED IN FEEDING IN FISHES AND CONCLUDING REMARKS

Our current knowledge on the endocrine regulation of feeding in fishes and the influences of intrinsic and extrinsic factors only allows us to build a simplified and relatively incomplete model of the regulation of appetite in fishes (Table 9.1, Figure 9.1). However, it appears that the general scheme of feeding regulation in fishes is similar to that of other vertebrates in that appetite is regulated by central feeding centers that are themselves influenced by hormonal factors arising from both the brain or from the periphery. As in mammals, the fish model shows redundancy, with the presence of several orexigenic or anorexigenic factors with apparently the same function, and a high degree of interaction between endocrine systems. Evidence gathered to date seems to indicate that fish appetite-regulating neuropeptides closely interact with each other and act in parallel, rather than in a hierarchical fashion. Long- and short-term regulators have been defined in mammals, the latter consisting mostly of satiety signals that act immediately following a meal, the former sensing the body energy status and regulating feeding according to fat mass/energy stores. To date, most of the studies performed on appetite-regulating hormones in fishes have been relatively short-term studies using acute treatments, so that such a distinction between short- and long-term feeding-related factors is difficult to establish.

Differences appear to exist between mammals and fishes with regards to the sites of synthesis, structure and function of some of the known endocrine regulators of appetite. Increasing evidence suggests that brain areas producing feeding-related hormones and those acting as targets for these hormones differ between fishes and mammals – and also between fishes themselves – which is not surprising, given the major differences in brain organization between groups (Cerdá-Reverter and Canosa, Chapter 1 of this volume). For example, OXs, which are predominantly produced by the lateral hypothalamus in mammals, display a widespread distribution in fish brains, as demonstrated by mRNA expression studies. Regarding structure, although the amino acid composition of a hormone and of its encoding cDNA might differ greatly between fishes and mammals, it might still have a conserved function. For example, although mammalian and fish leptins display pronounced structural differences (less than 10% identity at the protein level) and are produced by different tissues (adipose tissue and liver), they might act as regulators of feeding and metabolism in both groups. Conversely, hormones might have well-conserved structures but somewhat differ in function. For example, although GLPs have a similar structure in fishes and mammals and affect feeding in both groups, GLP affects gastric motility in mammals but does not appear to have such an effect in fishes. NPY is also structurally relatively well conserved between the two groups, but where it is considered the major orexigenic peptide in mammals, its expression is not affected by fasting in some species (e.g., cod), suggesting that its relative importance might be species specific.

The presence of specific fish tissues not present in mammals (e.g., caudal neurosecretory organ) and additional gene-duplication events in ray-finned fishes (leading to more genes than in land vertebrates and thus multiple peptide forms as opposed to a single mammalian form) might hint at the presence of additional endocrine players with distinct physiological roles in fishes. To add further complexity to the piscine model, fish usually display a higher degree of heterogeneity than mammals with regards to habitats and environmental conditions, which translate into species-specific physiological adaptations, including changes in feeding behavior. The endocrine mechanisms regulating these adaptations are to date poorly understood.

In conclusion, although significant advances have been made in recent years regarding the endocrine control of feeding and energy balance in fishes, lack of in-depth knowledge and controversies still exist and the puzzle is far from being solved. Further characterization of fish homologues to mammalian hormones might reveal group-specific differences in structure and might lead to the discovery of new functions for mammalian homologues in fishes. The list of appetite-regulating factors in fishes is also bound to increase in the years to come and might include fish-specific hormones that have not been isolated in mammals.

REFERENCES

Abbott, C. R., Rossi, M., Wren, A. M., Murphy, K. G., Kennedy, A. R., Stanley, S. A., Zollner, A. N., Morgan, D. G., Morgan, I., Ghatei, M. A., Small, C. J., and Bloom, S. R. (2001). Evidence of an orexigenic role for cocaine- and amphetamine-regulated transcript after administration into discrete hypothalamic nuclei. *Endocrinology* **142,** 3457–3463.

Albalat, A., Gutierrez, J., and Navarro, I. (2005). Regulation of lipolysis in isolated adipocytes of rainbow trout (*Oncorhynchus mykiss*): The role of insulin and glucagon. *Comp. Biochem. Physiol. A* **142,** 347–354.

Albalat, A., Saera-Vila, A., Capilla, E., Gutierrez, J., Perez-Sanchez, J., and Navarro, I. (2007). Insulin regulation of lipoprotein lipase (LPL) activity and expression in gilthead sea bream (*Sparus aurata*). *Comp. Biochem. Physiol. B* **148,** 151–159.

Aldegunde, M., and Mancebo, M. (2006). Effects of neuropeptide Y on food intake and brain biogenic amines in the rainbow trout (*Oncorhynchus mykiss*). *Peptides* **27,** 719–727.

Alderman, S. L., and Bernier, N. J. (2007). Localization of corticotropin-releasing factor, urotensin I, and CRF-binding protein gene expression in the brain of the zebrafish, *Danio rerio*. *J. Comp. Neurol.* **502,** 783–793.

Aldman, G., Jonsson, A. C., Jensen, J., and Holmgren, S. (1989). Gastrin/CCK-like peptides in the spiny dogfish, *Squalus acanthias*; concentrations and actions in the gut. *Comp. Biochem. Physiol. C* **92,** 103–108.

Amiya, N., Amano, M., Oka, Y., Iigo, M., Takahashi, A., and Yamamori, K. (2007). Immunohistochemical localization of orexin/hypocretin-like immunoreactive peptides and melanin-concentrating hormone in the brain and pituitary of medaka. *Neurosci. Lett.* **427,** 16–21.

Amole, N., and Unniappan, S. (2009). Fasting induces preproghrelin mRNA expression in the brain and gut of zebrafish. *Danio rerio. Gen. Comp. Endocrinol.* **161**(1), 133–137.

Andoh, T. (2007). Amino acids are more important insulinotropins than glucose in a teleost fish, barfin flounder (*Verasper moseri*). *Gen. Comp. Endocrinol.* **151,** 308–317.

Ayson, F. G., de Jesus-Ayson, E. G. T., and Takemura, A. (2007). mRNA expression patterns for GH, PRL, SL, IGF-I and IGF-II during altered feeding status in rabbitfish, *Siganus guttatus*. *Gen. Comp. Endocrinol.* **150,** 196–204.

Baker, D. M., Larsen, D. A., Swanson, P., and Dickhoff, W. W. (2000). Long-term peripheral treatment of immature coho salmon (*Oncorhynchus kisutch*) with human leptin has no clear physiologic effect. *Gen. Comp. Endocrinol.* **118,** 134–138.

Banos, N., Baro, J., Castejon, C., Navarro, I., and Gutierrez, J. (1998). Influence of high-carbohydrate enriched diets on plasma insulin levels and insulin and IGF-I receptors in trout. *Regul. Pept.* **77,** 55–62.

Bechtold, D. A., and Luckman, S. M. (2007). The role of RFamide peptides in feeding. *J. Endocrinol.* **192,** 3–15.

Bendiksen, E., Jobling, M., and Arnesen, A. (2002). Feed intake of Atlantic salmon parr *Salmo salar* L. in relation to temperature and feed composition. *Aquac. Res.* **33,** 525–532.

Bermudez, R., Vigliano, F., Quiroga, M. I., Nieto, J. M., Bosi, G., and Domeneghini, C. (2007). Immunohistochemical study on the neuroendocrine system of the digestive tract of turbot, *Scophthalmus maximus* (L.), infected by *Enteromyxum scophthalmi* (Myxozoa). *Fish Shellfish Immunol.* **22,** 252–263.

Bernier, N. J. (2006). The corticotropin-releasing factor system as a mediator of the appetite-suppressing effects of stress in fish. *Gen. Comp. Endocrinol.* **146,** 45–55.

Bernier, N. J., and Craig, P. M. (2005). CRF-related peptides contribute to stress response and regulation of appetite in hypoxic rainbow trout. *Am. J. Physiol. Regul. Integr. Comp. Physiol.* **289,** R982–R990.

Bernier, N. J., and Peter, R. E. (2001). Appetite-suppressing effects of urotensin I and corticotropin-releasing hormone in goldfish (*Carassius auratus*). *Neuroendocrinology* **73,** 248–260.

Bernier, N. J., Bedard, N., and Peter, R. E. (2004). Effects of cortisol on food intake, growth, and forebrain neuropeptide Y and corticotropin-releasing factor gene expression in goldfish. *Gen. Comp. Endocrinol.* **135,** 230–240.

Bjenning, C., Farrell, A., and Holmgren, S. (1991). Bombesin-like immunoreactivity in skates and the *in vitro* effect of bombesin on coronary vessels from the longnose skate, *Raja rhina*. *Regul. Pept.* **35,** 207–219.

Boeuf, G., and Le Bail, P. Y. (1999). Does light have an influence on fish growth? *Aquaculture* **177,** 129–152.

Boeuf, G., and Payan, P. (2001). How should salinity influence fish growth? *Comp. Biochem. Physiol. C* **130,** 411–423.

Bosi, G., Di Giancamillo, A., Arrighi, S., and Domeneghini, C. (2004). An immunohistochemical study on the neuroendocrine system in the alimentary canal of the brown trout, *Salmo trutta*, L., 1758. *Gen. Comp. Endocrinol.* **138,** 166–181.

Boujard, T., Ramezi, J., Vandeputte, M., Labbe, L., and Mambrini, M. (2007). Group feeding behavior of brown trout is a correlated response to selection for growth shaped by the environment. *Behav. Genet.* **37,** 525–534.

Bowen, L., Werner, I., and Johnson, M. L. (2006). Physiological and behavioral effects of zinc and temperature on coho salmon (*Oncorhynchus kisutch*). *Hydrobiologia* **559,** 161–168.
Brightman, M. W., and Broadwell, R. D. (1976). The morphological approach to the study of normal and abnormal brain permeability. *Adv. Exp. Med. Biol.* **69,** 41–54.
Buentello, J. A., Gatlin, D. M., and Neill, W. H. (2000). Effects of water temperature and dissolved oxygen on daily feed consumption, feed utilization and growth of channel catfish (*Ictalurus punctatus*). *Aquaculture* **182,** 339–352.
Canavate, J. P., Zerolo, R., and Fernandez-Diaz, C. (2006). Feeding and development of Senegal sole (*Solea senegalensis*) larvae reared in different photoperiods. *Aquaculture* **258,** 368–377.
Canosa, L. F., and Peter, R. E. (2004). Effects of cholecystokinin and bombesin on the expression of preprosomatostatin-encoding genes in goldfish forebrain. *Regul. Pept.* **121,** 99–105.
Canosa, L. F., Unniappan, S., and Peter, R. E. (2005). Periprandial changes in growth hormone release in goldfish: Role of somatostatin, ghrelin, and gastrin-releasing peptide. *Am. J. Physiol. Regul. Integr. Comp. Physiol.* **289,** R125–133.
Canosa, L. F., Chang, J. P., and Peter, R. E. (2007). Neuroendocrine control of growth hormone in fish. *Gen. Comp. Endocrinol.* **151,** 1–26.
Cerdá-Reverter, J., Martinez-Rodriguez, G., Zanuy, S., Carrillo, M., and Larhammar, D. (2000a). Molecular evolution of the neuropeptide Y (NPY) family of peptides: Cloning of three NPY-related peptides from the sea bass (*Dicentrarchus labrax*). *Regul. Pept.* **95,** 25–34.
Cerdá-Reverter, J. M., Martinez-Rodriguez, G., Anglade, I., Kah, O., and Zanuy, S. (2000b). Peptide YY (PYY) and fish pancreatic peptide Y (PY) expression in the brain of the sea bass (*Dicentrarchus labrax*) as revealed by in situ hybridization. *J. Comp. Neurol.* **426,** 197–208.
Cerdá-Reverter, J., Schioth, H., and Peter, R. (2003). The central melanocortin system regulates food intake in goldfish. *Regul. Pept.* **115,** 101–113.
Cerdá-Reverter, J. M., Haitina, T., Schioth, H. B., and Peter, R. E. (2005). Gene structure of the goldfish agouti-signaling protein: A putative role in the dorsal-ventral pigment pattern of fish. *Endocrinology* **146,** 1597–1610.
Cerdá-Reverter, J. M., Canosa, L. F., and Peter, R. E. (2006). Regulation of the hypothalamic melanin-concentrating hormone neurons by sex steroids in the goldfish: Possible role in the modulation of luteinizing hormone secretion. *Neuroendocrinol.* **84,** 364–377.
Chan, C. B., Leung, P. K., Wise, H., and Cheng, C. H. (2004). Signal transduction mechanism of the seabream growth hormone secretagogue receptor. *FEBS Lett.* **577,** 147–153.
Chang, C. L., Roh, J., and Hsu, S. Y. (2004). Intermedin, a novel calcitonin family peptide that exists in teleosts as well as in mammals: A comparison with other calcitonin/intermedin family peptides in vertebrates. *Peptides* **25,** 1633–1642.
Chiba, A., Sohn, Y. C., and Honma, Y. (1996). Distribution of neuropeptide Y and gonadotropin-releasing hormone immunoreactivities in the brain and hypophysis of the ayu, *Plecoglossus altivelis* (Teleostei). *Arch. Histol. Cytol.* **59,** 137–148.
Conlon, J. M., and Larhammar, D. (2005). The evolution of neuroendocrine peptides. *Gen. Comp. Endocrinol.* **142,** 53–59.
Craig, P. M., Al-Timimi, H., and Bernier, N. J. (2005). Differential increase in forebrain and caudal neurosecretory system corticotropin-releasing factor and urotensin I gene expression associated with seawater transfer in rainbow trout. *Endocrinology* **146,** 3851–3860.
Davis, K. B., and McEntire, M. (2006). Effect of photoperiod on feeding, intraperitoneal fat, and insulin-like growth factor-I in sunshine bass. *J. World Aquac. Soc.* **37,** 431–436.
De Boeck, G., Vlaeminck, A., Van der Linden, A., and Blust, R. (2000). The energy metabolism of common carp (*Cyprinus carpio*) when exposed to salt stress: An increase in energy expenditure or effects of starvation? *Physio. Bioch. Zool.* **73,** 102–111.
De Pedro, N., Alonso-Gomez, A., Gancedo, B., Delgado, M., and Alonso-Bedate, M. (1993). Role of corticotropin-releasing factor (CRF) as a food intake regulator in goldfish. *Physiol. Behav.* **53,** 517–520.

De Pedro, N., Cespedes, M. V., Delgado, M. J., and Alonsobedate, M. (1995a). The galanin-induced feeding stimulation is mediated via alpha(2)-adrenergic receptors in goldfish. *Regul. Pept.* **57,** 77–84.

De Pedro, N., Gancedo, B., Alonsogomez, A. L., Delgado, M. J., and Alonsobedate, M. (1995b). Alterations in food-intake and thyroid-tissue content by corticotropin-releasing factor in *Tinca-tinca*. *Rev. Esp. Fisiol.* **51,** 71–75.

De Pedro, N., Alonso-Gomez, A. L., Gancedo, B., Valenciano, A. I., Delgado, M. J., and Alonso-Bedate, M. (1997). Effect of alpha-helical-CRF[9-41] on feeding in goldfish: Involvement of cortisol and catecholamines. *Behav. Neurosci.* **111,** 398–403.

De Pedro, N., Pinillos, M. L., Valenciano, A. I., Alonso-Bedate, M., and Delgado, M. J. (1998). Inhibitory effect of serotonin on feeding behavior in goldfish: Involvement of CRF. *Peptides* **19,** 505–511.

De Pedro, N., Martinez-Alvarez, R., and Delgado, M. J. (2006). Acute and chronic leptin reduces food intake and body weight in goldfish (*Carassius auratus*). *J. Endocrinol.* **188,** 513–520.

De Pedro, N., Martinez-Alvarez, R. M., and Delgado, M. J. (2008). Melatonin reduces body weight in goldfish (*Carassius auratus*): Effects on metabolic resources and some feeding regulators. *J. Pineal Res.* **45,** 32–39.

del Sol Novoa, M., Capilla, E., Rojas, P., Baro, J., Gutierrez, J., and Navarro, I. (2004). Glucagon and insulin response to dietary carbohydrate in rainbow trout (*Oncorhynchus mykiss*). *Gen. Comp. Endocrinol.* **139,** 48–54.

Delicio, H., and Vicentini-Paulino, M. (1993). 2-deoxyglucose-induced food-intake by Nile tilapia, *Oreochromis-niloticus* (L). *Braz. J. Med. Biol. Res.* **26,** 327–331.

Despres, J. P. (2007). The endocannabinoid system: A new target for the regulation of energy balance and metabolism. *Crit. Pathw. Cardiol.* **6,** 46–50.

Diez, A., Menoyo, D., Perez-Benavente, S., Calduch-Giner, J. A., Vega-Rubin de Celis, S., Obach, A., Favre-Krey, L., Boukouvala, E., Leaver, M. J., Tocher, D. R., Perez-Sanchez, J., Krey, G., *et al.* (2007). Conjugated linoleic acid affects lipid composition, metabolism, and gene expression in gilthead sea bream (*Sparus aurata* L). *J. Nutr.* **137,** 1363–1369.

Dockray, G. J. (2004). The expanding family of -RFamide peptides and their effects on feeding behaviour. *Exp. Physiol.* **89,** 229–235.

Ehrman, M., Melroe, G., Moore, C., Kittilson, J.D, and Sheridan, M (2002). Nutritional regulation of somatostatin expression in rainbow trout, Oncorhynchus mykiss. *Fish Physiol. Biochem.* **26,** 309–314.

Eilertson, C. D., Carneiro, N. M., Kittilson, J. D., Comley, C., and Sheridan, M. A. (1996). Cholecystokinin, neuropeptide Y and galanin modulate the release of pancreatic somatostatin-25 and somatostatin-14 *in vitro*. *Regul. Pept.* **63,** 105–112.

Eilertson, C. D., and Sheridan, M. A. (1993). Differential effects of somatostatin-14 and somatostatin-25 on carbohydrate and lipid metabolism in rainbow trout *Oncorhynchus mykiss*. *Gen. Comp. Endocrinol.* **92,** 62–70.

Ekblad, E. (2006). CART in the enteric nervous system. *Peptides* **27,** 2024–2030.

Elphick, M. R., and Egertova, M. (2001). The neurobiology and evolution of cannabinoid signalling. *Philos. Trans. R. Soc. Lond. B. Biol. Sci.* **356,** 381–408.

Faraco, J. H., Appelbaum, L., Marin, W., Gaus, S. E., Mourrain, P., and Mignot, E. (2006). Regulation of hypocretin (orexin) expression in embryonic zebrafish. *J. Biol. Chem.* **281,** 29753–29761.

Forgan, L. G., and Forster, M. E. (2007). Effects of potential mediators of an intestinal brake mechanism on gut motility in Chinook salmon (*Oncorhynchus tshawytscha*). *Comp. Biochem. Physiol. C* **146,** 343–347.

Forlano, P. M., and Cone, R. D. (2007). Conserved neurochemical pathways involved in hypothalamic control of energy homeostasis. *J. Comp. Neurol.* **505,** 235–248.

Fu, C., Li, D., Hu, W., Wang, Y., and Zhu, Z. (2007). Fast-growing transgenic common carp mounting compensatory growth. *J. Fish Biol.* **71,** 174–185.

Gabillard, J. C., Weil, C., Rescan, P. Y., Navarro, I., Gutierrez, J., and Le Bail, P. Y. (2005). Does the GH/IGF system mediate the effect of water temperature on fish growth? A review. *Cybium* **29,** 107–117.

Gao, Q., and Horvath, T. L. (2007). Neurobiology of feeding and energy expenditure. *Annu. Rev. Neurosci.* **30,** 367–398.

Gelineau, A., and Boujard, T. (2001). Oral administration of cholecystokinin receptor antagonists increase feed intake in rainbow trout. *J. Fish Biol.* **58,** 716–724.

Godby, N. A., Rutherford, E. S., and Mason, D. M. (2007). Diet, feeding rate, growth, mortality, and production of juvenile steelhead in a Lake Michigan tributary. *N. Am. J. Fish. Man.* **27,** 578–592.

Gorissen, M. H. A. G., Flik, G., and Huising, M. O. (2006). Peptides and proteins regulating food intake: A comparative view. *Animal Biol.* **56,** 447–473.

Gregory, T. R., and Wood, C. M. (1999). The effects of chronic plasma cortisol elevation on the feeding behaviour, growth, competitive ability, and swimming performance of juvenile rainbow trout. *Physiol. Biochem. Zool.* **72,** 286–295.

Guijarro, A. I., Delgado, M. J., Pinillos, M. L., Lopez-Patino, M. A., Alonso-Bedate, M., and De Pedro, N. (1999). Galanin and beta-endorphin as feeding regulators in cyprinids: Effect of temperature. *Aquac. Res.* **30,** 483–489.

Haitina, T., Klovins, J., Takahashi, A., Lowgren, M., Ringholm, A., Enberg, J., Kawauchi, H., Larson, E. T., Fredriksson, R., and Schioth, H. B. (2007). Functional characterization of two melanocortin (MC) receptors in lamprey showing orthology to the MC1 and MC4 receptor subtypes. *BMC Evol. Biol.* **7,** 101.

Hallennan, E. M., McLean, E., and Fleming, I. A. (2007). Effects of growth hormone transgenes on the behavior and welfare of aquacultured fishes: A review identifying research needs. *Appl. Anim. Behav. Sci.* **104,** 265–294.

Hamoutene, D., Payne, J. F., and Volkoff, H. (2008). Effects of tebufenozide on some aspects of lake trout (*Salvelinus namaycush*) immune response. *Ecotoxicol. Environ. Saf.* **69,** 173–179.

Heino, M., and Kaitala, V. (1999). Evolution of resource allocation between growth and reproduction in animals with indeterminate growth. *J. Evol. Biol.* **12,** 423–429.

Himick, B. A., and Peter, R. E. (1994a). Bombesin acts to suppress feeding behavior and alter serum growth hormone in goldfish. *Physiol. Behav.* **55,** 65–72.

Himick, B. A., and Peter, R. E. (1994b). CCK/gastrin-like immunoreactivity in brain and gut, and CCK suppression of feeding in goldfish. *Am. J. Physiol.* **267,** R841–851.

Himick, B. A., Golosinski, A. A., Jonsson, A. C., and Peter, R. E. (1993). CCK/gastrin-like immunoreactivity in the goldfish pituitary: Regulation of pituitary hormone secretion by CCK-like peptides *in vitro*. *Gen. Comp. Endocrinol.* **92,** 88–103.

Himick, B. A., Vigna, S. R., and Peter, R. E. (1996). Characterization of cholecystokinin binding sites in goldfish brain and pituitary. *Am. J. Physiol.* **271,** R137–143.

Holmberg, A., Schwerte, T., Pelster, B., and Holmgren, S. (2004). Ontogeny of the gut motility control system in zebrafish *Danio rerio* embryos and larvae. *J. Exp. Biol.* **207,** 4085–4094.

Hoskins, L. J., Xu, M., and Volkoff, H. (2008). Interactions between gonadotropin-releasing hormone (GnRH) and orexin in the regulation of feeding and reproduction in goldfish (*Carassius auratus*). *Horm. Behav.* **54,** 379–385.

Hoyle, C. H. V. (1999). Neuropeptide families and their receptors: Evolutionary perspectives. *Brain Res.* **848,** 1–25.

Huising, M. O., Geven, E. J., Kruiswijk, C. P., Nabuurs, S. B., Stolte, E. H., Spanings, F. A., Verburg-van Kemenade, B. M., and Flik, G. (2006a). Increased leptin expression in common carp (*Cyprinus carpio*) after food intake but not after fasting or feeding to satiation. *Endocrinology* **147,** 5786–5797.

Huising, M. O., Kruiswijk, C. P., and Flik, G. (2006b). Phylogeny and evolution of class-I helical cytokines. *J. Endocrinol.* **189,** 1–25.

Irwin, D. (2004). A second insulin gene in fish genomes. *Gen. Comp. Endocrinol.* **135,** 150–158.

Irwin, D. M., and Wong, K. (2005). Evolution of new hormone function: Loss and gain of a receptor. *J. Heredity* **96,** 205–211.

Jadhao, A. G., and Meyer, D. L. (2000). Sexually dimorphic distribution of galanin in the preoptic area of red salmon, *Oncorhynchus nerka.*. *Cell Tissue Res.* **302,** 199–203.

Jadhao, A., and Pinelli, C. (2001). Galanin-like immunoreactivity in the brain and pituitary of the "four-eyed" fish, *Anableps anableps*. *Cell Tissue Res.* **306,** 309–318.

Jensen, H., Rourke, I. J., Moller, M., Jonson, L., and Johnsen, A. H. (2001). Identification and distribution of CCK-related peptides and mRNAs in the rainbow trout, *Oncorhynchus mykiss*. *Biochim. Biophys. Acta* **1517,** 190–201.

Johansson, V., Winberg, S., and Bjornsson, B. T. (2005). Growth hormone-induced stimulation of swimming and feeding behaviour of rainbow trout is abolished by the D-1 dopamine antagonist SCH23390. *Gen. Comp. Endocrinol.* **141,** 58–65.

Johnsen, A. H., Jonson, L., Rourke, I. J., and Rehfeld, J. F. (1997). Elasmobranchs express separate cholecystokinin and gastrin genes. *Proc. Natl. Acad. Sci. U S A* **94,** 10221–10226.

Johnson, R. M., Johnson, T. M., and Londraville, R. L. (2000). Evidence for leptin expression in fishes. *J. Exp. Zool.* **286,** 718–724.

Johnsson, J. I., and Bjornsson, B. T. (1994). Growth hormone increases growth rate, appetite and dominance in juvenile rainbow trout, *Oncorhynchus mykiss*. *Anim. Behav.* **48,** 177–186.

Jonsson, E., Forsman, A., Einarsdottir, I. E., Egner, B., Ruohonen, K., and Bjornsson, B. T. (2006). Circulating levels of cholecystokinin and gastrin-releasing peptide in rainbow trout fed different diets. *Gen. Comp. Endocrinol.* **148,** 187–194.

Jonsson, E., Forsman, A., Einarsdottir, I. E., Kaiya, H., Ruohonen, K., and Bjornsson, B. T. (2007). Plasma ghrelin levels in rainbow trout in response to fasting, feeding and food composition, and effects of ghrelin on voluntary food intake. *Comp. Biochem. Physiol. A* **147,** 1116–1124.

Kaiya, H., Miyazato, M., Kangawa, K., Peter, R. E., and Unniappan, S. (2008). Ghrelin: A multifunctional hormone in non-mammalian vertebrates. *Comp. Biochem. Physiol. A* **149,** 109–128.

Kamiji, M. M., and Inui, A. (2007). Neuropeptide Y receptor selective ligands in the treatment of obesity. *Endocr. Rev.* **28,** 664–684.

Kamisaka, Y., Fujii, Y., Yamamoto, S., Kurokawa, T., Ronnestad, I., Totland, G. K., Tagawa, M., and Tanaka, M. (2003). Distribution of cholecystokinin-immunoreactive cells in the digestive tract of the larval teleost, ayu, *Plecoglossus altivelis*. *Gen. Comp. Endocrinol.* **134,** 116–121.

Kamisaka, Y., Drivenes, O., Kurokawa, T., Tagawa, M., Ronnestad, I., Tanaka, M., and Helvik, J. V. (2005). Cholecystokinin mRNA in Atlantic herring, *Clupea harengus* – molecular cloning, characterization, and distribution in the digestive tract during the early life stages. *Peptides* **26,** 385–393.

Kawakoshi, A., Kaiya, H., Riley, L. G., Hirano, T., Grau, E. G., Miyazato, M., Hosoda, H., and Kangawa, K. (2007). Identification of a ghrelin-like peptide in two species of shark, *Sphyrna lewini* and *Carcharhinus melanopterus*. *Gen. Comp. Endocrinol.* **151,** 259–268.

Kehoe, A. S., and Volkoff, H. (2007). Cloning and characterization of neuropeptide Y (NPY) and cocaine and amphetamine regulated transcript (CART) in Atlantic cod (*Gadus morhua*). *Comp. Biochem. Physiol. A* **146,** 451–461.

Kehoe, A. S., and Volkoff, H. (2008). The effects of temperature on feeding and expression of two appetite-related factors, neuropeptide Y and cocaine- and amphetamine-regulated transcript, in Atlantic cod, *Gadus morhua*. *J. World Aquac. Soc.* **39,** 790–796.

Kelly, S. P., and Peter, R. E. (2006). Prolactin-releasing peptide, food intake, and hydromineral balance in goldfish. *Am. J. Physiol. Regul. Integr. Comp. Physiol.* **291,** R1474–1481.

Kim, E. J., Kim, C. H., Seo, J. K., Go, H. J., Lee, S., Takano, Y., Chung, J. K., Hong, Y. K., and Park, N. G. (2005). Structure-activity relationship of neuropeptide gamma derived from mammalian and fish. *J. Pept. Res.* **66,** 395–403.

Kinoshita, M., Morita, T., Toyohara, H., Hirata, T., Sakaguchi, M., Ono, M., Inoue, K., Wakamatsu, Y., and Ozato, K. (2001). Transgenic medaka overexpressing a melanin-concentrating hormone exhibit lightened body color but no remarkable abnormality. *Marine Biotechnology* **3,** 536–543.

Klein, S. E., and Sheridan, M. A. (2008). Somatostatin signaling and the regulation of growth and metabolism in fish. *Mol. Cell Endocrinol.* **286,** 148–154.

Klovins, J., and Schioth, H. B. (2005). Agouti-related proteins (AGRPs) and agouti-signaling peptide (ASIP) in fish and chicken. *Ann. N. Y. Acad. Sci.* **1040,** 363–367.

Kobayashi, Y., Peterson, B. C., and Waldbieser, G. C. (2008a). Association of cocaine- and amphetamine-regulated transcript (CART) messenger RNA level, food intake, and growth in channel catfish. *Comp. Biochem. Physiol. A* **151,** 219–225.

Kobayashi, Y., Tsuchiya, K., Yamanome, T., Schioth, H. B., Kawauchi, H., and Takahashi, A. (2008b). Food deprivation increases the expression of melanocortin-4 receptor in the liver of barfin flounder, *Verasper moseri*. *Gen. Comp. Endocrinol.* **155,** 280–287.

Korczynski, W., Ceregrzyn, M., Matyjek, R., Kato, I., Kuwahara, A., Wolinski, J., and Zabielski, R. (2006). Central and local (enteric) action of orexins. *J. Physiol. Pharmacol.* **57**(Suppl 6), 17–42.

Krueckl, S. L., Fradinger, E. A., and Sherwood, N. M. (2003). Developmental changes in the expression of growth hormone-releasing hormone and pituitary adenylate cyclase-activating polypeptide in zebrafish. *J. Comp. Neurol.* **455,** 396–405.

Kurokawa, T., Suzuki, T., and Hashimoto, H. (2003). Identification of gastrin and multiple cholecystokinin genes in teleost. *Peptides* **24,** 227–235.

Kurokawa, T., Uji, S., and Suzuki, T. (2005). Identification of cDNA coding for a homologue to mammalian leptin from pufferfish, *Takifugu rubripes*. *Peptides* **26,** 745–750.

Kurokawa, T., Murashita, K., Suzuki, T., and Uji, S. (2008). Genomic characterization and tissue distribution of leptin receptor and leptin receptor overlapping transcript genes in the pufferfish, *Takifugu rubripes*. *Gen. Comp. Endocrinol.* **158,** 108–114.

Kuz'mina, V. (2005). Regulation of the fish alimentary behavior: Role of humoral component. *J. Evol. Biochem. Physiol.* **41,** 282–295.

Kuz'mina, V., and Garina, D. (2001). Glucose, insulin, and adrenaline effects on some aspects of fish feeding behavior. *J. Evol. Biochem. Physiol.* **37,** 154–160.

Kuz'mina, V. V., Garina, D. V., and Gerasimov, Y. V. (2002). The role of glucose in regulation of feeding behavior of fish. *J. Ichthyol.* **42,** 210–215.

Lang, R., Gundlach, A. L., and Kofler, B. (2007). The galanin peptide family: Receptor pharmacology, pleiotropic biological actions, and implications in health and disease. *Pharmacol. Ther.* **115,** 177–207.

Leal, E., Sanchez, E., Muriach, B., and Cerdá-Reverter, J. M. (2009). Sex steroid-induced inhibition of food intake in sea bass (*Dicentrarchus labrax*). *J. Comp. Physiol. B* **179,** 77–86.

Liu, L., Conlon, J. M., Joss, J. M., and Burcher, E. (2002). Purification, characterization, and biological activity of a substance P-related peptide from the gut of the Australian lungfish, *Neoceratodus forsteri*. *Gen. Comp. Endocrinol.* **125,** 104–112.

Londraville, R. L., and Duvall, C. S. (2002). Murine leptin injections increase intracellular fatty acid-binding protein in green sunfish (*Lepomis cyanellus*). *Gen. Comp. Endocrinol.* **129,** 56–62.

Lopez-Olmeda, J. F., Madrid, J. A., and Sanchez-Vazquez, F. J. (2006). Melatonin effects on food intake and activity rhythms in two fish species with different activity patterns: Diurnal (goldfish) and nocturnal (tench). *Comp. Biochem. Physiol. A* **144,** 180–187.

Lopez-Patino, M. A., Guijarro, A. I., Isorna, E., Delgado, M. J., Alonso-Bedate, M., and De Pedro, N. (1999). Neuropeptide Y has a stimulatory action on feeding behavior in goldfish (*Carassius auratus*). *Eur. J. Pharmacol.* **377,** 147–153.

Lutz, T. A. (2006). Amylinergic control of food intake. *Physiol. Behav.* **89,** 465–471.

MacDonald, E., and Volkoff, H. (2009). Neuropeptide Y (NPY), cocaine- and amphetamine-regulated transcript (CART) and cholecystokinin (CCK) in winter skate (*Raja ocellata*): CDNA cloning, tissue distribution and mRNA expression responses to fasting. *Gen. Comp. Endocrinol.* **161,** 252–261.

Magalhaes, G. S., Junqueira-de-Azevedo, I. L., Lopes-Ferreira, M., Lorenzini, D. M., Ho, P. L., and Moura-da-Silva, A. M. (2006). Transcriptome analysis of expressed sequence tags from the venom glands of the fish *Thalassophryne nattereri*. *Biochimie* **88,** 693–699.

Mambrini, M., Medale, F., Sanchez, M. P., Recalde, B., Chevassus, B., Labbe, L., Quillet, E., and Boujard, T. (2004). Selection for growth in brown trout increases feed intake capacity without affecting maintenance and growth requirements. *J. Anim. Sci.* **82,** 2865–2875.

Mandiki, S. N. M., Babiak, I., Bopopi, J. M., Leprieur, F., and Kestemont, P. (2005). Effects of sex steroids and their inhibitors on endocrine parameters and gender growth differences in Eurasian perch (*Perca fluviatilis*) juveniles. *Steroids* **70,** 85–94.

Manning, A. J., Murray, H. M., Gallant, J. W., Matsuoka, M. P., Radford, E., and Douglas, S. E. (2008). Ontogenetic and tissue-specific expression of preproghrelin in the Atlantic halibut, *Hippoglossus hippoglossus* L. *J. Endocrinol.* **196,** 181–192.

Martinez-Alvarez, R. M., Volkoff, H., Cueto, J. A., and Delgado, M. J. (2008). Molecular characterization of calcitonin gene-related peptide (CGRP) related peptides (CGRP, amylin, adrenomedullin and adrenomedullin-2/intermedin) in goldfish (*Carassius auratus*): Cloning and distribution. *Peptides* **29,** 1534–1543.

Martinez-Alvarez, R. M., Volkoff, H., Munoz-Cueto, J. A., and Delgado, M. J. (2009). Effect of calcitonin gene-related peptide (CGRP), adrenomedullin and adrenomedullin-2/intermedin on food intake in goldfish (*Carassius auratus*). *Peptides* **30,** 803–807.

Maruyama, K., Konno, N., Ishiguro, K., Wakasugi, T., Uchiyama, M., Shioda, S., and Matsuda, K. (2008). Isolation and characterisation of four cDNAs encoding neuromedin U (NMU) from the brain and gut of goldfish, and the inhibitory effect of a deduced NMU on food intake and locomotor activity. *J. Neuroendocrinol.* **20,** 71–78.

Matsuda, K., and Maruyama, K. (2007). Regulation of feeding behavior by pituitary adenylate cyclase-activating polypeptide (PACAP) and vasoactive intestinal polypeptide (VIP) in vertebrates. *Peptides* **28,** 1761–1766.

Matsuda, K., Maruyama, K., Miura, T., Uchiyama, M., and Shioda, S. (2005a). Anorexigenic action of pituitary adenylate cyclase-activating polypeptide (PACAP) in the goldfish: feeding-induced changes in the expression of mRNAs for PACAP and its receptors in the brain, and locomotor response to central injection. *Neurosci. Lett.* **386,** 9–13.

Matsuda, K., Nagano, Y., Uchiyama, M., Takahashi, A., and Kawauchi, H. (2005b). Immunohistochemical observation of pituitary adenylate cyclase-activating polypeptide (PACAP) and adenohypophysial hormones in the pituitary of a teleost, *Uranoscopus japonicus*. *Zool. Sci.* **22,** 71–76.

Matsuda, K., Miura, T., Kaiya, H., Maruyama, K., Shimakura, S., Uchiyama, M., Kangawa, K., and Shioda, S. (2006a). Regulation of food intake by acyl and des-acyl ghrelins in the goldfish. *Peptides* **27,** 2321–2325.

Matsuda, K., Shimakura, S., Maruyama, K., Miura, T., Uchiyama, M., Kawauchi, H., Shioda, S., and Takahashi, A. (2006b). Central administration of melanin-concentrating hormone (MCH) suppresses food intake, but not locomotor activity, in the goldfish, *Carassius auratus*. *Neurosci. Lett.* **399,** 259–263.

Matsuda, K., Shimakura, S., Miura, T., Maruyama, K., Uchiyama, M., Kawauchi, H., Shioda, S., and Takahashi, A. (2007). Feeding-induced changes of melanin-concentrating hormone (MCH)-like immunoreactivity in goldfish brain. *Cell Tissue Res.* **328,** 375–382.

Matsuda, K., Kojima, K., Shimakura, S., Wada, K., Maruyama, K., Uchiyama, M., Kikuyama, S., and Shioda, S. (2008a). Corticotropin-releasing hormone mediates alpha-melanocyte-stimulating hormone-induced anorexigenic action in goldfish. *Peptides* **29,** 1930–1936.

Matsuda, K., Kojima, K., Shimakura, S. I., Miura, T., Uchiyama, M., Shioda, S., Ando, H., and Takahashi, A. (2008b). Relationship between melanin-concentrating hormone- and neuropeptide Y-containing neurons in the goldfish hypothalamus. *Comp. Biochem. Physiol. A. Mol. Integr. Physiol.* In press.

Matsuda, K., Nakamura, K., Shimakura, S., Miura, T., Kageyama, H., Uchiyama, M., Shioda, S., and Ando, H. (2008c). Inhibitory effect of chicken gonadotropin-releasing hormone II on food intake in the goldfish, *Carassius auratus*. *Horm. Behav.* **54,** 83–89.

Mazumdar, M., Lal, B., Sakharkar, A. J., Deshmukh, M., Singru, P. S., and Subhedar, N. (2006). Involvement of neuropeptide Y Y1 receptors in the regulation of LH and GH cells in the pituitary of the catfish, *Clarias batrachus*: An immunocytochemical study. *Gen. Comp. Endocrinol.* **149,** 190–196.

Mazumdar, M., Sakharkar, A. J., Singru, P. S., and Subhedar, N. (2007). Reproduction phase-related variations in neuropeptide Y immunoreactivity in the olfactory system, forebrain, and pituitary of the female catfish, *Clarias batrachus* (Linn.). *J. Comp. Neurol.* **504,** 450–469.

McCoy, J. G., and Avery, D. D. (1990). Bombesin: Potential integrative peptide for feeding and satiety. *Peptides* **11,** 595–607.

Mclean, E., Devlin, R. H., Byatt, J. C., Clarke, W. C., and Donaldson, E. M. (1997). Impact of a controlled release formulation of recombinant bovine growth hormone upon growth and seawater adaptation in coho (*Oncorhynchus kisutch*) and chinook (*Oncorhynchus tshawytscha*) salmon. *Aquaculture* **156,** 113–128.

McPartland, J. M., Glass, M., Matias, I., Norris, R. W., and Kilpatrick, C. W. (2007). A shifted repertoire of endocannabinoid genes in the zebrafish (*Danio rerio*). *Mol. Genet. Genomics* **277,** 555–570.

Metz, J. R., Peters, J. J., and Flik, G. (2006). Molecular biology and physiology of the melanocortin system in fish: A review. *Gen. Comp. Endocrinol.* **148,** 150–162.

Millington, G. W. (2007). The role of proopiomelanocortin (POMC) neurones in feeding behaviour. *Nutr. Metab. (Lond)* **4,** 18.

Millot, S., Begout, M. L., Ruyet, J. P. L., Breuil, G., Di-Poi, C., Fievet, J., Pineau, P., Roue, M., and Severe, A. (2008). Feed demand behavior in sea bass juveniles: Effects on individual specific growth rate variation and health (inter-individual and inter-group variation). *Aquaculture* **274,** 87–95.

Miura, T., Maruyama, K., Shimakura, S., Kaiya, H., Uchiyama, M., Kangawa, K., Shioda, S., and Matsuda, K. (2006). Neuropeptide Y mediates ghrelin-induced feeding in the goldfish, *Carassius auratus*. *Neurosci. Lett.* **407,** 279–283.

Miura, T., Maruyama, K., Shimakura, S., Kaiya, H., Uchiyama, M., Kangawa, K., Shioda, S., and Matsuda, K. (2007). Regulation of food intake in the goldfish by interaction between ghrelin and orexin. *Peptides* **28,** 1207–1213.

Miura, T., Maruyama, K., Kaiya, H., Miyazato, M., Kangawa, K., Uchiyama, M., Shioda, S., and Matsuda, K. (2008). Purification and properties of ghrelin from the intestine of the goldfish. *Carassius auratus*. *Peptides* **30,** 758–765.

Montserrat, N., Gabillard, J. C., Capilla, E., Navarro, M. I., and Gutierrez, J. (2007). Role of insulin, insulin-like growth factors, and muscle regulatory factors in the compensatory growth of the trout (*Oncorhynchus mykiss*). *Gen. Comp. Endocrinol.* **150,** 462–472.

Moriyama, S., Ito, T., Takahashi, A., Amano, M., Sower, S. A., Hirano, T., Yamamori, K., and Kawauchi, H. (2002). A homolog of mammalian PRL-releasing peptide (fish arginyl-phenylalanyl-amide peptide) is a major hypothalamic peptide of PRL release in teleost fish. *Endocrinology* **143,** 2071–2079.

Moriyama, S., Kasahara, M., Amiya, N., Takahashi, A., Amano, M., Sower, S. A., Yamamori, K., and Kawauchi, H. (2007). RFamide peptides inhibit the expression of melanotropin and growth hormone genes in the pituitary of an Agnathan, the sea lamprey, *Petromyzon marinus*. *Endocrinology* **148,** 3740–3749.

Murashita, K., Fukada, H., Hosokawa, H., and Masumoto, T. (2006). Cholecystokinin and peptide Y in yellowtail (*Seriola quinqueradiata*): Molecular cloning, real-time quantitative RT-PCR, and response to feeding and fasting. *Gen. Comp. Endocrinol.* **145,** 287–297.

Murashita, K., Fukada, H., Hosokawa, H., and Masumoto, T. (2007). Changes in cholecystokinin and peptide Y gene expression with feeding in yellowtail (*Seriola quinqueradiata*): Relation to pancreatic exocrine regulation. *Comp. Biochem. Physiol. B* **146,** 318–325.

Murashita, K., Uji, S., Yamamoto, T., Ronnestad, I., and Kurokawa, T. (2008). Production of recombinant leptin and its effects on food intake in rainbow trout (*Oncorhynchus mykiss*). *Comp. Biochem. Physiol. B* **150,** 377–384.

Muruzabal, F. J., Fruhbeck, G., Gomez-Ambrosi, J., Archanco, M., and Burrell, M. A. (2002). Immunocytochemical detection of leptin in non-mammalian vertebrate stomach. *Gen. Comp. Endocrinol.* **128,** 149–152.

Mustonen, A. M., Nieminen, P., and Hyvarinen, H. (2002). Leptin, ghrelin, and energy metabolism of the spawning burbot (*Lota lota*, L.). *J. Exp. Zool.* **293,** 119–126.

Nagasaka, R., Okamoto, N., and Ushio, H. (2006). Increased leptin may be involved in the short life span of ayu (*Plecoglossus altivelis*). *J. Exp. Zool.* **305,** 507–512.

Nakamachi, T., Matsuda, K., Maruyama, K., Miura, T., Uchiyama, M., Funahashi, H., Sakurai, T., and Shioda, S. (2006). Regulation by orexin of feeding behaviour and locomotor activity in the goldfish. *J. Neuroendocrinol.* **18,** 290–297.

Nakata, M., and Yada, T. (2007). PACAP in the glucose and energy homeostasis: Physiological role and therapeutic potential. *Curr. Pharm. Des.* **13,** 1105–1112.

Narnaware, Y. K., and Peter, R. E. (2001). Effects of food deprivation and refeeding on neuropeptide Y (NPY) mRNA levels in goldfish. *Comp. Biochem. Physiol. B* **129,** 633–637.

Narnaware, Y. K., and Peter, R. E. (2002). Influence of diet composition on food intake and neuropeptide Y (NPY) gene expression in goldfish brain. *Regul. Pept.* **103,** 75–83.

Narnaware, Y. K., Peyon, P. P., Lin, X., and Peter, R. E. (2000). Regulation of food intake by neuropeptide Y in goldfish. *Am. J. Physiol. Regul. Integr. Comp. Physiol.* **279,** R1025–1034.

Navarro, I., Leibush, B., Moon, T., Plisetskaya, E., Banos, N., Mendez, E., Planas, J., and Gutierrez, J. (1999). Insulin, insulin-like growth factor-I (IGF-I) and glucagon: The evolution of their receptors. *Com. Biochem. Physiol. B* **122,** 137–153.

Navarro, M. H., Lozano, M. T., and Agulleiro, B. (2006). Ontogeny of the endocrine pancreatic cells of the gilthead sea bream, *Sparus aurata* (Teleost). *Gen. Comp. Endocrinol.* **148,** 213–226.

Nelson, L. E., and Sheridan, M. A. (2006). Gastroenteropancreatic hormones and metabolism in fish. *Gen. Comp. Endocrinol.* **148,** 116–124.

Nieminen, P., Mustonen, A. M., and Hyvarinen, H. (2003). Fasting reduces plasma leptin-and ghrelin-immunoreactive peptide concentrations of the burbot (*Lota lota*) at 2 degrees C but not at 10 degrees C. *Zoolog. Sci.* **20,** 1109–1115.

Nishino, S. (2007). The hypothalamic peptidergic system, hypocretin/orexin and vigilance control. *Neuropeptides* **41,** 117–133.

Noble, C., Mizusawa, K., and Tabata, M. (2005). Does light intensity affect self-feeding and food wastage in group-held rainbow trout and white-spotted charr? *J. Fish Biol* **66,** 1387–1399.

Noble, C., Kadri, S., Mitchell, D. F., and Huntingford, F. A. (2007). The impact of environmental variables on the feeding rhythms and daily feed intake of cage-held 1+ Atlantic salmon parr (*Salmo salar* L.). *Aquaculture* **269,** 290–298.

Nordgarden, U., Bjornsson, B. T., and Hansen, T. (2007). Developmental stage of Atlantic salmon parr regulates pituitary GH secretion and parr-smolt transformation. *Aquaculture* **264,** 441–448.

Novak, C. M., Jiang, X., Wang, C., Teske, J. A., Kotz, C. M., and Levine, J. A. (2005). Caloric restriction and physical activity in zebrafish (*Danio rerio*). *Neurosci. Lett.* **383,** 99–104.

Ohno, K., and Sakurai, T. (2007). Orexin neuronal circuitry: Role in the regulation of sleep and wakefulness. *Front. Neuroendocrinol.* **29,** 70–87.

Oliver, A. S., and Vigna, S. R. (1996). CCK-X receptors in the endothermic mako shark (*Isurus oxyrinchus*). *Gen. Comp. Endocrinol.* **102,** 61–73.

Olsson, C., Holbrook, J. D., Bompadre, G., Jonsson, E., Hoyle, C. H., Sanger, G. J., Holmgren, S., and Andrews, P. L. (2008). Identification of genes for the ghrelin and motilin receptors and a novel related gene in fish, and stimulation of intestinal motility in zebrafish (*Danio rerio*) by ghrelin and motilin. *Gen. Comp. Endocrinol.* **155,** 217–226.

Olszewski, P. K., Schioth, H. B., and Levine, A. S. (2008). Ghrelin in the CNS: From hunger to a rewarding and memorable meal? *Brain Res. Rev.* **58,** 160–170.

Parhar, I. S., Sato, H., and Sakuma, Y. (2003). Ghrelin gene in cichlid fish is modulated by sex and development. *Biochem. Biophys. Res. Commun.* **305,** 169–175.

Patel, Y. C. (1999). Somatostatin and its receptor family. *Front. Neuroendocrinol.* **20,** 157–198.

Pauls, S., Zecchin, E., Tiso, N., Bortolussi, M., and Argenton, F. (2007). Function and regulation of zebrafish nkx2.2a during development of pancreatic islet and ducts. *Develop. Biol.* **304,** 875–890.

Peddu, S. C., Breves, J. P., Kaiya, H., Gordon Grau, E., and Riley, L. G., Jr. (2009). Pre- and postprandial effects on ghrelin signaling in the brain and on the GH/IGF-I axis in the mozambique tilapia (*Oreochromis mossambicus*). *Gen. Comp. Endocrinol.* **161,** 412–418.

Pedroso, F. L., de Jesus-Ayson, E. G. T., Cortado, H. H., Hyodo, S., and Ayson, F. G. (2006). Changes in mRNA expression of grouper (*Epinephelus coioides*) growth hormone and insulin-like growth factor I in response to nutritional status. *Gen. Comp. Endocrinol.* **145,** 237–246.

Pepels, P. P. L. M., Bonga, S. E. W., and Balm, P. H. M. (2004). Bacterial lipopolysaccharide (LPS) modulates corticotropin-releasing hormone (CRH) content and release in the brain of juvenile and adult tilapia (*Oreochromis mossambicus*; Teleostei). *J. Exp. Biol.* **207,** 4479–4488.

Peter, R.E (1979). The brain and feeding behavior. *In* " Fish Physiology," (Hoar, W. S., Randall, D. J., and Brett, J. R., Eds.), Vol. VIII, pp. 121–159. Academic Press, New York, NY.

Peterson, B. C., and Small, B. C. (2005). Effects of exogenous cortisol on the GH/IGF-I/IGFBP network in channel catfish. *Domest. Anim. Endocrinol.* **28,** 391–404.

Peyon, P., Lin, X. W., Himick, B. A., and Peter, R. E. (1998). Molecular cloning and expression of cDNA encoding brain preprocholecystokinin in goldfish. *Peptides* **19,** 199–210.

Peyon, P., Saied, H., Lin, X., and Peter, R. E. (1999). Postprandial, seasonal and sexual variations in cholecystokinin gene expression in goldfish brain. *Brain Res. Mol. Brain Res.* **74,** 190–196.

Peyon, P., Saied, H., Lin, X., and Peter, R. E. (2000). Preprotachykinin gene expression in goldfish brain: Sexual, seasonal, and postprandial variations. *Peptides* **21,** 225–231.

Pinillos, M. L., De Pedro, N., Alonso-Gomez, A. L., Alonso-Bedate, M., and Delgado, M. J. (2001). Food intake inhibition by melatonin in goldfish (*Carassius auratus*). *Physiol. Behav.* **72,** 629–634.

Pinuela, C., and Northcutt, R. G. (2007). Immunohistochemical organization of the forebrain in the white sturgeon, *Acipenser transmontanus. Brain Behav. Evol.* **69,** 229–253.

Pissios, P., Bradley, R. L., and Maratos-Flier, E. (2006). Expanding the scales: The multiple roles of MCH in regulating energy balance and other biological functions. *Endocr. Rev.* **27,** 606–620.

Prober, D. A., Rihel, J., Onah, A. A., Sung, R. J., and Schier, A. F. (2006). Hypocretin/orexin overexpression induces an insomnia-like phenotype in zebrafish. *J. Neurosci.* **26,** 13400–13410.

Quillet, E., Le Guillou, S., Aubin, J., Labbe, L., Fauconneau, B., and Medale, F. (2007). Response of a lean muscle and a fat muscle rainbow trout (*Oncorhynchus mykiss*) line on growth, nutrient utilization, body composition and carcass traits when fed two different diets. *Aquaculture* **269,** 220–231.

Rao, P. D., Murthy, C. K., Cook, H., and Peter, R. E. (1996). Sexual dimorphism of galanin-like immunoreactivity in the brain and pituitary of goldfish, *Carassius auratus. J. Chem. Neuroanat.* **10,** 119–135.

Raven, P. A., Devlin, R. H., and Higgs, D. A. (2006). Influence of dietary digestible energy content on growth, protein and energy utilization and body composition of growth hormone transgenic and non-transgenic coho salmon (*Oncorhynchus kisutch*). *Aquaculture* **254,** 730–747.

Raybould, H. E. (2007). Mechanisms of CCK signaling from gut to brain. *Curr. Opin. Pharmacol.* **7,** 570–574.

Reddy, P. K., and Leatherland, J. F. (2003). Influences of photoperiod and alternate days of feeding on plasma growth hormone and thyroid hormone levels in juvenile rainbow trout. *J. Fish Biol.* **63,** 197–212.

Riley, L. G., Fox, B. K., Kaiya, H., Hirano, T., and Grau, E. G. (2005). Long-term treatment of ghrelin stimulates feeding, fat deposition, and alters the GH/IGF-I axis in the tilapia, *Oreochromis mossambicus. Gen. Comp. Endocrinol.* **142,** 234–240.

Ripley, J. L., and Foran, C. M. (2007). Influence of estuarine hypoxia on feeding and sound production by two sympatric pipefish species (Syngnathidae). *Mar. Environ. Res.* **63,** 350–367.

Rodriguez-Gomez, F. J., Rendon-Unceta, C., Sarasquete, C., and Munoz-Cueto, J. A. (2001). Distribution of neuropeptide Y-like immunoreactivity in the brain of the Senegalese sole (*Solea senegalensis*). *Anat. Rec.* **262,** 227–237.

Ronnestad, I., Kamisaka, Y., Conceicao, L. E. C., Morais, S., and Tonheim, S. K. (2007). Digestive physiology of marine fish larvae: Hormonal control and processing capacity for proteins, peptides and amino acids. *Aquaculture* **268,** 82–97.

Rousseau, K., and Dufour, S. (2007). Comparative aspects of GH and metabolic regulation in lower vertebrates. *Neuroendocrinol.* **86,** 165–174.

Rubio, V. C., Sanchez-Vazquez, F. J., and Madrid, J. A. (2004). Oral administration of melatonin reduces food intake and modifies macronutrient selection in European sea bass (*Dicentrarchus labrax*, L.). *J. Pineal Res.* **37,** 42–47.

Russell, N. R., Fish, J. D., and Wootton, R. J. (1996). Feeding and growth of juvenile sea bass: The effect of ration and temperature on growth rate and efficiency. *J. Fish Biol.* **49,** 206–220.

Sakharkar, A. J., Singru, P. S., Sarkar, K., and Subhedar, N. K. (2005). Neuropeptide Y in the forebrain of the adult male cichlid fish *Oreochromis mossambicus*: Distribution, effects of castration and testosterone replacement. *J. Comp. Neurol.* **489,** 148–165.

Salaneck, E., Larsson, T. A., Larson, E. T., and Larhammar, D. (2008). Birth and death of neuropeptide Y receptor genes in relation to the teleost fish tetraploidization. *Gene* **409,** 61–71.

Saper, C. B. (2006). Staying awake for dinner: Hypothalamic integration of sleep, feeding, and circadian rhythms. *Hypoth. Integ. Energy Metabol.* **153,** 243–252.

Sawada, K., Ukena, K., Satake, H., Iwakoshi, E., Minakata, H., and Tsutsui, K. (2002). Novel fish hypothalamic neuropeptide. *Eur. J. Biochem.* **269,** 6000–6008.

Schioth, H. B., Haitina, T., Fridmanis, D., and Klovins, J. (2005). Unusual genomic structure: Melanocortin receptors in Fugu. *Ann. N. Y. Acad. Sci.* **1040,** 460–463.

Seale, A. P., Itoh, T., Moriyama, S., Takahashi, A., Kawauchi, H., Sakamoto, T., Fujimoto, M., Riley, L. G., Hirano, T., and Grau, E. G. (2002). Isolation and characterization of a homologue of mammalian prolactin-releasing peptide from the tilapia brain and its effect on prolactin release from the tilapia pituitary. *Gen. Comp. Endocrinol.* **125,** 328–339.

Selz, Y., Braasch, I., Hoffmann, C., Schmidt, C., Schultheis, C., Schartl, M., and Volff, J. N. (2007). Evolution of melanocortin receptors in teleost fish: The melanocortin type 1 receptor. *Gene* **401,** 114–122.

Shearer, K. D., Silverstein, J. T., and Plisetskaya, E. M. (1997). Role of adiposity in food intake control of juvenile chinook salmon (*Oncorhynchus tshawytscha*). *Comp. Biochem. Physiol. A* **118,** 1209–1215.

Shepherd, B. S., Johnson, J. K., Silverstein, J. T., Parhar, I. S., Vijayan, M. M., McGuire, A., and Weber, G. M. (2007). Endocrine and orexigenic actions of growth hormone secretagogues in rainbow trout (*Oncorhynchus mykiss*). *Comp. Biochem. Physiol. A* **146,** 390–399.

Sheridan, M. A., and Kittilson, J. D. (2004). The role of somatostatins in the regulation of metabolism in fish. *Comp. Biochem. Physiol. B* **138,** 323–330.

Shimakura, S., Kojima, K., Nakamachi, T., Kageyama, H., Uchiyama, M., Shioda, S., Takahashi, A., and Matsuda, K. (2008). Neuronal interaction between melanin-concentrating hormone- and alpha-melanocyte-stimulating hormone-containing neurons in the goldfish hypothalamus. *Peptides* **29,** 1432–1440.

Silverstein, J. T. (2002). Using genetic variation to understand control of feed intake in fish. *Fish Physiol. Biochem.* **27,** 173–178.

Silverstein, J. T., and Plisetskaya, E. M. (2000). The effects of NPY and insulin on food intake regulation in fish. *Amer. Zool.* **40,** 296–308.

Silverstein, J. T., Shearer, K. D., Dickhoff, W. W., and Plisetskaya, E. M. (1999a). Regulation of nutrient intake and energy balance in salmon. *Aquaculture* **177,** 161–169.

Silverstein, J. T., Wolters, W. R., and Holland, M. (1999b). Evidence of differences in growth and food intake regulation in different genetic strains of channel catfish. *J. Fish Biol.* **54,** 607–615.

Silverstein, J. T., Bondareva, V. M., Leonard, J. B., and Plisetskaya, E. M. (2001). Neuropeptide regulation of feeding in catfish, *Ictalurus punctatus*: A role for glucagon-like peptide-1 (GLP-1)? *Comp. Biochem. Physiol. B* **129,** 623–631.

Singru, P. S., Mazumdar, M., Sakharkar, A. J., Lechan, R. M., Thim, L., Clausen, J. T., and Subhedar, N. K. (2007). Immunohistochemical localization of cocaine- and amphetamine-regulated transcript peptide in the brain of the catfish, *Clarias batrachus* (Linn.). *J. Comp. Neurol.* **502,** 215–235.

Singru, P. S., Mazumdar, M., Barsagade, V., Lechan, R. M., Thim, L., Clausen, J. T., and Subhedar, N. (2008). Association of cocaine- and amphetamine-regulated transcript and neuropeptide Y in the forebrain and pituitary of the catfish, *Clarias batrachus*: A double immunofluorescent labeling study. *J. Chem. Neuroanat.* **36,** 239–250.

Small, B. C. (2005). Effect of fasting on nychthemeral concentrations of plasma growth hormone (GH), insulin-like growth factor I (IGF-1), and cortisol in channel catfish (*Ictalurus punctatus*). *Comp. Biochem. Physiol. B* **142,** 217–223.

Soengas, J., and Aldegunde, M. (2004). Brain glucose and insulin: Effects on food intake and brain biogenic amines of rainbow trout. *J. Comp. Physiol. A* **190,** 641–649.

Song, Y., and Cone, R. D. (2007). Creation of a genetic model of obesity in a teleost. *Faseb J.* **21,** 2042–2049.

Song, Y., Golling, G., Thacker, T. L., and Cone, R. D. (2003). Agouti-related protein (AGRP) is conserved and regulated by metabolic state in the zebrafish, *Danio rerio*. *Endocrine* **22,** 257–265.

Stevens, E. D., and Devlin, R. H. (2005). Gut size in GH-transgenic coho salmon is enhanced by both the GH transgene and increased food intake. *J. Fish Biol.* **66,** 1633–1648.

Sueiro, C., Carrera, I., Ferreiro, S., Molist, P., Adrio, F., Anadon, R., and Rodriguez-Moldes, I. (2007). New insights on Saccus vasculosus evolution: A developmental and immunohistochemical study in elasmobranchs. *Brain Behav. Evol.* **70,** 187–204.

Sundstrom, G., Larsson, T. A., Brenner, S., Venkatesh, B., and Larhammar, D. (2005). Ray-fin fish tetraploidization gave rise to pufferfish duplicates of NPY and PYY, but zebrafish NPY duplicate was lost. *Ann. N. Y. Acad. Sci.* **1040,** 476–478.

Sunuma, T., Amano, M., Yamanome, T., Furukawa, K., and Yamamori, K. (2007). Self-feeding activity of a pleuronectiform fish, the barfin flounder. *Aquaculture* **270,** 566–569.

Suzuki, H., Miyoshi, Y., and Yamamoto, T. (2007). Orexin-A (hypocretin 1)-like immunoreactivity in growth hormone-containing cells of the Japanese seaperch (*Lateolabrax japonicus*) pituitary. *Gen. Comp. Endocrinol.* **150,** 205–211.

Sze, K. H., Zhou, H., Yang, Y., He, M., Jiang, Y., and Wong, A. O. (2007). Pituitary adenylate cyclase-activating polypeptide (PACAP) as a growth hormone (GH)-releasing factor in grass carp: II. Solution structure of a brain-specific PACAP by nuclear magnetic resonance spectroscopy and functional studies on GH release and gene expression. *Endocrinology* **148,** 5042–5059.

Tachibana, T., Takagi, T., Tomonaga, S., Ohgushi, A., Ando, R., Denbow, D. M., and Furuse, M. (2003). Central administration of cocaine- and amphetamine-regulated transcript inhibits food intake in chicks. *Neurosci. Lett.* **337,** 131–134.

Takahashi, A., Tsuchiya, K., Yamanome, T., Amano, M., Yasuda, A., Yamamori, K., and Kawauchi, H. (2004). Possible involvement of melanin-concentrating hormone in food intake in a teleost fish, barfin flounder. *Peptides* **25,** 1613–1622.

Terova, G., Rimoldi, S., Chini, V., Gornati, R., Bernardini, G., and Saroglia, M. (2007). Cloning and expression analysis of insulin-like growth factor I and II in liver and muscle of sea bass (*Dicentrarchus labrax*, L.) during long-term fasting and refeeding. *J. Fish Biol.* **70,** 219–233.

Terova, G., Rimoldi, S., Bernardini, G., Gornati, R., and Saroglia, M. (2008). Sea bass ghrelin: Molecular cloning and mRNA quantification during fasting and refeeding. *Gen. Comp. Endocrinol.* **155,** 341–351.

Thavanathan, R., and Volkoff, H. (2006). Effects of amylin on feeding of goldfish: Interactions with CCK. *Regul. Pept.* **133,** 90–96.

Thorndyke, M. C., Reeve, J. R., Jr., and Vigna, S. R. (1990). Biological activity of a bombesin-like peptide extracted from the intestine of the ratfish, *Hydrolagus colliei*. *Comp. Biochem. Physiol. C* **96,** 135–140.

Todd, A. S., McKnight, D. M., Jaros, C. L., and Marchitto, T. M. (2007). Effects of acid rock drainage on stocked rainbow trout (*Oncorhynchus mykiss*): An in-situ, caged fish experiment. *Environmental Monitoring and Assessment* **130,** 111–127.

Tucker, B. J., Booth, M. A., Allan, G. L., Booth, D., and Fielder, D. S. (2006). Effects of photoperiod and feeding frequency on performance of newly weaned Australian snapper *Pagrus auratus*. *Aquaculture* **258,** 514–520.

Unniappan, S., and Peter, R. E. (2005). Structure, distribution and physiological functions of ghrelin in fish. *Comp. Biochem. Physiol. A* **140,** 396–408.

Unniappan, S., Lin, X., and Peter, R. (2003). Characterization of complementary deoxyribonucleic acids encoding preprogalanin and its alternative splice variants in the goldfish. *Mol. Cell Endocrinol.* **200,** 177–187.

Unniappan, S., Canosa, L. F., and Peter, R. E. (2004a). Orexigenic actions of ghrelin in goldfish: Feeding-induced changes in brain and gut mRNA expression and serum levels, and responses to central and peripheral injections. *Neuroendocrinology* **79,** 100–108.

Unniappan, S., Cerdá-Reverter, J. M., and Peter, R. E. (2004b). In situ localization of prepro-galanin mRNA in the goldfish brain and changes in its expression during feeding and starvation. *Gen. Comp. Endocrinol.* **136,** 200–207.

Valassi, E., Scacchi, M., and Cavagnini, F. (2008). Neuroendocrine control of food intake. *Nutr. Metab. Cardiovasc. Dis.* **18,** 158–168.

Valenti, M., Cottone, E., Martinez, R., De Pedro, N., Rubio, M., Viveros, M. P., Franzoni, M. F., Delgado, M. J., and Di Marzo, V. (2005). The endocannabinoid system in the brain of *Carassius auratus* and its possible role in the control of food intake. *J. Neurochem.* **95,** 662–672.

van Ginneken, V. J. T., and Maes, G. E. (2005). The european eel (*Anguilla anguilla*, Linnaeus), its lifecycle, evolution and reproduction: A literature review. *Rev. Fish. Biol. Fisheries* **15,** 367–398.

Vegusdal, A., Sundvold, H., Gjoen, T., and Ruyter, B. (2003). An *in vitro* method for studying the proliferation and differentiation of Atlantic salmon preadipocytes. *Lipids* **38,** 289–296.

Very, N. M., Knutson, D., Kittilson, J. D., and Sheridan, M. A. (2001). Somatostatin inhibits growth of rainbow trout. *J. Fish Biol.* **59,** 157–165.

Vicentic, A., and Jones, D. C. (2007). The CART (cocaine- and amphetamine-regulated transcript) system in appetite and drug addiction. *J. Pharmacol. Exp. Ther.* **320,** 499–506.

Volff, J. N. (2004). Genome evolution and biodiversity in teleost fish. *Heredity* **94,** 280–294.

Volkoff, H., and Peter, R. (2000). Effects of CART peptides on food consumption, feeding and associated behaviors in the goldfish, *Carassius auratus*: Actions on neuropeptide Y- and orexin A-induced feeding. *Brain Res.* **887,** 125–133.

Volkoff, H., and Peter, R. (2001a). Characterization of two forms of cocaine- and amphetamine-regulated transcript (CART) peptide precursors in goldfish: Molecular cloning and distribution, modulation of expression by nutritional status, and interactions with leptin. *Endocrinology* **142,** 5076–5088.

Volkoff, H., and Peter, R. E. (2001b). Interactions between orexin A, NPY and galanin in the control of food intake of the goldfish, *Carassius auratus*. *Regul. Pept.* **101,** 59–72.

Volkoff, H., and Peter, R. E. (2004). Effects of lipopolysaccharide treatment on feeding of goldfish: Role of appetite-regulating peptides. *Brain Res.* **998,** 139–147.

Volkoff, H., Bjorklund, J. M., and Peter, R. E. (1999b). Stimulation of feeding behavior and food consumption in the goldfish, *Carassius auratus*, by orexin-A and orexin-B. *Brain Res.* **846,** 204–209.

Volkoff, H., Peyon, P., Lin, X., and Peter, R. (2000). Molecular cloning and expression of cDNA encoding a brain bombesin/gastrin-releasing peptide-like peptide in goldfish. *Peptides* **21,** 639–648.

Volkoff, H., Eykelbosh, A. J., and Peter, R. E. (2003). Role of leptin in the control of feeding of goldfish *Carassius auratus*: Interactions with cholecystokinin, neuropeptide Y and orexin A, and modulation by fasting. *Brain Res.* **972,** 90–109.

Volkoff, H., Canosa, L. F., Unniappan, S., Cerdá-Reverter, J. M., Bernier, N. J., Kelly, S. P., and Peter, R. E. (2005). Neuropeptides and the control of food intake in fish. *Gen. Comp. Endocrinol.* **142,** 3–19.

Volkoff, H., Hamoutene, D., and Payne, J. F. (2007). Potential effects of tebufenozide on feeding and metabolism of lake trout (*Salvelinus namaycush*). *Can. Tech. Rep. Fish. Aquat. Sci.* **2777,** iv + 19.

Walton, K. M., Chin, J. E., Duplantier, A. J., and Mather, R. J. (2006). Galanin function in the central nervous system. *Curr. Opin. Drug Discov. Devel.* **9,** 560–570.

Westermark, G. T., Falkmer, S., Steiner, D. F., Chan, S. J., Engstrom, U., and Westermark, P. (2002). Islet amyloid polypeptide is expressed in the pancreatic islet parenchyma of the teleostean fish, *Myoxocephalus (cottus) scorpius*. *Comp. Biochem. Physiol. B* **133,** 119–125.

Westring, C. G., Ando, H., Kitahashi, T., Bhandari, R. K., Ueda, H., Urano, A., Dores, R. M., Sher, A. A., and Danielson, P. B. (2008). Seasonal changes in CRF-I and urotensin I transcript levels in masu salmon: Correlation with cortisol secretion during spawning. *Gen. Comp. Endocrinol.* **155,** 126–140.

Wong, A. O., Li, W., Leung, C. Y., Huo, L., and Zhou, H. (2005). Pituitary adenylate cyclase-activating polypeptide (PACAP) as a growth hormone (GH)-releasing factor in grass carp. I. Functional coupling of cyclic adenosine 3′,5′-monophosphate and Ca2+/calmodulin-dependent signaling pathways in PACAP-induced GH secretion and GH gene expression in grass carp pituitary cells. *Endocrinology* **146,** 5407–5424.

Wong, M. M., Yu, R. M., Ng, P. K., Law, S. H., Tsang, A. K., and Kong, R. Y. (2007). Characterization of a hypoxia-responsive leptin receptor (omLepR(L)) cDNA from the marine medaka (*Oryzias melastigma*). *Mar. Pollut. Bull.* **54,** 797–803.

Woo, N. Y. S., Chung, A. S. B., and Ng, T. B. (1993). Influence of oral-administration of estradiol-17-beta and testosterone on growth, digestion, food conversion and metabolism in the underyearling red-sea bream, *Chrysophrys major*. *Fish Physiol. Biochem.* **10,** 377–387.

Xing, Y., Wensheng, L., and Haoran, L. (2005). Polygenic expression of somatostatin in orange-spotted grouper (*Epinephelus coioides*): Molecular cloning and distribution of the mRNAs encoding three somatostatin precursors. *Mol. Cell Endocrinol.* **241,** 62–72.

Xu, M., and Volkoff, H. (2007). Molecular characterization of prepro-orexin in Atlantic cod (*Gadus morhua*): Cloning, localization, developmental profile and role in food intake regulation. *Mol. Cell Endocrinol.* **271,** 28–37.

Xu, M., and Volkoff, H. (2008). Cloning, tissue distribution and effects of food deprivation on pituitary adenylate cyclase activating polypeptide (PACAP)/PACAP-related peptide (PRP) and preprosomatostatin 1 (PPSS 1) in Atlantic cod (*Gadus morhua*). *Peptides* **30,** 766–776.

Xu, M., and Volkoff, H. (2009). Molecular characterization of ghrelin and gastrin-releasing peptide in Atlantic cod (*Gadus morhua*): Cloning, localization, developmental profile and role in food intake regulation. *Gen. Comp. Endocrinol.* **160,** 250–258.

Yada, T., Moriyama, S., Suzuki, Y., Azuma, T., Takahashi, A., Hirose, S., and Naito, N. (2002). Relationships between obesity and metabolic hormones in the "cobalt" variant of rainbow trout. *Gen. Comp. Endocrinol.* **128,** 36–43.

Yaghoubian, S., Filosa, M. F., and Youson, J. H. (2001). Proteins immunoreactive with antibody against a human leptin fragment are found in serum and tissues of the sea lamprey, *Petromyzon marinus* L. *Comp. Biochem. Physiol. B* **129,** 777–785.

Yokogawa, T., Marin, W., Faraco, J., Pezeron, G., Appelbaum, L., Zhang, J., Rosa, F., Mourrain, P., and Mignot, E. (2007). Characterization of sleep in zebrafish and insomnia in hypocretin receptor mutants. *PLoS Biol* **5,** e277.

10

THE NEURONAL AND ENDOCRINE REGULATION OF GUT FUNCTION

SUSANNE HOLMGREN
CATHARINA OLSSON

1. Introduction
2. Anatomy of the Gut Neuronal and Endocrine Systems
   2.1. Gut Innervation
   2.2. The Gut as an Endocrine Organ
3. Neurotransmitters and Hormones of the Gut
   3.1. Acetylcholine
   3.2. Amines, Amino Acids and Purines
   3.3. Nitric Oxide
   3.4. Peptides
4. Development of Gut Innervation and Neuroendocrine System
5. Control of Gut Motility
   5.1. Gut Muscles and Spontaneous Activity
   5.2. Effects of Individual Signal Substances
   5.3. Control of Propagating Activity
   5.4. Control of Non-propagating Activity
   5.5. Control of Gallbladder Motility
   5.6. Development of the Control of Gut Motility
6. Control of Secretion and Digestion
   6.1. Gastric Acid Secretion
   6.2. Pepsinogen/Pepsin Secretion
   6.3. Pancreatic Secretion
   6.4. Secretion from the Rectal Gland of Elasmobranchs
7. Control of Nutrient Absorption
8. Control of Water and Ion Transport
9. Control of Splanchnic Circulation
   9.1. Reduction in Flow – Catecholamines
   9.2. Increase in Flow
   9.3. Variable Effects
10. Concluding Remarks

Fish Neuroendo crinology: Volume 28
FISH PHYSIOLOGY

DOI: 10.1016/S1546-5098(09)28010-1

The chapter summarizes current knowledge on the neuronal and endocrine control of gut functions such as motility, secretion and absorption in the fish gut. Most knowledge is on elasmobranch and teleost species, but what little is known from the other groups is included when relevant. The anatomy of the gut innervation and the endocrine system of the gut is outlined. Many studies have concerned the identity and distribution of neurotransmitters and gut hormones, and this is summarized in one section. Most functional studies concern different aspects of motility control, and this comprises a major part of the chapter, but the more scarce knowledge on control of gut circulation, secretion, water and ion transport, and absorptive processes are also included. Recent studies dealing with the development of the fish gut nervous and endocrine control systems are reported. Comparisons are made between fish species and groups, and, when relevant, with vertebrates in general.

1. INTRODUCTION

This chapter will focus on the neuronal and endocrine regulation of various processes involved in digestion and absorption of food. Other functions of the gut, such as the control of the local immune defense, are outside the scope of this text. Secretory processes, motility patterns and blood flow through the gut are central mechanisms for food processing, controlled mainly by nerves and endocrine cells acting in elaborated reflex pathways together with muscle cells, interstitial cells of Cajal (ICCs) and secretory cells. Reflexes are either enteric, i.e., local to the gut, or may involve extrinsic pathways. Both types result in an integrated response to a certain stimulus.

In many contexts, the neuronal and endocrine systems is an interface only between a primary stimulus and adjustments of gut functions. Food itself is the most important primary stimulus. Studies in mammals show that food composition and properties modulate gut activity to optimize digestion and absorption (e.g., Schneeman, 2002). Sensory stimuli, and even anticipation of food, act on the gut via the central nervous system (CNS) and the vagus nerve. Local reflexes and hormone release may be triggered by distension of the gut wall, by components in the food, or by a changed pH in the lumen.

2. ANATOMY OF THE GUT NEURONAL AND ENDOCRINE SYSTEMS

The gut neuronal and endocrine systems in fish follows the general vertebrate plan, but some species deviate from this plan, showing anatomical specialities. For example, cyclostomes, holocephalans and some teleosts lack a stomach, i.e., acid secretion is lacking and the entrance of the bile duct is found in the gut region immediately distal to the esophagus. The size and number of pyloric ceca, when present, varies between species. Elasmobranchs, holocephalans and chondrosteans often have a spiral intestine, i.e., a short, wide middle intestine with elaborate spiral foldings of mucosa (Stevens and Hume, 1995). The fish pancreas is often diffuse, with large species variations in structure (Section 2.2). All this is reflected in the distribution of enteric nerves and endocrine cells.

2.1. Gut Innervation

The gastrointestinal (GI) tract is a densely innervated organ. Gut smooth muscle, blood vessels, endocrine cells, etc., are all controlled by autonomic nerves. Many of these nerves are located within the gut wall, forming the enteric nervous system (ENS). The gut also receives an extrinsic cranial and spinal autonomic innervation, via the vagus and the splanchnic nerves, respectively. In some stomach-less species, e.g., zebrafish, the vagus innervates the entire gut from the esophagus (see, e.g., Olsson *et al.*, 2008b). In most other species, the vagal innervation includes part of the esophagus, the stomach and the proximal intestine, whereas in some species vagal fibers only reach the esophagus (Burnstock, 1969; Nilsson, 1983). Both sensory and motor fibers run in the vagi but few studies have considered the proportion of sensory versus motor fibers in fish. Tracing studies in the Atlantic cod, *Gadus morhua,* indicate peripheral sensory nerve endings in the gut wall (Karila, 1997).

The spinal innervation in cyclostomes mainly involves the hindgut. In elasmobranchs, a pair of anterior splanchnic nerves may innervate the posterior stomach and anterior intestine. Mid and posterior splanchnic nerves innervate the spiral intestine and rectum. In teleosts, an anterior splanchnic nerve innervates the posterior stomach and most of the intestine. In elasmobranchs and teleosts, the most anterior spinal outflow often joins the vagus, forming a vago-sympathetic trunk running to the gut (Nilsson, 1983).

The ENS contains a large number of nerve cells in all vertebrates (Nilsson, 1983). In fish, cell bodies are located primarily in the myenteric plexus, between the longitudinal and circular muscle layers. In cyclostomes, a well-developed enteric plexus of scattered nerve cell bodies is found along the straight intestinal tube (e.g., Kirtisinghe, 1940; Baumgarten *et al.*, 1973). Similarly in teleosts, myenteric neurons are fairly evenly dispersed along the gut, only rarely forming microganglia containing more than two to three cells (Kirtisinghe, 1940; Olsson and Karila, 1995; Figure 10.1A). The exact distribution, however, varies between species as well as between regions. Still, the overall density of myenteric nerve cell bodies in teleosts is similar to that in small mammals. In elasmobranchs, nerve cell bodies usually occupy nodes in the myenteric plexus (Kirtisinghe, 1940; Olsson and Karila, 1995).

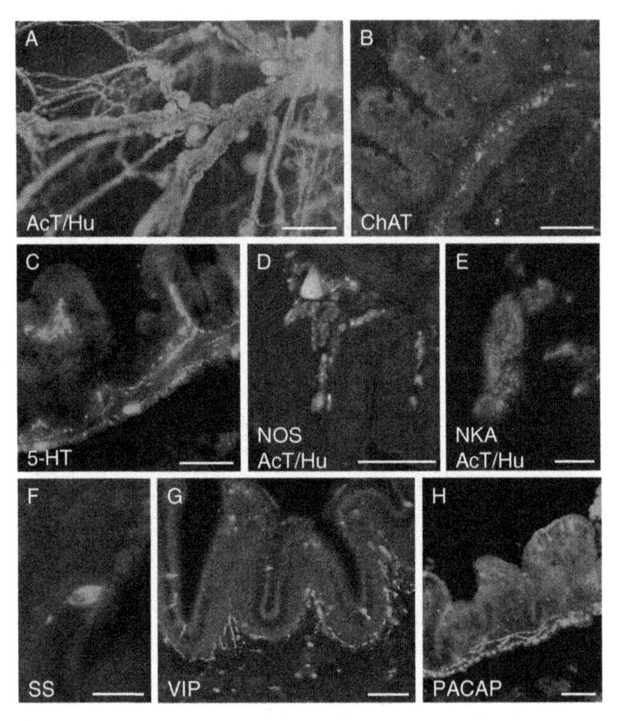

Fig. 10.1. Morphology and examples of signal substances immunoreactivity (ir) in the fish gut. (A) Whole-mount preparation of the sculpin (*Myoxocephalus scorpius*) gut; (B–C, G–H) sections of zebrafish (*Danio rerio*) gut; and (D–F) sections of sculpin gut. (A) Nerve cell bodies and nerve bundles in the myenteric plexus of the cardiac stomach. (B) ChAT ir in nerve fibers and cell bodies of the intestinal bulb, indicating cholinergic innervation. (C) 5-HT ir in nerve fibers and cell bodies in the mid-intestine. (D) Double-labeling against NOS (green) and AT/Hu (red) in the cardiac stomach, showing nitrergic nerve cell body in the myenteric plexus. (E) Double-labeling against NKA (red) and AT/Hu (green) in the cardiac stomach, showing NKA-ir in nerve cell body in the myenteric plexus. (F) SS-ir in endocrine cell of the mucosa of the cardiac stomach. (G) VIP-ir in nerve fibers and endocrine cells of the mucosa of the pyloric stomach. (H) PACAP-ir in nerve fibers of the myenteric plexus and circular muscle layer of the mid-intestine. AcT, acetylated tubulin; ChAT, choline acetyl transferase; 5-HT, serotonin; Hu, human neuronal protein C/D; NKA, neurokinin A; NOS, nitric oxide synthase; PACAP, pituitary adenylate cyclase-activating polypeptide; SS, somatostatin; VIP, vasoactive intestinal polypeptide. Scale bars: 50 μm (A–D, G–H), 25 μm (E–F). (See Color Insert.)

Other cell types, like ICCs and enteric glial cells that may play part in gut motility control, are little investigated in fish. Kirtisinghe (1940) described ICCs in fish gut, while results from more recent studies, using mammalian ICC markers, have given contradictory results (e.g., Mellgren and Johnson, 2005; Rich *et al.*, 2007).

2.2. The Gut as an Endocrine Organ

The gut has an extensive endocrine function. Numerous endocrine cells along the whole length of the gut mucosa and in the pancreas form the so-called "gastroenteropancreatic (GEP) endocrine system." Some signal substances are unique to endocrine cells, while others are identical or similar to neurotransmitters (Section 3). A population of endocrine and enzyme-releasing exocrine cells of the intestinal mucosa, which release the same substances as pancreatic cells, are proposed to correspond to a primitive early stage in pancreatic evolution (e.g., Van Noorden and Falkmer, 1980; Youson and Al-Mahrouki, 1999).

The anatomy of the pancreas varies markedly among fish species. In cyclostomes, exocrine pancreatic tissue is present in the wall of the proximal intestine, while scattered endocrine islet organs are found in the mesenterium (e.g., Van Noorden, 1990). In elasmobranchs and lungfishes, the pancreas is a compact organ, with distinct endo- and exocrine parts (e.g., Van Noorden and Falkmer, 1980). Amongst teleosts, smaller or larger islets are spread out in the mesenterium near the intestine and the pyloric ceca. Exocrine cells may comprise a thin layer surrounding endocrine tissue, as in the Brockmann bodies in rainbow trout, *Oncorhynchus mykiss,* or form a lobulated organ, separate from the endocrine tissue. In both elasmobranchs and teleosts, the four major types of endocrine cells [insulin, glucagon, somatostatin (SS) and pancreatic polypeptide (PP)] are present, as well as minor populations of cells secreting other peptides, but the occurrence varies between species (e.g., Youson and Al-Mahrouki, 1999). Autonomic nerves innervate exocrine and endocrine tissue in both elasmobranchs and teleosts (Van Noorden and Patent, 1980; Jönsson, 1991, 1993).

3. NEUROTRANSMITTERS AND HORMONES OF THE GUT

Signal substances may be synthesized in both endocrine cells and nerves. Many substances interact with more than one receptor, and the actual effect of the transmitter depends on the receptor subtype. Data regarding the distribution of receptors in fish are scarce, but several genes that correlate to mammalian receptor subtypes have been identified in fish. For full references of identified genes mentioned below, see, e.g., the National Center for Biotechnology Information (NCBI; www.ncbi.nlm.nih.gov).

Below follows an overview of the most investigated signal substances in fish gut. Their identity and presence will be exemplified by selected references, and overall patterns and exceptions to these patterns will be discussed. Table 10.1 shows a scan of the shorthorn sculpin, *Myoxocephalus scorpius,*

Table 10.1

Relative density of nerve fibers (+), nerve cell bodies (x) and mucosal endocrine cells (o) in the gut of sculpin (*Myoxocephalus scorpius*) using immunohistochemistry

	CCK/ Gastrin[a]	CGRP	GAL	GRL	GRP[a]	NPY	SS	TK	VIP	PACAP	5-HT	NOS
CS	o	-	+++ooo	-	-	++	ooo	+++xooo	+++	+++	ooo	+++x
PS	o	+	+ooo	o	-	++	ooo	+++xooo	+++	+++oo	ooo	+++
PI	ooo	+++	(+)ooo	(o)	ooo	+++ooo	-	+++ ooo	+++o	+++ooo	+++x	+++
MI	ooo	weak	(+)ooo	(o)	oo	+++ooo	-	+++ ooo	+++o	+++oo	+++x	+++x
DI	ooo	weak	(+)ooo	(o)	oo	+++ooo	-	+++ ooo	+++oo	+++oo	+++	+++x
R	?	weak	(+)ooo	(o)	oo	?	-	+++ ooo	+++oo	+++oo	+++	weak

CS, cardiac stomach; PS, pyloric stomach; PI, proximal intestine; MI, mid intestine; DI, distal intestine; R, rectum; CCK, cholecystokinin; CGRP, calcitonin gene-related peptide; GAL, galanin; GRL, ghrelin; GRP, gastrin-releasing peptide; NPY, neuropeptide Y; SS, somatostatin; TK, tachykinin (substance P/neurokinin A); VIP, vasoactive intestinal polypeptide; PACAP, pituitary adenylate cyclase-activating polypeptide; 5-HT, 5-hydroxytryptamine/serotonin; NOS, nitric oxide synthase.

[a]Bjenning and Holmgren (1988) found gastrin/CCK-like-ir nerve fibers in the stomach and intestine, and bombesin-like-ir (= GRP-like- ir) fibers in the stomach.

gut as a detailed example of the distribution of some neuroendocrine substances along the gut. Extensive tables and reference lists of earlier literature can be found in, e.g., Nilsson and Holmgren (1994). Possible coexistence of signal substances in nerves or endocrine cells is usually not considered since surprisingly little is known about this in fish. Only a few combinations of antisera have been tested, giving a scattered picture of the overall chemical coding. Several factors are likely to affect the distribution and abundance of neurotransmitters and hormones in the gut, e.g., feeding status, type of food, and seasonal variations, but studies in this field are only in their infancy. Also infections and inflammation can change the expression (e.g., Bosi *et al.*, 2005).

3.1. Acetylcholine

Pharmacological and physiological studies suggest a dominating role of acetylcholine (ACh) in the control of gut functions in fish as in other vertebrates (Section 5.2). This is supported by histochemical findings, although scarce, of the synthesizing enzyme, choline acetyl transferase (ChAT) or the vesicular ACh transporter (VAChT) in gut nerves, e.g., in myenteric nerve cell bodies in the cod (Karila *et al.*, 1998). It was suggested that cholinergic nerves include descending (projecting anally) interneurons and ascending (projecting orally) motorneurons in the cod. ChAT-immunoreactive (ir) nerves are also demonstrated in zebrafish (Figure 10.1B; Olsson *et al.*, 2008b). A vagal cholinergic innervation is present in most teleosts, but may be lacking in some teleosts and in elasmobranchs (Campbell and Burnstock, 1968; Campbell, 1975; Holmgren and Nilsson, 1981; Nilsson, 1983). Muscarinic M3-like receptors may be the most important subtype of cholinergic receptors in the control of rainbow trout gut motility (Aronsson and Holmgren, 2000).

3.2. Amines, Amino Acids and Purines

Amines used as signal substances in the fish gut include catecholamines, histamine and serotonin (5-hydroxytryptamine; 5-HT). *Catecholamines* [adrenaline (AD), noradrenaline (NA) and dopamine (DA)] are released from adrenergic nerves and chromaffin tissue. The gut of most classes of fish is densely innervated by adrenergic nerves, surrounding myenteric nerve cell bodies and innervating gut vessels (e.g., Baumgarten *et al.*, 1973; Anderson, 1983; Jensen and Holmgren, 1985; Karila *et al.*, 1997). The origin of these nerve fibers is believed to be mainly extrinsic (spinal autonomic). Our unpublished data, however, indicate the presence of the synthesizing enzyme tyrosine hydroxylase (TH) in enteric nerve cell bodies of several teleost species. TH catalyses the first step in catecholamine synthesis and TH-positive cells may thus be adrenergic, noradrenergic or dopaminergic cells. It is suggested that there are both dopaminergic and adrenergic nerve cell bodies in the gut of river lamprey, *Lampetra fluviatilis* (Baumgarten *et al.*, 1973), and some fluorescent nerves in the teleost intestine may contain DA (Nilsson, 1983). In elasmobranchs, as in mammals, birds and reptiles, NA is the dominating catecholamine in gut adrenergic nerves, while AD is the predominating form in teleosts and holosteans (von Euler and Fänge, 1961; Burnstock, 1969; Abrahamsson and Nilsson, 1976).

Histamine is found in mucosal endocrine cells of the fish stomach and intestine (Reite, 1972; Håkanson *et al.*, 1986; D'Este and Renda, 1995; own unpublished observation). The presence in gut nerves and mast cells is more uncertain. Careful studies showed histamine in mast cells only in lungfish and perciform fish, the most advanced teleosts, and it was hypothesized that this expression first developed in reptiles and independently at a later stage in the two fish groups (Reite, 1972; Mulero *et al.*, 2007). Pharmacological studies suggest that H1-like receptors mediate motility control in the gut (see Section 5.2.1), while H2 receptors control gastric acid secretion (see Section 6.1).

5-HT represents an interesting case. In cyclostomes, it is found in nerve cells only (Baumgarten *et al.*, 1973; Goodrich *et al.*, 1980). In elasmobranchs, nerves are sparse, while endocrine cells containing 5-HT are common. Sturgeons (Chondrostei) have 5-HT in both endocrine and nerve cells (Salimova and Fehér, 1982). Teleosts show large interspecies variation in the relative proportion of 5-HT-containing nerves vs. endocrine cells (Anderson, 1983). In cyprinids, 5-HT seems to be present in nerves only (e.g., Pan and Fang, 1993; Pederzoli *et al.*, 2004; Olsson et al., 2008b; Figure 10.1C). In the European eel, *Anguilla anguilla*, 5-HT is found only in endocrine cells (Domeneghini *et al.*, 2000), while in a closely related species both endocrine cells and nerves contain 5-HT (Anderson and Campbell, 1988). The staining of neurons, however, was weak. Anderson and Campbell (1988) argue that the endocrine cells only express 5-HT if no 5-HT nerve fibers reach the mucosa. At least seven classes of 5-HT receptors exist and members of most of the classes have been isolated from different fish species, although little is known about receptor distribution in the gut.

The amino acid *γ-amino butyric acid (GABA)* is present in the CNS but so far there are few reports from fish gut. Gábriel *et al.* (1990) demonstrated GABA-ir nerve cell bodies in common carp, *Cyprinus carpio*. Occasional GABA-ir myenteric nerve cells are found in rainbow trout (own unpublished data). GABA binds three types of receptors; two (A and B) have been sequenced in teleost species but nothing is known about the distribution in the gut.

Adenosine and ATP are purine derivates that may have transmitter functions when released in high enough concentrations from synaptic vesicles in nerve endings. The action is mediated by a large number of receptor subtypes, belonging either to the P1-receptor group stimulated by adenosine, or the P2-receptors interacting with ATP and other nucleotides. Both P1- and P2-receptors are found in fish gut (Lennard and Huddart, 1989; Knight and Burnstock, 1993).

3.3. Nitric Oxide

Nitric oxide (NO) is synthesized by the enzyme nitric oxide synthase (NOS), and the neuronal isoform has been isolated and sequenced from several fish species, e.g., zebrafish, rainbow trout and fugu (NCBI, 2008). The presence of NOS is the most commonly used indicator of nitrergic nerves (Figure 10.1D). Li and Furness (1993) first found NOS-containing nerve cell bodies in the myenteric plexus throughout the gut of rainbow trout. The cells sent fibers mainly to the muscle layers. Afterwards, nitrergic neurons were demonstrated in several fish species. Close to or just over half the population of myenteric nerve cells are NOS-containing in Atlantic cod and spiny dogfish, *Squalus acanthias*, while they are absent from the intestine of hagfish, *Myxine glutinosa* (Olsson and Karila, 1995). In goldfish *Carassius auratus*, a concentration of nitrergic neurons is found in all sphincters of the gut (Brüning *et al.*, 1996).

3.4. Peptides

Signal peptides in the gut consist of only a few to up to about 40 amino acids. They form families based on structural and genetic similarities. Changes in amino acid sequence of a peptide reflect evolutionary distance between species. An extra gene duplication has occurred in teleosts, increasing the possibility of the evolution of unique signal substances (e.g., Conlon and Larhammar, 2005). Below follows a presentation of the most investigated peptides involved in neuroendocrine control of gut function. However, this is not a complete list of the signal peptides present in the GI tract.

The cholecystokinin (CCK)/gastrin family includes several peptides of variable length, characterized by the four common C-terminal amino acids (Johnsen, 1998; Kurokawa *et al.*, 2003). In most species, CCK/gastrin-like material is found in endocrine cells in both stomach and intestine and it is not uncommon with additional immunoreactive nerve fibers (Van Noorden and Pearse, 1974; Holmgren and Nilsson, 1983a; Burkhardt-Holm and Holmgren, 1989; Yui *et al.*, 1990; Holmgren *et al.*, 1994). Few immunohistochemical studies have unequivocally distinguished between CCK and gastrin, but following the general vertebrate situation, it is most likely that gastric endocrine cells contain gastrin, while intestinal endocrine cells and gut nerves contain a CCK peptide. One type of CCK/gastrin receptor has been demonstrated in fish (Oliver and Vigna, 1996). This CCK-X receptor is believed to be the ancestral form of the mammalian CCK-A and CCK-B receptors. The central function of CCK in fish (larval) digestion has been reviewed in detail by Rønnestad *et al.* (2000).

Calcitonin gene-related peptide (CGRP) is a 37-amino acid peptide obtained by alternative splicing of the calcitonin/CGRP genes (Wimalawansa, 1997). It exists in two similar isoforms, in both teleosts (Ogoshi *et al.*, 2006) and mammals. In Atlantic cod, CGRP is expressed in nerve cell bodies and fibers in the intestine while no cell bodies were seen in the stomach (Karila, 1997; Shahbazi *et al.*, 1998). Similarly in zebrafish, vagal CGRP fibers innervate the proximal intestine, and there seem to be both extrinsic and intrinsic nerves in the distal part (Olsson *et al.*, 2008b). CGRP nerves are also found, e.g., in Arctic lamprey, *Lampetra japonica* (Yui *et al.*, 1988) and Australian lungfish, *Neoceratodus forsteri* (Holmgren *et al.*, 1994). Endocrine cells in the mucosa are reported from, e.g., cod and lungfish. A CGRP receptor is expressed in the gut of Japanese flounder, *Paralichthys olivaceus* (Suzuki *et al.*, 2000).

Galanin (GAL) is a 29-amino acid peptide isolated and sequenced from ancient fish groups as well as teleosts and elasmobranchs (e.g., Habu *et al.*, 1994; Wang and Conlon, 1994; Wang *et al.*, 1999). The primary structure is highly conserved in particular in the N-terminal. With the exception of the lamprey, where GAL is found also in endocrine cells (Bosi *et al.*, 2004), reports are almost exclusively of a neuronal occurrence in enteric and perivascular fibers (Karila *et al.*, 1993; Holmgren *et al.*, 1994; Preston *et al.*, 1995). Most enteric fibers are found in the mucosa and submucosa, but also some in the muscle layers, indicating effects on both secretion and motility. Nerve cell bodies are present in the myenteric plexus showing that at least part of the innervation is intrinsic to the gut (Bosi *et al.*, 2007), and in the vagus nerve, projecting either towards the CNS or the gut (Karila *et al.*, 1993). The density of innervation often decreases from stomach to rectum.

Ghrelin (GRL) is a recent addition to the list of gut hormones that are identified and sequenced in several fish species. The length varies between species, from 19 amino acids in goldfish to 25 in shark. In some species more than one isoform is expressed (see Kaiya *et al.*, 2008). GRL is present in endocrine cells in rainbow trout (Sakata *et al.*, 2004) and eel stomach (Kaiya *et al.*, 2006). However, GRL-ir endocrine cells are also present in the intestine of stomach-less species like zebrafish and common carp (Kono *et al.*, 2008; Olsson *et al.*, 2008a). GRL binds to a receptor related to the motilin receptor. In addition to GRL and motilin receptors, a third receptor has been found in the pufferfish genome (Olsson *et al.*, 2008a). The affinity for GRL to this receptor is not known.

Gastrin-releasing peptide (GRP) and bombesin (BBS) belong to a family of several peptides sharing the same eight C-terminal amino acids. GRP has been isolated from the GI tract of dogfish (Conlon *et al.*, 1987) and rainbow trout (Jensen and Conlon, 1992a), and the corresponding cDNA has been sequenced from zebrafish and goldfish (NCBI, 2008; Volkoff *et al.*, 2000). A neuromedin B-like sequence is also present in zebrafish (NCBI, 2008). In rainbow trout, the presence of a shorter GRP is indicated, while extracts contained no BBS (Jensen and Conlon, 1992a). On the other hand, immunohistochemistry suggests a peptide more like BBS than GRP in the gut of an elasmobranch and a holocephalan ratfish (Cimini *et al.*, 1985; Thorndyke *et al.*, 1990).

GRP/BBS-like-ir nerves are present in the gut of lampreys (Yui *et al.*, 1988). In various elasmobranch and teleost species, both nerves and endocrine cells are found along the entire GI tract, with a variable density among species and gut regions (e.g., Cimini *et al.*, 1985; Bjenning and Holmgren, 1988; Tagliafierro *et al.*, 1988). Nerve cell bodies occur in the myenteric plexus of, e.g., spiny dogfish and shorthorn sculpin indicating the intrinsic enteric nature of the innervation (Holmgren and Nilsson, 1983a; Bjenning and Holmgren, 1988). At least in elasmobranchs, perivascular nerve fibers of gut wall vessels as well as major vessels to the gut express a GRP/BBS-like peptide (Holmgren and Nilsson, 1983a; Tagliafierro *et al.*, 1988; Bjenning *et al.*, 1990).

The evolution of the *neuropeptide Y (NPY)* family of peptides has received considerable interest over the years. In a recent careful analysis by Sundström *et al.* (2008), it is established that the (teleost) fish genome, after the extra gene duplication, has evolved two forms of NPY and two forms of peptide YY (PYY), with slightly different expression in different species. True PP is missing. A number of studies have reported NPY/PYY-like material in the fish gut and pancreas but the exact identity of the peptide is seldom equivocally established. In general, PYY is the predominating form in nerves and endocrine cells throughout the body, including the gut (Sundström *et al.*, 2008).

In cyclostomes, NPY/PYY-like material appears in endocrine cells only (Yui *et al.*, 1988), while in the gut of elasmobranchs and teleosts, nerve fibers are present in the myenteric plexus and muscle layers and some surround blood vessels (e.g., Bjenning *et al.*, 1989; Burkhardt-Holm and Holmgren, 1989; Preston *et al.*, 1998; Shahbazi *et al.*, 2002). PP-like material is consistently found in the endocrine pancreas, except in cyclostomes (e.g., Van Noorden and Patent, 1980), as well as in intestinal endocrine cells and in a few cases in the stomach (Langer *et al.*, 1979; Cimini *et al.*, 1985; Yui *et al.*, 1990). At least seven NPY-receptor types are present in vertebrates. Species-dependent variations are common amongst fish (Salaneck *et al.*, 2008), but little is known of the presence of individual receptor types in the fish gut.

Several forms of *somatostatin (SS)*, usually 14 or around 28 amino acids long, occur in fish. SS-14 (I) is highly conserved, being identical in, e.g., hagfish, lungfish, trout and mammals. Other SS-14s as well as longer forms such as SS-28 vary more between species, and some forms appear unique to fish (see, e.g., Nelson and Sheridan, 2005). Besides a general occurrence in pancreatic cells, SS is present in gut endocrine cells of all fish investigated (Figure 10.1F). Neuronal SS is common in the elasmobranch gut, but more rarely found in teleosts. The SS distribution differs: in spiny dogfish, fibers are most common in the rectum; in greater spotted dogfish, *Scyliorhinus stellaris*, they are found only in the stomach (Holmgren and Nilsson, 1983a; Cimini *et al.*, 1985); and in the teleost *Barbus conchonius* (rosy barb) only in the intestine (Rombout *et al.*, 1986). Five groups of SS receptors (SSR_{1-5}) have been identified, and representatives of all except SSR_4 have been found in fish (see Nelson and Sheridan, 2005). Subtypes 1A, 1B and 2 are present in the liver and pancreas. The receptor expression is affected by hormone levels and feeding status. For example, in rainbow trout, insulin-like growth factor-I (IGF-I) down-regulates and fasting up-regulates all three subtypes (Slagter *et al.*, 2005).

Tachykinins (TK) constitute a large peptide family of which substance P and neurokinin A (NKA) are the most studied. Substance P contains 11 amino acids, and elasmobranch and teleost sequences differ in 3–4 positions compared to mammals. NKA comprises ten amino acids of which the last five are usually identical to the substance P C-terminal (see Holmgren and Jensen, 2001; Severini *et al.*, 2002). In addition, several related fish peptides have been isolated, such as scyliorhinin I and II from dogfish (Conlon *et al.*, 1986). Tachykininergic nerves are frequent in all gut regions of all fish groups except cyclostomes. The nerve fibers are most prominent in the myenteric plexus but are also present in the other layers of the gut wall (e.g., Jensen and Holmgren, 1985, 1991; Jensen *et al.*, 1993a). Enteric nerve cell bodies are common, indicating an intrinsic origin of at least a part of the fibers (Figure 10.1E; Karila *et al.*, 1998). Fibers innervating vessels to the gut in rainbow trout and spiny dogfish are identified using antisera raised against fish TK (Kågström *et al.*, 1996a, Kågström and Holmgren, 1998). In addition, endocrine cells expressing TK are found in all gut regions (e.g., Jensen and Holmgren, 1991). TK act through at least three distinct receptors, neurokinin (NK) 1–3 receptors, all found in fish. Functional data suggest the presence of NK1 receptors in fish gut (Jensen *et al.*, 1993b).

Vasoactive intestinal polypeptide (VIP) and *pituitary adenylate cyclase-activating polypeptide (PACAP)* are two closely related peptides containing 28 amino acids, and 38 or 27 amino acids, respectively. VIP and PACAPs have been sequenced in several fish and show large similarities with mammalian peptides (see Hoyle, 1998; Holmgren and Jensen, 2001). Except for cyclostomes, where it is only present in endocrine cells (Reinecke *et al.*, 1981; Van Noorden, 1990), neuronal VIP has been found in most species examined. VIP fibers innervate all layers of the gut wall (Figure 10.1G). Although few labeled nerve cell bodies are found, it is still believed that most of the VIP innervation is of intrinsic origin. Large arteries to the gut receive a relatively sparse VIP innervation (e.g., Langer *et al.*, 1979; Holmgren and Nilsson, 1983a,b; Jensen and Holmgren, 1985; Yui *et al.*, 1990; Holmgren *et al.*, 1994; Kågström and Holmgren, 1997). VIP-like material is also regularly found in endocrine cells in many fish (e.g., Falkmer *et al.*, 1980; Reinecke *et al.*, 1981; Rajjo *et al.*, 1989; Olsson and Holmgren, 1994). However, some species may not contain such cells (e.g., Domeneghini *et al.*, 2000; Lee *et al.*, 2004). The discrepancies could be genuine species differences, but could also be due to different specificity of the antisera used.

PACAP has been demonstrated in gut nerves in some species (Olsson and Holmgren, 1994; Holmberg *et al.*, 2004; Figure 10.1H). It is colocalized with VIP to 100% in enteric neurons in cod, rainbow trout and spiny dogfish (Olsson and Holmgren, 1994; own unpublished data). Among endocrine cells, usually a subpopulation of PACAP cells is VIP-negative. PACAP and VIP interact with three main types of receptors, VPAC1, VPAC2 and PAC1, with some species differences in distribution. Three pufferfish species express VPAC2 and PAC1 in the gut (Cardoso *et al.*, 2004). In goldfish, only VPAC1 is found in the intestine (Chow, 1997) while one PAC1 isoform exists in zebrafish intestine (Fradinger *et al.*, 2005).

4. DEVELOPMENT OF GUT INNERVATION AND NEUROENDOCRINE SYSTEM

Most knowledge on the early development of the fish nervous system comes from studies on zebrafish (e.g., Raible *et al.*, 1992; Shepherd *et al.*, 2001; Holmberg *et al.*, 2003, 2004, 2006, 2007), although some studies involve, e.g., turbot, *Scophthalmus maximus*, and green swordtail, *Xiphophorus helleri* (Sadaghiani and Vielkind, 1990; Reinecke *et al.*, 1997). The autonomic innervation of the gut derives from the neural crest (Lamers *et al.*, 1981; Sadaghiani and Vielkind, 1990; Raible *et al.*, 1992). Vagal neural crest cells migrate to the gut under the influence of various chemical cues. In zebrafish, the first nerve cells appear in the gut within 48 hours after fertilization (Bisgrove *et al.*, 1997; Holmberg *et al.*, 2003). Exogenous feeding begins around day 4–5 but already at 3 days post fertilization (dpf) the intestine is richly innervated and the nerves have begun to express various signal substances (Holmberg *et al.*, 2004, 2006; Olsson *et al.*, 2008b). However, the most proximal part (intestinal bulb) does not receive much innervation until later; around day 11–13 there is a dense nerve fiber network in most of the gut (Olsson *et al.*, 2008b). Gut endocrine cells derive from the endoderm in the primitive gut (see Wallace and Pack, 2003). In zebrafish, the endocrine cells differentiate and mature comparably late, around 4 dpf (Ng *et al.*, 2005).

Many transmitters and hormones are expressed before the onset of exogenous feeding. The exact timing and order of first appearance of signal substances varies between species. Further, whether a substance that can be expressed in both nerves and endocrine cells first appears in one or the other cell type also varies. In some species occurrence in nerves precedes that in endocrine cells, the opposite or an appearance in nerves and endocrine cells at the same stage is true for other species. See Table 10.2 for a more detailed comparison for some species. In addition, in larval Atlantic halibut, *Hippoglossus hippoglossus*, GRL is expressed before onset of exogenous feeding and reaches its maximum during metamorphosis (Manning *et al.*, 2008). In contrast, CCK appears after first feeding in halibut endocrine cells (Kamisaka *et al.*, 2001) but one day before first feeding in Japanese flounder (Kurokawa *et al.*, 2000). Different substances also show different regional

Table 10.2

First occurrence (using immunohistochemistry) of some transmitters and hormones in the gut of different fish species during development

	Sea bass[1]		Turbot[2]		Zebrafish[3]		Dogfish[4]		Dogfish[5]	
	ec	*nf*	*ec*	*nf*	*ec*	*nf*	*ec*	*nf*	*ec*	*nf*
CCK			11						4–5	–
NPY			8	24			2.5	4		
TK	4	12	11	8	3	3				
VIP/PACAP	18	4	24	5	4	3			4	7–8
5-HT			10	15	–	4	3	–	4–5	–
NOS	–	8[a]			–	3				
	dph		dph		dpf		months[b]		months	

ec, endocrine cell; *nf*, nerve fiber; CCK, cholecystokinin; NPY, neuropeptide Y; TK, tachykinin (substance P/neurokinin A); VIP, vasoactive intestinal polypeptide; PACAP, pituitary adenylate cyclase-activating polypeptide; 5-HT, 5-hydroxytryptamine/serotonin; NOS, nitric oxide synthase.

dpf, days post fertilization; dph, days post hatching.

[1]Pederzoli *et al.*, 2004, 2007 (*Dientrarchus labrax*); [2]Reinecke *et al.*, 1997 (*Scophthalmus maximus*); [3]Holmberg *et al.*, 2004, 2006; Olsson *et al.*, 2008b (*Danio rerio*); [4]Chiba, 1998 (*Scyliorhinus torazame*); [5]Tagliafierro *et al.*, 1989 (*Scyliorhinus stellaris*).

[a]Not tested at earlier stage; [b] hatching around 8–9 months.

development. In sea bass, a TK is first expressed in the rectum, and subsequently in the stomach while VIP shows the opposite pattern (Pederzoli *et al.*, 2004). In cloudy dogfish, *Scylirhinus torazame,* 5-HT is first expressed in endocrine cells in the midgut then spreads to the stomach with occasional cells in the hindgut (Chiba, 1998).

Some substances may be transitionally expressed during development. In larval lampreys, there are SS- or CCK/gastrin-ir endocrine cells in the intestine; neither of these is found in adults (Yui *et al.*, 1988). In greater spotted dogfish, glucagon-like material is expressed in endocrine cells of the pyloric stomach during the middle embryonic stages only (but they are present in stomach and spiral intestine throughout embryonic stages and birth) (Tagliafierro *et al.*, 1989).

5. CONTROL OF GUT MOTILITY

The general aim of gut motility is to mix and propel gut contents to optimize digestion and absorption along the GI tract. The underlying mechanisms involve contraction and relaxation of gut smooth muscles in response to chemical and mechanical stimulation by the ingested food. Motility also occurs in fasted animals. Motility patterns look different in different regions, and the control pathways vary with the situation.

5.1. Gut Muscles and Spontaneous Activity

GI smooth muscles display cyclic patterns of electrical activity, so-called "slow waves" that depend on spontaneous depolarizations and repolarizations. Slow waves start in ICCs and spread passively throughout the ICC network as well as to the muscle cells that act in synchrony. The ICCs set the frequency of slow waves and several pacemaker regions can exist along the GI tract. Neuronal or hormonal input converts slow wave activity to muscle contractions. These events are little investigated in fish.

Experimentally, contractile activity is often referred to as "spontaneous" if it occurs prior to any treatment of the preparation such as drug exposure or electrical stimulation, although it is most likely initiated by ICCs and/or local nerves or hormones. Spontaneous contractions in preparations of the fish gut were demonstrated early in both elasmobranchs and teleosts (e.g., Young, 1933; Burnstock, 1958a,b). For example, in the Atlantic cod intestine, these contractions are nerve-dependent since they are reduced or abolished by tetrodotoxin (TTX). Cholinergic and serotonergic nerves maintain a stimulatory tonus, counter-balanced by an inhibitory nitrergic tonus (Jensen and Holmgren, 1985; Karila and Holmgren, 1995; Olsson and Holmgren, 2000).

Other preparations may be insensitive to nerve blockade, or change pattern of activity after the blockade (own unpublished observations).

5.2. Effects of Individual Signal Substances

Gut motility in every instance depends on interacting nerves and hormones. The basis for our understanding of the integrated functions is to elucidate the action of individual substances. The number of pharmacological studies on gut motility in fish is in many cases limited, and most studies use isolated preparations *in vitro*. Below we have summarized the effects of the most well studied transmitters on gut motility (see also Figure 10.2). There are always exceptions, but ACh, histamine, 5-HT and TK are in general excitatory, NO and VIP/PACAP are inhibitory, while effects of catecholamines, purinergic substances and CCK may differ in different species or even different parts of the gut.

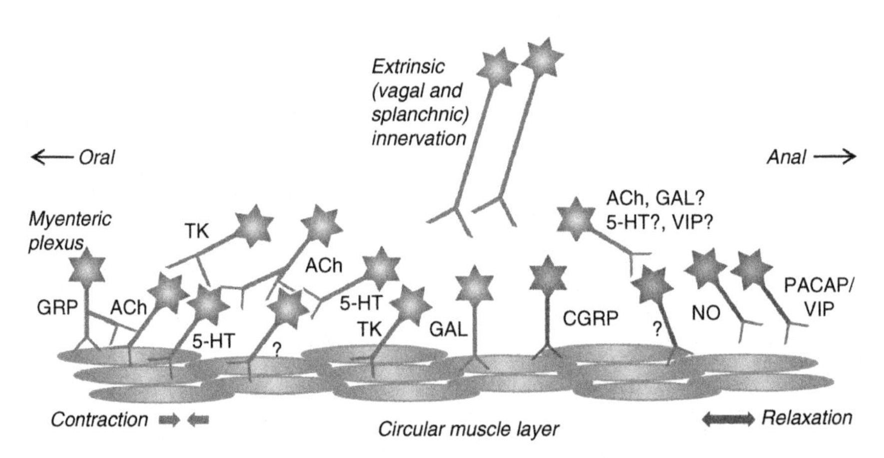

Fig. 10.2. Neuron types (demonstrated by immunohistochemistry) in the myenteric plexus of teleost fish including predominant projections and effects on gut smooth muscles. The picture is a tentative summary of different gut regions and species examined. Additional signal substances, released from either nerves or endocrine cells, are likely to be involved. A demonstrated direct effect on smooth muscle is shown by nerves contacting the circular muscle layer. If the nature of the innervation has not been demonstrated, this is indicated by nerves not making contact with other cells. Vertical nerve indicates no projection studies have been performed. Red = excitation; blue = inhibition. 5-HT, 5-hydroxytryptamine (serotonin); ACh, acetylcholine; CGRP, calcitonin gene-related peptide; GAL, galanin; GRP, gastrin-releasing peptide; NO, nitric oxide; PACAP, pituitary adenylate cyclase-activating polypeptide; TK, tachykinins; VIP, vasoactive intestinal polypeptide. (See Color Insert.)

5.2.1. Stimulatory

Acetylcholine (ACh) is a potent excitatory transmitter in the gut of most teleost species (see Nilsson, 1983; Jensen and Holmgren, 1994). The cholinergic nerves can be vagal extrinsic, or intrinsic (see Section 3.1). The cholinergic contractions of the cod intestine are reduced by atropine but not affected by TTX, suggesting a direct effect on muscarinic receptors on smooth muscles (Jensen and Holmgren, 1985). Furthermore, *in vitro* studies on isolated intestine indicate the presence of orally projecting, cholinergic, intrinsic motor neurons (Karila and Holmgren, 1995). Atropine also reduces reflex contractions in response to distension of rainbow trout stomach (Grove and Holmgren, 1992a).

Exogenous ACh causes contraction of the elasmobranch and hagfish gut (e.g., von Euler and Östlund, 1957; Holmgren and Fänge, 1981; Nilsson and Holmgren, 1983), but little is known about the cholinergic control *in vivo*. An extrinsic excitatory (cholinergic) innervation may be lacking in elasmobranchs (Nilsson, 1983) suggesting that ACh is released mainly from enteric neurons. Also, stimulation of the vagus nerve in hagfish gives weak and inconclusive responses (Patterson and Fair, 1933).

Histamine has weak or no effect on gut smooth muscle in a number of fish species (see Reite, 1972), but clearly and consistently contracts intestinal muscle in many other species (von Euler and Östlund, 1957; Valette and Augeraux, 1958). Burnstock (1958a) found that low concentrations of histamine had no effect, while high (unphysiological?) concentrations could elicit a contraction. Histamine contracts intestinal smooth muscle in *Sparus aurata* (gilthead sea bream), a perciform species with histamine contained in mast cells. The effects of specific agonists and antagonists suggest that the action is mediated by H2-like receptors, but again, the concentrations used are very high (10^{-5}–10^{-2} M; Mulero *et al.*, 2007). In contrast, experiments on isolated preparations of rainbow trout stomach give classical responses with a pD_2 (i.e., $-\log EC_{50}$) of 5.65 and, notably, in this case specific drugs suggest an effect via H1 receptors (Manera *et al.*, 2008).

5-HT stimulates contractile activity in the fish gut (Young, 1980a,b, 1983; Nilsson and Holmgren, 1983; Jensen and Holmgren, 1994). As discussed above, the source of 5-HT is either endocrine cells or enteric nerves. For example, in *Platycephalus bassensis* (sand flathead), which lacks endocrine serotonergic cells, there is a spontaneous release of 5-HT from enteric nerves, giving a serotonergic tonus (Anderson *et al.*, 1991).

The contractile effect of 5-HT on rainbow trout gut is not affected by atropine or TTX, excluding a cholinergic link and indicating a direct effect on smooth muscles (Holmgren *et al.*, 1985; Burka *et al.*, 1989; Grove and Holmgren, 1992a). In the cod intestine, TTX reduces 5-HT-induced contractions, suggesting an at least partly indirect effect, most likely involving release of ACh (Jensen and Holmgren, 1985) and the same has been concluded for several other teleosts (e.g., Grove and Campbell, 1979b; Kiliaan *et al.*, 1989). In addition to its stimulating effect on muscle, 5-HT is suggested to be released from descending interneurons in the cod intestine (Karila, 1997).

Tachykinins (TK) generally stimulate contractions. There are several studies on the effect of substance P and NKA on gut motility in fish, some using endogenous peptides (Jensen *et al.*, 1993b; Karila *et al.*, 1998). In a study on isolated strip preparations from species representing most major fish groups, substance P and NKA induced or enhanced contractions in starry ray, South American lungfish (*Lepidosiren paradoxa*), gray bichir (*Polypterus senegalensis*) and rainbow trout while no effect was seen in hagfish and river lamprey (Jensen and Holmgren, 1991). Both substance P and NKA also stimulate motility in perfused cod stomach and intestine (Jensen and Holmgren, 1985; Jensen *et al.*, 1987, 1993a). Using native peptides in cod and rainbow trout intestine, substance P and NKA are equally potent while substance P is the more potent in trout stomach (Jensen *et al.*, 1993b; Karila *et al.*, 1998), suggesting the presence of different receptors in stomach and intestine.

The effect of TK can be either directly on the gut muscle, as demonstrated by the lack of effect of application of TTX, or indirect. In most species, there is a direct effect of substance P (Andrews and Young, 1988a; Jensen and Holmgren, 1991; Jensen *et al.*, 1993a). In addition, in the intestine of teleosts there is an indirect effect via cholinergic neurons (Kitazawa *et al.*, 1986a; Jensen and Holmgren, 1991). Furthermore, there is an indirect effect via serotonergic neurons in rainbow trout stomach and intestine and in cod intestine (Holmgren *et al.*, 1985; Jensen and Holmgren, 1991). The physiological significance of three different modes of action as in the rainbow trout and cod intestine, i.e., if they are involved in different reflexes, has not been elucidated.

Ghrelin (GRL) is a hunger signal, a peptide released from the mammalian stomach wall which also has effects on gut motility. In zebrafish, GRL causes a small but significant increase in basal tension in intestinal strip preparations (Olsson *et al.*, 2008a).

Galanin (GAL) has a weak excitatory direct effect on *in vitro* preparations of Atlantic cod stomach (Karila *et al.*, 1993). From its distribution and the coexistence with the cholinergic enzyme ChAT in gut nerves, it has been suggested that GAL in the fish gut probably functions as a modulator of cholinergic activity (Bosi *et al.*, 2007).

Cod *NPY* (Jensen and Conlon, 1992b) contracts ring preparations of cod intestine but only at high concentrations, and it is suggested that the endogenous material found using immunohistochemistry resembles PYY rather than NPY (Shahbazi *et al.*, 2002).

5.2.2. Inhibitory

Nitric oxide (NO) is inhibitory on GI motility. Most work on fish and NO has been on cod intestine. Application of the NO donor sodium nitroprusside to isolated strip preparations abolishes spontaneous contractions (Olsson and Holmgren, 2000). Further, NOS inhibitors like L-NAME (NG-nitro-L-arginine methyl ester) increase amplitude of spontaneous contractions and sometimes also basal tension suggesting an endogenous tonic release of NO in the preparations. The reduced relaxation, anal to the site of stimulation, in the intestine after L-NAME indicates that NO is involved in the descending inhibitory pathway, presumably released from motor neurons (Karila and Holmgren, 1995).

NO also reduces electrically induced contractions in rainbow trout stomach (Green and Campbell, 1994). Furthermore, the inhibitory component of vagal stimulation was markedly reduced by NOS inhibitors. There seems to be no endogenous tonic release of NO in the trout stomach, since basal activity was not altered by L-NAME (Olsson *et al.*, 1999). Additional studies show that NO inhibits contractions also in cod stomach and rainbow trout intestine (own unpublished data) as well as zebrafish intestine (Holmberg *et al.*, 2007).

VIP and PACAP have a predominantly inhibitory effect on gut smooth muscles in fish. In Atlantic cod stomach, VIP abolishes spontaneous contractions *in vivo* (Jensen *et al.*, 1991). Similarly, VIP reduces distension-induced contractions in perfused stomach or spontaneous contractions of gastric strip preparations from rainbow trout stomach (Holmgren, 1983; Grove and Holmgren, 1992a). The effect of VIP in the rainbow trout stomach is blocked by methysergide, suggesting an effect via inhibition of serotonergic neurons (Grove and Holmgren, 1992a). In contrast, both mammalian and cod VIP are without effect on cod intestine (Jensen and Holmgren, 1985; Olsson and Holmgren, 2000). PACAP, on the other hand, reduces spontaneous contractions in both cod and rainbow trout intestinal strip preparations (Olsson and Holmgren, 2000; own unpublished data). Similarly, PACAP inhibits activity in zebrafish intestine and Japanese stargazer, *Uranoscopus japonicus*, rectum (Matsuda *et al.*, 2000; Holmberg *et al.*, 2004). In skate, no effect of VIP was seen in either part of the GI tract (Andrews and Young, 1988a) while VIP inhibits activity in spiny dogfish rectum (Lundin *et al.*, 1984).

CGRP inhibits spontaneous contractions in isolated cod intestine (Shahbazi *et al.*, 1998). The effect was not affected by TTX, indicating a mainly direct effect on smooth muscles. CGRP was suggested to play a role in the descending inhibitory pathways. In mammals, CGRP is often a transmitter in sensory neurons, but whether CGRP found in extrinsic nerves to the fish gut is involved in afferent signaling has not been determined.

Somatostatin (SS) inhibits activity induced by distension of the stomach in cod and rainbow trout, possibly through an inhibition of cholinergic neurons as in mammals (Grove and Holmgren, 1992a,b). In the cod intestine, an initial relaxation on exposure to SS is sometimes followed by a contraction (Jensen and Holmgren, 1985).

5.2.3. Variable Effects

Catecholamines. The effect of catecholamines varies between species and regions of the gut (Nilsson, 1983; Jensen and Holmgren, 1994). In hagfish, responses are weak and irregular (Holmgren and Fänge, 1981). In elasmobranchs, AD usually contracts the stomach and intestine by an action on α-adrenoceptors (AR), while it relaxes the rectum (Young, 1980b, 1983, 1988; Nilsson and Holmgren, 1983). The teleost stomach contracts in response to splanchnic nerve stimulation and to AD and NA in some species, e.g., salmon (*Salmo* spp.) and eel, and shows mixed responses or relaxes in other species such as cod, anglerfish (*Lophius piscatorius*) and stargazer. The dominating response of the intestine to stimulation of the splanchnic nerve supply is a relaxation due to release of catecholamines (Grove and Campbell, 1979b). Contraction is mediated by α-AR and relaxation may be mediated by either α- or β-AR, or both (e.g., Burnstock, 1958a,b; Grove and Campbell, 1979a,b; Young, 1936, 1980b). Relaxation of the carp intestinal bulb can be due to stimulation of presynaptic α_2-AR on cholinergic neurons, which inhibit release of ACh (Kitazawa *et al.*, 1986b). Although the pharmacological evidence is not clear-cut, it is likely that the AR in the rainbow trout stomach also is of the α_2 variety (Kitazawa *et al.*, 1986a, 1988). It is generally believed that the adrenergic control in fish is mainly exerted by extrinsic splanchnic nerves.

DA increased the electrical activity of the stomach wall in *Raja radiata* (starry ray) and cod, and pharmacological evidence suggests an effect on specific DA receptors (Groisman and Shparkovskii, 1989).

Adenosine and ATP. The scattered information on purinergic mechanisms in the fish gut indicates diverse and often mixed effects depending on gut region and species. ATP stimulates stomach strips from *Raja clavata* (thornback ray; Young, 1983) but causes inhibition of longitudinal muscle from the rectum, a direct effect on the smooth muscle (Young, 1988). Similarly, in a teleost, the anglerfish, ATP stimulates the stomach but causes inhibition of rhythmic activity in the intestine, in both cases mimicking vagal stimulation (Young, 1980a). In rainbow trout stomach strips, both adenosine and ATP caused contraction of longitudinal muscle, but relaxation of circular muscle (Holmgren, 1983). Both substances relax the perfused intestine of cod (Jensen and Holmgren, 1985). The results suggest a large variety of purinergic receptors also in the fish gut. The two main types of purinergic receptors, P1 and P2, are present in the intestine of the European flounder, *Platichthys flesus*.

P1 receptors, probably of two different subtypes, mediate an inhibitory effect of adenosine in the intestine and an excitatory effect in the rectum. P2 receptors mediate excitatory effects of ATP in the intestine (Lennard and Huddart, 1989).

CCK has an inhibitory effect on gastric emptying in rainbow trout, significantly increasing the amount of food left in the stomach after 20 h (Olsson *et al.*, 1999). CCK decreases the amplitude of spontaneous contractions in the intestine while at the same time increasing resting tension (Olsson *et al.*, 1999). In contrast, CCK has a stimulatory effect on isolated strip preparations from Atlantic cod (Jönsson *et al.*, 1987). The effect on stomach motility *in vivo* is concentration-dependent, decreasing frequency and amplitude at lower concentrations and inducing contractions at higher concentrations (Olsson *et al.*, 1999). CCK and related peptides also have an excitatory effect on skate and dogfish gut, but the responses are weak and irregular (Andrews and Young, 1988a; Aldman *et al.*, 1989).

GRP or BBS stimulates stomach motility in fish (Holmgren, 1983; Holmgren and Jönsson, 1988). Similarly, BBS stimulates both basic tone and rhythmic activity of the rectal smooth muscle from spiny dogfish (Lundin *et al.*, 1984). In contrast, BBS is weakly inhibitory on cod intestine (Jensen and Holmgren, 1985; Holmgren and Jönsson, 1988). TTX did not affect the response to BBS in cod or rainbow trout stomach indicating a direct effect. Interestingly, BBS also potentiates the effect of ACh (Thorndyke and Holmgren, 1990).

In cod stomach, BBS is more potent than mammalian GRP (Bjenning and Holmgren, 1988). To our knowledge, no studies have used species-specific endogenous peptides, hence the exact potency of GRP-like peptides in fish *in vivo* has not been determined. Neither have the receptor types involved in gut motility control been examined.

5.3. Control of Propagating Activity

Propagating activity typically moves gut contents along the gut. The speed, frequency and distance vary with type of food, species and also region of the gut. Burnstock (1958a) recorded propagating contractions in brown trout, both in stomach and intestine, when the gut was distended. The average frequency was one per 2 minutes. The contractions were nerve-dependent since they were blocked by hexamethonium, and involved an extrinsic component.

Peristalsis involves the combined action of ascending and descending reflex pathways, including both excitatory and inhibitory enteric nerves. Sensory neurons are activated by mechanical or chemical stimuli from the food, either directly or by release of 5-HT from endocrine cells. Such cells are found in many fish species, but not all (see Section 3.2). The sensory nerves simultaneously activate ascending and descending interneurons that in turn activate excitatory or inhibitory motor neurons, respectively. Comparatively few studies of peristaltic movement in fish *in vivo* exist; again most of our knowledge comes from *in vitro* or *in situ* experiments with extrapolations to mammalian studies. Little is also known about intrinsic sensory nerves in fish gut, their distribution, transmitter content and function, but it is suggested that the overall scheme is similar to mammals (Olsson and Holmgren, 2001). An ascending excitatory pathway and a descending inhibitory pathway is demonstrated in Atlantic cod (Karila and Holmgren, 1995, 1997; Karila *et al.*, 1998). The ascending pathway involves cholinergic motor neurons and a serotonergic mechanism, possibly by serotonergic interneurons acting on the cholinergic neurons (Karila and Holmgren, 1995). A TK is also involved in the excitatory phase – either acting directly on the smooth muscle, or indirectly via other neurons (Jensen *et al.*, 1987; Jensen and Holmgren, 1994), while NO exerts an inhibitory tonus on the ascending reflex (Karila and Holmgren, 1995). The descending pathway clearly involves NO, and possibly VIP (Karila and Holmgren, 1995, 1997; Olsson and Karila, 1995; Olsson and Holmgren, 2000), and serotonergic and possibly cholinergic neurons act as excitatory interneurons in the descending inhibition (Karila *et al.*, 1998).

Migrating motor complexes (MMCs) that occur in the interdigestive state constitute a different type of propagating contraction. They are slower than postprandial contraction waves and usually propagate for a longer distance. In mammals, MMCs contain three distinct phases with phase III being the most prominent with rhythmic contractions. Phase II is more irregular while phase I is more or less silent. A full MMC, containing all three phases, has not been demonstrated in fish so far but propagating contractions in fasted fish have been suggested to be correlated to phase III. In fasted rainbow trout, contractions occurred in the stomach *in vivo* at a frequency of approximately 0.6 per minute (Olsson *et al.*, 1999), i.e., similar to the frequency in brown trout. Likewise, in cod stomach and intestine, contractions with an average frequency of ca. 0.5 per minute have been recorded *in vitro* (Jensen *et al.*, 1991; Karila and Holmgren, 1995). In the cod intestine, the contractions propagated at a speed similar to mammalian MMCs (ca. 3.5 cm min^{-1}) (Karila and Holmgren, 1995).

Retrograde peristalsis, i.e., contraction waves propagating in an anal-to-oral direction, has been observed in fish. In developing zebrafish, video films demonstrate the occurrence of frequent retrograde peristaltic waves both in the upper intestine and in the rectal region (Holmberg *et al.*, 2003). Presumably these contractions will aid in mixing the chyme once the larvae start feeding. Similarly, retrograde peristalsis observed in halibut larvae is suggested to be important for mixing chyme with enzymes in the pyloric region (Rønnestad *et al.*, 2000). In adult *Scyliorhinus canicula*, retrograde peristalsis has been observed in connection with vomiting. The peristaltic wave can be initiated by stimulation of the splanchnic nerve, probably by release of 5-HT (Andrews and Young, 1993).

5.4. Control of Non-propagating Activity

Non-propagating motility patterns include, e.g., gastric accommodation, gastric emptying, and segmenting contractions for mixing of food. *Gastric accommodation*, i.e., relaxation in response to distension of the stomach wall, may involve both central and local reflex pathways. In fish, this accommodation occurs even after the vagal and splanchnic inputs are cut, suggesting a mainly local enteric control mechanism (Grove and Holmgren, 1992a,b).

Gastric emptying occurs when the pressure in the stomach exceeds that in the proximal intestine. Peristaltic contractions push chyme towards the pyloric sphincter that is under tonic contraction and hence normally closed. Gastric emptying in fish has been studied using a variety of methods, from *post mortem* examination of gut contents to X-ray (e.g., Fänge and Grove, 1979). The half-emptying time in rainbow trout is approximately 24 h (Olsson *et al.*, 1999). The lag phase before gastric emptying is usually between zero and 5 h, but is mainly dependent on the dryness of the food (Bucking and Wood, 2006). Chyme components entering the intestine act as a feedback signal for the rate of emptying. In particular, fat and acid stimulate release of CCK from endocrine cells in the proximal intestine, which acts via extrinsic and/or intrinsic nerves to reduce gastric emptying – the so-called "intestinal brake." In rainbow trout, CCK inhibits gastric emptying (Olsson *et al.*, 1999).

Segmentation of the gut for mixing of gut contents is achieved by local standing contractions in both the stomach and intestine. Little is known about the integrated control of this activity beyond the effects of individual transmitters as reported above (Section 5.2). Distension of the stomach, besides causing accommodation, also initiates contractile activity. In rainbow trout, the distension activates stimulatory cholinergic and serotonergic pathways. These may be modulated (inhibited) by SS released from local endocrine cells and by VIP neurons, respectively (Grove and Holmgren, 1992a). In cod, the VIP/cholinergic link appears to be missing (Grove and Holmgren, 1992b).

5.5. Control of Gallbladder Motility

The gallbladder of the hagfish is innervated by vagal cholinergic fibers which contract the gallbladder via muscarinic receptors (Holmgren and Fänge, 1981), and pharmacological evidence suggest that this is also the case in, e.g., rainbow trout (Aldman and Holmgren, 1987). CCK, presumably released from duodenal endocrine cells *in vivo*, contracts the gallbladder in teleost, elasmobranch and holostean species (e.g., Aldman and Holmgren, 1987; Andrews and Young, 1988b; Rajjo *et al.*, 1988). Duodenal acidification or infusion of fat or amino acids similarly increases gallbladder motility in fish – in mammals this is mediated by CCK (Aldman *et al.*, 1992; Aldman and Holmgren, 1995). Careful pharmacological studies, *in vivo*, suggest that CCK in low concentrations acts via stimulation of cholinergic neurons, while high concentrations have a direct effect on the gallbladder muscle (Aldman and Holmgren, 1995). Inhibition of the rainbow trout gallbladder motility is elicited by AD, or VIP acting via a β-adrenergic pathway (Aldman and Holmgren, 1987, 1992).

5.6. Development of the Control of Gut Motility

Zebrafish larvae display uncoordinated contractions at 3 dpf (Holmberg *et al.*, 2003). Propagating contraction waves spread both anally and orally from a region between the proximal and middle intestine at 4 dpf. Exogenous feeding in zebrafish starts around 5–6 dpf and there is an increase in neuronal control between 4 and 7 dpf (Holmberg *et al.*, 2007). Propagating contractions have also been observed in newly metamorphosed juvenile halibut *in vivo* (Rønnestad *et al.*, 2000). A number of signal substances expressed in the young larvae may affect the contractions (Holmberg *et al.*, 2004, 2006). At 4 dpf, ACh can be released and act on muscarinic receptors, increasing the frequency of contraction waves from the basal activity. Endogenously released NO creates an inhibitory tonus. Peptides like PACAP and NKA seem to be effective from 5 dpf.

6. CONTROL OF SECRETION AND DIGESTION

Gastric acid secretion is almost the only secretory mechanism of the gut where mechanistic studies have been performed to some extent in fish. Information on the neuronal and endocrine control of pepsinogen secretion or secretion and/or release of pancreatic juices and bile is scattered and the control of mucus secretion even more so. Gut secretion also comprises the activity of endocrine cells in the gut mucosa – in this case there is both the control of the endocrine cells as well as the control that the endocrine cells in turn exert on other cells to be considered.

6.1. Gastric Acid Secretion

Gastric acid is secreted from so-called "oxynticopeptic" cells, in fish as in many other non-mammalian vertebrates. The oxynticopeptic cells are usually gathered in gastric glands in the stomach mucosa, and they secrete both acid and pepsinogen (Bishop and Odense, 1966; Mattisson and Holstein, 1980; Ezeasor, 1981). An *in situ* hybridization study in winter flounder, *Pseudopleuronectes americanus*, confirms the existence of cells expressing both pepsinogen and proton pumps in this species (Gawlicka *et al.*, 2001). However, oxynticopeptic cells are not a feature of all fish species, and species differences commonly occur even between closely related species (Rebolledo and Vial, 1979; Michelangeli *et al.*, 1988). In some species purely oxyntic and/or peptic cells are found along with the oxynticopeptic cells (Smolka *et al.*, 1994). Interestingly, in winter flounder, a proton pump is found on mucous cells suggesting acid secretion also from these cells (Gawlicka *et al.*, 2001).

Gastric glands show large species variations in distribution. Often, they are restricted to the cardiac part of the stomach (Darias *et al.*, 2005), but sometimes they are found in all regions of the stomach (Arellano *et al.*, 2001). In some species, such as the rainbow trout, the acid secreting mucosa continues into the distal part of the esophagus (Ezeasor, 1984). Stomach-less fish usually lack an acid secreting mucosa (e.g., Koelz, 1992). Fish have a basal secretion of gastric acid, even if the stomach is not digesting. In Atlantic cod, *in vivo*, the basal secretion is controlled largely by a vagal tonus, since it is almost abolished after vagotomy (Holstein and Cederberg, 1980). Food intake increases gastric acid secretion. This was observed in an elasmobranch as early as 1905 (Sullivan, 1905–6) and has been repeatedly confirmed in several species (e.g., Norris *et al.*, 1973; Maier and Tullis, 1984). At least part of the stimulus is through distension of the stomach wall (Smit, 1968).

Several neuronal or endocrine signal substances are involved in the control of the oxynticopeptic cells (Figure 10.3). As in mammals, gastrin, histamine, ACh and SS play central roles in the control of acid secretion. Histamine and ACh stimulate secretion in several elasmobranchs and teleosts (e.g., Holstein, 1976; Smit, 1968). The effect of histamine is presumably mediated by H2 receptors in Atlantic cod since some H2 agonists stimulate and H2 receptor antagonists block or reduce the acid secretion (Bomgren and Jönsson, 1996; Holstein, 1976, 1986). In addition, VIP and SS inhibit histamine-induced secretion in the cod (Holmgren *et al.*, 1986; Holstein, 1983). A feedback mechanism reduces the secretory rate when luminal pH is lowered, but the mediating mechanism is not identified (Bomgren and Jönsson, 1996). ACh is presumably released from vagal pathways in cod *in vivo*, and acts on muscarinic receptors, since vagotomy almost abolishes gastric acid secretion, and atropine blocks the effect of cholinergic drugs (Holstein, 1977; Holstein and Cederberg, 1980). In addition, an H2 antagonist blocks the effect of ACh, suggesting an effect via histamine release (Holstein, 1976).

Gastrin has an unusual effect in Atlantic cod, being inhibitory on acid secretion (Holstein, 1982). In the common stingray, *Dasyatis pastinaca*, pentagastrin stimulates gastric secretion but there is no change in

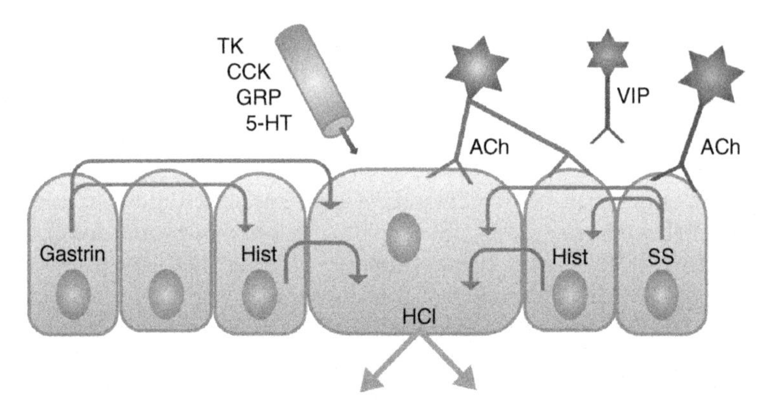

Fig. 10.3. Simplified summary of paracrine, endocrine and neuronal control of gastric acid secretion in fish (based on Bomgren, 2001). Red = stimulation; blue = inhibition. 5-HT, 5-hydroxytryptamine (serotonin); ACh, acetylcholine; CCK, cholecystokinin; GRP, gastrin-releasing peptide; HCl, hydrochloride acid; Hist, histamine; SS, somatostatin; TK, tachykinins; VIP, vasoactive intestinal polypeptide. (See Color Insert.)

intraluminal pH (Zaks *et al.*, 1975). It is possible that during evolution, changes have led to different mechanisms in fish compared to tetrapods (Vigna, 1983). However, mammalian gastrins were used for these studies, and it is possible that the slightly different amino acid sequence compared to fish gastrins makes them antagonistic at the fish receptors. BBS in low doses stimulates gastric acid secretion in the Atlantic cod (Holstein and Humphrey, 1980). In mammals, low doses of BBS (or the native GRP) act via release of (stimulatory) gastrin. However, if the inhibitory effect of mammalian gastrin in cod reflects the effect of native fish gastrin, the stimulation caused by BBS probably does not involve gastrin. This is supported by the fact that no changes in plasma levels of gastrin were found after BBS treatment (Holstein and Humphrey, 1980). Instead, inhibition of an inhibitory VIP tonus is suggested to mediate at least part of the effect of BBS, since BBS lowers plasma levels of VIP (Holstein and Humphrey, 1980; Holstein, 1983).

6.2. Pepsinogen/Pepsin Secretion

In addition to gastric acid secretion, oxynticopeptic cells synthesize, store and secrete pepsinogen, which is converted to the active enzyme pepsin when secreted (Smit, 1968). Pepsinogen-ir cells have been found in both teleost and elasmobranch species (Yasugi *et al.*, 1988). In Atlantic cod, *in vivo*, TK and 5-HT strongly stimulate pepsinogen secretion. Unusually, in comparison with mammals, ACh and histamine have weak effects. Gastric acid and pepsinogen secretion show different sensitivities to the stimulatory agents.

For example, TK have strong stimulatory effects on pepsinogen secretion but weak effects on acid secretion, while histamine and carbachol have the opposite effects. This means that acid and pepsinogen may be separately released from the oxynticopeptic cells (Holstein and Cederberg, 1986).

6.3. Pancreatic Secretion

The fish endocrine pancreas has been extensively described by, e.g., Youson and Al-Mahrouki (1999) and Youson *et al.* (2006), but very little is known of the neuronal and endocrine control of the release of pancreatic hormones in fish. Innervation of islet tissue includes CCK-, GAL-, NPY-, oxytocin- and VIP-ir fibers (Noe *et al.*, 1986; McDonald *et al.*, 1987; Jönsson, 1991). CCK, GAL and NPY affect the release of SSs from Brockmann bodies in rainbow trout, with separate effects on SST-14 and SST-35 (Eilertson *et al.*, 1996).

Exocrine pancreatic tissue secretes bicarbonate and all types of enzymes needed for digestion of nutrients. Feeding and components of the food stimulate secretion/activity of pancreatic enzymes (e.g., Krogdahl and Bakke-McKellep, 2005; Murashita *et al.*, 2007). Several studies indicate that CCK is an important mediator of this effect. Increased levels of CCK or its mRNA after feeding have been measured in rainbow trout (Jönsson *et al.*, 2006) and yellowtail, *Seriola quinqueradiata* (Murashita *et al.*, 2007). Intraperitoneal injection *in vivo* of mammalian CCK, mimicking a hormonal effect, stimulates release of trypsin and chymotrypsin in Atlantic salmon, *Salmo salar* (Einarsson *et al.*, 1997). Murashita *et al.* (2007) also show that mRNA levels of peptide Y (PY) follow the opposite trend compared to CCK, i.e., a transient decrease after feeding, and suggest that PY is antagonistic to CCK in control of enzyme secretion from fish pancreas. Secretin might be lacking in fish (Cardoso *et al.*, 2004).

6.4. Secretion from the Rectal Gland of Elasmobranchs

The rectal gland of elasmobranchs is a salt (NaCl) secreting exocrine gland, with its duct opening into the rectum. The gland gives elasmobranchs an extra capacity to deal with the hyperosmolarity caused by the urea in their blood. The function and neuronal and endocrine control of the rectal gland has been carefully reviewed by Olson (1999). In short, an increase in cAMP activity increases secretion. The most potent stimulator is scyliorhinin II (initially called rectin before it was fully sequenced) (Shuttleworth and Thorndyke, 1984; Anderson *et al.*, 1995). In spiny dogfish, secretion from the gland can also be stimulated by VIP (Stoff *et al.*, 1979), however, this effect could possibly be attributed to the increase in blood flow through the gland caused by VIP (Thorndyke *et al.*, 1989). Adenosine acting through adenosine A1 receptors is inhibitory on secretion from the rectal gland in spiny dogfish (Kelley *et al.*, 1991). Similarly, NPY as well as BBS interacting with SS cause an inhibition (Silva *et al.*, 1990, 1993). Immunohistochemistry supports the physiological studies (Holmgren and Nilsson, 1983a; Stoff *et al.*, 1988). It has also been speculated that peptides related to atrial natriuretic peptide (ANP) exert a hormonal control (Olson, 1999).

7. CONTROL OF NUTRIENT ABSORPTION

Transporters in the apical and basolateral cell membranes of mucosal enterocytes move amino acids and monosaccharides from the chyme into the blood. The absorption rate can be regulated by an increase or decrease in number or activity of the appropriate transporter. The diet composition may change the relative uptake of glucose and amino acids, and hence the population of transporters, also in fish (Buddington *et al.*, 1987). Several hormones affect the transporters, but although enteric nerves are present in the mucosal layer of the gut, little is known of whether and how they affect the absorption rate.

Reports on hormonal effects on absorption in fish are also few and sometimes circumstantial. Calcitonin is present in endocrine cells of goldfish intestinal mucosa. Diet affects the number of these cells, and it is speculated that calcitonin release may inhibit absorption of nutrients by a paracrine effect (Okuda *et al.*, 1999). Local gut hormones known from mammals to affect nutrient absorption, such as enteroglucagon, are found in fish (Rombout, 1977). The most solid evidence comes from studies of circulating hormones, which may affect the rate of absorption. For example, methyltestosterone rapidly stimulates glucose uptake, probably by increasing the insertion of stored glucose transporters into the cell membrane (Hazzard and Ahearn, 1992). Cortisol and growth hormone increase and AD decreases the uptake of proline in Coho salmon, *Oncorhynchus kisutch* (Collie and Stevens, 1985). Sex steroids increase uptake of leucine in rainbow trout intestine (Habibi and Ince, 1984).

8. CONTROL OF WATER AND ION TRANSPORT

Local and circulating hormones, and possibly to a certain extent mucosal nerves, control ion transport over the gut mucosa. The mechanisms are tightly integrated with the maintenance of homeostasis and the control of osmoregulation. Cortisol and prolactin thus play important roles in the control of water and ion fluxes, particularly when fish adapt to a change in salinity (e.g., Santos *et al.*, 2001; Takahashi *et al.*, 2006). SS controls Na^+, Cl^- and water fluxes in the eel gut (Uesaka *et al.*, 1994). ACh and NA (from nerves) decrease transcellular ion transport while 5-HT increases this transport. It is also hypothesized that NA decreases while 5-HT increases paracellular conductance for anions (Trischitta *et al.*, 1999). NO modulates paracellular pathways directly by changing tight junction permeability and transcellular transport indirectly by inhibiting bicarbonate ion entry into the endothelial cell (Trischitta *et al.*, 2007). Stanniocalcin (released from the corpuscles of Stannius) controls intestinal uptake of Ca^{2+} in teleost fish (e.g., Sundell *et al.*, 1992). A detailed presentation of the control of osmoregulation in the gut is beyond the scope of the current chapter, and the reader is referred to excellent reviews in the field such as that of McCormick (2001).

9. CONTROL OF SPLANCHNIC CIRCULATION

The fish gut is supplied by blood through arteries branching off the dorsal aorta, forming the splanchnic circulation (see, e.g., Farrell *et al.*, 2001). To optimize gut functions, blood can be redistributed to the gut from other parts of the body. The flow to the gut is determined by the general blood pressure and the resistance of the GI vascular bed relative to the resistance of the remaining (somatic) vascular bed. Local resistance depends on a number of factors, including vascular innervation, hormonal tonus, local oxygen levels, metabolites released from gut tissues, and the level of compression caused by smooth muscle in the surrounding gut wall (see Farrell *et al.*, 2001). The scope of this chapter is limited to the control exerted by (local) neurons and hormones, but it must be kept in mind that changes in general cardiovascular performance, circulating vasoactive hormones as well as other local factors, have profound effects on gut blood flow.

Nerves innervating the arteries and arterioles to the gut usually form a network of varicose fibers in the adventitia with a concentration to the medioadventitial border, and only a few fibers penetrate the media. Hormones released from nearby endocrine/paracrine cells, and neurotransmitters thus have effects "from the outside," and predominantly directly on the vascular smooth muscle. Circulating hormones or vascular active substances (angiotensin, bradykinin, etc.) affect vessel smooth muscle from the inside in two ways, either directly or, probably more common, indirectly through release of endothelial factors.

Intake of food is normally associated with an increase in blood flow to the gut, the so-called "postprandial hyperemia." This has been verified by direct measurements in several teleosts (Axelsson *et al.*, 1989, 2000; Axelsson and

Fritsche, 1991), while data from other fish groups are more circumstantial. Exercise and/or stress decrease blood flow to the gut.

The increase in blood flow may depend on decreased resistance of the GI vascular bed, and/or a simultaneous increase in resistance of the somatic vascular bed, both leading to a shunting of the blood to the gut. A number of neuronal and endocrine substances have been found to relax/dilate vessels or increase the flow to the fish gut, and are likely to be more or less important mediators of the postprandial hyperemia (Figure 10.4). Increase in resistance of the splanchnic vasculature is mainly due to an adrenergic activity, but other substances such as natriuretic peptides and endothelins may have supplementary effects (Evans, 2001).

9.1. Reduction in Flow – Catecholamines

Adrenergic mechanisms dominate the vasoconstrictor control in the gut. Adrenergic nerves innervate gut wall vessels and major vessels leading to the gut in several elasmobranch and teleost species (e.g., Anderson, 1983; Holmgren and Nilsson, 1983a), and in addition, levels of circulating catecholamines increase during voluntary exercise and stress (Abrahamsson, 1979; Axelsson, 1988). α-AR mediate an adrenergic contraction of the major gut vessels (e.g., Holmgren and Nilsson, 1974; Nilsson *et al.*, 1975), and *in vivo* studies show the presence of an α-adrenergic tonus in unfed fish (Holmgren *et al.*, 1992a; Axelsson *et al.*, 2000). There is a species difference in how much the postprandial increase in gut blood flow is due to a release of this tonus (Axelsson *et al.*, 1989; Axelsson and Fritsche, 1991).

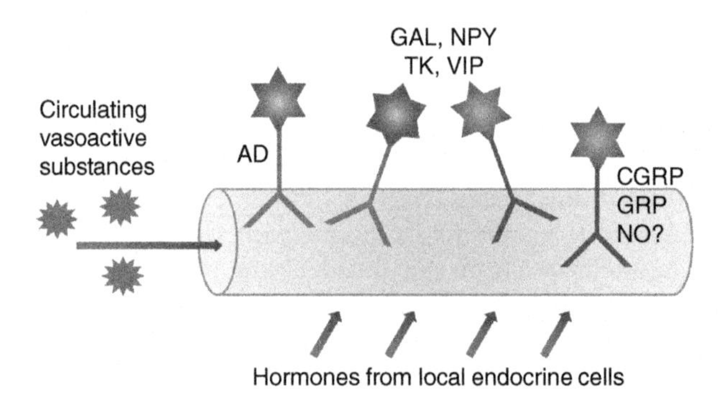

Fig. 10.4. Hormonal and neuronal control of gut blood flow in teleosts. Red = reduced flow; blue = increased flow; AD, adrenaline; GAL, galanin; CGRP, calcitonin gene-related peptide; GRP, gastrin-releasing peptide; NO, nitric oxide; NPY, neuropeptide Y; TK, tachykinins; VIP, vasoactive intestinal polypeptide. (See Color Insert.)

Injections, *in vivo*, of catecholamines mimic the decrease in blood flow to the gut caused by exercise or fright. In spiny dogfish, both AD and NA increase the resistance of the celiac artery vascular bed, decreasing the flow to the gut (Holmgren *et al.*, 1992a; Axelsson *et al.*, 2000). The relative importance of adrenergic nerves versus circulating catecholamines is not clear, but the effects of circulating catecholamines may be dominating.

9.2. Increase in Flow

CGRP causes vasorelaxation in isolated rainbow trout celiac arteries by receptors on the smooth muscle cells (Kågström and Holmgren, 1998). Similarly, an endothelium-independent, CGRP-induced relaxatory mechanism is reported for the dorsal aorta from the holocephalan elephant fish, *Callorhinchus milii* (Jennings *et al.*, 2007).

GRP/BBS release from the gut is increased when food is ingested (Jönsson *et al.*, 2006); the source may be both gut endocrine cells and enteric neurons. BBS and/or GRP may increase the flow to the gut by dilation of gut vessels, as in the perfused gut from spiny dogfish (Bjenning *et al.*, 1990), but the dominating effect is likely to be a general increase in somatic resistance, shunting blood to the gut (Holmgren *et al.*, 1992b).

NO may control gut vessels only in some species. Since addition of substrate (L-arginine) or NOS antagonists had no effect on gut vessels from the rainbow trout, it was concluded that these vessels are not controlled by a nitrergic mechanism, although the NO donor sodium nitroprusside induces vasodilatation (Olson and Villa, 1991). However, subsequent studies in other species suggested that NO derived from perivascular nerves causes vasodilatation (Donald and Broughton, 2005; Jennings *et al.*, 2007). Interestingly, fish may use neuronal NOS contained in perivascular nerve fibers for control of blood vessels, rather than endothelial NOS stimulated by circulating agents as in mammals.

9.3. Variable Effects

Galanin (GAL) contracts isolated celiac and mesenteric arteries from Atlantic cod and pancreatic-mesenteric arteries of elasmobranchs (Karila *et al.*, 1993; Preston *et al.*, 1995), but decreases resistance *in vivo* of the celiac artery in lungfish (Holmgren *et al.*, 1994). Vasoconstriction in the cod was independent of adrenergic mechanisms. The decrease in vascular resistance in the lungfish could involve a nitrergic link, as has been shown for GAL-induced decrease in general blood pressure in trout (Le Mevel *et al.*, 1998).

NPY, co-released with NA from sympathetic nerves, is considered an important factor in the control of vascular beds in mammals. NPY causes an increase in blood flow to the gut by a reduction in local vascular resistance in spiny dogfish (Holmgren *et al.*, 1992b). This is in contrast to its most common effect in mammals. However, in another three elasmobranch species, Preston *et al.* (1998) found that isolated gut arteries contract in response to NPY. It is not clear whether this is a true species difference or depends on the experimental setups. Similarly to spiny dogfish arteries, both untreated and precontracted celiac arteries from Atlantic cod relax in response to cod NPY. This is partly due to a direct effect on the vascular smooth muscle and partly due to release of prostaglandin, but not through release of NO or some endothelium derived factor (Shahbazi *et al.*, 2002). NPY is present in about 10% of the adrenergic neurons projecting to enteric blood vessels and the submucosa (Karila *et al.*, 1997), but further studies are needed to explain the physiological significance of releasing a stimulatory and an inhibitory substance simultaneously.

Tachykinins (TKs) are vasodilatory in most vascular beds in mammals, acting via an endothelium-dependent mechanism, but vasoconstrictory, by a direct action on the smooth muscle, in some vessels (see, e.g., Severini *et al.*, 2002). The effects in fish vary between species, but also within one species due to an action via several receptor types. Thus, mammalian substance P increases blood flow to the gut in the spiny dogfish and reduces it in the Australian lungfish (Holmgren *et al.*, 1992a, 1994). In cod, *in vivo*, mammalian substance P produces a mixed response; a larger initial increase in blood flow is transiently interrupted by a small superimposed cholinergic decrease (Jensen *et al.*, 1991). Subsequent studies using a number of native fish TKs in spiny dogfish and rainbow trout show that TKs may cause shunting of blood to the stomach, and also that several types of TK receptors are present, some mediating excitatory and some mediating inhibitory responses (Kågström *et al.*, 1996a,b). For example, native trout substance P and NKA occasionally induced vasoconstriction in isolated vessels from the rainbow trout, but *in vivo,* they induced an early increase in celiac artery blood flow followed by a decrease (Kågström *et al.*, 1996b; Kågström and Holmgren, 1998).

Using crude *VIP* extracts from the gut of trout and catfish, *Ictalurus melas,* Holder *et al.* (1983) demonstrated a vasodilatation in a perfused intestinal loop of the catfish gut. In Atlantic cod, *in vivo*, it was later found that VIP causes an increase in flow both in the celiac and the mesenteric arteries, but while the increase in the celiac artery depends on both lowered vascular resistance and an elevated cardiac output, the increased flow in the less densely innervated mesenteric artery is due to an increase in cardiac output only (Jensen *et al.*, 1991). Kågström and Holmgren (1997), using small vessels from the intestine of the rainbow trout, established that VIP-induced relaxation of the vessel wall is independent of the endothelium and a nitrergic link (in contrast to many mammalian vessels), but involves prostaglandin synthesis. In spiny dogfish, VIP increases vascular resistance, thereby reducing blood flow to the gut (Holmgren *et al.*, 1992a), which is an unusual effect, and in contrast to the vasodilatation combined with increased glandular secretion (as in mammalian exocrine glands) obtained in the rectal gland from the same species (Thorndyke *et al.*, 1989).

10. CONCLUDING REMARKS

Fish follow the general vertebrate gut basic anatomy, with some notable exceptions such as the lack of a stomach in several families and the occurrence of a spiral intestine in many elasmobranchs. Similarly, the neuronal and endocrine systems shows the same trends, with a general vertebrate structure combined with some notable exceptions. There is an abundance of immunohistochemical studies, which are getting increasingly more reliable concurrently with increased availability of sequenced material. They show us that fish are clearly as varied as other vertebrates in the occurrence of signal substances in the gut neuronal and endocrine systems, if not more. At the same time, there are few signal substances found as yet that are completely unique to fish.

There is some knowledge on the function of individual neurohormonal substances, but systematic studies comparing the situation in different fish groups are mainly lacking. Above all, there is very little known of the integrated function of gut neurohormonal mechanisms in fish, an issue that is of growing importance with the increasing effects of climate and other environmental changes. Here, although many mechanisms in general follow the vertebrate plan, there is much more variation in details, with examples of opposite effects or different interactions between signal substances between fish groups or in comparison with mammals. Many more comparative studies are needed to establish whether these exceptions are the results of an evolutionary coincidence or the results of an evolutionary pressure.

REFERENCES

Abrahamsson, T. (1979). Phenylethanolamine-N-methyl transferase (PNMT) activity and catecholamine storage and release from chromaffin tissue of the spiny dogfish, *Squalus acanthias. Comp. Biochem. Physiol.* **64C,** 169–172.

Abrahamsson, T., and Nilsson, S. (1976). Phenylethanolamine-N-methyl transferase (PNMT) activity and catecholamine content in chromaffin tissue and sympathetic neurons in the cod, *Gadus morhua. Acta Physiol. Scand.* **96,** 94–99.

Aldman, G., and Holmgren, S. (1987). Control of gallbladder motility in the rainbow trout, *Salmo gairdneri. Fish Physiol. Biochem.* **4,** 143–155.

Aldman, G., and Holmgren, S. (1992). VIP inhibits CCK-induced gallbladder contraction involving a β-adrenoceptor mediated pathway in the rainbow trout, *Oncorhynchus mykiss, in vivo*. *Gen. Comp. Endocrinol.* **88,** 287–291.

Aldman, G., and Holmgren, S. (1995). Intraduodenal fat and amino acids activate gallbladder motility in the rainbow trout, *Oncorhynchus mykiss*. *Gen. Comp. Endocrinol.* **100,** 27–32.

Aldman, G., Jönsson, A. C., Jensen, J., and Holmgren, S. (1989). Gastrin/CCK-like peptides in the spiny dogfish, *Squalus acanthias*; concentrations and actions in the gut. *Comp. Biochem. Physiol. C* **92,** 103–108.

Aldman, G., Grove, D. J., and Holmgren, S. (1992). Duodenal acidification and intra-arterial injection of CCK8 increase gallbladder motility in the rainbow trout *Oncorhynchus mykiss*. *Gen. Comp. Endocrinol.* **86,** 20–25.

Anderson, C. (1983). Evidence for 5-HT-containing intrinsic neurons in the teleost intestine. *Cell Tissue Res.* **230,** 377–386.

Anderson, C., and Campbell, G. (1988). Immunohistochemical study of 5-HT-containing neurons in the teleost intestine: Relationship to the presence of enterochromaffin cells. *Cell Tissue Res.* **254,** 553–559.

Anderson, C. R., Campbell, G., O'Shea, F., and Payne, M. (1991). The release of neuronal 5-HT from the intestine of a teleost fish, *Platycephalus bassensis*. *J. Auton. Nerv. Syst.* **33,** 239–246.

Anderson, W. G., Conlon, J. M., and Hazon, N. (1995). Characterization of the endogenous intestinal peptide that stimulates the rectal gland of *Scyliorhinus canicula*. *Am. J. Physiol.* **268,** R1359–R1364.

Andrews, P. L., and Young, J. Z. (1988a). The effect of peptides on the motility of the stomach, intestine and rectum in the skate (*Raja*). *Comp. Biochem. Physiol. C* **89,** 343–348.

Andrews, P. L., and Young, J. Z. (1988b). A pharmacological study of the control of motility in the gallbladder of the skate. *Comp. Biochem. Physiol. C* **89,** 349–354.

Andrews, P. L. R., and Young, J. Z. (1993). Gastric motility patterns for digestion and vomiting evoked by sympathetic nerve stimulation and 5-hydroxytryptamine in the dogfish *Scyliorhinus canicula*. *Phil. Trans. R Soc. Lond.* **342,** 363–380.

Arellano, J. M., Storch, V., and Sarasquete, C. (2001). Histological and histochemical observations in the stomach of the Senegal sole, *Solea senegalensis*. *Histol. Histopathol.* **16,** 511–521.

Aronsson, U., and Holmgren, S. (2000). Muscarinic M3-like receptors, cyclic AMP and L-type calcium channels are involved in the contractile response to cholinergic agents in gut smooth muscle of the rainbow trout, *Oncorhynchus mykiss*. *Fish Physiol. Biochem.* **23,** 353–361.

Axelsson, M. (1988). The importance of nervous and humoral mechanisms in the control of cardiac performance in the Atlantic cod *Gadus morhua* at rest and during non-exhaustive exercise. *J. Exp. Biol.* **137,** 287–302.

Axelsson, M., and Fritsche, R. (1991). Effects of exercise, hypoxia and feeding on the gastrointestinal blood flow in the Atlantic cod *Gadus morhua*. *J. Exp. Biol.* **158,** 181–198.

Axelsson, M., Driedzic, W. R., Farrell, A. P., and Nilsson, S. (1989). Regulation of cardiac output and gut flow in the sea raven, *Hemitripterus americanus*. *Fish Physiol. Biochem.* **196,** 2–12.

Axelsson, M., Thorarensen, H., Nilsson, S., and Farrell, A. P. (2000). Gastrointestinal blood flow in the red Irish lord, *Hemilepidotus hemilepidotus*: long-term effects of feeding and adrenergic control. *J. Comp. Physiol. B* **170,** 145–152.

Baumgarten, H. G., Björklund, A., Lachenmayer, L., Nobin, A., and Rosengren, E. (1973). Evidence for the existence of serotonin-, dopamine-, and noradrenaline-containing neurons in the gut of *Lampetra fluviatilis*. *Z. Zellforsch. Mikrosk. Anat.* **141,** 33–54.

Bisgrove, B. W., Raible, D. W., Walter, V., Eisen, J. S., and Grunwald, D. J. (1997). Expression of c-ret in the zebrafish embryo: Potential roles in motoneuronal development. *J. Neurobiol.* **33,** 749–768.

Bishop, C., and Odense, P. H. (1966). Morphology of the digestive tract of the Atlantic cod, *Gadus morhua*. *J. Fish. Res. Bd. Canada* **23,** 1607–1615.

Bjenning, C., and Holmgren, S. (1988). Neuropeptides in the fish gut. An immunohistochemical study of evolutionary patterns. *Histochemistry* **88,** 155–163.

Bjenning, C., Driedzic, W. R., and Holmgren, S. (1989). Neuropeptide Y-like immunoreactivity in the cardiovascular nerve plexus of the elasmobranchs *Raja erinacea* and *Raja radiata*. *Cell Tiss. Res.* **255,** 481–486.

Bjenning, C., Jönsson, A.-C., and Holmgren, S. (1990). Bombesin-like immunoreactive material in the gut, and the effect of bombesin on the stomach circulatory system of an elasmobranch fish, *Squalus acanthias*. *Regul. Pept.* **28,** 57–69.

Bomgren, P. (2001). "Gastrin and histamine in amphibians and fish." PhD thesis, University of Gothenburg, Göteborg, Sweden.

Bomgren, P., and Jönsson, A.-C. (1996). Basal, H2-receptor stimulated and pH-dependent gastric acid secretion from an isolated stomach mucosa preparation of the Atlantic cod, *Gadus morhua*, studied using a modified pH-static titration method. *Fish Physiol. Biochem.* **15,** 275–285.

Bosi, G., Shinn, A. P., Giari, L., Arrighi, S., and Domeneghini, C. (2004). The presence of a galanin-like peptide in the gut neuroendocrine system of *Lampetra fluviatilis* and *Acipenser transmontanus*: an immunohistochemical study. *Tissue Cell* **36,** 283–292.

Bosi, G., Shinn, A. P., Giari, L., Simoni, E., Pironi, F., and Dezfuli, B. S. (2005). Changes in the neuromodulators of the diffuse endocrine system of the alimentary canal of farmed rainbow trout, *Oncorhynchus mykiss* (Walbaum), naturally infected with *Eubothrium crassum (Cestoda)*. *J. Fish. Dis.* **28,** 703–711.

Bosi, G., Bermudez, R., and Domeneghini, C. (2007). The galaninergic enteric nervous system of *Pleuronectiformes (Pisces, Osteichthyes)*: An immunohistochemical and confocal laser scanning immunofluorescence study. *Gen. Comp. Endocrinol.* **152,** 22–29.

Brüning, G., Hattwig, K., and Mayer, B. (1996). Nitric oxide synthase in the peripheral nervous system of the goldfish, *Carassius auratus*. *Cell Tissue Res.* **284,** 87–98.

Bucking, C., and Wood, C. M. (2006). Water dynamics in the digestive tract of the freshwater rainbow trout during the processing of a single meal. *J. Exp. Biol.* **209,** 1883–1893.

Buddington, R. K., Chen, J. W., and Diamond, J. (1987). Genetic and phenotypic adaptation of intestinal nutrient transport to diet in fish. *J. Physiol.* **393,** 261–281.

Burka, J. F., Blair, R. M. J., and Hogan, J. E. (1989). Characterization of the muscarinic and serotonergic receptors of the intestine of the rainbow trout, *Salmo gairdneri*. *Can. J. Physiol. Pharmacol.* **67,** 477–482.

Burkhardt-Holm, P., and Holmgren, S. (1989). A comparative study of neuropeptides in the intestine of two stomachless teleosts (*Poecilia reticulata, Leuciscus idus melanotus)* under conditions of feeding and starvation. *Cell Tissue Res.* **255,** 245–254.

Burnstock, G. (1958a). The effect of drugs on spontaneous motility and on response to stimulation of the extrinsic nerves of the gut of a teleostean fish. *Br. J. Pharmacol. Chemother.* **13,** 216–226.

Burnstock, G. (1958b). Reversible inactivation of nervous activity in a fish gut. *J. Physiol.* **141,** 35–45.

Burnstock, G. (1969). Evolution of the autonomic innervation of visceral and cardiovascular systems in vertebrates. *Pharmacol. Rev.* **21,** 247–324.

Campbell, G. (1975). Inhibitory vagal innervation of the stomach in fish. *Comp. Biochem. Physiol.* **50C,** 169–170.

Campbell, G., and Burnstock, G. (1968). Comparative physiology of gastrointestinal motility. *In* "Handbook of Physiology" (C. F., Code, Ed.), pp. 2213–2266. American Physiological Society, Washington DC.

Cardoso, J. C., Power, D. M., Elgar, G., and Clark, M. S. (2004). Duplicated receptors for VIP and PACAP (VPAC1R and PAC1R) in a teleost fish, *Fugu rubripes*. *J. Mol. Endocrinol.* **33,** 411–428.

Chiba, A. (1998). Ontogeny of serotonin-immunoreactive cells in the gut epithelium of the cloudy dogfish, *Scyliorhinus torazame*, with reference to coexistence of serotonin and neuropeptide Y. *Gen. Comp. Endocrinol.* **111,** 290–298.

Chow, B. K. C. (1997). The goldfish vasoactive intestinal polypeptide receptor: Functional studies and tissue distribution. *Fish Physiol. Biochem.* **17,** 213–222.

Cimini, V., Van Noorden, S., Giordano-Lanza, G., Nardini, V., McGregor, G. P., Bloom, S. R., and Polak, J. M. (1985). Neuropeptides and 5-HT immunoreactivity in the gastric nerves of the dogfish (*Scyliorhinus stellaris*). *Peptides* **6**(Suppl 3), 373–377.

Collie, N. L., and Stevens, J. J. (1985). Hormonal effects on L-proline transport in coho salmon (*Oncorhynchus kisutch*) intestine. *Gen. Comp. Endocrinol.* **59,** 399–409.

Conlon, J. M., and Larhammar, D. (2005). The evolution of neuroendocrine peptides. *Gen. Comp. Endocrinol.* **142,** 53–59.

Conlon, J. M., Deacon, C. F., O'Toole, L., and Thim, L. (1986). Scyliorhinin I and II: two novel tachykinins from dogfish gut. *FEBS Lett.* **200,** 111–116.

Conlon, J. M., Henderson, I. W., and Thim, L. (1987). Gastrin-releasing peptide from the intestine of the elasmobranch fish, *Scyliorhinus canicula* (common dogfish). *Gen. Comp. Endocrinol.* **68,** 415–420.

D'Este, L., and Renda, T. (1995). Phylogenetic study on distribution and chromogranin/secretogranin content of histamine immunoreactive elements in the gut. *Ital. J. Anat. Embryol.* **100**(Suppl 1), 403–410.

Darias, M. J., Murray, H. M., Martínez-Rodrígueza, G., Cárdenasc, S., and Yúfera, M. (2005). Gene expression of pepsinogen during the larval development of red porgy (*Pagrus pagrus*). *Aquaculture* **248,** 245–252.

Domeneghini, C., Radaelli, G., Arrighi, S., Mascarello, F., and Veggetti, A. (2000). Neurotransmitters and putative neuromodulators in the gut of *Anguilla anguilla* (L.). Localizations in the enteric nervous and endocrine systems. *Eur. J. Histochem.* **44,** 295–306.

Donald, J. A., and Broughton, B. R. (2005). Nitric oxide control of lower vertebrate blood vessels by vasomotor nerves. *Comp. Biochem. Physiol.* **142A,** 188–197.

Eilertson, C. D., Carneiro, N. M., Kittilson, J. D., Comley, C., and Sheridan, M. A. (1996). Cholecystokinin, neuropeptide Y and galanin modulate the release of pancreatic somatostatin-25 and somatostatin-14 *in vitro*. *Regul. Pept.* **63,** 105–112.

Einarsson, S., Davies, P. S., and Talbot, C. (1997). Effect of exogenous cholecystokinin on the discharge of the gallbladder and the secretion of trypsin and chymotrypsin from the pancreas of the Atlantic salmon, *Salmo salar* L. *Comp. Biochem. Physiol.* **117C,** 63–67.

Evans, D. H. (2001). Vasoactive receptors in abdominal blood vessels of the dogfish shark, *Squalus acanthias*. *Physiol. Biochem. Zool.* **74,** 120–126.

Ezeasor, D. N. (1981). The fine structure of the gastric epithelium on the rainbow trout, *Salmo gairdneri*, Richardson. *J. Fish Biol.* **19,** 611–627.

Ezeasor, D. N. (1984). Light and electron microscopic studies on the oesophageal epithelium of the rainbow trout, *Salmo gairdneri*. *Anat. Anz.* **155,** 71–83.

Falkmer, S., Fahrenkrug, J., Alumets, J., Håkanson, R., and Sundler, F. (1980). Vasoactive intestinal polypeptide (VIP) in epithelial cells of the gut mucosa of an elasmobranchian cartilaginous fish, the ray. *Endocrinol. Japon* **1,** 31–35.

Fänge, R., and Grove, D. J. (1979). Digestion. *In* "Fish Physiology" (Hoar, W. S., and Randall, W. S., Eds.), pp. 161–260. Academic Press, London.

Farrell, A. P., Thorarensen, H., Axelsson, M., Crocker, C. E., Gamperl, A. K., and Cech, J. J., Jr. (2001). Gut blood flow in fish during exercise and severe hypercapnia. *Comp. Biochem. Physiol. A Mol. Integr. Physiol.* **128,** 551–563.

Fradinger, E. A., Tello, J. A., Rivier, J. E., and Sherwood, N. M. (2005). Characterization of four receptor cDNAs: PAC1, VPAC1, a novel PAC1 and a partial GHRH in zebrafish. *Mol. Cell. Endocrinol.* **231,** 49–63.

Gábriel, R., Halasy, K., Fekete, É., Eckert, M., and Benedeczki, I. (1990). Distribution of GABA-like immunoreactivity in myenteric plexus of carp, frog and chicken. *Histochemistry* **94,** 323–328.

Gawlicka, A., Leggiadro, C. T., Gallant, J. W., and Douglas, S. E. (2001). Cellular expression of the pepsinogen and the gastric proton pump genes in the stomach of winter flounder as determined by in situ hybridization. *J. Fish. Biol.* **58,** 529–536.

Goodrich, J. T., Bernd, P., Sherman, D., and Gershon, M. D. (1980). Phylogeny of enteric serotonergic neurons. *J. Comp. Neurol.* **190,** 15–28.

Green, K., and Campbell, G. (1994). Nitric oxide formation is involved in vagal inhibition of the stomach of the trout *(Salmo gairdneri)*. *J. Auton. Nerv. Syst.* **50,** 221–229.

Groisman, S. D., and Shparkovskii, I. A. (1989). Effect of dopamine and DOPA on electrical activity of stomach muscles in the skate *Raja radiata* and cod *Gadus morhua*. *Zh. Evolyut. Biokhim. Fiziol.* **25,** 505–511.

Grove, D. J., and Campbell, G. (1979a). Effects of extrinsic nerve stimulation on the stomach of the flathead, *Platychephalus bassensis* Cuvier and Valenciennes. *Comp. Biochem. Physiol. C* **63C,** 373–380.

Grove, D. J., and Campbell, G. (1979b). The role of extrinsic and intrinsic nerves in the coordination of gut motility in the stomachless flatfish *Rhombosolea tapirina* and *Ammotretis rostrata* Guenther. *Comp. Biochem. Physiol.* **63C,** 143–159.

Grove, D. J., and Holmgren, S. (1992a). Intrinsic mechanisms controlling cardiac stomach volume of the rainbow trout (*Oncorhynchus mykiss*) following gastric distension. *J. Exp. Biol.* **163,** 33–48.

Grove, D. J., and Holmgren, S. (1992b). Mechanisms controlling stomach volume of the Atlantic cod *Gadus morhua* following gastric distension. *J. Exp. Biol.* **163,** 49–63.

Habibi, H. R., and Ince, B. W. (1984). A study of androgen-stimulated L-leucine transport by the intestine of rainbow trout (*Salmo gairdneri* Richardson) *in vitro*. *Comp. Biochem. Physiol. A* **79,** 143–149.

Habu, A., Ohishi, T., Mihara, S., Ohkubo, R., Hong, Y.-M., Mochizuki, T., and Yanaihara, N. (1994). Isolation and sequenc determination of galanin from the pituitary of yellowfin tuna. *Biomed Res – Tokyo* **15,** 357–362.

Håkanson, R., Böttcher, G., Ekblad, E., Panula, P., Simonsson, M., Dohlsten, M., Hallberg, T., and Sundler, F. (1986). Histamine in endocrine cells in the stomach. A survey of several species using a panel of histamine antibodies. *Histochemistry* **86,** 5–17.

Hazzard, C. E., and Ahearn, G. A. (1992). Rapid stimulation of intestinal D-glucose transport in teleosts by 17 alpha-methyltestosterone. *Am. J. Physiol.* **262,** R412–R418.

Holder, F. C., Vincent, B., Ristori, M. T., and Laurent, P. (1983). Vascular perfusion of an intestinal segment in the catfish (*Ictalurus melas*, R): demonstration of the vasoactive effects of mammalian VIP and of gastrointestinal extracts from teleost fish. *C. R. Seances Acad. Sci. III* **296,** 783–788.

Holmberg, A., Schwerte, T., Fritsche, R., Pelster, B., and Holmgren, S. (2003). Ontogeny of intestinal motility in correlation to neuronal development in zebrafish embryos and larvae. *J. Fish Biol.* **63,** 318–331.

Holmberg, A., Schwerte, T., Pelster, B., and Holmgren, S. (2004). Ontogeny of the gut motility control system in zebrafish *Danio rerio* embryos and larvae. *J. Exp. Biol.* **207,** 4085–4094.

Holmberg, A., Olsson, C., and Holmgren, S. (2006). The effects of endogenous and exogenous nitric oxide on gut motility in zebrafish *Danio rerio* embryos and larvae. *J. Exp. Biol.* **209,** 2472–2479.

Holmberg, A., Olsson, C., and Hennig, G. W. (2007). TTX-sensitive and TTX-insensitive control of spontaneous gut motility in the developing zebrafish (*Danio rerio*) larvae. *J. Exp. Biol.* **210,** 1084–1091.

Holmgren, S. (1983). The effects of putative non-adrenergic, non-cholinergic autonomic transmitters on isolated strips from the stomach of the rainbow trout *Salmo gairdneri. Comp. Biochem. Physiol. C* **74,** 229–238.

Holmgren, S., and Fänge, R. (1981). Effects of cholinergic drugs on the intestine and gallbladder of the hagfish, *Myxine glutinosa* L., with a report on the inconsistent effects of catecholamines. *Mar. Biol. Lett.* **2,** 265–277.

Holmgren, S., and Jensen, J. (2001). Evolution of vertebrate neuropeptides. *Brain Res. Bull.* **55,** 723–735.

Holmgren, S., and Jönsson, A.-C. (1988). Occurrence and effects on motility of bombesin related peptides in the gastrointestinal tract of the Atlantic cod, *Gadus morhua. Comp. Biochem. Physiol.* **89C,** 249–256.

Holmgren, S., and Nilsson, S. (1974). Drug effects on isolated artery strips from two teleosts, *Gadus morhua* and *Salmo gairdneri. Acta Physiol. Scand.* **90,** 431–437.

Holmgren, S., and Nilsson, S. (1981). On the non-adrenergic, non-cholinergic innervation of the rainbow trout stomach. *Comp. Biochem. Physiol.* **70C,** 65–69.

Holmgren, S., and Nilsson, S. (1983a). Bombesin-, gastrin/CCK-, 5-hydroxytryptamine-, neurotensin-, somatostatin-, and VIP-like immunoreactivity and catecholamine fluorescence in the gut of the elasmobranch, *Squalus acanthias. Cell Tissue Res.* **234,** 595–618.

Holmgren, S., and Nilsson, S. (1983b). VIP-, bombesin- and neurotensin-like immunoreactivity in neurons of the gut of the holostean fish, *Lepisosteus platyrhincus. Acta Zool. (Stockh.)* **64,** 25–32.

Holmgren, S., Grove, D. J., and Nilsson, S. (1985). Substance P acts by releasing 5-hydroxytryptamine from enteric neurons in the stomach of the rainbow trout, *Salmo gairdneri. Neuroscience* **14,** 683–693.

Holmgren, S., Jönsson, A.-C., and Holstein, B. (1986). Gastrointestinal peptides in fish. *In* "Fish Physiology – Recent Advances" (Nilsson, S., and Holmgren, S., Eds.), pp. 119–139. Croom Helm Ltd, London.

Holmgren, S., Axelsson, M., and Farrell, A. P. (1992a). The effect of catecholamines, substance P and vasoactive intestinal polypeptide on blood flow to the gut in the dogfish *Squalus acanthias. J. Exp. Biol.* **168,** 161–175.

Holmgren, S., Axelsson, M., and Farrell, A. P. (1992b). The effects of neuropeptide Y and bombesin on blood flow to the gut in dogfish *Squalus acanthias. Regul. Pept.* **40,** 169.

Holmgren, S., Fritsche, R., Karila, P., Gibbins, I., Axelsson, M., Franklin, C., Grigg, G., and Nilsson, S. (1994). Neuropeptides in the Australian lungfish *Neoceratodus forsteri*: effects *in vivo* and presence in autonomic nerves. *Am. J. Physiol.* **266,** R1568–R1577.

Holstein, B. (1976). Effect of the H2-receptor antagonist metiamide on carbachol- and histamine-induced gastric acid secretion in the Atlantic cod, *Gadus morhua. Acta Physiol. Scand.* **97,** 189–195.

Holstein, B. (1977). Effect of atropine and SC-15396 on stimulated gastric acid secretion in the Atlantic cod, *Gadus morhua. Acta Physiol. Scand.* **101,** 185–193.

Holstein, B. (1982). Inhibition of gastric acid secretion inthe Atlantic cod, *Gadus morhua,* by sulphated and desulphated gastrin, caerulein, and CCK-octapeptide. *Acta Physiol. Scand.* **114,** 453–459.

Holstein, B. (1983). Effect of vasoactive intestinal polypeptide on gastric acid secretion and mucosal blood flow in the Atlantic cod, *Gadus morhua. Gen. Comp. Endocrinol.* **52,** 471–473.

Holstein, B. (1986). Characterization with agonists of the histamine receptors mediating stimulation of gastric acid secretion in the Atlantic cod, *Gadus morhua. Agents Actions* **19,** 42–47.

Holstein, B., and Cederberg, C. (1980). Effect of vagotomy and glucose administration on gastric acid secretion in the Atlantic cod, *Gadus morhua*. *Acta Physiol. Scand.* **109,** 37–44.

Holstein, B., and Cederberg, C. (1986). Effects of tachykinins on gastric acid and pepsin secretion and on gastric outflow in the Atlantic cod, *Gadus morhua*. *Am. J. Physiol.* **250,** G309–G315.

Holstein, B., and Humphrey, C. S. (1980). Stimulation of gastric acid secretion and suppression of VIP-like immunoreactivity by bombesin in the Atlantic codfish, *Gadus morhua*. *Acta Physiol. Scand.* **109,** 217–223.

Hoyle, C. H. (1998). Neuropeptide families: evolutionary perspectives. *Regul. Pept.* **73,** 1–33.

Jennings, B. L., Bell, J. D., Hyodo, S., Toop, T., and Donald, J. A. (2007). Mechanisms of vasodilation in the dorsal aorta of the elephant fish, *Callorhinchus milii (Chimaeriformes: Holocephali)*. *J. Comp. Physiol. B* **177,** 557–567.

Jensen, J., and Conlon, J. M. (1992a). Isolation and primary structure of gastrin-releasing peptide from a teleost fish, the trout (*Oncorhynchus mykiss*). *Peptides* **13,** 995–999.

Jensen, J., and Conlon, J. M. (1992b). Substance-P-related and neurokinin-A-related peptides from the brain of the cod and trout. *Eur. J. Biochem.* **206,** 659–664.

Jensen, J., and Holmgren, S. (1985). Neurotransmitters in the intestine of the Atlantic cod, *Gadus morhua*. *Comp. Biochem. Physiol. C* **82,** 81–89.

Jensen, J., and Holmgren, S. (1991). Tachykinins and intestinal motility in different fish groups. *Gen. Comp. Endocrinol.* **83,** 388–396.

Jensen, J., and Holmgren, S. (1994). The gastrointestinal canal. *In* "Comparative Physiology and Evolution of the Autonomic Nervous System" (Nilsson, S., and Holmgren, S., Eds.), pp. 119–167. Harwood Academic Publishers, Chur, Switzerland.

Jensen, J., Holmgren, S., and Jönsson, A.-C. (1987). Substance P-like immunoreactivity and the effects of tachykinins in the intestine of the Atlantic cod, *Gadus morhua*. *J. Auton. Nerv. Syst.* **20,** 25–33.

Jensen, J., Axelsson, M., and Holmgren, S. (1991). Effects of substance P and VIP on the gastrointestinal blood flow in the Atlantic cod, *Gadus morhua*. *J. Exp. Biol.* **156,** 361–373.

Jensen, J., Karila, P., Jönsson, A.-C., Aldman, G., and Holmgren, S. (1993a). Effects of substance P and distribution of substance P-like immunoreactivity in nerves supplying the stomach of the cod, *Gadus morhua*. *Fish Physiol. Biochem.* **12,** 237–247.

Jensen, J., Olson, K. R., and Conlon, J. M. (1993b). Primary structures and effects on gastrointestinal motility of tachykinins from the rainbow trout. *Am. J. Physiol.* **265,** R804–R810.

Johnsen, A. H. (1998). Phylogeny of the cholecystokinin/gastrin family. *Front. Neuroendocrinol.* **19,** 73–99.

Jönsson, A.-C. (1991). Regulatory peptides in the pancreas of two species of elasmobranchs and in the Brockmann bodies of four teleost species. *Cell Tissue Res.* **266,** 163–172.

Jönsson, A.-C. (1993). Co-localization of peptides in the Brockmann bodies of the cod (*Gadus morhua*) and the rainbow trout (*Oncorhynchus mykiss*). *Cell Tissue Res.* **273,** 547–555.

Jönsson, A.-C., Holmgren, S., and Holstein, B. (1987). Gastrin/CCK-like immunoreactivity in endocrine cells and nerves in the gastrointestinal tract of the cod, *Gadus morhua,* and the effect of peptides of the gastrin/CCK family on cod gastrointestinal smooth muscle. *Gen. Comp. Endocrinol.* **66,** 190–202.

Jönsson, E., Forsman, A., Einarsdottir, I. E., Egnér, B., Ruohonen, K., and Björnsson, B. T. (2006). Circulating levels of cholecystokinin and gastrin-releasing peptide in rainbow trout fed different diets. *Gen. Comp. Endocrinol.* **148,** 187–194.

Kågström, J., and Holmgren, S. (1997). VIP-induced relaxation of small arteries of the rainbow trout, *Oncorhynchus mykiss*, involves prostaglandin synthesis but not nitric oxide. *J. Auton. Nerv. Syst.* **63,** 68–76.

Kågström, J., and Holmgren, S. (1998). Calcitonin gene-related peptide (CGRP), but not tachykinins, causes relaxation of small arteries from the rainbow trout gut. *Peptides* **19,** 577–584.

Kågström, J., Axelsson, M., Jensen, J., Farrell, A. P., and Holmgren, S. (1996a). Vasoactivity and immunoreactivity of fish tachykinins in the vascular system of the spiny dogfish. *Am. J. Physiol. – Reg. Int. Comp. Phys.* **39,** R585–R593.

Kågström, J., Holmgren, S., Olson, K. R., Conlon, J. M., and Jensen, J. (1996b). Vasoconstrictive effects of native tachykinins in the rainbow trout, *Oncorhynchus mykiss*. *Peptides* **17,** 39–45.

Kaiya, H., Tsukada, T., Yuge, S., Mondo, H., Kangawa, K., and Takei, Y. (2006). Identification of eel ghrelin in plasma and stomach by radioimmunoassay and histochemistry. *Gen. Comp. Endocrinol.* **148,** 375–382.

Kaiya, H., Miyazato, M., Kangawa, K., Peter, R. E., and Unniappan, S. (2008). Ghrelin: A multifunctional hormone in non-mammalian vertebrates. *Comp. Biochem. Physiol. A Mol. Integr. Physiol.* **149,** 109–128.

Kamisaka, Y., Totland, G. K., Tagawa, M., Kurokawa, T., Suzuki, T., Tanaka, M., and Rønnestad, I. (2001). Ontogeny of cholecystokinin-immunoreactive cells in the digestive tract of Atlantic halibut, *Hippoglossus hippoglossus*, larvae. *Gen. Comp. Endocrinol.* **123,** 31–37.

Karila, P. (1997). "Nervous Control of Gastrointestinal Motility in the Atlantic Cod, *Gadus morhua*." PhD thesis, University of Gothenburg, Göteborg, Sweden.

Karila, P., and Holmgren, S. (1995). Enteric reflexes and nitric oxide in the fish intestine. *J. Exp. Biol.* **198,** 2405–2411.

Karila, P., and Holmgren, S. (1997). Anally projecting neurons exhibiting immunoreactivity to galanin, nitric oxide synthase and vasoactive intestinal peptide, detected by confocal laser scanning microscopy, in the intestine of the Atlantic cod, *Gadus morhua*. *Cell Tissue Res.* **287,** 525–533.

Karila, P., Jönsson, A.-C., Jensen, J., and Holmgren, S. (1993). Galanin-like immunoreactivity in extrinsic and intrinsic nerves to the gut of the Atlantic cod, *Gadus morhua*, and the effect of galanin on the smooth muscle of the gut. *Cell Tissue Res.* **271,** 537–544.

Karila, P., Messenger, J., and Holmgren, S. (1997). Nitric oxide synthase- and neuropeptide Y-containing subpopulations of sympathetic neurons in the coeliac ganglion of the Atlantic cod, *Gadus morhua*, revealed by immunohistochemistry and retrograde tracing from the stomach. *J. Auton. Nerv. Syst.* **66,** 35–45.

Karila, P., Shahbazi, F., Jensen, J., and Holmgren, S. (1998). Projections and actions of tachykininergic, cholinergic, and serotonergic neurones in the intestine of the Atlantic cod. *Cell Tissue Res.* **291,** 403–413.

Kelley, G. G., Aassar, O. S., and Forrest, J. N. J. (1991). Endogenous adenosine is an autacoid feedback inhibitor of chloride transport in the shark rectal gland. *J. Clin. Invest.* **88,** 1933–1939.

Kiliaan, A. J., Joosten, H. W. J., Bakker, R., Dekker, K., and Grooth, J. A. (1989). Serotonergic neurons in the intestine of two teleosts, *Carassius auratus* and *Oreochromis mossambicus,* and the effect of serotonin on trans-epithelial ion-selectivity and muscle tension. *Neuroscience* **31,** 817–824.

Kirtisinghe, P. (1940). The myenteric nerve plexus in some lower chordates. *Quart. J. Microscop. Sci.* **81,** 521–539.

Kitazawa, T., Kondo, H., and Temma, K. (1986a). Alpha 2-adrenoceptor-mediated contractile response to catecholamines in smooth muscle strips isolated from rainbow trout stomach (*Salmo gairdneri*). *Br. J. Pharmacol.* **89,** 259–266.

Kitazawa, T., Temma, K., and Kondo, H. (1986b). Presynaptic alpha-adrenoceptor mediated inhibition of the neurogenic cholinergic contraction of the isolated intestinal bulb of the carp (*Cyprinus carpio*). *Comp. Biochem. Physiol.* **83C,** 271–277.

Kitazawa, T., Miyashita, N., Chugun, A., Temma, K., and Kondo, H. (1988). Antagonist-like action of synthetic alpha2-adrenoceptor agonist on contractile response to catecholamines in smooth muscle strips isolated from rainbow trout stomach (*Salmo gairdneri*). *Comp. Biochem. Physiol.* **91C,** 585–588.

Knight, G. E., and Burnstock, G. (1993). Identification of purinoceptors in the isolated stomach and intestine of the three-spined stickleback *Gasterosteus aculeatus* L. *Comp. Biochem. Physiol. C* **106,** 71–78.

Koelz, H. R. (1992). Gastric acid in vertebrates. *Scand. J. Gastroenterol. Suppl.* **193,** 2–6.

Kono, T., Kitao, Y., Sonoda, K., Nomoto, R., Mekata, T., and Sakai, M. (2008). Identification and expression analysis of ghrelin gene in common carp *Cyprinus carpio*. *Fisheries Science* **74,** 603–612.

Krogdahl, A., and Bakke-McKellep, A. M. (2005). Fasting and refeeding cause rapid changes in intestinal tissue mass and digestive enzyme capacities of Atlantic salmon (*Salmo salar* L.). *Comp. Biochem. Physiol. A Mol. Integr. Physiol.* **141,** 450–460.

Kurokawa, T., Suzuki, T., and Andoh, T. (2000). Development of cholecystokinin and pancreatic polypeptide endocrine systems during the larval stage of Japanese flounder, *Paralichthys olivaceus*. *Gen. Comp. Endocrinol.* **120,** 8–16.

Kurokawa, T., Suzuki, T., and Hashimoto, H. (2003). Identification of gastrin and multiple cholecystokinin genes in teleost. *Peptides* **24,** 227–235.

Lamers, C. H., Rombout, J. W., and Timmermans, L. P. (1981). An experimental study on neural crest migration in *Barbus conchonius* (*Cyprinidae, Teleostei*), with special reference to the origin of the enteroendocrine cells. *J. Embryol. Exp. Morphol.* **62,** 309–323.

Langer, M., Van Noorden, S., Polak, J. M., and Pearse, A. G. (1979). Peptide hormone-like immunoreactivity in the gastrointestinal tract and endocrine pancreas of eleven teleost species. *Cell Tissue Res.* **199,** 493–508.

Le Mevel, J. C., Mabin, D., Hanley, A. M., and Conlon, J. M. (1998). Contrasting cardiovascular effects following central and peripheral injections of trout galanin in trout. *Am. J. Physiol.* **275,** R1118–R1126.

Lee, J. H., Ku, S. K., Park, K. D., and Lee, H. S. (2004). Immunohistochemical study of the gastrointestinal endocrine cells in the Korean aucha perch. *J. Fish Biol.* **65,** 170–181.

Lennard, R., and Huddart, H. (1989). Purinergic modulation in the flounder gut. *Gen. Pharmacol.* **20,** 849–853.

Li, Z. S., and Furness, J. B. (1993). Nitric oxide synthase in the enteric nervous system of the rainbow trout, *Salmo gairdneri*. *Arch. Histol. Cytol.* **56,** 185–193.

Lundin, K., Holmgren, S., and Nilsson, S. (1984). Peptidergic functions in the dogfish rectum. *Acta Physiol. Scand.* **121,** 46A.

Maier, K. J., and Tullis, R. E. (1984). The effects of diet a digestive cycle on the gastrointestinal tract pH values in the goldfish, *Carassius auratus* L., Mozambique tilapia, *Oreochromis mossambicus* (Peters), and channel catfish, *Ictalurus punctatus* (Rafinesque). *J. Fish Biol.* **25,** 151–165.

Manera, M., Giammarino, A., Perugini, M., and Amorena, M. (2008). *In vitro* evaluation of gut contractile response to histamine in rainbow trout (*Oncorhynchus mykiss* Walbaum, 1792). *Res. Vet. Sci.* **84,** 126–131.

Manning, A. J., Murray, H. M., Gallant, J. W., Matsuoka, M. P., Radford, E., and Douglas, S. E. (2008). Ontogenetic and tissue-specific expression of preproghrelin in the Atlantic halibut, *Hippoglossus hippoglossus* L. *J. Endocrinol.* **196,** 181–192.

Matsuda, K., Kashimoto, K., Higuchi, T., Yoshida, T., Uchiyama, M., Shioda, S., Arimura, A., and Okamura, T. (2000). Presence of pituitary adenylate cyclase-activating polypeptide (PACAP) and its relaxant activity in the rectum of a teleost, the stargazer, *Uranoscopus japonicus*. *Peptides* **21,** 821–827.

Mattisson, A., and Holstein, B. (1980). The ultrastructure of the gastric glands and its relation to induced secretory activity of cod, *Gadus morhua* (Day). *Acta Physiol. Scand.* **109,** 51–59.

McCormick, S. (2001). Endocrine control of osmoregulation in teleost fish. *Amer. Zool.* **41,** 781–794.

McDonald, J. K., Greiner, F., Bauer, G. E., Elde, R. P., and Noe, B. D. (1987). Separate cell types that express two different forms of somatostatin in anglerfish islets can be immunocytochemically differentiated. *J. Histochem. Cytochem.* **35,** 155–162.

Mellgren, E. M., and Johnson, S. L. (2005). kitb, a second zebrafish ortholog of mouse Kit. *Dev. Genes Evol.* **215,** 470–477.

Michelangeli, F., Ruiz, M. C., Dominguez, M. G., and Parthe, V. (1988). Mammalian-like differentiation of gastric cells in the shark *Hexanchus griseus*. *Cell Tissue Res.* **251,** 225–227.

Mulero, I., Sepulcre, M. P., Meseguer, J., Garcia-Ayala, A., and Mulero, V. (2007). Histamine is stored in mast cells of most evolutionarily advanced fish and regulates the fish inflammatory response. *Proc. Natl. Acad. Sci. USA* **104,** 19434–19439.

Murashita, K., Fukada, H., Hosokawa, H., and Masumoto, T. (2007). Changes in cholecystokinin and peptide Y gene expression with feeding in yellowtail (*Seriola quinqueradiata*): Relation to pancreatic exocrine regulation. *Comp. Biochem. Physiol. B Biochem. Mol. Biol.* **146,** 318–325.

NCBI (National Center for Biotechnology Information) www.ncbi.nlm.nih.gov.

Nelson, L. E., and Sheridan, M. A. (2005). Regulation of somatostatins and their receptors in fish. *Gen. Comp. Endocrinol.* **142,** 117–133.

Ng, A. N., de Jong-Curtain, T. A., Mawdsley, D. J., White, S. J., Shin, J., Appel, B., Dong, P. D., Stainier, D. Y., and Heath, J. K. (2005). Formation of the digestive system in zebrafish: III. Intestinal epithelium morphogenesis. *Dev. Biol.* **286,** 114–135.

Nilsson, S. (1983). "Autonomic Nerve Function in the Vertebrates." Springer Verlag, Berlin.

Nilsson, S., and Holmgren, S. (1983). Splanchnic nervous control of the stomach of the spiny dogfish, *Squalus acanthias*. *Comp. Biochem. Physiol. C* **76,** 271–276.

Nilsson, S., and Holmgren, S. (1994). "Comparative Physiology and Evolution of the Autonomic Nervous System." Harwood Academic Publishers, Chur, Switzerland.

Nilsson, S., Holmgren, S., and Grove, D. J. (1975). Effects of drugs and nerve stimulation on the spleen and arteries of two species of dogfish, *Scyliorhinus canicula* and *Squalus acanthias*. *Acta Physiol. Scand.* **95,** 219–230.

Noe, B. D., McDonald, J. K., Greiner, F., and Wood, J. D. (1986). Anglerfish islets contain NPY immunoreactive nerves and produce the NPY analog aPY. *Peptides* **7,** 147–154.

Norris, J. S., Norris, D. O., and Windell, J. T. (1973). Effect of simulated meal size of gastric acid and pepsin secretory rates in bluegill (*Lepomis macrochirus*). *J. Fisheries. Res. Board Canada* **30,** 201–204.

Ogoshi, M., Inoue, K., Naruse, K., and Takei, Y. (2006). Evolutionary history of the calcitonin gene-related peptide family in vertebrates revealed by comparative genomic analyses. *Peptides* **27,** 3154–3164.

Okuda, R., Sasayama, Y., Suzuki, N., Kambegawa, A., and Srivastav, A. K. (1999). Calcitonin cells in the intestine of goldfish and a comparison of the number of cells among saline-fed, soup-fed, or high Ca soup-fed fishes. *Gen. Comp. Endocrinol.* **113,** 267–273.

Oliver, A. S., and Vigna, S. R. (1996). CCK-X receptors in the endothermic mako shark (*Isurus oxyrinchus*). *Gen. Comp. Endocrinol.* **102,** 61–73.

Olson, K. R. (1999). Chapter 13. Rectal gland and volume homeostasis. *In* "Sharks, Skates and Rays: The Biology of Elasmobranch Fishes" (Hamlett, W. C., Ed.), pp. 329–352. The Johns Hopkins University Press, Baltimore, Maryland.

Olson, K. R., and Villa, J. (1991). Evidence against nonprostanoid endothelium-derived relaxing factor(s) in trout vessels. *Am. J. Physiol.* **260,** R925–R933.

Olsson, C., and Holmgren, S. (1994). Distribution of PACAP (pituitary adenylate cyclase-activating polypeptide)-like and helospectin-like peptides in the teleost gut. *Cell Tissue Res.* **277,** 539–547.

Olsson, C., and Holmgren, S. (2000). PACAP and nitric oxide inhibit contractions in proximal intestine of the Atlantic cod, *Gadus morhua*. *J. Exp. Biol.* **203,** 575–583.

Olsson, C., and Holmgren, S. (2001). The control of gut motility. *Comp. Biochem. Physiol. A Mol. Integr. Physiol.* **128,** 481–503.

Olsson, C., and Karila, P. (1995). Coexistence of NADPH-diaphorase and vasoactive intestinal polypeptide in the enteric nervous system of the Atlantic cod (*Gadus morhua*) and the spiny dogfish (*Squalus acanthias*). *Cell Tissue Res.* **280,** 297–305.

Olsson, C., Aldman, G., Larsson, A., and Holmgren, S. (1999). Cholecystokinin affects gastric emptying and stomach motility in the rainbow trout *Oncorhynchus mykiss*. *J. Exp. Biol.* **202,** 161–170.

Olsson, C., Holbrook, J. D., Bompadre, G., Jönsson, E., Hoyle, C. H., Sanger, G. J., Holmgren, S., and Andrews, P. L. (2008a). Identification of genes for the ghrelin and motilin receptors and a novel related gene in fish, and stimulation of intestinal motility in zebrafish (*Danio rerio*) by ghrelin and motilin. *Gen. Comp. Endocrinol.* **155,** 217–226.

Olsson, C., Holmberg, A., and Holmgren, S. (2008b). Development of enteric and vagal innervation of the zebrafish (*Danio rerio*) gut. *J. Comp. Neurol.* **508,** 756–770.

Pan, Q. S., and Fang, Z. P. (1993). An immunocytochemical study of endocrine cells in the gut of a stomachless teleost fish, grass carp, *Cyprinidae*. *Cell Transplant.* **2,** 419–427.

Patterson, T. L., and Fair, E. (1933). The action of the vagus on the stomach-intestine of the hagfish. Comparative studies. VIII. *J. Cell. Comp. Physiol.* **3,** 113–199.

Pederzoli, A., Bertacchi, I., Gambarelli, A., and Mola, L. (2004). Immunolocalisation of vasoactive intestinal peptide and substance P in the developing gut of *Dicentrarchus labrax* (L.). *Eur. J. Histochem.* **48,** 179–184.

Pederzoli, A., Conte, A., Tagliazucchi, D., Gambarelli, A., and Mola, L. (2007). Occurrence of two NOS isoforms in the developing gut of sea bass *Dicentrarchus labrax* (L.). *Histol. Histopathol.* **22,** 1057–1064.

Preston, E., McManus, C. D., Jönsson, A. C., and Courtice, G. P. (1995). Vasoconstrictor effects of galanin and distribution of galanin containing fibres in three species of elasmobranch fish. *Regul. Pept.* **58,** 123–134.

Preston, E., Jonsson, A. C., McManus, C. D., Conlon, J. M., and Courtice, G. P. (1998). Comparative vascular responses in elasmobranchs to different structures of neuropeptide Y and peptide YY. *Regul. Pept.* **78,** 57–67.

Raible, D. W., Wood, A., Hodsdon, W., Henion, P. D., Weston, J. A., and Eisen, J. S. (1992). Segregation and early dispersal of neural crest cells in the embryonic zebrafish. *Dev. Dyn.* **195,** 29–42.

Rajjo, I. M., Vigna, S. R., and Crim, J. W. (1988). Actions of cholecystokinin-related peptides on the gallbladder of bony fishes *in vitro*. *Comp. Biochem. Physiol. C* **90,** 267–273.

Rajjo, I. M., Vigna, S. R., and Crim, J. W. (1989). Immunocytochemical localization of vasoactive intestinal polypeptide in the digestive tracts of a holostean and a teleostean fish. *Comp. Biochem. Physiol.* **94C,** 411–418.

Rebolledo, I. M., and Vial, J. D. (1979). Fine structure of the oxynticopeptic cell in the gastric glands of an elasmobranch species (*Halaelurus chilensis*). *Anat. Rec.* **193,** 805–822.

Reinecke, M., Schluter, P., Yanaihara, N., and Forssmann, W. G. (1981). VIP immunoreactivity in enteric nerves and endocrine cells of the vertebrate gut. *Peptides* **2**(Suppl 2), 149–156.

Reinecke, M., Muller, C., and Segner, H. (1997). An immunohistochemical analysis of the ontogeny, distribution and coexistence of 12 regulatory peptides and serotonin in endocrine cells and nerve fibers of the digestive tract of the turbot, *Scophthalmus maximus* (*Teleostei*). *Anat. Embryol. (Berl)* **195,** 87–101.

Reite, O. B. (1972). Comparative physiology of histamine. *Physiol. Rev.* **52,** 778–819.

Rich, A., Leddon, S. A., Hess, S. L., Gibbons, S. J., Miller, S., Xu, X., and Farrugia, G. (2007). Kit-like immunoreactivity in the zebrafish gastrointestinal tract reveals putative ICC. *Dev. Dyn.* **236,** 903–911.

Rombout, J. H. (1977). Enteroendocrine cells in the digestive tract of *Barbus conchonius (Teleostei, Cyprinidae)*. *Cell Tissue Res.* **185,** 435–450.

Rombout, J. H., van der Grinten, C. P., Binkhorst, F. M., Taverne-Thiele, J. J., and Schooneveld, H. (1986). Immunocytochemical identification and localization of peptide hormones in the gastro-entero-pancreatic (GEP) endocrine system of the mouse and a stomachless fish, *Barbus conchonius*. *Histochemistry* **84,** 471–483.

Rønnestad, I., Rojas-Garcia, C. R., and Skadal, J. (2000). Retrograde peristalsis, a possible mechanism for filling the pyloric cecae? *J. Fish. Biol.* **56,** 216–218.

Sadaghiani, B., and Vielkind, J. R. (1990). Distribution and migration pathways of HNK-1-immunoreactive neural crest cells in teleost fish embryos. *Development* **110,** 197–209.

Sakata, I., Mori, T., Kaiya, H., Yamazaki, M., Kangawa, K., Inoue, K., and Sakai, T. (2004). Localization of ghrelin-producing cells in the stomach of the rainbow trout (*Oncorhynchus mykiss*). *Zoolog. Sci.* **21,** 757–762.

Salaneck, E., Larsson, T. A., Larson, E. T., and Larhammar, D. (2008). Birth and death of neuropeptide Y receptor genes in relation to the teleost fish tetraploidization. *Gene* **409,** 61–71.

Salimova, N., and Fehér, E. (1982). Innervation of the alimentary tract in Chondrostean fish (*Acipenseridae*). A histochemical, microspectro-fluorimetric and ultrastructural study. *Acta Morphol. Acad. Sci. Hung.* **30,** 213–222.

Santos, C. R., Ingleton, P. M., Cavaco, J. E., Kelly, P. A., Edery, M., and Power, D. M. (2001). Cloning, characterization, and tissue distribution of prolactin receptor in the sea bream (*Sparus aurata*). *Gen. Comp. Endocrinol.* **121,** 32–47.

Schneeman, B. O. (2002). Gastrointestinal physiology and functions. *Br. J. Nutr.* **88**(Suppl 2), S159–S163.

Severini, C., Improta, G., Falconieri-Erspamer, G., Salvadori, S., and Erspamer, V. (2002). The tachykinin peptide family. *Pharmacol. Rev.* **54,** 285–322.

Shahbazi, F., Karila, P., Olsson, C., Holmgren, S., and Jensen, J. (1998). Primary structure, distribution, and effects on motility of CGRP in the intestine of the cod *Gadus morhua*. *Am. J. Physiol.* **275,** R19–R28.

Shahbazi, F., Holmgren, S., Larhammar, D., and Jensen, J. (2002). Neuropeptide Y effects on vasorelaxation and intestinal contraction in the Atlantic cod *Gadus morhua*. *Am. J. Physiol. Regul. Integr. Comp. Physiol.* **282,** R1414–R1421.

Shepherd, I. T., Beattie, C. E., and Raible, D. W. (2001). Functional analysis of zebrafish GDNF. *Dev. Biol.* **231,** 420–435.

Shuttleworth, T. J., and Thorndyke, M. C. (1984). An endogenous peptide stimulates secretory activity in the elasmobranch rectal gland. *Science* **225,** 319–321.

Silva, P., Lear, S., Reichlin, S., and Epstein, F. H. (1990). Somatostatin mediates bombesin inhibition of chloride secretion by rectal gland. *Am. J. Physiol.* **258,** R1459–1463.

Silva, P., Epstein, F. H., Karnaky, K. J., Jr., Reichlin, S., and Forrest, J. N., Jr. (1993). Neuropeptide Y inhibits chloride secretion in the shark rectal gland. *Am. J. Physiol.* **265,** R439–R446.

Slagter, B. J., Kittilson, J., and Sheridan, M. A. (2005). Expression of somatostatin receptor mRNAs is regulated *in vivo* by growth hormone, insulin, and insulin-like growth factor-I in rainbow trout (*Oncorhynchus mykiss*). *Regul. Pept.* **128,** 27–32.

Smit, H. (1968). Gastric secretion in the lower vertebrates and birds. *In* "Handbook of Physiology, sect 6: Alimentary canal, volume V: Bile, digestion, ruminal physiology" (Code, C. F., Ed.), pp. 2791–2805. American Physiological Society, Washington DC.

Smolka, A. J., Lacy, E. R., Luciano, L., and Reale, E. (1994). Identification of gastric H,K-ATPase in an early vertebrate, the Atlantic stingray *Dasyatis sabina*. *J. Histochem. Cytochem.* **42,** 1323–1332.

Stevens, C. E., and Hume, I. D. (1995). "Comparative Physiology of the Vertebrate Digestive System." Cambridge University Press, Cambridge.

Stoff, J. S., Rosa, R., Hallac, R., Silva, P., and Epstein, F. H. (1979). Hormonal regulation of active chloride transport in the dogfish rectal gland. *Am. J. Physiol.* **237,** F138–F144.

Stoff, J. S., Silva, P., Lechan, R., Solomon, R., and Epstein, F. H. (1988). Neural control of shark rectal gland. *Am. J. Physiol.* **255,** R212–R216.

Sullivan, M. X. (1905–1906). The physiology of the digestive tract of elasmobranchs. *Am. J. Physiol.* **15,** 42–45.

Sundell, K., Björnsson, B. T., Itoh, H., and Kawauchi, H. (1992). Chum salmon (*Oncorhynchus keta*) stanniocalcin inhibits *in vitro* intestinal calcium uptake in Atlantic cod (*Gadus morhua*). *J. Comp. Physiol. B* **162,** 489–495.

Sundström, G., Larsson, T. A., Brenner, S., Venkatesh, B., and Larhammar, D. (2008). Evolution of the neuropeptide Y family: New genes by chromosome duplications in early vertebrates and in teleost fishes. *Gen. Comp. Endocrinol.* **155,** 705–716.

Suzuki, N., Suzuki, T., and Kurokawa, T. (2000). Cloning of a calcitonin gene-related peptide receptor and a novel calcitonin receptor-like receptor from the gill of flounder, *Paralichthys olivaceus*. *Gene* **244,** 81–88.

Tagliafierro, G., Zaccone, G., Bonini, E., Faraldi, G., Farina, L., Fasulo, S., and Rossi, G. G. (1988). Bombesin-like immunoreactivity in the gastrointestinal tract of some lower vertebrates. *Ann. NY Acad. Sci.* **547,** 458–460.

Tagliafierro, G., Rossi, G. G., Bonini, E., Faraldi, G., and Farina, L. (1989). Ontogeny and differentiation of regulatory peptide- and serotonin-immunoreactivity in the gastrointestinal tract of an elasmobranch. *J. Exp. Zool.* **252,** 165–174.

Takahashi, H., Sakamoto, T., Hyodo, S., Shepherd, B. S., Kaneko, T., and Grau, E. G. (2006). Expression of glucocorticoid receptor in the intestine of a euryhaline teleost, the Mozambique tilapia (*Oreochromis mossambicus*): Effect of seawater exposure and cortisol treatment. *Life Sciences* **78,** 2329–2335.

Thorndyke, M., and Holmgren, S. (1990). Bombesin potentiates the effect of acetylcholine on isolated strips of fish stomach. *Regul. Pept.* **30,** 125–135.

Thorndyke, M. C., Riddell, J. H., Thwaites, D. T., and Dimaline, R. (1989). Vasoactive intestinal polypeptide and its relatives – biochemistry, distributions and functions. *Biol. Bull.* **177,** 183–186.

Thorndyke, M. C., Reeve, J. R., Jr., and Vigna, S. R. (1990). Biological activity of a bombesin-like peptide extracted from the intestine of the ratfish, *Hydrolagus colliei*. *Comp. Biochem. Physiol. C* **96,** 135–140.

Trischitta, F., Denaro, M. G., and Faggio, C. (1999). Effects of acetylcholine, serotonin and noradrenaline on ion transport in the middle and posterior part of *Anguilla anguilla* intestine. *J. Comp. Physiol. B* **169,** 370–376.

Trischitta, F., Pidalà, P., and Faggio, C. (2007). Nitric oxide modulates ionic transport in the isolated intestine of the eel, *Anguilla anguilla*. *Comp. Biochem. Physiol.* **148A,** 368–373.

Uesaka, T., Yano, K., Yamasaki, M., Nagashima, K., and Ando, M. (1994). Somatostatin-related peptides isolated from the eel gut: Effects on ion and water absorption across the intestine of the seawater eel. *J. Exp. Biol.* **188,** 205–216.

Valette, G., and Augeraux, P. (1958). Réactivité des muscles lisses des poissons a l'histamine et a d'autres agents contracturants (5-hydroxytryptamine, acetylcholine et chlorure de baryum). *J. Physiol. (Paris)* **50,** 1067–1074.

Van Noorden, S. (1990). Gut hormones in cyclostomes. *Fish Physiol. Biochem.* **8,** 399–408.

Van Noorden, S., and Falkmer, S. (1980). Gut-islet endocrinology – some evolutionary aspects. *Invest. Cell. Pathol.* **3,** 21–35.

Van Noorden, S., and Patent, G. J. (1980). Vasoactive intestinal polypeptide-like immunoreactivity in nerves of the pancreatic islet of the teleost fish, *Gillichthys mirabilis*. *Cell Tissue Res.* **212,** 139–146.

Van Noorden, S., and Pearse, A. G. (1974). Immunoreactive polypeptide hormones in the pancreas and gut of the lamprey. *Gen. Comp. Endocrinol.* **23,** 311–324.

Vigna, S. R. (1983). Evolution of endocrine regulation of gastrointestinal function in lower vertebrates. *Amer. Zool.* **23,** 512–520.

Volkoff, H., Peyon, P., Lin, X., and Peter, R. E. (2000). Molecular cloning and expression of cDNA encoding a brain bombesin/gastrin-releasing peptide-like peptide in goldfish. *Peptides* **21,** 639–648.

von Euler, U. S., and Fänge, R. (1961). Catecholamines in nerves and organs of *Myxine glutinosa*, *Squalus acanthias* and *Gadus callarias*. *Gen. Comp. Endocrinol.* **1,** 191–194.

von Euler, U. S., and Östlund, E. (1957). Effects of certain biologically occurring substances on the isolated intestine of fish. *Acta Physiol. Scand.* **38,** 364–372.

Wallace, K. N., and Pack, M. (2003). Unique and conserved aspects of gut development in zebrafish. *Dev. Biol.* **255,** 12–29.

Wang, Y., and Conlon, J. M. (1994). Purification and characterization of galanin from the phylogenetically ancient fish, the bowfin (*Amia calva*) and dogfish (*Scyliorhinus canicula*). *Peptides* **15,** 981–986.

Wang, Y., Barton, B. A., Thim, L., Nielsen, P. F., and Conlon, J. M. (1999). Purification and characterization of galanin and scyliorhinin I from the hybrid sturgeon, *Scaphirhynchus platorynchus x Scaphirhynchus albus (Acipenseriformes)*. *Gen. Comp. Endocrinol.* **113,** 38–45.

Wimalawansa, S. J. (1997). Amylin, calcitonin gene-related peptide, calcitonin, and adrenomedullin: A peptide superfamily. *Crit. Rev. Neurobiol.* **11,** 167–239.

Yasugi, S., Matsunaga, T., and Mizuno, T. (1988). Presence of pepsinogens immunoreactive to anti-embryonic chicken pepsinogen antiserum in fish stomachs: Possible ancestor molecules of chymosin of higher vertebrates. *Comp. Biochem. Physiol. A* **91,** 565–569.

Young, J. Z. (1933). The autonomic nervous system of selachians. *Q. J. Microsc. Sci.* **75,** 571–624.

Young, J. Z. (1936). The innervation and reactions to drugs of the viscera of teleostean fish. *Proc. R. Soc. London, Ser. B, Biol. Sci* **120,** 303–318.

Young, J. Z. (1980a). Nervous control of gut movements in *Lophius*. *J. Mar. Biol. Assoc. UK* **60,** 19–30.

Young, J. Z. (1980b). Nervous control of stomach movements in dogfishes and rays. *J. Mar. Biol. Assoc. UK* **60,** 1–17.

Young, J. Z. (1983). Control of movements of the stomach and spiral intestine of *Raja* and *Scyliorinus*. *J. Mar. Biol. Assoc. UK* **63,** 557–574.

Young, J. Z. (1988). Sympathetic innervation of the rectum and bladder of the skate and parallel effects of ATP and adrenaline. *Comp. Biochem. Physiol. C* **89,** 101–107.

Youson, J. H., and Al-Mahrouki, A. A. (1999). Ontogenetic and phylogenetic development of the endocrine pancreas (islet organ) in fish. *Gen. Comp. Endocrinol.* **116,** 303–335.

Youson, J. H., Al-Mahrouki, A. A., Amemiya, Y., Graham, L. C., Montpetit, C. J., and Irwin, D. M. (2006). The fish endocrine pancreas: review, new data, and future research directions in ontogeny and phylogeny. *Gen. Comp. Endocrinol.* **148,** 105–115.

Yui, R., Nagata, Y., and Fujita, T. (1988). Immunocytochemical studies on the islet and the gut of the arctic lamprey, *Lampetra japonica*. *Arch. Histol. Cytol.* **51,** 109–119.

Yui, R., Shimada, M., and Fujita, T. (1990). Immunohistochemical studies on peptide- and amine-containing endocrine cells and nerves in the gut and rectal gland of the ratfish *Chimaera monstrosa*. *Cell Tissue Res.* **260,** 193–201.

Zaks, M. G., Gazhala, E. M., Gzgzian, D. M., Kuzina, M. M., and Pesennikova, D. I. (1975). Comparative physiological characteristics of the action of pentagastrin on gastric secretion in fish, frogs, turtles and chickens. *Zh. Evol. Biokhim. Fiziol.* **11,** 594–600.

INDEX

A

Acetylated β-endorphins, 270
Acetylcholine (ACh), 324, 473, 483–484, 491
Activator protein 1 (AP-1), 153
Adaptive immune response, 318–320
Adenohypophysis, fish
  hormone synthesis, 7–9
  regionalization, nomenclatures, 6–7
Adrenaline (AD), 473, 494
Adrenergic receptor agonists, on immune parameters, 326
Adrenocorticotropic hormone (ACTH), 7, 199, 237, 265–266, 283, 394. *See also* Corticotropic axis
  contribution of MCH in regulation of, 287
  releasing activity in fish, 244
  releasing efficacy, 243
  secretion in fish, 245
  secretion in teleost fish, factors affecting, 239–242
Adrenomedullins (AMs), 379–380, 438
African catfish (*Clarias gariepinus*), 119, 162
African cichlid fish (*Haplochromis burtoni*), 18
AGGT(C/A)A consensus sequence, 280
Agouti-related protein (AgRP), 41, 443
Agouti-signaling peptide (ASP), 268, 443
Aldosterone, 399
Alpha-2-macroglobulin (α2M), 317
Amago Salmon (Oncorhynchus rhodurus), 271
Androgen receptors (AR), in fish brain, 89–91
Angiotensin-converting enzyme (ACE), 384
Angiotensin II (ANG II), 376–379. *See also* Drinking in Fish, hormonal regulation
  dipsogenic effect, 377, 381
  induced drinking, 387
  in mammals, 387
  for sustaining body fluid balance, 394–397
Angiotensins, 244
ANP and VNP, antidipsogenic hormones in eel, 381
Anterior tuberal nucleus (NAT), 17
Anteroventral periventricular nucleus (AVPV), 87
Appetite-regulating factors, 430
Appetite regulators, 430
  brain signals
    galanin, 433–434
    neuropeptide Y family of peptides, 430–432
    orexins, 432–433
  gender and reproductive status, 445–446
  genetic influence, 446
  ontogeny, 445
  peripheral signals, ghrelin, 434–435
  satiation signals, 435
    amylin, 439–440
    bombesin/GRP, 439
    calcitonin gene-related peptide, 438
    cholecystokinin/gastrin, 438–439
    cocaine-and amphetamine-regulated transcript, 435–436
    corticotropin-releasing factor-related neuropeptides, 436
    endocannabinoids, 437
    glucagon/glucagon-like peptide-1, 440–441
    insulin/insulin-like growth factors, 440–441
    leptin, 440
    neuromedin U, 437–438
    RFamides, 436–437
    tachykinins, 437

Appetite regulatory endocrine factors, 423–428
Arctic charr (*Salvelinus alpinus*), 283
Arctic grayling (*Thymallus arcticus*), 168
Arginine-vasopressin (AVP), 22, 243, 390–394
Arginine-vasotocin (AVT), 22–25, 243, 385, 391, 393–394
Aromatase B expression, in fish brain, 79–80
Atlantic cod (*Gadus morhua*), 214, 432, 469
Atlantic croaker (*Micropogonias undulatus*), 54, 84, 122, 162, 217
Atlantic halibut (*Hippoglossus hippoglossus*), 157, 217, 480
Atlantic salmon (*Salmo salar*), 136, 157, 202, 270
Atrial natriuretic peptide (ANP), 172, 494
Atrial NP (ANP), 214
Autocrine/paracrine signals in pituitary, in GH release, 177–179
AVT and IST expression, in teleost brain, 243–244
AVT-ir neurons, 243
AVT-stimulated ACTH release, in rainbow trout pituitary, 243
AVT V2 receptor, 392
Ayu (*Plecoglossus altivelis*), 44

B

Barfin flounder (*Verasper moseri*), 444
Black porgy (*Acanthopagrus schlegeli*), 79, 181
Black sea bream (*Acanthopagrus schlegeli*), 201
Blood–brain barrier (BBB), 377
Bluehead wrasse (*Thalassoma bifasciatum*), 80, 131
B-lymphocyte, 319
Body fluid regulation, 370–372
Body fluids, fish, 366
Bombesin (BBS), 36, 439, 477
Bradykinin, 384
Brain–pituitary–gonadal (BPG) axis, 341
Brown ghost knifefish (*Apteronotus leptorhynchus*), 13, 15, 237
Brown trout (*Salmo trutta*), 284
Brown trout (*Salmo trutta fario*), 35, 97, 173
B-type natriuretic peptide (BNP), 172
B-type NP (BNP), 214
Budget of ions and water, 367
Bullshark (*Carcharhinus leucas*), 368, 389
Burton's mouthbrooder (*Haplochromis burtoni*), 273

C

Calcitonin gene-related peptide (CGRP), 438, 476, 485
cAMP response element (CRE), 153
cAMP signaling cascade, 265
CART mRNA expression, in goldfish, 31
Catecholamines. *See also* Immune modulation, in teleost fish
  effect on innate and adaptive immunity, 325
  in gut motility, 486
  and receptors, in immune system, 323–324
Catecholamines, definition, 205. *See also* Hypothalamic peptides, prolactin secretion
Catfish (*Clarias batrachus*), 13, 131
Catfish (*Clarias gariepinus*), 44
Catfish (*Ictalurus melas*), 498
Catfish (*Ictalurus punctatus*), 215, 284
Caudal neurosecretory system (CNSS), 273, 398
Caudal pars distalis (CPD), 89
CCK/gastrin-ir perikarya and fibers, in goldfish, 30–31
Central nervous system (CNS), 468
Central nucleus of the inferior lobe (NCLI), 19
ChAT-immunoreactive (ir) nerves, 473
Chemokines and chemokine receptors, 322–324
Chemotaxis, 322
Cherry salmon (*Oncorhynchus masou*), 119
Chinook salmon (*Oncorhynchus tshawytscha*), 34, 199, 264
Cholecystokinin (CCK), 30–31, 170–171, 438, 475, 482, 487, 489–490. *See also* Hypophysiotropic peptides, in teleost fish
Choline acetyl transferase (ChAT), 473
Chum salmon (*Oncorhynchus keta*), 28, 157, 208
Cichlid (*Astatotilapia burtoni*), 120
Clawed toad (*Xenopus laevis*), 245
Climbing perch (*Anabas testudineys*), 203

Cloudy dogfish (*Scylirhinus torazame*), 481
Cobia (*Rachycentron canadun*), 129
Cocaine-and amphetamine-regulated transcript (CART), 31–32, 435. *See also* Hypophysiotropic peptides, in teleost fish
Cod (*Gadus morhua*), 31
co-gonadotropin (co-GTH), 154
Coho salmon (*Oncorhynchus kisutch*), 120, 171, 203, 494
Colony-stimulating factors (CSFs), 322
Common carp (*Cyprinus carpio*), 92, 199, 277, 474
Common stingray (*Dasyatis pastinaca*), 491
Copeptine peptide, 214. *See also* Prolactin (PRL) hormone
Corticoid receptors, 91
  glucocorticoid receptors
    chemically identified, 94–95
    distribution of, 92–94
  mineralocorticoid receptors, 92
Corticosteroid receptor (CR), in fish brain, 91, 273
Corticotropes
  hypothalamic regulation, 237
    inhibitory factors, 244–245
    stimulatory factors, 237–238, 243–244
  secretion, targets and functions, 264–265
    ACTH and non-acetylated βEND, 265–267
Corticotropic axis, 282–283, 286
Corticotropin-like intermediate peptide (CLIP), 42, 264
Corticotropin-releasing factor (CRF), 32–35, 170, 394. *See also* Hypophysiotropic peptides, in teleost fish
Corticotropin-releasing factor (CRF)-related neuropeptides, 436
Cortisol, 236, 272, 394
  DOC and, 272
  effect on immune responses, 328–329
  functions, 274–275
  in GH release, 175–176
  negative feedback, 275–276
  $P450_{c11}$, role, 266
  receptor types in immune cells, 326–328
  roles in control of water and ion fluxes, 494
  vs. T4, 341
CRF-BP gene expression, 286
CRF-expressing neurons, in zebrafish preoptic area, 32–33
CRF gene expression, 275–276, 286, 448
CRF-related peptides, 286, 288, 448
CRF/ UI, in thyrotropic axis regulation, 286
CRH-induced ACTH release, in rainbow trout, 253
C-terminal peptide of isotocin precursor (cNpCp), 214
C-type natriuretic peptide (CNP), 172
C type NPs (CNP), 214, 380
CXC chemokine, 323–324
CXCL12/CXCR4 signaling, 323
cyp19a1b gene, 80
Cyprinid (*Spinibarbus denticulatus*), 90–91
Cystic fibrosis transmembrane conductance regulator (CFTR), 275
Cytochrome P450, 265
Cytochrome P450c17 (CYP17), 78–79
Cytokines, 314, 317
  classification and receptors, 321–322
  family members, in teleost fish species, 318
  and phylogenetic relationship, 321

D

5′-Deiodinases D1 and D2, 277
Deiodination. *See also* Extrathyroidal deiodination
  T4 50-deiodination activities, 277
  of T4's phenolic ring, 282
11-Deoxycorticosterone (DOC), 272, 399. *See also* Mineralocorticoid receptors (MR)
  role in spermiation, 275
11-deoxycorticosterone (DOC), 91
Diffuse nucleus of inferior lobe (NDLI), 19
Dihydroxyphenylacetic acid (DOPAC), 168
Dogfish (*Scyliorhinus canicula*), 375, 488
Dogfish (*Triakis scyllia*), 215, 375
Dopamine (DA), 205–207, 473. *See also* Hypothalamic neurotransmitters
  functions of, 117
  in GH release, 167–168
  inhibiting basal ACTH secretion, 245
  inhibitory effects and endocrine stress response, 288

Dopamine (DA) (*continued*)
neurotransmitters, 52–53
role as gonadotropin release inhibitory factor, 124–126
stimulatory role in stressed fish, 253
Dopaminergic neurons, in fish brain, 88
Drinking in fish, hormonal regulation, 376
central *vs.* peripheral hormones, 386–387
hormones inducing drinking, 376
adrenomedullins, 379–380
angiotensin II, 376–379
hormones inhibiting drinking, 380
bradykinin, 384–385
ghrelin, 383–384
natriuretic peptides, 380–382
hormones regulating, in fishes and mammals, 386
neural mechanisms to elicit drinking, 387–388
sodium appetite, regulation of, 388–389

E

Eel (*Anguilla anguilla*), 24, 88
Eel (*Anguilla japonica*), 99, 214
Eelpout (*Zoarces viviparous*), 89
Elasmobranchs fish, neurohypophysis, 5
Elephant fish (*Callorhinchus milii*), 497
Endocannabinoids (ECs), 437
Endocrine disrupting compounds (EDCs), 342
Endocrine–immune interactions, in teleosts, 203–204
Endocrine pancreas, fish, 493
Endogenous UI in fish, 243
β-Endorphin (β–END), 237, 267
secretion in teleost fish, factors affecting, 250
Endorphins, role, 266–267
English sole (*Parophrys vetulus*), 217
Enteric nervous system (ENS), 469
Erythropoietin, 322
17β-Estradiol (E2), 123, 342
Estrogen/androgen receptors, 342
Estrogen receptors (ERs), 83–89
Estrogen responsive element (ERE), 83
European eel (*Anguilla anguilla*), 40, 119, 213
European plaice (*Pleuronectes platessa*), 25
European sea bass (*Dicentrarchus labrax*), 87, 119, 220
Extant teleosts, stress responses, 236
Extracellular signal-regulated kinase (ERK), 167
Extrathyroidal deiodination, 276–279

F

Fathead minnow (*Pimephales promelas*), 119, 173
Feeding activity and effect of photoperiod, 447
Feeding, in vertebrates, 422
Feeding regulation
fishes as models for study, 429
mammalian models, comparison, 429
methods for studying, 429–430
Fish brain, estrogen receptors distribution, 85–87
Fish, growth hormone
autocrine/paracrine signals in pituitary, GH release, 177–179
biological actions, 154–155
and growth hormone receptors, 152–153
at hypothalamic and pituitary levels, 179–183
hypothalamic signals from CNS, GH release
inhibitors, 157–163
neuroendocrine regulators, GH, 171–172
stimulators, 163–171
peripheral organs/tissues signals, in GH release
inhibitors, 172–174
stimulators, 174–177
secretion and synthesis regulation, 155–157
Flounder (*Platichthys flesus*), 24–25, 286
Fluid balance, regulation
renal and extrarenal mechanisms, 389–394
prolactin, growth hormone and cortisol, 397–399
renin–angiotensin system, 394–397
urotensin I, 399–403
Fluid exchange and balancing mechanisms, 366–370
elasmobranchs, 368
for fish in FW, major challenges, 366

for marine cyclostome and teleosts, 366–367
roles for claudins, occludin and aquaporins, 368–370
Fluid intake, regulation, 372–373
regulatory mechanism in fish, 373
acceleration *vs.* inhibition of drinking, 374
reflex drinking *vs.* thirst-motivated drinking, 373–374
volemic *vs.* osmotic factors, 375–376
fOat protein, 280
Follicle-stimulating hormone (FSH), in teleost fish, 7, 264. *See also* Gonadotropin-releasing hormone (GnRH)
dopamine in release of, 125
gonadal steroids regulation, 121–122
negative feedback effects, 122–123
positive feedback effects, 123–124
Four eyed fish (*Anableps anableps*), 35
Free T4 in Senegal sole, 284
Freshwater, fish in, 366
FSHβ mRNA expression, measurement of, 120

G

Galanin (GAL), 432, 434–435, 476, 484, 497
in GH release, 171
Galanin-ir cell populations, in teleost fish, 35
Galanin peptide, 35–36. *See also* Hypophysiotropic peptides, in teleost fish
Gamma-amino butyric acid (GABA), 51–52, 123, 245, 385, 474. *See also* Hypothalamic neurotransmitters
in GH release, 162–163
neurons, in fish brain, 88
neurotransmitter, in LH and FSH neuroendocrine regulation, 128
Gar (*Lepisosteus oculatus*), 44
Gastric acid, 490, 492
Gastric inhibitory peptide (GIP), 440
Gastrin, effect in Atlantic cod, 491
Gastrin-releasing peptide (GRP), 36–37, 171, 439, 477, 492. *See also* Hypophysiotropic peptides, in teleost fish
Gastroenteropancreatic (GEP) endocrine system, 471
Gastrointestinal (GI) tract, 469
GH/IGF-I axis, 398
Ghrelin (GRL), 213, 434–435, 476–477, 480, 484
in fish, structure and function of, 174–175
GHRH and PACAP fibers, in pituitary zone, 41
GHRH-like peptides, 442
Gill claudin 3-and 4-like immunoreactive proteins, 369
Gilthead sea bream (*Sparus auratus*), 199, 398
Glandular nerve cells, role, 4
Glossopharyngeal–vagal motor complex (GVC), 388
Glucagon-like peptide-1 (GLP-1), 441
Glucagon treatment, 441
Glucocorticoid receptor (GR), 273, 399
Glucocorticoid-responsive element (GRE), 92
Glucocorticoids receptors, in fish brain, 91
chemically identified, 94–95
distribution of, 92–94
Glutamate, in GH release, 171–172
Glutamate neurotransmitters, 51–52. *See also* Hypothalamic neurotransmitters
Glutamic acid decarboxylase (GAD), 52
Glutamic acid decarboxylase 67 (GAD67), 126
gnrh1 gene, role, 121
GnRH-induced LH synthesis, DA inhibitory control, 124–125
GnRH neurons, in fish brain, 87–88
Goldfish (*Carassius auratus*), 13, 46, 85, 119, 200, 475
Gonadal growth and reproduction seasonality, GnRH and GnRH receptors in, 121
Gonadal steroid hormones, 77–78
steroid production in fish brain, 78–82
steroid receptors, 82–83
androgen receptors, 89–91
estrogen receptors, 83–89
progesterone receptors, 91
Gonadectomy, in LH and FSH secrete evaluation, 122
Gonadotropin-releasing hormone (GnRH), 37–39, 80, 210–212, 446. *See also*

Hypophysiotropic peptides, in teleost fish
actions of, 120–121
in GH release, 163–167
gonadal steroids regulation, 121–122
negative feedback effects, 122–123
positive feedback effects, 123–124
multiplicity, 117–119
receptors, 119–120
Gonadotropins hormone (GTH), 7, 78
G-protein coupled receptors (GPCR), 48, 130, 199
Grass carp (*Ctenopharyngodon idellus*), 156
Green molly (*Poecilia latipina*), 25
Green swordtail (*Xiphophorus helleri*), 479
Grey mullet (*Mugil cephalus*), 41, 121
Growth hormone factor-1 (GHF-1), 153
Growth hormone (GH), 7, 198, 322, 397–398, 441
autocrine/paracrine signals in pituitary, in GH release, 177–179
biological actions, 154–155
and growth hormone receptors, 152–153
at hypothalamic and pituitary levels, 179–183
hypothalamic signals from CNS, in GH release
inhibitors, 157–163
neuroendocrine regulators of GH, 171–172
stimulators, 163–171
peripheral organs/tissues signals, in GH release
inhibitors, 172–174
stimulators, 174–177
in regulation of feeding in fishes, 442
secretion and synthesis regulation, 155–157
Growth hormone-releasing hormone (GHRH), 39–41, 155, 163, 442. *See also* Hypophysiotropic peptides, in teleost fish
Gulf toadfish (*Opsanus beta*), 25
Guppy (*Poecilia reticulata*), 208
Gut as endocrine organ, 471
Gut blood flow, in teleosts, 496
Gut innervation, 469
development, neuroendocrine system and, 479–481
in elasmobranchs, nerve cell bodies, 469
In fish, cell bodies, 469
ICCs in fish gut, 470
spinal innervation in cyclostomes, 469
Gut motility control, 481
development, 490
gallbladder motility, control, 489–490
gut muscles and spontaneous activity, 481–482
individual signal substances, effects, 482
acetylcholine, 483
adenosine and ATP, 486–487
catecholamines, 486
CCK, 487
CGRP, 485
galanin, 484
ghrelin, 484
GRP/ BBS, 487
histamine, 483
5-HT, 483–484
nitric oxide, 485
NPY, 484
somatostatin, 486
tachykinins, 484
VIP and PACAP, 485
non-propagating activity, control, 489
propagating activity, control
migrating motor complexes, 488
peristalsis, 487–488
retrograde peristalsis, 488–489
Gut neuronal and endocrine systems in fish, 468

H

Hagfish (*Myxine glutinosa*), 366, 475
Hawaiian parrotfish (*Scarus dubius*), 271
Hematopoietic growth factors, 322
Herring (*Clupea harengus*), 119, 445
Histamine, 474, 483
Horseradish peroxidase (HRP), 19
HPI axis in fish, 244
5-Hydroxyindoleacetic acid (5-HIAA), 54
3β-Hydroxysteroid dehydrogenase/D4–D5 isomerase (3βHSD), 78–79

5-Hydroxytryptamine (5-HT), 53–54, 126, 473–474, 482–483, 487
Hypocretins. *See* Orexins peptide
Hypophysiotropic factors to stress response, 285–288
Hypophysiotropic peptides, in teleost fish
  cholecystokinin, 30–31
  cocaine-and amphetamine-regulated transcript, 31–32
  corticotropin-releasing factor and related peptides, 32–35
  galanin, 35–36
  gastrin-releasing peptide, 36–37
  GHRH and PACAP, 39–41
  gonadotropin-releasing hormone, 37–39
  melanocortin system, 41–43
  neuropeptide tyrosine family of peptides, 43–46
  orexins, 46–47
  RF-amide peptides, 47–49
  somatostatin peptides, 49–50
  thyrotropin-releasing hormone, 50–51
Hypophysiotropic territories
  cytoarchitecture of
    hypothalamus, 15–19
    preoptic area, 13–15
    telencephalon, 9–12
  in fish brain, 19, 22
Hypothalamic MCH, orexigenic role, 253
Hypothalamic neurotransmitters
  amino acid neurotransmitters, 51–52
  dopamine, 52–53
  serotonin, 53–54
Hypothalamic peptides, prolactin secretion, 204–205. *See also* Prolactin (PRL) hormone
  inhibiting factors, 205–207
  stimulating peptides
    GnRH, 210–212
    GRL, 213
    NpCp and NPs, 214
    PACAP, 213
    PrRP, 207–210
    TRH, 212
    VIP, 212–213
Hypothalamic–pituitary–adrenal (HPA), 155
Hypothalamic–pituitary–interrenal (HPI) axis, 236
Hypothalamic–pituitary–thyroid (HPT) axis, 236, 285, 341
Hypothalamic signals from CNS, in GH release. *See also* Growth hormone
  inhibitors
    5-HT and GABA in, 162–163
    somatostatin, 157–162
  stimulators
    dopamine, 167–168
    GHRH and GnRHs, 163–167
    GRP and Galanin, 171
    NPY, 169–170
    PACAP, 168–169
    TRH, CRF and CCK, 170–171
Hypothalamo-hypophysial system, 4
Hypothalamo-pituitary complex, steroid effects, 78–82
Hypothalamus, cytoarchitecture, 15–19
Hypoxia, appetite suppressive effects, 448

I

icv OX-A injections, 446
IL-1 family and receptors, 322
Immune modulation, in teleost fish
  BPG axis, 341–342
  by cytokines, 342
    cytokine-induced activation, 345–346
    peripheral immune signals, 344–345
    from periphery to brain, 343
    production in brain, 345
    stress response, 343
  HPA axis, 325
  HPT axis, 341
  opioid system, 330
  sympathetic innervation, 324–325
  type I cytokine family, 336
    effects of PRL, GH and leptin, 339–340
    expression in lymphoid tissues of fish, 338
    production of GH, PRL, SL and leptin, 337
    receptors in lymphoid tissues of fish, 339
    receptors on leukocytes, 337–339
Immunity, 315
  adaptive immune response, 318–320
  cell types for innate immune system, 316
  immune organs and communication, 316
  immune response, termination of, 320
  innate immune response, 317–318

Immunity (*continued*)
and interaction with neuroendocrine system, 320–321
and pathogen recognition, 316–317
Immunocytochemical staining
in rainbow trout, 48
in sea bass, 35
Immunoprotective effects, of GH, 155
Immunostaining studies, in plainfin midshipman, 23
Indian catfish (*Clarias batrachus*), 170
Indian catfish (*Heteropneustes fossilis*), 122, 162
Innate immune response, 317–318
Inositol triphosphate (IP3), 211
Insulin and insulin-like growth factor, 97–98
Insulin-like growth factor (IGF), 75, 97–98, 172–173, 198
Insulin-like growth factor I (IGF-I), 132–133, 398
Interferon gamma (IFN-γ), 317–318
Interleukin 1 beta (IL-1β), 317
Interleukin 6 (IL-6), 317
Interleukin 12 (IL-12), 317–318
Intermedin (IM), 438
Internal cell layer (ICL), 45
Interstitial cells of Cajal (ICCs), 468, 470, 481
Isotocin peptide (IST), 25, 28, 243

J

Janus kinase (JAK), 198, 314
Janus kinase 2 (JAK 2), 153
Japanese conger eel (*Conger myriaster*), 99
Japanese crucian carp (*Carassius auratus langsdorfi*), 208
Japanese crucian carp (*Carassius cuvieri*), 48
Japanese flounder (*Paralichthys olivaceus*), 201, 438
Japanese stargazer (*Uranoscopus japonicus*), 485

K

Kallikrein–kinin system (KKS), 377, 384
Ketanserin, role, 126
Killifish (*Fundulus heteroclitus*), 13, 202, 259, 399
kiss1 gene, 129
KISS neurons, in fish brain, 87
Kisspeptins, fish reproductive function of, 129–130

L

Lake trout (*Salvelinus namaycush*), 448
Lamprey (*Lampetra fluviatilis*), 375, 389, 474
Lateral forebrain bundle (LFB), 11
Lateral hypothalamic nucleus (LH), 15
Lateral recess (LR), 17
Lateral tuberal nucleus (NLT), 18
Leptin
in GH release, 177
and leptin receptors, 96–97
Ligand-binding domain (LBD), 83
Lipopolysaccharide (LPS), 316
β-Lipotrophic hormone (β-LPH), 264
Lungfish (*Protopterus aethiopicus*), 200
Lungfish (*Protopterus annectens*), 135
Luteinizing hormone (LH), in teleost fish, 7, 264. *See also* Gonadotropin-releasing hormone (GnRH)
gonadal steroids regulation, 121–122
negative feedback effects, 122–123
positive feedback effects, 123–124

M

Major histocompatibility complex (MHC), 318–319
Mammalian genes, positively regulated by T3, 281
Marine cyclostome species, plasma osmolality, 366
Marine goby (*Trimma okinawae*), 25
Masu salmon (*Oncorhynchus masu*), 98, 211, 215, 286, 446
MCH genes, 444
MC4R-and MC5R-expressing neurons, 266
MCT proteins, 280
Medaka (*Oryzias latipes*), 129, 215
Medaka (*Oryzias melastigma*), 87, 440

Melanin-concentrating hormone (MCH), 18, 28–30, 88, 132, 244, 443–444
Melanocortin (MC) system, 443
Melanocortin system, in teleost fish, 41–43. *See also* Hypophysiotropic peptides, in teleost fish
α–melanocyte-stimulating hormone (α–MSH), 96, 245, 268–270, 276, 285. *See also* Melanotropic axis
  contribution of MCH in regulation, 287
  secretion in teleost fish, factors affecting, 246–249
Melanocyte-stimulating hormone (MSH), 7
Melanotropes
  hypothalamic regulation of, 245, 251
    hypothalamic factors TRH and CRF, 251–252
    MCH and DA, 252–254
  secretion, targets and functions of, 264–265
    MSH and *N*-acetylated β-ENDs, 268–270
Melanotropic axis, 236, 283–284
  CRF/ UI contribute to regulation of, 286
  factors affecting activity, in teleosts, 251
Melatonin, in LH and FSH neuroendocrine regulation, 127–128. *See also* Monoamine and amino acid neurotransmitters, fish reproductive function
Metabolic hormones, 95
  insulin and insulin-like growth factor, 97–98
  leptin and leptin receptors, 96–97
  thyroid hormone receptors, 98–99
Metabolic signals/energy reserves, 444–445
  hyperglycemia and feeding latency time, 444
  linoleic acid and food intake, 445
  lipostatic control of food intake, 444
Min brain and hormone relationships, in fish hormonal milieu maintenance, 77
Mineralocorticoid receptors (MR), 91, 272–273
Mitogen-activated protein kinase (MAPK), 84, 173, 281
Molly (*Poecilia latipinna*), 207
Monoamine and amino acid neurotransmitters, fish reproductive function. *See also* Gonadotropin-releasing hormone (GnRH)
  dopamine, 124–126
  γ-aminobutyric acid, 128–129
  noradrenaline and melatonin, 127–128
  serotonin, 126–127
Monocarboxylate transporters (MCT), 280
Mozambique tilapia (*Oreochromis mossambicus*), 199, 253, 259–260, 432
Mudskipper (*Periophthalmus modestus*), 207
Mudsucker (*Gillichthys mirabilis*), 50
Muscarinic M3-like receptors, 473

N

Na and Cl retention, 397
N-Acetylated βEND, 252, 265, 267, 270
Na/K-ATPase, 275, 281, 367
Na/K/2Cl cotransporter (NKCC), 367
Natriuretic peptides (NPs), 172, 214, 380–382
Natural killer (NK) cells, 316
Neuroendocrine control of GH release, 156–157
Neuroendocrine–immune interaction, 314
Neuroendocrine pathways, in LH and FSH regulation, 116–117, 134–137
  future development, 137–140
  gonadotropin-releasing hormone
    actions of, 120–121
    gonadal steroids regulation of, 121–124
    multiplicity of, 117–119
    receptors of, 119–120
  monoamine and amino acid neurotransmitters
    dopamine, 124–126
    γ-aminobutyric acid, 128–129
    noradrenaline and melatonin, 127–128
    serotonin, 126–127
  neuropeptides
    kisspeptins, 129–130
    neuropeptides and neuropeptide Y, 131–132
    pituitary adenylate cyclase-activating polypeptide, 130–131
  protein hormones, 132–134
Neuroendocrinology, definition, 76
Neurohypophysial hormones
  arginine-vasotocin peptide, 22–25
  isotocin, 25–28

Neurohypophysial hormones (*continued*)
melanin-concentrating hormone, 28–30
Neurohypophysis structure, 4–5
Neurointermediate lobe (NIL), 7
Neurokinin A (NKA), 478, 484
Neurokinin (NK) 1–3 receptors, 478
Neuromedin U (NMU), 437–438
Neuronal cell bodies, in producing MCH, 29
Neuronal tract tracing and AVT immunohistochemical studies, in Atlantic salmon, 24
Neuropeptides, fish reproductive function of
kisspeptins, 129–130
neuropeptides and neuropeptide Y, 131–132
pituitary adenylate cyclase-activating polypeptide, 130–131
Neuropeptide tyrosine peptides (NPY peptides), 43–46, 125, 131, 169–170, 205
Neuropeptide Y (NPY), 430–432, 477, 484, 497
Neurotransmitters and hormones of gut, 471–473
acetylcholine (ACh), for gut functions in fish, 473
amines, as signal substances in fish gut, 473–474
adenosine and ATP, 475
histamine, 5-HT and GABA, 474
nitric oxide, 475
signal peptides in gut, 475
calcitonin gene-related peptide, 476
cholecystokinin/gastrin family, 475
forms of somatostatin, 478
galanin, 476
gastrin-releasing peptide and bombesin, 477
ghrelin, 476–477
GRP/BBS-like-ir nerves, 477
neuropeptide Y family, 477
NPY/PYY-like material, 477
pituitary adenylate cyclaseactivating polypeptide, 478–479
tachykinins, 478
vasoactive intestinal polypeptide, 478
Nile perch (*Lates niloticus*), 38, 211
Nile tilapia (*Oreochromis niloticus*), 120, 201, 271
Nissl's staining, in CNS cytoarchitecture investigation, 9
N-methyl-D-aspartate (NMDA), 171
Non-acetylated βEND, 266–268
Non-specific cytotoxic cells (NCC), 316
Non-teleost bony ray-finned fish, neurohypophysis, 5
Noradrenaline (NA), 127, 473. *See also* Monoamine and amino acid neurotransmitters, fish reproductive function
NOS-containing in Atlantic cod and spiny dogfish, 475
NP family, in Telsost fish, 380
NPO CRF gene expression, 275
NP receptors (NPRs), 214
NPY genes, 446
Nuclear estrogen receptors, 84
Nucleus lateralis tuberis (NLT), 236
Nucleus of the saccus vasculosus (NSV), 17
Nucleus posterioris periventricularis (NPPv), 129
Nucleus preopticus (NPO), 236, 252, 273, 287
Nucleus preopticus periventricularis (NPP), 124
Nucleus ventral tuberis (NVT), 130
Nutrient absorption, control, 494

O

OATP family, mediating thyroid hormone, 280
Occludin expression, 370
Opioid, 330
opioid receptor agonists, on immune parameters, 334–335
peptide synthesis, in immune cells, 330–331
prohormones, for opioid peptides and hormones, 331
receptors and function, in immune system, 331–336
receptor types and ligands in mammals, 332
Orexin A (OX-A), 432–433
Orexins peptide, 46–47. *See also* Hypophysiotropic peptides, in teleost fish

Organic anion transporter polypeptides (OATP), 280
Organon vasculosum of the lamina terminalis (OVLT), 377
Ostariophysan fish, GH of, 152
Oxyntomodulin, 440–441
Oxytocin, 243
Oyster toadfish (*Opsanus tau*), 89

P

PACAP-related peptide (PRP), 39–40
Pancreas, fish, 469
Pancreatic polypeptide (PP), 471
Paraventricular organ (PVO), 18
Pars distalis (PD), 7
Pars intermedia (PI), 7, 198
Pars nervosa (PN), 7
Pathogen-associated molecular patterns (PAMPs), 316
Pejerrey (*Odontesthes bonariensis*), 80, 165, 218
Pepsinogen/pepsin secretion, 492–493
Periodic acid-Schiff (PAS), 214
Peripheral organs/tissues, signals from
  inhibitors
    insulin-like growth factor, 172–173
    noradrenaline, 173–174
  stimulators, 174–177
Phosphatidylinositol 3-kinase (PI3K), 84
Phosphoenolpyruvate carboxykinase (PEPCK), 281
Phospholipase C (PLC), 211
Photoperiod and feeding activity in fish, 447
Pituitary adenylate cyclase-activating polypeptide (PACAP), 39–41, 130–131, 168–169, 442, 490. *See also* Hypophysiotropic peptides, in teleost fish
Pituitary cells, hypothalamic regulation
  corticotropes in, 237
    inhibitory factors, 244–245
    stimulatory factors, 237–238, 243–244
  melanotropes in, 245, 251
    hypothalamic factors TRH and CRF, 251–252
    MCH and DA, 252–254
  thyrotropes in, 254
    hypothalamic factors, 261, 264
    thyrotropin-releasing hormone, 254–255, 259–261
Pituitary
  estrogen receptor in, 88–89
  structure, 5
Plainfin midshipman (*Porichthys notatus*), 5, 80
Plasma Na^+ concentration, 371
Platyfish (*Xhiphophorus maculates*), 35, 85
Pollutants and health status, 448
Posterior recess nucleus (NRP), 17
Posterior tuberal nucleus (NPT), 17
Postprandial hyperemia, 495
Preoptic area. *See also* Hypophysiotropic territories, cytoarchitecture
  cytoarchitecture of, 13–15
  in teleost pituitary, 22
Preoptic area (POA), 243
Prepro-SS peptides (PSSs), 159
PRL_{188} gene transcript, in tilapia larvae, 203
Proconvertase 1 (PC1), 42
Progesterone receptors (PRs), in fish brain, 91
Prohormone convertases (PCs), 264
Prolactin (PRL) hormone, 7, 322, 385, 397, 494
  neuroendocrine regulation of prolactin secretion
    functions of, 202–204
    hypothalamic peptides, 204–214
    PRL and receptor, 199–202
Prolactin-releasing peptide (PrRP), 47–48, 198, 205, 207–210, 397
Pro-opiomelanocortin (POMC), 7, 96, 237, 264, 283, 442–443
Pro-opiomelanocortin-related peptides, 329–330
Protein hormones, fish reproductive function, 132–134
Protein kinase C (PKC), 84
Proximal pars distalis (PPD), 7, 198
PrRP-immunoreactive (tPrRP-ir), 208
PSS-I and PSS-III, in preoptic area, 50
Pufferfish (*Fugu ruprides*), 119
Pufferfish (*Takifugu rubripes*), 31, 201, 266, 432
Pufferfish (*Teraodon nigrovidis*), 119, 438

R

Rabbitfish (*Siganus guttatus*), 156
Radial cells, in embryonic neurogenesis, 80
Rainbow trout (*Oncorhynchus mykiss*), 24, 79, 120, 199, 243, 379, 470–471
Ray-finned fish, telencephalon topology, 10
Reactive oxygen species (ROS), 317
Reciprocal interactions, 76–77
Rectal gland of elasmobranchs, 493
Red drum (*Sciaenops ocellatus*), 217
Red sea bream (*Pagrus major*), 121
Renin-angiotensin system (RAS), 377, 394–397
Reproduction
  FSH and LH in, 116
  reproductive functions, GH receptors, 154
Reverse transcriptase polymerase chain reaction (RT-PCR), 200
RF-amide peptides, 436
  in GH release, 172
  in teleost fish, 47–49 (*see also* Hypophysiotropic peptides, in teleost fish)
RF-amide-related peptide (RFRP), 47
RNA polymerase, 279
Roach (*Rutilus rutilus*), 119
Rostral pars distalis (RPD), 7, 198, 244–245

S

Saddleback wrasse (*Thalassoma duperrey*), 135
Salinity and feeding, 448
Sculpin (*Myoxocephalus octodecimspinosus*), 40
Sculpin (*Myoxocephalus scorpius*), 470–471
Sea bass (*Dicentrarchus labrax*), 11, 207, 432
Sea bream (*Sparus aurata*), 33, 153
Sea lamprey (*Petromyzon marinus*), 389
Sea perch (*Lateolabrax japonicus*), 47, 433
Secretion and digestion, control
  gastric acid secretion, 490–492
  pancreatic secretion, 493
  pepsinogen/pepsin secretion, 492–493
  rectal gland of elasmobranchs, secretion, 493–494
Secretogranin-II (Sg-II), goldfish LH release regulation, 132
Secretoneurin (SN), goldfish LH release regulation, 132
Senegalese sole (*Solea senegalensis*), 36, 284
Serotonin
  neurotransmitters, 53–54 (*see also* Hypothalamic neurotransmitters)
  in teleost fish, 126–127, 162 (*see also* Dopamine (DA))
Serum amyloid A (SAA), 317
Sex steroids, in GH release, 176–177
Signal transducer and activator of transcription (STAT), 315
Signal transducers and activators of transcription (STAT), 198
Sockeye salmon (*Oncorhynchus nerka*), 36, 132, 167, 220
Sodium appetite, (preference), 388
Somatolactin receptor (SLR), 215–216
Somatolactin (SL) hormone, 7, 198
  neuroendocrine regulation of somatolactin secretion
    functions of, 216–218
    hypothalamic control of, 218–220
    somatolactin and receptor, 214–216
Somatostatin gene, 205, 207
Somatostatin (SS) peptide, in teleosts, 49–50, 157–162, 442, 471, 478
South American cichlid (*Cichlasoma dimerus*), 38
Spadefish (*Chaetodipterus faber*), 135
Spiny dogfish (*Squalus acanthias*), 268, 438, 475
Splanchnic circulation, control, 495–496
  increase in flow, 497
  reduction in flow, 496–497
  variable effects, 497–499
Spotted dogfish (*Scyliorhinus stellaris*), 478
Spotted sea trout (*Cynoscion nebulosus*), 91
Spotted snakehead (*Channa punctatus*), 267
SS-or CCK/gastrin-ir endocrine cells, 481
SS receptors (SST), 207
Stargazer (*Uranoscopus japonicus*), 40, 168, 213, 218
Steroid effects, on hypothalamo-pituitary complex, 78–82
Steroid hormone replacement, in LH and FSH secrete evaluation, 122

Steroid membrane receptors, 82–83
androgen receptors, 89–91
estrogen receptors, 83–89
progesterone receptors, 91
Stingray (*Dasyatis akajei*), 168
Striped bass (*Morone saxatilus*), 119
Sturgeon (*Acipenser gueldenstaedtii*), 200
SW-adapting hormones in teleosts, 385
SW-induced drinking, 387
Swordtail (*Xyphophorus helleii*), 47

T

Tachykinins (TKs), 437, 478, 482, 498
T3 and T4 levels, in catfish, 284
Target genes, for T3 in teleost species, 281
T-cell receptor (TCR), 319
Telencephalon, development of, 9–12. *See also* Hypophysiotropic territories, cytoarchitecture
Teleost CXC cytokines, 323
Teleostean growth hormone, 152
Teleostean hypothalamus, cytoarchitecture, 15–19. *See also* Hypophysiotropic territories, cytoarchitecture
Temperature and food intake, relationship, 447
Testosterone, in LH and FSH synthesis, 123
Tetrodotoxin (TTX), 481, 485
TGF beta (TGF-β), 320
T4 hormone, 259–260, 270, 277. *See also* Thyroid hormone
levels, in rainbow trout, 284
Three-spined sticklebacks (*Gasterosteus aculeatus*), 137, 204
Thyroid hormone
biologically active metabolites, 281–282
in GH release, 175
metabolism, 278
negative feedback, 282
signaling, 279
targets and functions, 281
Thyroid hormone receptors (TRs), 98–99, 279
Thyroid peroxidase (TPO), 271
Thyroid response element (TRE), 279
Thyroid-stimulating hormone (TSH), 7, 259–260, 282
production within immune system, 341
targets and functions of, 270
receptor, 271–272
structure and functions, 270–271
Thyrotropes
hypothalamic regulation of, 254
hypothalamic factors, 261, 264
thyrotropin-releasing hormone, 254–255, 259–261
secretion, targets and functions of, 264–265
Thyrotropic axis, 284–286
Thyrotropin. *See* Thyroid-stimulating hormone (TSH)
Thyrotropin-releasing hormone (TRH), 50–51, 212, 244, 254–255, 281, 287
effects on thyrotropin secretion, in teleost fish, 256–258
in GH release, 170
Thyrotropin secretion in teleost fish, factors affecting, 262–263
Thyroxine, 276–277
and extrathyroidal deiodination, 276–277, 279 (*see also* Thyroid hormone, metabolism)
factors affecting secretion, 255
Tilapia (*Astatotilapia burtoni*), 89–90, 95
Tilapia larvae, PRL_{188} gene transcript in, 203
Tilapia (*Oreochromis mossambicus*), 29, 94, 119
TLR activation, 317
T-lymphocytes, 318–319, 336
Toadfish (*Opsanus tau*), 368
Toll-like receptors (TLRs), 316–317
Torpedo fish (*Torpedo marmorata*), 162
T3-responsive genes, 280–281
TRH-ir fibers, 244
3,5,3′-Triiodo-L-thyronine (T3), 254, 259, 277, 279, 282
TR–RXR–TRE complex, 279
TSH receptor (TSHR), 271
TSHR gene, 272
T3-target genes, 280
Tuberal hypothalamus, in teleost pituitary, 22
Tumor necrosis factor alpha (TNF-α), 317–318
Turbot (*Psetta maxima*), 159
Turbot (*Scophthalmus maximus*), 479

Type I cytokines and linkage to immune system, 336–337
Type II helical cytokines (HCII), 322
Tyrosine hydroxylase (TH), 52, 473–474

U

UI-ir nerve fibers, 243
Ultrastructural studies
- in goldfish, 24
- in green molly, 25, 28

Upper esophageal sphincter muscle (UES), 388
Urotensin II sequences, 400
Urotensins I (UI) and II (UII), 398

V

Valley pupfish (*Cyprinodon nevadensis*), 24
Vasoactive intestinal peptide (VIP), 39, 205, 212–213, 440, 478
Vasotocin V1 receptor, 393
Ventricular NP (VNP), 214, 380
Vertebrates, feeding, 422
Vesicular ACh transporter (VAChT), 473
Volemic *vs.* osmotic mechanisms, 370–372
- blood volume, determinant of blood pressure in trout, 371
- euryhaline eels, 371
- SW fish, 371
- teleost fish in FW, 370
- terrestrial animals, to maintain plasma Na^+, 372

Voltage-sensitive Ca^{2+} channels (VSCC), 165
V1-type vasopressin receptor, 243

W

Water and ion transport, control, 494–495
Whitefish (*Coregonus clupeaformis*), 168
White sea bream (*Diplodus sargus*), 25
White sturgeon (*Ascipenser transmontanus*), 169
White sucker (*Catostomus commersoni*), 32, 157, 243
Winter flounder (*Pseudopleuronectes americanus*), 217, 280, 491

Y

Yellow snapper (*Lutjanus argentiventris*), 54
Yellowtail flounder (*Pleuronectes ferrugineus*), 168
Yellowtail (*Seriola quinqueradiata*), 432, 493

Z

Zebrafish (*Danio rerio*), 17, 46–47, 79, 119, 201, 243

OTHER VOLUMES IN THE FISH PHYSIOLOGY SERIES

VOLUME 1 Excretion, Ionic Regulation, and Metabolism
Edited by W. S. Hoar and D. J. Randall

VOLUME 2 The Endocrine System
Edited by W. S. Hoar and D. J. Randall

VOLUME 3 Reproduction and Growth: Bioluminescence, Pigments, and Poisons
Edited by W. S. Hoar and D. J. Randall

VOLUME 4 The Nervous System, Circulation, and Respiration
Edited by W. S. Hoar and D. J. Randall

VOLUME 5 Sensory Systems and Electric Organs
Edited by W. S. Hoar and D. J. Randall

VOLUME 6 Environmental Relations and Behavior
Edited by W. S. Hoar and D. J. Randall

VOLUME 7 Locomotion
Edited by W. S. Hoar and D. J. Randall

VOLUME 8 Bioenergetics and Growth
Edited by W. S. Hoar, D. J. Randall, and J. R. Brett

VOLUME 9A Reproduction: Endocrine Tissues and Hormones
Edited by W. S. Hoar, D. J. Randall, and E. M. Donaldson

VOLUME 9B Reproduction: Behavior and Fertility Control
Edited by W. S. Hoar, D. J. Randall, and E. M. Donaldson

VOLUME 10A Gills: Anatomy, Gas Transfer, and Acid-Base Regulation
Edited by W. S. Hoar and D. J. Randall

VOLUME 10B Gills: Ion and Water Transfer
Edited by W. S. Hoar and D. J. Randall

VOLUME 11A The Physiology of Developing Fish: Eggs and Larvae
Edited by W. S. Hoar and D. J. Randall

VOLUME 11B The Physiology of Developing Fish: Viviparity and Posthatching Juveniles
Edited by W. S. Hoar and D. J. Randall

VOLUME 12A The Cardiovascular System
Edited by W. S. Hoar, D. J. Randall, and A. P. Farrell

VOLUME 12B The Cardiovascular System
Edited by W. S. Hoar, D. J. Randall, and A. P. Farrell

VOLUME 13 Molecular Endocrinology of Fish
Edited by N. M. Sherwood and C. L. Hew

VOLUME 14 Cellular and Molecular Approaches to Fish Ionic Regulation
Edited by Chris M. Wood and Trevor J. Shuttleworth

VOLUME 15 The Fish Immune System: Organism, Pathogen, and Environment
Edited by George Iwama and Teruyuki Nakanishi

VOLUME 16 Deep Sea Fishes
Edited by D. J. Randall and A. P. Farrell

VOLUME 17 Fish Respiration
Edited by Steve F. Perry and Bruce Tufts

VOLUME 18 Muscle Growth and Development
Edited by Ian A. Johnson

VOLUME 19 Tuna: Physiology, Ecology, and Evolution
Edited by Barbara A. Block and E. Donald Stevens

VOLUME 20 Nitrogen Excretion
Edited by Patricia A. Wright and Paul M. Anderson

VOLUME 21 The Physiology of Tropical Fishes
Edited by Adalberto L. Val, Vera Maria F. De Almeida-Val, and David J. Randall

VOLUME 22 The Physiology of Polar Fishes
Edited by Anthony P. Farrell and John F. Steffensen

VOLUME 23 Fish Biomechanics
Edited by Robert E. Shadwick and George V. Lauder

VOLUME 24 Behaviour and Physiology of Fish
Edited by Katherine A. Sloman, Rod W. Wilson, and Sigal Balshine

VOLUME 25 Sensory Systems Neuroscience
Edited by Toshiaki J. Hara and Barbara S. Zielinski

VOLUME 26 Primitive Fishes
Edited by David J. McKenzie, Anthony P. Farrell, and Colin J. Brauner

VOLUME 27 Hypoxia
Edited by Jeffrey G. Richards, Anthony P. Farrell, and Colin J. Brauner

VOLUME 28 Fish Neuroendocrinology
Edited by Nicholas J. Bernier, Glen Van Der Kraak, Anthony P. Farrell, and Colin J. Brauner

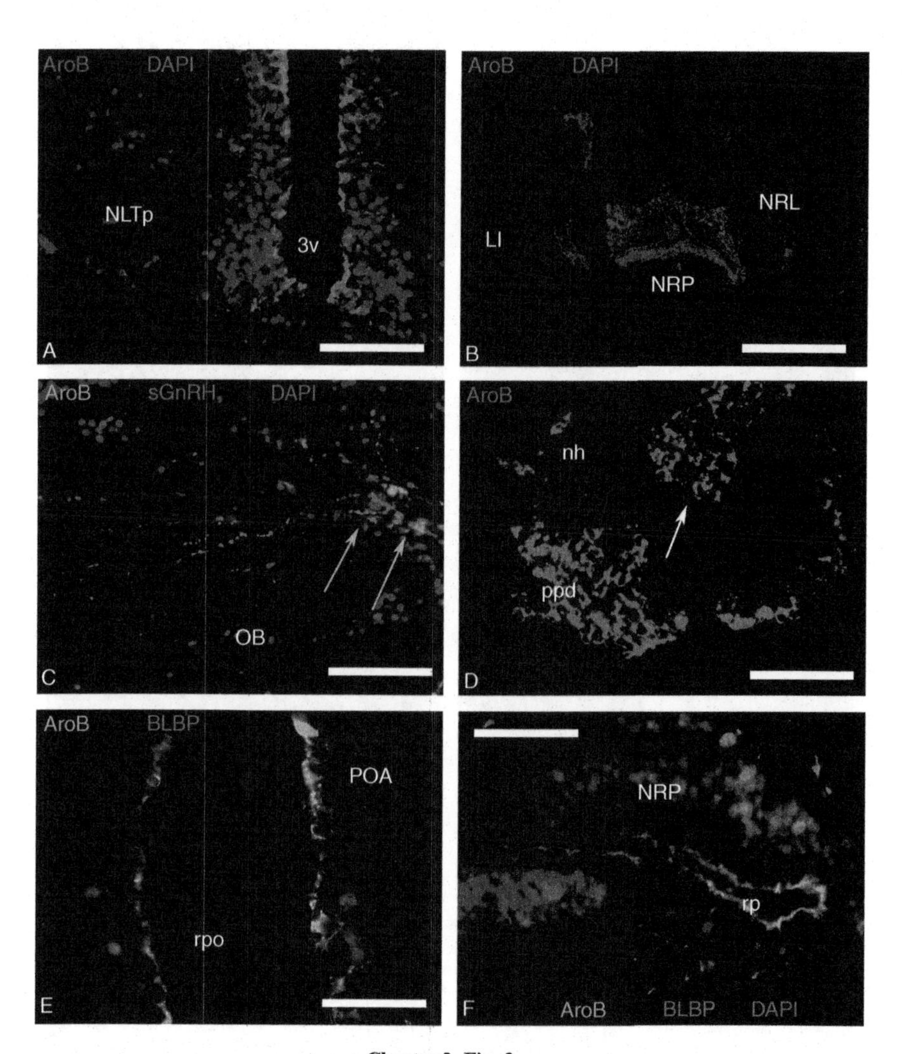

Chapter 2, Fig. 2.

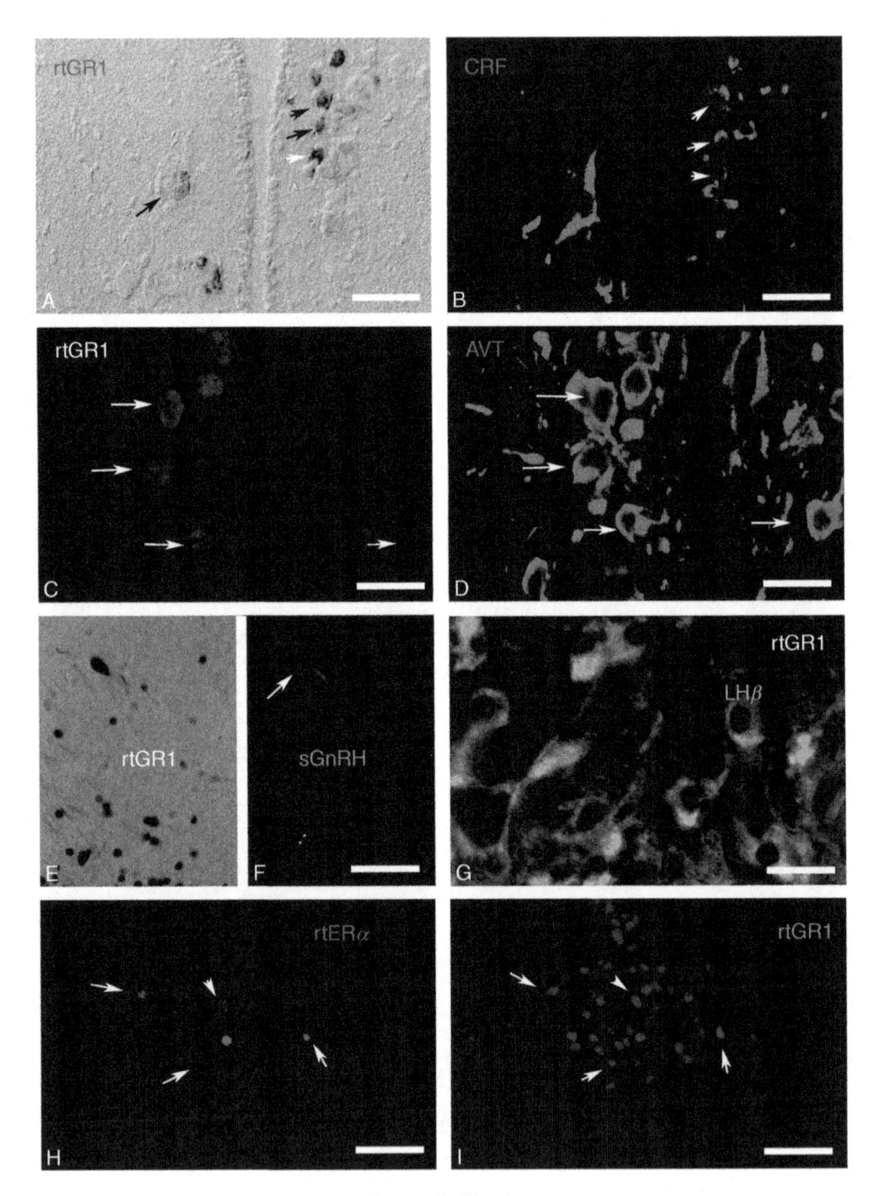

Chapter 2, Fig. 4.

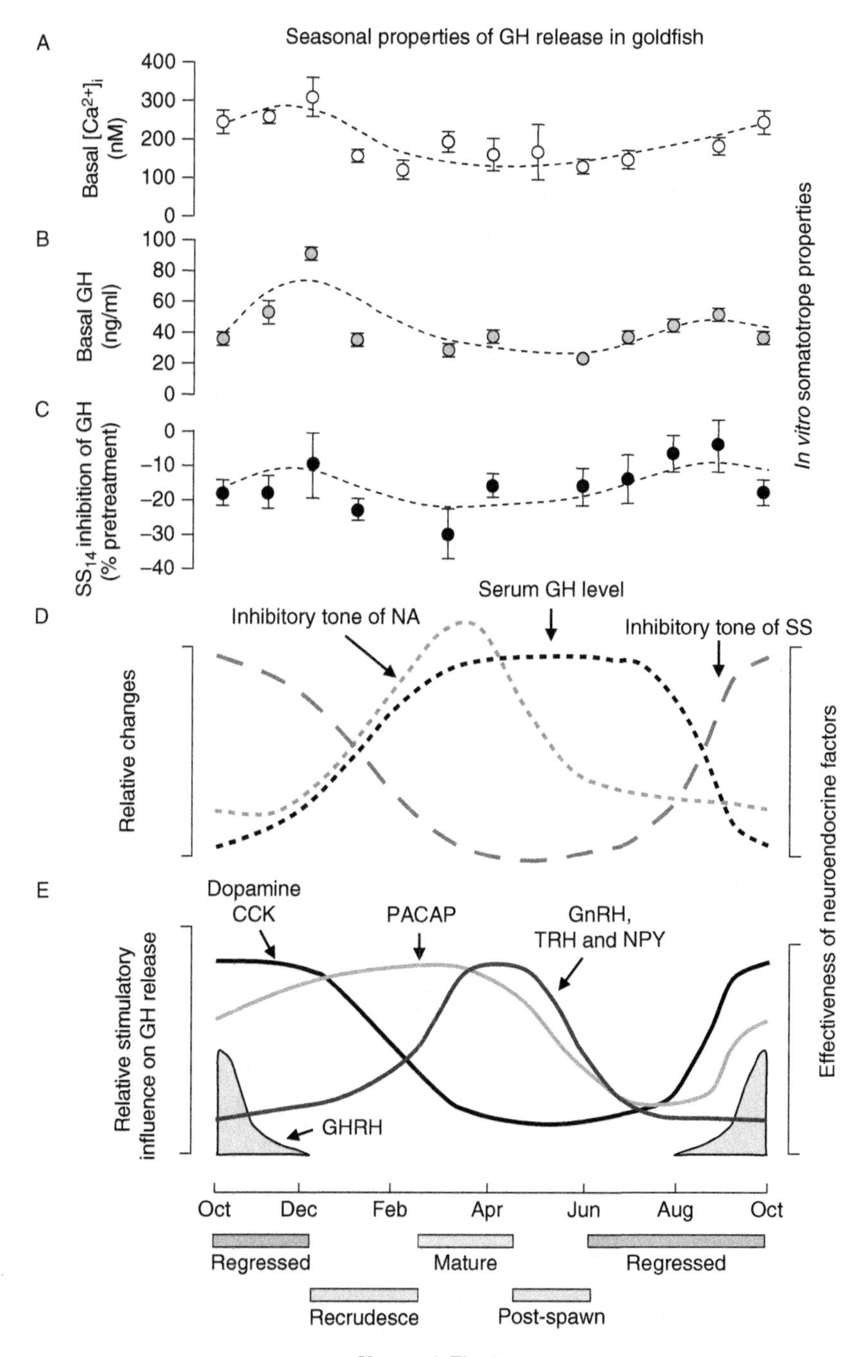

Chapter 4, Fig. 1.

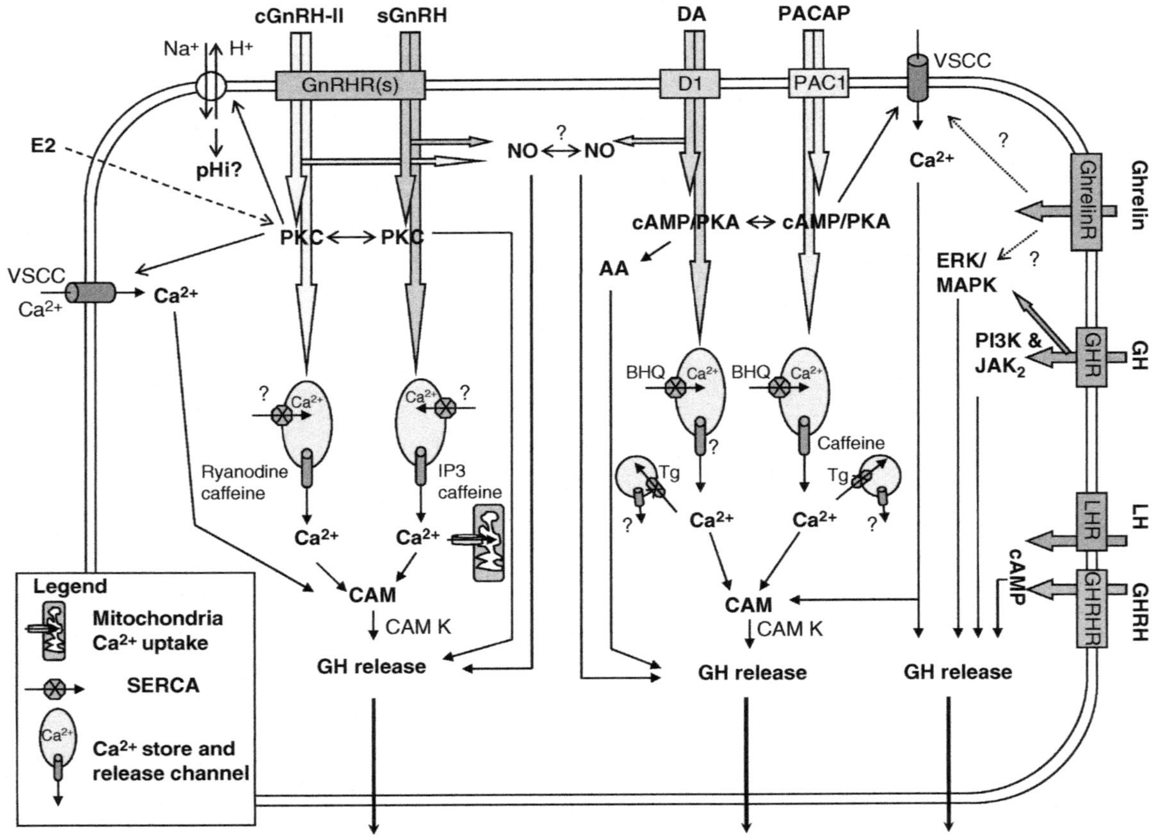

Chapter 4, Fig. 2.

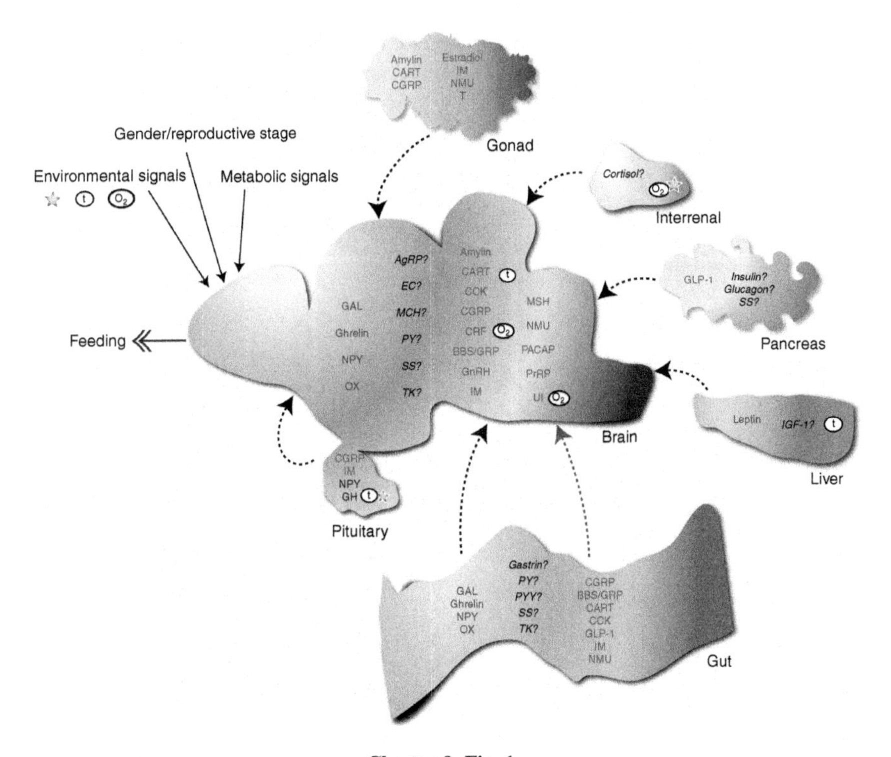

Chapter 9, Fig. 1.

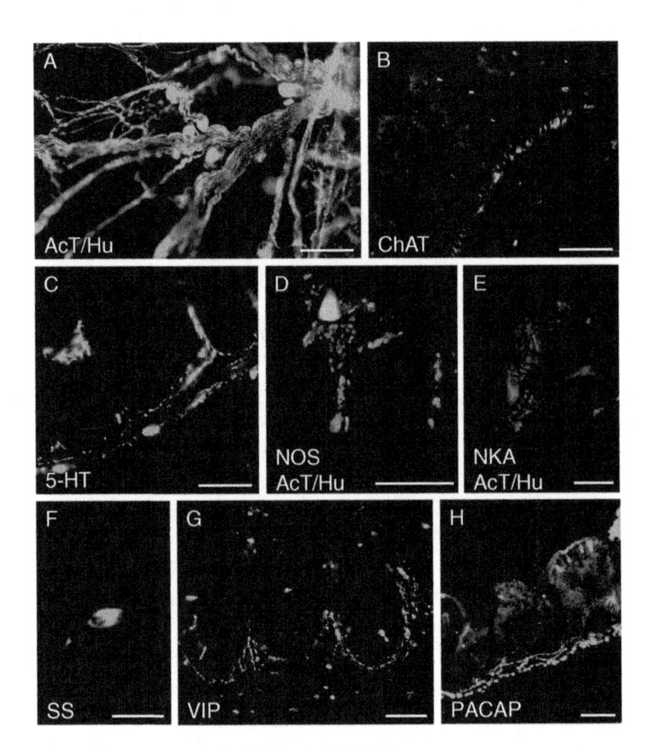

Chapter 10, Fig. 1.

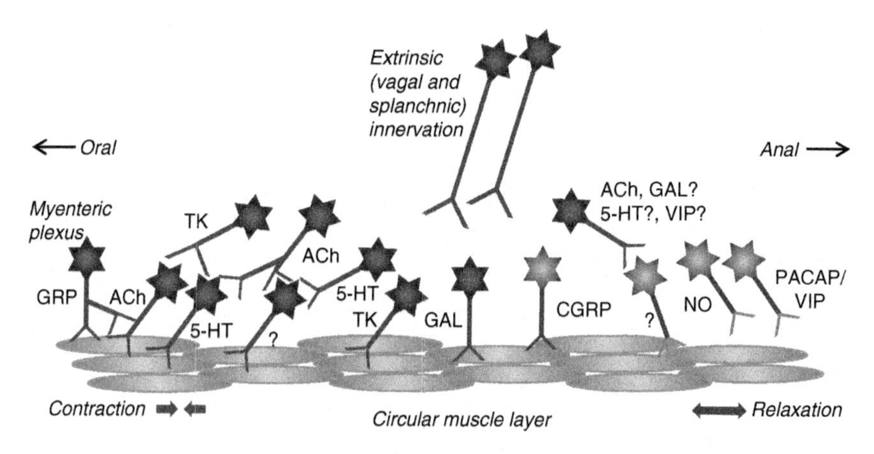

Chapter 10, Fig. 2.

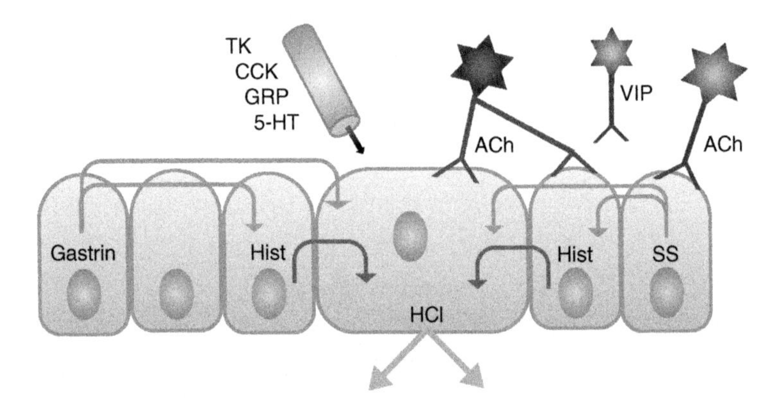

Chapter 10, Fig. 3.

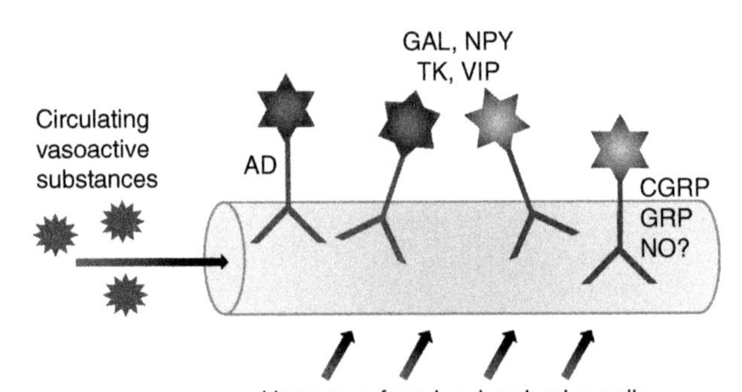

Chapter 10, Fig. 4.

Printed in the United States
By Bookmasters